Signals and Systems
Using MATLAB®

Signals and Systems Using MATLAB®

Third Edition

Luis F. Chaparro
Department of Electrical and Computer Engineering
University of Pittsburgh
Pittsburgh, PA, USA

Aydin Akan
Department of Biomedical Engineering
Izmir Katip Celebi University
Izmir, Turkey

ACADEMIC PRESS

An imprint of Elsevier

Academic Press is an imprint of Elsevier
125 London Wall, London EC2Y 5AS, United Kingdom
525 B Street, Suite 1650, San Diego, CA 92101, United States
50 Hampshire Street, 5th Floor, Cambridge, MA 02139, United States
The Boulevard, Langford Lane, Kidlington, Oxford OX5 1GB, United Kingdom

Library of Congress Cataloging-in-Publication Data
A catalog record for this book is available from the Library of Congress

British Library Cataloguing-in-Publication Data
A catalogue record for this book is available from the British Library

ISBN: 978-0-12-814204-2

For information on all Academic Press publications
visit our website at https://www.elsevier.com/books-and-journals

Working together
to grow libraries in
developing countries

www.elsevier.com • www.bookaid.org

Publisher: Katey Birtcher
Sr. Acquisitions Editor: Steve Merken
Sr. Content Development Specialist: Nate McFadden
Production Project Manager: Sreejith Viswanathan
Designer: Matthew Limbert

Typeset by VTeX

To our families and students

Contents

PART 1 INTRODUCTION

PART 2 THEORY AND APPLICATIONS OF CONTINUOUS-TIME SIGNALS AND SYSTEMS

PART 3 THEORY AND APPLICATIONS OF DISCRETE-TIME SIGNALS AND SYSTEMS

Preface

... in this book I have only made up a bunch of other men's flowers, providing of my own only the string that ties them together.

French Essayist M. de Montaigne (1533–1592)

This third edition of *Signals and Systems Using MATLAB®* is the result of adding new material, and of rewriting and reorganizing the material from the past two editions. Many of the changes resulted from helpful comments from students and faculty who have used the book in their courses.

As indicated in the past editions, even though it is not possible to keep up with advances in technology, innovation in engineering is possible only through a solid understanding of basic principles. The theory of signals and systems constitutes one of those fundamentals, and it will be the foundation of much research and development in engineering for years to come. Moreover, in the near future not only engineers will need to know about signals and systems—to some degree everybody will, as the pervasiveness of digital technologies will require it.

Learning as well as teaching the theory of signals and systems is complicated by the required combination of mathematical abstraction and of concrete engineering applications. A course in signals and systems must be designed to nurture not only the students' interest in applications, but also to make them appreciate the significance of the concepts and of the mathematical tools used. The aim of this textbook is to serve the students' needs in learning the theory of signals and systems as well as to facilitate for faculty the teaching of the material. The approach followed in the book is to present the material as clearly and as simply as possible, and to enhance the learning experience by using MATLAB—an essential tool in the practice of engineering. MATLAB not only helps to illustrate the theoretical results but also to make students aware of the computational issues that engineers face in implementing them.

Book level

The material in the past two editions was aimed to be used in junior- and senior-level courses in signals and systems in electrical and computer engineering, or in biomedical or mechanical engineering. However, in this third edition material on two-dimensional signal processing has been added and as a result the third part of the book could be also used for an introductory graduate course in digital signal processing—providing students the benefit of having the first two parts for remedial background. The book could be of interest to students in applied mathematics and, given its reader-friendly nature, also to practicing engineers interested in learning or in reviewing the basic principles of signals and systems on their own. The material is organized so that the reader not only gets a solid understanding of the theory—enhanced by analytic examples and software examples using MATLAB to learn about applications, but also develop confidence and proficiency in the material by working on analytic and computational problems.

The organization of the material in the book follows the assumption that the reader has been exposed to the theory of linear circuits, differential equations, and linear algebra, and that the course using this material will be followed by courses in control, communications, or digital image processing. The content is guided by the goal of nurturing the interest of students in applications, and of assisting them in becoming more mathematically sophisticated.

To help the reader to learn the book's material, a set of solved problems is made available on the web with this edition. This set consists of analytic and MATLAB-based problems covering the different topics in the text. These problems would be useful to the reader by providing additional examples of problems similar to those at the end of each of the chapters.

Book approach and content

The material is divided in three parts: introduction, theory and applications of continuous-time signals and systems, and theory and applications of discrete-time signals and systems. Material in the first part is intended to help the reader to understand the connection between continuous- and discrete-time signals and systems, and between infinitesimal and finite calculus, and to learn why complex numbers and functions are used in the study of signals and systems. MATLAB is introduced here as a tool for numeric as well as symbolic computations. An overview of the rest of the chapters is weaved through-out this chapter, and motivational practical applications are provided. Significantly, this introductory chapter is named Chapter 0, to serve as an analogy to the ground floor of the building formed by the other chapters.

The core of the material is presented in the second and third parts of the book. In the second part we cover the basics of continuous-time signals and systems and illustrate their applications, and we do similarly in the third part for discrete-time signals and systems. The treatment of continuous- and discrete-time signals and systems is done separately. A concerted effort is made in the second and third parts of the book to show students not only the relevance but the connections of each of the tools used in the analysis of signals and systems. In particular, we emphasize that the transformations used should be seen as a progression rather than as disconnected methods.

Although the treatment of the continuous- and the discrete-time signals and systems is very similar, as can be seen when comparing the contents of parts two and three, it is important to recognize that from the learning point of view putting them together may be expeditious but confusing. It is thus that in the book we introduce first the processing in time and in frequency of continuous-time signals that are familiar to students from their previous courses in linear circuits and differential equations. Building on the experience and mathematical sophistication acquired in the second part we then consider the theory and applications of discrete-time signals and systems. Moreover, the discussion of the tools needed to represent and process signals is done in such a way as to display their connection. Thus, for instance, the concept of the convolution integral—of great significance in the theory of linear time-invariant systems—follows the generic representation of signals by impulse signals and the characterization of systems as linear and time-invariant. Similarly, the presentation of the Fourier analysis connects not only the representations of periodic and aperiodic signals but also the more general Laplace transform representations. It is important that these relations be clearly seen to allow the reader to achieve a holistic understanding of the material. To complement these theoretical results, however, it is necessary to show where they apply. It is thus that after the chapters introducing signals and systems, their applications in control and communications systems are shown.

Using sampling theory as a bridge, the third part of the book covers the theory and illustrates the applications of discrete-time signals and systems. Despite the great similarity in the treatment of continuous and discrete signals and systems there are important differences that need to be pointed out.

The discrete nature of the analysis provides a simpler structure than in the continuous case, thus making some of the concepts more esoteric. This is one more reason for which the treatment of continuous- and discrete-time signals and systems is done separately. For instance, the concept that the discrete frequency is not measurable and that it depends on the sampling period used to discretize the signal is not easily absorbed when it is known that the continuous frequency is measurable and independent of time. Moreover, introducing the Z-transform without connecting it with the Laplace transform misses the connection of continuous and discrete signals. To complement the theoretical concepts, the application of sampling and discrete processing for control and communications is given. A more advanced, but interesting, theory is obtained by extending the one-dimensional material to two dimensions and applying it to the processing of images. After achieving a good understanding of the one-dimensional theory, this should constitute a challenging but rewarding experience for the reader. Just as with the control and communication examples, the introduction of two-dimensional processing is done to illustrate applications that the reader could engage in later at a more appropriate depth.

A great deal of effort has been put into making the text reader friendly. To make sure the reader does not miss important issues presented on a topic, we have inserted well-thought-out remarks intended to minimize common misunderstandings we have observed with our students in the past. Plenty of analytic examples, with different levels of complexity, are given to illustrate important issues. Each chapter has a set of examples using MATLAB, illustrating topics presented in the text or special issues that the reader should know. The MATLAB code is given so that the reader can learn by example from it. To help the reader follow the mathematical derivations, we give additional steps whenever necessary and do not skip steps that are basic in the understanding of a derivation. Summaries of important issues are boxed and concepts and terms are bolded to help the reader to grasp the main points and terminology.

Without any doubt, learning the material in signals and systems requires working analytical as well as computational problems. Thus we consider it important to provide problems of different levels of complexity to exercise not only basic problem-solving skills, but to achieve a level of proficiency in the subject and mathematical sophistication. The basic problems are designed to provide a way for the reader to achieve mastery of conceptual issues, while the MATLAB problems are designed to deepen the conceptual understanding as they are applied. To encourage the readers to work on their own, partial or complete answers are provided for most of the problems at the end of each chapter. Also the new set of solved problems available on the website https://textbooks.elsevier.com/web/product_details.aspx?isbn=9780128142042 should help in understanding the material and in solving problems at the end of each of the chapters.

Two additional features should be beneficial to the reader. One is the inclusion of quotations and footnotes to present interesting ideas, remarks or historical comments, and the other is the inclusion of sidebars that attempt to teach historical facts that the reader should be aware of. Indeed, the theory of signal and systems clearly connects with mathematics and a great number of mathematicians have contributed to it. Likewise, there are a large number of engineers who have contributed significantly to the development and application of signals and systems. All of them should be recognized for their contributions, and we should learn from their experiences.

Teaching using this book

The material in this text could be used for a two-term sequence in signals and systems: one on continuous-time signals and systems, followed by a second course in discrete-time signals and systems with a lab component using MATLAB. These two courses would cover most of the chapters in the text with various degrees of depth, depending on the emphasis the faculty would like to give to the course. As indicated, Chapter 0 was written as a necessary introduction to the rest of the material, but it does not need to be covered in great detail—students can refer to it as needed. Applications in either control, communications, filtering, or two-dimensional processing are to be chosen depending on the emphasis wanted in the course.

A second alternative for teaching a one-term course using this text is to cover the material in Chapters 1 to 5 and Chapters 8 and 9—omitting some relatively less important issues in these chapters. Material from Chapters 6 and 7 could be used depending on the emphasis given to the course.

The material in the third part of the text, as a third alternative, could be used for an introductory course in digital signal processing with the material in the first two parts serving as remedial material for students not familiar with it, in particular the theory of analog filtering.

To readers in general and to students in particular

As user of this book, it is important for you to understand the features of the book so that you can take advantage of them in learning the material. In particular:

1. Refer as often as necessary to the material in Chapter 0 and to the appendix to review, recall or learn the mathematical background; to visualize the overall structure of the material; or to review or learn MATLAB as it applies to signal processing.
2. As you will experience, the complexity of the material grows as it develops. So keep up with your studying of it, and develop a connection of the material with applications in your own areas of interest to motivate you.
3. To help you learn the material, clear and concise results are emphasized in each part by putting them in boxes. Justification for these results are given right before or right after, complemented with remarks regarding issues that need additional clarification, and illustrated with plenty of analytic and computational examples. Also important terms are emphasized throughout the text, and they have been indexed so you can find them easily. Special notation and names of relevant MATLAB functions also appear in the index. Tables provide a good summary of properties and formulas.
4. Most of the problems at the end of each chapter have been given full or partial answers, aimed at helping and encouraging you to solve them on your own. Some of these problems have been used by the authors in exams. The MATLAB problems have headings indicating how they relate to specific topics. Use the solved problems given in the website https://textbooks.elsevier.com/web/product_details.aspx?isbn=9780128142042 for a better understanding of the material and to become more proficient in solving the problems at the end of the chapters.
5. One of the objectives of this text is to help you learn MATLAB, as it applies to signals and systems, on your own. This is done by providing the soft introduction to MATLAB in Chapter 0, and then by showing you the code in each of the chapter examples. You will notice that in the first two parts the MATLAB code is more complete than in the latter part of the book. The assumption is that you will be by then very proficient in the use of MATLAB and could easily supply the missing code.
6. Finally, notice the footnotes, the vignettes, and the historical sidebars, which have been included to provide you with a glance at the background in which the theory and practice of signals and systems developed.

Acknowledgments

This third edition has been made possible by the support, cooperation and help from several individuals. The support and gentle push from Stephen Merken, senior editor in Elsevier, convinced the first author of the need for a revision of the material. This and the cooperation of Professor Aydin Akan from the department of Biomedical Engineering at Izmir Katip Celebi University, Turkey, who gracefully agreed to become co-author, made the new edition a reality. Our thanks go to our development editor Nate McFadden for his continual help in getting the book material reviewed and processed. Thanks also to the book reviewers who provided new insights into the material and ideas on how to improve it. In particular, most sincere thanks go to Professor Danijela Cabric from the University of California in Los Angeles (UCLA), who has taught from our book and who made very significant suggestions that have been incorporated into this edition. Our thanks go to those faculty and students who have used the previous editions and have made suggestions and pointed out unclear material. Thanks also to our families who have endured the long process of writing and revising that is involved in writing a book. Finally, thanks to our students, who have unknowingly contributed to the impetus to write a book that would make the teaching of signal and system theory more accessible and fun to future students.

INTRODUCTION

FROM THE GROUND UP!

CONTENTS

Signals and Systems Using MATLAB®. https://doi.org/10.1016/B978-0-12-814204-2.00009-0

In theory there is no difference between theory and practice. In practice there is.

New York Yankees Baseball Player Lawrence "Yogi" Berra (1925)

0.1 INTRODUCTION

In our modern world, signals of all kinds emanate from different types of devices—radios and TVs, cell phones, global positioning systems (GPSs), radars and sonars. These systems allow us to communicate messages, to control processes, and to sense or measure signals. In the last 70 years, with the advent of the transistor, of the digital computer and of the theoretical fundamentals of digital signal processing the trend has been towards digital representation and processing of data, which in many applications is in analog form. Such a trend highlights the importance of learning how to represent signals in analog as well as in digital forms and how to model and design systems capable of dealing with different types of signals.

The year 1948 is considered the year when technologies and theories responsible for the spectacular advances in communications, control, and biomedical engineering since then were born [58]. Indeed, in June of that year, Bell Telephone Laboratories announced the invention of the transistor. Later that month, a prototype computer built at Manchester University in the United Kingdom became the first operational stored-program computer. Also in that year fundamental theoretical results were published: Claude Shannon's mathematical theory of communications, Richard W. Hamming's theory on error-correcting codes, and Norbert Wiener's *Cybernetics* comparing biological systems to communication and control systems.

Digital signal processing advances have gone hand-in-hand with progress in electronics and computers. In 1965, Gordon Moore, one of the founders of Intel, envisioned that the number of transistors on a chip would double about every two years [38]. It is these advances in digital electronics and in computer engineering that have permitted the proliferation of digital technologies. Today, digital hardware and software process signals from cell phones, high-definition television (HDTV) receivers, digital radio, radars and sonars, just to name a few. The use of digital signal processors (DSPs) and of field-programmable gate arrays (FPGAs) have been replacing the use of application-specific integrated circuits (ASICs) in industrial, medical and military applications.

It is clear that digital technologies are here to stay. The abundance of algorithms for processing digital signals, and the pervasive presence of DSPs and FPGAs in thousands of applications make digital signal processing theory a necessary tool not only for engineers but for anybody who would be dealing with digital data—soon, that will be everybody! This book serves as an introduction to the theory of signals and systems—a necessary first step in the road towards understanding digital signal processing.

0.2 EXAMPLES OF SIGNAL PROCESSING APPLICATIONS

With the availability of digital technologies for processing signals, it is tempting to believe there is no need to understand their connection with analog technologies. That it is precisely the opposite is illustrated by considering the following three interesting applications: the compact-disc (CD) player, software-defined and cognitive radio, and computer-controlled systems.

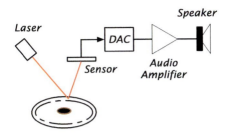

FIGURE 0.1

When playing a CD, the CD-player follows the tracks in the disc, focusing a laser on them, as the CD is spun. The laser shines a light which is reflected by the pits and bumps put on the surface of the disc and corresponding to the coded digital signal from an acoustic signal. A sensor detects the reflected light and converts it into a digital signal, which is then converted into an analog signal by the digital-to-analog converter (DAC). When amplified and fed to the speakers such a signal sounds like the originally recorded acoustic signal.

0.2.1 COMPACT-DISC (CD) PLAYER

Compact discs [11] were first produced in Germany in 1982. Recorded voltage variations over time due to an acoustic sound is called an *analog signal* given its similarity with the differences in air pressure generated by the sound waves over time. Audio CDs and CD-players illustrate best the conversion of a binary signal—unintelligible—into an intelligible analog signal. Moreover, the player is a very interesting control system.

To store an analog audio signal, e.g., voice or music, on a CD the signal must be first sampled and converted into a sequence of binary digits—a digital signal—by an analog-to-digital (A/D) converter and then especially encoded to compress the information and to avoid errors when playing the CD. In the manufacturing of a CD, pits and bumps—corresponding to the ones and zeros from the quantization and encoding processes—are impressed on the surface of the disc. Such pits and bumps will be detected by the CD-player and converted back into an analog signal that approximates the original signal when the CD is played. The transformation into an analog signal uses a digital-to-analog (D/A) converter.

As we will see in Chapter 8, an audio signal is sampled at a rate of about 44,000 samples/s (corresponding to a maximum frequency around 22 kHz for a typical audio signal) and each of these samples is represented by a certain number of bits (typically 8 bits/sample). The need for stereo sound requires that two channels be recorded. Overall, the number of bits representing the signal is very large and needs to be compressed and especially encoded. The resulting data, in the form of pits and bumps impressed on the CD surface, are put into a spiral track that goes from the inside to the outside of the disc.

Besides the binary-to-analog conversion, the CD-player exemplifies a very interesting control system (see Fig. 0.1). Indeed, the player must: (i) rotate the disc at different speeds depending on the location of the track within the CD being read; (ii) focus a laser and a lens system to read the pits and bumps on the disc, and (iii) move the laser to follow the track being read. To understand the exactness required, consider that the width of the track is typically less than a micrometer (10^{-6} meters or 3.937×10^{-5} inches), and the height of the bumps is about a nanometer (10^{-9} meters or 3.937×10^{-8} inches).

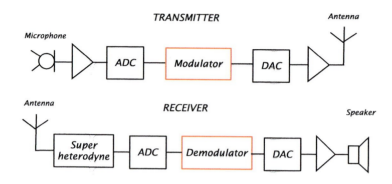

FIGURE 0.2

Schematics of a voice SDR mobile two-way radio. Transmitter: the voice signal is inputted by means of a microphone, amplified by an audio amplifier, converted into a digital signal by an analog-to-digital converter (ADC) and then modulated using software, before being converted into analog by a digital-to-analog converter (DAC), amplified and sent as a radio frequency signal via an antenna. Receiver: the signal received by the antenna is processed by a superheterodyne front-end, converted into a digital signal by an ADC before being demodulated and converted into an analog signal by a DAC, amplified and fed to a speaker. The modulator and demodulator blocks indicate software processing.

0.2.2 SOFTWARE-DEFINED RADIO AND COGNITIVE RADIO

Software-defined and cognitive radio are important emerging technologies in wireless communications [48]. In a software-defined radio (SDR), some of the radio functions typically implemented in hardware are converted into software [69]. By providing smart processing to SDRs, cognitive radio (CR) will provide the flexibility needed to more efficiently use the radio frequency spectrum and to make available new services to users. In the United States the Federal Communication Commission (FCC), and likewise in other parts of the world the corresponding agencies, allocates the bands for different users of the radio spectrum (commercial radio and TV, amateur radio, police, etc.). Although most bands have been allocated, implying a scarcity of spectrum for new users, it has been found that locally at certain times of the day the allocated spectrum is not being fully utilized. Cognitive radio takes advantage of this.

Conventional radio systems are composed mostly of hardware, and as such cannot easily be reconfigured. The basic premise in SDR as a wireless communication system is its ability to reconfigure by changing the software used to implement functions typically done by hardware in a conventional radio. In an SDR transmitter, software is used to implement different types of modulation procedures, while analog-to-digital converters (ADCs) and digital-to-analog converter (DACs) are used to change from one type of signal into another. Antennas, audio amplifiers and conventional radio hardware are used to process analog signals. Typically, an SDR receiver uses an ADC to change the analog signals from the antenna into digital signals that are processed using software on a general-purpose processor. See Fig. 0.2.

Given the need for more efficient use of the radio spectrum, cognitive radio (CR) uses SDR technology while attempting to dynamically manage the radio spectrum. A cognitive radio monitors locally the radio spectrum to determine regions that are not occupied by their assigned users and transmits in

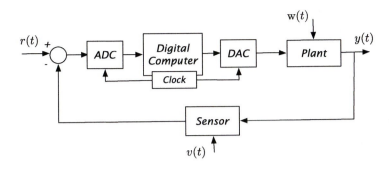

FIGURE 0.3

Computer-control system for an analog plant (e.g., cruise control for a car). The reference signal is $r(t)$ (e.g., desired speed) and the output is $y(t)$ (e.g., car speed). The analog signals are converted to digital signals by an analog-to-digital converter (ADC), while the digital signal from the computer is converted into an analog signal (an actuator is probably needed to control the car) by a digital-to-analog converter (DAC). The signals $v(t)$ and $w(t)$ are disturbances or noise in the plant and the sensor (e.g., electronic noise in the sensor and undesirable vibration in the car).

those bands. If the primary user of a frequency band recommences transmission, the CR either moves to another frequency band, or stays in the same band but decreases its transmission power level or modulation scheme to avoid interference with the assigned user. Moreover, a CR will search for network services that it can offer to its users. Thus SDR and CR are bound to change the way we communicate and use network services.

0.2.3 COMPUTER-CONTROL SYSTEMS

The application of computer-control ranges from controlling simple systems such as a heater (e.g., keeping a room temperature comfortable while reducing energy consumption) or cars (e.g., controlling their speed), to that of controlling rather sophisticated machines such as airplanes (e.g., providing automatic flight control), or chemical processes in very large systems such as oil refineries. A significant advantage of computer control is the flexibility computers provide—rather sophisticated control schemes can be implemented in software and adapted for different control modes.

Typically, control systems are feedback systems where the dynamic response of a system is changed to make it follow a desirable behavior. As indicated in Fig. 0.3, the plant is a system, such as a heater, a car, an airplane, or a chemical process in need of some control action so that its output (it is also possible for a system to have several outputs) follows a reference signal (or signals). For instance, one could think of a cruise control system in a car that attempts to keep the speed of the car at a certain value by controlling the gas pedal mechanism. The control action will attempt to have the output of the system follow the desired response, despite the presence of disturbances either in the plant (e.g., errors in the model used for the plant) or in the sensor (e.g., measurement error). By comparing the reference signal with the output of the sensor, and using a control law implemented in the computer, a control action is generated to change the state of the plant and attain the desired output.

To use a computer in a control application it is necessary to transform analog signals into digital signals so that they can be inputted into the computer, while it is also necessary that the output of the computer be converted into an analog signal to drive an actuator (e.g., an electrical motor) to provide an action capable of changing the state of the plant. This can be done by means of ADCs and DACs. The sensor should act as a transducer whenever the output of the plant is of a different type than the reference. Such would be the case, for instance, if the plant output is a temperature while the reference signal is a voltage.

0.3 IMPLEMENTATION OF DIGITAL SIGNAL PROCESSING ALGORITHMS

Continuous-time signals are typically processed using analog systems composed of electrical circuit components such as resistors, capacitors, and inductors together with semiconductor electronic components such as diodes, transistors, and operational amplifiers, among others. Digital signals, on the other hand, are sequences of numbers and processing them requires numerical manipulation of these sequences. Simple addition, multiplication, and delay operations are enough to implement many discrete-time systems. Thus, digital signal processing systems are easier to design, develop, simulate, test, and implement than analog systems by using flexible, reconfigurable, and reliable software and hardware tools. Digital signal processing systems are employed these days in many applications such as cell phones, household appliances, cars, ships and airplanes, smart home applications, and many other consumer electronic devices. The fast development of digital technology has enabled high-capacity processing hardware tools at reasonable costs available for real-time applications. Refer to [44,54] for in-depth details.

A digital signal processing system may be used to perform a task on an analog signal $x(t)$, or on an inherently discrete-time signal $x[n]$. In the former case, the analog signal is first converted into digital form by using an analog-to-digital converter which performs sampling of the analog signal, quantization of the samples, and encoding the amplitude values using a binary representation. A digital signal processing system may be represented by a mathematical equation defining the output signal as a function of the input by using arithmetic operations. Designing these systems requires the development of an algorithm that implements arithmetic operations.

A general-purpose computer may be used to develop and test these algorithms. Algorithm development, debugging and testing steps are generally done by using a high-level programming tool such as MATLAB or C/C++. Upon successful development of the algorithm, and after running simulations on test signals, the algorithm is ready to be implemented on hardware. Digital signal processing applications often require heavy arithmetic operations, e.g., repeated multiplications and additions, and as such dedicated hardware is required. Possible implementations for a real-time implementation of the developed algorithms are:

- General-purpose microprocessors (μPs) and micro-controllers (μCs).
- General-purpose digital signal processors (DSPs).
- Field-programmable gate arrays (FPGAs).

Selecting the best implementation hardware depends on the requirements of the application such as performance, cost, size, and power consumption.

0.3.1 MICROPROCESSORS AND MICRO-CONTROLLERS

With increasing clock frequencies (for processing fast changing signals) and lower costs, general-purpose microprocessors and micro-controllers have become capable of handling many digital signal processing applications. However, complex operations such as multiplication and division are time consuming for general-purpose microprocessors since they need a series of operations. These processors do not have the best architecture or on chip facilities required for efficient digital signal processing operations. Moreover, they are usually not cost effective or power efficient for many applications.

Micro-controllers are application-specific micro-computers that contain built-in hardware components such as central processing unit (CPU), memory, and input/output (I/O) ports. As such, they are referred to as embedded controllers. A variety of consumer and industrial electronic products such as home appliances, automotive control applications, medical devices, space and military applications, wireless sensor networks, smart phones, and games are designed using micro-controllers. They are preferred in many applications due to their small size, low cost, and providing processor, memory, and random-access memory (RAM) components, all together in one chip.

A very popular micro-controller platform is the Arduino electronic board with an on-board micro-controller necessary and input/output ports. Arduino is an open-source and flexible platform that offers a very simple way to design a digital signal processing application. The built-in micro-controller is produced in an architecture having a powerful arithmetic logic unit that enables very fast execution of the operations. User-friendly software development environment is available for free, and it makes it very easy to design digital signal processing systems on Arduino boards.

0.3.2 DIGITAL SIGNAL PROCESSORS

A digital signal processor is a fast special-purpose microprocessor with architecture and instruction set designed specifically for efficient implementation of digital signal processing algorithms. Digital signal processors are used for a wide range of applications, from communications and control to speech and image processing. Applications embedded digital signal processors are often used in consumer products such as mobile phones, fax/modems, disk drives, radio, printers, medical and health care devices, MP3 players, high-definition television (HDTV), and digital cameras. These processors have become a very popular choice for a wide range of consumer applications, since they are very cost effective. Software development for digital signal processors has been facilitated by especially designed software tools. DSPs may be reprogrammed in the field to upgrade the product or to fix any software bugs, with useful built-in software development tools including a project build environment, a source code editor, a C/C++ compiler, a debugger, a profiler, a simulator, and a real-time operating system. Digital signal processors provide the advantages of microprocessors, while being easy to use, flexible, and lower cost.

0.3.3 FIELD PROGRAMMABLE GATE ARRAYS

Another way to implement a digital signal processing algorithm is using field-programmable gate arrays (FPGAs) which are field-programmable logic elements, or programmable devices that contain fields of small logic blocks (usually NAND gates) and elements. The logic block size in the field-programmable logic elements is referred to as the "granularity" which is related to the effort required to complete the wiring between the blocks. There are three main granularity classes:

- Fine granularity or Pilkington (sea of gates) architecture
- Medium granularity
- Large granularity (Complex Programmable Logic Devices)

Wiring or linking between the gates is realized by using a programming tool. The field-programmable logic elements are produced in various memory technologies that allow the device to be reprogrammable, requiring short programming time and protection against unauthorized use. For many high-bandwidth signal processing applications such as wireless, multimedia, and satellite communications, FPGA technology provides a better solution than digital signal processors.

0.4 CONTINUOUS OR DISCRETE?

Infinitesimal Calculus, or just plain *Calculus*, deals with functions of one or more continuously changing variables. Based on the representation of these functions, the concepts of *derivative* and of *integral* are developed to measure the rate of change of functions and the areas under the graphs of these functions, or their volumes. Ordinary differential equations are then introduced to characterize dynamic systems.

Finite Calculus, on the other hand, deals with sequences. Thus derivatives and integrals are replaced by differences and summations, while ordinary differential equations are replaced by difference equations. Finite Calculus makes possible the computations of Calculus by means of a combination of digital computers and numerical methods—thus Finite Calculus becomes the more concrete mathematics.[1] Numerical methods applied to these sequences permit us to approximate derivatives, integrals, and the solution of differential equations.

In engineering, as in many areas of science, the inputs and outputs of electrical, mechanical, chemical and biological processes are measured as functions of time with amplitudes expressed in terms of voltage, current, torque, pressure, etc. These functions are called *continuous-time signals*, and to process them with a computer they must be converted into binary sequences—or a string of ones and zeros that is understood by the computer. Such a conversion is done in such a way as to preserve as much as possible the information contained in the original signal. Once in binary form, signals can be processed using algorithms (coded procedures understood by computers and designed to obtain certain desired information from the signals or to change them) in a computer or in a dedicated piece of hardware.

In a digital computer, differentiation and integration can be done only approximately, and the solution of ordinary differential equations requires a discretization process as we will illustrate later in this chapter. Not all signals are functions of a continuous parameter—there exist inherently discrete-time signals that can be represented as sequences, converted into binary form and processed by computers. For these signals the Finite Calculus is the natural way of representing and processing them.

---

[1]The use of *concrete*, rather than abstract, mathematics was coined by Graham, Knuth and Patashnik the authors of the book *Concrete Mathematics—a Foundation for Computer Science* [31]. Professor Donald Knuth, from Stanford University, is the inventor of the TeX and Metafont typesetting systems that are the precursors of LaTeX, the document layout system in which the original manuscript of this book was done.

Continuous-time signals are converted into binary sequences by means of an analog-to-digital converter which, as we will see, compresses the data by converting the continuous-time signal into a discrete-time signal or sequence of samples, and each sample is represented by a string of ones and zeros giving a binary signal. Both time and signal amplitude are made discrete in this process. Likewise, digital signals can be transformed into continuous-time signals by means of a digital-to-analog converter that uses the reverse process of an analog-to-digital converter. These converters are commercially available, and it is important to learn how they work so that digital representation of continuous-time signals is obtained with minimal information loss. Chapters 1, 8, and 9 will provide the necessary information as regards continuous-time and discrete-time signals, and show how to convert one into the other and back. The Sampling Theory in Chapter 8 is the backbone of digital signal processing.

0.4.1 CONTINUOUS AND DISCRETE REPRESENTATIONS

There are significant differences between continuous-time and discrete-time signals and in their processing. A discrete-time signal is a sequence of measurements typically made at uniform times, while a continuous-time signal depends continuously on time. Thus, a discrete-time signal $x[n]$ and the corresponding continuous-time signal $x(t)$ are related by a sampling process:

$$x[n] = x(nT_s) = x(t)|_{t=nT_s}. \tag{0.1}$$

That is, the signal $x[n]$ is obtained by sampling $x(t)$ at times $t = nT_s$, where n is an integer and T_s is the **sampling period** or the time between samples. This results in a sequence

$$\{\cdots x(-T_s) \quad x(0) \quad x(T_s) \quad x(2T_s) \cdots\}$$

according to the sampling times, or equivalently

$$\{\cdots x[-1] \quad x[0] \quad x[1] \quad x[2] \cdots\}$$

according to the ordering of the samples (as referenced to time 0). This process is called **sampling or discretization** of a continuous-time signal.

Clearly, by choosing a small value for T_s we could make the continuous- and the discrete-time signals look very similar—almost indistinguishable—which is good but at the expense of memory space required to keep the numerous samples. If we make the value of T_s large, we improve the memory requirements but at the risk of losing information contained in the original signal. For instance, consider a sinusoid obtained from a signal generator:

$$x(t) = 2\cos(2\pi t)$$

for $0 \le t \le 10$ s. If we sample it every $T_{s1} = 0.1$ s, the analog signal becomes the following sequence:

$$x_1[n] = x(t)|_{t=0.1n} = 2\cos(2\pi n/10) \qquad 0 \le n \le 100,$$

providing a very good approximation of the original signal. If on the other hand, we let $T_{s2} = 1$ s, then the discrete-time signal is

$$x_2[n] = x(t)|_{t=n} = 2\cos(2\pi n) = 2 \qquad 0 \le n \le 10.$$

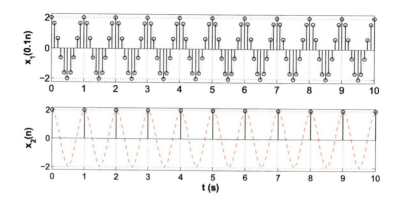

FIGURE 0.4

Sampling a sinusoid $x(t) = 2\cos(2\pi t)$, $0 \le t \le 10$, with two different sampling periods $T_{s1} = 0.1$ s (top) and $T_{s2} = 1$ s (bottom) giving $x_1(0.1n) = x_1[n]$ and $x_2(n) = x_2[n]$. The sinusoid is shown by dashed lines. Notice the similarity between the discrete-time signal and the analog signal when $T_{s1} = 0.1$ s, while they are very different when $T_{s2} = 1$ s, indicating loss of information.

See Fig. 0.4. Although by using T_{s2} the number of samples is considerably reduced, the representation of the original signal is very poor—it appears as if we had sampled a constant signal; we have thus lost information! This indicates that it is necessary to develop a way to choose T_s so that sampling provides not only a reasonable number of samples, but more important that it guarantees that the information in the continuous- and the discrete-time signals remains the same.

As indicated before, not all signals are analog; there are some that are naturally discrete. Fig. 0.5 displays the weekly average of the stock price of a fictitious company ACME. Thinking of it as a signal, it is naturally discrete-time as it does not result from the discretization of a continuous-time signal.

We have shown in this section the significance of the sampling period T_s in the transformation of a continuous-time signal into a discrete-time signal without losing information. Choosing the sampling period requires knowledge of the frequency content of the signal—a consequence of the inverse relation between time and frequency to be discussed in Chapters 4 and 5 where the Fourier representation of periodic and non-periodic signals is given. In Chapter 8 we consider the problem of sampling, and we will then use this relation to determine appropriate values for the sampling period.

0.4.2 DERIVATIVES AND FINITE DIFFERENCES

Differentiation is an operation that is approximated in Finite Calculus. The **derivative operator**

$$D[x(t)] = \frac{dx(t)}{dt} = \lim_{h \to 0} \frac{x(t+h) - x(t)}{h} \tag{0.2}$$

measures the rate of change of a signal $x(t)$. In Finite Calculus the **forward finite-difference operator**

$$\Delta[x(nT_s)] = x((n+1)T_s) - x(nT_s) \tag{0.3}$$

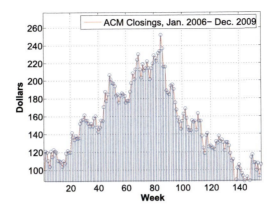

FIGURE 0.5

Weekly closings of ACM stock for 160 weeks in 2006 to 2009. ACM is the trading name of the stock of the imaginary company ACME Inc. makers of everything you can imagine.

measures the change in the signal from one sample to the next. If we let $x[n] = x(nT_s)$, for a known T_s, the forward finite-difference operator becomes a function of n

$$\Delta[x[n]] = x[n+1] - x[n]. \qquad (0.4)$$

The forward finite-difference operator measures the difference between two consecutive samples: one in the future $x((n+1)T_s)$ and the other in the present $x(nT_s)$ (see the problems at the end of the chapter for definition of the **backward finite-difference operator**). The symbols D and Δ are called operators as they operate on functions to give other functions. The derivative and the finite-difference operators are clearly not the same. In the limit, we have

$$\frac{dx(t)}{dt}\Big|_{t=nT_s} = \lim_{T_s \to 0} \frac{\Delta[x(nT_s)]}{T_s}. \qquad (0.5)$$

Depending on the signal and the chosen value of T_s, the finite-difference operation can be a crude or an accurate approximation to the derivative multiplied by T_s.

Intuitively, if a signal does not change very fast with respect to time, the finite-difference approximates well the derivative for relatively large values of T_s, but if the signal changes very fast one needs very small values of T_s. The concept of frequency of a signal can help us understand this. We will learn that the frequency content of a signal depends on how fast the signal varies with time, thus a constant signal has zero frequency, while a noisy signal that changes rapidly has high frequencies. Consider a constant signal, $x_0(t) = 2$ having a derivative of zero (i.e., such a signal does not change at all with respect to time, i.e., a zero-frequency signal). If we convert this signal into a discrete-time signal using a sampling period $T_s = 1$ (or any other positive value), then $x_0[n] = 2$ and so

$$\Delta[x_0[n]] = 2 - 2 = 0$$

coinciding with the derivative. Consider then a signal that changes faster than $x_0(t)$ such as $x_1(t) = t^2$. Sampling $x_1(t)$ with $T_s = 1$, we have $x_1[n] = n^2$ and its forward finite difference is given by

$$\Delta[x_1[n]] = \Delta[n^2] = (n+1)^2 - n^2 = 2n + 1$$

and resulting in an approximation $\Delta[x_1[n]]/T_s = 2n + 1$ to the derivative. The derivative of $x_1(t)$ is $2t$, which at zero equals 0, and at 1 equals 2. On the other hand, $\Delta[n^2]/T_s$ equals 1 and 3 at $n = 0$ and $n = 1$, different values from those of the derivative. Suppose then we choose $T_s = 0.01$, so that $x_1[n] = x_1(nT_s) = (0.01n)^2 = 0.0001n^2$. If we compute the difference for this signal we get

$$\Delta[x_1(0.01n)] = \Delta[(0.01n)^2] = (0.01n + 0.01)^2 - 0.0001n^2 = 10^{-4}(2n+1),$$

which gives as approximation to the derivative $\Delta[x_1(0.01n)]/T_s = 10^{-2}(2n + 1)$, or 0.01 when $n = 0$ and 0.03 when $n = 1$, which are a lot closer to the actual values of

$$\frac{dx_1(t)}{dt}\Big|_{t=0.01n} = 2t\big|_{t=0.01n} = 0.02n$$

as the error now is 0.01 for each case instead of 1 as in the case when $T_s = 1$. Thus, whenever the rate of change of the signal is fast the difference gets closer to the derivative by making T_s smaller.

It becomes clear that the faster the signal changes, the smaller the sampling period T_s should be in order to get a better approximation of the signal and its derivative. As we will learn in Chapters 4 and 5 the frequency content of a signal depends on the signal variation over time. A constant signal has zero frequency, while a signal that changes very fast over time has high frequencies. Clearly, the higher the frequencies in a signal, the more samples would be needed to represent it with no loss of information thus requiring that T_s be small.

0.4.3 INTEGRALS AND SUMMATIONS

Integration is the opposite of differentiation. To see this, suppose $I(t)$ is the integration of a continuous signal $x(t)$ from some time t_0 to t ($t_0 < t$),

$$I(t) = \int_{t_0}^{t} x(\tau)d\tau \tag{0.6}$$

or the area under $x(t)$ from t_0 to t. Notice that the upper bound of the integral is t so the integrand depends on τ, a dummy variable.[2] The derivative of $I(t)$ is

$$\frac{dI(t)}{dt} = \lim_{h \to 0} \frac{I(t) - I(t-h)}{h} = \lim_{h \to 0} \frac{1}{h} \int_{t-h}^{t} x(\tau)d\tau \approx \lim_{h \to 0} \frac{x(t) + x(t-h)}{2} = x(t)$$

---

[2] The integral $I(t)$ is a function of t and as such the integrand needs to be expressed in terms of a so-called **dummy variable** τ that takes values from t_0 to t in the integration. It would be confusing to let the integration variable be t. The variable τ is called a **dummy variable** because it is not crucial to the integration, any other variable could be used with no effect on the integration.

where the integral is approximated as the area of a trapezoid with sides $x(t)$ and $x(t - h)$ and height h. Thus, for a continuous signal $x(t)$

$$\frac{d}{dt} \int_{t_0}^{t} x(\tau)d\tau = x(t) \tag{0.7}$$

or if we use the derivative operator $D[.]$, then its inverse $D^{-1}[.]$ should be the integration operator, i.e., the above equation can be written

$$D[D^{-1}[x(t)]] = x(t). \tag{0.8}$$

We will see in Chapter 3 a similar relation between derivative and integral. The Laplace transform operators s and $1/s$ (just like D and $1/D$) imply differentiation and integration in the time domain.

Computationally, integration is implemented by sums. Consider, for instance, the integral of $x(t) = t$ from 0 to 10, which we know is equal to

$$\int_{0}^{10} t \, dt = \frac{t^2}{2} \Big|_{t=0}^{10} = 50 \,,$$

i.e., the area of a triangle of base of 10 and height of 10. For $T_s = 1$, suppose we approximate the signal $x(t)$ by an aggregation of pulses $p[n]$ of width $T_s = 1$ and height $nT_s = n$, or pulses of area n for $n = 0, \cdots, 9$. See Fig. 0.6. This can be seen as a lower-bound approximation to the integral, as the total area of these pulses gives a result smaller than the integral. In fact, the sum of the areas of the pulses is given by

$$\sum_{n=0}^{9} p[n] = \sum_{n=0}^{9} n = 0.5 \left[\sum_{n=0}^{9} n + \sum_{k=9}^{0} k \right] = 0.5 \left[\sum_{n=0}^{9} n + \sum_{n=0}^{9} (9 - n) \right] = \frac{9}{2} \sum_{n=0}^{9} 1 = \frac{10 \times 9}{2} = 45.$$

The approximation of the area using $T_s = 1$ is very poor (see Fig. 0.6). In the above, we used the fact that the sum is not changed whether we add the numbers from 0 to 9 or backwards from 9 to 0, and that doubling the sum and dividing by 2 would not change the final answer. The above sum can thus be generalized to

$$\sum_{n=0}^{N-1} n = \frac{1}{2} \left[\sum_{n=0}^{N-1} n + \sum_{n=0}^{N-1} (N - 1 - n) \right] = \frac{1}{2} \sum_{n=0}^{N-1} (N - 1) = \frac{N \times (N - 1)}{2},$$

a result that Gauss figured out when he was a preschooler![3]

To improve the approximation of the integral we use $T_s = 10^{-3}$, which gives a discretized signal nT_s for $0 \le nT_s < 10$ or $0 \le n \le (10/T_s) - 1$. The area of each of the pulses is nT_s^2 and the approximation

---

[3]Carl Friedrich Gauss (1777–1855), German mathematician, considered one of the most accomplished mathematicians of all times [5]. He was seven years old when he amazed his teachers with his trick for adding the numbers from 1 to 100 [9].

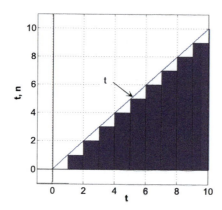

FIGURE 0.6

Approximation of area under $x(t) = t, t \geq 0, 0$ otherwise, by stacking pulses of width 1 and height nT_s, where $T_s = 1$ and $n = 0, 1, \cdots, 9$.

to the integral is then

$$\sum_{n=0}^{10^4-1} p[n] = \sum_{n=0}^{10^4-1} n10^{-6} = \frac{10^4 \times (10^4 - 1)}{10^6 \times 2} = 49.995.$$

A lot better result. In general we see that the integral can be computed quite accurately using a very small value of T_s, indeed

$$\sum_{n=0}^{(10/T_s)-1} p[n] = \sum_{n=0}^{(10/T_s)-1} nT_s^2 = T_s^2 \frac{(10/T_s) \times ((10/T_s) - 1)}{2} = \frac{10 \times (10 - T_s)}{2}$$

which for very small values of T_s (so that $10 - T_s \approx 10$) gives $100/2 = 50$, as desired.

Derivatives and integrals take us into the processing of signals by systems. Once a mathematical model for a dynamic system is obtained, typically ordinary differential equations characterize the relation between input and output variable or variables of the system. A significant subclass of systems (used as a valid approximation in some way to actual systems) is given by linear ordinary differential equations with constant coefficients. As we will see in Chapter 3, the solution of these equations can be efficiently found by means of the Laplace transform which converts them into algebraic equations that are much easier to solve.

0.4.4 DIFFERENTIAL AND DIFFERENCE EQUATIONS

An ordinary differential equation characterizes the dynamics of a continuous-time system, or the way the system responds to inputs over time. There are different types of ordinary differential equations, corresponding to different systems. Most systems are characterized by non-linear, time-dependent co-efficient ordinary differential equations. The analytic solution of these equations is rather complicated.

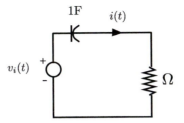

FIGURE 0.7

RC circuit.

To simplify the analysis, these equations are locally approximated as linear constant-coefficient ordinary differential equations.

Solution of ordinary differential equations can be obtained by means of analog and digital computers. An **analog computer** consists of operational amplifiers (op-amps), resistors, capacitors, voltage sources and relays. Using the linearized model of the op-amps, and resistors and capacitors it is possible to realize integrators to solve the ordinary differential equation. Relays are used to set the initial conditions in the capacitors, and the voltage sources give the input signal. Although this arrangement permits the solution of ordinary differential equations, its drawback is the storage of the solution which can be seen with an oscilloscope but is difficult to record. Hybrid computers were suggested as a solution—the analog computer is assisted by a digital component that stores the data. Both analog and hybrid computers have gone the way of the dinosaurs, and it is digital computers aided by numerical methods that are used now to solve ordinary differential equations.

Before going into the numerical solution provided by digital computers, let us consider why integrators are needed in the solution of ordinary differential equations. A first-order (the highest derivative present in the equation), linear (no non-linear functions of the input or the output are present) with constant coefficients ordinary differential equation obtained from a simple RC circuit (Fig. 0.7) with a constant voltage source $v_i(t)$ as input and with resistor $R = 1\ \Omega$, and capacitor $C = 1$ F (with huge plates!) connected in series is given by

$$v_i(t) = v_c(t) + \frac{dv_c(t)}{dt} \qquad t \geq 0 \qquad (0.9)$$

with an initial voltage $v_c(0)$ across the capacitor.

Intuitively, in this circuit the capacitor starts with an initial charge of $v_c(0)$, and it will continue charging until it reaches saturation at which time no more charge will flow (i.e., the current across the resistor and the capacitor is zero) so that the voltage across the capacitor is equal to the voltage source; the capacitor is acting as an open circuit given that the source is constant.

Suppose, ideally, that we have available devices that can perform differentiation, there is then the tendency to propose that the above ordinary differential equation be solved following the **block diagram** shown on the left in Fig. 0.8. Although nothing is wrong analytically, the problem with this approach is that in practice most signals are noisy (each device produces electronic noise) and the noise present in the signal may cause large derivative values given its rapidly changing amplitudes. Thus the realization of the ordinary differential equation using differentiators is prone to being very

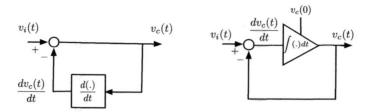

FIGURE 0.8

Realization of first-order ordinary differential equation using differentiators (left) and integrators (right).

noisy (i.e., not good). Instead, as proposed years ago by Lord Kelvin[4] instead of using differentiators we need to smooth out the process by using integrators, so that the voltage across the capacitor $v_c(t)$ is obtained by integrating both sides of Equation (0.9). Assuming that the source is switched on at time $t = 0$ and that the capacitor has an initial voltage $v_c(0)$, using the inverse relation between derivatives and integrals gives

$$v_c(t) = \int_0^t [v_i(\tau) - v_c(\tau)]d\tau + v_c(0) \qquad t \geq 0, \tag{0.10}$$

which is represented by the block diagram in the right of Fig. 0.8. Notice that the integrator also provides a way to include the initial condition, which in this case is the initial voltage across the capacitor, $v_c(0)$. Different from the accentuating effect of differentiators on noise, integrators average the noise thus reducing its effects.

Block diagrams like the ones shown in Fig. 0.8 allow us to visualize the system much better, and they are commonly used. Integrators can be efficiently implemented using operational amplifiers with resistors and capacitors.

How to Solve Ordinary Differential Equations

Let us then show how the above ordinary differential equation can be solved using integration and its approximation, resulting in a difference equation. Using Equation (0.10) at $t = t_1$ and $t = t_0$ for $t_1 > t_0$ we have the difference:

$$v_c(t_1) - v_c(t_0) = \int_{t_0}^{t_1} v_i(\tau)d\tau - \int_{t_0}^{t_1} v_c(\tau)d\tau.$$

If we let $t_1 - t_0 = \Delta t$ where $\Delta t \to 0$, i.e. is a very small time interval, the integrals can be seen as the area of small trapezoids of height Δt and bases $v_i(t_1)$ and $v_i(t_0)$ for the input source and $v_c(t_1)$ and $v_c(t_0)$ for the voltage across the capacitor (see Fig. 0.9). Using the formula for the area of a trapezoid

---

[4]William Thomson, Lord Kelvin, proposed in 1876 the **differential analyzer**, a type of analog computer capable of solving ordinary differential equations of order 2 and higher. His brother James designed one of the first differential analyzers [80].

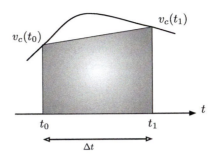

FIGURE 0.9

Approximation of area under the curve by a trapezoid.

we get an approximation for the above integrals so that

$$v_c(t_1) - v_c(t_0) = [v_i(t_1) + v_i(t_0)]\frac{\Delta t}{2} - [v_c(t_1) + v_c(t_0)]\frac{\Delta t}{2},$$

from which we obtain

$$v_c(t_1)\left[1 + \frac{\Delta t}{2}\right] = [v_i(t_1) + v_i(t_0)]\frac{\Delta t}{2} + v_c(t_0)\left[1 - \frac{\Delta t}{2}\right].$$

Assuming $\Delta t = T$, and letting $t_1 = nT$ and $t_0 = (n-1)T$ the above equation can be written as

$$v_c(nT) = \frac{T}{2+T}[v_i(nT) + v_i((n-1)T)] + \frac{2-T}{2+T}v_c((n-1)T) \qquad n \geq 1 \qquad (0.11)$$

and initial condition $v_c(0) = 0$. This is a first-order linear difference equation with constant coefficients approximating the ordinary differential equation characterizing the RC circuit. Letting the input be $v_i(t) = 1$ for $t > 0$, and zero otherwise, we have

$$v_c(nT) = \begin{cases} 0 & n = 0, \\ \frac{2T}{2+T} + \frac{2-T}{2+T}v_c((n-1)T) & n \geq 1. \end{cases} \qquad (0.12)$$

The advantage of the difference equation is that it can be solved for increasing values of n using previously computed values of $v_c(nT)$. Such a solution is called a **recursive solution**. For instance, letting $T = 10^{-3}$, $v_i(t) = 1$, for $t \geq 0$, zero otherwise, and defining $M = 2T/(2+T)$, $K = (2-T)/(2+T)$ we obtain

$$
\begin{aligned}
n &= 0 & v_c(0) &= 0 \\
n &= 1 & v_c(T) &= M \\
n &= 2 & v_c(2T) &= M + KM = M(1 + K) \\
n &= 3 & v_c(3T) &= M + K(M + KM) = M(1 + K + K^2) \\
n &= 4 & v_c(4T) &= M + KM(1 + K + K^2) = M(1 + K + K^2 + K^3)
\end{aligned}
$$

$$\vdots$$

The value $M = 2T/(2+T) \approx T = 10^{-3}$, $K = (2-T)/(2+T) < 1$ and $1 - K = M$. The response increases from the zero initial condition to a constant value (which is the effect of the dc source—the capacitor eventually becomes an open circuit, so that the voltage across the capacitor equals that of the input). Extrapolating from the above results it seems that in the steady state (i.e., when $nT \to \infty$) we have[5]

$$v_c(nT) = M \sum_{m=0}^{\infty} K^m = \frac{M}{1-K} = 1.$$

Even though this is a very simple example, it clearly illustrates that very good approximations to the solution of ordinary differential equations can be obtained using numerical methods which are appropriate for implementation in digital computers.

The above example shows how to solve an ordinary differential equation using integration and approximation of the integrals to obtain a difference equation that a computer can easily solve. The integral approximation used above is the **trapezoidal rule** method which is one among many numerical methods used to solve ordinary differential equations. Also we will see in Chapter 12 that the above results in the **bilinear transformation** which connects the Laplace s variable with the z variable of the Z-transform and that will be used in the design of discrete filters from analog filters.

0.5 COMPLEX OR REAL?

Most of the theory of signals and systems is based on functions of complex variables. Clearly signals are functions of real variables either time or space (if the signal is two-dimensional, like an image) or both (like video), thus why would one need complex numbers in processing signals? As we will see in Chapter 3, continuous-time signals can be equivalently characterized by means of frequency and damping. These two characteristics are given by the complex variable $s = \sigma + j\Omega$ (where σ is the damping factor and Ω is the frequency) in the representation of continuous-time signals using the Laplace transform. In the representation of discrete-time signals in Chapter 10, the Z-transform uses a complex variable $z = re^{j\omega}$ where r is the damping factor and ω is the discrete frequency.

The other reason for using complex variables is due to the response of linear systems to sinusoids. We will see that such result is fundamental in the analysis and synthesis of signals and systems. Thus a solid grasp of what is meant by complex variables and what a function of these is all about is needed. In this section, complex variables will be connected to vectors and phasors (commonly used in the sinusoidal steady state analysis of linear circuits).

---

[5]The infinite sum converges if $|K| < 1$, which is satisfied in this case. If we multiply the sum by $(1 - K)$ we get

$$(1-K) \sum_{m=0}^{\infty} K^m = \sum_{m=0}^{\infty} K^m - \sum_{m=0}^{\infty} K^{m+1} = 1 + \sum_{m=1}^{\infty} K^m - \sum_{\ell=1}^{\infty} K^{\ell} = 1$$

where we changed the variable in the second sum of the lower equation to $\ell = m + 1$. This explains why the sum equals $1/(1-K)$.

0.5.1 COMPLEX NUMBERS AND VECTORS

A complex number z represents any point (x, y) in a two-dimensional plane by $z = x + jy$ where $x = \mathcal{R}e[z]$ (real part of z) is the coordinate in the x-axis and $y = \mathcal{I}m[z]$ (imaginary part of z) is the coordinate in the y-axis. The symbol $j = \sqrt{-1}$ just indicates that z needs two components to represent a point in the two-dimensional plane. Interestingly, a vector $\vec{z}$ that emanates from the origin of the complex plane $(0, 0)$ to the point (x, y) with a length

$$|\vec{z}| = \sqrt{x^2 + y^2} = |z| \tag{0.13}$$

and an angle

$$\theta = \angle\vec{z} = \angle z \tag{0.14}$$

also represents the point (x, y) in the plane and has the same attributes as the complex number z. The couple (x, y) is therefore equally representable by the vector $\vec{z}$ or by a complex number z, which can be equivalently written as

$$
\begin{aligned}
z &= x + jy & \text{rectangular representation} \\
&= |z|e^{j\theta} & \text{polar representation}
\end{aligned} \tag{0.15}
$$

where the magnitude $|z|$ and the phase θ are defined in Equations (0.13) and (0.14).

Remarks

1. It is important to understand that a rectangular complex plane is identical to a polar complex plane despite the different representations (although equivalent) of each point in the plane.
2. When adding or subtracting complex numbers the rectangular form is the appropriate one, while when multiplying or dividing complex numbers the polar form is more advantageous. Thus, if complex numbers $z = x + jy = |z|e^{j\angle z}$ and $v = p + jq = |v|e^{j\angle v}$ are added analytically we obtain

$$z + v = (x + p) + j(y + q)$$

which is geometrically equivalent to the addition of vectors (see Fig. 0.10B). On the other hand, the multiplication of z and v is easily done using their polar forms as

$$zv = |z|e^{j\angle z}|v|e^{j\angle v} = |z||v|e^{j(\angle z + \angle v)}$$

but it requires more operations if done in the rectangular form, i.e.,

$$zv = (x + jy)(p + jq) = (xp - yq) + j(xq + yp).$$

Geometrically, the multiplication in the polar form is seen as changing the amplitude $|z|$ by a value $|v|$ (decreasing it when $|v| < 1$, increasing it when $|v| > 1$ and keeping it the same when $|v| = 1$) while adding $\angle v$, the angle of v, to the angle $\angle z$ of z (see Fig. 0.10C). No geometric interpretation is possible in general for the multiplication using a rectangular form (is there a case when it is possible?)

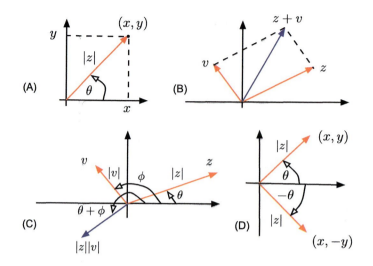

FIGURE 0.10

(A) Representation of a complex number z by a vector; (B) addition of complex numbers z and v; (C) multiplication of complex numbers $z = |z|e^{j\theta}$ and $v = |v|e^{j\phi}$; (D) complex conjugation of z.

3. One operation that is possible with complex numbers but not with real numbers is **complex conjugation**. Given a complex number $z = x + jy = |z|e^{j\angle z}$ its complex conjugate is $z^* = x - jy = |z|e^{-j\angle z}$, i.e., we negate the imaginary part of z or reflect its angle (see Fig. 0.10D). This operation gives for $z = x + jy = |z|e^{j\angle z}$

$$(i) \quad z + z^* = 2x \quad \text{or} \quad \mathcal{R}e[z] = 0.5[z + z^*]$$
$$(ii) \quad z - z^* = 2jy \quad \text{or} \quad \mathcal{I}m[z] = -0.5j[z - z^*]$$
$$(iii) \quad zz^* = |z|^2 \quad \text{or} \quad |z| = \sqrt{zz^*}$$
$$(iv) \quad \frac{z}{z^*} = e^{j2\angle z} \quad \text{or} \quad \angle z = -0.5j[\log(z) - \log(z^*)]$$
$$(v) \quad \frac{1}{z} = \frac{z^*}{|z|^2} = \frac{1}{|z|}e^{-j\angle z} \tag{0.16}$$

- Notice that $\log(.)$ used above is the natural or Naperian logarithm in base $e \approx 2.71828$; if the base is different from e it is indicated in the log—for instance, $\log_{10}$ is the logarithm in base 10.
- Equation (v) shows that complex conjugation provides a different approach to the inversion or division of complex numbers in rectangular form. This is done by making the denominator a positive real number by multiplying both numerator and denominator by the complex conjugate of the denominator. If $z = x + jy$ and $u = v + jw$, then

$$\frac{z}{u} = \frac{zu^*}{|u|^2} = \frac{(x + jy)(v - jw)}{v^2 + w^2} = \frac{|z|e^{j(\angle z - \angle u)}}{|u|} \tag{0.17}$$

The conversion of complex numbers from rectangular to polar needs to be done with care, especially when computing the angles. Consider the conversion of the following complex numbers from rectangular to polar form:

$$z = 3 + 4j, \quad u = -3 + j, \quad w = -4 - 3j, \quad v = 1 - j.$$

A good way to avoid errors in the conversion is to plot the complex number and then figure out how to obtain its magnitude and angle. Determining the quadrant of the complex plane in which the complex number is helps in finding the angle. For instance the complex number $z = 3 + 4j$ is in the first quadrant, and its magnitude and angle are

$$|z| = \sqrt{25} = 5, \quad \angle z = \tan^{-1}\left(\frac{4}{3}\right) = 0.927 \text{ (rad)}.$$

It is important to realize that the resulting angle is in radians, and that it should be multiplied by $180°/\pi$ if we want it to be in degrees so that equivalently $\angle z = 53.13°$. The magnitude of $u = -3 + j$, in the second quadrant, is $|u| = \sqrt{10}$ and its angle is

$$\angle u = \pi - \tan^{-1}\left(\frac{1}{3}\right) = 2.820 \text{ (rad)} = 161.56°.$$

An equivalent angle is obtained by subtracting this angle from 2π or $360°$ and negating it, or

$$\angle u = -\left[2\pi - \left(\pi - \tan^{-1}\left(\frac{1}{3}\right)\right)\right] = -\pi - \tan^{-1}\left(\frac{1}{3}\right) = -3.463 \text{ (rad)} = -198.43°.$$

The number $w = -4 - 3j$ is in the third quadrant, its magnitude is $|w| = 5$, while its angle is

$$\angle w = \pi + \tan^{-1}\left(\frac{3}{4}\right) = 3.785 \text{ (rad)} \quad \text{or} \quad \angle w = -\left[\pi - \tan^{-1}\left(\frac{3}{4}\right)\right] = -2.498 \text{ (rad)}$$

or equivalent angles measured in the positive and in the negative directions. For $v = 1 - j$ in the fourth quadrant, its magnitude is $v = \sqrt{2}$, its angle is

$$\angle v = -\tan^{-1}(1) = -\pi/4 \text{ (rad)} = -45°.$$

An equivalent angle is $7\pi/4$ or $315°$. Conventionally, angles are positive when measured in a counterclockwise direction with respect to the positive real axis, otherwise they are negative. The above results are shown in Fig. 0.11.

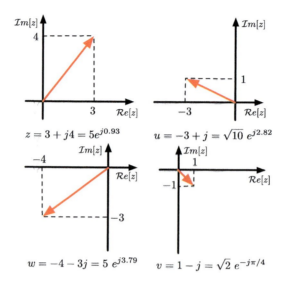

$$z = 3 + j4 = 5e^{j0.93}$$

$$u = -3 + j = \sqrt{10}\ e^{j2.82}$$

$$w = -4 - 3j = 5\ e^{j3.79}$$

$$v = 1 - j = \sqrt{2}\ e^{-j\pi/4}$$

FIGURE 0.11

Conversion of rectangular into polar representation for complex numbers.

Euler's Identity

One of the most famous equations of all times[6] is

$$1 + e^{j\pi} = 1 + e^{-j\pi} = 0,$$

due to one of the most prolific mathematicians of all times, Euler.[7] The above equation can easily be understood by establishing Euler's identity which connects the complex exponential and sinusoids:

$$e^{j\theta} = \cos(\theta) + j\sin(\theta). \tag{0.18}$$

One way to verify this identity is to consider the polar representation of the complex number $\cos(\theta) + j\sin(\theta)$, which has a unit magnitude since $\sqrt{\cos^2(\theta) + \sin^2(\theta)} = 1$ given the trigonometric identity $\cos^2(\theta) + \sin^2(\theta) = 1$. The angle of this complex number is

$$\psi = \tan^{-1}\left[\frac{\sin(\theta)}{\cos(\theta)}\right] = \theta.$$

---

[6]A reader poll done by *Mathematical Intelligencer* in 1990 named Euler's identity the most beautiful equation in mathematics. Another poll by *Physics World* in 2004 named Euler's identity the "greatest equation ever," together with Maxwell's equations. The book *Dr. Euler's Fabulous Formula* by Paul Nahin, is devoted to Euler's identity. The book states that the identity sets "the gold standard for mathematical beauty" [57].

[7]Leonard Euler (1707–1783) was a Swiss mathematician and physicist. Student of John Bernoulli and advisor of Joseph Lagrange. We owe Euler the notation $f(x)$ for functions, e for the base of natural logs, $i = \sqrt{-1}$, π for pi and Σ for sum, the finite-difference notation Δ and many more!

Thus the complex number can be written

$$\cos(\theta) + j\sin(\theta) = 1e^{j\theta},$$

which is Euler's identity. Now in the case when $\theta = \pm\pi$ the identity implies that $e^{\pm j\pi} = -1$, explaining the famous Euler's equation.

The relation between the complex exponentials and the sinusoidal functions is of great importance in signals and systems analysis. Using Euler's identity the cosine can be expressed as

$$\cos(\theta) = \mathcal{R}e[e^{j\theta}] = \frac{e^{j\theta} + e^{-j\theta}}{2}, \tag{0.19}$$

while the sine is given by

$$\sin(\theta) = \mathcal{I}m[e^{j\theta}] = \frac{e^{j\theta} - e^{-j\theta}}{2j}. \tag{0.20}$$

Indeed, we have

$$e^{j\theta} = \cos(\theta) + j\sin(\theta), \quad e^{-j\theta} = \cos(\theta) - j\sin(\theta)$$

and adding them we get the above expression for the cosine and subtracting the second from the first we get the given expression for the sine. The variable θ is in radians, or in the corresponding angle in degrees.

Applications of Euler's Identity

- **Polar to rectangular conversion.** Euler's identity is used to find the real and imaginary parts of a complex number in the polar form. If the vector corresponding to the complex number is in the first or the fourth quadrant of the complex plane, or equivalently the phase of the polar representation θ is $-\pi/2 \leq \theta \leq \pi/2$, Euler's identity gives the real and the imaginary parts directly. When the vector corresponding to the complex number is in the second or third quadrant, or the phase of the polar representation θ is $\pi/2 \leq \theta \leq 3\pi/2$, we express the phase of the complex number as $\theta = \pi \pm \phi$, where ϕ is the angle formed by the vector with respect to the negative axis, and then use $e^{j\pi} = -1$ to obtain

$$e^{j\theta} = e^{j(\pi \pm \phi)} = -e^{\pm j\phi}.$$

Euler's identity is then used to find the real and imaginary parts. To illustrate this conversion consider the following complex numbers in polar form:

$$z = \sqrt{2}e^{j\pi/4}, \quad u = \sqrt{2}e^{-j\pi/4}, \quad v = 3e^{-j190°}, \quad w = 5e^{j190°}.$$

The vectors corresponding to z and u are in the first and fourth quadrants and so direct application of Euler's identity is needed to obtain their rectangular forms:

$$
\begin{aligned}
z &= \sqrt{2}e^{j\pi/4} = \sqrt{2}\cos(\pi/4) + j\sqrt{2}\sin(\pi/4) = 1 + j, \\
u &= \sqrt{2}e^{-j\pi/4} = \sqrt{2}\cos(-\pi/4) + j\sqrt{2}\sin(-\pi/4) \\
&= \sqrt{2}\cos(\pi/4) - j\sqrt{2}\sin(\pi/4) = 1 - j.
\end{aligned}
$$

On the other hand, the vectors corresponding to v and w are in the second and third quadrant, and so we express their angles in terms of π (or 180°) and the angle formed by the vector with respect to the negative real axis:

$$
\begin{aligned}
v &= 3e^{-j190°} = 3e^{-j180°}e^{-j10°} \\
&= -3\cos(-10°) - j3\sin(-10°) = -3\cos(10°) + j3\sin(10°) = -2.95 + j0.52, \\
w &= 5e^{j190°} = 5e^{j180°}e^{j10°} \\
&= -5\cos(10°) - j5\sin(10°) = -4.92 - j0.87.
\end{aligned}
$$

- **Roots of special polynomials.** A problem of interest is finding the roots of complex polynomials of the form $F(z) = z^n + a$, for integer $n > 0$. Euler's identity can be used for that. For instance, given the polynomial $F(z) = z^3 + 1$ we want to find its three roots.[8] The roots are found as follows:

$$
z^3 + 1 = 0 \;\Rightarrow\; z_k^3 = -1 = e^{j(2k+1)\pi}, \; k = 0, 1, 2 \;\Rightarrow\; z_k = e^{j(2k+1)\pi/3}, \; k = 0, 1, 2,
$$
$$
z_0 = e^{j\pi/3}, \; z_1 = e^{j\pi} = -1, \; z_2 = e^{j(6-1)\pi/3} = e^{j2\pi}e^{-j\pi/3} = e^{-j\pi/3}.
$$

Notice that one of the roots is real (z_1) and the other two are complex conjugate pairs ($z_2 = z_0^*$). See left figure in Fig. 0.12. In general, the roots of $F(z) = z^n + a$, for integer $n > 0$ and a real are obtained according to

$$
z^n + a = 0 \;\Rightarrow\; z_k^n = -a = |a|e^{j((2k+1)\pi + \angle a)},
$$
$$
z_k = |a|^{(1/n)}e^{j((2k+1)\pi + \angle a)/n}, \; k = 0, 1, \cdots, n - 1.
$$

- **Powers of complex numbers.** Euler's identity is also useful in finding integer as well as non-integer powers of a complex number. For instance, consider finding the integer powers of $j = \sqrt{-1}$ for $n \geq 0$:

$$
j^n = (-1)^{n/2} = \begin{cases} (-1)^{(n/2)} & n \text{ even}, \\ j(-1)^{(n-1)/2} & n \text{ odd}. \end{cases}
$$

So that $j^0 = 1$, $j^1 = j$, $j^2 = -1$, $j^3 = -j$, and so on. Letting $j = 1e^{j\pi/2}$, we can see that the increasing powers of $j^n = 1e^{jn\pi/2}$ are vectors with angles of 0 when $n = 0$; $\pi/2$ when $n = 1$; π when $n = 2$; and $3\pi/2$ when $n = 3$. The angles repeat for the next 4 values, and the 4 after that, and so on. See right figure in Fig. 0.12.

---

[8] The fundamental theorem of algebra indicates that an n-order polynomial (n a positive integer) has n roots. These roots can be real, complex or a combination of real and complex. If the coefficients of the polynomial are real, then the complex roots must be complex conjugate pairs.

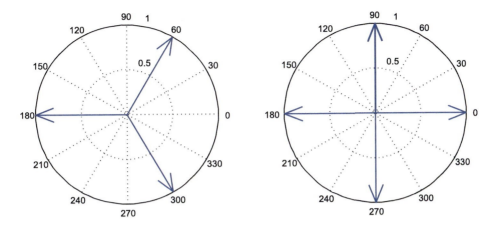

FIGURE 0.12

Left: roots of $z^3 + 1 = 0$; right: integer powers of j, periodic of period 4, with values for a period of 1, j, -1 and $-j$ as indicated by the arrows.

- **Trigonometric identities.** Trigonometric identities can be obtained using Euler's identity. For instance,

$$\sin(-\theta) = \frac{e^{-j\theta} - e^{j\theta}}{2j} = -\sin(\theta),$$

$$\cos(\pi + \theta) = e^{j\pi} \frac{e^{j\theta} + e^{-j\theta}}{2} = -\cos(\theta),$$

$$\cos^2(\theta) = \left[\frac{e^{j\theta} + e^{-j\theta}}{2}\right]^2 = \frac{1}{4}[2 + e^{j2\theta} + e^{-j2\theta}] = \frac{1}{2} + \frac{1}{2}\cos(2\theta),$$

$$\sin(\theta)\cos(\theta) = \frac{e^{j\theta} - e^{-j\theta}}{2j} \frac{e^{j\theta} + e^{-j\theta}}{2} = \frac{e^{j2\theta} - e^{-j2\theta}}{4j} = \frac{1}{2}\sin(2\theta).$$

0.5.2 FUNCTIONS OF A COMPLEX VARIABLE

Just like real-valued functions, functions of a complex variable can be defined. For instance, the natural logarithm of a complex number can be written as

$$v = \log(z) = \log(|z|e^{j\theta}) = \log(|z|) + j\theta$$

by using the inverse connection between the exponential and the logarithmic functions.

It is important to mention that complex variables as well as functions of complex variables are more general than real variables and real-valued functions. The above definition of the logarithmic function is valid when $z = x$, x a real value, and when $z = jy$ a purely imaginary value.

In particular, hyperbolic functions—which will be useful in the filter design in Chapters 7 and 12—can be obtained from the definition of the sinusoids with imaginary arguments. We have

$$\cos(j\alpha) = \frac{e^{-\alpha} + e^{\alpha}}{2} = \cosh(\alpha) \qquad \text{hyperbolic cosine,} \tag{0.21}$$

$$-j\sin(j\alpha) = \frac{e^{\alpha} - e^{-\alpha}}{2} = \sinh(\alpha) \qquad \text{hyperbolic sine,} \tag{0.22}$$

from which the other hyperbolic functions are defined. Notice that the hyperbolic functions are defined in terms of real, rather than complex, exponentials. As such, we obtain the following expression for the real-valued exponential by subtracting the expressions for the hyperbolic cosine and sine:

$$e^{-\alpha} = \cosh(\alpha) - \sinh(\alpha). \tag{0.23}$$

The theory of differentiation, integration and properties of functions of one or more complex variables is more general than the corresponding theory for real-valued functions and requires a more precise coverage than we can provide here. We thus encourage the reader to consult one or more books in complex analysis [52,1].

0.5.3 PHASORS AND SINUSOIDAL STEADY STATE

A sinusoid $x(t)$ is a periodic signal represented by

$$x(t) = A\cos(\Omega_0 t + \psi) \qquad -\infty < t < \infty \tag{0.24}$$

where A is the amplitude, $\Omega_0 = 2\pi f_0$ is the frequency in rad/s and ψ is the phase in radians. The signal $x(t)$ is defined for all values of t, and it repeats periodically with a period $T_0 = 1/f_0$ (s), where f_0 is the frequency in cycles/s or in Hertz (Hz) (in honor of H.R. Hertz[9]). Given that the units of Ω_0 is rad/s, then $\Omega_0 t$ has as units (rad/s) $\times$ (s) $=$ (rad), which coincides with the units of the phase ψ, and permits the computation of the cosine. If $\psi = 0$, then $x(t)$ is a cosine, and if $\psi = -\pi/2$ then $x(t)$ is a sine.

If one knows the frequency Ω_0 (rad/s) in Equation (0.24), the cosine is characterized by its amplitude and phase. This permits us to define **phasors**[10] as complex numbers characterized by the amplitude and the phase of a cosine of a certain frequency Ω_0. That is, for a voltage signal $v(t) = A\cos(\Omega_0 t + \psi)$ the corresponding **phasor** is

$$V = Ae^{j\psi} = A\cos(\psi) + jA\sin(\psi) = A\angle\psi \tag{0.25}$$

and such that

$$v(t) = \mathcal{R}e[Ve^{j\Omega_0 t}] = \mathcal{R}e[Ae^{j(\Omega_0 t + \psi)}] = A\cos(\Omega_0 t + \psi). \tag{0.26}$$

---

[9]Heinrich Rudolf Hertz was a German physicist known for being the first to demonstrate the existence of electromagnetic radiation in 1888.

[10]In 1883, Charles Proteus Steinmetz (1885–1923), German–American mathematician and engineer, introduced the concept of phasors for alternating current analysis. In 1902, Steinmetz became Professor of Electrophysics at Union College in Schenectady, New York.

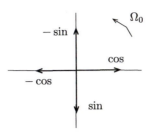

FIGURE 0.13

Generation of sinusoids from phasors of a frequency Ω_0 shown at their initial position at $t = 0$.

One can thus think of the voltage signal $v(t)$ as the projection of the phasor V onto the real axis as it turns counter-clockwise at Ω_0 radians per second. At time $t = 0$ the angle of the phasor is ψ. Clearly, the phasor definition is only valid for one frequency, in this case Ω_0, and it is connected to a cosine function.

Interestingly enough, the angle ψ can be used to differentiate cosines and sines. For instance when $\psi = 0$, the phasor V points right (at $t = 0$) and moves around at a rate of Ω_0 to generate as a projection on the real axis the signal $A \cos(\Omega_0 t)$. On the other hand, when $\psi = -\pi/2$ the phasor V points down (at $t = 0$) and moves around again at a rate of Ω_0 to generate a sinusoid $A \sin(\Omega_0 t) = A \cos(\Omega_0 t - \pi/2)$ as it is projected onto the real axis. This indicates that the sine lags the cosine by $\pi/2$ radians or $90°$, or equivalently that the cosine leads the sine by $\pi/2$ radians or $90°$. Thus the generation and relation of sines and cosines can easily be obtained using the plot in Fig. 0.13.

Computationally phasors can be treated as vectors rotating in the positive direction (counter-clockwise) at some frequency Ω_0 (rad/s). To illustrate the phasor computation, consider a current source

$$i(t) = A \cos(\Omega_0 t) + B \sin(\Omega_0 t)$$

to be expressed as

$$i(t) = C \cos(\Omega_0 t + \gamma)$$

where C and γ are to be determined using the sum of the two phasors corresponding to the sinusoidal components of $i(t)$. To accomplish that, the sinusoidal components of $i(t)$ must depend on a unique frequency, if that is not the case the concept of phasors would not apply as they would be rotating at different rates. To obtain the equivalent representation, we first obtain the phasor corresponding to $A \cos(\Omega_0 t)$, which is $I_1 = A e^{j0} = A$, and for $B \sin(\Omega_0 t)$ the corresponding phasor is $I_2 = B e^{-j\pi/2}$ so that

$$i(t) = \mathcal{R}e[(I_1 + I_2)e^{j\Omega_0 t}] = \mathcal{R}e[I e^{j\Omega_0 t}].$$

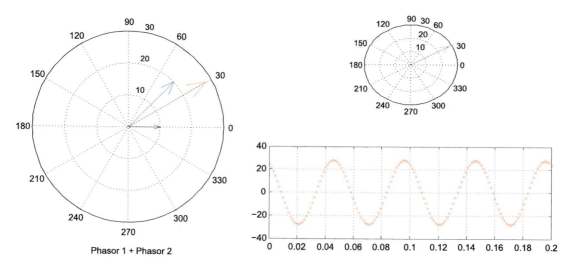

Phasor 1 + Phasor 2

FIGURE 0.14

Sum of phasors $I_1 = 10e^{j0}$ and $I_2 = 20e^{j\pi/4}$ with the result in the top-left and the corresponding sinusoid (bottom-right).

Thus the problem has been transformed into the addition of two vectors I_1 and I_2 which gives a vector

$$I = \sqrt{A^2 + B^2}\, e^{-j\, \tan^{-1}(B/A)}$$

and as such

$$i(t) = \mathcal{R}e[Ie^{j\Omega_0 t}] = \mathcal{R}e[\sqrt{A^2 + B^2}\, e^{-j\, \tan^{-1}(B/A)} e^{j\Omega_0 t}] = \sqrt{A^2 + B^2} \cos(\Omega_0 t - \tan^{-1}(B/A))$$

or an equivalent source with amplitude $C = \sqrt{A^2 + B^2}$ and phase $\gamma = -\tan^{-1}(B/A)$ and frequency Ω_0.

In Fig. 0.14 we display the result of adding two phasors I_1 and I_2, with frequency $f_0 = 20$ Hz, and the sinusoid that is generated by their sum $I = I_1 + I_2 = 27.98e^{j30.4°}$.

0.5.4 THE PHASOR CONNECTION

The fundamental property of a circuit made up of constant-valued resistors, capacitors and inductors is that its steady-state response to a sinusoid is also a sinusoid of the same frequency [23,21,14]. The effect of the circuit upon the input sinusoid is on its magnitude and phase and it depends on the frequency of the input sinusoid. This is due to the **linear** and **time-invariant** nature of the circuit. As we will see in Chapters 3, 4, 5, 10 and 11, this behavior can be generalized to more complex continuous-time as well as discrete-time systems.

To illustrate the connection of phasors with dynamic systems consider the RC circuit ($R = 1\ \Omega$ and $C = 1$ F) in Fig. 0.7. If the input to the circuit is a sinusoidal voltage source $v_i(t) = A\cos(\Omega_0 t)$ and the voltage across the capacitor $v_c(t)$ is the output of interest, the circuit can easily be represented by the first-order ordinary differential equation

$$v_i(t) = \frac{dv_c(t)}{dt} + v_c(t).$$

Assume that the steady-state response of this circuit (i.e., $v_c(t)$ as $t \to \infty$) is also a sinusoid

$$v_c(t) = C\cos(\Omega_0 t + \psi)$$

of the same frequency as the input, but with amplitude C and phase ψ to be determined. Since this response must satisfy the ordinary differential equation, we have

$$
\begin{aligned}
v_i(t) &= \frac{dv_c(t)}{dt} + v_c(t), \\
A\cos(\Omega_0 t) &= -C\Omega_0 \sin(\Omega_0 t + \psi) + C\cos(\Omega_0 t + \psi) \\
&= C\Omega_0 \cos(\Omega_0 t + \psi + \pi/2) + C\cos(\Omega_0 t + \psi) \\
&= C\sqrt{1 + \Omega_0^2}\cos(\Omega_0 t + \psi + \tan^{-1}\Omega_0).
\end{aligned}
$$

Comparing the two sides of the above equation gives

$$C = \frac{A}{\sqrt{1 + \Omega_0^2}}, \qquad \psi = -\tan^{-1}(\Omega_0),$$

and for a steady-state response

$$v_c(t) = \frac{A}{\sqrt{1 + \Omega_0^2}}\cos(\Omega_0 t - \tan^{-1}(\Omega_0)).$$

Comparing the steady-state response $v_c(t)$ with the input sinusoid $v_i(t)$, we see that they both have the same frequency Ω_0, but the amplitude and phase of the input are changed by the circuit depending on the frequency Ω_0. Since at each frequency the circuit responds differently, obtaining the **frequency response** of the circuit is useful not only in analysis but in design of circuits.

The above sinusoidal steady-state response can also be obtained using phasors. Expressing the steady-state response of the circuit as

$$v_c(t) = \mathcal{R}e\left[V_c e^{j\Omega_0 t}\right]$$

where $V_c = Ce^{j\psi}$ is the corresponding phasor for $v_c(t)$ we find that

$$\frac{dv_c(t)}{dt} = \frac{d\mathcal{R}e[V_c e^{j\Omega_0 t}]}{dt} = \mathcal{R}e\left[V_c \frac{de^{j\Omega_0 t}}{dt}\right] = \mathcal{R}e\left[j\Omega_0 V_c e^{j\Omega_0 t}\right].$$

Replacing $v_c(t)$, $dv_c(t)/dt$, obtained above, and

$$v_i(t) = \mathcal{R}e\left[V_i e^{j\Omega_0 t}\right] \quad \text{where } V_i = Ae^{j0}$$

in the ordinary differential equation we obtain

$$\mathcal{R}e\left[V_c(1+j\Omega_0)e^{j\Omega_0 t}\right] = \mathcal{R}e\left[Ae^{j\Omega_0 t}\right],$$

so that

$$V_c = \frac{A}{1+j\Omega_0} = \frac{A}{\sqrt{1+\Omega_0^2}}e^{-j\tan^{-1}(\Omega_0)} = Ce^{j\psi}$$

and the sinusoidal steady-state response is

$$v_c(t) = \mathcal{R}e\left[V_c e^{j\Omega_0 t}\right] = \frac{A}{\sqrt{1+\Omega_0^2}}\cos(\Omega_0 t - \tan^{-1}(\Omega_0)),$$

which coincides with the response obtained above. The ratio of the output phasor V_c to the input phasor V_i

$$\frac{V_c}{V_i} = \frac{1}{1+j\Omega_0}$$

gives the response of the circuit at frequency Ω_0. If the frequency of the input is a generic Ω, changing Ω_0 above for Ω gives the frequency response for all possible frequencies.

The concepts of **linearity** and **time-invariance** will be used in both continuous as well as discrete-time systems, along with the Fourier representation of signals in terms of sinusoids or complex exponentials, to simplify the analysis and to allow the design of systems. Thus, transform methods such as Laplace and the Z-transform will be used to solve differential and difference equations in an algebraic setup. Fourier representations will provide the frequency perspective. This is a general approach for both continuous and discrete-time signals and systems. The introduction of the concept of transfer function will provide tools for the analysis as well as the design of linear time-invariant systems. The design of analog and discrete filters is the most important application of these concepts. We will look into this topic in Chapters 5, 7 and 12.

0.6 SOFT INTRODUCTION TO MATLAB

MATLAB, which stands for MATrix LABoratory, is a computing language based on vectorial computations.[11] MATLAB provides a convenient environment for numeric and symbolic computations in signal processing, control, communications and many other related areas.

---

[11] The MatWorks, the developer of MATLAB, was founded in 1984 by Jack Little, Steve Bangert and Cleve Moler. Moler, a math professor at the University of New Mexico, developed the first version of MATLAB in Fortran in the late 1970s. It only had 80 functions, no M-files or toolboxes. Little and Bangert reprogrammed it in C and added M-files, toolboxes and more powerful graphics [56].

The following instructions are intended for users who have no background in MATLAB, but who are interested in using it in signal processing. Once you get the basic information on how to use the language you will be able to progress on your own.

1. There are two types of programs in MATLAB: one is the **script** which consists of a list of commands using MATLAB functions or your own functions, and the other is the **function** which is a program that can be called from scripts; it has certain inputs and provides corresponding outputs. We will show examples of both.
2. Create a directory where you will put your work, and from where you will start MATLAB. This is important because when executing a script, MATLAB will search for it in the current directory and if it is not present in there MATLAB gives an error indicating that it cannot find the desired script.
3. Once you start MATLAB, you will see three windows: the **command window**, where you will type commands, the **command history** which keeps a list of commands that have been used, and the **workspace**, where the variables used are kept.
4. Your first command on the command window should be to change to the directory where you keep your work. You can do this by using the command *cd* (change directory) followed by the name of the desired directory. It is also important to use the command **clear all** and **clf** to clear all previous variables in memory and all figures. If at any time you want to find in which directory you are, use the command *pwd* which will print the working directory which you are in.
5. Help is available in several forms in MATLAB. Just type *helpwin, helpdesk* or *demo* to get started. If you know the name of the function, *help* followed by the name of the function will give you the necessary information on that particular function, and it will also indicate a list of related functions. Typing 'help' on the command window will give all the HELP topics. Use *help* to find more about the functions listed in Table 0.1 for numeric MATLAB and in Table 0.2 for symbolic MATLAB.
6. To type your scripts or functions you can use the editor provided by MATLAB, simply type *edit* and the name of your file. You can also use any text editor to create scripts or functions, which need to be saved with the .m extension.

0.6.1 NUMERICAL COMPUTATIONS

The following examples of scripts and functions are intended to illustrate different features of numerical computations—numerical data in and numerical data out—using MATLAB. Please consult Table 0.1 for a basic set of variables and functions, and use *help* to provide details. Examples and more explanation on how to use MATLAB are provided by excellent reference books, such as [4].

MATLAB as a Calculator

In many respects MATLAB can be considered a sophisticated calculator. To illustrate this consider the following script where computations with complex numbers, conversion of complex numbers from one form into another, and plotting of complex numbers are done. It is started with the commands *clear all* and *clf* in order to clear all the existing variables, and all the figures.

Table 0.1 Basic numeric MATLAB functions

Special variables	ans	default name for result
	pi	π value
	inf, NaN	infinity, not-a-number error, e.g., 0/0
	i, j	$i = j = \sqrt{-1}$
	FUNCTION(S)	OPERATION
Mathematical	abs, angle	magnitude, angle of complex number
	acos, asine, atan	inverse cosine, sine, tangent
	acosh, asinh, atanh	inverse cosh, sinh, tanh
	cos, sin, tan	cosine, sine, tangent
	cosh, sinh, tanh	hyperbolic cosine, sine, tangent
	conj, imag, real	complex conjugate, imaginary, real parts
	exp, log, log10	exponential, natural and base 10 logarithms
Special operations	ceil, floor	round up, round down to integer
	fix, round	round toward zero, to nearest integer
	.*, ./	entry-by-entry multiplication, division
	.^	entry-by-entry power
	x', A'	transpose of vector x, matrix A
Array operations	x=*first:increment:last*	row vector x from *first* to *last* by *increment*
	x=linspace(*first,last,n*)	row vector x with *n* elements from *first* to *last*
	A=[x1;x2]	A matrix with rows x1, x2
	ones(N,M), zeros(N,M)	$N \times M$ ones and zeros arrays
	A(i,j)	(i,j) entry of matrix A
	A(i,:), A(:,j)	i-row (j-column) and all columns (rows) of A
	whos	display variables in workspace
	size(A)	(number rows, number of colums) of A
	length(x)	number rows (colums) of vector x
Control flow	for, elseif	for loop, else-if-loop
	while	while loop
	pause, pause(n)	pause and pause n seconds
Plotting	plot, stem	continuous, discrete plots
	figure	figure for plotting
	subplot	subplots
	hold on, hold off	hold plot on or off
	axis, grid	axis, grid of plots
	xlabel, ylabel, title, legend	labeling of axes, plots, and subplots
Saving and loading	save, load	saving and loading data
Information and managing	help	help
	clear, clf	clear variables from memory, clear figures
Operating system	cd, pwd	change directory, present working directory

```
%%
% Example 0.1---Computation with complex numbers
%%
clear all; clf
z=8+j*3
v=9-j*2
a=real(z)+imag(v)       % real and imaginary
b=abs(z+conj(v))        % absolute value, conjugate
c=abs(z*v)
d=angle(z)+angle(v)     % angle
d1=d*180/pi             % conversion to degrees
e=abs(v/z)
f=log(j*imag(z+v))      % natural log
```

Results

```
z = 8.0000 + 3.0000i
v = 9.0000 - 2.0000i
a = 6
b =  17.7200
c = 78.7718
d = 0.1401
d1 = 8.0272
e = 1.0791
f = 0 + 1.5708i
```

Notice that each of the commands are not followed by a semicolon (;) and as such when we execute this script MATLAB provides the corresponding answer—if we want to avoid this the semicolon suppresses the answer from MATLAB. Also, each of the statements are followed by a comment initiated after the % symbol—text in these comments are ignored by MATLAB. This simple script illustrates the large number of functions available to perform different calculations, some of them having names similar to the actual operations: *real* and *imag* find the real and imaginary parts while *abs* and *angle* compute the magnitude and the phase of a complex number. As expected, the phase is given in radians which can be transformed into degrees. The values of $j = \sqrt{-1}$ (*i* in MATLAB) and π do not need to be predefined.

The following script shows the conversion from one form to another of a complex number. The function *cart2pol* is used to transform a complex number in rectangular (or cartesian) form into polar form, and as expected the function *pol2cart* does the opposite. The help for *cart2pol* indicates that the real part X and the imaginary part Y of the complex number $X + jY$ are converted into an angle TH and a radius R to obtain the corresponding polar form, i.e.

$$[TH,R] = \text{cart2pol}(X,Y)$$

The function *pol2cart* has the reverse input/output of *cart2pol*. Using the vectors defined in the previous script we have the following computations.

```
%%
% Example 0.2---Rectangular--polar conversion
%%
m=z+v
[theta,r]=cart2pol(real(m),imag(m));    % rectangular to polar
disp(' magnitude of m');   r           % display text in '  ' and r
disp(' phase of m');   theta
[x,y]=pol2cart(theta,r)                 % polar to rectangular
```

which gives as results

```
        m = 17.0000 + 1.0000i
        magnitude of m
        r = 17.0294
        phase of m
        theta = 0.0588
        x = 17
        y = 1.0000
```

To see the functions *pol2cart* type in the command window

type pol2cart

In it you will see the general structure of a function:

> function [output(s)] = function_name(input(s))
> % function description
> % comments
> % author, revision
> commands

The information given in the function description is the information displayed when you type *help function_name*. The input(s) and output(s) can be one or more or none depending on what the function does. Once the function is created and saved (the name of the function followed by the extension .m), MATLAB will include it as a possible function than can be executed within a script. The function by itself cannot be executed.

   The following script illustrates the graphing capabilities of MATLAB. It will show graphically the generation of a sinusoid from a phasor. It is interactive and as such it asks for the frequency, the amplitude, and the phase of the sinusoid to be generated

```
%%
% Example 0.3---Interactive graphing example
%%
clear all;  clf
f=input(' frequency in Hz  >>  ')
A=input(' amplitude (>0) >>   ')
theta=input(' phase in degrees >>  ')
omega=2*pi*f;                    % frequency rad/s
```

```
tmax=1/f;                          % one period
time=[ ];  n=0;                    % initialization
figure(1)
for t=0:tmax/36:tmax              % loop
   z=A*exp(j*(omega*t+theta*pi/180));
   x=real(z); y=imag(z);          %  projection
   time=[time t];                 %  sequence
   subplot(121)                   % subplot 1 of 2
   compass(x,y);                  % plot vector
   axis('square')                 % square axis
   subplot(122)
   plot(n*tmax/36,x,'*r')         % plot x point in red '*'
   axis('square');
   axis([0 tmax -1.1*A  1.1*A]); grid    % bounds; grid
   hold on                        % hold current plot
   if n==0                        %  execute next statement when n=0
   pause(1)                       % pause
   else                           % if n=0 not satisfied go next statement
   pause(0.1)
   end                            % end of conditional
   n=n+1;                         % increment
end                               % end loop
hold off                          % dismiss hold on
```

This script displays the initial location of the phasor at $t = 0$ using *compass* for 1 s, and then computes the real and imaginary part of the phasor multiplied by $e^{j\Omega_0 t}$, and use them to plot the phasor with *compass* at consecutive times. The projection on the real axis (given by the real values x as functions of time) gives one period of the desired sinusoid. The results are plotted in a figure numbered 1 with two subplots organized as a row; *subplot(121)* corresponds to the first subplot (last number) of one row (first number) and two columns (middle number). To plot the resulting sinusoid point by point we use *plot* with each point represented by a red asterisk. We use *axis* to make the plots square, and define the range of the plot values using again *axis*. The functions *pause, hold on and hold off* are used to slow the graphical presentation and to keep and finally dismiss the plots being generated at each step of the for loop.

The initial location of the phasor is highlighted by pausing the plot for 3 s, and then the pausing is reduced to 0.1 s. The function *input* is used to enter the parameters of the desired sinusoid. The graphical display is similar to the one shown in the right of Fig. 0.14. In this example MATLAB works as an interactive graphical calculator (run this script to see what we mean).

MATLAB as a Signal Generator

Numerically the generation of a signal in MATLAB consists in creating either a vector, if the signal is one-dimensional, or a matrix if the signal is two-dimensional, or a sequence of matrices if three dimensions are needed. These vectors and matrices would correspond to sampled signals depending on time, to signals that depend on two space variables—like images—or to a video signal which is represented by a sequence of matrices that change with time. It is possible, however, that these vectors and matrices do not have any connection with a continuous signal as illustrated by the weekly closing of the ACM shown in Fig. 0.5.

MATLAB provides data files for experimentation that you can load and work with using the function *load*. For instance, the file *train.mat* is the recording of a train whistle, sampled at the rate of F_s samples/s (which accompanies the sampled signal $y(n)$). To be able to work with this file you can just type the following commands to load the data file (with extension *.mat*) into the work space:

```
%%
% Example 0.4a---Loading of test signal train
%%
    clear all
    load train
    whos
```

Information as regards the file is given by the function *whos* and shown as

Name	Size	Bytes	Class
Fs	1x1	8	double array
y	12880x1	103040	double array

The sampling frequency is $F_s = 8192$ samples/s, and the sampled train sequence is a column vector with 12,880 samples. One could then listen to this signal using the function *sound* with two input parameters, the sampled signal y and F_s, and then plot it:

```
%%
% Example 0.4b---Listening to/plotting train signal
%%
    sound(y,Fs)
    t=0:1/Fs:(length(y)-1)/Fs;
    figure(2); plot(t,y'); grid
    ylabel('y[n]'); xlabel('n')
```

To plot the signal, MATLAB provides the functions *plot* and *stem*. The function *plot* gives an interpolated version of the sequence, making the plot look continuous (see left figure in Fig. 0.15). The function *stem* plots the sequence of samples (see right Fig. 0.15). To plot 200 samples of the train signal with *stem* we use the following script. Notice the way MATLAB gets samples 100 to 299 from the signal.

```
%%
% Example 0.4c---Using stem to plot 200 samples of train
%%
figure(3)
n=100:299;
stem(n,y(100:299)); xlabel('n');ylabel('y[n]')
title('Segment of train signal')
axis([100 299 -0.5 0.5])
```

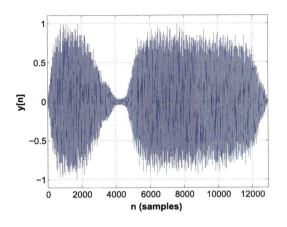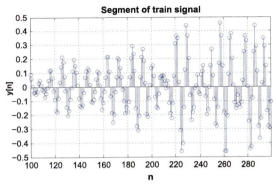

FIGURE 0.15

Left: whole train signal plotted using *plot*; right: segment of train signal plotted using *stem*.

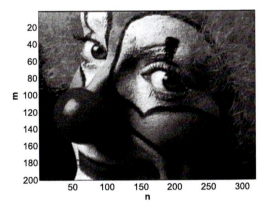

FIGURE 0.16

Clown in gray levels.

MATLAB also provides test images, such as *clown.mat* (with 200×320 pixels). It can be loaded using the script

```
clear all
load clown
```

To plot this image in gray levels (see Fig. 0.16) we use the following script:

```
colormap('gray')
imagesc(X)
```

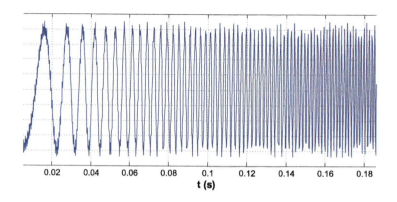

FIGURE 0.17

Generated signal composed of sinusoid added to chirp.

It is possible to generate your own signals by using many of the functions available in MATLAB. The following script illustrates how to generate a wave audio file and then reading it back and listening to it.

```
%%
% Example 0.5---Creating a WAV file from scratch and reading it back
%%
clear all
Fs=5000;                                    % sampling rate
t=0:1/Fs: 5;                                % time parameter
y=0.1*cos(2*pi*2000*t)-0.8*cos(2*pi*2000*t.^2); % sinusoid and chirp
%% writing chirp.wav file
audiowrite('chirp.wav',y, Fs)
%% reading chirp.wav back into MATLAB as y1 and listening to it
[y1, Fs] = audioread('chirp.wav');
sound(y1, Fs)                               % sound generated by y1
figure(4)
plot(t(1:1000), y1(1:1000))
```

The signal we create with this script is a sinusoid added to a chirp (a sinusoid with varying frequency). The signal has a duration of 5 s, and it is sampled at a rate of 5000 samples per second, so the signal has a total of 25,001 samples. The function *wavwrite* converts the signal generated by MATLAB into a wav audio file called 'chirp.wav'. This file is then read back by *wavread* and the function *sound* plays the sound file. Fig. 0.17 shows a segment of the generated sound file.

Saving and Loading Data

In many situations you would like to either save some data or load some data. The following illustrates a way to do it. Suppose you want to build and save a table of sine values for angles between zero and 360 degrees in intervals of 3 degrees. The following script generates the table using our function *singable* shown below. To save the generated values in a file sine.mat we use the function **save** (use **help save** to learn more about it).

```
%%
% Example 0.6a---Computing and saving angles and corresponding sine values
%%
theta=0:3:360; % angles from 0 to 360 degrees in intervals of 3 degrees
thetay=sintable(theta);
save sine.mat thetay

function xy= sintable(theta)
% This function generates sine values of input theta
% theta: angle in degrees, converted into radians
% xy: array with values for theta and y
y=sin(theta*pi/180);       % sine computes the argument in radians
xy=[theta' y'];            % array with 2 columns: theta' and y'
```

To load the table, we use the function **load** with the name given to the saved table *sine*, the extension **.mat* is not needed. The following script illustrates this:

```
%%
% Example 0.6b---Loading and checking size of sine.mat
%%
    clear all
    load sine
    whos
```

where we use **whos** to check the size of *thetay*:

```
    Name      Size              Bytes   Class
    thetay    121x2             1936    double array
```

This indicates that the array thetay has 121 rows and 2 columns, the first column corresponding to theta, the degree values, and the second column the sine values y. Verify this and plot the values by using

```
%%
% Example 0.6c---Plotting angles and corresponding sine values
%%
    x=thetay(:,1);
    y=thetay(:,2);
    figure(5); stem(x,y)
```

0.6.2 SYMBOLIC COMPUTATIONS

We have considered the numerical capabilities of MATLAB by which numerical data is transformed into numerical data. There will be many situations when we would like to do algebraic or calculus

Table 0.2 Basic symbolic MATLAB functions

	FUNCTION	OPERATION
Calculus	diff$^{(*)}$	differentiate
	int$^{(*)}$	integrate
	limit	limit
	taylor	Taylor series
	symsum	summation
Simplification	simplify	simplify
	expand	expand
	factor$^{(*)}$	factor
	simple	find shortest form
	subs	symbolic substitution
Solving equations	solve	solve algebraic equations
	dsolve	solve differential equations
Transforms	fourier	Fourier transform
	ifourier	inverse Fourier transform
	laplace	Laplace transform
	ilaplace	inverse Laplace transform
	ztrans	Z-transform
	iztrans	inverse Z-transform
Symbolic operations	sym	create symbolic objects
	syms	create symbolic objects
	pretty	make pretty expression
Special functions	dirac	Dirac or delta function
	heaviside	unit-step function
Plotting	fplot, ezplot	function plotter
	ezpolar	polar coordinate plotter
	ezcontour	contour plotter
	ezsurf	surface plotter
	ezmesh	mesh (surface) plotter

$^{(*)}$ *To get help on these symbolic function use 'help sym/xxx' where xxx is the name of either of these functions.*

operations resulting in terms of variables rather than numerical data. For instance, we might want to find a formula to solve quadratic algebraic equations, or find an integral, or obtain the Laplace or the Fourier transform of a signal. For those cases MATLAB provides the **Symbolic Math Toolbox**. In this section we give an introduction to symbolic computations by means of examples, and hope to get you interested into learning more on your own. The basic symbolic functions are given by Table 0.2.

Derivatives and Differences

The following script compares symbolic with numeric computations of the derivative of a chirp signal (a sinusoid with changing frequency) $y(t) = \cos(t^2)$:

$$z(t) = \frac{dy(t)}{dt} = -2t \sin(t^2).$$

```
%%
% Example 0.7---Derivatives and differences
% %
clf; clear all
% SYMBOLIC
    syms t y z            % define the symbolic variables
    y=cos(t^2)            % chirp signal -- notice no . before ^ since
                              % t is no vector
    z=diff(y)             % derivative
    figure(1)
    subplot(211)
    fplot(y,[0,2*pi]);grid    % plotting for symbolic y between 0 and 2*pi
    hold on
    subplot(212)
    fplot(z,[0,2*pi]);grid
    hold on
% NUMERIC
    Ts=0.1;                   % sampling period
    t1=0:Ts:2*pi;             % sampled time
    y1=cos(t1.^2);            % sampled signal -- notice difference with y
    z1=diff(y1)./diff(t1);    % difference  -- approximation to derivative
    figure(1)
    subplot(211)
    stem(t1,y1,'k');axis([0 2*pi 1.1*min(y1) 1.1*max(y1)])
    subplot(212)
    stem(t1(1:length(y1)-1),z1,'k');axis([0 2*pi 1.1*min(z1) 1.1*max(z1)])
    legend('Derivative (black)','Difference (blue)')
    hold off
```

The symbolic function **syms** defines the symbolic variables (use *help syms* to learn more). The signal $y(t)$ is written differently than $y_1(t)$ in the numeric computation. Since t_1 is a vector, squaring it requires a dot before the symbol, that is not the case for t which is not a vector but a variable. The results of using **diff** to compute the derivative of $y(t)$ is given in the same form as you would have obtained doing the derivative by hand, i.e.,

```
y =cos(t^2)
z =-2*t*sin(t^2)
```

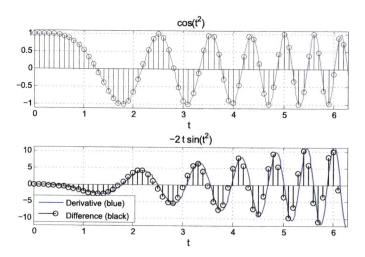

FIGURE 0.18

Symbolic and numeric computation of the derivative of the chirp $y(t) = \cos(t^2)$. Top plot is $y(t)$ and the sampled signal $y(nT_s)$, $T_s = 0.1$ s. The bottom plot displays the exact derivative (continuous line) and the approximation of the derivative at samples nT_s (samples). Better approximation to the derivative can be obtained by using a smaller value of T_s.

The symbolic toolbox provides its own graphic routines (use **help** to learn about the different **ez**-routines). For plotting $y(t)$ and $z(t)$, we use the function **fplot** (**ezplot** could also be used) which plots the above two functions for $t \in [0, 2\pi]$ and titles the plots with these functions.

The numeric computations differ from the symbolic in that vectors are being processed, and we are obtaining an approximation to the derivative $z(t)$. We sample the signal with $T_s = 0.1$ and use again the function **diff** to approximate the derivative (the denominator *diff(t1)* is the same as T_s). Plotting the exact derivative (continuous line) with the approximated one (samples) using **stem** clarifies that the numeric computation is an approximation at nT_s values of time. See Fig. 0.18, where we have superposed the numeric (samples) and the symbolic (continuous) results.

The Sinc Function and Integration

The sinc function is very significant in the theory of signals and systems, it is defined as

$$y(t) = \frac{\sin \pi t}{\pi t} \qquad -\infty < t < \infty.$$

It is symmetric with respect to the origin. The value of $y(0)$ (which is zero divided by zero) can be found using L'Hopital's rule to be unity. We will see later (Parseval's result in Chapter 5) that the integral of $y^2(t)$ is equal to 1. In the following script we are combining numeric and symbolic computations to show this.

In the symbolic part, after defining the variables we use the symbolic function **int** to compute the integral of the squared sinc function, with respect to t, from 0 to integer values $1 \le k \le 10$. We then

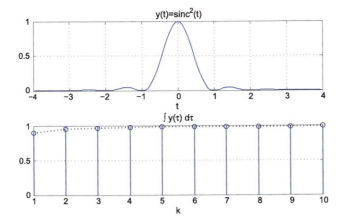

FIGURE 0.19

Computation of the integral of the squared sinc function shown in top plot. Bottom plot illustrates that the area under the curve of this function, or its integral, is unity. Using the symmetry of the function only the integral for $t \geq 0$ needs to be computed.

use the function **subs** to convert the symbolic results into a numerical array zz. The numeric part of the script defines a vector y to have the values of the sinc function for 100 time values equally spaced between $[-4, 4]$ obtained using the function **linspace**. We then use **plot** and **stem** to plot the sinc and the values of the integrals, which as seen in Fig. 0.19 reach a value close to unity in less than 10 steps.

```
%%
% Example 0.8---Integration of sinc squared
%%
clf; clear all
% SYMBOLIC
   syms t z
   for k=1:10,
       z=int(sinc(t)^2,t,0,k);     % integral of sinc^2 from 0 to k
       zz(k)=subs(2*z);            % substitution to numeric value zz
   end
% NUMERIC
   t1=linspace(-4, 4);     % 100 equally spaced points in [-4,4]
   y=sinc(t1).^2;          % numeric definition of the squared sinc function

   n=1:10;
   figure(1)
   subplot(211)
   plot(t1,y);grid; title('y(t)=sinc^2(t)');
   xlabel('t')
   subplot(212)
```

```
stem(n(1:10),zz(1:10)); hold on
plot(n(1:10),zz(1:10),'r');grid;title('\int y(\tau) d\tau'); hold off
xlabel('k')
```

Fig. 0.19 shows the squared sinc function and the values of the integral

$$2 \int_0^k \text{sinc}^2(t)dt = 2 \int_0^k \left[\frac{\sin(\pi t)}{\pi t} \right]^2 dt \qquad k = 1, \cdots, 10,$$

which quickly reaches the final value of unity. In computing the integral from $(-\infty, \infty)$ we are using the symmetry of the function and thus multiplication by 2.

Chebyshev Polynomials and Lissajous Figures

In this section we give two more illustrations of the use of symbolic MATLAB, while introducing results that will be useful later. The Chebyshev polynomials are used for the design of filters. They can be obtained by plotting two cosines functions as they change with time t, one of fix frequency and the other with increasing frequency:

$$
\begin{aligned}
x(t) &= \cos(2\pi t), \\
y(t) &= \cos(2\pi k t) \qquad k = 1, \cdots, N.
\end{aligned}
$$

The $x(t)$ gives the x-axis coordinate and $y(t)$ the y-axis coordinate at each value of t. If we solve for t in the upper equation we get

$$t = \frac{1}{2\pi} \cos^{-1}(x(t)),$$

which then replaced in the second equation gives

$$y(t) = \cos\left[k \cos^{-1}(x(t)) \right] \qquad k = 1, \cdots, N$$

as an expression for the Chebyshev polynomials (we will see in the Chapter 7 that these equations can be expressed as regular polynomials). Fig. 0.20 shows the Chebyshev polynomials for $N = 4$. The following script is used to compute and plot these polynomials.

```
%%
% Example 0.9---Chebyshev polynomials
%%
  clear all;clf
  syms x y t
  x=cos(2*pi*t);theta=0;
  figure(1)
  for k=1:4,
     y=cos(2*pi*k*t+theta);
    if k==1, subplot(221)
      elseif k==2, subplot(222)
```

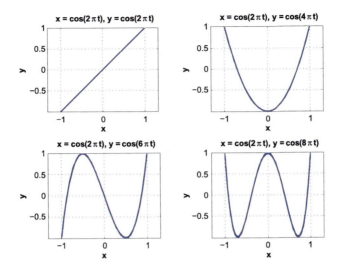

FIGURE 0.20

The Chebyshev polynomials for $n = 1, 2, 3, 4$. First (top-left) to fourth (bottom-right) polynomials.

```
        elseif k==3, subplot(223)
        else subplot(224)
      end
        fplot(x,y);grid;hold on
    end
    hold off
```

The Lissajous figures we consider next are a very useful extension of the above plotting of sinusoids in the x-axis and the y-axis. These figures are used to determine the difference between a sinusoidal input and its corresponding sinusoidal steady state. In the case of linear systems, which we will formally define in Chapter 2, for a sinusoidal input the output of the systems is also a sinusoid of the same frequency, but it differs with the input in the amplitude and phase. To find the ratio of the amplitudes of the input and the output sinusoids, or to determine the phase different between the input and the output sinusoids the Lissajous figures are used.

The differences in amplitude and phase can be measured using an oscilloscope for which we put the input in the horizontal sweep and the output in the vertical sweep, giving figures from which we can find the differences in amplitude and phase. Two situations are simulated in the following script, one where there is no change in amplitude but the phase changes from zero to $3\pi/4$, while in the other case the amplitude decreases as indicated and the phase changes in the same way as before. The plots, or Lissajous figures, indicate such changes. The difference between the maximum and the minimum of each of the figures in the x-axis gives the amplitude of the input, while the difference between the maximum and the minimum in the y-axis gives the amplitude of the output. The orientation of the

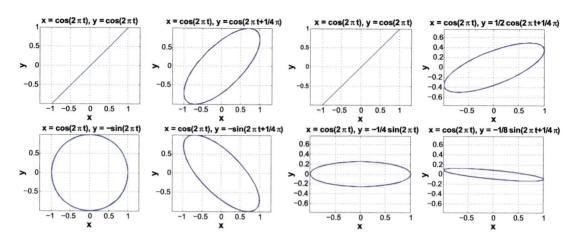

FIGURE 0.21

Lissajous figures: case 1 (left) input and output of same amplitude ($A = 1$) but phase differences of 0, $\pi/4$, $\pi/2$, and $3\pi/4$; case 2 (right) input has unit amplitude but output has decreasing amplitudes and same phase differences as in case 1.

ellipse provides the difference in phase with respect to that of the input. The following script is used to obtain the Lissajous figures in these cases. Fig. 0.21 displays the results.

```
%%
% Example 0.10---Lissajous figures
%%
clear all;clf
syms x y t
x=cos(2*pi*t);  % input of unit amplitude and frequency 2*pi
A=1;figure(1)   % amplitude of output in case 1
for i=1:2,
for k=0:3,
    theta=k*pi/4;      % phase of output
    y=A^k*cos(2*pi*t+theta);
  if k==0,subplot(221)
     elseif k==1,subplot(222)
     elseif k==2,subplot(223)
     else subplot(224)
  end
  fplot(x,y);grid;hold on
end
A=0.5; figure(2) % amplitude of output in case 2
end
```

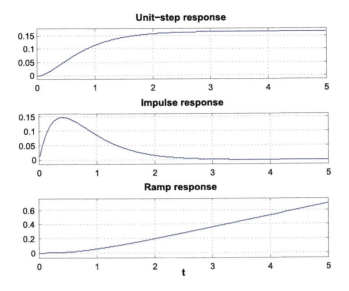

FIGURE 0.22

Response of a second order system represented by an ordinary differential equation for input the unit-step signal, its derivative or the impulse signal and the ramp signal which is the integral of the unit-step input.

Ramp, Unit-Step and Impulse Responses

To close this soft introduction to symbolic computations we illustrate the response of a linear system represented by an ordinary differential equation

$$\frac{d^2 y(t)}{dt^2} + 5\frac{dy(t)}{dt} + 6y(t) = x(t)$$

where $y(t)$ is the output and $x(t)$ the input. The input is a constant $x(t) = 1$ for $t \geq 0$ and zero otherwise (MATLAB calls this function Heaviside, we will call it the unit-step signal). We then let the input be the derivative of $x(t)$, which is a signal that we will call "impulse," and finally we let the input be the integral of $x(t)$, which is what we will call the "ramp" signal. The following script is used to find the responses, which are displayed in Fig. 0.22.

```
%%
% Example 0.11---Impulse, unit-step and ramp responses
%%
clear all; clf
syms y(t)
eqn=diff(y,t,2)+5*diff(y,t)+6*y==1
Dy=diff(y,t);
cond=[y(0)==0, Dy(0)==0];
y(t)=dsolve(eqn,cond)
Dy=diff(y,t)
```

```
Iy=int(y,t)
figure(1)
subplot(311)
fplot(y,[0,5]);title('Unit-step response');grid
subplot(312)
fplot(Dy,[0,5]);title('Impulse response');grid
subplot(313)
fplot(Iy,[0,5]);title('Ramp response');grid
```

This example illustrates the intuitive appeal of linear systems. When the input is a constant value (or a unit-step signal or a heaviside signal) the output tries to follow the input after some initial inertia and it ends up being constant. The impulse signal (obtained as the derivative of the unit-step signal) is a signal of very short duration equivalent to shocking the system with a signal that disappears very fast, different from the unit-step, which is like a dc source. Again the output tries to follow the input, eventually disappearing as t increases (no energy from the input!), and the ramp which is the integral of the unit-step grows with time providing more and more energy to the system as time increases thus the response we obtained. The function **dsolve** solves ordinary differential equations explicitly given (D stands for the derivative operator, so D is the first derivative and D2 is the second derivative). A second order system requires two initial conditions, the output and its derivative at $t = 0$.

We hope this soft introduction to MATLAB has provided you with the necessary background to understand the basic way MATLAB operates, and shown you how to continue increasing your knowledge of it. Your best source of information is the *help* command. Explore the different modules that MATLAB has and you will become quickly convinced that these modules provide a great number of computational tools for many areas of engineering and mathematics. Try it!

0.7 PROBLEMS
0.7.1 BASIC PROBLEMS

0.1 Let $z = 8 + j3$ and $v = 9 - j2$.

(a) Find

$$(i) \quad Re(z)+Im(v), \quad (ii) \quad |z+v|, \quad (iii) \quad |zv|,$$
$$(iv) \quad \angle z+\angle v, \quad (v) \quad |v/z|, \quad (vi) \quad \angle(v/z).$$

(b) Find the trigonometric and polar forms of

$$(i) \ z+v, \quad (ii) \ zv, \quad (iii) \ z^*, \quad (iv) \ zz^*, \quad (v) \ z-v.$$

Answers: (a) $Re(z)+Im(v)=6$; $|v/z|=\sqrt{85}/\sqrt{73}$; (b) $zz^*=|z|^2=73$.

0.2 Use Euler's identity to

(a) show that

$$(i) \ \cos(\theta - \pi/2) = \sin(\theta), \quad (ii) \ -\sin(\theta - \pi/2) = \cos(\theta),$$
$$(iii) \ \cos(\theta) = \sin(\theta + \pi/2);$$

(b) to find

$$(i) \ \int_0^1 \cos(2\pi t) \sin(2\pi t) dt, \quad (ii) \ \int_0^1 \cos^2(2\pi t) dt.$$

Answers: (b) 0 and 1/2.

0.3 Use Euler's identity to

(a) show the identities

$$(i) \quad \cos(\alpha + \beta) = \cos(\alpha) \cos(\beta) - \sin(\alpha) \sin(\beta),$$
$$(ii) \quad \sin(\alpha + \beta) = \sin(\alpha) \cos(\beta) + \cos(\alpha) \sin(\beta);$$

(b) find an expression for $\cos(\alpha) \cos(\beta)$, and for $\sin(\alpha) \sin(\beta)$.
Answers: $e^{j\alpha} e^{j\beta} = \cos(\alpha + \beta) + j \sin(\alpha + \beta) = [\cos(\alpha) \cos(\beta) - \sin(\alpha) \sin(\beta)] + j[\sin(\alpha) \cos(\beta) + \cos(\alpha) \sin(\beta)]$.

0.4 Consider the calculation of roots of the equation $z^N = \alpha$ where $N \geq 1$ is an integer and $\alpha = |\alpha| e^{j\phi}$ a nonzero complex number.

(a) First verify that there are exactly N roots for this equation and that they are given by $z_k = r e^{j\theta_k}$ where $r = |\alpha|^{1/N}$ and $\theta_k = (\phi + 2\pi k)/N$ for $k = 0, 1, \cdots, N-1$.

(b) Use the above result to find the roots of the following equations:

$$(i) \ z^2 = 1, \quad (ii) \ z^2 = -1, \quad (iii) \ z^3 = 1, \quad (iv) \ z^3 = -1,$$

and plot them in a polar plane (i.e., indicating their magnitude and phase). Explain how the roots are distributed in the polar plane.
Answers: Roots of $z^3 = -1 = 1e^{j\pi}$ are $z_k = 1e^{j(\pi + 2\pi k)/3}$, $k = 0, 1, 2$, equally spaced around circle of radius r.

0.5 Consider a function of $z = 1 + j1$, $w = e^z$.

(a) Find (i) $\log(w)$, (ii) $\mathcal{R}e(w)$, (iii) $\mathcal{I}m(w)$.

(b) What is $w + w^*$, where w^* is the complex conjugate of w?

(c) Determine $|w|$, $\angle w$ and $|\log(w)|^2$?

(d) Express $\cos(1)$ in terms of w using Euler's identity.
Answers: $\log(w) = z$; $w + w^* = 2\mathcal{R}e[w] = 2e \cos(1)$.

0.6 A phasor can be thought of as a vector, representing a complex number, rotating around the polar plane at a certain frequency in rad/s. The projection of such a vector onto the real axis gives a cosine with a certain amplitude and phase. This problem will show the algebra of phasors which would help you with some of the trigonometric identities that are hard to remember.

(a) When you plot $y(t) = A\sin(\Omega_0 t)$ you notice that it is a cosine $x(t) = A\cos(\Omega_0 t)$ shifted in time, i.e.,

$$y(t) = A\sin(\Omega_0 t) = A\cos(\Omega_0(t - \Delta_t)) = x(t - \Delta_t).$$

How much is this shift Δ_t? Better yet, what is $\Delta_\theta = \Omega_0\Delta_t$ or the shift in phase? One thus only need to consider cosine functions with different phase shifts instead of sines and cosines.

(b) From above, the phasor that generates $x(t) = A\cos(\Omega_0 t)$ is Ae^{j0} so that $x(t) = \mathcal{R}e[Ae^{j0}e^{j\Omega_0 t}]$. The phasor corresponding to the sine $y(t)$ should then be $Ae^{-j\pi/2}$. Obtain an expression for $y(t)$ similar to the one for $x(t)$ in terms of this phasor.

(c) From the above results, give the phasors corresponding to $-x(t) = -A\cos(\Omega_0 t)$ and $-y(t) = -\sin(\Omega_0 t)$. Plot the phasors that generate cos, sin, $-\cos$ and $-\sin$ for a given frequency. Do you see now how these functions are connected? How many radians do you need to shift in positive or negative direction to get a sine from a cosine, etc.

(d) Suppose then you have the sum of two sinusoids, for instance $z(t) = x(t) + y(t) = A\cos(\Omega_0 t) + A\sin(\Omega_0 t)$, adding the corresponding phasors for $x(t)$ and $y(t)$ at some time, e.g., $t = 0$, which is just a sum of two vectors, you should get a vector and the corresponding phasor. For $x(t)$, $y(t)$, obtain their corresponding phasors and then obtain from them the phasor corresponding to $z(t) = x(t) + y(t)$.

(e) Find the phasors corresponding to

$$(i)\ 4\cos(2t + \pi/3), \quad (ii)\ -4\sin(2t + \pi/3), \quad (iii)\ 4\cos(2t + \pi/3) - 4\sin(2t + \pi/3).$$

Answers: $\sin(\Omega_0 t) = \cos(\Omega_0(t - T_0/4)) = \cos(\Omega_0 t - \pi/2)$ since $\Omega_0 = 2\pi/T_0$; $z(t) = \sqrt{2}A\cos(\Omega_0 t - pi/4)$; (e) (i) $4e^{j\pi/3}$; (iii) $4\sqrt{2}e^{j7\pi/12}$.

0.7 To get an idea of the number of bits generated and processed by a digital system consider the following applications:

(a) A compact disc (CD) is capable of storing 75 minutes of "CD quality" stereo (left and right channels are recorded) music. Calculate the number of bits that are stored in the CD as raw data.
Hint: find out what 'CD quality' means in the binary representation of each sample.

(b) Find out what the vocoder in your cell phone is used for. To attaining "telephone quality" voice you use a sampling rate of 10,000 samples/s, and that each sample is represented by 8 bits. Calculate the number of bits that your cell phone has to process every second that you talk. Why would you then need a vocoder?

(c) Find out whether text messaging is cheaper or more expensive than voice. Explain how the text messaging works.

(d) Find out how an audio CD and an audio DVD compare. Find out why it is said that a vinyl long-play record reproduces sounds much better. Are we going backwards with digital technology in music recording? Explain.

(e) To understand why video streaming in the internet is many times of low quality, consider the amount of data that needs to be processed by a video compressor every second. Assume the size of a video frame, in pixels, is 352×240, and that an acceptable quality for the image is obtained by allocating 8 bits/pixel and to avoid jerking effects we use 60 frames/s.

- How many pixels need to be processed every second?
- How many bits would be available for transmission every second?
- The above is raw data, compression changes the whole picture (literally), find out what some of the compression methods are.

Answers: (a) About 6.4 Gbs; vocoder (short for voice encoder) reduces number of transmitted bits while keeping voice recognizable.

0.8 The geometric series

$$S = \sum_{n=0}^{N-1} \alpha^n$$

will be used quite frequently in the next chapters so let us look at some of its properties.

(a) Suppose $\alpha = 1$; what is S equal to?

(b) Suppose $\alpha \neq 1$ show that

$$S = \frac{1 - \alpha^N}{1 - \alpha}.$$

Verify that $(1 - \alpha)S = (1 - \alpha^N)$. Why do you need the constraint that $\alpha \neq 1$? Would this sum exist if $\alpha > 1$? Explain.

(c) Suppose now that $N = \infty$; under what conditions will S exist? if it does, what would S be equal to? Explain.

(d) Suppose again that $N = \infty$ in the definition of S. The derivative of S with respect to α is

$$S_1 = \frac{dS}{d\alpha} = \sum_{n=0}^{\infty} n\alpha^{n-1};$$

obtain a rational expression to find S_1.

Answers: $S = N$ when $\alpha = 1$, $S = (1 - \alpha^N)/(1 - \alpha)$ when $\alpha \neq 1$.

0.7.2 PROBLEMS USING MATLAB

0.9 Derivative and finite difference—Let $y(t) = dx(t)/dt$, where $x(t) = 4\cos(2\pi t)$, $-\infty < t < \infty$. Find $y(t)$ analytically and determine a value of T_s for which $\Delta[x(nT_s)]/T_s = y(nT_s)$ (consider as possible values $T_s = 0.01$ and $T_s = 0.1$). Use the MATLAB function *diff* or create your own to compute the finite difference. Plot the finite difference in the range [0, 1] and compare it with the actual derivative $y(t)$ in that range. Explain your results for the given values of T_s.

Answers: $y(t) = -8\pi \sin(2\pi t)$ has same sampling period as $x(t)$, $T_s \leq 0.5$; $T_s = 0.01$ gives better results.

0.10 Backward difference—Another definition for the finite difference is the backward difference:

$$\Delta_1[x(nT_s)] = x(nT_s) - x((n-1)T_s)$$

$(\Delta_1[x(nT_s)]/T_s$ approximates the derivative of $x(t)$).

(a) Indicate how this new definition connects with the finite difference defined earlier in this chapter.

(b) Solve Problem 0.9 with MATLAB using this new finite difference and compare your results with the ones obtained there.

(c) For the value of $T_s = 0.1$, use the average of the two finite differences to approximate the derivative of the analog signal $x(t)$. Compare this result with the previous ones. Provide an expression for calculating this new finite difference directly.

Answers: $\Delta_1[x(n+1)] = x(n+1) - x(n) = \Delta[x(n)]$; $0.5\{\Delta_1[x(n)] + \Delta[x(n)]\} = 0.5[x(n+1) - x(n-1)]$.

0.11 Sums and Gauss—Three laws in the computation of sums are

$$\begin{aligned}
\text{Distributive:} \quad & \sum_k ca_k = c\sum_k a_k \\
\text{Associative:} \quad & \sum_k (a_k + b_k) = \sum_k a_k + \sum_k b_k \\
\text{Commutative:} \quad & \sum_k a_k = \sum_{p(k)} a_{p(k)}
\end{aligned}$$

for any permutation $p(k)$ of the set of integers k in the summation.

(a) Explain why the above rules make sense when computing sums. To do that consider

$$\sum_k a_k = \sum_{k=0}^{2} a_k, \text{ and } \sum_k b_k = \sum_{k=0}^{2} b_k.$$

Let c be a constant, and choose any permutation of the values $[0, 1, 2]$ for instance $[2, 1, 0]$ or $[1, 0, 2]$.

(b) The trick that Gauss played when he was a preschooler can be explained by using the above rules. Suppose you want to find the sum of the integers from 0 to 10,000 (Gauss did it for integers between 0 and 100 but he was then just a little boy, and we can do better!). That is, we want to find S where

$$S = \sum_{k=0}^{10,000} k = 0 + 1 + 2 + \cdots + 10,000.$$

To do so consider

$$2S = \sum_{k=0}^{10,000} k + \sum_{k=10,000}^{0} k$$

and apply the above rules to find then S. Come up with a MATLAB function of your own to do this sum.

(c) Find the sum of an arithmetic progression

$$S_1 = \sum_{k=0}^{N} (\alpha + \beta k)$$

for constants α and β, using the given three rules.

(d) Find out if MATLAB can do these sums symbolically, i.e., without having numerical values. Use the found symbolic function to calculate the sum in the previous item when $\alpha = \beta = 1$ and $N = 100$.

Answers: $N = 10{,}000$, $S = N(N+1)/2$; $S_1 = \alpha(N+1) + \beta(N(N+1))/2$.

0.12 Integrals and sums—Suppose you wish to find the area under a signal $x(t)$ using sums. You will need the following result found above:

$$\sum_{n=0}^{N} n = \frac{N(N+1)}{2}$$

(a) Consider first $x(t) = t$, $0 \leq t \leq 1$, and zero otherwise. The area under this signal is 0.5. The integral can be approximated from above and below as

$$\sum_{n=1}^{N-1} (nT_s)T_s < \int_0^1 t\,dt < \sum_{n=1}^{N} (nT_s)T_s$$

where $NT_s = 1$ (i.e., we divide the interval $[0, 1]$ into N intervals of width T_s). Graphically show for $N = 4$ that the above equation makes sense by showing the right and left bounds as approximations for the area under $x(t)$.

(b) Let $T_s = 0.001$, use the symbolic function *symsum* to compute the left and right bounds for the above integral. Find the average of these results and compare it with the actual value of the integral.

(c) Verify the symbolic results by finding the sums on the left and the right of the above inequality using the summation given at the beginning of the problem. What happens when $N \to \infty$.

(d) Write a MATLAB script to compute the area under the signal $y(t) = t^2$ from $0 \leq t \leq 1$. Let $T_s = 0.001$. Compare the average of the lower and upper bounds to the value of the integral.

Answer: For $T_s = 1/N$

$$\left[\frac{(N-1)(N-2) + 2(N-1)}{2N^2} \right] \leq \frac{1}{2} \leq \left[\frac{(N-1)(N-2) + 2(N-1)}{2N^2} \right] + \frac{1}{N}.$$

0.13 Exponentials—The exponential $x(t) = e^{at}$ for $t \geq 0$ and zero otherwise is a very common continuous-time signal. Likewise, $y(n) = \alpha^n$ for integers $n \geq 0$ and zero otherwise is a very common discrete-time signal. Let us see how they are related. Do the following using MATLAB:

(a) Let $a = -0.5$, plot $x(t)$.

(b) Let $a = -1$, plot the corresponding signal $x(t)$. Does this signal go to zero faster than the exponential for $a = -0.5$?

(c) Suppose we sample the signal $x(t)$ using $T_s = 1$ what would be $x(nT_s)$ and how can it be related to $y(n)$, i.e., what is the value of α that would make the two equal?

(d) Suppose that a current $x(t) = e^{-0.5t}$ for $t \geq 0$ and zero otherwise is applied to a discharged capacitor of capacitance $C = 1$ F at $t = 0$. What would be the voltage in the capacitor at $t = 1$ s?

(e) How would you obtain an approximate result to the above problem using a computer? Explain.

Answers: $0 < e^{-\alpha t} < e^{-\beta t}$ for $\alpha > \beta \geq 0$; $v_c(1) = 0.79$.

0.14 Algebra of complex numbers—Consider complex numbers $z = 1 + j$, $w = -1 + j$, $v = -1 - j$ and $u = 1 - j$. You may use MATLAB *compass* to plot vectors corresponding to complex numbers to verify your analytic results.

(a) In the complex plane, indicate the point (x, y) that corresponds to z and then show a vector $\vec{z}$ that joins the point (x, y) to the origin. What is the magnitude and the angle corresponding to z or $\vec{z}$?

(b) Do the same for the complex numbers w, v and u. Plot the four complex numbers and find their sum $z + w + v + u$ analytically and graphically.

(c) Find the ratios z/w, w/v, and u/z. Determine the real and imaginary parts of each, as well as their magnitudes and phases. Using the ratios find u/w.

(d) The phase of a complex number is only significant when the magnitude of the complex number is significant. Consider z and $y = 10^{-16}z$, compare their magnitudes and phases. What would you say about the phase of y?

Answers: $|w| = \sqrt{2}$, $\angle w = 3\pi/4$, $|v| = \sqrt{2}$, $\angle v = 5\pi/4$, $|u| = \sqrt{2}$, $\angle u = -\pi/4$.

THEORY AND APPLICATIONS OF CONTINUOUS-TIME SIGNALS AND SYSTEMS

CONTINUOUS-TIME SIGNALS

CONTENTS

A journey of a thousand miles begins with a single step.

Chinese philosopher Lao-Tzu (604–531 BC)

1.1 INTRODUCTION

Starting in this chapter we will concentrate on the representation and processing of continuous-time signals. Such signals are familiar to us. Voice, music as well as images and video coming from radios, cell phones, iPads, and MP3 players exemplify these signals. Clearly each of these signals has some type of information, but what is not clear is how we could capture, represent and perhaps modify these signals and their information content.

To process signals we need to understand their nature—to classify them—so as to clarify the limitations of our analysis and our expectations. Several realizations could then come to mind. One could be that almost all signals vary randomly and continuously with time. Consider a voice signal. If you are able to capture such a signal, by connecting a microphone to your computer and using the hardware and software necessary to display it, you realize that when you speak into the microphone a rather complicated signal that changes in unpredictable ways is displayed. You would then ask yourself how is that your spoken words are converted into this signal, and how it could be represented mathematically to allow you to develop algorithms to change it. In this book we consider the representation of deterministic—rather than random—signals, just a first step in the long process of answering these significant questions. A second realization could be that to input signals into a computer such signals must be in binary form. But how to convert the continuously changing voltage signal generated by the microphone into a binary form? This requires that we compress the information in such a way that it would permit us to get it back, as when we wish to listen to the voice signal stored in the computer. One more realization could be that the generation and processing of signals require systems. In our example, one could think of the human vocal system as a system, of the microphone also as a system that converts differences in air pressure into a voltage signal, likewise of the computer which together with special hardware and software displays and processes the signal. Signals and systems go together. We will consider the interaction of signals and systems in the next chapter.

The mathematical representation of signals means how to think of a signal as a function of either time (e.g., music and voice signals), space (e.g., images), or of time and space (e.g., videos). Using practical characteristics of signals we offer a classification of signals that is connected with the way a signal is stored, processed or both. In this chapter we start the representation and analysis of continuous-time signals and systems. That is, what it means to delay or advance a signal, to reflect it or to find its odd or even components, to determine if a signal is periodic or not, if it is finite energy or not. These are signal operations that will help us in their representation and processing. We show also that any signal can be represented using basic signals. This will permit us to highlight certain characteristics of signals and to simplify finding their corresponding outputs when applied to systems. In particular, the representation in terms of sinusoids is of great interest as it allows the development of the so-called Fourier representation—essential in the development of the theory of linear time-invariant systems to be considered in the next chapter.

1.2 CLASSIFICATION OF TIME-DEPENDENT SIGNALS

Considering signals functions of time carrying information, there are many ways to classify them according to:

1. *The predictability of their behavior*: signals can be *random* or *deterministic*. While a deterministic signal can be represented by a formula or a table of values, random signals can only be approached probabilistically. In this book we will only consider deterministic signals.
2. *The variation of their time variable and their amplitude*: signals can be either *continuous-time* or *discrete-time*—with *analog* or *discrete amplitude*—or *digital*, which are discrete in time with each sample represented by a binary number. This classification relates to the way signals are either processed, stored or both.
3. *Whether the signals exhibit repetitive behavior or not* as *periodic* or *aperiodic* signals.
4. *Their energy content*: signals can be characterized as *finite-energy*, *finite-power* signals or infinite energy and infinite power.

5. *Their symmetry with respect to the time origin*, signals can be *even, odd* or neither.
6. *The dimension of their support*: signals can be said of *finite* or of *infinite* support. Support can be understood as the time interval of the signal outside of which the signal is always zero.

1.3 CONTINUOUS-TIME SIGNALS

That some signals are functions of time carrying information is easily illustrated by means of a recorded voice signal. Such a signal can be thought of as a continuously varying voltage, originally generated by a microphone, that can be transformed into an audible acoustic signal—providing the voice information—by means of an amplifier and speakers. Thus, the voice signal is represented by a function of time

$$v(t), \quad t_b \le t \le t_f \tag{1.1}$$

where t_b is the time at which this signal starts and t_f the time at which it ends. The function $v(t)$ varies continuously with time, and its amplitude can take any possible value. This signal obviously carries the information provided by the voice message.

Not all signals are functions of time. A digital image stored in a computer provides visual information. The intensity of the illumination of the image depends on its location within the image. Thus, a digital image can be represented as a function of space variables (m, n) that vary discretely, creating an array of intensity values called *picture elements* or *pixels*. The visual information in the image is thus provided by the signal $p(m, n)$ where $0 \le m \le M - 1$ and $0 \le n \le N - 1$ for an image of size $M \times N$ pixels. Each pixel value can be represented, for instance, by 256 gray-level values when 8 bits/pixel are used. Thus, the signal $p(m, n)$ varies discretely in space and in amplitude. A video, as a sequence of images in time, is accordingly a function of time and of two space variables. How their time or space variables, and their amplitudes vary characterize signals.

For a time-dependent signal, time and amplitude varying continuously or discretely characterize different signals. Thus according to how the independent variable time, t, varies we have **continuous-time** or **discrete-time** signals, i.e., t takes an innumerable or a finite set of values. Likewise, the amplitude of either a continuous-time or a discrete-time signal can vary continuously or discretely. Continuous-time signals with continuous amplitude, are called **analog signals** as they resemble the pressure variations caused by an acoustic signal. If the signal amplitude varies discretely, but continuously in time, the signal is called **multilevel signal**. A continuous-amplitude, discrete-time signal is called a **discrete-time signal**. A **digital signal** has discrete time and discrete amplitude. Each of the samples of a digital signal is represented by a binary number.

A good way to illustrate the above signal classification is to consider the steps needed to process $v(t)$ in (1.1) with a computer—or the conversion of an analog signal into a digital signal. As indicated before, the independent time variable t in $v(t)$ varies continuously between t_b and t_f, and the amplitude of $v(t)$ also varies continuously and we assume it could take any possible real value, i.e., $v(t)$ is an analog signal. As such, $v(t)$ cannot be processed with a computer as it would require one to store an innumerable number of signal values (even when t_b is very close to t_f). Moreover, for an accurate representation of the possible amplitude values of $v(t)$, we would need a large number of bits. Thus, it is necessary to reduce the amount of data without losing the information provided by the signal. To

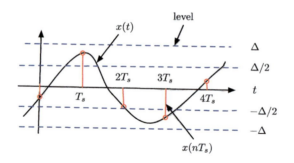

FIGURE 1.1

Discretization in time and in amplitude of an analog signal using as parameters the sampling period T_s and the quantization level Δ. In time, samples are taken at uniform times $\{nT_s\}$, and in amplitude the range of amplitudes is divided into a finite number of levels so that each sample value is approximated by one of them.

accomplish that, we sample the signal $v(t)$ by taking signal values at equally spaced times nT_s, where n is an integer and T_s is the **sampling period** which is appropriately chosen for this signal (in Chapter 8 we will discuss how to chose T_s). As a result of the sampling—thereby reducing the number of values in time—we obtain the discrete-time signal

$$v(nT_s) = v(t)|_{t=nT_s} \qquad 0 \le n \le N \tag{1.2}$$

where $T_s = (t_f - t_b)/N$ and we have taken samples at times $t_b + nT_s$. Clearly, this discretization of the time variable reduces the number of values to enter into the computer, but the amplitudes of these samples still can take innumerable values. Now, to represent each of the $v(nT_s)$ values with a certain number of bits, we also discretize the amplitude of the samples. To do so, a set of levels equally representing the positive as well as the negative values of the signal amplitude are used. A sample value falling within one of these levels is allocated a unique binary code according to some approximation scheme. For instance, if we want each sample to be represented by 8 bits we have 2^8 or 256 possible levels balanced to represent, by rounding or truncating, positive as well as negative values of the sample amplitudes. Assigning a unique binary code to each of the levels we convert an analog signal into a digital signal, in this case a binary signal. These last two operations are called **quantization and coding**.

Given that many of the signals we encounter in practical applications are analog, if it is desirable to process such signals with a computer the above procedure needs to be done regularly. The device that converts an analog signal into a digital signal is called an **analog-to-digital converter** or **A/D converter** and it is characterized by the number of samples it takes per second (**sampling rate** $1/T_s$) and by the number of bits that it allocates to each sample. To convert a digital signal into an analog signal a **digital-to-analog converter** or **D/A converter** is used. Such a device inverts the A/D converter process: binary values are converted into a multilevel signal, or pulses with amplitudes approximating those of the original samples, which are then smoothed out resulting in an analog signal. We will discuss in Chapter 8 the sampling, the binary representation, and the reconstruction of analog signals.

Fig. 1.1 shows how the discretization of an analog signal in time and in amplitude can be understood, while Fig. 1.2 illustrates the sampling and quantization of a segment of speech.

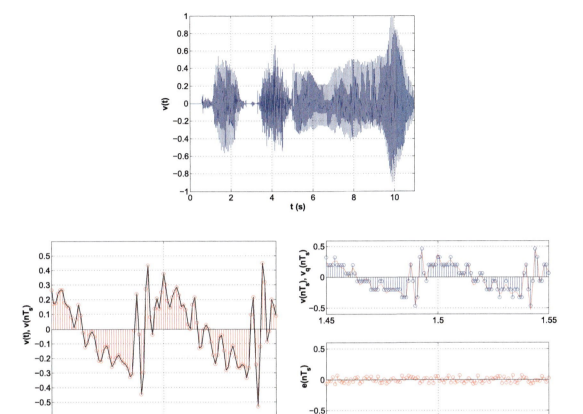

FIGURE 1.2

A segment of the voice signal shown on top is sampled and quantized. The bottom-left figure displays the voice segment (continuous line) and the sampled signal (vertical samples) using a sampling period $T_s = 10^{-3}$ s. The bottom-right figures show the sampled and the quantized signal at the top, while the quantization error (difference between the sampled and the quantized signals) is displayed at the bottom.

A **continuous-time** signal can be thought of as a real-valued, or complex-valued, function of time:

$$x(.) : \mathcal{R} \to \mathcal{R} \ (\mathcal{C})$$
$$t \qquad x(t) \tag{1.3}$$

Thus, the independent variable is time t, and the value of the function at some time t_0, $x(t_0)$, is a real (or a complex) value. (Although in practice signals are real, it is useful in theory to have the option of complex-valued signals.) It is assumed that both time t and signal amplitude $x(t)$ can vary continuously, if needed from $-\infty$ to ∞. The amplitude of $x(t)$ can vary either continuously (**analog signal**) or discretely (**multilevel signal**).

Example 1.1. Characterize the sinusoidal signal $x(t) = \sqrt{2}\cos(\pi t/2 + \pi/4)$, $-\infty < t < \infty$.

Solution: The signal $x(t)$ is

- deterministic, as the value of the signal can be obtained for any possible value of t;
- analog, as there is a continuous variation of the time variable t from $-\infty$ to ∞, and of the amplitude of the signal between $-\sqrt{2}$ to $\sqrt{2}$; and
- of infinite support, as the signal does not become zero outside any finite interval.

The amplitude of the sinusoid is $\sqrt{2}$, its frequency is $\Omega = \pi/2$ (rad/s), and its phase is $\pi/4$ rad (notice that Ωt has radians as units so that it can be added to the phase). Because of the infinite support this signal does not exist in practice, but we will see that sinusoids are extremely important in the representation and processing of signals. ☐

Example 1.2. A complex signal $y(t)$ is defined as $y(t) = (1 + j)e^{j\pi t/2}$, $0 \le t \le 10$, and zero otherwise. Express $y(t)$ in terms of the signal $x(t)$ in Example 1.1. Characterize $y(t)$.

Solution: Since $1 + j = \sqrt{2}e^{j\pi/4}$, using Euler's identity,

$$y(t) = \sqrt{2}e^{j(\pi t/2 + \pi/4)} = \sqrt{2}\left[\cos(\pi t/2 + \pi/4) + j\sin(\pi t/2 + \pi/4)\right] \qquad 0 \le t \le 10$$

and zero otherwise. Thus, the real and imaginary parts of this signal are

$$\mathcal{R}e[y(t)] = \sqrt{2}\cos(\pi t/2 + \pi/4),$$
$$\mathcal{I}m[y(t)] = \sqrt{2}\sin(\pi t/2 + \pi/4),$$

for $0 \le t \le 10$ and zero otherwise. The signal $y(t)$ can be written using $x(t)$ as

$$y(t) = x(t) + jx(t - 1) \qquad 0 \le t \le 10$$

and zero otherwise. Notice that

$$x(t - 1) = \sqrt{2}\cos(\pi(t - 1)/2 + \pi/4) = \sqrt{2}\cos(\pi t/2 - \pi/2 + \pi/4) = \sqrt{2}\sin(\pi t/2 + \pi/4).$$

The signal $y(t)$ is

- analog of finite support, i.e., the signal is zero outside the interval $0 \le t \le 10$,
- complex, composed of two sinusoids of frequency $\Omega = \pi/2$ rad/s, phase $\pi/4$ in rad, and amplitude $\sqrt{2}$ in $0 \le t \le 10$ and zero outside that time interval, i.e., finite support. ☐

Example 1.3. Consider the pulse signal $p(t) = 1$, $0 \le t \le 10$ and zero elsewhere. Characterize this signal, and use it along with $x(t)$ in Example 1.1, to represent $y(t)$ in Example 1.2.

Solution: The analog signal $p(t)$ is of finite support and real-valued. We have

$$\mathcal{R}e[y(t)] = x(t)p(t), \quad \mathcal{I}m[y(t)] = x(t - 1)p(t)$$

so that

$$y(t) = [x(t) + jx(t - 1)]p(t).$$

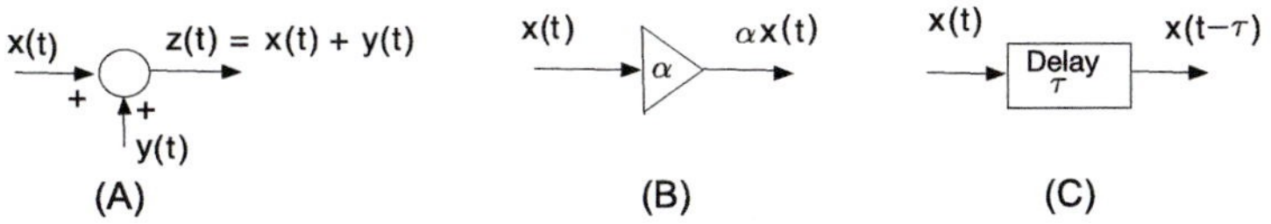

FIGURE 1.3

Diagrams of basic signal operations: (A) adder, (B) constant multiplier, and (C) delay.

The multiplication by $p(t)$ makes $x(t)p(t)$ and $x(t-1)p(t)$ finite-support signals. This operation is called **time windowing** as the signal $p(t)$ only provides information on $x(t)$ within the range where $p(t)$ is unity, ignoring the rest of the signal outside the support of $p(t)$. ☐

The above three examples not only illustrate how different types of signal can be related to each other, but also how signals can be defined in more precise forms. Although the representations for $y(t)$ in Examples 1.2 and 1.3 are equivalent, the one in Example 1.3 is shorter and easier to visualize by the use of the pulse $p(t)$.

1.3.1 ADDITION, CONSTANT MULTIPLICATION, TIME SHIFTING AND REFLECTION

Adding two signals, multiplying a signal by a constant, as well as time shifting and reflection of a signal are basic operations needed in the representation and processing of continuous-time signals. These operations can be described as follows:

- *Signal addition*—two signals $x(t)$ and $y(t)$ are added to obtain their sum $z(t) = x(t) + y(t)$. An **adder** is used to implement this operation.
- *Constant multiplication*—a signal $x(t)$ is multiplied by a constant α to obtain $y(t) = \alpha x(t)$. A **constant multiplier** is used.
- *Time shifting*—The signal $x(t)$ is **delayed** τ seconds to get $x(t - \tau)$, **advanced** by τ to get $x(t + \tau)$. A time delay, or shift to the right, of a signal is implemented by a **delay**.
- *Time reflection*—The time variable of a signal $x(t)$ is negated to give the **reflected** signal $x(-t)$.

In Fig. 1.3 we show the diagrams used for the implementation of these operations. These operations will be used in the block diagrams for system in the next chapters.

Given the simplicity of the first two operations we will only discuss the other two. Special operations useful in communications, control and filtering are considered later.

Remarks

1. It is important to understand that advancing or reflecting a signal cannot be implemented in **real-time**, i.e., as the signal is being processed. Delays can be implemented in real-time, but advancing and reflection require the signal be saved or recorded. Thus an acoustic signal recorded on tape can be delayed or advanced with respect to an initial time, or played back faster or slower, but it can only be delayed if the signal is coming from a live microphone.
2. We will see later in this chapter that shifting in frequency is also possible. The operation is called **signal modulation** and it is of great significance in communications. We will also show later that reflection is a special scaling of the time variable.

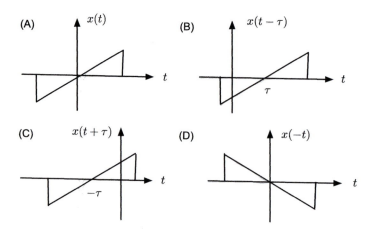

FIGURE 1.4

Continuous-time signal (A), and its delayed (B), advanced (C), and reflected (D) versions.

For a positive value τ,

- a signal $x(t - \tau)$ is the original signal $x(t)$ **shifted to the right or delayed** τ seconds, as illustrated in Fig. 1.4B for the signal shown in Fig. 1.4A. That the original signal has been shifted to the right can be verified by finding that the $x(0)$ value of the original signal appears in the delayed signal at $t = \tau$ (which results from making $t - \tau = 0$).
- Likewise, a signal $x(t + \tau)$ is the original signal $x(t)$ **shifted left or advanced** by τ seconds as illustrated in Fig. 1.4C for the signal shown in Fig. 1.4A. The original signal is now shifted to the left, i.e., the value $x(0)$ of the original signal occurs now earlier (i.e., it has been advanced) at time $t = -\tau$.
- **Reflection** consists in negating the time variable. Thus the reflection of $x(t)$ is $x(-t)$. This operation can be visualized as flipping the signal about the origin. See Fig. 1.4D for the signal shown in Fig. 1.4A.

Given an analog signal $x(t)$ and $\tau > 0$ we have with respect to $x(t)$

1. $x(t - \tau)$ is **delayed** or **shifted right** τ seconds,
2. $x(t + \tau)$ is **advanced** or **shifted left** τ seconds,
3. $x(-t)$ is **reflected**,
4. $x(-t - \tau)$ is **reflected, and advanced (or shifted left)** τ seconds, while $x(-t + \tau)$ is **reflected, and delayed (or shifted right)** τ seconds.

Remarks

Whenever we combine the delaying or advancing with reflection, delaying and advancing are swapped. Thus, $x(-t + 1)$ is $x(t)$ reflected and delayed, or shifted to the right, by 1. Likewise, $x(-t - 1)$ is $x(t)$ reflected and advanced, or shifted to the left by 1. Again, the value $x(0)$ of the original signal is found in $x(-t + 1)$ at $t = 1$, and in $x(-t - 1)$ at $t = -1$.

Example 1.4. Consider an analog pulse

$$x(t) = \begin{cases} 1 & 0 \le t \le 1, \\ 0 & \text{otherwise.} \end{cases}$$

Find mathematical expressions for $x(t)$ delayed by 2, advanced by 2, and for the reflection $x(-t)$.

Solution: An expression for the delayed signal $x(t-2)$ can be found mathematically by replacing the variable t by $t-2$ in the definition of $x(t)$ so that

$$x(t-2) = \begin{cases} 1 & 0 \le t-2 \le 1 \text{ or } 2 \le t \le 3, \\ 0 & \text{otherwise.} \end{cases}$$

The value $x(0)$ (which in $x(t)$ occurs at $t=0$) in $x(t-2)$ now occurs when $t=2$, so that the signal $x(t)$ has been shifted to the right 2 units of time, and since the values are occurring later, the signal $x(t-2)$ is said to be "delayed" by 2 with respect to $x(t)$.

Likewise, we have

$$x(t+2) = \begin{cases} 1 & 0 \le t+2 \le 1 \text{ or } -2 \le t \le -1, \\ 0 & \text{otherwise.} \end{cases}$$

The signal $x(t+2)$ can be seen to be the advanced version of $x(t)$, as it is this signal shifted to the left by 2 units of time. The value $x(0)$ for $x(t+2)$ now occurs at $t=-2$ which is ahead of $t=0$.

Finally, the signal $x(-t)$ is given by

$$x(-t) = \begin{cases} 1 & 0 \le -t \le 1 \text{ or } -1 \le t \le 0, \\ 0 & \text{otherwise.} \end{cases}$$

This signal is a mirror image of the original: the value $x(0)$ still occurs at the same time, but $x(1)$ occurs when $t=-1$. □

Example 1.5. When shifting and reflection are considered together the best approach to visualize the operation is to make a table calculating several values of the new signal and comparing these with those from the original signal. Consider the signal

$$x(t) = \begin{cases} t & 0 \le t \le 1, \\ 0 & \text{otherwise.} \end{cases}$$

Determine if $x(-t+2)$ is reflected and advanced or delayed.

Solution: Although one can see that this signal is reflected, it is not clear whether it is advanced or delayed by 2. By computing a few values:

t	$x(-t+2)$
-1	$x(3) = 0$
0	$x(2) = 0$
1	$x(1) = 1$
1.5	$x(0.5) = 0.5$
2	$x(0) = 0$

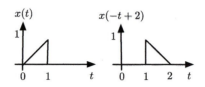

FIGURE 1.5

Signal $x(t)$ and its reflected and delayed version $x(-t+2)$.

it is clear now that $x(-t+2)$ is reflected and "delayed" by 2 (see Fig. 1.5). In fact, as indicated above, whenever the signal is a function of $-t$, i.e., reflected, the $-t+\tau$ operation becomes reflection and "delay", and $-t-\tau$ becomes reflection and "advancing".   □

Remark

When computing the convolution integral later on, we will consider the signal $x(t-\tau)$ as a function of τ for different values of t. As indicated in the above example, this signal is a reflected version of $x(\tau)$ being shifted to the right t seconds. To see this, consider $t=0$ then $x(t-\tau)|_{t=0}=x(-\tau)$, the reflected version and $x(0)$ occurs at $\tau=0$. When $t=1$ then $x(t-\tau)|_{t=1}=x(1-\tau)$ and $x(0)$ occurs at $\tau=1$, so that $x(1-\tau)$ is $x(-\tau)$ shifted to the right by 1, and so on.

1.3.2 EVEN AND ODD SIGNALS

Symmetry with respect to the origin differentiates signals and will be useful in their Fourier analysis. An analog signal $x(t)$ is called

- **even** when $x(t)$ coincides with its reflection $x(-t)$. Such a signal is symmetric with respect to the time origin,
- **odd** when $x(t)$ coincides with $-x(-t)$, i.e., the negative of its reflection. Such a signal is antisymmetric with respect to the time origin.

Even and odd signals are defined as follows:

$$x(t) \quad \textbf{even}: \quad x(t)=x(-t), \qquad\qquad (1.4)$$
$$x(t) \quad \textbf{odd}: \quad x(t)=-x(-t). \qquad\qquad (1.5)$$

Even and odd decomposition: Any signal $y(t)$ is representable as a sum of an even component $y_e(t)$ and an odd component $y_o(t)$

$$y(t)=y_e(t)+y_o(t) \qquad\qquad (1.6)$$

where

$$y_e(t) = 0.5\big[y(t)+y(-t)\big],$$
$$y_o(t) = 0.5\big[y(t)-y(-t)\big].$$

Using the definitions of even and odd signals, any signal $y(t)$ can be decomposed into the sum of an even and an odd function. Indeed, the following is an identity:

$$y(t) = \frac{1}{2}\left[y(t) + y(-t)\right] + \frac{1}{2}\left[y(t) - y(-t)\right]$$

where the first term is the even component $y_e(t)$ and the second is the odd component $y_o(t)$ of $y(t)$. It can easily be verified that $y_e(t)$ is even and that $y_o(t)$ is odd.

Example 1.6. Consider the analog signal $x(t) = \cos(2\pi t + \theta)$, $-\infty < t < \infty$. Determine the value of θ for which $x(t)$ is even and for which it is odd. If $\theta = \pi/4$ is $x(t) = \cos(2\pi t + \pi/4)$, $-\infty < t < \infty$, even or odd?

Solution: The reflection of $x(t)$ is $x(-t) = \cos(-2\pi t + \theta)$, then:
(i) $x(t)$ is even if $x(t) = x(-t)$ or

$$\cos(2\pi t + \theta) = \cos(-2\pi t + \theta) = \cos(2\pi t - \theta)$$

or $\theta = -\theta$, that is, when $\theta = 0$, or π. Thus, $x_1(t) = \cos(2\pi t)$ as well as $x_2(t) = \cos(2\pi t + \pi) = -\cos(2\pi t)$ are even.
(ii) For $x(t)$ to be odd, we need that $x(t) = -x(-t)$ or

$$\cos(2\pi t + \theta) = -\cos(-2\pi t + \theta) = \cos(-2\pi t + \theta \pm \pi) = \cos(2\pi t - \theta \mp \pi),$$

which can be obtained with $\theta = -\theta \mp \pi$ or equivalently when $\theta = \mp\pi/2$. Indeed, $\cos(2\pi t - \pi/2) = \sin(2\pi t)$ and $\cos(2\pi t + \pi/2) = -\sin(2\pi t)$ both are odd.
When $\theta = \pi/4$, $x(t) = \cos(2\pi t + \pi/4)$ is neither even nor odd according to the above.   □

Example 1.7. Consider the signal

$$x(t) = \begin{cases} 2\cos(4t) & t > 0, \\ 0 & \text{otherwise.} \end{cases}$$

Find its even and odd decomposition. What would happen if $x(0) = 2$ instead of 0, i.e., when we let $x(t) = 2\cos(4t)$ at $t \geq 0$, and zero otherwise? Explain.

Solution: The signal $x(t)$ is neither even nor odd given that its values for $t \leq 0$ are zero. For its even–odd decomposition, the even component is given by

$$x_e(t) = 0.5[x(t) + x(-t)] = \begin{cases} \cos(4t) & t > 0, \\ \cos(4t) & t < 0, \\ 0 & t = 0, \end{cases}$$

and the odd component is given by

$$x_o(t) = 0.5[x(t) - x(-t)] = \begin{cases} \cos(4t) & t > 0, \\ -\cos(4t) & t < 0, \\ 0 & t = 0, \end{cases}$$

which when added give $x(t)$. See Fig. 1.6.

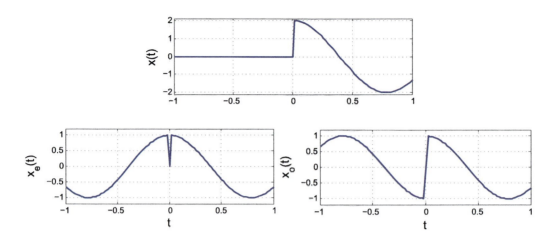

FIGURE 1.6

Even and odd components of $x(t)$ $(x(0) = 0)$. Notice that in this case $x_e(0) = 0$. The value $x_o(0)$ is always zero.

If $x(0) = 2$, we have

$$x_e(t) = 0.5[x(t) + x(-t)] = \begin{cases} \cos(4t) & t > 0, \\ \cos(4t) & t < 0, \\ 2 & t = 0, \end{cases}$$

while the odd component is the same. The even component has a discontinuity at $t = 0$ in both cases.   □

1.3.3 PERIODIC AND APERIODIC SIGNALS

A useful characterization of signals is whether they are **periodic** or **aperiodic** (non-periodic).

A continuous-time signal $x(t)$ is **periodic** if the following two conditions are satisfied:

1. it is defined for all possible values of t, $-\infty < t < \infty$, and
2. there is a positive real value T_0, the **fundamental period** of $x(t)$, such that

$$x(t + kT_0) = x(t) \tag{1.7}$$

for any integer k.

The **fundamental period** of $x(t)$ is the smallest $T_0 > 0$ that makes the periodicity possible.

Remarks

1. The infinite support and the uniqueness of the fundamental period make periodic signals non-existent in practical applications. Despite this, periodic signals are of great significance in the

Fourier signal representation and processing as we will see later. Indeed, the representation of aperiodic signals is obtained from that of periodic signals, and the response of systems to sinusoids is fundamental in the theory of linear systems.

2. Although seemingly redundant, the first part of the definition of a periodic signal indicates that it is not possible to have a nonzero periodic signal with a finite support (i.e., the signal is zero outside an interval $t \in [t_1, t_2]$) or any support different from the infinite support $-\infty < t < \infty$. This first part of the definition is needed for the second part to make sense.

3. It is exasperating to find the period of a constant signal $x(t) = A$. Visually $x(t)$ is periodic but its fundamental period is not clear. Any positive value could be considered the period, but none will be taken. The reason is that $x(t) = A\cos(0t)$ or a cosine of zero frequency, and as such its period $(2\pi/\text{frequency})$ is not determined since we would have to divide by zero—which is not permitted. Thus, a constant signal is periodic of non-definable fundamental period!

Example 1.8. Consider the sinusoid $x(t) = A\cos(\Omega_0 t + \theta)$, $-\infty < t < \infty$. Determine the fundamental period of this signal, and indicate for what frequency Ω_0 the fundamental period of $x(t)$ is not defined.

Solution: The analog frequency is $\Omega_0 = 2\pi/T_0$ so $T_0 = 2\pi/\Omega_0$ is the fundamental period. Whenever $\Omega_0 > 0$ these sinusoids are periodic. For instance, consider $x(t) = 2\cos(2t - \pi/2)$, $-\infty < t < \infty$; its fundamental period is found from its analog frequency $\Omega_0 = 2 = 2\pi f_0$ (rad/s), or a Hz frequency of $f_0 = 1/\pi = 1/T_0$, so that $T_0 = \pi$ is the fundamental period in s. That this is so can be seen using the fact that $\Omega_0 T_0 = 2T_0 = 2\pi$ and as such

$$
\begin{aligned}
x(t + NT_0) &= 2\cos(2(t + NT_0) - \pi/2) = 2\cos(2t + 2\pi N - \pi/2) \\
&= 2\cos(2t - \pi/2) = x(t) \qquad \text{integer } N,
\end{aligned}
$$

since adding $2\pi N$ (a multiple of 2π) to the angle of the cosine gives the original angle. If $\Omega_0 = 0$, i.e., dc frequency, the fundamental period cannot be defined because of the division by zero when finding $T_0 = 2\pi/\Omega_0$. □

Example 1.9. Consider a periodic signal $x(t)$ of fundamental period T_0, determine whether the following signals are periodic and if so find their corresponding fundamental periods:

1. $y(t) = A + x(t)$, where A is either a positive, negative or zero value,
2. $z(t) = x(t) + v(t)$ where $v(t)$ is periodic of fundamental period $T_1 = NT_0$, where N is a positive integer, i.e., a multiple of T_0,
3. $w(t) = x(t) + s(t)$ where $s(t)$ is periodic of fundamental period T_1, not a multiple of T_0.

Solution: (1) Adding a constant to a periodic signal does not change the periodicity. Thus, $y(t)$ is periodic of fundamental period T_0, i.e., for an integer k, $y(t + kT_0) = A + x(t + kT_0) = A + x(t)$ since $x(t)$ is periodic of fundamental period T_0.

(2) The fundamental period $T_1 = NT_0$ of $v(t)$ is also a period of $x(t)$ and so $z(t)$ is periodic of fundamental period T_1 since for any integer k

$$
z(t + kT_1) = x(t + kT_1) + v(t + kT_1) = x(t + kNT_0) + v(t) = x(t) + v(t)
$$

given that $v(t + kT_1) = v(t)$, and that kN is an integer so that $x(t + kNT_0) = x(t)$. The periodicity can be visualized by considering that in one period of $v(t)$ we can include N periods of $x(t)$.

(3) The condition for $w(t)$ to be periodic is that the ratio of the periods of $x(t)$ and of $s(t)$ be

$$\frac{T_1}{T_0} = \frac{N}{M}$$

where N and M are positive integers not divisible by each other, or that MT_1 periods of $s(t)$ can be exactly included into NT_0 periods of $x(t)$. Thus, $MT_1 = NT_0$ becomes the fundamental period of $w(t)$, or

$$w(t + MT_1) = x(t + MT_1) + s(t + MT_1) = x(t + NT_0) + s(t + MT_1) = x(t) + s(t). \quad \square$$

Example 1.10. Let $x(t) = e^{j2t}$ and $y(t) = e^{j\pi t}$, consider their sum $z(t) = x(t) + y(t)$, and their product $w(t) = x(t)y(t)$. Determine if $z(t)$ and $w(t)$ are periodic, and if so their fundamental periods. Is $p(t) = (1 + x(t))(1 + y(t))$ periodic?

Solution: According to Euler's identity

$$x(t) = \cos(2t) + j\sin(2t), \quad y(t) = \cos(\pi t) + j\sin(\pi t)$$

indicating $x(t)$ is periodic of fundamental period $T_0 = \pi$ (the frequency of $x(t)$ is $\Omega_0 = 2 = 2\pi/T_0$) and $y(t)$ is periodic of fundamental period $T_1 = 2$ (the frequency of $y(t)$ is $\Omega_1 = \pi = 2\pi/T_1$). For $z(t) = x(t) + y(t)$ to be periodic T_1/T_0 must be a rational number, which is not the case as $T_1/T_0 = 2/\pi$ is irrational due to π. So $z(t)$ is not periodic. The product, $w(t) = x(t)y(t) = e^{j(2+\pi)t} = \cos(\Omega_2 t) + j\sin(\Omega_2 t)$ where $\Omega_2 = 2 + \pi = 2\pi/T_2$. Thus, $w(t)$ is periodic of fundamental period $T_2 = 2\pi/(2 + \pi)$. The terms $1 + x(t)$ and $1 + y(t)$ are periodic of fundamental period $T_0 = \pi$ and $T_1 = 2$, respectively, and from the case of the product above giving $w(t)$, one would hope their product $p(t) = (1 + x(t))(1 + y(t))$ would be periodic. But since $p(t) = 1 + x(t) + y(t) + x(t)y(t)$ and as shown above $x(t) + y(t) = z(t)$ is not periodic, $p(t)$ is not periodic. $\quad \square$

---

1. A sinusoid of frequency $\Omega_0 > 0$ is periodic of fundamental period $T_0 = 2\pi/\Omega_0$. If $\Omega_0 = 0$ the fundamental period is not defined.
2. The sum of two periodic signals $x(t)$ and $y(t)$, of periods T_1 and T_2, is periodic if the ratio of the periods T_1/T_2 is a rational number N/M, with N and M non-divisible integers. The fundamental period of the sum is $MT_1 = NT_2$.

1.3.4 FINITE-ENERGY AND FINITE-POWER SIGNALS

Another possible classification of signals is based on their energy and power. The concepts of energy and power introduced in circuit theory can be extended to any signal. Recall that for a resistor of unit resistance its **instantaneous power** is given by

$$p(t) = v(t)i(t) = i^2(t) = v^2(t)$$

where $i(t)$ and $v(t)$ are the current and the voltage in the resistor. The **energy** in the resistor for an interval $[t_0, t_1]$, is the accumulation of instantaneous power over that time interval

$$E_T = \int_{t_0}^{t_1} p(t)dt = \int_{t_0}^{t_1} i^2(t)dt = \int_{t_0}^{t_1} v^2(t)dt.$$

The **power** in the interval $[t_0, t_1]$ is the average energy

$$P_T = \frac{E_T}{T} = \frac{1}{T}\int_{t_0}^{t_1} i^2(t)dt = \frac{1}{T}\int_{t_0}^{t_1} v^2(t)dt \qquad T = t_1 - t_0,$$

corresponding to the heat dissipated by the resistor (and for which you pay the electric company). The energy and power concepts can thus easily be generalized as follows.

The **energy** and the **power** of a continuous-time signal $x(t)$ are defined for either finite- or infinite-support signals by

$$E_x = \int_{-\infty}^{\infty} |x(t)|^2 dt, \quad P_x = \lim_{T \to \infty} \frac{1}{2T} \int_{-T}^{T} |x(t)|^2 dt. \tag{1.8}$$

A signal $x(t)$ is then said to be **finite-energy**, or **square integrable**, whenever

$$E_x < \infty. \tag{1.9}$$

A signal $x(t)$ is said to be **finite-power** if

$$P_x < \infty. \tag{1.10}$$

Remarks

1. The above definitions of energy and power are valid for any signal of finite or infinite support, since a finite-support signal is zero outside its support.
2. In the formulas for energy and power we are considering the possibility that the signal might be complex and so we are squaring its magnitude. If the signal being considered is real this simply is equivalent to squaring the signal.
3. According to the above definitions, a finite-energy signal has zero power. Indeed, if the energy of the signal is some constant $E_x < \infty$ then

$$P_x = \left[\lim_{T \to \infty} \frac{1}{2T}\right]\left[\int_{-\infty}^{\infty} |x(t)|^2 dt\right] = \lim_{T \to \infty} \frac{E_x}{2T} = 0.$$

4. A signal $x(t)$ is said to be **absolutely integrable**, if $x(t)$ satisfies the condition

$$\int_{-\infty}^{\infty} |x(t)|dt < \infty. \tag{1.11}$$

Example 1.11. Consider an aperiodic signal $x(t) = e^{-at}$, $a > 0$, for $t \geq 0$ and zero otherwise. Find the energy and the power of this signal and determine whether the signal is finite-energy, finite-power or both.

Solution: The energy of $x(t)$ is given by

$$E_x = \int_0^\infty e^{-2at} dt = \frac{1}{2a} < \infty.$$

The power of $x(t)$ is then zero. Thus $x(t)$ is a finite-energy and finite-power signal. □

Example 1.12. Find the energy and the power of
(a) the complex signal $y(t) = (1 + j)e^{j\pi t/2}$, $0 \leq t \leq 10$, and zero otherwise, and
(b) the pulse $z(t) = 1$ for $0 \leq t \leq 10$, and zero otherwise.
Determine whether these signals are finite-energy, finite-power or both.

Solution: The energy in these signals is computed as follows:

$$E_y = \int_0^{10} |(1 + j)e^{j\pi t/2}|^2 dt = 2\int_0^{10} dt = 20, \quad E_z = \int_0^{10} dt = 10,$$

where we used $|(1 + j)e^{j\pi t/2}|^2 = |1 + j|^2 |e^{j\pi t/2}|^2 = |1 + j|^2 = 2$. Thus, $y(t)$ and $z(t)$ are finite-energy signals, and the power of $y(t)$ and $z(t)$ are zero because they have finite energy. □

Example 1.13. Consider the periodic signal $x(t) = \cos(\pi t/2)$, $-\infty < t < \infty$ which can be written as $x(t) = x_1(t) + x_2(t)$ where

$$x_1(t) = \begin{cases} \cos(\pi t/2) & t \geq 0, \\ 0 & \text{otherwise,} \end{cases} \quad \text{and} \quad x_2(t) = \begin{cases} \cos(-\pi t/2) & t < 0, \\ 0 & \text{otherwise.} \end{cases}$$

The signal $x_1(t)$ is called a "causal" sinusoid because it is zero for $t < 0$. It is a signal that you would get from a signal generator that is started at a certain initial time, in this case 0, and that continues until the signal generator is switched off (in this case possibly infinity). The signal $x_2(t)$ is called "anticausal." It cannot be obtained from a generator as it starts at $-\infty$ and ends at 0. Determine the power of $x(t)$ and relate it to the power of $x_1(t)$ and of $x_2(t)$.

Solution: Clearly, $x(t)$ as well as $x_1(t)$ and $x_2(t)$ are infinite-energy signals:

$$E_x = \int_{-\infty}^\infty x^2(t) dt = \underbrace{\int_{-\infty}^0 \cos^2(-\pi t/2) dt}_{E_{x_2} \to \infty} + \underbrace{\int_0^\infty \cos^2(\pi t/2) dt}_{E_{x_1} \to \infty} \to \infty.$$

The power of $x(t)$ can be calculated by using the symmetry of the signal squared and letting $T = NT_0$, with T_0 the fundamental period of $x(t)$ and N a positive integer:

$$
\begin{aligned}
P_x &= \lim_{T \to \infty} \frac{1}{2T} \int_{-T}^{T} x^2(t)dt = \lim_{T \to \infty} \frac{2}{2T} \int_{0}^{T} \cos^2(\pi t/2)dt = \lim_{N \to \infty} \frac{1}{NT_0} \int_{0}^{NT_0} \cos^2(\pi t/2)dt \\
&= \lim_{N \to \infty} \frac{1}{NT_0} \left[N \int_{0}^{T_0} \cos^2(\pi t/2)dt \right] = \frac{1}{T_0} \int_{0}^{T_0} \cos^2(\pi t/2)dt.
\end{aligned}
$$

Using the trigonometric identity $\cos^2(\theta) = 0.5 + 0.5\cos(2\theta)$ we have

$$
\cos^2(\pi t/2) = \frac{1}{2}[\cos(\pi t) + 1]
$$

and, since the fundamental period of $x(t)$ is $T_0 = 4$,

$$
P_x = \frac{1}{T_0} \int_{0}^{T_0} \cos^2(\pi t/2)dt = \frac{1}{8} \int_{0}^{4} \cos(\pi t)dt + \frac{1}{8} \int_{0}^{4} dt = 0 + 0.5 = 0.5
$$

where the area of the sinusoid over two of its periods is zero. So we see that $x(t)$ is a finite-power but infinite-energy signal.

Now although $x_1(t)$ is not periodic it repeats periodically for $t \geq 0$ and its power is the same as the power of $x_2(t)$. Indeed,

$$
\begin{aligned}
P_{x_2} &= \lim_{T \to \infty} \frac{1}{2T} \int_{-T}^{0-} \cos^2(-\pi t/2)dt \\
&= \lim_{T \to \infty} \frac{1}{2T} \int_{0}^{T} \cos^2(\pi \tau/2)d\tau = P_{x_1}
\end{aligned}
$$

where we used a change of variable $\tau = -t$, and that the integral from 0 or 0− is the same. Since $x_2(t)$ repeats every $T_0 = 4$ s, its power is

$$
\begin{aligned}
P_{x_2} &= \lim_{T \to \infty} \frac{1}{2T} \int_{-T}^{T} x_2^2(t)dt = \lim_{T \to \infty} \frac{1}{2T} \int_{0}^{T} x_2^2(t)dt \\
&= \lim_{N \to \infty} \frac{1}{2NT_0} \left[N \int_{0}^{T_0} x_2^2(t)dt \right] = \frac{1}{2T_0} \int_{0}^{T_0} \cos^2(\pi t/2)dt = 0.25
\end{aligned}
$$

i.e., it is half the power of $x(t)$. So $P_x = P_{x_1} + P_{x_2}$.   □

The power of a periodic signal $x(t)$ of fundamental period T_0 is

$$
P_x = \frac{1}{T_0} \int_{t_0}^{t_0+T_0} x^2(t)dt \tag{1.12}
$$

for any value of t_0, i.e., the average energy in a period of the signal.

The above result is similar to the computation of the power of a sinusoid. Letting $T = NT_0$, for an integer $N > 0$ we have

$$P_x = \lim_{T \to \infty} \frac{1}{2T} \int_{-T}^{T} x^2(t)dt = \lim_{N \to \infty} \frac{1}{2NT_0} \int_{-NT_0}^{NT_0} x^2(t)dt$$

$$= \lim_{N \to \infty} \frac{1}{2NT_0} \left[N \int_{-T_0}^{T_0} x^2(t)dt \right] = \frac{1}{2T_0} \int_{-T_0}^{T_0} x^2(t)dt = \frac{1}{T_0} \int_{t_0}^{t_0+T_0} x^2(t)dt.$$

Superposition of Power

As we will see later in the Fourier Series representation, any periodic signal is representable as a possibly infinite sum of sinusoids of frequencies multiples of the frequency of the periodic signal being represented. These frequencies are said to be **harmonically related**. The power of a periodic signal is shown to be the sum of the power of each of its sinusoidal components, i.e., there is superposition of the power. Although this superposition is still possible when a sum of sinusoids form a non-periodic signal, the power calculation needs to be done for each of the components. This is illustrated in the following example.

Example 1.14. Consider the signals $x(t) = \cos(2\pi t) + \cos(4\pi t)$ and $y(t) = \cos(2\pi t) + \cos(2t)$, $-\infty < t < \infty$. Determine if these signals are periodic and if so find their fundamental periods. Compute the power of these signals.

Solution: The sinusoids $\cos(2\pi t)$ and $\cos(4\pi t)$ have fundamental periods $T_1 = 1$ and $T_2 = 1/2$ so $x(t)$ is periodic since $T_1/T_2 = 2$ and its fundamental period is $T_0 = T_1 = 2T_2 = 1$. The two frequencies are harmonically related. On the other hand $\cos(2t)$ has as fundamental period $T_3 = \pi$ and thus the ratio of the fundamental periods of the sinusoidal components of $y(t)$ is $T_1/T_3 = 1/\pi$ which is not rational and so $y(t)$ is not periodic, and the frequencies 2π and 2 are not harmonically related.

Using the trigonometric identities

$$\cos^2(\theta) = \frac{1}{2}[1 + \cos(2\theta)],$$

$$\cos(\alpha)\cos(\beta) = \frac{1}{2}[\cos(\alpha + \beta) + \cos(\alpha - \beta)],$$

we have

$$x^2(t) = \cos^2(2\pi t) + \cos^2(4\pi t) + 2\cos(2\pi t)\cos(4\pi t)$$

$$= 1 + \frac{1}{2}\cos(4\pi t) + \frac{1}{2}\cos(8\pi t) + \cos(6\pi t) + \cos(2\pi t).$$

Thus, we have for $T_0 = 1$

$$P_x = \frac{1}{T_0} \int_0^{T_0} x^2(t)dt = 1$$

or the integral for the constant term, since the other integrals are zero. In this case we used the periodicity of $x(t)$ to calculate the power directly.

The power calculation is more involved for $y(t)$, because it is not periodic with harmonically related frequencies so we have to consider each of its components. Indeed, letting

$$y(t) = \underbrace{\cos(2\pi t)}_{y_1(t)} + \underbrace{\cos(2t)}_{y_2(t)}$$

we have the power of $y(t)$ is

$$P_y = \lim_{T\to\infty} \frac{1}{2T} \int_{-T}^{T} [\cos(2\pi t) + \cos(2t)]^2 dt$$

$$= \lim_{T\to\infty} \frac{1}{2T} \int_{-T}^{T} \cos^2(2\pi t)dt + \lim_{T\to\infty} \frac{1}{2T} \int_{-T}^{T} \cos^2(2t)dt + \lim_{T\to\infty} \frac{1}{2T} \int_{-T}^{T} 2\cos(2\pi t)\cos(2t)dt$$

$$= P_{y_1} + P_{y_2} + \lim_{T\to\infty} \frac{1}{2T} \int_{-T}^{T} [\cos(2(\pi + 1)t) + \cos(2(\pi - 1))t]dt$$

$$= P_{y_1} + P_{y_2} + 0 = 0.5 + 0.5 = 1.$$

So that the power of $y(t)$ is the sum of the powers of its components $y_1(t)$ and $y_2(t)$.

Likewise if we let $x(t) = \cos(2\pi t) + \cos(4\pi t) = x_1(t) + x_2(t)$ then $P_{x_1} = P_{x_2} = 0.5$ and

$$P_x = P_{x_1} + P_{x_2} = 1.$$

Thus the power of a signal composed of sinusoids is the sum of the power of each of the components, whether or not the frequencies of the sinusoids are harmonically related.   □

The power of a sum of sinusoids,

$$x(t) = \sum_k A_k \cos(\Omega_k t) = \sum_k x_k(t) \tag{1.13}$$

with harmonically or non-harmonically related frequencies $\{\Omega_k\}$, is the sum of the power of each of the sinusoidal components,

$$P_x = \sum_k P_{x_k} \tag{1.14}$$

1.4 REPRESENTATION OF CONTINUOUS-TIME SIGNALS USING BASIC SIGNALS

A fundamental idea in signal processing is to represent signals in terms of basic signals which we know how to process. In this section we consider some of these basic signals (complex exponentials, sinusoids, impulse, unit step and ramp) which will be used to represent signals and for which we are able to obtain their responses in a simple way, as it will be shown in the next chapter.

1.4.1 COMPLEX EXPONENTIALS

A **complex exponential** is a signal of the form

$$
\begin{aligned}
x(t) &= Ae^{at} \\
&= |A|e^{rt}\left[\cos(\Omega_0 t + \theta) + j\sin(\Omega_0 t + \theta)\right] \quad -\infty < t < \infty,
\end{aligned} \tag{1.15}
$$

where $A = |A|e^{j\theta}$, and $a = r + j\Omega_0$ are complex numbers.

Using Euler's identity, and the definitions of A and a, we have $x(t) = Ae^{at}$ equaling

$$
x(t) = |A|e^{j\theta}e^{(r+j\Omega_0)t} = |A|e^{rt}e^{j(\Omega_0 t + \theta)} = |A|e^{rt}\left[\cos(\Omega_0 t + \theta) + j\sin(\Omega_0 t + \theta)\right].
$$

We will see later that complex exponentials are fundamental in the Fourier representation of signals.

Remarks

1. Suppose that A and a are real, then

$$
x(t) = Ae^{at} \qquad -\infty < t < \infty
$$

is a decaying exponential if $a < 0$, and a growing exponential if $a > 0$. The top left figure in Fig. 1.7 illustrates the case of a decaying real exponential $e^{-0.5t}$, $a = -0.5 < 0$, while the figure on the top right the case of a growing exponential $e^{0.5t}$, $a = 0.5 > 0$.
2. If A is real, but $a = j\Omega_0$ then we have

$$
x(t) = Ae^{j\Omega_0 t} = A\cos(\Omega_0 t) + jA\sin(\Omega_0 t)
$$

where the real part of $x(t)$ is $\mathcal{R}e[x(t)] = A\cos(\Omega_0 t)$ and the imaginary part of $x(t)$ is $\mathcal{I}m[x(t)] = A\sin(\Omega_0 t)$.
3. If $A = |A|e^{j\theta}$, and $a = r + j\Omega_0$, $x(t) = Ae^{at}$, $-\infty < t < \infty$, is a complex signal and we need to consider separately its real and its imaginary parts. The real part function is

$$
f(t) = \mathcal{R}e[x(t)] = |A|e^{rt}\cos(\Omega_0 t + \theta)
$$

while the imaginary part function is

$$
g(t) = \mathcal{I}m[x(t)] = |A|e^{rt}\sin(\Omega_0 t + \theta).
$$

The envelope of $f(t)$ can be found by considering that $-1 \le \cos(\Omega_0 t + \theta) \le 1$ and that when multiplied by $|A|e^{rt} > 0$ we have $-|A|e^{rt} \le |A|e^{rt}\cos(\Omega_0 t + \theta) \le |A|e^{rt}$, so that

$$
-|A|e^{rt} \le f(t) \le |A|e^{rt}.
$$

Whenever $r < 0$ the $f(t)$ signal is a damped sinusoid, and when $r > 0$ then $f(t)$ grows. The bottom-left figure in Fig. 1.7 illustrates the case of a decaying modulated exponential $e^{-0.5t}\cos(2\pi t)$, as $r = -0.5 < 0$, while the bottom-right figure shows the case of a growing modulated exponential $e^{0.5t}\cos(2\pi t)$, as $r = 0.5 > 0$.

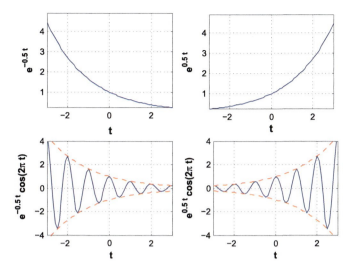

FIGURE 1.7

Analog exponentials: decaying exponential (top left), growing exponential (top right), and modulated exponential decaying and growing (bottom left and right).

Sinusoids

A sinusoid is of the general form

$$A\cos(\Omega_0 t + \theta) = A\sin(\Omega_0 t + \theta + \pi/2) \qquad -\infty < t < \infty \qquad (1.16)$$

where A is the amplitude of the sinusoid, $\Omega_0 = 2\pi f_0$ (rad/s) is its analog frequency, and θ its phase shift. The fundamental period T_0 of the above sinusoid is inversely related to the frequency:

$$\Omega_0 = 2\pi f_0 = \frac{2\pi}{T_0}.$$

The cosine and the sine signals, as indicated above, are out of phase by $\pi/2$ radians. The frequency of a sinusoid $\Omega_0 = 2\pi f_0$ is in radians per second, so that f_0 is in Hz (or 1/s units). The fundamental period of a sinusoid is found by the relation $f_0 = 1/T_0$ (it is important to point out the inverse relation between time and frequency shown here, which will be important in the representation of signals later on).

Recall from Chapter 0 that Euler's identity provides the relation of the sinusoids with the complex exponential

$$e^{j\Omega_0 t} = \cos(\Omega_0 t) + j\sin(\Omega_0 t) \qquad (1.17)$$

so that we will be able to represent in terms of sines and cosines any signal that is represented in terms of complex exponentials. Likewise, Euler's identity also permits us to represent sines and cosines in

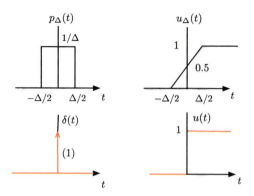

FIGURE 1.8

Generation of $\delta(t)$ and $u(t)$ from the limit as $\Delta \to 0$ of a pulse $p_\Delta(t)$ and its integral $u_\Delta(t)$.

terms of complex exponentials, since

$$
\cos(\Omega_0 t) = \frac{1}{2}(e^{j\Omega_0 t} + e^{-j\Omega_0 t}),
$$

$$
\sin(\Omega_0 t) = \frac{1}{2j}(e^{j\Omega_0 t} - e^{-j\Omega_0 t}). \tag{1.18}
$$

1.4.2 UNIT STEP, UNIT IMPULSE AND RAMP SIGNALS

Consider a rectangular pulse of duration Δ and unit area

$$
p_\Delta(t) = \begin{cases} \frac{1}{\Delta} & -\Delta/2 \le t \le \Delta/2, \\ 0 & t < -\Delta/2 \text{ and } t > \Delta/2. \end{cases} \tag{1.19}
$$

Its integral from $-\infty$ to t is

$$
u_\Delta(t) = \int_{-\infty}^{t} p_\Delta(\tau)d\tau = \begin{cases} 1 & t > \Delta/2, \\ \frac{1}{\Delta}(t + \frac{\Delta}{2}) & -\Delta/2 \le t \le \Delta/2, \\ 0 & t < -\Delta/2. \end{cases} \tag{1.20}
$$

The pulse $p_\Delta(t)$ and its integral $u_\Delta(t)$ are shown in Fig. 1.8.

Suppose that $\Delta \to 0$, then:

- the pulse $p_\Delta(t)$ still has a unit area but is an extremely narrow pulse. We will call the limit the **unit-impulse signal**

$$
\delta(t) = \lim_{\Delta \to 0} p_\Delta(t), \tag{1.21}
$$

which is zero for all values of t except at $t = 0$ when its value is not defined,

- the integral $u_\Delta(t)$, as $\Delta \to 0$, has a left-side limit of $u_\Delta(-\epsilon) \to 0$ and a right-side limit of $u_\Delta(\epsilon) \to 1$, for an infinitesimal $\epsilon > 0$, and at $t = 0$ it is $1/2$. Thus the limit is

$$\lim_{\Delta \to 0} u_\Delta(t) = \begin{cases} 1 & t > 0, \\ 1/2 & t = 0, \\ 0 & t < 0. \end{cases} \tag{1.22}$$

Ignoring the value at $t = 0$ we define the **unit-step signal** as

$$u(t) = \begin{cases} 1 & t > 0, \\ 0 & t < 0. \end{cases}$$

You can think of the $u(t)$ as the switching at $t = 0$ of a dc signal generator from off to on, while $\delta(t)$ is a very strong pulse of very short duration. We can then summarize the above definitions as follows.

The **unit-impulse signal** $\delta(t)$

- is zero everywhere except at the origin where its value is not well defined, i.e., $\delta(t) = 0, t \neq 0$, undefined at $t = 0$,
- has an integral

$$\int_{-\infty}^{t} \delta(\tau)d\tau = \begin{cases} 1 & t > 0, \\ 0 & t < 0, \end{cases} \tag{1.23}$$

so that the area under the impulse is unity.

The **unit-step** signal is

$$u(t) = \begin{cases} 1 & t > 0, \\ 0 & t < 0. \end{cases}$$

The $\delta(t)$ and $u(t)$ are related as follows:

$$u(t) = \int_{-\infty}^{t} \delta(\tau)d\tau, \tag{1.24}$$

$$\delta(t) = \frac{du(t)}{dt}. \tag{1.25}$$

According to the fundamental theorem of calculus, indicating that indefinite integration can be reversed by differentiation, we see that if

$$u_\Delta(t) = \int_{-\infty}^{t} p_\Delta(\tau)d\tau \quad \text{then} \quad p_\Delta(t) = \frac{du_\Delta(t)}{dt}.$$

Letting $\Delta \to 0$ we obtain the relation between $u(t)$ and $\delta(t)$ from the above equations.

Remarks

1. Since $u(t)$ is not a continuous function, it jumps from 0 to 1 instantaneously at $t = 0$, from the calculus point of view it should not have a derivative. That $\delta(t)$ is its derivative must be taken with suspicion, which makes the $\delta(t)$ signal also suspicious. Such signals can, however, be formally defined using the theory of distributions [10].

2. Although it is impossible to physically generate the impulse signal $\delta(t)$ it characterizes very brief pulses of any shape. It can be derived using functions different from the rectangular pulse we considered (see Equation (1.19)). In the problems at the end of the chapter, we consider how it can be derived from either a triangular pulse or a sinc function of unit area.

3. Signals with jump discontinuities can be represented as the sum of a continuous signal and unit-step signals. This is useful in computing the derivative of these signals.

The Ramp Signal

The **ramp** signal is defined as

$$r(t) = t\, u(t). \tag{1.26}$$

The relation between the ramp, the unit-step, and the unit-impulse signals is given by

$$\frac{dr(t)}{dt} = u(t), \tag{1.27}$$

$$\frac{d^2r(t)}{dt^2} = \delta(t). \tag{1.28}$$

The ramp is a continuous function and its derivative is given by

$$\frac{dr(t)}{dt} = \frac{dtu(t)}{dt} = u(t) + t\frac{du(t)}{dt} = u(t) + t\,\delta(t) = u(t) + 0\,\delta(t) = u(t).$$

Example 1.15. Consider the discontinuous signals

$$x_1(t) = \cos(2\pi t)[u(t) - u(t-1)], \quad x_2(t) = u(t) - 2u(t-1) + u(t-2).$$

Represent each of these signals as the sum of a continuous signal and unit-step signals, and find their derivatives.

Solution: The signal $x_1(t)$ is a period of $\cos(2\pi t)$, from 0 to 1, and zero otherwise. It displays discontinuities at $t = 0$ and at $t = 1$. Subtracting $u(t) - u(t-1)$ from $x_1(t)$ we obtain a continuous signal $x_{1a}(t)$, but to compensate we must add a unit pulse $x_{1b}(t)$ between $t = 0$ and $t = 1$, giving

$$x_1(t) = \underbrace{(\cos(2\pi t) - 1)[u(t) - u(t-1)]}_{x_{1a}(t)} + \underbrace{[u(t) - u(t-1)]}_{x_{1b}(t)}.$$

See Fig. 1.9 to visualize this decomposition.

The derivative is

$$\begin{aligned}
\frac{dx_1(t)}{dt} &= -2\pi \sin(2\pi t)[u(t) - u(t-1)] + (\cos(2\pi t) - 1)[\delta(t) - \delta(t-1)] + \delta(t) - \delta(t-1) \\
&= -2\pi \sin(2\pi t)[u(t) - u(t-1)] + \delta(t) - \delta(t-1) \tag{1.29}
\end{aligned}$$

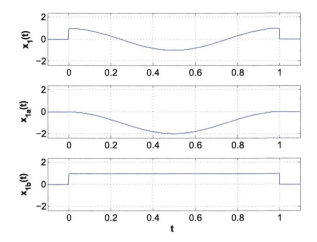

FIGURE 1.9

Decomposition of $x_1(t) = \cos(2\pi t)[u(t) - u(t-1)]$ (top) into a continuous (middle) and a discontinuous signal (bottom).

since

$$
\begin{aligned}
(\cos(2\pi t) - 1)[\delta(t) - \delta(t-1)] &= (\cos(2\pi t) - 1)\delta(t) - (\cos(2\pi t) - 1)\delta(t-1) \\
&= (\cos(0) - 1)\delta(t) - (\cos(2\pi) - 1)\delta(t-1) \\
&= 0\delta(t) - 0\delta(t-1) = 0.
\end{aligned}
$$

The term $\delta(t)$ in the derivative (see Equation (1.29)) indicates that there is a jump from 0 to 1 in $x_1(t)$ at $t=0$, and the term $-\delta(t-1)$ indicates that there is a jump of -1 (from 1 to zero) at $t=1$.

The signal $x_2(t)$, Fig. 1.10, has jump discontinuities at $t=0$, $t=1$, and $t=2$ and we can think of it as completely discontinuous so that its continuous component is 0. Its derivative is

$$
\frac{dx_2(t)}{dt} = \delta(t) - 2\delta(t-1) + \delta(t-2).
$$

The values of 1, -2 and 1 associated with $\delta(t)$, $\delta(t-1)$ and $\delta(t-2)$ coincide with the jump from 0 to 1 (positive jump of 1) at $t=0$, from 1 to -1 (negative jump of -2) at $t=1$, and finally from -1 to 0 (positive jump of 1) at $t=2$.   □

Signal Generation With MATLAB

In the following examples we illustrate how to generate continuous-time signals using MATLAB. This is done by either approximating continuous-time signals by discrete-time signals or by using symbolic MATLAB. The function *plot* uses an interpolation algorithm that makes the plots of discrete-time signals look like continuous-time signals.

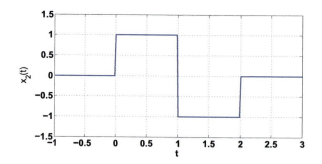

FIGURE 1.10

The signal $x_2(t)$ has discontinuities at $t = 0$, 1 and 2.

Example 1.16. Write a script and the necessary functions to generate a signal

$$y(t) = 3r(t+3) - 6r(t+1) + 3r(t) - 3u(t-3)$$

plot it and verify analytically that the obtained figure is correct.

Solution: The functions *ramp* and *unitstep* shown below generate ramp and unit-step signals for obtaining a numeric approximation of the signal $y(t)$. The following script shows how these functions are used to generate $y(t)$. The arguments of *ramp* determine the support of the signal, the slope and the shift (for advance a positive number and for delay a negative number). For *unitstep* we need to provide the support and the shift.

```
%%
% Example 1.16---Signal generation
%%
clear all; clf
Ts=0.01; t=-5:Ts:5;              % support of signal
% ramps with support [-5, 5]
y1=ramp(t,3,3);                  % slope of 3 and advanced by 3
y2=ramp(t,-6,1);                 % slope of -6 and advanced by 1
y3=ramp(t,3,0);                  % slope of 3
% unit-step signal with support [-5,5]
y4=-3*unitstep(t,-3);           % amplitude -3 and delayed by 3
y=y1+y2+y3+y4;
plot(t,y,'k'); axis([-5 5 -1 7]); grid
```

Analytically,

- $y(t) = 0$ for $t < -3$ and for $t > 3$, as we will see later, so the chosen support $-5 \le t \le 5$ displays the signal in a region where the signal is not zero;
- for $-3 \le t \le -1$, $y(t) = 3r(t+3) = 3(t+3)$ which is 0 at $t = -3$ and 6 at $t = -1$;

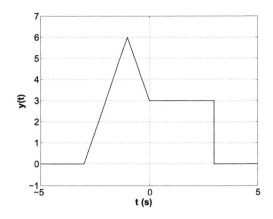

FIGURE 1.11

Generation of $y(t) = 3r(t+3) - 6r(t+1) + 3r(t) - 3u(t-3)$, $-5 \leq t \leq 5$, and zero otherwise.

- for $-1 \leq t \leq 0$, $y(t) = 3r(t+3) - 6r(t+1) = 3(t+3) - 6(t+1) = -3t + 3$ which is 6 at $t = -1$ and 3 at $t = 0$;
- for $0 \leq t \leq 3$, $y(t) = 3r(t+3) - 6r(t+1) + 3r(t) = -3t + 3 + 3t = 3$, and finally
- for $t \geq 3$ the signal $y(t) = 3r(t+3) - 6r(t+1) + 3r(t) - 3u(t-3) = 3 - 3 = 0$,

coinciding with the signal shown in Fig. 1.11.

Our functions *ramp* and *unitstep* are

```
function y=ramp(t,m,ad)
% ramp generation
% t: time support
% m: slope of ramp
% ad : advance (positive), delay (negative) factor
% USE: y=ramp(t,m,ad)
N=length(t);
y=zeros(1,N);
for i=1:N,
    if t(i)>=-ad,
    y(i)=m*(t(i)+ad);
    end
end

function y=unitstep(t,ad)
% generation of unit step
% t: time
% ad : advance (positive), delay (negative)
% USE y=unitstep(t,ad)
N=length(t);
```

```
y=zeros(1,N);
for i=1:N,
    if t(i)>=-ad,
    y(i)=1;
    end
end
```

☐

Example 1.17. Consider the following script that uses the functions *ramp* and *unitstep* to generate a signal $y(t)$. Obtain analytically the formula for the signal $y(t)$. Write a function to compute and to plot the even and odd components of $y(t)$.

```
%%
% Example 1.17---Signal generation
%%
    clear all; clf
    t=-5:0.01:5;
    y1=ramp(t,2,2.5);
    y2=ramp(t,-5,0);
    y3=ramp(t,3,-2);
    y4=unitstep(t,-4);
    y=y1+y2+y3+y4;
    plot(t,y,'k'); axis([-5 5 -3 5]); grid
```

The signal $y(t) = 0$ for $t < -5$ and $t > 5$.

Solution: The signal $y(t)$ displayed on the left of Fig. 1.12 is given analytically by

$$y(t) = 2r(t + 2.5) - 5r(t) + 3r(t - 2) + u(t - 4).$$

Indeed, we have

t	$y(t)$
$t \leq -2.5$	0
$-2.5 \leq t \leq 0$	$2(t + 2.5)$, so $y(-2.5) = 0$, $y(0) = 5$
$0 \leq t \leq 2$	$2(t + 2.5) - 5t = -3t + 5$, so $y(0) = 5$, $y(2) = -1$
$2 \leq t \leq 4$	$2(t + 2.5) - 5t + 3(t - 2) = -3t + 5 + 3t - 6 = -1$
$t \geq 4$	$2(t + 2.5) - 5t + 3(t - 2) + 1 = -1 + 1 = 0$

Clearly, $y(t)$ is neither even nor odd, to find its even and odd components we use the function *evenodd*, shown below, with inputs the signal and its support and outputs the even and odd components. The decomposition of $y(t)$ is shown on the right side of Fig. 1.12. Adding these two signals gives back the original signal $y(t)$. The script used is

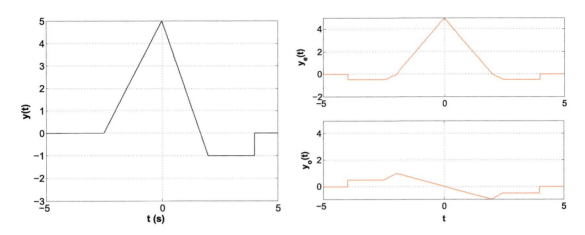

FIGURE 1.12

Signal $y(t) = 2r(t + 2.5) - 5r(t) + 3r(t - 2) + u(t - 4)$ (left), even component $y_e(t)$ (right-top), odd component $y_o(t)$ (right-bottom).

```
%%
% Example 1.17---Even and odd decomposition
%%
    [ye, yo]=evenodd(t,y);
    subplot(211)
    plot(t,ye,'r')
    grid
    axis([min(t) max(t) -2 5])
    subplot(212)
    plot(t,yo,'r')
     grid
    axis([min(t) max(t) -1 5])


    function [ye,yo]=evenodd(t,y)
    % even/odd decomposition
    % t: time
    % y: analog signal
    % ye, yo: even and odd components
    % USE [ye,yo]=evenodd(t,y)
    yr=fliplr(y);
    ye=0.5*(y+yr);
    yo=0.5*(y-yr);
```

The MATLAB function *fliplr* reverses the values of the vector y giving the reflected signal.   □

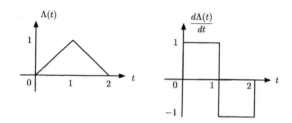

FIGURE 1.13

The triangular signal $\Lambda(t)$ and its derivative.

Example 1.18. Consider a triangular pulse

$$\Lambda(t) = \begin{cases} t & 0 \le t \le 1, \\ -t+2 & 1 < t \le 2, \\ 0 & \text{otherwise,} \end{cases}$$

represent it using ramp signals $r(t)$ and determine its derivative $d\Lambda(t)/dt$ in terms of unit-step signals.

Solution: The triangular signal can be rewritten as

$$\Lambda(t) = r(t) - 2r(t-1) + r(t-2). \tag{1.30}$$

In fact, since $r(t-1)$ and $r(t-2)$ have values different from 0 for $t \ge 1$ and $t \ge 2$, respectively, $\Lambda(t) = r(t) = t$ in $0 \le t \le 1$. For $1 \le t \le 2$, $\Lambda(t) = r(t) - 2r(t-1) = t - 2(t-1) = -t+2$. Finally, for $t > 2$ the three ramp signals are different from zero so $\Lambda(t) = r(t) - 2r(t-1) + r(t-2) = t - 2(t-1) + (t-2) = 0$ and by definition $\Lambda(t)$ is zero for $t < 0$. So the given expression for $\Lambda(t)$ in terms of the ramp functions is identical to given mathematical definition of it.

Using the mathematical definition of the triangular signal, its derivative is given by

$$\frac{d\Lambda(t)}{dt} = \begin{cases} 1 & 0 \le t \le 1, \\ -1 & 1 < t \le 2, \\ 0 & \text{otherwise.} \end{cases}$$

Using the representation in Equation (1.30) this derivative is also given by

$$\frac{d\Lambda(t)}{dt} = u(t) - 2u(t-1) + u(t-2),$$

which are two unit square pulses as shown in Fig. 1.13.   □

Example 1.19. Consider a full-wave rectified signal $x(t) = |\cos(2\pi t)|$, $-\infty < t < \infty$, of fundamental period $T_0 = 0.5$ (see Fig. 1.14). Obtain a representation for a period of $x(t)$ between 0 and 0.5, and use it to represent $x(t)$ in terms of shifted versions of it. A full-wave rectified signal is used in designing dc sources. Obtaining this signal is a first step in converting an alternating voltage into a dc voltage.

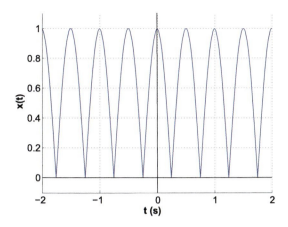

FIGURE 1.14

Eight periods of full-wave rectified signal $x(t) = |\cos(2\pi t)|$, $-\infty < t < \infty$.

Solution: The period between 0 and 0.5 can be expressed as

$$p(t) = x(t)[u(t) - u(t - 0.5)] = |\cos(2\pi t)|[u(t) - u(t - 0.5)].$$

Since $x(t)$ is a periodic signal of period $T_0 = 0.5$, we have

$$x(t) = \sum_{k=-\infty}^{\infty} p(t - kT_0). \quad \Box$$

Example 1.20. Generate a causal train of pulses $\rho(t)$ that repeats every 2 units of time using as first period $s(t) = u(t) - 2u(t - 1) + u(t - 2)$. Find the derivative of the train of pulses.

Solution: Considering that $s(t)$ is the first period of the train of pulses of period 2 then

$$\rho(t) = \sum_{k=0}^{\infty} s(t - 2k)$$

is the desired signal. Notice that $\rho(t)$ equals zero for $t < 0$, thus it is causal. Given that the derivative of a sum of signals is the sum of the derivative of each of the signals, the derivative of $\rho(t)$ is

$$\frac{d\rho(t)}{dt} = \sum_{k=0}^{\infty} \frac{ds(t - 2k)}{dt} = \sum_{k=0}^{\infty} [\delta(t - 2k) - 2\delta(t - 1 - 2k) + \delta(t - 2 - 2k)],$$

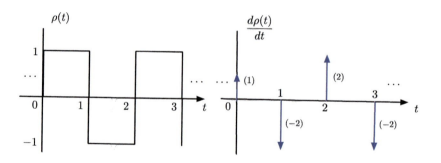

FIGURE 1.15

Causal train of pulses $\rho(t)$ and its derivative. The number enclosed in ( ) is the area of the corresponding delta signal and it indicates the jump at the particular discontinuity, positive when increasing and negative when decreasing.

which can be simplified to

$$\frac{d\rho(t)}{dt} = [\delta(t) - 2\delta(t-1) + \delta(t-2)] + [\delta(t-2) - 2\delta(t-3) + \delta(t-4)] + [\delta(t-4)\cdots$$

$$= \delta(t) + 2\sum_{k=1}^{\infty} \delta(t-2k) - 2\sum_{k=1}^{\infty} \delta(t-2k+1)$$

where $\delta(t)$, $2\delta(t-2k)$ and $-2\delta(t-2k+1)$ for $k \geq 1$ occur at $t = 0$, $t = 2k$ and $t = 2k - 1$ for $k \geq 1$, or the times at which the discontinuities of $\rho(t)$ occur. The value associated with the $\delta(t)$ corresponds to the jump of the signal from the left to the right. Thus $\delta(t)$ indicates there is a discontinuity of 1 in $\rho(t)$ at zero as it jumps from 0 to 1, while the discontinuities at 2, 4, $\cdots$ have a jump of 2 from -1 to 1, increasing. The discontinuities indicated by $\delta(t - 2k - 1)$ occurring at 1, 3, 5, $\cdots$ are from 1 to -1, i.e., decreasing so the value of -2. See Fig. 1.15. □

Example 1.21. Let us determine if the unit-impulse, the unit-step and the ramp functions are of finite energy and finite power or not.

Solution: Computing the energy of a pulse

$$p(t) = \begin{cases} 1/\Delta & -0.5\Delta \leq t \leq 0.5\Delta, \\ 0 & \text{otherwise,} \end{cases}$$

gives

$$E_p = \int_{-0.5\Delta}^{0.5\Delta} \frac{1}{\Delta^2} dt = \frac{1}{\Delta}$$

and as $\Delta \to 0$ then $p(t) \to \delta(t)$ and the energy of the resulting unit-impulse is infinite. However, the power of $\delta(t)$ is

$$P_\delta = \lim_{T \to \infty} \frac{1}{T} \int_{-T/2}^{T/2} \delta^2(t)dt = \lim_{T \to \infty, \Delta \to 0} \frac{E_p}{T},$$

i.e., undetermined since it would be a ratio of infinite quantities.

The energies of $u(t)$ and $r(t)$ are infinite, indeed

$$E_u = \int_0^\infty u(t)dt \to \infty, \qquad E_r = \int_0^\infty r(t)dt \to \infty$$

and their powers are

$$P_u = \lim_{T \to \infty} \frac{E_u}{T}, \qquad P_r = \lim_{T \to \infty} \frac{E_r}{T},$$

which as ratios of infinite quantities they are undetermined.   □

1.4.3 GENERIC REPRESENTATION OF SIGNALS

Consider the integral

$$\int_{-\infty}^\infty f(t)\delta(t)dt.$$

The product of $f(t)$ and $\delta(t)$ gives zero everywhere except at the origin where we get an impulse of area $f(0)$, that is, $f(t)\delta(t) = f(0)\delta(t)$ (let $t_0 = 0$ in Fig. 1.16); therefore

$$\int_{-\infty}^\infty f(t)\delta(t)dt = \int_{-\infty}^\infty f(0)\delta(t)dt = f(0) \int_{-\infty}^\infty \delta(t)dt = f(0), \qquad (1.31)$$

since the area under the impulse is unity. This property of the impulse function is appropriately called the **sifting property**. The aim of this integration is to sift out $f(t)$ for all t except for $t = 0$, where $\delta(t)$ occurs.

The above result can be generalized. If we delay or advance the $\delta(t)$ function in the integrand, the result is that all values of $f(t)$ are sifted out except for the value corresponding to the location of the delta function, that is,

$$\int_{-\infty}^\infty f(t)\delta(t - \tau)dt = \int_{-\infty}^\infty f(\tau)\delta(t - \tau)dt = f(\tau) \int_{-\infty}^\infty \delta(t - \tau)dt = f(\tau) \qquad \text{for any } \tau,$$

since the last integral is unity. Fig. 1.16 illustrates the multiplication of a signal $f(t)$ by an impulse signal $\delta(t - t_0)$, located at $t = t_0$.

> By the sifting property of the impulse function $\delta(t)$, any signal $x(t)$ can be represented by the following **generic representation**:
>
> $$x(t) = \int_{-\infty}^\infty x(\tau)\delta(t - \tau)d\tau. \qquad (1.32)$$

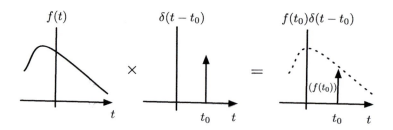

FIGURE 1.16

Multiplication of a signal $f(t)$ by an impulse signal $\delta(t - t_0)$.

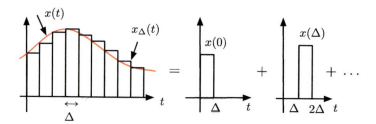

FIGURE 1.17

Generic representation of $x(t)$ as an infinite sum of pulses of height $x(k\Delta)$ and width Δ. When $\Delta \to 0$ the sum becomes and integral of weighted unit-impulse signals.

This generic representation can be visualized as in Fig. 1.17. Equation (1.32) basically indicates that any signal can be viewed as a stacking of pulses $x_\Delta(t - k\Delta) = x(k\Delta)p_\Delta(t - k\Delta)\Delta$ of width Δ and height $x(k\Delta)$ (i.e., pulse $p_\Delta(t - k\Delta)$ has height $1/\Delta$ but unit area and it is shifted by $k\Delta$). Thus an approximation of $x(t)$ is given by

$$x_\Delta(t) = \sum_{k=-\infty}^{\infty} x_\Delta(t - k\Delta) = \sum_{k=-\infty}^{\infty} x(k\Delta)p_\Delta(t - k\Delta)\Delta.$$

In the limit as $\Delta \to 0$ these pulses become impulses, separated by an infinitesimal value, or

$$\lim_{\Delta \to 0} x_\Delta(t) \quad \to \quad x(t) = \int_{-\infty}^{\infty} x(\tau)\delta(t - \tau)d\tau.$$

Equation (1.32) provides a generic representation of a signal in terms of basic signals, in this case impulse signals. As we will see in the next chapter, once we determine the response of a system to an impulse we will use the generic representation to find the response of the system to any signal.

1.5 TIME SCALING, MODULATION, WINDOWING AND INTEGRATION

In communications, control and filtering there is the need for more specific signal operations than those presented in subsection 1.3.1 such as time scaling, modulation, windowing, and integration that are described as follows:

Given a signal $x(t)$, and real values $\alpha \neq 0$ or 1, and $\phi > 0$, we have

1. **Time scaling**: $x(\alpha t)$ is said to be

   - **reflected** if $\alpha = -1$,
   - **contracted** if $|\alpha| > 1$, and if $\alpha < 0$ it is also **reflected**,
   - **expanded** if $|\alpha| < 1$, and if $\alpha < 0$ it is also **reflected**.

2. **Modulation**:

   - $x(t)e^{j\phi t}$ is said to be **modulated or shifted in frequency** by ϕ radians.

3. **Windowing**:

   - For a **window signal** $w(t)$, the **time-windowed** signal $x(t)w(t)$ displays $x(t)$ within the support of $w(t)$.

4. **Integration**:

   - The **integral** of $x(t)$ is the signal

$$y(t) = \int_{-\infty}^{t} x(\tau)d\tau.$$

Time scaling is an operation that changes the characteristics of the original signal, in time as well as in frequency. For instance, consider a signal $x(t)$ with a finite support $t_0 \leq t \leq t_1$, assume that $\alpha > 1$ then $x(\alpha t)$ is defined in $t_0 \leq \alpha t \leq t_1$ or $t_0/\alpha \leq t \leq t_1/\alpha$, a smaller support than the original one. Thus, if $\alpha = 2$, $t_0 = 2$ and $t_1 = 4$ then the support of $x(2t)$ is $1 \leq t \leq 2$ while the support of $x(t)$ is $2 \leq t \leq 4$. If $\alpha = -2$ then $x(-2t)$ is not only contracted but also reflected. On the other hand, $x(0.5t)$ would have a support $2t_0 \leq t \leq 2t_1$, larger than the original support. The change in frequency can be illustrated with a periodic signal $x(t) = \cos(2\pi t)$ of fundamental period $T_o = 1$ s, while the contracted signal $x_1(t) = x(2t) = \cos(4\pi t)$ has a fundamental period $T_1 = 1/2$ and the expanded signal $x_2(t) = x(t/2) = \cos(\pi t)$ a fundamental period $T_2 = 2$, and the original frequency of 1 Hz of the original signal $x(t)$ is changed to a frequency of 2 Hz for $x_1(t)$ and 1/2 Hz for $x_2(t)$, i.e., they are different from the original signal.

Example 1.22. An acoustic signal $x(t)$ has a duration of 3.3 minutes and a radio station would like to use the signal for a 3 minute segment. Indicate how to make it possible.

Solution: We need to contract the signal by a factor of $\alpha = 3.3/3 = 1.1$, so that $x(1.1t)$ can be used in the 3-min piece. If the signal is recorded on tape, the tape player can be ran 1.1 times faster than the recording speed. This would change the voice or music on the tape, as the frequencies in $x(1.1t)$ are increased with respect to the original frequencies in $x(t)$. □

Multiplication by a complex exponential shifts the frequency of the original signal. To illustrate this consider the case of an exponential $x(t) = e^{j\Omega_0 t}$ of frequency Ω_0, if we multiply $x(t)$ by an exponential $e^{j\phi t}$ then

$$x(t)e^{j\phi t} = e^{j(\Omega_0 + \phi)t} = \cos((\Omega_0 + \phi)t) + j\sin((\Omega_0 + \phi)t)$$

so that the frequency of the new exponential is greater than Ω_0 if $\phi > 0$ or smaller if $\phi < 0$. Thus, we have shifted the frequency of $x(t)$. If we have a sum of exponentials (they do not need to be harmonically related as in the Fourier series we will consider later)

$$x(t) = \sum_k A_k e^{j\Omega_k t} \quad \text{then} \quad x(t)e^{j\phi t} = \sum_k A_k e^{j(\Omega_k + \phi)t}$$

so that each of the frequency components of $x(t)$ has been shifted to new frequencies $\{\Omega_k + \phi\}$. This shifting of the frequency is significant in the development of amplitude modulation, and as such this frequency-shift process is called **modulation**, i.e., the signal $x(t)$ modulates the complex exponential and $x(t)e^{j\phi t}$ is the modulated signal.

A sinusoid is characterized by its amplitude, frequency and phase. When we allow these three parameters to be functions of time, or

$$A(t)\cos(\Omega(t)t + \theta(t)),$$

different types of modulation systems in communications are obtained:

- **Amplitude modulation or AM:** the amplitude $A(t)$ changes according to the message, while the frequency and the phase are constant.
- **Frequency modulation or FM:** the frequency $\Omega(t)$ changes according to the message, while the amplitude and phase are constant.
- **Phase modulation or PM:** the phase $\theta(t)$ varies according to the message and the other parameters are kept constant.

We can thus summarize the above as follows:

---

1. If $x(t)$ is periodic of fundamental period T_0 then the **time-scaled signal** $x(\alpha t)$, $\alpha \neq 0$, is also periodic of fundamental period $T_0/|\alpha|$.

2. The frequencies present in a signal can be changed by **modulation**, i.e., multiplying the signal by a complex exponential or, equivalently, by sines and cosines. The frequency change is also possible by expansion and compression of the signal.

3. Reflection is a special case of time scaling with $\alpha = -1$.

Example 1.23. Let $x_1(t)$, $0 \le t \le T_0$, be a period of a periodic signal $x(t)$ of fundamental period T_0. Represent $x(t)$ in terms of advanced and delayed versions of $x_1(t)$. What would be $x(2t)$?

Solution: The periodic signal $x(t)$ can be written

$$
\begin{aligned}
x(t) &= \cdots + x_1(t + 2T_0) + x_1(t + T_0) + x_1(t) + x_1(t - T_0) + x_1(t - 2T_0) + \cdots \\
&= \sum_{k=-\infty}^{\infty} x_1(t - kT_0)
\end{aligned}
$$

and the contracted signal $x(2t)$ is then

$$
x(2t) = \sum_{k=-\infty}^{\infty} x_1(2t - kT_0)
$$

and periodic of fundamental period $T_0/2$.   □

Example 1.24. Use symbolic MATLAB to generate a damped sinusoid signal $y(t) = e^{-t} \cos(2\pi t)$, and its envelope. Let then $f(t) = y(t)u(t)$ and find using symbolic MATLAB $f(0.5t - 1)$, $f(2t - 2)$ and $f(1.2t - 5)$ and plot them.

Solution: The following script, using symbolic MATLAB, generates $y(t)$.

```
%%
% Example 1.24---Damped sinusoid, scaling and shifting
%%
% Generation y(t) and its envelope
t=sym('t');
y=exp(-t)*cos(2*pi*t);
ye=exp(-t);
figure(1)
fplot(y,[-2,4]); grid
hold on
fplot(ye,[-2,4])
hold on
fplot(-ye,[-2,4]); axis([-2 4 -8 8])
hold off
xlabel('t'); ylabel('y(t)'); title('')
```

The desired scaled and shifted damped sinusoids are obtained using the following symbolic script:

```
%%
% Scaled and shifted damped sinusoid
%%
for n=1:3
    if n==1
    a=0.5;b=1;
    elseif n==2
    a=2;b=2
```

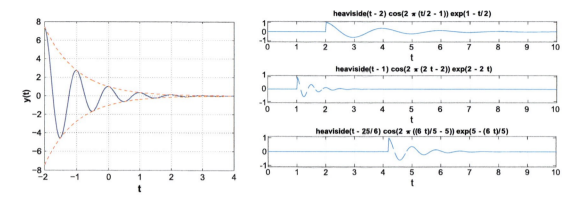

FIGURE 1.18

Damped sinusoid $y(t)$ and its envelope (left); scaled and shifted damped $f(t)$.

```
    else
    a=1.2; b=5
end
figure(2)
syms t f
f(t)=cos(2*pi*(a*t-b))*exp(-(a*t-b))*heaviside(a*t-b)
subplot(3,1,n)
ezplot(f,[0 10]);grid
axis([0 10 -1.1 1.1])
end
```

The plots of the damped sinusoid and the scaled and shifted $f(t)$ are given in Fig. 1.18.   □

Example 1.25. Use the *ramp* and *unitstep* functions given before to generate a signal

$$y(t) = 2r(t+2) - 4r(t) + 3r(t-2) - r(t-3) - u(t-3)$$

to modulate a so-called carrier signal $x(t) = \sin(5\pi t)$ to give the AM modulated signal $z(t) = y(t)x(t)$. Obtain a script to generate the AM signal, and to plot it. Scale and shift the AM signal to obtain compressed $z(2.5t)$ and compressed and delayed signal $z(2.5t - 3)$ as well as expanded $z(0.35t)$ and expanded and advanced signal $z(0.35t + 2)$. (See Fig. 1.19.)

Solution: The signal $y(t)$ analytically equals

$$y(t) = \begin{cases} 0 & t < -2, \\ 2(t+2) = 2t+4 & -2 \le t < 0, \\ 2t+4-4t = -2t+4 & 0 \le t < 2, \\ -2t+4+3(t-2) = t-2 & 2 \le t < 3, \\ t-2-(t-3)-1 = 0 & 3 \le t. \end{cases}$$

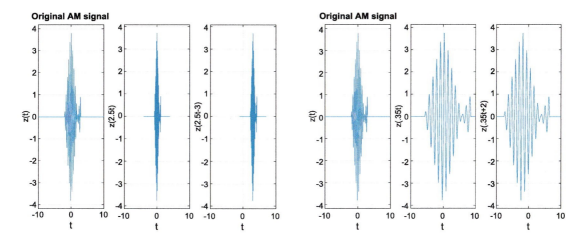

FIGURE 1.19

Original AM signal $z(t)$, compressed $z(2.5t)$ and compressed and delayed $z(2.5t - 3)$ (left); original $z(t)$, expanded $z(.35t)$ and expanded and advanced $z(.35t + 2)$ signals (right).

The following script is used to generate the message signal $y(t)$ and the AM signal $z(t)$, and the corresponding plots. The MATLAB function *sound* is used produce the sound corresponding to $100z(t)$.

```
%%
% Example 1.25---AM signal
%%
clear all;clf
    % Message signal
    T=5;Ts=0.001; t=-T:Ts:T;
    y1=ramp(t,2,2); y2=ramp(t,-4,0); y3=ramp(t,3,-2); y4=ramp(t,-1,-3);
    y5=unitstep(t,-3);
    y=y1+y2+y3+y4-y5;                          % message
    % AM modulation
    x=sin(5*pi*t);                             % carrier
    z=y.*x;
    sound(100*z,1000)
    figure(1)
    plot(t,z,'k'); hold on
    plot(t,y,'r',t,-y,'r'); axis([-5 5 -5 5]); grid
    hold off
    xlabel('t'); ylabel('z(t)'); title('AM signal')
    % scaling and shifting
       for n=1:2
       if n==1,
```

```
        gamma=2.5; sht=3;
        [z1,t1,t2]=scale_shift(z,gamma,sht,T,Ts); % compression, delay
        figure(2)
        subplot(131)
        plot(t,z); axis([-T T 1.1*min(z)  1.1*max(z)]); grid
        xlabel('t'); ylabel('z(t)'); title('Original AM signal')
        hold on
        subplot(132)
        plot(t1,z1);grid; axis([-T T 1.1*min(z1) 1.1*max(z1)])
        xlabel('t'); ylabel('z(2.5t)')
        hold on
        subplot(133)
        plot(t2,z1);grid; axis([-T T 1.1*min(z1) 1.1*max(z1)])
        xlabel('t'); ylabel('z(2.5t-3)')
        hold off
        else
        gamma=.35; sht=-2;
        [z1,t1,t2]=scale_shift(z,gamma,sht,T,Ts); % expansion, advance
        figure(3)
        subplot(131)
        plot(t,z); axis([-T T 1.1*min(z)  1.1*max(z)]); grid
        xlabel('t'); ylabel('z(t)'); title('Original AM signal')
        hold on
        subplot(132)
        plot(t1,z1);grid; axis([-T T 1.1*min(z1) 1.1*max(z1)])
        xlabel('t'); ylabel('z(.35t)')
        hold on
        subplot(133)
        plot(t2,z1);grid; axis([-T T 1.1*min(z1) 1.1*max(z1)])
        xlabel('t'); ylabel('z(.35t+2)')
        hold off
    end
end
```

Our function *scale_shift* was used to do the scaling and shifting.   □

```
    function [z3,t1,t2]=scale_shift (z,gamma,delay,T,Ts)
% performs scale and shift of digitized signal
% gamma positive real with two decimal
% shf positive real
% [-T T] range of signal
% Ts sampling period
    beta1=100;alpha1=round(gamma,2)*beta1;
    g=gcd(beta1,alpha1);beta=beta1/g;alpha=alpha1/g;
    z1=interp(z,beta);z2=decimate(z1,alpha);
```

```
t1=-T/gamma:Ts:T/gamma;
M=length(t1);
z3=z2(1:M);
t2=t1+delay;
```

Remark

It is important to understand the scaling and shifting of an analog signal using MATLAB:

- In Example 1.24, the scaling and shifting are done using symbolic MATLAB and so it appears as if we would have done the scaling and the shifting in theory. Recall that the function *heaviside* corresponds to the unit-step function.
- In Example 1.25, the analog signal is approximated by a discrete signal and in this case the scaling needs to be done using the discrete functions *interp* and *decimate*. Function *inter* expands the discrete signal by adding interpolated values into given samples, while *decimate* contracts the discrete signal by getting rid of samples. Both of these require integer scaling factors. If we wish to scale by a non-integer factor γ the discrete signal we need to express or approximate this factor by a rational number. Our function does this by rounding the scaling factor to the nearest hundredth and then finding a rational number for it. For instance if the scaling factor is $\gamma = 0.251$ it is rounded to 0.25 and then find the greatest common divisor to get an interpolation factor of $\beta = 4$ and a decimation factor of $\alpha = 1$.

Example 1.26. Consider the effect of time scaling on the unit-impulse, unit-step and ramp signals which as you recall are not regular functions. Let $u(t)$ be the unit-step function, what are $u(\alpha t)$ and $r(\alpha t)$ where $r(t)$ is the ramp function? What does the derivative $du(\alpha t)/dt$ say about $\delta(\alpha t)$?

Solution: We have for any $\alpha \neq 0$

$$u(\alpha t) = \begin{cases} 1 & t > 0/\alpha \text{ or } t > 0, \\ 0 & t \leq 0, \end{cases}$$

so $u(\alpha t) = u(t)$. Likewise, for the scaled ramp function we have

$$r(\alpha t) = \int_0^{\alpha t} u(\tau)d\tau = \begin{cases} \alpha t & t \geq 0, \\ 0 & t < 0, \end{cases}$$

or $r(\alpha t) = \alpha r(t)$. Thus the scaling does not seem to affect the unit step but it does affect the ramp. Consider then the derivative of $u(\alpha t)$:

$$\frac{du(\alpha t)}{dt} = \frac{du(\alpha t)}{d\alpha t} \frac{d\alpha t}{dt} = \alpha \delta(\alpha t),$$

but since $u(\alpha t) = u(t)$, and $\delta(t) = du(t)/dt$, according to the above $\delta(t) = \alpha \delta(\alpha t)$, so that

$$\delta(\alpha t) = \frac{1}{\alpha} \delta(t)$$

so it seems that the scaling affects the unit-impulse function.[1] Notice that, for $\alpha = -1$, we get $\delta(-t) = -\delta(t)$, but since the unit-impulse is only defined at 0, $\delta(-t) = \delta(t)$ or an even function.   □

In filtering applications one could like to consider the representation of a signal only in a finite time or frequency support and in such cases the concept of time-windowing or frequency-filtering are useful. Multiplying an infinite-support signal $x(t)$ by a time-window $w(t)$ of finite support gives a view of the original signal only in the support of the window. Likewise, if we are interested in what the signal looks like in a finite frequency support we could get by means of a filter, an almost-finite-support window in frequency. As the previous operations, windowing does change not only the time but also the frequency characteristic of the original signal. Finally, integration is a very important operation given that differential equations are used to represent systems. As indicated in Chapter 0, the solution of these equations uses integration.

Example 1.27. An approximation of an analog signal $x(t)$ is given by

$$x_{\Delta/M}(t) = \sum_{k=-\infty}^{\infty} x(k\Delta/M)p(Mt - k\Delta/M)$$

where the pulse $p(t)$ is given by

$$p(t) = \begin{cases} 1 & 0 \leq t \leq \Delta, \\ 0 & \text{otherwise.} \end{cases}$$

Δ has a small positive value and $M = 2^n, n = 0, 1, 2, \cdots$. Determine the definite integral

$$\int_{-\infty}^{\infty} x_{\Delta/M}(t)dt$$

for values of $\Delta/M \to 0$, i.e., for a fixed Δ, $M \to \infty$. Suppose $x(t) = t$ for $0 \leq t \leq 10$ and zero otherwise, how does the approximation for $M = 2^n, n = 0, 1, \cdots$ of the definite integral compares with the actual area. For some $0 < t \leq 10$, what is

$$y(t) = \int_0^t x(t)dt \ ?$$

Solution: The compressed pulse $p(Mt)$ is

$$p(Mt) = \begin{cases} 1 & 0 \leq t \leq \Delta/M, \\ 0 & \text{otherwise,} \end{cases}$$

---

[1] This is not a very intuitive result given that the support of the impulse function is zero, and the scaling does not change that. However, if we use the same pulse $p_\Delta(t)$ as we used in the derivation of $\delta(t)$ then $p_\Delta(\alpha t)$ is a pulse of width Δ/α and height $1/\Delta$, i.e., of area $1/\alpha$. Thus when $\Delta \to 0$ we get $\delta(\alpha t) = \delta(t)/\alpha$.

with an area equal to Δ/M and thus the integral becomes

$$\int_{-\infty}^{\infty} x_{\Delta/M}(t)dt = \sum_{k=-\infty}^{\infty} x(k\Delta/M) \int_{-\infty}^{\infty} p(Mt - k\Delta/M)dt = \sum_{k=-\infty}^{\infty} x(k\Delta/M)\frac{\Delta}{M}. \quad (1.33)$$

As $M \to \infty$, the above approximates the area under $x(t)$.

For the special case of $x(t) = t$ for $0 \le t \le 10$ and zero otherwise, we know $x(t)$ has an area of 50. The sum in the approximation goes in general from zero to $K/M = 10 - 1/M$, for an integer K, or $K = 10M - 1$ so letting $\Delta = 1$ we have

$$\sum_{k=0}^{10M-1} x(k/M)\frac{1}{M} = \frac{1}{M}\sum_{k=0}^{10M-1}\frac{k}{M} = \frac{1}{M^2}\frac{(10M)(10M-1)}{2} = 50 - \frac{5}{M}$$

and thus as M increases the result approaches the exact area of 50. For $M = 1, 2,$ and 4 the integral (1.33) equals 45, 47.5 and 48.75. If instead of 10 we let it have a value $t > 0$, then the undefined integral is given by

$$y(t) = \lim_{M\to\infty} \sum_{k=0}^{tM-1} \frac{x(k/M)}{M} = \lim_{M\to\infty}\left[\frac{t^2}{2} - \frac{t}{2M}\right] = \frac{t^2}{2},$$

coinciding with the actual value of the integral. □

Example 1.28. Given that the unit-impulse is not defined at $t = 0$, solving ordinary differential equations with input a unit-impulse and zero initial conditions is not very intuitive. This solution is called the impulse response of the system represented by a differential equation and it is very important. One way to resolve this is by using the relation that exists between the unit-step and the unit-impulse signals. Consider the first order differential equation representing a system

$$\frac{y(t)}{dt} + y(t) = x(t), \qquad t > 0,$$

where $x(t)$ is the input, $y(t)$ the output, and the initial condition is zero (i.e., $y(0) = 0$), find the impulse response, i.e., the output when the input is $\delta(t)$.

Solution: Calling the impulse response $h(t)$, the equation we would like to solve is

$$\frac{dh(t)}{dt} + h(t) = \delta(t), \qquad h(0) = 0. \quad (1.34)$$

The difficulty in solving this equation is that the input is undefined and possibly very large at $t = 0$ and makes it hard to understand how the system will display a zero initial response. Thus, instead if we let the input be $u(t)$ and call the corresponding output $z(t)$, then the differential equation for that case is

$$\frac{dz(t)}{dt} + z(t) = u(t), \qquad z(0) = 0. \quad (1.35)$$

In this case the input is still not defined at zero but is possibly not very large, so the response is more feasible. The derivative with respect to t of the two sides of this equation gives after replacing the connection between $u(t)$ and $\delta(t)$:

$$\frac{d^2z(t)}{dt^2} + \frac{dz(t)}{dt} = \frac{du(t)}{dt} = \delta(t),$$

which when compared to (1.34) indicates that $dz(t)/dt = h(t)$ or equivalently the integral of the impulse response $h(t)$ is

$$\int_{-\infty}^{t} h(\tau)d\tau = \int_{-\infty}^{t} \frac{dz(\tau)}{d\tau} d\tau = z(t).$$

Thus solving (1.35) for $z(t)$ will permit us to find $h(t)$. The solution $z(t)$ is[2]:

$$z(t) = (1 - e^{-t})u(t)$$

and the impulse response is

$$h(t) = \frac{dz(t)}{dt} = \delta(t) + e^{-t}u(t) - e^{-t}\delta(t) = e^{-t}u(t) + \delta(t) - \delta(t) = e^{-t}u(t).$$

The zero initial condition $z(0) = 0$ is satisfied as $1 - e^0 = 0$, and $z(0) = 0u(0) = 0$ independent of what value $u(0)$ takes. On the other hand, $h(0) = 1\, u(0)$ is not defined; however, assuming $h(0 - \epsilon) = h(0) = 0$ when $\epsilon \to 0$, i.e., the state of a system cannot change in an infinitesimal interval, we say that the zero initial condition is satisfied. None of this would be necessary if we use the relation between $u(t)$ and $\delta(t)$ and their corresponding responses. $\square$

1.6 SPECIAL SIGNALS—THE SAMPLING AND THE SINC SIGNALS

Two signals of great significance in the sampling of continuous-time signals and in their reconstruction are the sampling and the sinc signals. Sampling a continuous-time signal consists in taking samples of the signal at uniform times. One can think of this process as the multiplication of a continuous-time signal $x(t)$ by a periodic train of very narrow pulses of fundamental period T_s (the sampling period).

---

[2]The solution of (1.35) can be obtained as follows. Let the derivative operator $D() = d()/dt$, then the derivative of (1.35) becomes

$$D(D + 1)[z(t)] = 0, \qquad t > 0$$

given that $Du(t) = 0$ when $t > 0$ where $u(t) = 1$. The solution of this homogeneous equation is obtained using the roots of the characteristic polynomial $D(D + 1) = 0$ or 0 and -1 and thus the solution of this equation for $t \geq 0$ is $z(t) = Ae^{-t} + B$ where A and B are constants to be determined. The zero initial condition gives $z(0) = A + B = 0$ and for $t > 0$

$$\frac{dz(t)}{dt} + z(t) = [-Ae^{-t}] + [Ae^{-t} + B] = B = 1$$

so $B = 1$ and $A = -1$, giving $z(t) = (1 - e^{-t})u(t)$.

For simplicity, if the width of the pulses is much smaller than T_s the train of pulses can be approximated by a periodic train of impulses of fundamental period T_s, i.e., the **sampling signal**:

$$\delta_{T_s}(t) = \sum_{n=-\infty}^{\infty} \delta(t - nT_s). \tag{1.36}$$

The sampled signal $x_s(t)$ is then

$$x_s(t) = x(t)\delta_{T_s}(t) = \sum_{n=-\infty}^{\infty} x(nT_s)\delta(t - nT_s) \tag{1.37}$$

or a sequence of uniformly shifted impulses with amplitude the value of the signal $x(t)$ at the time when the impulse occurs.

A fundamental result in sampling theory is the recovery of the original signal, under certain constraints, by means of an interpolation using **sinc signals**. Moreover, we will see that the sinc is connected with ideal low-pass filters. The sinc function is defined as

$$S(t) = \frac{\sin \pi t}{\pi t} \qquad -\infty < t < \infty. \tag{1.38}$$

This signal has the following characteristics:
(1) It is an even function of t, as

$$S(-t) = \frac{\sin(-\pi t)}{-\pi t} = \frac{-\sin(\pi t)}{-\pi t} = S(t). \tag{1.39}$$

(2) At $t = 0$ the numerator and the denominator of the sinc are zero, thus the limit as $t \to 0$ is found using L'Hopital's rule, i.e.,

$$\lim_{t \to 0} S(t) = \lim_{t \to 0} \frac{d\sin(\pi t)/dt}{d\pi t/dt} = \lim_{t \to 0} \frac{\pi \cos(\pi t)}{\pi} = 1. \tag{1.40}$$

(3) $S(t)$ is bounded, i.e., since $-1 \le \sin(\pi t) \le 1$ for $t > 0$

$$\frac{-1}{\pi t} \le S(t) = \frac{\sin(\pi t)}{\pi t} \le \frac{1}{\pi t} \tag{1.41}$$

and given that $S(t)$ is even, it is equally bounded for $t < 0$. And since at $t = 0$ $S(0) = 1$, $S(t)$ is bounded for all t. As $t \to \pm\infty$, $S(t) \to 0$.
(4) The zero-crossing times of $S(t)$ are found by letting the numerator equal zero, i.e., when $\sin(\pi t) = 0$, the zero-crossing times are such that $\pi t = k\pi$, or $t = k$ for a nonzero integer k or $k = \pm 1, \pm 2, \cdots$.
(5) A property which is not obvious and that requires the frequency representation of $S(t)$ is that the integral

$$\int_{-\infty}^{\infty} |S(t)|^2 dt = 1. \tag{1.42}$$

Recall that we showed this in Chapter 0 using numeric and symbolic MATLAB.

Table 1.1 Basic signals

Signal	Definition/Properties
Damped complex exponential	$\|A\|e^{rt}\left[\cos(\Omega_0 t + \theta) + j\sin(\Omega_0 t + \theta)\right]$ $-\infty < t < \infty$
Sinusoid	$A\cos(\Omega_0 t + \theta) = A\sin(\Omega_0 t + \theta + \pi/2)$ $-\infty < t < \infty$
Unit-impulse	$\delta(t) = 0, t \neq 0$, undefined at $t = 0$, $\int_{-\infty}^{t} \delta(\tau)d\tau = 1, t > 0$, $$\int_{-\infty}^{\infty} f(\tau)\delta(t - \tau)d\tau = f(t)$$
Unit step	$u(t) = \begin{cases} 1 & t > 0 \\ 0 & t < 0 \end{cases}$
Ramp	$r(t) = tu(t) = \begin{cases} t & t > 0 \\ 0 & t \leq 0 \end{cases}$ $$\delta(t) = du(t)/dt$$ $$u(t) = \int_{-\infty}^{t} \delta(\tau)d\tau$$ $$r(t) = \int_{-\infty}^{t} u(\tau)d\tau$$
Rectangular pulse	$p(t) = A\left[u(t) - u(t-1)\right] = \begin{cases} A & 0 \leq t \leq 1 \\ 0 & \text{otherwise} \end{cases}$
Triangular pulse	$\Lambda(t) = A[r(t) - 2r(t-1) + r(t-2)] = \begin{cases} At & 0 \leq t \leq 1 \\ A(2-t) & 1 < t \leq 2 \\ 0 & \text{otherwise} \end{cases}$
Sampling signal	$\delta_{T_s}(t) = \sum_k \delta(t - kT_s)$
Sinc	$S(t) = \sin(\pi t)/(\pi t)$ $$S(0) = 1$$ $$S(k) = 0, \quad k \text{ integer} \neq 0$$ $$\int_{-\infty}^{\infty} S^2(t)dt = 1$$

1.7 WHAT HAVE WE ACCOMPLISHED? WHERE DO WE GO FROM HERE?

We have taken another step in our long journey. In this chapter you have learned the main classification of signals and have started the study of deterministic, continuous-time signals. We have discussed important characteristics of signals such as periodicity, energy, power, evenness and oddness and learned basic signal operations which will be useful as we will see in the next chapters. Interestingly, we also began to see how some of these operations lead to practical applications, such as amplitude, frequency, and phase modulations, which are basic in the theory of communication. Very importantly, we have also begun to represent signals in terms of basic signals, which in future chapters will allow us to simplify the analysis and will give us flexibility in the synthesis of systems. These basic signals are used as test signals in control systems. A summary of the different basic signals is given in Table 1.1. Our next step is to connect signals with systems. We are particularly interested in developing a theory that can be used to approximate, to some degree, the behavior of most systems of interest in engineering. After that we will consider the analysis of signals and systems in the time and in the frequency domains.

Excellent books in signals and systems are available [28,34,66,63,43,53,46,64,71]. Some of these are classical books in the area that readers should look into for a different perspective of the material presented in this part of the book.

1.8 PROBLEMS
1.8.1 BASIC PROBLEMS

1.1 Consider the following continuous-time signal:

$$x(t) = \begin{cases} 1 - t & 0 \leq t \leq 1, \\ 0 & \text{otherwise.} \end{cases}$$

Carefully plot $x(t)$ and then find and plot the following signals:
  (a) $x(t + 1)$, $x(t - 1)$ and $x(-t)$,
  (b) $0.5[x(t) + x(-t)]$ and $0.5[x(t) - x(-t)]$,
  (c) $x(2t)$ and $x(0.5t)$,
  (d) $y(t) = dx(t)/dt$ and

$$z(t) = \int_{-\infty}^{t} y(\tau)d\tau.$$

Answers: $x(t + 1)$ is $x(t)$ shifted left by 1; $0.5[x(t) + x(-t)]$ discontinuous at $t = 0$.

1.2 The following problems relate to the symmetry of the signal:
  (a) Consider a causal exponential $x(t) = e^{-t}u(t)$.
    i.   Plot $x(t)$ and explain why it is called causal. Is $x(t)$ an even or an odd signal?
    ii. Is it true that $0.5e^{-|t|}$ is the even component of $x(t)$? Explain.
  (b) Using Euler's identity $x(t) = e^{jt} = \cos(t) + j\sin(t)$, find the even $x_e(t)$ and the odd $x_o(t)$ components of $x(t)$.
  (c) A signal $x(t)$ is known to be even, and not exactly zero for all time; explain why

$$\int_{-\infty}^{\infty} x(t)\sin(\Omega_0 t)dt = 0.$$

  (d) Is it true that

$$\int_{-\infty}^{\infty} [x(t) + x(-t)]\sin(\Omega_0 t)dt = 0$$

  for any signal $x(t)$?
  Answer: (a) (ii) yes, it is true; (b) $x_e(t) = \cos(t)$; (c) the integrand is odd; (d) $x(t) + x(-t)$ is even.

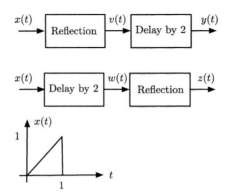

FIGURE 1.20

Problem 1.3.

1.3 Do reflection and the time shifting commute? That is, do the two block diagrams in Fig. 1.20 provide identical signals, i.e., is $y(t)$ equal to $z(t)$? To provide an answer to this consider the signal $x(t)$ shown in Fig. 1.20 is the input to the two block diagrams. Find $y(t)$ and $z(t)$, plot them and compare these plots. What is your conclusion? Explain.
Answers: The operations do not commute.

1.4 The following problems relate to the periodicity of signals:

   (a) Determine the frequency Ω_0 in rad/s, the corresponding frequency f_0 in Hz, and the fundamental period T_0 s of these signals defined in $-\infty < t < \infty$,

$$(i)\ \cos(2\pi t), \quad (ii)\ \sin(t-\pi/4), \quad (iii)\ \tan(\pi t).$$

   (b) Find the fundamental period T of $z(t) = 1 + \sin(t) + \sin(3t)$, $-\infty < t < \infty$.
   (c) If $x(t)$ is periodic of fundamental period $T_0 = 1$, determine the fundamental period of the following signals:

$$(i)\ y(t) = 2 + x(t), \quad (ii)\ w(t) = x(2t), \quad (iii)\ v(t) = 1/x(t).$$

   (d) What is the fundamental frequency f_0, in Hz, of

$$(i)\ x(t) = 2\cos(t), \quad (ii)\ y(t) = 3\cos(2\pi t + \pi/4), \quad (iii)\ c(t) = 1/\cos(t).$$

   (e) If $z(t)$ is periodic of fundamental period T_0, is $z_e(t) = 0.5[z(t) + z(-t)]$ also periodic? If so determine its fundamental period T_0. What about $z_o(t) = 0.5[z(t) - z(-t)]$?
Answers: (a) (iii) the frequency is $f_0 = 1/2$ Hz; (b) $T = 2\pi$; (c) $x(2t)$ has fundamental period $1/2$; (d) $c(t)$ has $f_0 = 1/(2\pi)$ Hz; (e) $z_e(t)$ is periodic of fundamental period T_0.

1.5 In the following problems find the fundamental period of signals and determine periodicity.

   (a) Find the fundamental period of the following signals, and verify it:

$$(i)\ x(t) = \cos(t + \pi/4), \quad (ii)\ y(t) = 2 + \sin(2\pi t), \quad (iii)\ z(t) = 1 + (\cos(t)/\sin(3t)).$$

(b) The signal $x_1(t)$ is periodic of fundamental period T_0, and the signal $y_1(t)$ is also periodic of fundamental period $10T_0$. Determine if the following signals are periodic, and if so give their fundamental periods:

$$(a)\ z_1(t) = x_1(t) + 2y_1(t) \quad (b)\ v_1(t) = x_1(t)/y_1(t) \quad (c)\ w_1(t) = x(t) + y_1(10t).$$

Answers: (a) Fundamental period of $y(t)$ is 1; (b) $v_1(t)$ periodic of fundamental period $10T_0$.

1.6 The following problems are about energy and power of signals.

(a) Plot the signal $x(t) = e^{-t}u(t)$ and determine its energy. What is the power of $x(t)$?

(b) How does the energy of $z(t) = e^{-|t|}$, $-\infty < t < \infty$, compare to the energy of $z_1(t) = e^{-t}u(t)$? Carefully plot the two signals.

(c) Consider the signal

$$y(t) = \text{sign}[x_i(t)] = \begin{cases} 1 & x_i(t) \geq 0, \\ -1 & x_i(t) < 0, \end{cases}$$

for $-\infty < t < \infty$, $i = 1, 2$. Find the energy and the power of $y(t)$ when

$$(a)\ x_1(t) = \cos(2\pi t), \quad (b)\ x_2(t) = \sin(2\pi t).$$

Plot $y(t)$ in each case.

(d) Given $v(t) = \cos(t) + \cos(2t)$.
 i. Compute the power of $v(t)$.
 ii. Determine the power of each of the components of $v(t)$, add them, and compare the result to the power of $v(t)$.

(e) Find the power of $s(t) = \cos(2\pi t)$ and of $f(t) = s(t)u(t)$. How do they compare?
Answer: (a) $E_x = 0.5$; (b) $E_z = 2E_{z_1}$; (c) $P_y = 1$; (d) $P_v = 1$.

1.7 Consider a circuit consisting of a sinusoidal source $v_s(t) = \cos(t)u(t)$ connected in series to a resistor R and an inductor L and assume they have been connected for a very long time.

(a) Let $R = 0$, $L = 1$ H, compute the instantaneous and the average powers delivered to the inductor.

(b) Let $R = 1\ \Omega$ and $L = 1$ H, compute the instantaneous and the average powers delivered to the resistor and the inductor.

(c) Let $R = 1\ \Omega$ and $L = 0$ H compute the instantaneous and the average powers delivered to the resistor.

(d) The complex power supplied to the circuit is defined as $P = \frac{1}{2}V_s I^*$ where V_s and I are the phasors corresponding to the source and the current in the circuit, and I^* is the complex conjugate of I. Consider the values of the resistor and the inductor given above, and compute the complex power and relate it to the average power computed in each case.
Answers: (a) $P_a = 0$; (b) $P_a = 0.25$; (c) $P_a = 0.5$.

1.8 Consider the periodic signal $x(t) = \cos(2\Omega_0 t) + 2\cos(\Omega_0 t)$, $-\infty < t < \infty$, and $\Omega_0 = \pi$. The frequencies of the two sinusoids are said to be harmonically related.

(a) Determine the period T_0 of $x(t)$. Compute the power P_x of $x(t)$ and verify that the power P_x is the sum of the power P_1 of $x_1(t) = \cos(2\pi t)$ and the power P_2 of $x_2(t) = 2\cos(\pi t)$.

(b) Suppose that $y(t) = \cos(t) + \cos(\pi t)$, where the frequencies are not harmonically related. Find out whether $y(t)$ is periodic or not. Indicate how you would find the power P_y of $y(t)$. Would $P_y = P_1 + P_2$ where P_1 is the power of $\cos(t)$ and P_2 that of $\cos(\pi t)$? Explain what is the difference with respect to the case of harmonic frequencies.

Answers: (a) $T_0 = 2$; $P_x = 2.5$; (b) $y(t)$ is not periodic, but $P_y = P_1 + P_2$.

1.9 A signal $x(t)$ is defined as $x(t) = r(t+1) - r(t) - 2u(t) + u(t-1)$.

(a) Plot $x(t)$ and indicate where it has discontinuities. Compute $y(t) = dx(t)/dt$ and plot it. How does it indicate the discontinuities? Explain.

(b) Find the integral

$$\int_{-\infty}^{t} y(\tau)d\tau$$

and give the values of the integral when $t = -1, 0, 0.99, 1.01, 1.99$, and 2.01. Is there any problem with calculating the integral at exactly $t = 1$ and $t = 2$? Explain.

Answers: $x(t)$ has discontinuities at $t = 0$ and at $t = 1$, indicated by delta functions in $dx(t)/dt$.

1.10 One of the advantages of defining the $\delta(t)$ function is that we are now able to find the derivative of discontinuous signals. Consider a periodic sinusoid defined for all times

$$x(t) = \cos(\Omega_0 t) \quad -\infty < t < \infty$$

and a causal sinusoid defined as $x_1(t) = \cos(\Omega_0 t)u(t)$, where the unit-step function indicates that the function has a discontinuity at zero, since for $t = 0+$ the function is close to 1 and for $t = 0-$ the function is zero.

(a) Find the derivative $y(t) = dx(t)/dt$ and plot it.

(b) Find the derivative $z(t) = dx_1(t)/dt$ (treat $x_1(t)$ as the product of two functions $\cos(\Omega_0 t)$ and $u(t)$) and plot it. Express $z(t)$ in terms of $y(t)$.

(c) Verify that the integral $\int_{-\infty}^{t} z(\tau)d\tau$ gives back $x_1(t)$.

Answers: (a) $y(t) = -\Omega_0 \sin(\Omega_0 t)$; (b) $z(t) = y(t)u(t) + \delta(t)$.

1.11 Let $x(t) = t[u(t) - u(t-1)]$, we would like to consider its expanded and compressed versions.

(a) Plot $x(2t)$ and determine if it is a compressed or expanded version of $x(t)$.

(b) Plot $x(t/2)$ and determine if it is a compressed or expanded version of $x(t)$.

(c) Suppose $x(t)$ is an acoustic signal, e.g., a music signal recorded in a magnetic tape, what would be a possible application of the expanding and compression operations? Explain.

Answers: (a) $x(2t) = 2t[u(t) - u(t - 0.5)]$, compressed.

1.12 Consider the signal $x(t)$ in Fig. 1.21.

(a) Plot the even–odd decomposition of $x(t)$, i.e., find and plot the even $x_e(t)$ and the odd $x_o(t)$ components of $x(t)$.

(b) Show that the energy of the signal $x(t)$ can be expressed as the sum of the energies of its even and odd components, i.e. that

$$\int_{-\infty}^{\infty} x^2(t)dt = \int_{-\infty}^{\infty} x_e^2(t)dt + \int_{-\infty}^{\infty} x_o^2(t)dt.$$

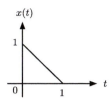

FIGURE 1.21

Problem 1.12.

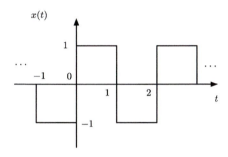

FIGURE 1.22

Problem 1.13.

(c) Verify that the energy of $x(t)$ is equal to the sum of the energies of $x_e(t)$ and $x_o(t)$.

Answers: $x_o(t) = -0.5(1+t)[u(t+1) - u(t)] + 0.5(1-t)[u(t) - u(t-1)]$.

1.13 A periodic signal can be generated by repeating a period.

(a) Find the function $g(t)$, defined in $0 \le t \le 2$ only, in terms of basic signals and such that when repeated using a period of 2 generates the periodic signal $x(t)$ shown in Fig. 1.22.

(b) Obtain an expression for $x(t)$ in terms of $g(t)$ and shifted versions of it.

(c) Suppose we shift and multiply by a constant the periodic signal $x(t)$ to get new signals $y(t) = 2x(t-2)$, $z(t) = x(t+2)$ and $v(t) = 3x(t)$ are these signals periodic?

(d) Let then $w(t) = dx(t)/dt$, and plot it. Is $w(t)$ periodic? If so, determine its period.

Answers: (a) $g(t) = u(t) - 2u(t-1) + u(t-2)$; (c) signals $y(t)$, $v(t)$ are periodic.

1.14 Consider a complex exponential signal $x(t) = 2e^{j2\pi t}$.

(a) Suppose $y(t) = e^{j\pi t}$, would the sum of these signals $z(t) = x(t) + y(t)$ also be periodic? If so, what is the fundamental period of $z(t)$?

(b) Suppose we then generate a signal $v(t) = x(t)y(t)$, with the $x(t)$ and $y(t)$ signals given before, is $v(t)$ periodic? If so, what is its fundamental period?

Answers: (b) $z(t)$ is periodic of period $T_1 = 2$; (c) $v(t)$ is periodic of period $T_3 = 2/3$.

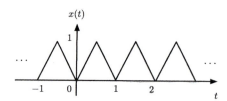

FIGURE 1.23

Problem 1.15.

1.15 Consider the triangular train of pulses $x(t)$ in Fig. 1.23.
  (a) Carefully plot the derivative of $x(t)$, $y(t) = dx(t)/dt$.
  (b) Can you compute

$$z(t) = \int_{-\infty}^{\infty} [x(t) - 0.5] dt?$$

If so, what is it equal to? If not, explain why not.
  (c) Is $x(t)$ a finite-energy signal? How about $y(t)$?
  Answers: (a) $y(t) = \sum_k [u(t-k) - 2u(t-0.5-k) + u(t-1-k)]$; (c) $x(t)$, $y(t)$ have infinite energy.

1.8.2 PROBLEMS USING MATLAB

1.16 Signal energy and RC circuit—The signal $x(t) = e^{-|t|}$ is defined for all values of t.
  (a) Plot the signal $x(t)$ and determine if this signal is of finite energy.
  (b) If you determine that $x(t)$ is absolutely integrable, or that the integral

$$\int_{-\infty}^{\infty} |x(t)| dt$$

is finite, could you say that $x(t)$ has finite energy? Explain why or why not. HINT: Plot $|x(t)|$ and $|x(t)|^2$ as functions of time.
  (c) From your results above, is it true the energy E_y of the signal

$$y(t) = e^{-t} \cos(2\pi t) u(t)$$

is less than half the energy of $x(t)$? Explain. To verify your result, use symbolic MATLAB to plot $y(t)$ and to compute its energy.
  (d) To discharge a capacitor of 1 mF charged with a voltage of 1 volt we connect it, at time $t = 0$, with a resistor R. When we measure the voltage in the resistor we find it to be $v_R(t) = e^{-t} u(t)$. Determine the resistance R. If the capacitor has a capacitance of 1 μF, what would be R? In general, how are R and C related?
  Answers: (a) $E_x = 1$; (c) $E_y = E_x/2$; (d) $R = 1/C$.

1.17 Periodicity of sum of sinusoids—

(a) Consider the periodic signals $x_1(t) = 4\cos(\pi t)$ and $x_2(t) = -\sin(3\pi t + \pi/2)$. Find the periods T_1 of $x_1(t)$ and T_2 of $x_2(t)$ and determine if $x(t) = x_1(t) + x_2(t)$ is periodic. If so, what is its fundamental period T_0?

(b) Two periodic signals $x_1(t)$ and $x_2(t)$ have periods T_1 and T_2 such that their ratio $T_1/T_2 = 3/12$, determine the fundamental period T_0 of $x(t) = x_1(t) + x_2(t)$.

(c) Determine whether $x_1(t) + x_2(t)$, $x_3(t) + x_4(t)$ are periodic when

- $x_1(t) = 4\cos(2\pi t)$ and $x_2(t) = -\sin(3\pi t + \pi/2)$,
- $x_3(t) = 4\cos(2t)$ and $x_4(t) = -\sin(3\pi t + \pi/2)$.

Use symbolic MATLAB to plot $x_1(t) + x_2(t)$, $x_3(t) + x_4(t)$ and confirm your analytic result about their periodicity or lack of periodicity.

Answers: (b) $T_0 = 4T_1 = T_2$; (c) $x_1(t)$ is periodic, $x_2(t)$ is non-periodic.

1.18 Impulse signal generation—When defining the impulse or $\delta(t)$ signal the shape of the signal used to do so is not important. Whether we use the rectangular pulse we considered in this chapter or another pulse, or even a signal that is not a pulse, in the limit we obtain the same impulse signal. Consider the following cases:

(a) The triangular pulse

$$\Lambda_\Delta(t) = \frac{1}{\Delta}\left(1 - \left|\frac{t}{\Delta}\right|\right)(u(t + \Delta) - u(t - \Delta)).$$

Carefully plot it, compute its area, and find its limit as $\Delta \to 0$. What do you obtain in the limit? Explain.

(b) Consider the signal

$$S_\Delta(t) = \frac{\sin(\pi t/\Delta)}{\pi t}.$$

Use the properties of the sinc signal $S(t) = \sin(\pi t)/(\pi t)$ to express $S_\Delta(t)$ in terms of $S(t)$. Then find its area, and the limit as $\Delta \to 0$. Use symbolic MATLAB to show that for decreasing values of Δ the $S_\Delta(t)$ becomes like the impulse signal.

Answers: $S_\Delta(0) = 1/\Delta$, $S_\Delta(t) = 0$ at $t = \pm k\Delta$.

1.19 Contraction and expansion and periodicity—Consider the periodic signal $x(t) = \cos(\pi t)$ of fundamental period $T_0 = 2$ s.

(a) Is the expanded signal $x(t/2)$ periodic? If periodic indicate its period.

(b) Is the compressed signal $x(2t)$ periodic? If periodic indicate its period.

(c) Use MATLAB to plot the above two signals and verify your analytic results.

Answers: (a) $x(t/2)$ is periodic of fundamental period 4.

1.20 Full-wave rectified signal—Consider the full-wave rectified signal

$$y(t) = |\sin(\pi t)| \qquad -\infty < t < \infty.$$

(a) As a periodic signal, $y(t)$ does not have finite energy but it has a finite power P_y. Find it.

(b) It is always useful to get a quick estimate of the power of a periodic signal by finding a bound for the signal squared. Find a bound for $|y(t)|^2$ and show that $P_y < 1$.

(c) Use symbolic MATLAB to check if the full-wave rectified signal has finite power and if that value coincides with the P_y you found above. Plot the signal and provide the script for the computation of the power. How does it coincide with the analytical result?

Answers: (a) $P_y = 0.5$.

1.21 Shifting and scaling a discretized analog signal—The discretized approximation of a pulse is given by

$$
w(nTs) = \begin{cases} 1 & -N/4 \le n \le -1, \\ -1 & 1 \le n \le (N/4) + 1, \\ 0 & \text{otherwise,} \end{cases}
$$

where $N = 10{,}000$ and $T_s = 0.001$ s.

(a) Obtain this signal and let the plotted signal using *plot* be the analog signal. Determine the duration of the analog signal.

(b) There are two possible ways to visualize the shifting of an analog signal. Since when advancing or delaying a signal the values of the signal remain the same, it is only the time values that change we could visualize a shift by changing the time scale to be a shifted version of the original scale. Using this approach, plot the shifted signal $w(t - 2)$.

(c) The other way to visualize the time shifting is to obtain the values of the shifted signal and plot it against the original time support. This way we could continue processing the signal while with the previous approach we can only visualize it. Using this approach, obtain $w(t - 2)$ and then plot it.

(d) Obtain the scaled and shifted approximations to $w(1.5t)$ and $w(1.5t - 2)$ using our function *scale_shift* and comment on your results.

Answers: The duration of the pulse is 5.001 s.

1.22 Windowing, scaling and shifting a discretized analog signal—We wish to obtain a discrete approximation to a sinusoid $x(t) = \sin(3\pi t)$ from 0 to 2.5 s. To do so a discretized signal $x(nT_s)$, with $T_s = 0.001$, is multiplied by a causal window $w(nT_s)$ of duration 2.5, i.e., $w(nT_s) = 1$ for $0 \le n \le 2500$ and zero otherwise. Use our *scale_shift* function to find $x(2t)$ and $x(2t - 5)$ for $-1 \le t \le 10$ and plot them.

1.23 Multipath effects—In wireless communications, the effects of *multipath* significantly affect the quality of the received signal. Due to the presence of buildings, cars, etc., between the transmitter and the receiver the sent signal does not typically go from the transmitter to the receiver in a straight path (called *line of sight*). Several copies of the signal, shifted in time and frequency as well as attenuated, are received—i.e., the transmission is done over multiple paths each attenuating and shifting the sent signal. The sum of these versions of the signal appears to be quite different from the original signal given that constructive as well as destructive effects may occur. In this problem we consider the time shift of an actual signal to illustrate the effects of attenuation and time shift. In the next problem we consider the effects of time and frequency shifting, and attenuation.

Assume that the MATLAB *handel.mat* signal is an analog signal $x(t)$ that it is transmitted over three paths, so that the received signal is

$$y(t) = x(t) + 0.8x(t - \tau) + 0.5x(t - 2\tau)$$

and let $\tau = 0.5$ s. Determine the number of samples corresponding to a delay of τ seconds by using the sampling rate Fs (samples per second) given when the file *handel* is loaded. To simplify matters just work with a signal of duration 1 s; that is, generate a signal from *handel* with the appropriate number of samples. Plot the segment of the original *handel* signal $x(t)$ and the signal $y(t)$ to see the effect of multipath. Use the MATLAB function *sound* to listen to the original and the received signals.

1.24 Multipath effects, Part 2—Consider now the Doppler effect in wireless communications. The difference in velocity between the transmitter and the receiver causes a shift in frequency in the signal, which is called the Doppler effect. Just like the acoustic effect of a train whistle as the train goes by. To illustrate the frequency-shift effect, consider a complex exponential $x(t) = e^{j\Omega_0 t}$, assume two paths one which does not change the signal while the other causes the frequency shift and attenuation, resulting in the signal

$$y(t) = e^{j\Omega_0 t} + \alpha e^{j\Omega_0 t} e^{j\phi t} = e^{j\Omega_0 t} \left[1 + \alpha e^{j\phi t} \right]$$

where α is the attenuation and ϕ is the Doppler frequency shift which is typically much smaller than the signal frequency. Let $\Omega_0 = \pi$, $\phi = \pi/100$, and $\alpha = 0.7$. This is analogous to the case where the received signal is the sum of the line of sight signal and an attenuated signal affected by Doppler.

(a) Consider the term $\alpha e^{j\phi t}$, a phasor with frequency $\phi = \pi/100$, to which we add 1. Use the MATLAB plotting function *compass* to plot the addition $1 + 0.7e^{j\phi t}$ for times from 0 to 256 s changing in increments of $T = 0.5$ s.

(b) If we write $y(t) = A(t)e^{j(\Omega_0 t + \theta(t))}$ give analytical expressions for $A(t)$ and $\theta(t)$, and compute and plot them using MATLAB for the times indicated above.

(c) Compute the real part of the signal

$$y_1(t) = x(t) + 0.7x(t - 100)e^{j\phi(t-100)},$$

i.e., the effects of time and frequency delays, put together with attenuation, for the times indicated in part (a). Use the function *sound* (let $F_s = 2000$ in this function) to listen to the different signals.

Answers: $A(t) = \sqrt{1.49 + 1.4\cos(\phi t)}$, $\theta(t) = \tan^{-1}(0.7\sin(\phi t)/(1 + 0.7\cos(\phi t)))$.

1.25 Beating or pulsation—An interesting phenomenon in the generation of musical sounds is beating or pulsation. Suppose NP different players try to play a pure tone, a sinusoid of frequency 160 Hz, and that the signal recorded is the sum of these sinusoids. Assume the NP players while trying to play the pure tone end up playing tones separated by Δ Hz, so that the recorded signal is

$$y(t) = \sum_{i=1}^{NP} 10\cos(2\pi f_i t)$$

where the f_i are frequencies from 159 to 161 separated by Δ Hz. Each player is playing a different frequency.

(a) Generate the signal $y(t)$ $0 \le t \le 200$ (s) in MATLAB. Let each musician play a unique frequency. Consider an increasing number of players, letting NP to go from 51 players, with $\Delta = 0.04$ Hz, to 101 players with $\Delta = 0.02$ Hz. Plot $y(t)$ for each of the different number of players.

(b) Explain how this is related with multipath and the Doppler effect discussed in the previous problems.

1.26 Chirps—Pure tones or sinusoids are not very interesting to listen to. Modulation and other techniques are used to generate more interesting sounds. Chirps, which are sinusoids with time-varying frequency, are some of those more interesting sounds. For instance, the following is a chirp signal:

$$y(t) = A\cos(\Omega_c t + s(t)).$$

(a) Let $A = 1$, $\Omega_c = 2$, and $s(t) = t^2/4$. Use MATLAB to plot this signal for $0 \le t \le 40$ s in steps of 0.05 s. Use *sound* to listen to the signal.

(b) Let $A = 1$, $\Omega_c = 2$, and $s(t) = -2\sin(t)$ use MATLAB to plot this signal for $0 \le t \le 40$ s in steps of 0.05 s. Use *sound* to listen to the signal.

(c) What is the frequency of a chirp? It is not clear. The instantaneous frequency IF(t) is the derivative with respect to t of the argument of the cosine. For instance, for a cosine $\cos(\Omega_0 t)$ the IF$(t) = d\Omega_0 t/dt = \Omega_0$, so that the instantaneous frequency coincides with the conventional frequency. Determine the instantaneous frequencies of the two chirps and plot them. Do they make sense as frequencies? Explain.

CONTINUOUS-TIME SYSTEMS

2

CONTENTS

Things should be made as simple as possible, but not any simpler.

Albert Einstein, Physicist, 1879–1955

2.1 INTRODUCTION

The concept of system is useful in dealing with actual devices or processes for purposes of analysis and of synthesis. A transmission line, for instance, is a system—even though physically is just wires connecting two terminals. Voltages and currents in this system are not just functions of time but also of space. It takes time for a voltage signal to "travel" from one point to another separated by miles—Kirchhoff's laws do not apply. Resistance, capacitance and inductance of the line are distributed over the length of the line, i.e., the line is modeled as a cascade of circuits characterized by values of resistance, capacitance and inductance per unit length. A less complicated system is one consisting of resistors, capacitors and inductors where ideal models are used to represent these elements and to perform analysis and synthesis. The word "ideal" indicates that the models only approximate the real behavior of actual resistors, capacitors, and inductors. A more realistic model for a resistor would need to consider possible changes in the resistance due to temperature, and perhaps other marginal effects present in the resistor. Although this would result in a better model, for most practical applications it would be unnecessarily complicated.

We initiate the characterization of systems, and propose the linear and time invariant (LTI) model as a mathematical idealization of the behavior of systems—a good starting point. It will be seen that most practical systems deviate from it, but despite that the behavior of many devices is approximated as linear and time invariant. A transistor, which is a non-linear device, is analyzed using linear models around an operating point. Although the vocal system is hardly time-invariant or linear, or even represented by an ordinary differential equation, in speech synthesis short intervals of speech are modeled as the output of linear time-invariant models. Finally, it will be seen that the LTI model is not appropriate to represent communication systems, that non-linear or time-varying systems are more appropriate.

The output of a LTI system due to any signal is obtained by means of the generic signal representation obtained in the previous chapter. The response due to an impulse, together with the linearity and time-invariance of the system gives the output as an integral. This convolution integral although difficult to compute, even in simple cases, has significant theoretical value. It allows us not only to determine the response of the system for very general cases, but also provides a way to characterize causal and stable systems. Causality relates to the cause and effect of the input and the output of the system, giving us conditions for real-time processing while stability characterizes useful systems. These two conditions are of great practical significance.

2.2 SYSTEM CONCEPT AND CLASSIFICATION

Although we view a **system** as a mathematical transformation of an input signal (or signals) into an output signal (or signals), it is important to understand that such transformation results from an idealized model of the physical device or process we are interested in.

For instance, in the connection of physical resistors, capacitors, and inductors, the model idealizes how to deal with the resistors, the capacitors and the inductors. For a simple circuit, we would ignore in the model, for instance, stray inductive and capacitive as well as environmental effects. Resistance, capacitance and inductance would be assumed localized in the physical devices and the wires would not have resistance, inductance or capacitance. We would then use the circuits laws to obtain an ordinary differential equation to characterize the interconnection. Now a wire that in this RLC model connects two elements, in a transmission line a similar wire is modeled as having capacitance, inductance and resistance distributed over the length of the line to realize the way voltage travels over it. In practice, the model and the mathematical representation are not unique.

A system can be considered a connection of subsystems. Thinking of the RLC circuit as a system, for instance, the resistor, the capacitor, the inductor and the source would be the subsystems.

In engineering, models are typically developed in the different areas. There will be however analogues as it is the case between mechanical and electrical systems. In such cases the mathematical equations are similar, or even identical, but their significance is very different.

According to general characteristics attributed to systems, they can be classified as:

- **Static or dynamic systems:** a dynamic system has the capability of storing energy, or remembering its state, while a static system does not. A battery connected to resistors is a static system, while the same battery connected to resistors, capacitors and inductors constitutes a dynamic system. The main difference is the capability of capacitors and inductors to store energy, to remember the state of the device, which resistors do not have. Most systems of interest are dynamic.

- **Lumped- or distributed-parameter systems.** This classification relates as to how the elements of the system are viewed. In the case of the RLC circuit, the resistance, capacitance and inductance are localized so that these physical elements are modeled as lumped elements. In the case of a transmission line resistance, capacitance, and inductance are modeled as distributed over the length of the line.
- **Passive or active systems.** A system is passive if it is not able to deliver energy to the outside world. A circuit composed of constant resistors, capacitors and inductors is passive, but if it also contains operational amplifiers the circuit is an active system.

Dynamic systems with lumped parameters, such as the RLC circuit, are typically represented by ordinary differential equations, while distributed-parameter dynamic systems like the transmission line are represented by partial differential equations. In the case of lumped systems only the time variation is of interest, while in the case of distributed systems we are interested in both time and space variations of the signals. In this book we consider only dynamic systems with lumped parameters, possibly changing with time, with a single input and a single output. These systems however can be passive or active.

A further classification of systems is obtained by considering the type of signals present at the input and the output of the system.

Whenever the input(s) and the output(s) of a system are both continuous-time, discrete-time or digital the corresponding systems are **continuous-time**, **discrete-time** or **digital**, respectively. It is also possible to have **hybrid** systems when the input(s) and the output(s) are not of the same type.

Of the systems presented in Chapter 0, the CD player is a hybrid system as it has a digital input (the bits stored on the disc) and a continuous-time output (the acoustic signal put out by the player). The SDR system, on the other hand, can be considered to have a continuous-time input (in the transmitter) and a continuous-time output (at the receiver) making it a continuous-time system, although having hybrid subsystems.

2.3 LINEAR TIME-INVARIANT (LTI) CONTINUOUS-TIME SYSTEMS

A continuous-time system is a system in which the signals at its input and at its output are continuous-time. In this book, we will consider only single-input single-output systems, which mathematically are represented as a transformation S that converts an input signal $x(t)$ into an output signal $y(t) = S[x(t)]$ (see Fig. 2.1):

$$x(t) \quad \Rightarrow \quad y(t) = S[x(t)] \qquad (2.1)$$

$$\text{Input} \qquad \text{Output}$$

When developing a mathematical model for a continuous-time system is important to contrast the accuracy of the model with its simplicity and practicality. The following are some of the characteristics of the model being considered:

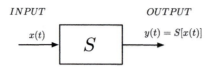

FIGURE 2.1

Continuous-time system S with input $x(t)$ and output $y(t)$.

- Linearity
- Time-invariance
- Causality
- Stability

Linearity between the input and the output, as well as the constancy of the system parameters (time-invariance) simplify the mathematical model. Causality, or non-anticipatory behavior of the system, relates to the cause-and-effect relationship between the system input and the system output. Causality is essential when the system is working under real-time situations, i.e., when there is limited time for the system to process signals coming into the system. Stability is required of practical systems: a stable system behaves well under reasonable inputs. Unstable systems are useless.

2.3.1 LINEARITY

A system represented by S is said to be **linear** if for inputs $x(t)$ and $v(t)$, and any constants α and β, **superposition** holds, that is

$$
\begin{aligned}
S[\alpha x(t) + \beta v(t)] &= S[\alpha x(t)] + S[\beta v(t)] \\
&= \alpha S[x(t)] + \beta S[v(t)].
\end{aligned} \tag{2.2}
$$

Thus when checking the linearity of a system we first need to check **scaling**, that is, if the output $y(t) = S[x(t)]$ for some input $x(t)$ is known, then for a scaled input $\alpha x(t)$ the output should be $\alpha y(t) = \alpha S[x(t)]$. If this condition is not satisfied, the system is non-linear. If the condition is satisfied, you would then test the **additivity** or that the response to the sum of inputs, $S[x(t) + v(t)]$, is the sum of the corresponding responses $S[x(t)] + S[v(t)]$. Equivalently, we can check these two properties of a linear system by checking **superposition**.

Notice that the scaling property of linear systems indicates that whenever the input of a linear system is zero its output is zero. Indeed, if the output corresponding to an input $x(t)$ is $y(t)$, then the output corresponding to $\alpha x(t)$ is $\alpha y(t)$ so that whenever $\alpha = 0$ then both input and output are zero.

Example 2.1. Consider a biased averager, i.e., the output $y(t)$ of such a system is given by

$$
y(t) = \frac{1}{T} \int_{t-T}^{t} x(\tau) d\tau + B
$$

for an input $x(t)$. The system finds the average over an interval $[t - T \ t]$, of duration T, and adds to it a constant bias B. Is this system linear? If not, is there a way to make it linear? Explain.

Solution: Let $y(t)$ be the system response corresponding to $x(t)$, assume then that we scale the input by a factor α so that the input is $\alpha x(t)$. The corresponding output is then

$$\frac{1}{T}\int_{t-T}^{t} \alpha x(\tau)d\tau + B = \frac{\alpha}{T}\int_{t-T}^{t} x(\tau)d\tau + B,$$

which is not equal to

$$\alpha y(t) = \frac{\alpha}{T}\int_{t-T}^{t} x(\tau)d\tau + \alpha B,$$

so the system is not linear. Notice that the difference is due to the term associated with the bias B, which is not affected at all by the scaling of the input. To make the system linear we would need that the bias be $B = 0$.

You can think of the biased averager as a system having two inputs: $x(t)$ and the bias B. The constant B is the response when the input is zero, $x(t) = 0$. Thus the response $y(t)$ can be seen as the sum of the response of a linear system

$$y_0(t) = \frac{1}{T}\int_{t-T}^{t} x(\tau)d\tau$$

and a zero-input response B, or

$$y(t) = y_0(t) + B.$$

Since when determining linearity we only change $x(t)$, the zero-input response does not change, and so the system is not linear. If $B = 0$, the system is clearly linear.

A spring illustrates this behavior quite well. Due to its elasticity, the spring when normally stretched returns to its original shape—it behaves like a linear system. However, if we stretch the spring beyond a certain point it loses its ability to return to its original shape and a permanent deformation or bias occurs, and from then on the spring behaves like a non-linear system. □

Example 2.2. Whenever the explicit relation between the input and the output of a system is given by a non-linear expression the system is non-linear. Considering the following input–output relations, show the corresponding systems are non-linear:

$$(i)\ y(t) = |x(t)|, \quad (ii)\ z(t) = \cos(x(t))\ \text{assuming}\ |x(t)| \le 1, \quad (iii)\ v(t) = x^2(t),$$

where $x(t)$ is the input and $y(t)$, $z(t)$ and $v(t)$ are the outputs.

Solution: Superposition is not satisfied by the first system, so the system is non-linear. Indeed, if the outputs for $x_1(t)$ and $x_2(t)$ are $y_1(t) = |x_1(t)|$ and $y_2(t) = |x_2(t)|$, respectively, the output for $x_1(t) + x_2(t)$ is not the sum of the responses since

$$y_{12}(t) = |x_1(t) + x_2(t)| \le \underbrace{|x_1(t)|}_{y_1(t)} + \underbrace{|x_2(t)|}_{y_2(t)}.$$

For the second system, if the response for $x(t)$ is $z(t) = \cos(x(t))$, the response for $-x(t)$ is not $-z(t)$ because the cosine is an even function of its argument. Thus for

$$-x(t) \to \cos(-x(t)) = \cos(x(t)) = z(t).$$

Therefore, the outputs for $x(t)$ and $-x(t)$ are the same. Thus the system is non-linear.

In the third system, if $x_1(t) \to v_1(t) = (x_1(t))^2$ and $x_2(t) \to v_2(t) = (x_2(t))^2$ are corresponding input–output pairs, then for

$$x_1(t) + x_2(t) \to (x_1(t) + x_2(t))^2 = (x_1(t))^2 + (x_2(t))^2 + 2x_1(t)x_2(t) \neq v_1(t) + v_2(t);$$

thus this system is non-linear. □

Example 2.3. Consider each of the components of an RLC circuit and determine under what conditions they are linear. In an analogous manner, the ideal components of a mechanical system are the dashpot, the mass and the spring. How can they be related to the ideal resistor, inductor and capacitor? Under what conditions are the ideal mechanical components considered linear?

Solution: A resistor R has a **voltage–current relation** $v(t) = Ri(t)$. If this relation is a straight line through the origin the resistor is linear, otherwise it is non-linear. A diode is an example of a non-linear resistor, as its voltage–current relation is not a straight line through the origin. The slope of a linear resistor, $dv(t)/di(t) = R > 0$, indicates the possibility of active elements to display negative resistance. If the voltage–current relation is a straight line of constant slope $R > 0$ considering the current the input, superposition is satisfied. Indeed, if we apply to the resistor a current $i_1(t)$ to get $Ri_1(t) = v_1(t)$ and get $Ri_2(t) = v_2(t)$ when we apply a current $i_2(t)$, then when a current $ai_1(t) + bi_2(t)$, for any constants a and b, is applied, the voltage across the resistor is $v(t) = R(ai_1(t) + bi_2(t)) = av_1(t) + bv_2(t)$, i.e., the resistor R is a linear system.

A capacitor is characterized by the **charge–voltage relation** $q(t) = Cv_c(t)$. If this relation is not a straight line through the origin, the capacitor is non-linear. A varactor is a diode whose capacitance depends non-linearly on the voltage applied, and thus it is a non-linear capacitor. When the charge–voltage relation is a straight line through the origin with a constant slope of $C > 0$, using the current–charge relation $i(t) = dq(t)/dt$ we get the ordinary differential equation characterizing the capacitor:

$$i(t) = Cdv_c(t)/dt \qquad t \geq 0$$

with initial condition $v_c(0)$. Letting $i(t)$ be the input, this ordinary differential equation can be solved by integration giving as output the voltage

$$v_c(t) = \frac{1}{C} \int_0^t i(\tau)d\tau + v_c(0), \tag{2.3}$$

which explains the way the capacitor works. For $t > 0$, the capacitor accumulates charge on its plates beyond the original charge due to an initial voltage $v_c(0)$ at $t = 0$. The capacitor is seen to be a linear system if $v_c(0) = 0$, otherwise it is not. In fact, when $v_c(0) = 0$ if the outputs corresponding to $i_1(t)$ and $i_2(t)$ are

$$v_{c1}(t) = \frac{1}{C} \int_0^t i_1(\tau)d\tau, \quad v_{c2}(t) = \frac{1}{C} \int_0^t i_2(\tau)d\tau,$$

respectively, and the output due to a combination $ai_1(t) + bi_2(t)$ is

$$\frac{1}{C} \int_0^t [ai_1(\tau) + bi_2(\tau)]d\tau = av_{c1}(t) + bv_{c2}(t).$$

Thus a capacitor is a linear system if not initially charged. When the initial condition is not zero, $v_c(0) \neq 0$, the capacitor is affected by the current input $i(t)$ as well as by the initial condition $v_c(0)$, and as such it is not possible to satisfy linearity as only the current input can be changed.

The inductor L is the dual of the capacitor—replacing currents by voltages and C by L in the above equations, we obtain the equations for the inductor. A constant linear inductor is characterized by the **magnetic flux–current relation** $\phi(t) = Li_L(t)$, being a straight line of slope $L > 0$. If the plot of $\phi(t)$ and $i_L(t)$ is not a line through the origin, the inductor is non-linear. According to Faraday's induction law, the voltage across the inductor is

$$v(t) = \frac{d\phi(t)}{dt} = L\frac{di_L(t)}{dt}.$$

Solving this ordinary differential equation for the current we obtain

$$i_L(t) = \frac{1}{L} \int_0^t v(\tau)d\tau + i_L(0). \tag{2.4}$$

Like the capacitor, the inductor is not a linear system unless the initial current in the inductor is zero. Notice that an explicit relation between the input and the output was necessary to determine linearity.

If force $f(t)$ and velocity $w(t)$ are considered the analogs of voltage $v(t)$ and current $i(t)$ we then have that a dashpot, approximated mechanically by a plunger inside a cylinder containing air or liquid, or the friction caused by a rough surface, the velocity is proportional to the applied force:

$$f(t) = Dw(t) \tag{2.5}$$

where D is the **damping** caused by the air or liquid in the cylinder or by the friction. The dashpot is clearly analogous to the resistor. Now, from Newton's law we know that the force applied to a **mass** M of a body on a frictionless surface is related to the resulting velocity by

$$f(t) = M\frac{dw(t)}{dt} \tag{2.6}$$

where the derivative of $w(t)$ is the acceleration. This equation looks very much like the relation defining a constant inductor L. Thus the analog of an inductor is a mass. Finally, according to Hooke's law the force applied to a spring is proportional to the displacement or as the integral of the velocity, i.e.,

$$f(t) = Kx(t) = K\int_0^t w(\tau)d\tau. \tag{2.7}$$

The constant K is the **stiffness** of the spring and it is analogous to the inverse of the capacitance. The inverse $1/K$ is called the **compliance**. Fig. 2.2 shows the representations of the dashpot, the mass, and the spring. $\qquad\qquad\qquad\square$

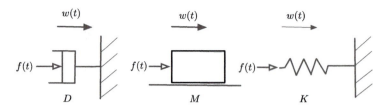

FIGURE 2.2

Mechanical system components: dashpot, mass and spring.

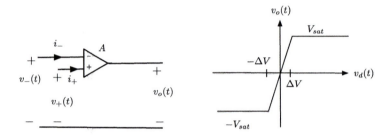

FIGURE 2.3

Operational amplifier: circuit diagram (left), and input–output voltage relation.

The Op-Amp

An excellent example of a device that can be used as either a linear or a non-linear system is the operational amplifier or **op-amp** [76,14,2]. It is modeled as a two-terminal circuit (see Fig. 2.3) with two voltage inputs: $v_-(t)$, in the *inverting terminal*, and $v_+(t)$, in the *non-inverting terminal*. The output voltage $v_0(t)$ is a non-linear function of the difference between the two inputs, i.e.,

$$v_o(t) = f[v_+(t) - v_-(t)] = f(v_d(t)).$$

The function $f(v_d(t))$ is approximately linear for a small range of values $[-\Delta V, \Delta V]$ of $v_d(t)$, in the order of millivolt, and it becomes constant beyond $\pm \Delta V$. The corresponding output voltage $v_0(t)$ is however in the order of volt, so that letting

$$v_o(t) = A v_d(t) \qquad -\Delta V \leq v_d(t) \leq \Delta V$$

be a line through the origin, its slope A is very large. If $|v_d(t)| > \Delta V$ the output voltage is a constant V_{sat}. That is, the gain of the amplifier saturates. Furthermore, the input resistance of the op-amp is large so that the currents into the negative and the positive terminals are very small. The op-amp output resistance is relatively small.

Thus depending on the dynamic range of the input signals, the op-amp operates in either a *linear* region or a *non-linear* region. Restricting the operational amplifier to operate in the linear region, simplifies the model. Assuming that the gain $A \to \infty$, and that the input resistance $R_{in} \to \infty$ then we obtain the following equations defining an **ideal operational amplifier**:

$$i_-(t) = i_+(t) = 0, \quad v_d(t) = v_+(t) - v_-(t) = 0. \tag{2.8}$$

These equations are called **the virtual short**, and are valid only if the output voltage of the operational amplifier is limited by the saturation voltage V_{sat}, i.e., when

$$-V_{sat} \le v_o(t) \le V_{sat}.$$

Later in the chapter we will consider ways to use the op-amp to get inverters, adders, buffers, and integrators.

2.3.2 TIME-INVARIANCE

A continuous-time system $\mathcal{S}$ is **time-invariant** if whenever for an input $x(t)$, with a corresponding output $y(t) = \mathcal{S}[x(t)]$, the output corresponding to a shifted input $x(t \mp \tau)$ (delayed or advanced) is the original output equally shifted in time, $y(t \mp \tau) = \mathcal{S}[x(t \mp \tau)]$ (delayed or advanced). Thus

$$\begin{aligned} x(t) &\Rightarrow & y(t) = \mathcal{S}[x(t)], \\ x(t \mp \tau) &\Rightarrow & y(t \mp \tau) = \mathcal{S}[x(t \pm \tau)]. \end{aligned} \tag{2.9}$$

That is, the system does not age—its parameters are constant.

A system that satisfies both the linearity and the time-invariance is called **Linear Time-Invariant or LTI**.

Remarks

1. It should be clear that linearity and time-invariance are independent of each other. Thus, one can have linear time-varying, or non-linear time-invariant systems.
2. Although most actual systems are, according to the above definitions, non-linear and time-varying, linear models are used to approximate (around an operating point) the non-linear behavior, and time-invariant models are used to approximate (in short segments) the system's time-varying behavior. For instance, in speech synthesis the vocal system is typically modeled as a linear time-invariant system for intervals of about 20 ms attempting to approximate the continuous variation in shape in the different parts of the vocal system (mouth, cheeks, nose, etc.). A better model for such a system is clearly a linear time-varying model.
3. In many cases time invariance can be determined by identifying—if possible—the input, the output and letting the rest represent the parameters of the system. If these parameters change with time, the system is time-varying. For instance, if the input $x(t)$ and the output $y(t)$ of a system are related by the equation $y(t) = f(t)x(t)$, the parameter of the system is the function $f(t)$, and if it is not constant, the system is time-varying. Thus, the system $y(t) = Ax(t)$, where A is a constant, is time invariant as can be easily verified. But the AM modulation system given by $y(t) = \cos(\Omega_0 t)x(t)$ is time-varying as the function $f(t) = \cos(\Omega_0 t)$ is not constant.

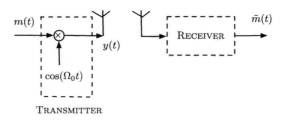

FIGURE 2.4

AM modulation: transmitter and receiver.

THE BEGINNINGS OF RADIO

The names of Nikola Tesla (1856–1943) and Reginald Fessenden (1866–1932) are linked to the invention of amplitude modulation and radio [65,78,7]. Radio was initially called "wireless telegraphy" and then "wireless" [26].

Tesla was a mechanical as well as an electrical engineer, but mostly an inventor. He has been credited with significant contributions to electricity and magnetism in the late 19th and early 20th century. His work is the basis of the alternating current (AC) power system, and the induction motor. His work on wireless communications using the "Tesla coils" was capable of transmitting and receiving radio signals. Although Tesla submitted a patent application for his basic radio before Guglielmo Marconi, it was Marconi who was initially given the patent for the invention of radio (1904). The Supreme Court in 1943 reversed the decision in favor of Tesla [6].

Fessenden has been called the "father of radio broadcasting." His early work on radio led to demonstrations on December 1906, of the capability of point-to-point wireless telephony, and what appears to be the first audio radio broadcasts of entertainment and music ever made to an audience (in this case shipboard radio operators in the Atlantic). Fessenden was a professor of electrical engineering at Purdue University and the first chairman of the electrical engineering department of the University of Pittsburgh in 1893.

AM Communication System

Amplitude modulation (AM) communication systems arose from the need to send an acoustic signal, a "message", over the airwaves using a reasonably sized antenna to radiate it. The size of the antenna depends inversely on the highest frequency present in the message, and voice and music have relatively low frequencies. A voice signal typically has frequencies in the range of 100 Hz to about 5 kHz (the frequencies needed to make a telephone conversation intelligible) while music typically displays frequencies up to about 22 kHz. The transmission of such signals with a practical antenna is impossible. To make the transmission possible, **modulation** was introduced, i.e., multiplying the message $m(t)$ by a periodic signal such as a cosine $\cos(\Omega_0 t)$, **the carrier**, with a frequency Ω_0 much larger than those in the acoustic signal. Amplitude modulation provided the larger frequencies needed to reduce the size of the antenna. Thus $y(t) = m(t)\cos(\Omega_0 t)$ is the signal to be transmitted, and the effect of this multiplication is to change the frequency content of the input.

The AM system is clearly linear, but time-varying. Indeed, if the input is $m(t - \tau)$, the message delayed τ seconds, the output would be $m(t - \tau)\cos(\Omega_0 t)$ which is not $y(t - \tau) = \cos(\Omega_0(t - \tau))m(t - \tau)$ as a time-invariant system would give. Fig. 2.4 illustrates the AM transmitter and receiver. The carrier continuously changes independent of the input and as such the system is time-varying.

FM Communication System

In comparison with an AM system, a frequency modulation (FM) system is non-linear and time-varying. An FM modulated signal $z(t)$ is given by

$$z(t) = \cos\left(\Omega_c t + \int_{-\infty}^{t} m(\tau)d\tau\right),$$

where $m(t)$ is the input message.

To show the FM system is non-linear, assume we scale the message to $\gamma m(t)$, for some constant γ, the corresponding output is given by

$$\cos\left(\Omega_c t + \gamma \int_{-\infty}^{t} m(\tau)d\tau\right)$$

which is not the previous output scaled, i.e., $\gamma z(t)$, thus FM is a non-linear system. Likewise, if the message is delayed or advanced, the output will not be equally delayed or advanced—thus the FM system is not time-invariant.

Vocal System

A remarkable system that we all have is the vocal system (see Fig. 2.5). The air pushed out from the lungs in this system is directed by the trachea through the vocal cords making them vibrate and create resonances similar to those from a musical wind instrument. The generated sounds are then muffled by the mouth and the nasal cavities resulting in an acoustic signal carrying a message. Given the length of the typical vocal system, on average for adult males about 17 cm and 14 cm for adult females, it is modeled as a distributed system and represented by partial differential equations. Due to the complexity of this model, it is the speech signal along with the understanding of the speech production that is used to obtain models of the vocal system. Speech processing is one of the most fascinating areas of electrical engineering.

A typical linear time-invariant model for speech production considers segments of speech of about 20 ms, and for each a low-order LTI system is developed. The input is either a periodic pulse for the generation of voiced sounds (e.g., vowels) or a noise-like signal for unvoiced sounds (e.g., the /sh/ sound). Processing these inputs gives speech-like signals. A linear time-varying model would take into consideration the variations of the vocal system with time and it would be thus more appropriate.

Example 2.4. Consider constant linear capacitors and inductors, represented by ordinary differential equations

$$\frac{dv_c(t)}{dt} = \frac{1}{C}i(t), \qquad \frac{di_L(t)}{dt} = \frac{1}{L}v(t),$$

with initial conditions $v_c(0) = 0$ and $i_L(0) = 0$, under what conditions are these time-invariant systems?

Solution: Given the duality of the capacitor and the inductor, we only need to consider one of these. Solving the ordinary differential equation for the capacitor we obtain (the initial voltage $v_c(0) = 0$):

$$v_c(t) = \frac{1}{C}\int_0^t i(\tau)d\tau.$$

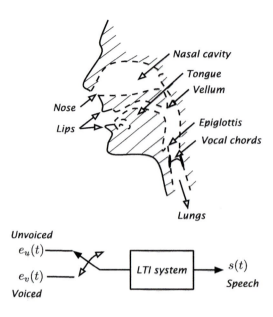

FIGURE 2.5

Vocal system: principal organs of articulation. Model for speech production.

Suppose then we delay the input current $i(t)$ by λ s. The corresponding output is given by

$$\frac{1}{C}\int_0^t i(\tau - \lambda)d\tau = \frac{1}{C}\int_{-\lambda}^0 i(\rho)d\rho + \frac{1}{C}\int_0^{t-\lambda} i(\rho)d\rho \tag{2.10}$$

by changing the integration variable to $\rho = \tau - \lambda$. For the above equation to equal the voltage across the capacitor delayed λ s, given by

$$v_c(t - \lambda) = \frac{1}{C}\int_0^{t-\lambda} i(\rho)d\rho,$$

we need that $i(t) = 0$ for $t < 0$, so that the first integral in the right expression in Equation (2.10) is zero. Thus, the system is time-invariant if the input current $i(t) = 0$ for $t < 0$. Putting this together with the condition on the initial voltage $v_c(0) = 0$, we can say that for the capacitor to be a linear and time-invariant system it should not be **initially energized**—i.e., the input $i(t) = 0$ for $t < 0$ and no initial voltage across the capacitor, $v(0) = 0$. Similarly, using duality, for the inductor to be linear and time-invariant, the input voltage across the inductor $v(t) = 0$ for $t < 0$ (to guarantee time-invariance) and the initial current in the inductor $i_L(0) = 0$ (to guarantee linearity).   □

Example 2.5. Consider the RLC circuit in Fig. 2.6 consisting of a series connection of a resistor R, an inductor L and a capacitor C. The switch has been open for a very long time and it is closed at $t = 0$, so that there is no initial energy stored in either the inductor or the capacitor and the voltage

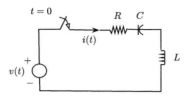

FIGURE 2.6

RLC circuit with a switch for setting the initial conditions to zero.

applied to the elements is zero for $t < 0$. Obtain an equation connecting the input $v(t)$ and output the current $i(t)$.

Solution: Because of the presence of the capacitor and the inductor, both capable of storing energy, this circuit is represented by a second order ordinary differential equation with constant coefficients. According to Kirchhoff's voltage law:

$$v(t) = Ri(t) + \frac{1}{C}\int_0^t i(\tau)d\tau + L\frac{di(t)}{dt}.$$

To get rid of the integral we find the derivative of $v(t)$ with respect to t:

$$\frac{dv(t)}{dt} = R\frac{di(t)}{dt} + \frac{1}{C}i(t) + L\frac{d^2i(t)}{dt^2},$$

a second order ordinary differential equation, with input the voltage source $v(t)$, and output the current $i(t)$. Because the circuit is not energized for $t < 0$, the circuit represented by the above ordinary differential equation is linear and time-invariant. □

Remark

An RLC circuit is represented by an ordinary differential equation of order equal to the number of independent inductors and capacitors in the circuit. If two or more capacitors (two or more inductors) are connected in parallel (in series), sharing the same initial voltage across them (same initial current) we can convert them into an equivalent capacitor with the same initial voltage (equivalent inductor with the same current).

A system represented by a linear ordinary differential equation, of any order N, having constant coefficients, and with input $x(t)$ and output $y(t)$:

$$a_0 y(t) + a_1 \frac{dy(t)}{dt} + \cdots + \frac{d^N y(t)}{dt^N} = b_0 x(t) + b_1 \frac{dx(t)}{dt} + \cdots + b_M \frac{d^M x(t)}{dt^M} \qquad t > 0 \qquad (2.11)$$

is linear time-invariant if the system is **not initially energized** (i.e., the initial conditions are zero, and the input $x(t)$ is zero for $t < 0$). If one or more of the coefficients $\{a_i, b_i\}$ are functions of time, the system is time-varying.

2.3.3 REPRESENTATION OF SYSTEMS BY ORDINARY DIFFERENTIAL EQUATIONS

In this section we present a system approach to the theory of linear ordinary differential equations with constant coefficients.

Given a dynamic system represented by a linear ordinary differential equation with constant coefficients,

$$a_0 y(t) + a_1 \frac{dy(t)}{dt} + \cdots + \frac{d^N y(t)}{dt^N} = b_0 x(t) + b_1 \frac{dx(t)}{dt} + \cdots + b_M \frac{d^M x(t)}{dt^M} \qquad t > 0 \tag{2.12}$$

with N initial conditions: $y(0)$, $d^k y(t)/dt^k|_{t=0}$, for $k = 1, \cdots, N-1$ and input $x(t) = 0$ for $t < 0$, the system's **complete response** $y(t)$, $t \geq 0$, has two components:

- the **zero-state response**, $y_{zs}(t)$, due exclusively to the input as the initial conditions are zero, and
- the **zero-input response**, $y_{zi}(t)$, due exclusively to the initial conditions as the input is zero.

Thus,

$$y(t) = y_{zs}(t) + y_{zi}(t). \tag{2.13}$$

Most continuous-time dynamic systems with lumped parameters are represented by **linear ordinary differential equations with constant coefficients**. By linear it is meant that there are no non-linear terms such as products of the input and the output, quadratic terms of the input or the output, etc., and if non-linear terms are present the system is non-linear. The coefficients are constant, and if they change with time the system is time-varying. The order of the differential equation equals the number of independent elements capable of storing energy.

Consider the dynamic system represented by the Nth-order linear ordinary differential equation with constant coefficients, and with $x(t)$ as the input and $y(t)$ as the output given by Equation (2.12). Defining the derivative operator as

$$D^n[y(t)] = \frac{d^n y(t)}{dt^n} \qquad n > 0, \text{ integer,}$$
$$D^0[y(t)] = y(t),$$

we rewrite (2.12) as

$$(a_0 + a_1 D + \cdots + D^N)[y(t)] = (b_0 + b_1 D + \cdots + b_M D^M)[x(t)] \qquad t > 0,$$
$$D^k[y(t)]_{t=0}, \qquad k = 0, \cdots, N-1.$$

As indicated, if the system is not energized for $t < 0$, i.e., the input and the initial conditions are zero, it is LTI. However, many LTI systems represented by ordinary differential equations have nonzero initial conditions. Considering the input signal $x(t)$ and the initial conditions two different inputs, using superposition we have that the **complete response** of the ordinary differential equation is composed of a **zero-input response**, due to the initial conditions when the input $x(t)$ is zero, and the **zero-state response** due to the input $x(t)$ with zero initial conditions. See Fig. 2.7.

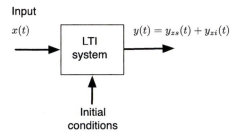

FIGURE 2.7

LTI system with $x(t)$ and initial conditions as inputs.

Thus to find the complete response of the system we need to solve the following two related ordinary differential equations:

$$(a_0 + a_1 D + \cdots + D^N)[y(t)] = 0 \tag{2.14}$$

with initial conditions $D^k[y(t)]_{t=0}, k = 0, \cdots, N-1$, and the ordinary differential equation

$$(a_0 + a_1 D + \cdots + D^N)[y(t)] = (b_0 + b_1 D + \cdots + b_M D^M)[x(t)] \tag{2.15}$$

with zero initial conditions. If $y_{zi}(t)$ is the solution of the zero-input ordinary differential equation (2.14), and $y_{zs}(t)$ is the solution of the ordinary differential equation (2.15), with zero initial conditions, we see that the complete response of the system is their sum

$$y(t) = y_{zi}(t) + y_{zs}(t).$$

Indeed, $y_{zi}(t)$ and $y_{zs}(t)$ satisfy their corresponding equations

$$(1 + a_1 D + \cdots + D^N)[y_{zi}(t)] = 0,$$
$$D^k[y_{zi}(t)]_{t=0}, \qquad k = 0, \cdots, N-1,$$
$$(a_0 + a_1 D + \cdots + D^N)[y_{zs}(t)] = (b_0 + b_1 D + \cdots + b_M D^M)[x(t)],$$

and adding these equations gives

$$(a_0 + a_1 D + \cdots + D^N)[y_{zi}(t) + y_{zs}(t)] = (b_0 + b_1 D + \cdots + b_M D^M)[x(t)]$$
$$D^k[y(t)]_{t=0}, \qquad k = 0, \cdots, N-1,$$

indicating that $y_{zi}(t) + y_{zs}(t)$ is the complete response.

To find the solution of the zero-input and the zero-state equations we need to factor out the derivative operator $a_0 + a_1 D + \cdots + D^N$. We can do so by replacing D by a complex variable s, as the roots will be either real or in complex-conjugate pairs, simple or multiple. The **characteristic polynomial**

$$a_0 + a_1 s + \cdots + s^N = \prod_k (s - p_k)$$

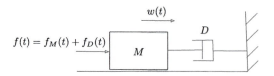

FIGURE 2.8

Mechanical system composed of a mass ($M = 1$) and a dashpot ($D = 1$) with an applied force $f(t)$ and a resulting velocity $w(t)$.

is then obtained. The roots of this polynomial are called the **natural frequencies** or **eigenvalues** and characterize the dynamics of the system as it is being represented by the ordinary differential equation. The solution of the zero-state can be obtained from a modified characteristic polynomial.

The solution of ordinary differential equations, with or without initial conditions, will be efficiently done using the Laplace transform in the next chapter. You will then find some similarity with the above approach.

Example 2.6. Consider the mechanical system shown in Fig. 2.8 where $M = 1$, $D = 1$, and the applied force is constant for $t > 0$ and zero otherwise, i.e., $f(t) = Bu(t)$, and the initial velocity is W_0. Obtain the differential equation that represents this system. Solve this equation for $B = 1$ and $B = 2$ when the initial condition W_0 is 1 and 0. Determine the zero-input and the zero-state responses, and use them to discuss the linearity and time invariance of the system.

Solution: Given that the applied force $f(t)$ is the sum of the forces in the mass and the dashpot, this mechanical system is analogous to a circuit connecting in series a voltage source, a resistor and an inductor. As such it is represented by the first-order ordinary differential equation

$$f(t) = w(t) + \frac{dw(t)}{dt}, \qquad t > 0, \quad w(0) = W_0. \tag{2.16}$$

Its complete solution,

$$w(t) = \begin{cases} W_0 & t = 0, \\ W_0 e^{-t} + B(1 - e^{-t}) & t > 0, \end{cases} \tag{2.17}$$

and zero for $t < 0$, satisfies the initial condition $w(0) = W_0$, as can easily be seen when $t = 0$, and the ordinary differential equation. In fact, when for $t > 0$ we replace in the ordinary differential equation (2.16) the input force by B, and $w(t)$ and $dw(t)/dt$ according to the solution (2.17) for $t > 0$ we get

$$\underbrace{B}_{f(t)} = \underbrace{[W_0 e^{-t} + B(1 - e^{-t})]}_{w(t)} + \underbrace{[Be^{-t} - W_0 e^{-t}]}_{dw(t)/dt} = B \qquad t > 0$$

or an identity, indicating $w(t)$ in (2.17) is the solution of the ordinary differential equation (2.16).

The zero-state response, i.e., the response due to $f(t) = Bu(t)$ and a zero initial condition, $W_0 = 0$, is

$$w_{zs}(t) = B(1 - e^{-t})u(t),$$

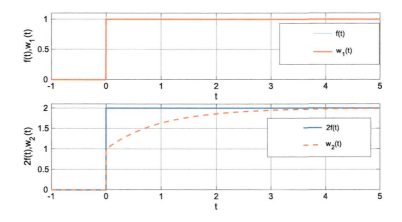

FIGURE 2.9

Non-linear behavior of mechanical system: (top) for $W_0 = 1$, $f(t) = u(t)$, the output velocity is $w_1(t) = 1$, $t \geq 0$ and zero otherwise; (bottom) for $W_0 = 1$, $f(t) = 2u(t)$ (doubled previous input) we obtain an output velocity $w_2(t) = 2 - e^{-t}$, $t \geq 0$ and zero otherwise so $w_2(t) \neq 2w_1(t)$.

which is obtained by letting $W_0 = 0$ in (2.17). The zero-input response, when $f(t) = 0$ and the initial condition is W_0, is

$$w_{zi}(t) = W_0 e^{-t}, \qquad t \geq 0,$$

and zero otherwise, obtained by subtracting the zero-state response from the complete response (2.17).

We consider next the effect of the initial conditions on the response of the system as it relates to the linearity and time-invariance of the system.

Initial condition different from zero: When $W_0 = 1$ and $B = 1$, the complete solution (2.17) becomes

$$w_1(t) = e^{-t} + (1 - e^{-t}) = 1 \qquad t \geq 0 \tag{2.18}$$

and zero for $t < 0$. Now if we let $B = 2$ (i.e., we double the original input) and keep $W_0 = 1$, the complete response is given by

$$w_2(t) = e^{-t} + 2(1 - e^{-t}) = 2 - e^{-t} \qquad t \geq 0$$

and zero for $t < 0$, which is different from $2w_1(t)$ the expected solution if the system were linear (see Fig. 2.9). Thus the system behaves non-linearly—when testing linearity we only consider the changes due to input or the zero-state response and not those due to the initial condition or the zero-input response.

Zero initial condition: For zero initial condition, $W_0 = 0$, and $B = 1$, i.e., $f(t) = u(t)$, the complete response is according to Equation (2.17):

$$w_1(t) = (1 - e^{-t}), \qquad t \geq 0,$$

and zero otherwise. For $W_0 = 0$ and $B = 2$ the complete response is

$$w_2(t) = 2(1 - e^{-t}) = 2w_1(t), \qquad t \geq 0,$$

and zero otherwise, which indicates the system is linear. In this case the response only depends on the input $f(t)$ as the initial condition is zero.

Time-invariance: Suppose now that $f(t) = u(t-1)$ and the initial condition is any value W_0. The complete response is

$$w_3(t) = \begin{cases} W_0 e^{-t} & 0 \leq t < 1, \\ W_0 e^{-t} + (1 - e^{-(t-1)}) & t \geq 1, \end{cases}$$

and zero otherwise. When $W_0 = 0$ then

$$w_3(t) = (1 - e^{-(t-1)})u(t-1),$$

which equals the zero-state response shifted by 1, indicating the system is behaving as time-invariant. On the other hand, when $W_0 = 1$ the complete response $w_3(t)$ is not the shifted zero-state response, and the system is behaving as time-varying. Again, the time-invariant behavior only considers the zero-state response. □

Example 2.7. A mechanical or electrical system is represented by a first-order differential equation

$$\frac{dy(t)}{dt} + a(t)y(t) = x(t)$$

where $a(t) = u(t)$, or a constant 1 for $t > 0$, the input is $x(t) = e^{-2t}u(t)$, and the initial condition is $y(0) = 0$. Find the output $y(t)$ for $t \geq 0$. Let the input be $x_1(t) = x(t-1)$ and the initial condition be zero at $t = 1$. Find the solution $y_1(t)$ of the corresponding differential equation, and determine if the system is time invariant.

Solution:
For input $x(t) = e^{-2t}u(t)$, coefficient $a(t) = u(t) = 1$ for $t > 0$, and initial condition $y(0) = 0$, the differential equation to solve (using the notation D for the derivative) is:

$$(D+1)y(t) = e^{-2t}, \qquad t > 0, \qquad y(0) = 0. \tag{2.19}$$

Noticing that for $t > 0$

$$(D+2)e^{-2t} = -2e^{-2t} + 2e^{-2t} = 0$$

we have

$$(D+2)(D+1)y(t) = (D+2)e^{-2t} = 0, \quad t > 0.$$

Replacing D by the complex variable s we obtain the characteristic polynomial $(s+2)(s+1) = 0$ having as roots $s = -2$ and $s = -1$. Thus the solution is

$$y(t) = Ae^{-2t} + Be^{-t}, \quad t \geq 0$$

where A and B are constants to be determined. For the above equation to be the desired solution it needs to satisfy the initial condition $y(0) = 0$, and the differential equation at $t = 0 + \epsilon = 0_+$ (where $\epsilon \to 0$):

$$y(0) = A + B = 0 \Rightarrow A + B = 0,$$
$$(D+1)y(t)|_{t=0_+} = -2Ae^{-2\,0_+} - Be^{-0_+} = e^{-2\,0_+} \Rightarrow -2A - B = 1,$$

using $e^{-2\,0_+} = e^{-0_+} = 1$. Solving for A and B in the above two equations, we get $A = -1$ and $B = 1$ so that the output, or solution of the differential equation, is

$$y(t) = [e^{-t} - e^{-2t}] \qquad t \geq 0,$$

which can easily be verified to satisfy THE initial condition and the differential equation.

Suppose now that we delay the input by 1, i.e., the input is now $x_1(t) = x(t - 1)$ for a new differential equation with output $y_1(t)$:

$$(D+1)y_1(t) = x_1(t) \tag{2.20}$$

with zero initial condition at time $t = 1$. To show the system is time-invariant let $y_1(t) = y(t - 1)$ (notice that the coefficient a for this equation has not changed, it is still $a = 1$). Since

$$D[y_1(t)] = D[y(t-1)] = \frac{dy(t-1)}{d(t-1)} \frac{d(t-1)}{dt} = D[y(t-1)]$$

we see that, replacing this $y_1(t)$ and $x_1(t)$ in (2.20), we get

$$(D+1)y(t-1) = x(t-1) \quad \text{initial condition: } y(t-1)|_{t=1} = 0.$$

Letting $\rho = t - 1$ (a shifted time variable) the above becomes

$$(D+1)y(\rho) + y(\rho) = x(\rho), \quad \rho > 0,$$

and the initial condition is $y(\rho = 0) = 0$, which is exactly the same initial differential equation as in (2.19), just computed using the shifted time ρ. Thus, $y_1(t) = y(t-1)$ when $x_1(t) = x(t-1)$ and as such the system is time invariant. ☐

Example 2.8. Finding the solution of linear but time-varying systems is more complicated. Consider now the system in the previous example with a time-varying coefficient $a(t)$, i.e.,

$$\frac{dy(t)}{dt} + a(t)y(t) = x(t)$$

where the time-varying coefficient $a(t) = (1 + e^{-0.1t})u(t)$ and the input $x(t) = e^{-2t}u(t)$. Given the time variation of the coefficient $a(t)$ the system is time-varying, although linear. To show the system is time-varying use symbolic MATLAB to find the solution of the differential equation for the given $a(t)$ and $x(t)$ and compare it to the solution for $t \geq 6$ when the input is $x(t - 6)$.

Solution:

The following script shows how to solve the differential equation when the input is $x(t)$ and $x(t-6)$. The initial conditions are zero for both cases, but the coefficient $a(t)$ changes from $t \geq 0$ to $t \geq 6$ units of time.

```
%%
% Example 2.8---Solution of a LTV system
%%
clear all; clf
syms y(t) a(t) x(t) y1(t) x1(t)
Dy=diff(y,t); x=exp(-2*t)*heaviside(t);a=(1+exp(-0.1*t))*heaviside(t);
Dy1=diff(y1,t);
% original input and coefficient
y=dsolve(Dy+a*y == x,y(0)==0)
figure(1)
subplot(311)
fplot(x,[0,6]);grid;title('input x(t)')
subplot(312)
fplot(a,[0,6]);grid;title('coefficient a(t)')
subplot(313)
fplot(y,[0,6]);grid; title('output y(t)')
% shifted input, coefficient
a=a*heaviside(t-6);x1=exp(-2*(t-6))*heaviside(t-6);
y1=dsolve(Dy1+a*y1 == x1,y1(1)==0)
figure(2)
subplot(311)
fplot(x1,[0,12]);grid; title('shifted input x_1(t)=x(t-6)')
subplot(312)
fplot(a,[0,12]);grid;title('coefficient a(t)u(t-6)')
subplot(313)
fplot(y1,[0,12]);grid;title('ouput y_1(t)')
% comparison of solutions
figure(3)
fplot(y,[0,6]);grid
hold on
fplot(y1,[6,12]);title('y(t) vs y_1(t)')
hold off
```

The results are shown in Fig. 2.10. Notice the change in the coefficient $a(t)$ and that although the outputs look similar they have different values indicating that the system is time-varying. ☐

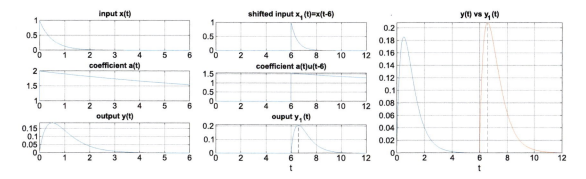

FIGURE 2.10

Input, coefficient, and output for $t \geq 0$ (left), $t \geq 6$ (middle) and comparison of outputs for $t \geq 0$ and $t \geq 6$ (right) for time-varying system.

2.4 THE CONVOLUTION INTEGRAL

The computation of the output of a LTI system is simplified when the input can be represented as a combination of signals for which we know their responses. This is done by applying superposition and time-invariance. This property of LTI systems will be of great importance in their analysis as you will soon learn.

> If $\mathcal{S}$ is the transformation corresponding to a LTI system, so that the response of the system is
>
> $$y(t) = \mathcal{S}[x(t)]$$
>
> to an input $x(t)$, then we have by superposition and time-invariance
>
> $$\mathcal{S}\left[\sum_k A_k x(t - \tau_k)\right] = \sum_k A_k \mathcal{S}[x(t - \tau_k)] = \sum_k A_k y(t - \tau_k),$$
>
> $$\mathcal{S}\left[\int g(\tau) x(t - \tau) d\tau\right] = \int g(\tau) \mathcal{S}[x(t - \tau)] d\tau = \int g(\tau) y(t - \tau) d\tau.$$

This property will allow us to find the response of a linear time-invariant system due to any signal, if we know the response to an impulse signal.

Example 2.9. The response of an RL circuit to a unit step source $u(t)$ is $i(t) = (1 - e^{-t})u(t)$. Find the response to a source $v(t) = u(t) - u(t - 2)$.

Solution: Using superposition and time-invariance, the output current due to the pulse $v(t) = u(t) - u(t - 2)$ V is

$$i(t) - i(t - 2) = 2(1 - e^{-t})u(t) - 2(1 - e^{(t-2)})u(t - 2)$$

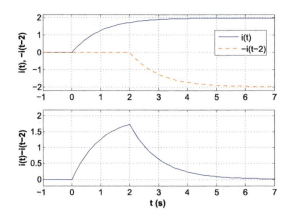

FIGURE 2.11

Response of an RL circuit to a pulse $v(t) = u(t) - u(t-2)$ using superposition and time-invariance.

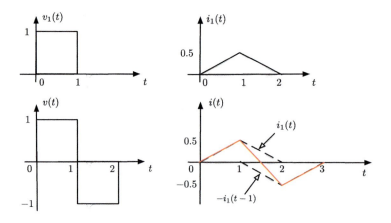

FIGURE 2.12

Application of superposition and time invariance to find the response of a LTI system.

Fig. 2.11 shows the responses to $u(t)$ and $u(t-2)$ and the overall response to $v(t) = u(t) - u(t-2)$. □

Example 2.10. Suppose we know that the response of a LTI system to a rectangular pulse $v_1(t)$ is the current $i_1(t)$ shown in Fig. 2.12. If the input voltage is a train of two pulses, $v(t)$, find the corresponding current $i(t)$.

Solution: Graphically, see Fig. 2.12, the response to $v(t)$ of the LTI system is given by $i(t)$ shown in the lower-right figure. □

2.4.1 IMPULSE RESPONSE AND CONVOLUTION INTEGRAL

In this section we consider the computation of the output of a continuous-time linear time-invariant (LTI) system due to any continuous-time input signal.

Recall the generic representation of a signal $x(t)$ in terms of shifted $\delta(t)$ signals found in the previous chapter:

$$x(t) = \int_{-\infty}^{\infty} x(\tau)\delta(t - \tau)d\tau \tag{2.21}$$

and define the **impulse response of a system** as follows:

> The **impulse response** of a continuous-time LTI system, $h(t)$, is the output of the system corresponding to an impulse $\delta(t)$ and initial conditions equal to zero.

If the input $x(t)$ in (2.21) is seen as an infinite sum of weighted and shifted impulses $x(\tau)\delta(t - \tau)$ then the output of a linear time-invariant (LTI) system is the superposition of the responses to each of these terms. We have the following.

> The response of a LTI system $\mathcal{S}$, represented by its impulse response $h(t)$, to any signal $x(t)$ is the **convolution integral**
>
> $$\begin{aligned} y(t) &= \int_{-\infty}^{\infty} x(\tau)h(t - \tau)d\tau = \int_{-\infty}^{\infty} x(t - \tau)h(\tau)d\tau \\ &= [x * h](t) = [h * x](t) \end{aligned} \tag{2.22}$$
>
> where the symbol $*$ stands for the convolution integral of the input signal $x(t)$ and the impulse response $h(t)$ of the system.

The above can be seen as follows:

(a) Assuming no energy is initially stored in the system (i.e., initial conditions are zero) the response to $\delta(t)$ is the impulse response $h(t)$.

(b) Given that the system is time invariant, the response to $\delta(t - \tau)$ is $h(t - \tau)$ and by linearity the response to $x(\tau)\delta(t - \tau)$ is $x(\tau)h(t - \tau)$ since $x(\tau)$ is not a function of time t.

(c) Thus the response of the system to the generic representation (2.21) is by superposition

$$y(t) = \int_{-\infty}^{\infty} x(\tau)h(t - \tau)d\tau$$

or equivalently

$$y(t) = \int_{-\infty}^{\infty} x(t - \sigma)h(\sigma)d\sigma$$

after letting $\sigma = t - \tau$. Notice that in the convolution integral the input and the impulse response commute, i.e., are interchangeable.

Remarks

1. We will see that the impulse response is fundamental in the characterization of linear time invariant systems.
2. Any system represented by the convolution integral is linear and time-invariant by the above construction. The convolution integral is a general representation of LTI systems, given that it was obtained from a generic representation of the input signal and assuming zero initial conditions (required to find $h(t)$). It is equivalent to the zero-state (i.e., zero initial conditions) response when the system is represented by an ordinary differential equation.

Example 2.11. Obtain the impulse response of a capacitor and use it to find its unit step response by means of the convolution integral. Let $C = 1$ F.

Solution:
For a capacitor with an initial voltage $v_c(0-) = 0$, we have

$$v_c(t) = \frac{1}{C} \int_{0-}^{t} i(\tau)d\tau \qquad t > 0.$$

The impulse response of a capacitor is then found by letting the input $i(t) = \delta(t)$ and the output $v_c(t) = h(t)$, giving

$$h(t) = \frac{1}{C} \int_{0-}^{t} \delta(\tau)d\tau = \frac{1}{C} \qquad t > 0$$

and zero if $t < 0$, or $h(t) = (1/C)u(t) = u(t)$ for $C = 1$ F. Notice that in the computation of $h(t)$ we assumed that the initial condition was at $0-$, right before zero, so that we could find the above integral by including $\delta(t)$ which has only a nonzero value at $t = 0$. This is equivalent to assuming that $v_c(0-) = v_c(0)$ or that the voltage in the capacitor cannot change instantaneously.

To compute the unit-step response of the capacitor we let the input be $i(t) = u(t)$, and $v_c(0) = 0$. The voltage across the capacitor using the convolution integral is

$$v_c(t) = \int_{-\infty}^{\infty} h(t-\tau)i(\tau)d\tau = \int_{-\infty}^{\infty} \frac{1}{C}u(t-\tau)u(\tau)d\tau = \int_{-\infty}^{\infty} u(t-\tau)u(\tau)d\tau$$

for $C = 1$ F. Since, as a function of τ, $u(t-\tau)u(\tau) = 1$ for $0 \le \tau \le t$ and zero otherwise, we have

$$v_c(t) = \int_{0}^{t} d\tau = t$$

for $t \ge 0$ and zero otherwise, or $v_c(t) = r(t)$, the ramp signal. This result makes physical sense since the capacitor is accumulating charge and the input is providing a constant charge, so that the response is a ramp signal. Notice that the impulse response is the derivative of the unit-step response. □

The relation between the impulse response, the unit-step and the ramp responses can be generalized for any system as follows:

The impulse response $h(t)$, the unit-step response $s(t)$, and the ramp response $\rho(t)$ are related by

$$h(t) = \begin{cases} ds(t)/dt, \\ d^2\rho(t)/dt^2. \end{cases} \tag{2.23}$$

This can be shown by first computing the ramp response $\rho(t)$ of a LTI system, represented by the impulse response $h(t)$, using the convolution integral

$$\rho(t) = \int_{-\infty}^{\infty} h(\tau) \underbrace{[(t-\tau)u(t-\tau)]}_{r(t-\tau)} d\tau = t \int_{-\infty}^{t} h(\tau) d\tau - \int_{-\infty}^{t} \tau\, h(\tau) d\tau$$

and its derivative is

$$\frac{d\rho(t)}{dt} = th(t) + \int_{-\infty}^{t} h(\tau) d\tau - th(t) = \int_{-\infty}^{t} h(\tau) d\tau,$$

corresponding to the unit-step response

$$s(t) = \int_{-\infty}^{\infty} u(t-\tau) h(\tau) d\tau = \int_{-\infty}^{t} h(\tau) d\tau.$$

The second derivative of $\rho(t)$ is

$$\frac{d^2\rho(t)}{dt^2} = \frac{d}{dt}\left[\int_{-\infty}^{t} h(\tau) d\tau\right] = h(t).$$

We will see in next chapter that using the Laplace transform the above relations are obtained more easily.

Example 2.12. The output $y(t)$ of an analog averager, a LTI system, is given by

$$y(t) = \frac{1}{T} \int_{t-T}^{t} x(\tau) d\tau,$$

which corresponds to the accumulation of area of $x(t)$ in a segment $[t-T, t]$ divided by its length T, or the average of $x(t)$ in $[t-T, t]$. Use the convolution integral to find the response of the averager to a ramp signal.

Solution:
To find the ramp response using the convolution integral we need first $h(t)$. The impulse response of an averager can be found by letting $x(t) = \delta(t)$ and $y(t) = h(t)$ or

$$h(t) = \frac{1}{T} \int_{t-T}^{t} \delta(\tau) d\tau.$$

If $t < 0$ or $t - T > 0$, or equivalently $t < 0$ or $t > T$, this integral is zero as in these two situations $t = 0$—where the delta function occurs—is not included in the limits. However, when $t - T < 0$ and $t > 0$, or $0 < t < T$, the integral is 1 as the origin $t = 0$, where $\delta(t)$ occurs, is included in this interval. Thus, the impulse response of the analog averager is

$$h(t) = \begin{cases} 1/T & 0 < t < T, \\ 0 & \text{otherwise.} \end{cases}$$

We then see that the output $y(t)$, for a given input $x(t)$, is given by the convolution integral

$$y(t) = \int_{-\infty}^{\infty} h(\tau)x(t - \tau)d\tau = \int_0^T \frac{1}{T}x(t - \tau)d\tau,$$

which can be shown to equal the definition of the averager by a change of variable. Indeed, let $\sigma = t - \tau$, so when $\tau = 0$ then $\sigma = t$, and when $\tau = T$ then $\sigma = t - T$. Moreover, we have $d\sigma = -d\tau$. The above integral is

$$y(t) = -\frac{1}{T} \int_t^{t-T} x(\sigma)d\sigma = \frac{1}{T} \int_{t-T}^t x(\sigma)d\sigma.$$

Thus we have

$$y(t) = \frac{1}{T} \int_0^t x(t - \tau)d\tau = \frac{1}{T} \int_{t-T}^t x(\sigma)d\sigma. \tag{2.24}$$

If the input is a ramp, $x(t) = tu(t)$, using the second integral in (2.24) the ramp response $\rho(t)$ is

$$\rho(t) = \frac{1}{T} \int_{t-T}^t x(\sigma)d\sigma = \frac{1}{T} \int_{t-T}^t \sigma u(\sigma)d\sigma.$$

If $t - T < 0$ and $t \geq 0$, or equivalently $0 \leq t < T$, the above integral is equal to

$$\rho(t) = \frac{1}{T} \int_{t-T}^0 0 \, d\sigma + \frac{1}{T} \int_0^t \sigma d\sigma = \frac{t^2}{2T} \qquad 0 \leq t < T,$$

but if $t - T \geq 0$, or $t \geq T$, we would then get

$$\rho(t) = \frac{1}{T} \int_{t-T}^t \sigma d\sigma = \frac{t^2 - (t - T)^2}{2T} = t - \frac{T}{2} \qquad t \geq T.$$

Therefore, the ramp response is

$$\rho(t) = \begin{cases} 0 & t < 0, \\ t^2/(2T) & 0 \leq t < T, \\ t - T/2 & t \geq T. \end{cases}$$

Notice that the second derivative of $\rho(t)$ is

$$\frac{d^2\rho(t)}{dt^2} = \begin{cases} 1/T & 0 \le t < T, \\ 0 & \text{otherwise,} \end{cases}$$

which is the impulse response of the averager found before.   □

Example 2.13. The impulse response of a LTI system is $h(t) = u(t) - u(t-1)$. Considering the inputs

$$x_2(t) = 0.5[\delta(t) + \delta(t-0.5)], \quad x_4(t) = 0.25[\delta(t) + \delta(t-0.25) + \delta(t-0.5) + \delta(t-0.75)]$$

find the corresponding outputs of the system. In general, let the input be

$$x_N(t) = \sum_{k=0}^{N-1} \delta(t - k\Delta\tau)\Delta\tau, \quad \text{where } \Delta\tau = \frac{1}{N};$$

write a MATLAB script to find the output when $N = 2^M$, $M = 1, 2, \cdots$. Show that as $N \to \infty$, $x_N(t) \to u(t) - u(t-1)$ and comment on the resulting output as N grows.

Solution:
Consider the response due to $x_2(t)$. If the input is $\delta(t)$ the corresponding output is $h(t)$ as $h(t)$ is the impulse response of the system. By the linearity of the system then the response due to $0.5\delta(t)$ is $0.5h(t)$, and by the time-invariance of the system the response to $0.5\delta(t-0.5)$ is $0.5h(t-0.5)$. The response to $x_1(t)$ is then the superposition of these responses or

$$y_2(t) = 0.5[h(t) + h(t-0.5)] = \begin{cases} 0.5 & 0 \le t < 0.5, \\ 1 & 0.5 \le t < 1, \\ 0.5 & 1 \le t < 1.5, \\ 0 & \text{otherwise.} \end{cases}$$

Equivalently, $y_2(t)$ is obtained from the convolution integral

$$y_2(t) = \int_{-\infty}^{\infty} x(\tau)h(t-\tau)d\tau = \int_{0-}^{0.5+} 0.5[\delta(\tau) + \delta(\tau-0.5)]h(t-\tau)d\tau = 0.5[h(t) + h(t-0.5)]$$

where the integral limits are set to include the delta functions. Similarly for $x_4(t)$, the output is

$$y_4(t) = 0.25[h(t) + h(t-0.25) + h(t-0.5) + h(t-0.75)]$$

by using the linearity, time-invariance of the system and that $h(t)$ is the impulse response.

Notice that the given two inputs can be obtained from the general expression for $x_N(t)$ by letting $N = 2$ and 4. Consider the following approximation of the integral defining the pulse

$$x(t) = u(t) - u(t-1) = \int_0^1 \delta(t-\tau)d\tau \approx \sum_{k=0}^{N-1} \delta(t - k\Delta\tau)\Delta\tau = \frac{1}{N}\left[\sum_{k=0}^{N-1} \delta(t - k/N)\right] = x_N(t)$$

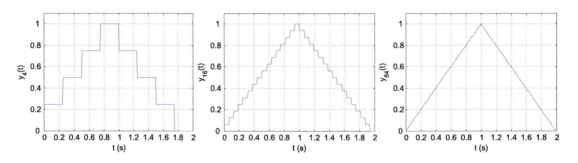

FIGURE 2.13

Outputs for inputs $x_N(t)$, $N = 4$, 16 and 64.

where we let $1 = N\Delta\tau$ or $\Delta\tau = 1/N$. The larger the value of N the better the approximation of the integral. So as $N \to \infty$ the input becomes $x(t) = u(t) - u(t - 1)$, which gives a triangular signal as output as we will show later.

The following script computes the response of the system for any $N = 2^M$, where $M \geq 1$ is an integer. Fig. 2.13 shows the cases when $N = 2$, 16, and 64. ☐

```
%%
% Example 2.13---Approximate convolution integral
%%
clear all; clf
Ts=0.01;Tend=2
t=0:Ts:Tend;M=6
h0=unitstep(t,0)-unitstep(t,-1);N=2^M
y=h0/N;
for k=1:N-1,
    y=y+(1/N)*(unitstep(t,-k/N)-unitstep(t,-1-k/N));
end
t1=0:Ts:(length(y)-1)*Ts;
figure(1)
plot(t1,y);axis([0 2 0 1.1]); grid; xlabel('t (s)'); ylabel('y_{64}(t)')
```

Example 2.14. Suppose that the input $x(t)$ and the output $y(t)$ of a LTI system are

$$x(t) = e^{-2t}u(t), \qquad y(t) = (e^{-t} - e^{-2t})u(t),$$

when the initial conditions are zero. Determine the impulse response $h(t)$ of the system. Then assume that $h(t)$ and $y(t)$ are given and find the input $x(t)$. These two problems are called **de-convolutions**.

Solution: These problems cannot be solved directly using the convolution integral, however the convolution integral indicates that whenever the input $x(t) = \delta(t)$ the output is $h(t)$. Indeed, we have

$$\int_{-\infty}^{t} x(\tau)h(t - \tau)d\tau = \int_{-\infty}^{t} \delta(\tau)h(t - \tau)d\tau = h(t)\int_{-\infty}^{t} \delta(\tau)d\tau = h(t), \qquad t > 0;$$

likewise when $h(t) = \delta(t)$ the convolution integral gives $x(t)$ as the output. Thus we need to convert the input and the impulse response into a delta function to solve the de-convolution problems. Consider the following result where D is the derivative operator[1]

$$(D+2)x(t) = (D+2)e^{-2t}u(t) = [-2e^{-2t}u(t) + e^{-2t}\delta(t)] + 2e^{-2t}u(t) = \delta(t).$$

Thus the derivative operator transforms $x(t)$ into a $\delta(t)$, and as a linear system the response will be $(D+2)y(t) = h(t)$, i.e., the desired impulse response. Thus

$$
\begin{aligned}
h(t) &= (D+2)[(e^{-t} - e^{-2t})u(t)] = -e^{-t}u(t) + \delta(t) + 2e^{-2t}u(t) - \delta(t) + 2e^{-t}u(t) - 2e^{-2t}u(t) \\
&= e^{-t}u(t).
\end{aligned}
$$

Now let us use $h(t) = e^{-t}u(t)$ and $y(t) = (e^{-t} - e^{-2t})u(t)$ to obtain the input $x(t)$. According to the convolution integral the output can be obtained by interchanging $x(t)$ and $h(t)$. Let us then transform $h(t)$, being considered the input to the system, to obtain an impulse. This can be done by using the derivative operator,[2] i.e.,

$$(D+1)h(t) = (D+1)e^{-t}u(t) = [-e^{-t}u(t) + e^{-t}\delta(t)] + e^{-t}u(t) = \delta(t).$$

The output is then for $t > 0$

$$\int_{-\infty}^{t} \delta(\tau)x(t-\tau)d\tau = x(t)\int_{-\infty}^{t} \delta(\tau)d\tau = x(t)$$

so that

$$
\begin{aligned}
x(t) &= (D+1)y(t) = (D+1)[(e^{-t} - e^{-2t})u(t)] \\
&= [-e^{-t} + 2e^{-2t}]u(t) + [e^{-t} - e^{-2t}]\delta(t) + [e^{-t} - e^{-2t}]u(t) \\
&= e^{-2t}u(t)
\end{aligned}
$$

coinciding with the given input.   □

---

[1] This result will be understood better when the Laplace transform is covered. We will see that

$$\mathcal{L}[e^{-2t}u(t)] = \frac{1}{s+2}, \qquad \mathcal{L}[\delta(t)] = 1 \;\Rightarrow\; (s+2)\mathcal{L}[e^{-2t}u(t)] = 1 = \mathcal{L}[\delta(t)]$$

where $\mathcal{L}[]$ is the Laplace transform. If we consider that s, the derivative operator in frequency, corresponds to D the derivative operator in time the above result gives

$$(D+2)e^{-2t}u(t) = \delta(t).$$

[2] Similar to the previous case we have the Laplace transforms:

$$\mathcal{L}[e^{-t}u(t)] = \frac{1}{s+1}, \qquad \mathcal{L}[\delta(t)] = 1 \;\Rightarrow\; (s+1)\mathcal{L}[e^{-t}u(t)] = 1 = \mathcal{L}[\delta(t)]$$

and in the time domain

$$(D+1)e^{-t}u(t) = \delta(t).$$

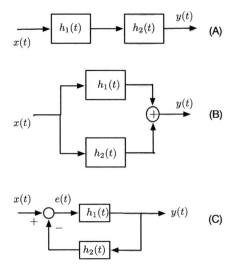

FIGURE 2.14

Block diagrams of the connection of two LTI systems with impulse responses $h_1(t)$ and $h_2(t)$ in (A) cascade, (B) parallel, and (C) negative feedback.

2.4.2 INTERCONNECTION OF SYSTEMS—BLOCK DIAGRAMS

Systems can be considered a connection of subsystems. In the case of LTI systems, to visualize the interaction of the different subsystems, each of the subsystems is represented by a block with the corresponding impulse response, or equivalently its Laplace transform as we will see in the next chapter. The flow of the signals is indicated by arrows, while the addition of signals is indicated by circles.

Of the block diagrams in Fig. 2.14, the **cascade** and the **parallel** connections, result from the properties of the convolution integral, while the **feedback** connection is found in many natural systems and has been replicated in engineering, especially in control. The concept of feedback is one of the greatest achievements of the 20th century.

Cascade Connection. When connecting LTI systems in cascade the impulse response of the overall system can be found using the convolution integral.

Two LTI systems with impulse responses $h_1(t)$ and $h_2(t)$ connected in **cascade** have as an overall impulse response

$$h(t) = [h_1 * h_2](t) = [h_2 * h_1](t)$$

where $h_1(t)$ and $h_2(t)$ commute (i.e., they can be interchanged).

In fact, if the input to the cascade connection is $x(t)$, the output $y(t)$ is found as

$$y(t) = [[x * h_1] * h_2](t) = [x * [h_1 * h_2]](t) = [x * [h_2 * h_1]](t)$$

where the last two equations show the *commutative property* of convolution. The impulse response of the cascade connection indicates that the order in which we connect LTI systems is not important: that we can interchange the impulse responses $h_1(t)$ and $h_2(t)$ with no effect in the overall response of the system (we will see later that this is true provided that the two systems do not load each other). When dealing with linear but time-varying systems, the order in which we connect the systems in cascade is, however, important.

Parallel Connection. In this connection the two systems have the same input, while the output is the sum of the outputs of the two systems.

> If we connect in parallel two LTI systems with impulse responses $h_1(t)$ and $h_2(t)$, the impulse response of the overall system is
>
> $$h(t) = h_1(t) + h_2(t).$$

In fact, the output of the parallel combination is

$$y(t) = [x * h_1](t) + [x * h_2](t) = [x * (h_1 + h_2)](t),$$

which is the *distributive property* of convolution.

Feedback Connection. In these connections the output of the system is fed back and compared with the input of the system. The feedback output is either added to the input giving a **positive feedback** system or subtracted from the input giving a **negative feedback** system. In most cases, especially in control systems, negative feedback is used. Fig. 2.14C illustrates the negative feedback connection.

> Given two LTI systems with impulse responses $h_1(t)$ and $h_2(t)$, a negative feedback connection (Fig. 2.14C) is such that the output is
>
> $$y(t) = [h_1 * e](t)$$
>
> where the error signal is
>
> $$e(t) = x(t) - [y * h_2](t).$$
>
> The overall impulse response $h(t)$, or the impulse response of the **closed-loop** system, is given by the following implicit expression
>
> $$h(t) = [h_1 - h * h_1 * h_2](t).$$
>
> If $h_2(t) = 0$, i.e., there is no feedback, the system is called **open-loop** and $h(t) = h_1(t)$.

Using the Laplace transform we will obtain later an explicit expression for the Laplace transform of $h(t)$. To obtain the above result we consider the output of the system as the overall impulse response $y(t) = h(t)$ due to an input $x(t) = \delta(t)$. Then $e(t) = \delta(t) - [h * h_2](t)$ and so when replaced in the expression for the output

$$h(t) = [e * h_1](t) = [(\delta - h * h_2) * h_1](t) = [h_1 - h * h_1 * h_2](t)$$

the implicit expression given above. When there is no feedback, $h_2(t) = 0$, then $h(t) = h_1(t)$.

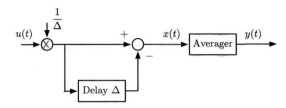

FIGURE 2.15

Block diagram of the cascading of two LTI systems one of them being an averager.

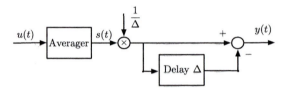

FIGURE 2.16

Equivalent block diagram of the cascading of two LTI systems one of them being an averager.

Example 2.15. Consider the block diagram in Fig. 2.15 with input a unit-step signal, $u(t)$. The averager is such that for an input $x(t)$ the output is

$$y(t) = \frac{1}{T} \int_{t-T}^{t} x(\tau)d\tau.$$

Determine what the system is doing as we let the delay $\Delta \to 0$. Consider that the averager and the system with input $u(t)$ and output $x(t)$ are LTI.

Solution:
Since it is not clear from the given block diagram what the system is doing, using the LTI of the two systems connected in series lets us reverse their order so that the averager is first (see Fig. 2.16), obtaining an equivalent block diagram.

The output of the averager is

$$s(t) = \frac{1}{T} \int_{t-T}^{t} u(\tau)d\tau = \begin{cases} 0 & t < 0, \\ t/T & 0 \leq t < T, \\ 1 & t \geq T, \end{cases}$$

as obtained before. The output $y(t)$ of the other system is given by

$$y(t) = \frac{1}{\Delta}[s(t) - s(t - \Delta)].$$

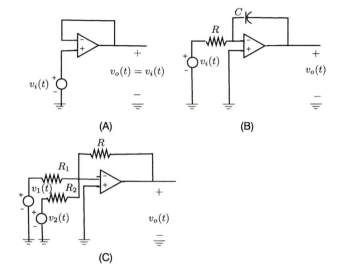

FIGURE 2.17

Operational amplifier circuits: (A) voltage follower, (B) inverting integrator, and (C) adder with inversion.

If we then let $\Delta \to 0$ we have that (recall that $ds(t)/dt = h(t)$ is the relation between the unit-step response $s(t)$ and the impulse response $h(t)$)

$$\lim_{\Delta \to 0} y(t) = \frac{ds(t)}{dt} = h(t) = \frac{1}{T}[u(t) - u(t - T)],$$

i.e., this system approximates the impulse response of the averager.   □

Example 2.16. In order to make the operational amplifier useful in the realization of filters, negative feedback is used. Consider the circuits obtained with an operational amplifier when we feed back its output with a wire, a capacitor, and a resistor (Fig. 2.17). Assume the linear model for the op-amp. The circuits in Fig. 2.17 are called a **voltage follower**, an **integrator**, and an **adder**. Obtain input/output equations for these circuits.

Solution:
Voltage follower circuit: Although the operational amplifier can operate as a linear element under certain conditions, its large gain A makes it not useful. Feedback is needed to make the op-amp useful. The **voltage follower** circuit (Fig. 2.17A) is a good example of a feedback system. Given that the voltage differential is assumed to be zero, then $v_-(t) = v_i(t)$, and therefore the output voltage

$$v_o(t) = v_i(t).$$

The input resistance of this circuit is $R_{in} = \infty$ and the output resistance is $R_{out} = 0$, so that the system behaves as an ideal voltage source. The voltage follower is used to isolate two circuits connected in cascade so that the connected circuit at either its input or its output port does not draw any current from the other, i.e., it does not load the other circuit. This is because of the infinite input resistance, or the behavior of the circuit as a voltage source ($R_{out} = 0$). This circuit is very useful in the implementation of analog filters.

Inverting integrator circuit: If we let the feedback element be a capacitor, we obtain the following equation from the virtual short equations. The current through the resistor R is $v_i(t)/R$ given that $v_-(t) = 0$ and $i_-(t) = 0$. Therefore, the output voltage is

$$v_o(t) = -v_c(t) = -\frac{1}{C}\int_0^t \frac{v_i(\tau)}{R}d\tau - v_c(0)$$

where $v_c(0)$ is the voltage across the capacitor at $t = 0$. If we let $v_c(0) = 0$ and $RC = 1$ the above equation is the negative of the integral of the voltage source. We thus have a circuit that realizes an integrator with a sign inversion. Again this circuit will be very useful in the implementation of analog filters.

Adder circuit: Since the circuit components are linear, the circuit is linear and we can use superposition. Letting $v_2(t) = 0$ the output voltage due to it is zero, and the output voltage due to $v_1(t)$ is $v_{o1}(t) = -v_1(t)R/R_1$. Similarly if we let $v_1(t) = 0$, its corresponding output is zero, and the output due to $v_2(t)$ is $v_{o2}(t) = -v_2(t)R/R_2$, so that when both $v_1(t)$ and $v_2(t)$ are considered the output is

$$v_o(t) = v_{o1}(t) + v_{o2}(t) = -v_1(t)\frac{R}{R_1} - v_2(t)\frac{R}{R_2}.$$

Using this circuit:
(1) when $R_1 = R_2 = R$ we have an **adder with a sign inversion**: $v_o(t) = -[v_1(t) + v_2(t)]$,
(2) when $R_2 \to \infty$ and $R_1 = R$, we get an **inverter** of the input $v_o(t) = -v_1(t)$,
(3) when $R_2 \to \infty$ and $R_1 = \alpha R$, we get a **constant multiplier with sign inversion**: $v_o(t) = -1/\alpha v_1(t)$. □

2.5 CAUSALITY

Causality refers to the cause-and-effect relationship between input and output of a system. In real-time processing, i.e., when it is necessary to process the signal as it comes into the system, causality of the system is required. In many situations the data can be stored and processed without the requirements of real-time processing—under such circumstances causality is not necessary. The following is the definition of causality of a system.

A continuous-time system S is called **causal** if

• whenever its input $x(t) = 0$, and there are no initial conditions, the output is $y(t) = 0$,
• the output $y(t)$ does not depend on future inputs.

For a value $\tau > 0$, when considering causality it is helpful to think of

- the time t (the time at which the output $y(t)$ is being computed) as the *present,*
- times $t - \tau$ as the *past,* and
- times $t + \tau$ as the *future.*

Causality is independent of the linearity and the time-invariance properties of a system. For instance, the system represented by the input/output equation $y(t) = x^2(t)$, where $x(t)$ is the input and $y(t)$ the output, is non-linear, time invariant, and according to the above definition causal. Likewise, a LTI system can be noncausal. Consider the following averager:

$$y(t) = \frac{1}{2T} \int_{t-T}^{t+T} x(\tau)d\tau,$$

which can be written as

$$y(t) = \frac{1}{2T} \int_{t-T}^{t} x(\tau)d\tau + \frac{1}{2T} \int_{t}^{t+T} x(\tau)d\tau.$$

At the present time t, $y(t)$ consists of the average of past and a present value, from $t - T$ to t, of the input, and of the average of future values of the signal from t to $t + T$. Thus this system is not causal. Finally, a biased averager,

$$y(t) = \frac{1}{T} \int_{t-T}^{t} x(\tau)d\tau + B,$$

is noncausal given that when the input is zero, $x(t) = 0$, the output is not zero but B.

A LTI system represented by its impulse response $h(t)$ is **causal** if

$$h(t) = 0 \qquad \text{for } t < 0. \tag{2.25}$$

The output of a causal LTI system with a causal input $x(t)$, i.e., $x(t) = 0$ for $t < 0$, is

$$y(t) = \int_{0}^{t} x(\tau)h(t - \tau)d\tau. \tag{2.26}$$

One can understand the above results by considering the following:

- The choice of the starting time as $t = 0$ is for convenience. It is purely arbitrary given that the system being considered is time-invariant, so that similar results are obtained for any other starting time.
- When computing the impulse response $h(t)$, the input $\delta(t)$ only occurs at $t = 0$ and there are no initial conditions, thus $h(t)$ should be zero for $t < 0$, since for $t < 0$ there is no input and there are no initial conditions.

- A causal LTI system is represented by the convolution integral

$$y(t) = \int_{-\infty}^{\infty} x(\tau)h(t-\tau)d\tau = \int_{-\infty}^{t} x(\tau)h(t-\tau)d\tau + \int_{t}^{\infty} x(\tau)h(t-\tau)d\tau$$

where the last integral is zero according to the causality of the system, i.e., $h(t-\tau)=0$ when $\tau > t$ since the argument of $h(.)$ becomes negative. Thus we obtain

$$y(t) = \int_{-\infty}^{t} x(\tau)h(t-\tau)d\tau. \tag{2.27}$$

If the input signal $x(t)$ is causal, i.e., $x(t)=0$ for $t<0$, we can simplify further the above equation. Indeed, when $x(t)=0$ for $t<0$ then Equation (2.27) becomes

$$y(t) = \int_{0}^{t} x(\tau)h(t-\tau)d\tau.$$

Thus the lower limit of the integral is set by the causality of the input signal, and the upper limit is set by the causality of the system. This equation clearly indicates that the system is causal, as the output $y(t)$ depends on present and past values of the input (considering the integral an infinite sum, the integrand depends continuously on $x(\tau)$, from $\tau = 0$ to $\tau = t$, which are past and present input values). Also if $x(t) = 0$ the output is also zero.

2.5.1 GRAPHICAL COMPUTATION OF CONVOLUTION INTEGRAL

The computation of the convolution integral is complicated even in the situation when the input as well as the system are causal. We will see that the convolution computation is much easier by using the Laplace transform, even in the case of noncausal inputs or systems.

Graphically, the computation of the convolution integral (2.26) for a causal input ($x(t) = 0$, $t < 0$) and a causal system ($h(t) = 0$, $t < 0$), consists in:

1. Choosing a time t_0 for which we want to compute $y(t_0)$.
2. Obtaining as functions of τ, the stationary $x(\tau)$ signal, and the reflected and delayed (shifted right) by t_0 impulse response $h(t_0 - \tau)$.
3. Obtaining the product $x(\tau)h(t_0 - \tau)$ and integrating it from 0 to t_0 to obtain $y(t_0)$.
4. Increasing the time value t_0 and moving from $-\infty$ to ∞.

The output is zero for $t < 0$ since the initial conditions are zero and the system is causal. In the above steps we can interchange the input and the impulse response and obtain identical results. It can be seen that the convolution integral is computationally very intensive as the above steps need to be done for each value of $t > 0$.

Example 2.17. An averager has an impulse response $h(t) = u(t) - u(t-1)$. Find graphically its unit-step response $y(t)$.

Solution: The input signal $x(\tau) = u(\tau)$, as a function of τ, and the reflected and delayed impulse response $h(t - \tau)$, also as a function of τ, for some value of $t < 0$ are shown in Fig. 2.18. Notice that

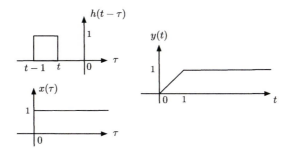

FIGURE 2.18

Graphical convolution for a unit-step input into an averager with $T = 1$.

when $t = 0$, $h(-\tau)$ is the reflected version of the impulse response, and for $t > 0$, then $h(t - \tau)$ is $h(-\tau)$ shifted by t to the right. As t goes from $-\infty$ to ∞, $h(t - \tau)$ moves from left to right, while $x(\tau)$ remains stationary. Interchanging the input and the impulse response (i.e., making $h(\tau)$ stationary and $x(t - \tau)$ the one that shifts linearly from left to right) does not change the final result—try it!

We then have the following results for different values of t:

- If $t < 0$, then $h(t - \tau)$ and $x(\tau)$ do not overlap and so the convolution integral is zero, or $y(t) = 0$ for $t < 0$. That is, the system for $t < 0$ has not yet been affected by the input.
- For $t \geq 0$ and $t - 1 < 0$, or equivalently $0 \leq t < 1$, $h(t - \tau)$ and $x(\tau)$ increasingly overlap and the integral increases linearly from 0 at $t = 0$ to 1 when $t = 1$. Therefore, $y(t) = t$ for $0 \leq t < 1$. That is, for these times the system is reacting slowly to the input.
- For $t \geq 1$, the overlap of $h(t - \tau)$ and $x(\tau)$ remains constant, and as such the integral is unity from then on, or $y(t) = 1$ for $t \geq 1$. The response for $t \geq 1$ has attained steady state. Thus the complete response is given as

$$y(t) = r(t) - r(t - 1)$$

where $r(t) = tu(t)$, the ramp function.
- Notice that the support of $y(t)$ can be considered the sum of the finite support of the impulse response $h(t)$ and the infinite support of the input $x(t)$. □

Example 2.18. Consider the graphical computation of the convolution integral of two pulses of the same duration (see Fig. 2.19).

Solution: In this case, $x(t) = h(t) = u(t) - u(t - 1)$. Again we plot $x(\tau)$ and $h(t - \tau)$ as functions of τ, for $-\infty < t < \infty$. The following remarks are relevant:

- While computing the convolution integral for t increasing from negative to positive values, $h(t - \tau)$ moves from left to right while $x(\tau)$ remains stationary, and that they only overlap on a finite support.
- For $t < 0$, $h(t - \tau)$ and $x(\tau)$ do not overlap, therefore $y(t) = 0$ for $t < 0$.

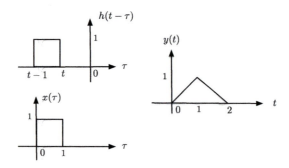

FIGURE 2.19

Graphical convolution of two equal pulses, i.e., a system with input $x(t) = u(t) - u(t-1)$ and impulse response $h(t) = x(t)$.

- As $h(t - \tau)$ and $x(\tau)$ overlap more and more for $0 \le t < 1$ the integral increases, and it decreases as $h(t - \tau)$ and $x(\tau)$ overlap less and less for $1 \le t < 2$. So that $y(t) = t$ for $0 \le t < 1$, and $y(t) = 2 - t$ for $1 \le t < 2$.
- For $t > 2$, there is no overlap and so $y(t) = 0$ for $t > 2$.

Thus the complete response is

$$y(t) = r(t) - 2r(t-1) + r(t-2)$$

where $r(t) = tu(t)$ is the ramp signal.

Notice in this example that:

(a) The result of the convolution of these two pulses, $y(t)$, is smoother than $x(t)$ and $h(t)$. This is because $y(t)$ is the continuous average of $x(t)$, as $h(t)$ is the impulse response of the averager in the previous example.

(b) The length of the support of $y(t)$ equals the sum of the lengths of the supports of $x(t)$ and $h(t)$. This is a general result that applies to any two signals $x(t)$ and $h(t)$. □

The length of the support of $y(t) = [x * h](t)$ is equal to the sum of the lengths of the supports of $x(t)$ and $h(t)$.

Example 2.19. Consider the following inputs and impulse responses of LTI systems.

$$(a) \ \ x_1(t) = u(t) - u(t-1), \qquad h_1(t) = u(t) - u(t-2),$$
$$(b) \ \ x_2(t) = h_2(t) = r(t) - 2r(t-1) + r(t-2),$$
$$(c) \ \ x_3(t) = e^{-t}u(t), \qquad h_3(t) = e^{-10t}u(t).$$

We want to compute the corresponding output using the convolution integral. Use the MATLAB functions *ramp* and *unistep* we developed in Chapter 1, as well as the MATLAB function *conv*.

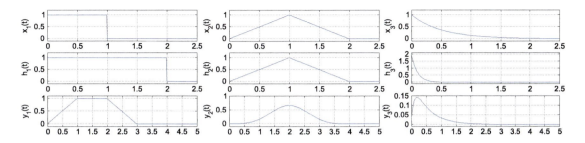

FIGURE 2.20

Results $y_i(t)$ (bottom) of convolution of $x_i(t)$ (top) with $h_i(t)$ (middle), $i = 1, 2, 3$, from left to right.

Solution: The computation of the convolution integral can be approximated using the function *conv*. This requires that the signals be discretized. The first two signals are generated using the functions *unitstep* and *ramp* created in Chapter 1. The results of the convolutions are shown in Fig. 2.20 for the three inputs. Notice that the length of the convolution is equal to the sum of the lengths of the signal and the impulse response. The following script illustrates the convolution with $x_1(t)$ as input, the other two can be found in a similar way.

```
%%
% Example 2.19---Convolution using MATLAB
%%
clear all; clf
Ts=0.01; delay=1; Tend=2.5; t=0:Ts:Tend;
for n=1:3,
    if n==1,
    x=unitstep(t,0)-unitstep(t,-delay);
    h=unitstep(t,0)-unitstep(t,-2*delay); figure(1)
    elseif n==2,
    x=ramp(t,1,0)+ramp(t,-2,-1)+ramp(t,1,-2);
    h=x; figure(2)
    else
    x=exp(-2*t);h=2*exp(-10*t); figure(3)
    end
    y=Ts*conv(x,h);
    t1=0:Ts:length(y)*Ts-Ts;
    subplot(311)
    plot(t,x); axis([0 2.5 -0.1 1.2]);grid;ylabel('x_3(t)');
    subplot(312)
    plot(t,h); axis([0 2.5 -0.1 2.1]);grid;ylabel('h_3(t)');
    subplot(313)
    plot(t1,y);; axis([0 5 -0.01 1.1*max(y)]);grid;ylabel('y_3(t)');
end
```

2.6 BOUNDED INPUT–BOUNDED OUTPUT STABILITY

Stability characterizes useful systems—an unstable system is useless. A stable system is such that well-behaved outputs are obtained for well-behaved inputs. Of the possible definitions of stability, we consider here bounded input–bounded output (BIBO) stability which is defined as follows.

> **Bounded input–bounded output (BIBO) stability** establishes that for a bounded (that is what is meant by well-behaved) input $x(t)$ the output of a BIBO stable system $y(t)$ is also bounded. This means that if there is a finite bound $M < \infty$ such that $|x(t)| \leq M$ (you can think of it as an envelope $[-M, M]$ inside which the input is in) the output is also bounded. That is, there is a bound $L < \infty$ such that $|y(t)| \leq L < \infty$.

In the case of a LTI system, if the input is bounded, i.e., $|x(t)| \leq M$, the system output $y(t)$—represented by the convolution integral—is also bounded under certain condition on the impulse response. We have

$$|y(t)| = \left| \int_{-\infty}^{\infty} x(t-\tau)h(\tau)d\tau \right| \leq \int_{-\infty}^{\infty} |x(t-\tau)||h(\tau)|d\tau \leq M \int_{-\infty}^{\infty} |h(\tau)|d\tau \leq MK < \infty$$

where the integral

$$\int_{-\infty}^{\infty} |h(\tau)|d\tau \leq K < \infty$$

or the impulse response is absolutely integrable.

> A LTI system is **bounded input–bounded output (BIBO) stable** provided that the system impulse response $h(t)$ is **absolutely integrable**, i.e.,
>
> $$\int_{-\infty}^{\infty} |h(t)|dt < \infty. \tag{2.28}$$

Remarks

- Testing the stability of a LTI system using the above criteria requires knowing the impulse response of the system and verifying that it is absolutely integrable. A simpler way to determine BIBO stability, using the Laplace transform, will be given in Chapter 3.
- Although BIBO stability refers to the input and the output in determining stability, stability of a system is an internal characteristic of it and as such independent of the input and the output. Thus, stability definitions that do not relate to the input or the output are more appropriate.

Example 2.20. Consider the BIBO stability and causality of RLC circuits. Consider for instance a series RL circuit where $R = 1 \ \Omega$ and $L = 1$ H, and a voltage source $v_s(t)$ which is bounded. Discus why such a system is causal and stable.

Solution:
RLC circuits are naturally stable. As you know, inductors and capacitors simply store energy and so CL circuits simply exchange energy between these elements. Resistors consume energy which is transformed into heat, and so RLC circuits spend the energy given to them. This characteristic is called **passivity**, indicating that RLC circuits can only use energy not generate it. Clearly, RLC circuits are also causal systems as one would not expect them to provide any output before they are activated.

According to Kirchhoff's voltage law, a series connection of a resistor $R = 1\ \Omega$, an inductor $L = 1$ H, and a source $v_s(t)$ is a circuit represented by a first order ordinary differential equation

$$v_s(t) = i(t)R + L\frac{di(t)}{dt} = i(t) + \frac{di(t)}{dt},$$

where $i(t)$ is the current in the circuit. To find the impulse response of this circuit we would need to solve the above equation with input $v_s(t) = \delta(t)$ and zero initial condition, $i(0) = 0$. In the next chapter, the Laplace domain will provide us an algebraic way to solve the ordinary differential equation and will confirm our intuitive solution. Intuitively, in response to a large and sudden impulse $v_s(t) = \delta(t)$ the inductor tries to follow it by instantaneously increasing its current. But as time goes by and the input is not providing any additional energy, the current in the inductor goes to zero. Thus, we conjecture that the current in the inductor is $i(t) = h(t) = e^{-t}u(t)$ when $v_s(t) = \delta(t)$ and initial conditions are zero $i(0) = 0$. It is possible to confirm that is the case. Replacing $v_s(t) = \delta(t)$, and $i(t) = e^{-t}u(t)$ in the ordinary differential equation we get

$$\underbrace{\delta(t)}_{v_s(t)} = \underbrace{e^{-t}u(t)}_{i(t)} + \underbrace{[e^{-t}\delta(t) - e^{-t}u(t)]}_{di(t)/dt} = e^0\delta(t) = \delta(t),$$

which is an identity, confirming that indeed our conjectured solution is the solution of the ordinary differential equation. The initial condition is also satisfied, from the response $i(0-) = 0$ and physically the inductor keeps that current for an extremely short time before reacting to the strong input.

Thus, the RL circuit where $R = 1\ \Omega$ and $L = 1$ H, has an impulse response $h(t) = e^{-t}u(t)$ indicating that it is causal since $h(t) = 0$ for $t < 0$, i.e., the circuit output is zero given that the initial conditions are zero and that the input $\delta(t)$ is also zero before 0. We can also show that the RL circuit is stable. In fact, $h(t)$ is absolutely integrable:

$$\int_{-\infty}^{\infty} |h(t)|dt = \int_0^{\infty} e^{-t}dt = 1. \quad \square$$

Example 2.21. Consider the causality and BIBO stability of an echo system (or a multi-path system). See Fig. 2.21. Let the output $y(t)$ be given by

$$y(t) = \alpha_1 x(t - \tau_1) + \alpha_2 x(t - \tau_2)$$

where $x(t)$ is the input, and α_i, $\tau_i > 0$, $i = 1,\ 2$, are attenuation factors and delays. Thus the output is the superposition of attenuated and delayed versions of the input. Typically, the attenuation factors are less than unity. Is this system causal and BIBO stable?

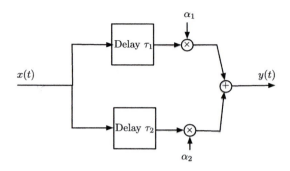

FIGURE 2.21

Echo system with two paths.

Solution:

Since the output depends only on past values of the input, the echo system is causal. To determine if the system is BIBO stable we consider a bounded input signal $x(t)$, and determine if the output is bounded. Suppose $x(t)$ is bounded by a finite value M, or $|x(t)| < M < \infty$, for all times, which means that the value of $x(t)$ cannot exceed an envelope $[-M, M]$, at all times. This would also hold when we shift $x(t)$ in time, so that

$$|y(t)| \leq |\alpha_1||x(t - \tau_1)| + |\alpha_2||x(t - \tau_2)| < [|\alpha_1| + |\alpha_2|]M$$

so the corresponding output is bounded. The system is BIBO stable.

We can also find the impulse response $h(t)$ of the echo system, and show that it satisfies the absolute integral condition of BIBO stability. Indeed, if we let the input of the echo system be $x(t) = \delta(t)$ the output is

$$y(t) = h(t) = \alpha_1 \delta(t - \tau_1) + \alpha_2 \delta(t - \tau_2)$$

and the integral

$$\int_{-\infty}^{\infty} |h(t)|dt \leq |\alpha_1| \int_{-\infty}^{\infty} \delta(t - \tau_1)dt + |\alpha_2| \int_{-\infty}^{\infty} \delta(t - \tau_2)dt = |\alpha_1| + |\alpha_2| < \infty. \quad \square$$

Example 2.22. Consider a positive feedback system created by a microphone close to a set of speakers that are putting out an amplified acoustic signal (see Fig. 2.22). The microphone (symbolized by the adder) picks up the input signal $x(t)$ as well as the amplified and delayed signal $\beta y(t - \tau)$, where $\beta \geq 1$ is the gain of the amplifier. The speakers provide the feedback. Find the equation that connects the input $x(t)$ and the output $y(t)$ and recursively from it obtain an expression for $y(t)$ in terms of past values of the input. Determine if the system is BIBO stable or not; use $x(t) = u(t)$, $\beta = 2$ and $\tau = 1$ in doing so.

Solution:

The input–output equation is

$$y(t) = x(t) + \beta y(t - \tau).$$

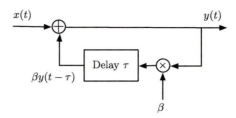

FIGURE 2.22

Positive feedback system: microphone picks up input signal $x(t)$ and amplified and delayed output signal $\beta y(t - \tau)$ making system unstable.

If we use this expression to obtain $y(t - \tau)$ we get

$$y(t - \tau) = x(t - \tau) + \beta y(t - 2\tau)$$

and replacing it in the input/output equation we get

$$y(t) = x(t) + \beta[x(t - \tau) + \beta y(t - 2\tau)] = x(t) + \beta x(t - \tau) + \beta^2 y(t - 2\tau).$$

Repeating the above scheme, we will obtain the following expression for $y(t)$ in terms of the input:

$$y(t) = x(t) + \beta x(t - \tau) + \beta^2 x(t - 2\tau) + \beta^3 x(t - 3\tau) + \cdots.$$

If we let $x(t) = u(t)$ and $\beta = 2$, the corresponding output is

$$y(t) = u(t) + 2u(t - 1) + 4u(t - 2) + 8u(t - 3) + \cdots,$$

which continuously grows as time increases. The output is clearly not a bounded signal, although the input is bounded. Thus the systems is unstable, and the screeching sound will prove it—you need to separate the speakers and the microphone to avoid it.   □

2.7 WHAT HAVE WE ACCOMPLISHED? WHERE DO WE GO FROM HERE?

By now you should have begun to see the forest for the trees. In this chapter we connected signals with systems. We have initiated the study of linear, time-invariant dynamic systems. As you will learn throughout your studies, this model is of great use in representing systems in many engineering applications. The appeal is its simplicity and mathematical structure. We have also presented practical properties of systems such as causality and stability. Simple yet significant examples of systems, ranging from the vocal system to simple RLC circuits illustrate the application of the LTI model and point to its practical application. At the same time, modulators show that more complicated systems need to be explored to be able to communicate wirelessly. Finally, you were given a system's approach to the theory of ordinary differential equations and shown some features that will come back when we apply transformations. Our next step is to do the analysis of systems with continuous-time signals by means of transformations. In the following chapter we will discuss the Laplace transform, which allows a transient as well as a steady-state analysis and which will convert the solution of differential equations into an algebraic problem. More important, it will provide the concept of transfer function which connects with the impulse response and the convolution integral covered in this chapter. The Laplace transform is very significant in the area of classical control.

2.8 PROBLEMS
2.8.1 BASIC PROBLEMS

2.1 The input–output equation characterizing an amplifier that saturates once the input reaches certain values is

$$y(t) = \begin{cases} 100x(t) & -10 \leq x(t) \leq 10, \\ 1000 & x(t) > 10, \\ -1000 & x(t) < -10, \end{cases}$$

where $x(t)$ is the input and $y(t)$ the output.
   (a) Plot the relation between the input $x(t)$ and the output $y(t)$. Is this a linear system? For what range of input values is the system linear, if any?
   (b) Suppose the input is a sinusoid $x(t) = 20\cos(2\pi t)u(t)$, carefully plot $x(t)$ and $y(t)$ for $t = -2$ to 4.
   (c) Let the input be delayed by 2 units of time, i.e., the input is $x_1(t) = x(t-2)$ find the corresponding output $y_1(t)$, and indicate how it relates to the output $y(t)$ due to $x(t)$ found above. Is the system time-invariant?
   Answers: If the input is always in $[-10 \ 10]$ the system behaves linearly; the system is time-invariant.

2.2 Consider an averager represented by the input/output equation

$$y(t) = \int_{t-1}^{t} x(\tau)d\tau + 2$$

where $x(t)$ is the input and $y(t)$ the output.
   (a) Let the input be $x_1(t) = \delta(t)$, find graphically the corresponding output $y_1(t)$ for $-\infty < t < \infty$. Let then the input be $x_2(t) = 2x_1(t)$, find graphically the corresponding output $y_2(t)$ for $-\infty < t < \infty$. Is $y_2(t) = 2y_1(t)$? Is the system linear?
   (b) Suppose the input is $x_3(t) = u(t) - u(t-1)$, graphically compute the corresponding output $y_3(t)$ for $-\infty < t < \infty$. If a new input is $x_4(t) = x_3(t-1) = u(t-1) - u(t-2)$, find graphically the corresponding output $y_4(t)$ for $-\infty < t < \infty$, and indicate if $y_4(t) = y_3(t-1)$. Accordingly, would this averager be time-invariant?
   (c) Is this averager a causal system? Explain.
   (d) If the input to the averager is bounded, would its output be bounded? Is the averager BIBO stable?
   Answers: $y_1(t) = 2 + [u(t) - u(t-1)]$; the system is non-linear, noncausal, and BIBO stable.

2.3 An RC circuit in series with a voltage source $x(t)$ is represented by an ordinary differential equation

$$\frac{dy(t)}{dt} + 2y(t) = 2x(t)$$

where $y(t)$ is the voltage across the capacitor. Assume $y(0)$ is the initial voltage across the capacitor.

   (a) If it is known that the resistor has a resistance R, and the capacitor $C = 1$ F. Draw the circuit that corresponds to the given ordinary differential equation.

   (b) For zero initial condition, and $x(t) = u(t)$, assess the following as possible output of the system:

$$y(t) = e^{-2t} \int_0^t e^{2\tau} d\tau.$$

   If so, find and plot $y(t)$.

Answers: $R = 0.5$; $y(t) = 0.5(1 - e^{-2t})u(t)$.

2.4 A time-varying capacitor is characterized by the charge–voltage equation $q(t) = C(t)v(t)$. That is, the capacitance is not a constant but a function of time.

   (a) Given that $i(t) = dq(t)/dt$, find the voltage–current relation for this time-varying capacitor.

   (b) Let $C(t) = 1 + \cos(2\pi t)$ and $v(t) = \cos(2\pi t)$ determine the current $i_1(t)$ in the capacitor for all t.

   (c) Let $C(t)$ be as above, but delay $v(t)$ by 0.25 s, determine $i_2(t)$ for all time. Is the system TI?

Answer: (b) $i_1(t) = -2\pi \sin(2\pi t)[1 + 2\cos(2\pi t)]$.

2.5 An analog system has the following input–output relation:

$$y(t) = \int_0^t e^{-(t-\tau)}x(\tau)d\tau \qquad t \geq 0,$$

and zero otherwise. The input is $x(t)$ and $y(t)$ is the output.

   (a) Is this system LTI? If so, can you determine without any computation the impulse response of the system? Explain.

   (b) Is this system causal? Explain.

   (c) Find the unit step response $s(t)$ of the given system and from it find the impulse response $h(t)$. Is this a BIBO stable system? Explain.

   (d) Find the response due to a pulse $x(t) = u(t) - u(t - 1)$.

Answers: Yes, LTI with $h(t) = e^{-t}u(t)$; causal and BIBO stable.

2.6 A fundamental property of linear time-invariant systems is that whenever the input of the system is a sinusoid of a certain frequency the output will also be a sinusoid of the same frequency but with an amplitude and phase determined by the system. For the following systems let the input be $x(t) = \cos(t)$, $-\infty < t < \infty$, find the output $y(t)$ and determine if the system is LTI:

$$(a)\ \ y(t) = |x(t)|^2, \qquad (b)\ \ y(t) = 0.5[x(t) + x(t-1)],$$
$$(c)\ \ y(t) = x(t)u(t), \qquad (d)\ \ y(t) = \tfrac{1}{2}\int_{t-2}^{t} x(\tau)d\tau.$$

Answers: (a) $y(t) = 0.5(1 + \cos(2t))$; (c) system is not LTI.

2.7 Consider the system where for an input $x(t)$ the output is $y(t) = x(t)f(t)$.
 (a) Let $f(t) = u(t) - u(t-10)$, determine whether the system with input $x(t)$ and output $y(t)$ is linear, time-invariant, causal and BIBO stable.
 (b) Suppose $x(t) = 4\cos(\pi t/2)$, and $f(t) = \cos(6\pi t/7)$ and both are periodic, is the output $y(t)$ also periodic? What frequencies are present in the output? Is this system linear? Is it time-invariant? Explain.
 (c) Let $f(t) = u(t) - u(t-2)$ and the input $x(t) = u(t)$, find the corresponding output $y(t)$. Suppose you shift the input so that it is $x_1(t) = x(t-3)$; what is the corresponding output $y_1(t)$? Is the system time-invariant? Explain.
Answer: (a) System is time-varying, BIBO stable.

2.8 The response of a first-order system is for $t \geq 0$

$$y(t) = y(0)e^{-t} + \int_{0}^{t} e^{-(t-\tau)}x(\tau)d\tau$$

and zero otherwise.
 (a) Consider $y(0) = 0$; is the system linear? If $y(0) \neq 0$, is the system linear? Explain.
 (b) If $x(t) = 0$, how is the response of the system called? If $y(0) = 0$ how is the response of the system to any input $x(t)$ called?
 (c) Let $y(0) = 0$, find the response due to $\delta(t)$. How is this response called?
 (d) When $y(0) = 0$, and $x(t) = u(t)$ call the corresponding output $s(t)$. Find $s(t)$ and calculate $ds(t)/dt$, what does this correspond to from the above results?
Answers: If $y(0) = 0$, $y(t) = [x * h](t)$ is zero-state response with $h(t) = e^{-t}u(t)$.

2.9 The impulse response of an LTI continuous-time system is $h(t) = u(t) - u(t-1)$.
 (a) If the input of this system is $x(t)$, assess if it is true that the system output is

$$y(t) = \int_{t-1}^{t} x(\tau)d\tau .$$

 (b) If the input is $x(t) = u(t)$, graphically determine the corresponding output $y(t)$ of the system.
 (c) Calculate the unit-step response, call it $s(t)$, and then figure out how to obtain the impulse response $h(t)$ of the system from it.
Answers: Yes, use convolution integral; if $x(t) = u(t)$, then $y(t) = r(t) - r(t-1)$.

2.10 The input of an LTI continuous-time system with impulse response $h(t) = u(t) - u(t-1)$ is

$$x(t) = \sum_{k=0}^{9} \delta(t - kT).$$

(a) Find the output $y(t)$ of the system using the convolution integral.
(b) If $T = 1$, obtain and plot the system output $y(t)$.
(c) If $T = 0.5$, obtain and plot the system output $y(t)$.
Answers: $y(t) = \sum_{k=0}^{9} h(t - kT)$.

2.11 The impulse response $h(t)$ of a causal, linear, time-invariant, continuous-time system is

$$h(t) = \sum_{k=0}^{\infty} h_1(t - 2k), \quad \text{where} \quad h_1(t) = u(t) - 2u(t-1) + u(t-2).$$

Assuming zero initial conditions, determine the outputs $y_i(t)$, $i = 1, 2$, of this system if the input is

$$(a) \ x_1(t) = u(t) - u(t-2) \qquad (b) \ x_2(t) = \delta(t) - \delta(t-2).$$

Answers: $y_1(t) = r(t) - 2r(t-1) + r(t-2)$; $y_2(t) = h_1(t)$.

2.12 A quadrature amplitude modulation (QAM) system is a communication system capable of transmitting two messages $m_1(t)$, $m_2(t)$ at the same time. The transmitted signal $s(t)$ is

$$s(t) = m_1(t)\cos(\Omega_c t) + m_2(t)\sin(\Omega_c t).$$

Carefully draw a block diagram for the QAM system.
(a) Determine if the system is time-invariant or not.
(b) Assume $m_1(t) = m_2(t) = m(t)$, i.e., we are sending the same message using two different modulators. Express the modulated signal in terms of a cosine with carrier frequency Ω_c, amplitude A, and phase θ. Obtain A and θ. Is the system linear? Explain.
Answer: $s(t) = \sqrt{2}m(t)\cos(\Omega_c t - \pi/4)$.

2.13 An echo system could be modeled using or not using feedback.
(a) Feedback systems are of great interest in control and in modeling of many systems. An echo is created as the sum of one or more delayed and attenuated output signals that are fed back into the present signal. A possible model for an echo system is

$$y(t) = x(t) + \alpha_1 y(t - \tau) + \cdots + \alpha_N y(t - N\tau)$$

where $x(t)$ is the present input signal and $y(t)$ is the present output and $y(t - k\tau)$ the previous delayed outputs, the $|\alpha_k| < 1$ values are attenuation factors. Carefully draw a block diagram for this system.
(b) Consider the echo model for $N = 1$ and parameters $\tau = 1$ and $\alpha_1 = 0.1$; is the resulting echo system LTI? Explain.

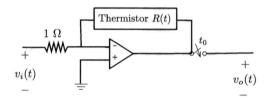

FIGURE 2.23

Problem 2.14.

(c)   Another possible model is given by a non-recursive system, i.e., one without feedback,

$$z(t) = x(t) + \beta_1 x(t - \tau) + \cdots + \beta_M x(t - M\tau)$$

where several present and past inputs, are delayed and attenuated, and added up to form the output. The parameters $|\beta_k| < 1$ are attenuation factors and τ a delay. Carefully draw a block diagram for the echo system characterized by the above equation. Does the above equation represent a LTI system? Explain.

Answers: (b) The input/output equation for the echo system is $y(t) = x(t) + 0.1y(t - 1)$; it is LTI.

2.8.2 PROBLEMS USING MATLAB

2.14 Temperature measuring system—The op-amp circuit shown in Fig. 2.23 is used to measure the changes of temperature in a system. The output voltage is given by

$$v_o(t) = -R(t)v_i(t).$$

Suppose that the temperature in the system changes cyclically after $t = 0$, so that

$$R(t) = [1 + 0.5 \cos(20\pi t)] u(t).$$

Let the input be $v_i(t) = 1$ V.
   (a)   Assuming that the switch closes at $t_0 = 0$ s, use MATLAB to plot the output voltage $v_o(t)$ for $0 \le t \le 0.2$ s in time intervals of 0.01 s.
   (b)   If the switch closes at $t_0 = 50$ ms, plot the output voltage $v_{o1}(t)$ for $0 \le t \le 0.2$ s in time intervals of 0.01 s.
   (c)   Use the above results to determine if this system is time invariant. Explain.
Answers: $v(0) = -1.5$; $v_{o1}(50 \times 10^{-3}) = -0.5$.

2.15 Zener diode—A Zener diode is one such that the output corresponding to an input $v_s(t) = \cos(\pi t)$ is a "clipped" sinusoid,

$$x(t) = \begin{cases} 0.5 & v_s(t) > 0.5, \\ -0.5 & v_s(t) < -0.5, \\ v_s(t) & \text{otherwise,} \end{cases}$$

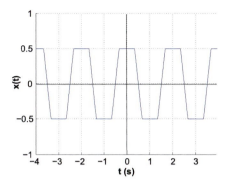

FIGURE 2.24

Problem 2.15.

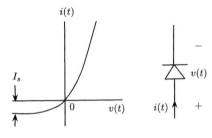

FIGURE 2.25

Problem 2.16: voltage–current characterization of p-n diode.

as shown in Fig. 2.24 for a few periods. Use MATLAB to generate the input and output signals and plot them in the same plot for $0 \leq t \leq 4$ at time intervals of 0.001.

(a) Is this system linear? Compare the output obtained from $v_s(t)$ with that obtained from $0.3v_s(t)$.

(b) Is the system time-invariant? Explain.

Answers: Zener diode is non-linear, time-invariant.

2.16 p-n diode—The voltage–current characterization of a p-n diode is given by (see Fig. 2.25)

$$i(t) = I_s(e^{qv(t)/kT} - 1)$$

where $i(t)$ and $v(t)$ are the current and the voltage in the diode (in the direction indicated in the diode) I_s is the reversed saturation current, and kT/q is a constant.

(a) Consider the voltage $v(t)$ as the input and the current $i(t)$ as the output of the diode. Is the p-n diode a linear system? Explain.

(b) An *ideal diode* is such that when the voltage is negative, $v(t) < 0$, the current is zero, i.e., open circuit; and when the current is positive, $i(t) > 0$, the voltage is zero or short circuit. Under what conditions does the p-n diode voltage–current characterization approximate

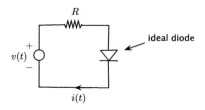

FIGURE 2.26

Problem 2.16: ideal diode circuit.

the characterization of the ideal diode? Use MATLAB to plot the current–voltage plot for a diode with $I_s = 0.0001$ and $kT/1 = 0.026$ and compare it to the ideal diode current–voltage plot. Determine if the ideal diode is linear.

(c) Consider the circuit using an ideal diode in Fig. 2.26, where the source is a sinusoid signal $v(t) = \sin(2\pi t)u(t)$ and the output is the voltage in the resistor $R = 1\ \Omega$ or $v_r(t)$. Plot $v_r(t)$. Is this system linear? Where would you use this circuit?

Answers: The diode is non-linear.

2.17 Convolution—The impulse response of a LTI is $h(t) = e^{-2t}u(t)$. Use MATLAB functions to approximate the convolution integral when the inputs of the system are

$$x_1(t) = \cos(2\pi t)[u(t) - u(t - 20)], \qquad x_2(t) = \sin(\pi t)e^{-20t}[u(t) - u(t - 20)],$$
$$x_3(t) = r(t) - 2r(t - 2) + r(t - 4).$$

Plot the impulse response, the input, and the corresponding output for each case.

2.18 Steady-state response of averager—An analog averager is given by

$$y(t) = \frac{1}{T}\int_{t-T}^{t} x(\tau)d\tau.$$

(a) Let $x(t) = u(t) - u(t - 1)$; find the average signal $y(t)$ using the above integral. Let $T = 1$. Carefully plot $y(t)$. Verify your result by graphically computing the convolution of $x(t)$ and the impulse response $h(t)$ of the averager.

(b) To see the effect of the averager, consider the signal to be averaged to be $x(t) = \cos(2\pi t/T_0)u(t)$, select the smallest possible value of T in the averager so that the steady-state response of the system, $y(t)$ as $t \to \infty$, will be 0.

(c) Use MATLAB to compute the output in part (b). Compute the output $y(t)$ for $0 \le t \le 2$ at intervals $T_s = 0.001$. Approximate the convolution integral using the function *conv* (use help to find about *conv*) multiplied by T_s.

Answers: $y(t) = t[u(t) - u(t - 1)] + u(t - 1)$.

2.19 AM envelope detector—Consider an *envelope detector* that is used to detect the message sent in the AM system shown in the examples. The envelope detector as a system is composed of two cascaded systems: one which computes the absolute value of the input (implemented with ideal diodes), and a second one that low-pass filters its input (implemented with an RC circuit). The following is an implementation of these operations in discrete time, so we can use numeric MATLAB. Let the input to the envelope detector be

$$x(t) = [p(t) + P]\cos(\Omega_0 t).$$

Use MATLAB to solve numerically this problem.

(a) Consider first $p(t) = 20[u(t) - u(t - 40)] - 10[u(t - 40) - u(t - 60)]$ and let $\Omega_0 = 2\pi$, $P = 1.1|\min(p(t))|$. Generate the signals $p(t)$ and $x(t)$ for $0 \le t \le 100$ with an interval of $T_s = 0.01$.

(b) Consider then the subsystem that computes the absolute value of the input $x(t)$. Plot $y(t) = |x(t)|$.

(c) Compute the low-pass filtered signal $(h * y)(t)$ by using an RC circuit with impulse response $h(t) = e^{-0.8t}u(t)$. To implement the convolution use *conv* function multiplied by T_s. Plot together the message signal $p(t)$, the modulated signal $x(t)$, the absolute value $y(t)$ and the envelope $z(t) = (h * y)(t) - P$. Does this envelope look like $p(t)$?

(d) Consider the message signal $p(t) = 2\cos(0.2\pi t)$, $\Omega_0 = 10\pi$ and $P = |\min(p(t))|$ and repeat the process. Scale the signal to get the original $p(t)$.

2.20 Frequency Modulation (FM)—*Frequency modulation*, or FM, uses a wider bandwidth that amplitude modulation, or AM, but it is not affected as much by noise as AM is. The output of an FM transmitter is of the form

$$y(t) = \cos\left(2\pi t + 2\pi v \int_0^t m(\tau)d\tau\right)$$

where $m(t)$ is the message and v is a factor in Hz/V if the units of the message are in volt.

(a) Create as message a signal

$$m(t) = \cos(t).$$

Find the FM signal $y(t)$ for $v = 10$ and then for $v = 1$. Use MATLAB to generate the different signals for times $0 \le t \le 10$ at intervals of $T_s = 0.01$. Plot $m(t)$ and the two FM signals (one for $v = 10$ and the other for $v = 1$) in the same plot. Is the FM transmitter a linear system? Explain.

(b) Create a message signal

$$m_1(t) = \begin{cases} 1 & \text{when } m(t) \ge 0, \\ -1 & \text{when } m(t) < 0. \end{cases}$$

Find the corresponding FM signal for $v = 1$.

THE LAPLACE TRANSFORM

CONTENTS

What we know is not much. What we do not know is immense.

Pierre-Simon marquis de Laplace (1749–1827), French Mathematician and Astronomer

3.1 INTRODUCTION

The material in the next three chapters is very significant in the analysis of continuous-time signals and systems. In this chapter, we begin the frequency domain analysis of continuous-time signals and systems using the Laplace transform, the most general of these transforms, to be followed by the analysis using a Fourier series and the Fourier transform. These transforms provide representations of signals complementary to their time-domain representation. The Laplace transform depends on a complex variable $s = \sigma + j\Omega$, composed of a damping factor σ and of a frequency Ω variable, while the Fourier transform considers only frequency Ω. The growth or decay of a signal—damping—as well as its repetitive nature—frequency—in the time domain are characterized in the Laplace domain by the location of the roots of the numerator and denominator, or zeros and poles, of the Laplace transform of the signal. The location of the poles and the zeros of the transfer function relates to the dynamic characteristics of the system.

The Laplace transform provides a significant algebraic characterization of continuous-time systems: the ratio of the Laplace transform of the output to that of the input—or the transfer function of the system. The transfer function concept unifies the convolution integral and the ordinary differential equations system representations in the LTI case. The concept of transfer function is not only useful in analysis, but also in design as we will see in Chapter 7.

Certain characteristics of continuous-time systems can only be studied via the Laplace transform. Such is the case of stability, transient and steady-state responses. This is a significant reason to study the Laplace analysis before the Fourier analysis which deals exclusively with the frequency characterization of continuous-time signals and systems. Stability and transients are important issues in classical control theory, thus the importance of the Laplace transform in this area. The frequency characterization of signals and the frequency response of systems—provided by the Fourier transform—are significant in communications.

Given the prevalence of causal signals (those that are zero for negative time) and of causal systems (those with impulse responses that are zero for negative time) the Laplace transform is typically known as "one-sided," but the "two-sided" transform also exists! The impression is that these are two different transforms, but in reality it is the Laplace transform applied to different types of signals and systems. We will show that by separating the signal into its causal and its anticausal components, we only need to apply the one-sided transform. Care should be exercised, however, when dealing with the inverse transform so as to get the correct signal.

Since the Laplace transform requires integration over an infinite domain, it is necessary to consider if and where this integration converges—or the "region of convergence" in the s-plane. Now, if such a region includes the $j\Omega$-axis of the s-plane, then the Laplace transform exists for $s = j\Omega$ and coincides with the Fourier transform of the signal as we will see in Chapter 5. Thus, the Fourier transform for a large class of functions can be obtained directly from their Laplace transforms—another good reason to study first the Laplace transform. In a subtle way, the Laplace transform is also connected with the Fourier series representation of periodic continuous-time signals. Such a connection reduces the computational complexity of the Fourier series by eliminating integration in cases when the Laplace transform of a period of the signal is already available.

Linear time-invariant systems respond to complex exponentials or eigenfunctions in a very special way: their output is the input complex exponential with its magnitude and phase changed by the response of the system. This provides the characterization of the system by the Laplace transform, when the exponent is the complex frequency s, and by the Fourier representation when the exponent is $j\Omega$. The eigenfunction concept is linked to phasors used to compute the steady-state response in circuits.

LAPLACE AND HEAVISIDE

The Marquis Pierre-Simon de Laplace (1749–1827) [5,9] was a French mathematician and astronomer. Although from humble beginnings he became royalty by his political abilities. As an astronomer, he dedicated his life to the work of applying the Newtonian law of gravitation to the entire solar system. He was considered an applied mathematician and, as a member of the French Academy of Sciences, knew other great mathematicians of the time such as Legendre, Lagrange and Fourier. Besides his work on celestial mechanics, Laplace did significant work in the theory of probability from which the Laplace transform probably comes. He felt that "... the theory of probabilities is only common sense expressed in numbers." Early transformations similar to Laplace's were used by Euler and Lagrange. It was, however, Oliver Heaviside (1850–1925) who used the Laplace transform in the solution of ordinary differential equations. Heaviside, an Englishman, was a self-taught electrical engineer, mathematician, and physicist [79].

3.2 THE TWO-SIDED LAPLACE TRANSFORM

Rather than giving the definitions of the Laplace transform and its inverse, let us see how they are obtained intuitively. As indicated before, a basic idea in characterizing signals—and their response when applied to LTI systems—is to consider them a combination of basic signals for which we can easily obtain a response. In the previous chapter, when considering the "time-domain" solutions, we represented the input as an infinite combination of impulses, weighted by the value of the signal, occurring at all possible times. The reason we did so is because the response due to an impulse is the impulse response of the LTI system which is fundamental in our studies. A similar approach will be followed when attempting to obtain the "frequency domain" representations of signals when applied to a LTI system. In this case, the basic functions used are complex exponentials or sinusoids. The concept of **eigenfunction**, discussed next, is somewhat abstract at the beginning, but after you see it applied here and in the Fourier representation later you will think of it as a way to obtain a representation analogous to the impulse representation.

3.2.1 EIGENFUNCTIONS OF LTI SYSTEMS

Consider as the input of a LTI system the complex signal $x(t) = e^{s_0 t}$, $s_0 = \sigma_0 + j\Omega_0$, $-\infty < t < \infty$, and let $h(t)$ be the impulse response of the system. (See Fig. 3.1.) According to the convolution integral, the output of the system is

$$y(t) = \int_{-\infty}^{\infty} h(\tau)x(t-\tau)d\tau = \int_{-\infty}^{\infty} h(\tau)e^{s_0(t-\tau)}d\tau = e^{s_0 t}\underbrace{\int_{-\infty}^{\infty} h(\tau)e^{-\tau s_0}d\tau}_{H(s_0)} = x(t)H(s_0). \quad (3.1)$$

Since the same exponential at the input appears at the output, $x(t) = e^{s_0 t}$ is called an **eigenfunction**[1] of the LTI system. The input $x(t)$ is changed at the output by the complex function $H(s_0)$, which is related to the system through the impulse response $h(t)$. In general, for any s, the eigenfunction at the

---

[1]David Hilbert (1862–1943), German mathematician, seems to have been the first to use the German word *eigen* to denote eigenvalues and eigenvectors in 1904. The word *eigen* means own, proper.

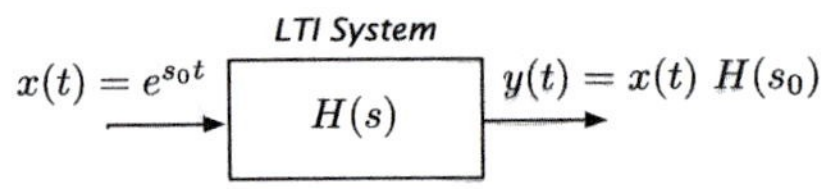

$$x(t) = e^{s_0 t} \quad \boxed{\begin{array}{c} \textit{LTI System} \\ H(s) \end{array}} \quad y(t) = x(t) \, H(s_0)$$

FIGURE 3.1

Eigenfunction property of LTI systems: The input of the system is $x(t) = e^{s_0 t} = e^{\sigma_0 t} e^{j \Omega_0 t}$ and the output of the system is the same input multiplied by the complex value $H(s_0)$ where $H(s) = \mathcal{L}[h(t)]$ or the Laplace transform of the impulse response $h(t)$ of the system.

output is modified by the complex function

$$H(s) = \int_{-\infty}^{\infty} h(\tau) e^{-\tau s} d\tau, \tag{3.2}$$

which corresponds to the Laplace transform of $h(t)$!

An input $x(t) = e^{s_0 t}$, $s_0 = \sigma_0 + j \Omega_0$, is called an **eigenfunction** of a LTI system with impulse response $h(t)$ if the corresponding output of the system is

$$y(t) = x(t) \int_{-\infty}^{\infty} h(t) e^{-s_0 t} dt = x(t) H(s_0)$$

where $H(s_0)$ is the Laplace transform of $h(t)$ computed at $s = s_0$. This property is only valid for LTI systems, it is not satisfied by time-varying or non-linear systems.

Remarks

1. One could think of $H(s)$ as an infinite combination of complex exponentials, weighted by the impulse response $h(t)$. A similar rationalization can be used for the Laplace transform of a signal.
2. Consider now the significance of applying the eigenfunction result. Suppose a signal $x(t)$ is expressed as a sum of complex exponentials in $s = \sigma + j\Omega$,

$$x(t) = \frac{1}{2\pi j} \int_{\sigma - j\infty}^{\sigma + j\infty} X(s) e^{st} ds,$$

i.e., an infinite sum of exponentials in s each weighted by the function $X(s)/(2\pi j)$ (this equation is connected with the inverse Laplace transform as we will see soon). Using the superposition property of LTI systems, and considering that for a LTI system with impulse response $h(t)$ the output due to e^{st} is $H(s)e^{st}$, then the output due to $x(t)$ is

$$y(t) = \frac{1}{2\pi j} \int_{\sigma - j\infty}^{\sigma + j\infty} X(s) \left[H(s) e^{st} \right] ds = \frac{1}{2\pi j} \int_{\sigma - j\infty}^{\sigma + j\infty} Y(s) e^{st} ds$$

where we let $Y(s) = X(s)H(s)$. But from the previous chapter we see that $y(t)$ is the convolution $y(t) = [x * h](t)$. Thus the following two expressions must be connected:

$$y(t) = [x * h](t) \quad \Leftrightarrow \quad Y(s) = X(s)H(s).$$

The expression on the left indicates how to compute the output in the time domain, and the one on the right how to compute the Laplace transform of the output in the frequency domain. This is the most important property of the Laplace transform: it reduces the complexity of the convolution integral in time to the multiplication of the Laplace transforms of the input, $X(s)$, and of the impulse response, $H(s)$.

Now we are ready for the proper definition of the direct and the inverse Laplace transforms of a signal or of the impulse response of a system.

The two-sided Laplace transform of a continuous-time function $f(t)$ is

$$F(s) = \mathcal{L}[f(t)] = \int_{-\infty}^{\infty} f(t)e^{-st}\,dt \qquad s \in \text{ROC} \qquad (3.3)$$

where the variable $s = \sigma + j\Omega$, with σ a damping factor and Ω frequency in rad/s. ROC stands for the region of convergence of $F(s)$, i.e., where the infinite integral exists.

The inverse Laplace transform is given by

$$f(t) = \mathcal{L}^{-1}[F(s)] = \frac{1}{2\pi j} \int_{\sigma-j\infty}^{\sigma+j\infty} F(s)e^{st}\,ds \qquad \sigma \in \text{ROC}. \qquad (3.4)$$

Remarks

1. The Laplace transform $F(s)$ provides a representation of $f(t)$ in the s-domain, which in turn can be converted back into the original time-domain function in a one-to-one manner using the region of convergence. Thus,[2]

$$F(s), \ \text{ROC} \quad \Leftrightarrow \quad f(t).$$

2. If $f(t) = h(t)$, the impulse response of a LTI system, then $H(s)$ is called the **system** or **transfer function** of the system and it characterizes the system in the s-domain just like $h(t)$ does in the time domain. On the other hand, if $f(t)$ is just a signal, then $F(s)$ is its Laplace transform.

3. The inverse Laplace transform equation (3.4) can be understood as the representation of $f(t)$ (whether it is a signal or an impulse response) by an infinite summation of complex exponentials weighted by $F(s)$. The computation of the inverse Laplace transform using Equation (3.4) requires complex integration. Algebraic methods will be used later to find the inverse Laplace transform thus avoiding the complex integration.

---

[2]This notation simply indicates that corresponding to the Laplace transform $F(s)$, with a certain ROC, there is a function of time $f(t)$. It does not mean that $F(s)$ equals $f(t)$, far from it—$F(s)$ and $f(t)$ are in completely different domains!

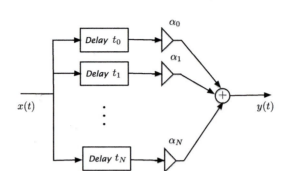

FIGURE 3.2

Block diagram of a wireless communication channel causing a multi-path effect on the sent message $x(t)$. The message $x(t)$ is delayed and attenuated when sent over $N + 1$ paths. The effect is similar to that of an echo in acoustic signals.

Example 3.1. A common problem in wireless communications is the so-called "multi-path" effect on the transmitted message. Consider the channel between the transmitter and the receiver a system like the one depicted in Fig. 3.2. The sent message $x(t)$ does not necessarily go from the transmitter to the receiver directly (line of sight) but it may take different paths, each with different length so that the signal in each path is attenuated and delayed differently.[3] At the receiver, these delayed and attenuated signals are added causing a fading effect—given the different phases of the incoming signals their addition at the receiver may result in a weak or a strong signal thus giving the sensation of the message fading back and forth. If $x(t)$ is the message sent from the transmitter, and the channel has N different paths with attenuation factors $\{\alpha_i\}$ and corresponding delays $\{t_i\}$, $i = 0, \cdots, N$, use the eigenfunction property to find the system function of the channel causing the multi-path effect.

Solution: The output of the transmission channel in Fig. 3.2 can be written as

$$y(t) = \alpha_0 x(t - t_0) + \alpha_1 x(t - t_1) + \cdots + \alpha_N x(t - t_N). \tag{3.5}$$

For $s_0 = \sigma_0 + j\Omega_0$, the response of the multi-path system to $e^{s_0 t}$ is $e^{s_0 t} H(s_0)$ so that for $x(t) = e^{s_0 t}$ we get two equivalent responses,

$$
\begin{aligned}
y(t) &= e^{s_0 t} H(s_0) \\
&= \alpha_0 e^{s_0(t-t_0)} + \cdots + \alpha_N e^{s_0(t-t_N)} = e^{s_0 t} \left[\alpha_0 e^{-s_0 t_0} + \cdots + \alpha_N e^{-s_0 t_N}\right].
\end{aligned}
$$

Thus the system function for the channel, in terms of the variable s, is

$$H(s) = \alpha_0 e^{-s t_0} + \cdots + \alpha_N e^{-s t_N}.$$

---

[3]Typically, there are three effects each path can have on the sent signal. The distance the signal needs to travel to get to the receiver (in each path this distance might be different, and is due to reflection or refraction on buildings, structures, cars, etc.) determines how much the signal is attenuated, and delayed with respect to a signal that goes directly to the receiver. The other effect is a frequency-shift—or Doppler effect—that is caused by the relative velocity between the transmitter and the receiver. The channel model in Fig. 3.2 does not include the Doppler effect.

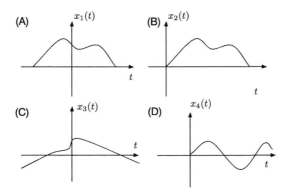

FIGURE 3.3

Examples of different types of signals: (A) noncausal finite-support signal $x_1(t)$, (B) causal finite-support signal $x_2(t)$, (C) noncausal infinite-support signal $x_3(t)$, and (D) causal infinite support $x_4(t)$.

Notice that the time-shifts in the input–output equation became exponentials in the Laplace domain, a property we will see later. □

3.2.2 REGION OF CONVERGENCE

Let us consider the different types of functions (either continuous-time signals or the impulse responses of continuous-time systems) we might be interested in calculating the Laplace transforms of:

- Finite-support functions:

$$f(t) = 0 \qquad \text{for } t \text{ not in a finite segment } t_1 \leq t \leq t_2$$

  for any finite, positive or negative, t_1 and t_2 as long as $t_1 < t_2$. We will see that the Laplace transform of these finite-support signals always exists and is of particular interest in the computation of the coefficients of the Fourier series of periodic signals.
- Infinite-support functions: In this case, $f(t)$ is defined over an infinite support, e.g., $t_1 < t < t_2$ where either t_1 or t_2 or both are infinite.

A finite, or infinite, support function $f(t)$ is called:

(i) **casual** if $f(t) = 0$ $\quad t < 0$,

(ii) **anticausal** if $f(t) = 0$ $\quad t > 0$,

(iii) **noncausal**: a combination of the above.

Fig. 3.3 illustrates the different types of signals.

Because the definition of $F(s)$, see Equation (3.3), requires integration over an infinite support in each of the above cases we need to consider the region in the s-plane where the transform exists—or its **region of convergence** (ROC). This is obtained by looking at the convergence of the transform.

For the Laplace transform $F(s)$ of $f(t)$ to exist we need that

$$\left| \int_{-\infty}^{\infty} f(t)e^{-st}\, dt \right| = \left| \int_{-\infty}^{\infty} f(t)e^{-\sigma t}e^{-j\Omega t}\, dt \right|$$

$$\leq \int_{-\infty}^{\infty} |f(t)e^{-\sigma t}|\, dt < \infty$$

or that $f(t)e^{-\sigma t}$ be absolutely integrable. This may be possible by choosing an appropriate σ even in the case when $f(t)$ is not absolutely integrable. The value chosen for σ determines the ROC of $F(s)$. The frequency Ω does not affect the ROC.

Poles and Zeros and the Region of Convergence

The region of convergence (ROC) is obtained from the conditions for the integral in the Laplace transform to exist. In most of the cases of interest, the Laplace transform is a ratio of a numerator polynomial $N(s)$ and a denominator polynomial $D(s)$. The roots of $N(s)$ are called zeros, and the roots of $D(s)$ are called poles. The ROC is related to the *poles* of the transform.

For a rational function $F(s) = \mathcal{L}[f(t)] = N(s)/D(s)$, its **zeros** are the values of s that make the function $F(s) = 0$, and its **poles** are the values of s that make the function $F(s) \to \infty$. Although only finite numbers of zeros and poles are considered, infinite numbers of zeros and poles are also possible.

Typically, $F(s)$ is rational, a ratio of two polynomials $N(s)$ and $D(s)$, or $F(s) = N(s)/D(s)$, and as such its zeros are the values of s that make the numerator polynomial $N(s) = 0$, while the poles are the values of s that make the denominator polynomial $D(s) = 0$. For instance, for

$$F(s) = \frac{2(s^2 + 1)}{s^2 + 2s + 5} = \frac{2(s + j)(s - j)}{(s + 1)^2 + 4} = \frac{2(s + j)(s - j)}{(s + 1 + 2j)(s + 1 - 2j)}$$

the zeros are $s_{1,2} = \pm j$, or the roots of $N(s) = 0$ so that $F(\pm j) = 0$; and the pair of complex conjugate poles $-1 \pm 2j$ are the roots of the equation $D(s) = 0$, so that $F(-1 \pm 2j) \to \infty$. Geometrically, zeros can be visualized as those values that make the function go to zero, and poles as those values that make the function approach infinity (looking like the main "pole" of a circus tent). See Fig. 3.4 for an alternative behavior of zeros and poles.

Not all Laplace transforms have poles or zeros or a finite number of them. Consider

$$P(s) = \frac{e^s - e^{-s}}{s}.$$

$P(s)$ seems to have a pole at $s = 0$, however, it is canceled by a zero at the same place. Indeed, the zeros of $P(s)$ are obtained by letting $e^s - e^{-s} = 0$, which when multiplied by e^s gives

$$e^{2s} = 1 = e^{j2\pi k}$$

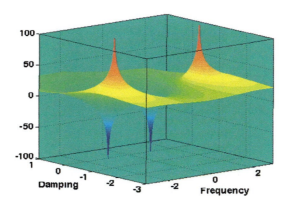

FIGURE 3.4

Three-dimensional plot of the logarithm of the magnitude of $F(s) = 2(s^2 + 1)/(s^2 + 2s + 5)$ as a function of damping σ and frequency Ω. The function $\log(F(s_i)) \to \infty$ when s_i is a pole (i.e., $s_{1,2} = -1 \pm 2j$) while $\log(F(s_i)) \to -\infty$ when s_i is a zero (i.e., $s_{1,2} = \pm j$).

for an integer $k = 0, \pm 1, \pm 2, \cdots$. Thus the zeros are $s_k = j\pi k$, $k = 0, \pm 1, \pm 2, \cdots$. Now, when $k = 0$, the zero at 0 cancels the pole at zero, therefore $P(s)$ has only zeros, an infinite number of them, $\{j\pi k$, $k = \pm 1, \pm 2, \cdots\}$. Likewise, the function $Q(s) = 1/P(s)$ has an infinite number of poles but no zeros.

Poles and Region of Convergence

The ROC consists of the values of σ for which $x(t)e^{-\sigma t}$ is absolutely integrable. Two general comments that apply to the Laplace transform of all types of signals when finding their ROCs are:

- No poles are included in the ROC, which means that, for the ROC to be the region where the Laplace transform is defined, the transform cannot become infinite at any point in it. So poles should not be present in the ROC.
- The ROC is a plane parallel to the $j\Omega$-axis. This means that it is the damping σ that defines the ROC, not the frequency Ω. This is because when we compute the absolute value of the integrand in the Laplace transform to test for convergence we let $s = \sigma + j\Omega$ and the term $|e^{j\Omega}| = 1$. Thus all the regions of convergence will contain $-\infty < \Omega < \infty$.

If $\{\sigma_i\}$ are the real parts of the poles of $F(s) = \mathcal{L}[f(t)]$, the region of convergence corresponding to different types of signals or impulse responses is determined from its poles as follows:

1. For a **causal** $f(t)$, $f(t) = 0$ for $t < 0$, the region of convergence of its Laplace transform $F(s)$ is a plane to the **right** of the poles,

$$\mathcal{R}_c = \{(\sigma, \Omega) : \sigma > \max\{\sigma_i\}, -\infty < \Omega < \infty\}.$$

2. For an **anticausal** $f(t)$, $f(t) = 0$ for $t > 0$, the region of convergence of its Laplace transform $F(s)$ is a plane to the **left** of the poles,

$$\mathcal{R}_{ac} = \{(\sigma, \Omega) : \sigma < \min\{\sigma_i\}, -\infty < \Omega < \infty\}.$$

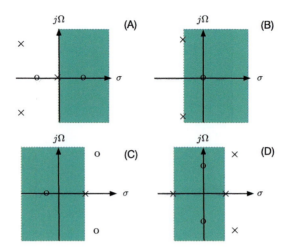

FIGURE 3.5

ROC (shaded region) for (A) causal signal with poles with $\sigma_{max} = 0$; (B) causal signal with poles with $\sigma_{max} < 0$; (C) anticausal signal with poles with $\sigma_{min} > 0$; (D) two-sided or noncausal signal where ROC is bounded by poles (poles on left-hand plane give causal component and poles on the right-hand s-plane give the anticausal component of the signal). The ROCs do not contain poles, but they can contain zeros.

3. For a **noncausal** $f(t)$, i.e., $f(t)$ defined for $-\infty < t < \infty$, the region of convergence of its Laplace transform $F(s)$ is the intersection of the regions of convergence corresponding to the causal component, $\mathcal{R}_c$, and $\mathcal{R}_{ac}$ corresponding to the anticausal component,

$$\mathcal{R}_c \bigcap \mathcal{R}_{ac}.$$

See Fig. 3.5 for examples illustrating how the ROCs connect with the poles and the type of signal.

Example 3.2. Find the Laplace transform of $\delta(t)$, $u(t)$ and a pulse $p(t) = u(t) - u(t-1)$ indicating their regions of convergence. Use MATLAB to verify the transforms.

Solution: Even though $\delta(t)$ is not a regular signal, its Laplace transform can easily be obtained:

$$\mathcal{L}[\delta(t)] = \int_{-\infty}^{\infty} \delta(t)e^{-st}dt = \int_{-\infty}^{\infty} \delta(t)e^{-s0}dt = \int_{-\infty}^{\infty} \delta(t)dt = 1.$$

Since there are no conditions for the integral to exist, we say that $\mathcal{L}[\delta(t)] = 1$ exists for all values of s, or that its ROC is the whole s-plane. This is also indicated by the fact that $\mathcal{L}[\delta(t)] = 1$ has no poles. The Laplace transform of $u(t)$ can be found as

$$U(s) = \mathcal{L}[u(t)] = \int_{-\infty}^{\infty} u(t)e^{-st}dt = \int_{0}^{\infty} e^{-st}dt = \int_{0}^{\infty} e^{-\sigma t}e^{-j\Omega t}dt$$

where we replaced the variable $s = \sigma + j\Omega$. Using Euler's equation, the above equation becomes

$$U(s) = \int_0^\infty e^{-\sigma t}[\cos(\Omega t) - j\sin(\Omega t)]dt$$

and since the sine and the cosine are bound, then we need to find a value for σ so that the exponential $e^{-\sigma t}$ does not grow as t increases. If $\sigma < 0$ the exponential $e^{-\sigma t}$ for $t \geq 0$ will grow and the integral will not converge. On the other hand, if $\sigma > 0$ the integral will converge as $e^{-\sigma t}$ for $t \geq 0$ will decay, and it is not clear what happens when $\sigma = 0$. Thus the integral exists in the region defined by $\sigma > 0$ and all frequencies $-\infty < \Omega < \infty$, (the frequency values do not interfere in the convergence), or the open right-hand s-plane. In the region of convergence, $\sigma > 0$, the integral is found to be

$$U(s) = \frac{e^{-st}}{-s}\Big|_{t=0}^\infty = \frac{1}{s}$$

where the limit for $t = \infty$ is zero since $\sigma > 0$. So the Laplace transform $U(s) = 1/s$ converges in the region defined by $\{(\sigma, \Omega) : \sigma > 0, -\infty < \Omega < \infty\}$, or the open (i.e., the $j\Omega$-axis is not included) right-hand s-plane. This ROC can also be obtained by considering that the pole of $U(s)$ is at $s = 0$ and that $u(t)$ is casual.

The pulse $p(t) = u(t) - u(t-1)$ is a finite-support signal and so its ROC is the whole s-plane. Its Laplace transform is

$$P(s) = \mathcal{L}[u(t) - u(t-1)] = \int_0^1 e^{-st}dt = \frac{-e^{-st}}{s}\Big|_{t=0}^1 = \frac{1}{s}[1 - e^{-s}].$$

That the ROC of $P(s)$ is the whole s-plane is due to a pole/zero cancellation. The zeros are the values of s that make $1 - e^{-s} = 0$ or $e^s = 1 = e^{j2\pi k}$, thus the zeros are $s_k = j2\pi k$, $k = 0, \pm1, \pm2, \cdots$. For $k = 0$, the zero is $s_0 = 0$ and it cancels the pole $s = 0$ so that

$$P(s) = \prod_{k=-\infty, k\neq 0}^\infty (s - j2\pi k)$$

has an infinite number of zeros but no poles, so that the ROC of $P(z)$ is the whole s-plane.

We can find the Laplace transform of the signals the MATLAB function *laplace* as shown by the following script:

```
%%
% Example 3.2---Laplace transform of unit-step, delta and pulse
%%
syms t u d p
% unit-step function
u=sym('heaviside(t)')
U=laplace(u)
% delta function
d=sym('dirac(t)')
D=laplace(d)
```

```
% pulse
p=heaviside(t)-heaviside(t-1)
P=laplace(p)

u = heaviside(t)
U = 1/s
d = dirac(t)
D =1
p = heaviside(t) - heaviside(t - 1)
P = 1/s - 1/(s*exp(s))
```

The naming of $u(t)$ and $\delta(t)$ as Heaviside and Dirac[4] functions is used in symbolic MATLAB. □

3.3 THE ONE-SIDED LAPLACE TRANSFORM

The one-sided Laplace transform is of significance given that most of the applications consider causal systems and causal signals—in which cases the two-sided transform is not needed—and that any signal or impulse response of a LTI system can be decomposed into causal and anticausal components requiring only the computation of one-sided Laplace transforms.

For any function $f(t)$, $-\infty < t < \infty$, its **one-sided Laplace transform** $F(s)$ is defined as

$$F(s) = \mathcal{L}[f(t)u(t)] = \int_{0-}^{\infty} f(t)e^{-st}dt, \qquad \text{ROC} \qquad (3.6)$$

or the two-sided Laplace transform of a causal or made-causal signal.

Remarks

1. The functions $f(t)$ above can be either a signal or the impulse response of a LTI system.
2. If $f(t)$ is causal, multiplying it by $u(t)$ is redundant—but harmless, but if $f(t)$ is not causal the multiplication by $u(t)$ makes $f(t)u(t)$ causal. When $f(t)$ is causal, the two-sided and the one-sided Laplace transforms of $f(t)$ coincide.
3. For a causal function $f(t)u(t)$ (notice $u(t)$ indicates the function is causal so it is an important part of the function) the corresponding Laplace transform is $F(s)$ with a certain region of convergence. This unique relation is indicated by the pair

$$f(t)u(t) \quad \leftrightarrow \quad F(s), \quad \text{ROC}$$

where the symbol $\leftrightarrow$ indicates a unique relation between a function in t with a function in s—it is not an equality, far from it!

---

[4]Paul Dirac (1902–1984) was an English electrical engineer, better known for his work in physics.

4. The lower limit of the integral in the one-sided Laplace transform is set to $0- = 0 - \varepsilon$, where $\varepsilon \to 0$, or a value on the left of 0. The reason for this is to make sure that an impulse function, $\delta(t)$, only defined at $t = 0$, is included when we are computing its Laplace transform. For any other function this limit can be taken as 0 with no effect on the transform.

5. An important use of the one-sided Laplace transform is for solving ordinary differential equations with initial conditions. The two-sided Laplace transform by starting at $t = -\infty$ (lower bound of the integral) ignores possible nonzero initial conditions at $t = 0$, and thus it is not useful in solving ordinary differential equations unless the initial conditions are zero.

The one-sided Laplace transform can be used to find the two-sided Laplace transform of any signal or impulse response.

The Laplace transform of

- a finite-support function $f(t)$, i.e., $f(t) = 0$ for $t < t_1$ and $t > t_2$, $t_1 < t_2$, is

$$F(s) = \mathcal{L}\big[f(t)[u(t-t_1) - u(t-t_2)]\big] \qquad \text{ROC: whole } s\text{-plane};\qquad (3.7)$$

- a causal function $g(t)$, i.e., $g(t) = 0$ for $t < 0$, is

$$G(s) = \mathcal{L}[g(t)u(t)] \qquad \mathcal{R}_c = \{(\sigma, \Omega) : \sigma > \max\{\sigma_i\}, -\infty < \Omega < \infty\} \qquad (3.8)$$

where $\{\sigma_i\}$ are the real parts of the poles of $G(s)$;
- an anticausal function $h(t)$, i.e., $h(t) = 0$ for $t > 0$, is

$$H(s) = \mathcal{L}[h(-t)u(t)]_{(-s)} \qquad \mathcal{R}_{ac} = \{(\sigma, \Omega) : \sigma < \min\{\sigma_i\}, -\infty < \Omega < \infty\} \qquad (3.9)$$

where $\{\sigma_i\}$ are the real parts of the poles of $H(s)$;
- a noncausal function $p(t)$, i.e., $p(t) = p_{ac}(t) + p_c(t) = p(t)u(-t) + p(t)u(t)$, is

$$P(s) = \mathcal{L}[p(t)] = \mathcal{L}[p_{ac}(t)u(t)]_{(-s)} + \mathcal{L}[p_c(t)u(t)] \qquad \mathcal{R}_c \cap \mathcal{R}_{ac}. \qquad (3.10)$$

The Laplace transform of a bounded function $f(t)$ of finite support $t_1 \leq t \leq t_2$, always exists and has the whole s-plane as ROC. Indeed, the integral defining the Laplace transform is bounded for any value of σ. If $A = \max(|f(t)|)$, then

$$|F(s)| \leq \int_{t_1}^{t_2} |f(t)||e^{-st}|dt \leq A \int_{t_1}^{t_2} e^{-\sigma t} dt = \begin{cases} A(e^{-\sigma t_1} - e^{-\sigma t_2})/\sigma & \sigma \neq 0, \\ A(t_2 - t_1) & \sigma = 0, \end{cases}$$

is less than infinity so that the integral converges for all σ.

For an anticausal function $h(t)$, so that $h(t) = 0$ for $t > 0$, its Laplace transform is obtained after the variable substitution $\tau = -t$ as

$$\begin{aligned} H(s) &= \mathcal{L}[h(t)u(-t)] = \int_{-\infty}^{0} h(t)u(-t)e^{-st} dt = -\int_{\infty}^{0} h(-\tau)u(\tau)e^{s\tau} d\tau \\ &= \int_{0}^{\infty} h(-\tau)u(\tau)e^{s\tau} d\tau = \mathcal{L}[h(-t)u(t)]_{(-s)}. \end{aligned}$$

That is, it is the Laplace transform of the causal function $h(-t)u(t)$ (the reflection of the anticausal function $h(t)$) with s replaced by $-s$.

As a result, for a noncausal function $p(t) = p_{ac}(t) + p_c(t)$ with $p_{ac}(t) = p(t)u(-t)$, the anticausal component, and $p_c(t) = p(t)u(t)$, the causal component, the Laplace transform of $p(t)$ is

$$P(s) = \mathcal{L}[p(-t)u(t)]_{(-s)} + \mathcal{L}[p(t)u(t)].$$

The ROC of $P(s)$ is the intersection of the ROCs of its anticausal and causal components.

Example 3.3. Find and use the Laplace transform of $e^{j(\Omega_0 t + \theta)}u(t)$ to obtain the Laplace transform of $x(t) = \cos(\Omega_0 t + \theta)u(t)$. Consider the special cases that $\theta = 0$ and $\theta = -\pi/2$. Determine the ROCs. Use MATLAB to plot the signals and the corresponding poles/zeros when $\Omega_0 = 2$, $\theta = 0$ and $\pi/4$.

Solution: The Laplace transform of the complex causal signal $e^{j(\Omega_0 t + \theta)}u(t)$ is found to be

$$
\begin{aligned}
\mathcal{L}[e^{j(\Omega_0 t + \theta)}u(t)] &= \int_0^\infty e^{j(\Omega_0 t + \theta)}e^{-st}\,dt = e^{j\theta}\int_0^\infty e^{-(s - j\Omega_0)t}\,dt \\
&= \frac{-e^{j\theta}}{s - j\Omega_0}e^{-\sigma t - j(\Omega - \Omega_0)t}\Big|_{t=0}^\infty = \frac{e^{j\theta}}{s - j\Omega_0} \qquad \text{ROC: } \sigma > 0.
\end{aligned}
$$

According to Euler's identity

$$\cos(\Omega_0 t + \theta) = \frac{e^{j(\Omega_0 t + \theta)} + e^{-j(\Omega_0 t + \theta)}}{2},$$

by the linearity of the integral and using the above result we get

$$
\begin{aligned}
X(s) &= \mathcal{L}[\cos(\Omega_0 t + \theta)u(t)] = 0.5\mathcal{L}[e^{j(\Omega_0 t + \theta)}u(t)] + 0.5\mathcal{L}[e^{-j(\Omega_0 t + \theta)}u(t)] \\
&= 0.5\frac{e^{j\theta}(s + j\Omega_0) + e^{-j\theta}(s - j\Omega_0)}{s^2 + \Omega_0^2} = \frac{s\cos(\theta) - \Omega_0\sin(\theta)}{s^2 + \Omega_0^2}
\end{aligned}
$$

and a region of convergence $\{(\sigma, \Omega) : \sigma > 0, -\infty < \Omega < \infty\}$ or the open right-hand s-plane. The poles of $X(s)$ are $s_{1,2} = \pm j\Omega_0$, and its zero is $s = (\Omega_0 \sin(\theta))/\cos(\theta) = \Omega_0\tan(\theta)$.

Now if we let $\theta = 0, -\pi/2$ in the above equation we have the following Laplace transforms:

$$\mathcal{L}[\cos(\Omega_0 t)u(t)] = \frac{s}{s^2 + \Omega_0^2}, \qquad \mathcal{L}[\sin(\Omega_0 t)u(t)] = \frac{\Omega_0}{s^2 + \Omega_0^2}$$

as $\cos(\Omega_0 t - \pi/2) = \sin(\Omega_0 t)$. The ROC of the above Laplace transforms is still $\{(\sigma, \Omega) : \sigma > 0, -\infty < \Omega < \infty\}$, or the open right-hand s-plane (i.e., not including the $j\Omega$-axis). See Fig. 3.6 for the pole–zero plots and the corresponding signals for $\theta = 0$, $\theta = \pi/4$ and $\Omega_0 = 2$. Notice that, for all the cases, the regions of convergence do not include the poles of the Laplace transforms, located on the $j\Omega$-axis. ☐

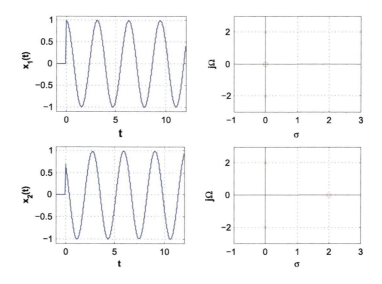

FIGURE 3.6

Location of the poles and zeros of $\mathcal{L}[\cos(2t + \theta)u(t)]$ for $\theta = 0$ (top figure) and for $\theta = \pi/4$ (bottom figure). Note that the zero in the top figure is moved to the right to 2 in the bottom figure because the zero of the Laplace transform of $x_2(t)$ is $s = \Omega_0 \tan(\theta) = 2\tan(\pi/4) = 2$.

Example 3.4. Use MATLAB symbolic computation to find the Laplace transform of a real exponential, $x(t) = e^{-t}u(t)$, and of $x(t)$ modulated by a cosine or $y(t) = e^{-t}\cos(10t)u(t)$. Plot the signals and the poles and zeros of their Laplace transforms.

Solution: The script shown below is used. The MATLAB function *laplace* is used for the computation of the Laplace transform and the function *fplot* allow us to do the plotting of the signal. For the plotting of the poles and zeros we use our function *splane*. When you run the script you obtain the Laplace transforms,

$$X(s) = \frac{1}{s+1},$$
$$Y(s) = \frac{s+1}{s^2 + 2s + 101} = \frac{s+1}{(s+1)^2 + 100};$$

$X(s)$ has a pole at $s = -1$, but no zeros, while $Y(s)$ has a zero at $s = -1$ and poles at $s_{1,2} = -1 \pm j10$. The results are shown in Fig. 3.7. Notice that

$$Y(s) = \mathcal{L}[e^{-t}\cos(10t)u(t)] = \int_0^\infty \cos(10t)e^{-(s+1)t}dt = \mathcal{L}[\cos(10t)u(t)]_{s'=s+1}$$
$$= \frac{s'}{(s')^2 + 1}\bigg|_{s'=s+1} = \frac{s+1}{(s+1)^2 + 1}$$

or a "frequency shift" of the original variable s. $\qquad\square$

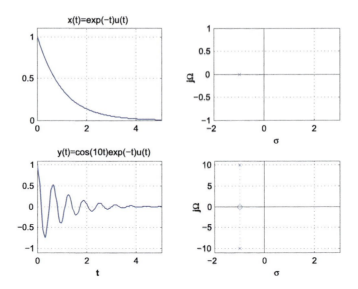

FIGURE 3.7

Poles and zeros of the Laplace transform of causal signal $x(t) = e^{-t}u(t)$ (top) and of the causal decaying signal $y(t) = e^{-t}\cos(10t)u(t)$.

```
%%
% Example 3.4---Laplace transform of exponential and modulated exponential
%%
syms t
x=exp(-t);
y=x*cos(10*t);
X=laplace(x)
Y=laplace(y)
% plotting of signals and poles/zeros
figure(1)
subplot(221)
fplot(x,[0,5]);grid
axis([0 5 0 1.1]);title('x(t)=exp(-t)u(t)')
numx=[0 1];denx=[1 1];
subplot(222)
splane(numx,denx)
subplot(223)
fplot(y,[-1,5]);grid
axis([0 5 -1.1 1.1]);title('y(t)=cos(10t)exp(-t)u(t)')
numy=[0 1 1];deny=[1 2 101];
subplot(224)
splane(numy,deny)
```

The function *splane* is used to plot the poles and zeros of the Laplace transforms.

```
function splane(num,den)
%
% function splane
% input: coefficients of numerator (num) and denominator (den) in
% decreasing order
% output: pole/zero plot
% use: splane(num,den)
%
z=roots(num); p=roots(den);
A1=[min(imag(z)) min(imag(p))];A1=min(A1)-1;
B1=[max(imag(z)) max(imag(p))];B1=max(B1)+1;
N=20;
D=(abs(A1)+abs(B1))/N;
im=A1:D:B1;
Nq=length(im);
re=zeros(1,Nq);
A=[min(real(z)) min(real(p))];A=min(A)-1;
B=[max(real(z)) max(real(p))];B=max(B)+1;
stem(real(z),imag(z),'o:')
hold on
stem(real(p),imag(p),'x:')
hold on
plot(re,im,'k');xlabel('\sigma');ylabel('j\Omega'); grid
axis([A 3 min(im) max(im)])
hold off
```

Example 3.5. In statistical signal processing, the autocorrelation function $c(\tau)$ of a random signal describes the correlation that exists between the random signal $x(t)$ and shifted versions of it, $x(t + \tau)$ for shifts $-\infty < \tau < \infty$. Typically, $c(\tau)$ is two-sided, i.e., nonzero for both positive and negative values of τ, and symmetric. Its two-sided Laplace transform is related to the power spectrum of the signal $x(t)$. Let $c(t) = e^{-a|t|}$, where $a > 0$ (we replaced the τ variable for t for convenience), find its Laplace transform indicating its region of convergence. Determine if it would be possible to compute $|C(\Omega)|^2$, which is called the power spectral density of the random signal $x(t)$.

Solution: The autocorrelation can be expressed as $c(t) = c(t)u(t) + c(t)u(-t) = c_c(t) + c_{ac}(t)$, where $c_c(t)$ is the causal component and $c_{ac}(t)$ the anticausal component of $c(t)$. The Laplace transform of $c(t)$ is then given by

$$C(s) = \mathcal{L}[c_c(t)] + \mathcal{L}[c_{ac}(-t)]_{(-s)}.$$

The Laplace transform for $c_c(t) = e^{-at}u(t)$ is

$$C_c(s) = \int_0^\infty e^{-at} e^{-st} dt = \frac{-e^{-(s+a)t}}{s+a}\Big|_{t=0}^\infty = \frac{1}{s+a}$$

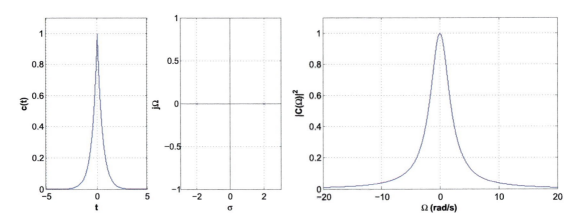

FIGURE 3.8

Two-sided autocorrelation function autocorrelation $c(t) = e^{-2|t|}$ and poles of $C(s)$ (left figure). The ROC of $C(s)$ is the region in between the poles which includes the $j\Omega$-axis. The power spectral density $|C(\Omega)|^2$ corresponding to $c(t)$ is shown on the right figure—it is the magnitude square of the Fourier transform of $c(t)$.

with a region of convergence $\{(\sigma, \Omega) : \sigma > -a, -\infty < \Omega < \infty\}$. The Laplace transform of the anti-causal part is

$$\mathcal{L}[c_{ac}(-t)u(t)]_{(-s)} = \frac{1}{-s + a}$$

and since it is anticausal and has a pole at $s = a$ its region of convergence is $\{(\sigma, \Omega) : \sigma < a, -\infty < \Omega < \infty\}$. We thus have

$$C(s) = \frac{1}{s + a} + \frac{1}{-s + a} = \frac{2a}{a^2 - s^2}$$

with a region of convergence the intersection of $\sigma > -a$ with $\sigma < a$ or

$$\{(\sigma, \Omega) : -a < \sigma < a, -\infty < \Omega < \infty\}.$$

This region contains the $j\Omega$-axis which will permit us to compute the distribution of the power over frequencies, or the power spectral density of the random signal, $|C(\Omega)|^2$ (shown in Fig. 3.8 for $a = 2$). □

Example 3.6. Consider a noncausal LTI system with impulse response

$$h(t) = e^{-t}u(t) + e^{2t}u(-t) = h_c(t) + h_{ac}(t).$$

Find the system function $H(s)$, its ROC, and indicate whether we could compute $H(j\Omega)$ from it.

Solution: The Laplace transform of the causal component, $h_c(t)$, is

$$H_c(s) = \frac{1}{s+1}$$

provided that $\sigma > -1$. For the anticausal component

$$\mathcal{L}[h_{ac}(t)] = \mathcal{L}[h_{ac}(-t)u(t)]_{(-s)} = \frac{1}{-s+2},$$

which converges when $\sigma - 2 < 0$ or $\sigma < 2$, or its region of convergence is $\{(\sigma, \Omega) : \sigma < 2, -\infty < \Omega < \infty\}$. Thus the system function is

$$H(s) = \frac{1}{s+1} + \frac{1}{-s+2} = \frac{-3}{(s+1)(s-2)}$$

with a region of convergence the intersection of $\{(\sigma, \Omega) : \sigma > -1, -\infty < \Omega < \infty\}$ and $\{(\sigma, \Omega) : \sigma < 2, -\infty < \Omega < \infty\}$, or

$$\{(\sigma, \Omega) : -1 < \sigma < 2, -\infty < \Omega < \infty\}$$

which is a sector of the s-plane that includes the $j\Omega$-axis. Thus $H(j\Omega)$ can be obtained from its Laplace transform. $\quad\square$

Example 3.7. Find the Laplace transform of the ramp function $r(t) = tu(t)$ and use it to find the Laplace of a triangular pulse $\Lambda(t) = r(t+1) - 2r(t) + r(t-1)$.

Solution: Notice that although the ramp is an ever increasing function of t, we still can obtain its Laplace transform

$$R(s) = \int_0^\infty te^{-st}dt = \frac{e^{-st}}{s^2}(-st-1)\Big|_{t=0}^{\infty} = \frac{1}{s^2}$$

where we let $\sigma > 0$ for the integral to exist. Thus $R(s) = 1/s^2$ with region of convergence

$$\{(\sigma, \Omega) : \sigma > 0, -\infty < \Omega < \infty\}.$$

The above integration can be avoided by noticing that if we find the derivative with respect to s of the Laplace transform of $u(t)$, or

$$\frac{d\,U(s)}{ds} = \int_0^\infty \frac{de^{-st}}{ds}dt = \int_0^\infty (-t)e^{-st}dt = -\underbrace{\int_0^\infty te^{-st}dt}_{R(s)}$$

where we assumed the derivative and the integral can be interchanged. We then have

$$R(s) = -\frac{d\,U(s)}{ds} = \frac{1}{s^2}.$$

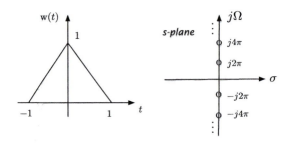

FIGURE 3.9

The Laplace transform of triangular signal $\Lambda(t)$ has as ROC the whole s-plane as it has no poles, but an infinite number of double zeros at $\pm j2\pi k$, for $k = \pm 1, \pm 2, \cdots$.

The Laplace transform of $\Lambda(t)$ can then be shown to be (try it!)

$$\Lambda(s) = \frac{1}{s^2}[e^s - 2 + e^{-s}].$$

The zeros of $\Lambda(s)$ are the values of s that make $e^s - 2 + e^{-s} = 0$ or multiplying by e^{-s},

$$1 - 2e^{-s} + e^{-2s} = (1 - e^{-s})^2 = 0,$$

or double zeros at

$$s_k = j2\pi k \qquad k = 0, \pm 1, \pm 2, \cdots.$$

In particular, when $k = 0$ there are two zeros at 0 which cancel the two poles at 0 resulting from the denominator s^2. Thus $\Lambda(s)$ has an infinite number of zeros but no poles given this pole–zero cancellation (see Fig. 3.9). Therefore, $\Lambda(s)$, as a signal of finite support, has the whole s-plane as its region of convergence, and can be calculated at $s = j\Omega$. $\qquad\square$

3.4 PROPERTIES OF THE ONE-SIDED LAPLACE TRANSFORM

We consider next basic properties of the one-sided Laplace transform. Three issues to observe in this section are:

- That time and frequency domain representations of continuous-time signals and systems are complementary—that is, certain characteristics of the signal or the system can be seen better in one domain than in the other.
- That given the inverse relationship between time and frequency a reverse relationship, or **duality**, exists between operations in the time and the frequency domains.
- That the properties of the Laplace transform of a signal equally apply to the Laplace transform of an impulse response of a system. Thus, we denote signals and impulse responses as functions.

3.4.1 LINEARITY

For functions $f(t)$ and $g(t)$, with Laplace transforms $F(s)$ and $G(s)$, and constants a and b we see that the Laplace transform is linear:

$$\mathcal{L}[af(t)u(t) + bg(t)u(t)] = aF(s) + bG(s).$$

The linearity of the Laplace transform is easily verified using integration properties:

$$\mathcal{L}[af(t)u(t) + bg(t)u(t)] = \int_0^\infty [af(t) + bg(t)]u(t)e^{-st}dt$$

$$= a\int_0^\infty f(t)u(t)e^{-st}dt + b\int_0^\infty g(t)u(t)e^{-st}dt$$

$$= a\mathcal{L}[f(t)u(t)] + b\mathcal{L}[g(t)(t)].$$

We will use the linearity property to illustrate the significance of the location of the poles of the Laplace transform of causal signals. As seen before, the Laplace transform of an exponential signal $f(t) = Ae^{-at}u(t)$ where a in general can be a complex number, is

$$F(s) = \frac{A}{s+a} \qquad \text{ROC: } \sigma > -|a|.$$

The location of the pole $s = -a$ closely relates to the signal. For instance, if $a = 5$, $f(t) = Ae^{-5t}u(t)$ is a decaying exponential and the pole of $F(s)$ is at $s = -5$ (in left-hand s-plane); if $a = -5$ we have an increasing exponential and the pole is at $s = 5$ (in right-hand s-plane). The larger the value of $|a|$ the faster the exponential decays (for $a > 0$) or increases (for $a < 0$), thus $Ae^{-10t}u(t)$ decays a lot faster that $Ae^{-5t}u(t)$, and $Ae^{10t}u(t)$ grows a lot faster than $Ae^{5t}u(t)$.

The Laplace transform $F(s) = 1/(s+a)$ of $f(t) = e^{-at}u(t)$, for any real value of a, has a pole on the real axis σ of the s-plane, and we have the following three cases:

- For $a = 0$, the pole at the origin $s = 0$ corresponds to the signal $f(t) = u(t)$ which is constant for $t \geq 0$, i.e., it does not decay.
- For $a > 0$ the signal $f(t) = e^{-at}u(t)$ is a decaying exponential, and the pole $s = -a$ of $F(s)$ is in the real axis σ of the left-hand s-plane. As the pole is moved away from the origin towards the left, the faster the exponential decays and as it moves towards the origin the slower the exponential decays.
- For $a < 0$, the pole $s = -a$ is on the real axis σ of the right-hand s-plane, and corresponds to a growing exponential. As the pole moves to the right the exponential grows faster and as it is moved towards the origin it grows at a slower rate—clearly this signal is not useful, as it grows continuously.

> The conclusion is that the σ-axis of the Laplace plane corresponds to damping, and that a single pole on this axis and in the left-hand s-plane corresponds to a decaying exponential and that a single pole on this axis and in the right-hand s-plane corresponds to a growing exponential.

Suppose then we consider

$$g(t) = A\cos(\Omega_0 t)u(t) = A\frac{e^{j\Omega_0 t}}{2}u(t) + A\frac{e^{-j\Omega_0 t}}{2}u(t)$$

and let $a = j\Omega_0$ to express $g(t)$ as

$$g(t) = 0.5[Ae^{at}u(t) + Ae^{-at}u(t)].$$

Then, by the linearity of the Laplace transform and the previous result, we obtain

$$G(s) = \frac{A}{2}\frac{1}{s - j\Omega_0} + \frac{A}{2}\frac{1}{s + j\Omega_0} = \frac{As}{s^2 + \Omega_0^2}, \tag{3.11}$$

with a zero at $s = 0$, and the poles are values for which

$$s^2 + \Omega_0^2 = 0 \;\Rightarrow\; s^2 = -\Omega_0^2 \;\; \text{or} \;\; s_{1,2} = \pm j\Omega_0,$$

which are located on the $j\Omega$-axis. The farther away from the origin of the $j\Omega$-axis the poles are, the higher the frequency Ω_0, and the closer the poles are to the origin the lower the frequency. Thus the $j\Omega$-axis corresponds to the frequency axis. Furthermore, notice that to generate the real-valued signal $g(t)$ we need two complex conjugate poles, one at $+j\Omega_0$ and the other at $-j\Omega_0$. Although frequency, as measured by frequency meters, is a positive value "negative" frequencies are needed to represent "real" signals (if the poles are not complex conjugate pairs, the inverse Laplace transform is complex—rather than real-valued).

> The conclusion is that a sinusoid has a pair of poles on the $j\Omega$-axis. For these poles to correspond to a real-valued signal they should be complex conjugate pairs, requiring negative as well as positive values of the frequency. Furthermore, when these poles are moved away from the origin of the $j\Omega$-axis, the frequency increases, and the frequency decreases whenever the poles are moved towards the origin.

Consider then the case of a signal $d(t) = Ae^{-\alpha t}\sin(\Omega_0 t)u(t)$ or a causal sinusoid multiplied (or modulated) by $e^{-\alpha t}$. According to Euler's identity

$$d(t) = A\left[\frac{e^{(-\alpha+j\Omega_0)t}}{2j}u(t) - \frac{e^{(-\alpha-j\Omega_0)t}}{2j}u(t)\right]$$

and as such we can again use linearity to get

$$D(s) = \frac{A}{2j}\left[\frac{1}{s + \alpha - j\Omega_0} - \frac{1}{s + \alpha + j\Omega_0}\right] = \frac{A\Omega_0}{(s + \alpha)^2 + \Omega_0^2}. \tag{3.12}$$

Notice the connection between Equations (3.11) and (3.12). Given $G(s)$ then $D(s) = G(s + \alpha)$, with $G(s)$ corresponding to $g(t) = A\cos(\Omega_0 t)$ and $D(s)$ to $d(t) = g(t)e^{-\alpha t}$. Multiplying a function $g(t)$ by an exponential $e^{-\alpha t}$, with α real or imaginary, shifts the transform to $G(s + \alpha)$, i.e., it is a **complex frequency-shift** property. The poles of $D(s)$ have as real part the damping factor $-\alpha$ and as imaginary part the frequencies $\pm\Omega_0$. The real part of the pole indicates decay (if $\alpha > 0$) or growth (if $\alpha < 0$) in the signal, while the imaginary part indicates the frequency of the cosine in the signal. Again, the poles will be complex conjugate pairs since the signal $d(t)$ is real-valued.

> The conclusion is that the location of the poles (and to some degree the zeros), as indicated in the previous two cases, determines the characteristics of the signal. Signals are characterized by their damping and frequency and as such can be described by the poles of its Laplace transform.

Finally, consider the case when we multiply the signal $d(t) = Ae^{-\alpha t}\sin(\Omega_0 t)u(t)$ by t to get $p(t) = Ate^{-\alpha t}\sin(\Omega_0 t)u(t)$. In the Laplace domain, this is equivalent to differentiating $D(s)$ with respect to s and multiplying it by -1, indeed assuming differentiation and integration can be interchanged we obtain

$$\frac{dD(s)}{ds} = \int_0^\infty d(t)\frac{de^{-st}}{ds}dt = \int_0^\infty [-td(t)]e^{-st}dt = \mathcal{L}[-td(t)].$$

The Laplace transform of $p(t)$ is then

$$P(s) = \mathcal{L}[td(t)] = -\frac{dD(s)}{ds} = -\frac{-2A\Omega_0(s + \alpha)}{[(s + \alpha)^2 + \Omega_0^2]^2} = \frac{2A\Omega_0(s + \alpha)}{[(s + \alpha)^2 + \Omega_0^2]^2}.$$

In general, we have

> The conclusion is that double poles result from the multiplication of a signal by t. In general, for a causal function $f(t)$ with Laplace transform $F(s)$ we have the pair
>
> $$t^n f(t)u(t) \quad \leftrightarrow \quad (-1)^n \frac{d^n F(s)}{ds^n}, \qquad n \text{ integer bigger or equal to } 1. \qquad (3.13)$$

which can be shown by computing an nth-order derivative of $F(s)$ or

$$\frac{d^n F(s)}{ds^n} = \int_0^\infty f(t)\frac{d^n e^{-st}}{ds^n}dt = \int_0^\infty f(t)(-t)^n e^{-st}dt.$$

If we were to add the different signals considered above, then the Laplace transform of the resulting signal would be the sum of the Laplace transform of each of the signals and the poles/zeros would be the aggregation of the poles/zeros from each. This observation will be important when finding the inverse Laplace transform, then we would like to do the opposite: to isolate poles or pairs of poles (when they are complex conjugate) and associate with each a general form of the signal with parameters that are found by using the zeros and the other poles of the transform. Fig. 3.10 provides an example illustrating the importance of the location of the poles, and the significance of the σ and $j\Omega$ axes.

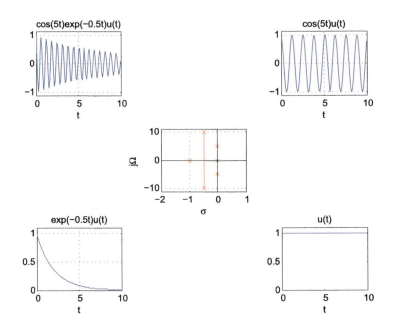

FIGURE 3.10

For the poles shown in the middle plot, possible signals are displayed around them: the pole $s = 0$ corresponds to a unit-step signal; the complex conjugate poles on the $j\Omega$-axis correspond to a sinusoid; the pair of complex conjugate poles with negative real part provides a sinusoid multiplied by an exponential; and the pole in the negative real axis gives a decaying exponential. The actual amplitudes and phases are determined by the zeros and by the other poles.

3.4.2 DIFFERENTIATION

For a signal $f(t)$, with Laplace transform $F(s)$, the one-sided Laplace transform of its first- and second-order derivatives are

$$\mathcal{L}\left[\frac{df(t)}{dt}u(t)\right] = sF(s) - f(0-), \tag{3.14}$$

$$\mathcal{L}\left[\frac{d^2 f(t)}{dt^2}u(t)\right] = s^2 F(s) - sf(0-) - \frac{df(t)}{dt}\Big|_{t=0-}. \tag{3.15}$$

In general, if $f^{(N)}(t)$ denotes the Nth-order derivative of a function $f(t)$ that has a Laplace transform $F(s)$, we have

$$\mathcal{L}[f^{(N)}(t)u(t)] = s^N F(s) - \sum_{k=0}^{N-1} f^{(k)}(0-)s^{N-1-k} \tag{3.16}$$

where $f^{(m)}(t) = d^m f(t)/dt^m$ is the mth-order derivative, $m > 0$, and $f^{(0)}(t) \triangleq f(t)$.

The Laplace transform of the derivative of a causal signal is

$$\mathcal{L}\left[\frac{df(t)}{dt}u(t)\right] = \int_{0-}^{\infty} \frac{df(t)}{dt}e^{-st}dt.$$

This integral is evaluated by parts. Let $w = e^{-st}$ then $dw = -se^{-st}dt$, and let $v = f(t)$ so that $dv = [df(t)/dt]dt$, and

$$\int wdv = wv - \int vdw.$$

We would then have

$$\int_{0-}^{\infty} \frac{df(t)}{dt}e^{-st}dt = e^{-st}f(t)\Big|_{0-}^{\infty} - \int_{0-}^{\infty} f(t)(-se^{-st})dt = -f(0-) + s\int_{0-}^{\infty} f(t)e^{-st}dt$$

$$= -f(0-) + sF(s)$$

where $e^{-st}f(t)|_{t=0-} = f(0-)$ and $e^{-st}f(t)|_{t\to\infty} = 0$ since the region of convergence guarantees it. For a second-order derivative we have

$$\mathcal{L}\left[\frac{d^2f(t)}{dt^2}u(t)\right] = \mathcal{L}\left[\frac{df^{(1)}(t)}{dt}u(t)\right] = s\mathcal{L}[f^{(1)}(t)] - f^{(1)}(0-)$$

$$= s^2F(s) - sf(0-) - \frac{df(t)}{dt}\Big|_{t=0-}$$

where we used the notation $f^{(1)}(t) = df(t)/dt$. This approach can be extended to any higher order to obtain the general result shown above.

Remarks

1. The derivative property for a signal $x(t)$ defined for all t is

$$\int_{-\infty}^{\infty} \frac{dx(t)}{dt}e^{-st}dt = sX(s).$$

This can be seen by computing the derivative of the inverse Laplace transform with respect to t, assuming that the integral and the derivative can be interchanged. Using Equation (3.4):

$$\frac{dx(t)}{dt} = \frac{1}{2\pi j}\int_{\sigma-j\infty}^{\sigma+j\infty} X(s)\frac{de^{st}}{dt}ds = \frac{1}{2\pi j}\int_{\sigma-j\infty}^{\sigma+j\infty} (sX(s))e^{st}ds$$

or that $sX(s)$ is the Laplace transform of the derivative of $x(t)$. Thus the two-sided transform does not include initial conditions. The above result can be generalized for the two-sided Laplace transform and any order of the derivative as

$$\mathcal{L}[d^Nx(t)/dt^N] = s^NX(s).$$

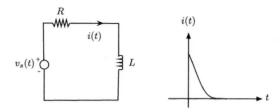

FIGURE 3.11

Impulse response $i(t)$ of an RL circuit with input $v_s(t)$.

2. Application of the linearity and the derivative properties of the Laplace transform makes solving differential equations an algebraic problem.

Example 3.8.   Find the impulse response of an RL circuit in series with a voltage source $v_s(t)$ (see Fig. 3.11). The current $i(t)$ is the output and the input is the voltage source $v_s(t)$.

Solution: To find the impulse response of the RL circuit we let $v_s(t) = \delta(t)$ and set the initial current in the inductor to zero. According to Kirchhoff's voltage law:

$$v_s(t) = L\frac{di(t)}{dt} + Ri(t) \qquad i(0-) = 0,$$

which is a first-order linear ordinary differential equation with constant coefficients, zero initial condition, and a causal input so that it is a linear time-invariant system, as discussed before.

   Letting $v_s(t) = \delta(t)$ and computing the Laplace transform of the above equation (using the linearity and the derivative properties of the transform and remembering the initial condition is zero) we obtain

$$\mathcal{L}[\delta(t)] \;=\; \mathcal{L}[L\frac{di(t)}{dt} + Ri(t)]$$
$$1 \;=\; sLI(s) + RI(s)$$

where $I(s)$ is the Laplace transform of $i(t)$. Solving for $I(s)$ we have

$$I(s) = \frac{1/L}{s + R/L},$$

which as we have seen is the Laplace transform of

$$i(t) = \frac{1}{L}e^{-(R/L)t}u(t).$$

Notice that $i(0-) = 0$ and that the response has the form of a decaying exponential trying to follow the input signal, a delta function.   □

Example 3.9.   Obtain from the Laplace transform of $x(t) = \cos(\Omega_0 t)u(t)$ the Laplace transform of $\sin(t)u(t)$ using the derivative property.

Solution: The causal sinusoid $x(t) = \cos(\Omega_0 t)u(t)$ has a Laplace transform

$$X(s) = \frac{s}{s^2 + \Omega_0^2}.$$

Then

$$
\begin{aligned}
\frac{dx(t)}{dt} &= u(t)\frac{d\cos(\Omega_0 t)}{dt} + \cos(\Omega_0 t)\frac{du(t)}{dt} = -\Omega_0 \sin(\Omega_0 t)u(t) + \cos(\Omega_0 t)\delta(t) \\
&= -\Omega_0 \sin(\Omega_0 t)u(t) + \delta(t).
\end{aligned}
$$

The presence of $\delta(t)$ indicates that $x(t)$ is discontinuous at $t = 0$, and the discontinuity is $+1$. Then the Laplace transform of $dx(t)/dt$ is given by

$$sX(s) - x(0-) = -\Omega_0 \mathcal{L}[\sin(\Omega_0 t)u(t)] + \mathcal{L}[\delta(t)];$$

thus, the Laplace transform of the sine is

$$\mathcal{L}[\sin(\Omega_0 t)u(t)] = -\frac{sX(s) - x(0-) - 1}{\Omega_0} = \frac{1 - sX(s)}{\Omega_0} = \frac{\Omega_0}{s^2 + \Omega_0^2}$$

since $x(0-) = 0$ and $X(s) = \mathcal{L}[\cos(\Omega_o T)]$.   □

Notice that whenever the signal is discontinuous at $t = 0$, as in the case of $x(t) = \cos(\Omega_0 t)u(t)$, its derivative will include a $\delta(t)$ signal due to the discontinuity. On the other hand, whenever the signal is continuous at $t = 0$, for instance $y(t) = \sin(\Omega_0 t)u(t)$, its derivative does not contain $\delta(t)$ signals. In fact,

$$\frac{dy(t)}{dt} = \Omega_0 \cos(\Omega_0 t)u(t) + \sin(\Omega_0 t)\delta(t) = \Omega_0 \cos(\Omega_0 t)u(t),$$

since the sine is zero at $t = 0$.

3.4.3 INTEGRATION

The Laplace transform of the integral of a causal signal $y(t)$ is given by

$$\mathcal{L}\left[\int_0^t y(\tau)d\tau \ u(t)\right] = \frac{Y(s)}{s}. \tag{3.17}$$

This property can be shown by using the derivative property. Call the integral

$$f(t) = \left[\int_0^t y(\tau)d\tau\right]u(t)$$

and so

$$\frac{df(t)}{dt} = y(t)u(t) + \delta(t)\int_0^t y(\tau)d\tau = y(t)u(t) + 0,$$

$$\mathcal{L}\left[\frac{df(t)}{dt}\right] = sF(s) - f(0) = Y(s);$$

since $f(0) = 0$ (the area over a point)

$$F(s) = \mathcal{L}\left[\int_0^t y(\tau)d\tau \; u(t)\right] = \frac{Y(s)}{s}.$$

Example 3.10. Suppose that

$$\int_0^t y(\tau)d\tau = 3u(t) - 2y(t),$$

Find the Laplace transform of $y(t)$, a causal signal.

Solution: Applying the integration property gives

$$\frac{Y(s)}{s} = \frac{3}{s} - 2Y(s)$$

so that solving for $Y(s)$ we obtain

$$Y(s) = \frac{3}{2(s+0.5)}$$

corresponding to $y(t) = 1.5e^{-0.5t}u(t)$. ☐

3.4.4 TIME-SHIFTING

If the Laplace transform of $f(t)u(t)$ is $F(s)$, the Laplace transform of the time-shifted signal $f(t-\tau)u(t-\tau)$ is

$$\mathcal{L}[f(t-\tau)u(t-\tau)] = e^{-\tau s}F(s). \tag{3.18}$$

This property is easily shown by a change of variable when computing the Laplace transform of the shifted signals. It indicates that when we delay (or advance) the signal to get $f(t-\tau)u(t-\tau)$ (or $f(t+\tau)u(t+\tau)$) its corresponding Laplace transform is $F(s)$ multiplied by $e^{-\tau s}$ (or $e^{\tau s}$). It should be emphasized that the property requires that the shift be done in both $f(t)$ and $u(t)$, if done in one of them the property does not apply directly.

Example 3.11. Suppose we wish to find the Laplace transform of the causal sequence of pulses $x(t)$ shown in Fig. 3.12. Let $x_1(t)$ denote the first pulse, i.e., for $0 \le t < 1$.

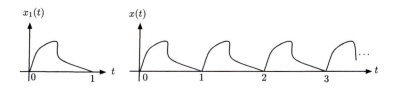

FIGURE 3.12

Generic causal pulse signal.

Solution: We have for $t \geq 0$

$$x(t) = x_1(t) + x_1(t-1) + x_1(t-2) + \cdots$$

and 0 for $t < 0$. According to the shifting and linearity properties we have[5]

$$X(s) = X_1(s)\left[1 + e^{-s} + e^{-2s} + \cdots\right] = X_1(s)\left[\frac{1}{1-e^{-s}}\right].$$

The poles of $X(s)$ are the poles of $X_1(s)$ and the roots of $1 - e^{-s} = 0$ (the s values such that $e^{-s} = 1$, or $s_k = \pm j2\pi k$ for any integer $k \geq 0$). Thus there are an infinite number of poles for $X(s)$, and the partial fraction expansion method that uses poles to invert Laplace transforms, presented later, will not be useful. The reason this example is presented in here, ahead of the inverse Laplace, is to illustrate that when we are finding the inverse of this type of Laplace function we need to consider the time-shift property, otherwise we would need to consider an infinite partial fraction expansion. ◻

Example 3.12. Consider the causal full-wave rectified signal shown in Fig. 3.13. Find its Laplace transform.

Solution: The first period of the full-wave rectified signal can be expressed as

$$x_1(t) = \sin(2\pi t)u(t) + \sin(2\pi(t-0.5))u(t-0.5)$$

and its Laplace transform is

$$X_1(s) = \frac{2\pi(1 + e^{-0.5s})}{s^2 + (2\pi)^2}.$$

The train of sinusoidal pulses

$$x(t) = \sum_{k=0}^{\infty} x_1(t - 0.5k)$$

---

[5]Notice that $1 + e^{-s} + e^{-2s} + \cdots = 1/(1 - e^{-s})$, which is verified by cross-multiplying:

$$[1 + e^{-s} + e^{-2s} + \cdots](1 - e^{-s}) = (1 + e^{-s} + e^{-2s} + \cdots) - (e^{-s} + e^{-2s} + \cdots) = 1.$$

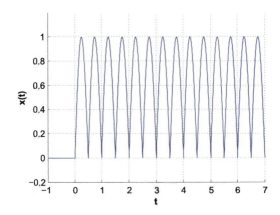

FIGURE 3.13

Full-wave rectified causal signal.

will then have the Laplace transform

$$X(s) = X_1(s)[1 + e^{-s/2} + e^{-s} + \cdots] = X_1(s)\frac{1}{1 - e^{-s/2}} = \frac{2\pi(1 + e^{-s/2})}{(1 - e^{-s/2})(s^2 + 4\pi^2)}. \quad \square$$

Example 3.13. Find the Laplace transform of the signal $x(t) = e^{-3t}u(t-4)$.

Solution: Since the two components of the signal are not equally delayed the shift property cannot be applied directly, however, an equivalent expression for $x(t)$ is

$$x(t) = [e^{-3(t-4)}u(t-4)]e^{12} \quad \Rightarrow \quad X(s) = e^{12}\frac{e^{-4s}}{s+3} = \frac{e^{-4(s-3)}}{s+3}. \quad \square$$

3.4.5 DUALITY

Because of the inverse relationship between time and frequency, operations in time have a dual in the frequency domain—or the Laplace domain in this case. This is illustrated for three special cases.

Duality in derivatives and integrals—For a causal function $f(t)$, such that $f(0-) = 0$, consider the following pairs:

$$
\begin{aligned}
f(t) \quad &\leftrightarrow \quad F(s), \\
df(t)/dt \quad &\leftrightarrow \quad sF(s), \\
tf(t) \quad &\leftrightarrow \quad -dF(s)/ds.
\end{aligned}
\tag{3.19}
$$

Notice that multiplying $F(s)$ by s in the Laplace domain is equivalent to finding the derivative of $f(t)$ in the time domain, while multiplying by t the function $f(t)$ is equivalent to finding the derivative of $-F(s)$. This inverse connection is what is referred to as duality in the derivatives in time and in frequency.

Similarly, we have the following duality in integrals for the same pair $f(t) \leftrightarrow F(s)$:

$$\int_{0-}^{t} f(\tau)d\tau \quad \leftrightarrow \quad F(s)/s,$$

$$f(t)/t \quad \leftrightarrow \quad \int_{-\infty}^{-s} F(-\rho)d\rho. \tag{3.20}$$

The first pair corresponds to the integration property. The second can be shown as follows:

$$\int_{-\infty}^{-s} F(-\rho)d\rho = \int_{-\infty}^{-s} \int_{0}^{\infty} f(t)e^{\rho t} dt \, d\rho$$

$$= \int_{0}^{\infty} f(t) \int_{-\infty}^{-s} e^{\rho t} d\rho \, dt = \int_{0}^{\infty} \frac{f(t)}{t} e^{-st} dt.$$

In this case, dividing $F(s)$ by s is equivalent to integrating $f(t)$ in the time domain, while dividing by t the function $f(t)$ is equivalent to integrating $F(s)$ as shown above.

Example 3.14. Duality is useful in calculating the Laplace transform without using the integral definition. For instance, for the ramp function $r(t) = tu(t)$ with $R(s) = 1/s^2$ we have

$$\mathcal{L}\left[\frac{dr(t)}{dt} = u(t)\right] = sR(s) - r(0-) = \frac{1}{s},$$

$$\mathcal{L}[tr(t) = t^2 u(t)] = -\frac{dR(s)}{ds} = -\frac{ds^{-2}}{ds} = \frac{2}{s^3},$$

$$\mathcal{L}\left[\int_{0}^{t} r(\tau)d\tau = \frac{t^2}{2}u(t)\right] = \frac{R(s)}{s} = \frac{1}{s^3},$$

$$\mathcal{L}\left[\frac{r(t)}{t} = u(t)\right] = \int_{-\infty}^{-s} R(-\rho)d\rho = \int_{-\infty}^{-s} \rho^{-2}d\rho = -\rho^{-1}|_{-\infty}^{-s} = \frac{1}{s}. \quad \square$$

Duality in time and frequency shifts—For a pair $f(t)u(t) \leftrightarrow F(s)$ we have

$$f(t-\alpha)u(t-\alpha) \quad \leftrightarrow \quad F(s)e^{-\alpha s},$$

$$f(t)e^{-\alpha t}u(t) \quad \leftrightarrow \quad F(s+\alpha), \tag{3.21}$$

indicating that shifting in one domain corresponds to multiplication by an exponential in the other. The second pair is shown as follows:

$$\mathcal{L}[f(t)e^{-\alpha t}u(t)] = \int_{0}^{\infty} f(t)e^{-\alpha t}e^{-st}dt = \int_{0}^{\infty} f(t)e^{-(s+\alpha)t}dt = F(s+\alpha).$$

Example 3.15. This property is useful in the calculation of Laplace transform without integration. For instance, $f(t) = e^{-\alpha t}u(t)$ having a Laplace transform $F(s) = 1/(s+\alpha)$ then

$$\mathcal{L}[f(t-\beta) = e^{-\alpha(t-\beta)}u(t-\beta)] = F(s)e^{-\beta s} = \frac{e^{-\beta s}}{s+\alpha},$$

$$\mathcal{L}[f(t)e^{-\beta t} = e^{-(\alpha+\beta)t}u(t)] = F(s+\beta) = \frac{1}{(s+\beta)+\alpha}.$$

The multiplying exponential does not have to be real, for instance, letting $f(t) = u(t)$ with Laplace transform $F(s) = 1/s$

$$
\begin{aligned}
\cos(\Omega_0 t)u(t) &= 0.5e^{j\Omega_0 t}u(t) + 0.5e^{-j\Omega_0 t}u(t) \\
&= \frac{0.5}{s - j\Omega_0} + \frac{0.5}{s + j\Omega_0} = \frac{s}{s^2 + \Omega_0^2},
\end{aligned}
$$

which coincides with the Laplace transform of $\cos(\Omega_0 t)$ obtained before.   □

Duality in time expansion, contraction and reflection—Consider again the Laplace pair $f(t)u(t) \leftrightarrow F(s)$ and $\alpha \neq 0$, then we have the following pairs:

$$
\begin{aligned}
f(\alpha t)u(t) &\quad\leftrightarrow\quad (1/|\alpha|)\, F(s/\alpha), \\
(1/|\alpha|)\, f(t/\alpha)u(t) &\quad\leftrightarrow\quad F(\alpha s).
\end{aligned}
\tag{3.22}
$$

Indeed, we have

$$
\mathcal{L}[f(\alpha t)u(t)] = \int_0^\infty f(\alpha t)e^{-st}\,dt = \int_0^\infty f(\tau)e^{-s\tau/\alpha}\frac{d\tau}{|\alpha|} = \frac{1}{|\alpha|}F(s/\alpha),
$$

$$
F(\alpha s) = \int_0^\infty f(t)e^{-s\alpha t}\,dt = \int_0^\infty f(\tau/\alpha)e^{-s\tau}\frac{d\tau}{|\alpha|} = \int_0^\infty \left[\frac{f(\tau/\alpha)}{|\alpha|}\right]e^{-s\tau}\,d\tau.
$$

These pairs clearly show, besides the duality, the inverse relationship between the time and frequency domains. Compression in time corresponds to expansion in frequency and vice versa. Moreover, they indicate that when $\alpha = -1$ we have

$$
\mathcal{L}[f(-t)u(t)] = F(-s),
$$

or that if $f(t)$ is anticausal its Laplace transform is that of its reflection $f(-t)u(t)$ with s changed to $-s$.

Example 3.16. Let $f_i(t)$, $i = 1, 2, 3$ be the unit-step, the ramp and unit impulse signals, respectively. Determine the Laplace transform of $f_i(\alpha t)$ for $\alpha > 0$. Consider then a pulse of unit length $x(t) = u(t) - u(t-1)$, what is the Laplace transform of $x(\alpha t)u(t)$ for $\alpha > 0$?

For the three functions we have for any $\alpha > 0$

$$
f_1(t) = u(t), \quad f_1(\alpha t) = u(\alpha t) = u(t) \;\Rightarrow\; \mathcal{L}[f_1(\alpha t)] = \frac{1}{\alpha}\frac{\alpha}{s} = \frac{1}{s},
$$

$$
f_2(t) = r(t) = tu(t), \quad f_2(\alpha t) = (\alpha t)u(\alpha t) = \alpha r(t) \;\Rightarrow\; \mathcal{L}[f_2(\alpha t)] = \frac{1}{\alpha}\frac{1}{(s/\alpha)^2} = \frac{\alpha}{s^2},
$$

$$
f_3(t) = \delta(t), \quad f_3(\alpha t) = \delta(\alpha t) = \frac{1}{\alpha}\delta(t) \;\Rightarrow\; \mathcal{L}[f_3(\alpha t)] = \frac{1}{\alpha},
$$

where the Laplace transforms are computed using the top equation in (3.22). Of the above three cases, the only thing that is surprising is that for $f_3(\alpha t)$ one expects that the scaling factor would not have any effect, but it does, as $\delta(\alpha t) = (1/\alpha)\delta(t)$.

According to the top equation in (3.22) given that $X(s) = (1 - e^{-s})/s$:

$$\mathcal{L}[x(\alpha t)u(t)] = \frac{1}{\alpha}X(s/\alpha) = \frac{1 - e^{-s/\alpha}}{\alpha s/\alpha} = \frac{1}{s} - \frac{e^{-s/\alpha}}{s}.$$

To verify this we express

$$x(\alpha T)u(t) = u(\alpha t) - u(\alpha t - 1) = u(t) - u(t - 1/\alpha)$$

having the above transform.   □

3.4.6 CONVOLUTION INTEGRAL

Because this is the most important property of the Laplace transform we will provide a more extensive coverage later, after considering the inverse Laplace transform.

> The Laplace transform of the convolution integral of a causal signal $x(t)$, with Laplace transforms $X(s)$, and a causal impulse response $h(t)$, with Laplace transform $H(s)$, is given by
>
> $$\mathcal{L}[(x * h)(t)] = X(s)H(s). \tag{3.23}$$

If the input of a LTI system is the causal signal $x(t)$ and the impulse response of the system is $h(t)$ then the output $y(t)$ can be written as

$$y(t) = \int_0^\infty x(\tau)h(t - \tau)d\tau \qquad t \geq 0$$

and zero otherwise. Its Laplace transform is

$$
\begin{aligned}
Y(s) &= \mathcal{L}\left[\int_0^\infty x(\tau)h(t - \tau)d\tau\right] = \int_0^\infty \left[\int_0^\infty x(\tau)h(t - \tau)d\tau\right]e^{-st}dt \\
&= \int_0^\infty x(\tau)\left[\int_0^\infty h(t - \tau)\ e^{-s(t-\tau)}\ dt\right]e^{-s\tau}d\tau = X(s)H(s)
\end{aligned}
$$

where the internal integral is shown to be $H(s) = \mathcal{L}[h(t)]$ (change variable to $v = t - \tau$) using the causality of $h(t)$. The remaining integral is the Laplace transform of $x(t)$.

> The **system function** or **transfer function** $H(s) = \mathcal{L}[h(t)]$, the Laplace transform of the impulse response $h(t)$ of a LTI system with input $x(t)$ and output $y(t)$, can be expressed as the ratio
>
> $$H(s) = \frac{\mathcal{L}[\text{output}]}{\mathcal{L}[\text{input}]} = \frac{\mathcal{L}[y(t)]}{\mathcal{L}[x(t)]} = \frac{Y(s)}{X(s)}. \tag{3.24}$$
>
> This function is called "transfer function" because it transfers the Laplace transform of the input to the output. Just as with the Laplace transform of signals, $H(s)$ characterizes a LTI system by means of its poles and zeros. Thus it becomes a very important tool in the analysis and synthesis of systems.

Table 3.1 Basic properties of one-sided Laplace transforms

Causal functions and constants	$\alpha f(t), \beta g(t)$	$\alpha F(s), \beta G(s)$
Linearity	$\alpha f(t) + \beta g(t)$	$\alpha F(s) + \beta G(s)$
Time-shifting	$f(t - \alpha)u(t - \alpha)$	$e^{-\alpha s} F(s)$
Frequency shifting	$e^{\alpha t} f(t)$	$F(s - \alpha)$
Multiplication by t	$t\, f(t)$	$-\dfrac{dF(s)}{ds}$
Derivative	$\dfrac{df(t)}{dt}$	$sF(s) - f(0-)$
Second derivative	$\dfrac{d^2 f(t)}{dt^2}$	$s^2 F(s) - sf(0-) - f^{(1)}(0)$
Integral	$\displaystyle\int_{0-}^{t} f(t')dt'$	$\dfrac{F(s)}{s}$
Expansion/contraction	$f(\alpha t), \alpha \neq 0$	$\dfrac{1}{\lvert\alpha\rvert} F\left(\dfrac{s}{\alpha}\right)$
Initial value	$f(0-) = \lim\limits_{s \to \infty} sF(s)$	
Derivative Duality	$\dfrac{df(t)}{dt}$	$sF(s)$
	$tf(t)$	$-\dfrac{dF(s)}{ds}$
Integration Duality	$\displaystyle\int_{0-}^{t} f(\tau)d\tau$	$F(s)/s$
	$f(t)/t$	$\displaystyle\int_{-\infty}^{-s} F(-\rho)d\rho$
Time and Frequency Duality	$f(t - \alpha)u(t - \alpha)$	$F(s)e^{-\alpha s}$
	$f(t)e^{-\alpha t}u(t)$	$F(s + \alpha)$
Time Scaling Duality	$f(\alpha t)u(t)$	$(1/\lvert\alpha\rvert)F(s/\alpha)$
	$(1/\lvert\alpha\rvert) f(t/\alpha)u(t)$	$F(\alpha s)$
Convolution	$[f * g](t)$	$F(s)G(s)$
Initial value	$f(0-) = \lim\limits_{s \to \infty} sF(s)$	

3.5 INVERSE LAPLACE TRANSFORM

Inverting the Laplace transform consists in finding a function (either a signal or an impulse response of a system) that has the given transform with the given region of convergence. We will consider three cases:

- inverse of one-sided Laplace transforms giving causal functions,
- inverse of Laplace transforms with exponentials,
- inverse of two-sided Laplace transforms giving anticausal or noncausal functions.

The given function $X(s)$ we wish to invert can be the Laplace transform of a signal or a transfer function, i.e., the Laplace transform of an impulse response. As a reference Table 3.1 provides the basic properties of the one-sided Laplace transform.

3.5.1 INVERSE OF ONE-SIDED LAPLACE TRANSFORMS

When we consider a causal function $x(t)$, the region of convergence of $X(s)$ is of the form

$$\{(\sigma, \Omega) : \sigma > \sigma_{max}, -\infty < \Omega < \infty\}$$

where σ_{max} is the maximum of the real parts of the poles of $X(s)$. Since in this section we only consider causal signals, the region of convergence will be assumed known and will not be shown with the Laplace transform.

The most common inverse Laplace method is the so-called **partial fraction expansion** which consists in expanding the given function in s into a sum of components whose inverse Laplace transforms can be found in a table of Laplace transform pairs. Assume the signal we wish to find has a rational Laplace transform, i.e.,

$$X(s) = \frac{N(s)}{D(s)} \tag{3.25}$$

where $N(s)$ and $D(s)$ are polynomials in s with real-valued coefficients. In order for the partial fraction expansion to be possible, it is required that $X(s)$ be **proper rational** which means that the degree of the numerator polynomial $N(s)$ is less than that of the denominator polynomial $D(s)$. If $X(s)$ is not proper then we need to do a long division until we obtain a proper rational function, i.e.,

$$X(s) = g_0 + g_1 s + \cdots + g_m s^m + \frac{B(s)}{D(s)} \tag{3.26}$$

where the degree of $B(s)$ is now less than that of $D(s)$, so that we can perform partial expansion for $B(s)/D(s)$. The inverse of $X(s)$ is then given by

$$x(t) = g_0 \delta(t) + g_1 \frac{d\delta(t)}{dt} + \cdots + g_m \frac{d^m \delta(t)}{dt^m} + \mathcal{L}^{-1}\left[\frac{B(s)}{D(s)}\right]. \tag{3.27}$$

The presence of $\delta(t)$ and its derivatives (called doublets, triplets, etc.) are very rare in actual signals and as such the typical rational function has a numerator polynomial which is of lower degree than the denominator polynomial.

Remarks

1. Things to remember before performing the inversion are:

- the poles of $X(s)$ provide the basic characteristics of the signal $x(t)$,
- if $N(s)$ and $D(s)$ are polynomials in s with real coefficients, then the zeros and poles of $X(s)$ are real and/or complex conjugate pairs, and can be simple or multiple, and
- in the inverse, $u(t)$ should be included since the result of the inverse is causal—the function $u(t)$ is an integral part of the inverse.

Table 3.2 One-sided Laplace transforms	
$\delta(t)$	1, whole s-plane
$u(t)$	$\dfrac{1}{s}$, $\mathcal{R}e[s] > 0$
$r(t)$	$\dfrac{1}{s^2}$, $\mathcal{R}e[s] > 0$
$e^{-at}u(t), a > 0$	$\dfrac{1}{s+a}$, $\mathcal{R}e[s] > -a$
$\cos(\Omega_0 t)u(t)$	$\dfrac{s}{s^2 + \Omega_0^2}$, $\mathcal{R}e[s] > 0$
$\sin(\Omega_0 t)u(t)$	$\dfrac{\Omega_0}{s^2 + \Omega_0^2}$, $\mathcal{R}e[s] > 0$
$e^{-at}\cos(\Omega_0 t)u(t), a > 0$	$\dfrac{s+a}{(s+a)^2 + \Omega_0^2}$, $\mathcal{R}e[s] > -a$
$e^{-at}\sin(\Omega_0 t)u(t), a > 0$	$\dfrac{\Omega_0}{(s+a)^2 + \Omega_0^2}$, $\mathcal{R}e[s] > -a$
$2A\, e^{-at}\cos(\Omega_0 t + \theta)u(t), a > 0$	$\dfrac{A\angle\theta}{s+a-j\Omega_0} + \dfrac{A\angle-\theta}{s+a+j\Omega_0}$, $\mathcal{R}e[s] > -a$
$\dfrac{1}{(N-1)!}\, t^{N-1}u(t)$	$\dfrac{1}{s^N}$ N an integer, $\mathcal{R}e[s] > 0$

2. The basic idea of the partial expansion is to decompose proper rational functions into a sum of rational components whose inverse transform can be found directly in tables. Table 3.2 displays the basic one-sided Laplace transform pairs.

3. Because when doing a partial fraction expansion it is possible to make simple algebraic errors, it is a good idea to check the final result. One of the tests is provided by the location of the poles, the generic form of the inverse should be deducted from the poles before doing the expansion. Another test is given by the so-called *initial-value theorem*, which can be used to verify that the initial value of the inverse $x(0-)$ is equal to the limit of $sX(s)$ as $s \to \infty$ or

$$\lim_{s\to\infty} sX(s) = x(0-), \qquad (3.28)$$

which is shown using the derivative property:

$$\mathcal{L}[dx(t)/dt] = \int_{0-}^{\infty} \frac{dx(t)}{dt}e^{-st}dt = sX(s) - x(0-),$$

$$\lim_{s\to\infty}\int_{0-}^{\infty} \frac{dx(t)}{dt}e^{-st}dt = 0 \;\; \Rightarrow \;\; \lim_{s\to\infty} sX(s) = x(0-),$$

as the exponential in the integral goes to 0.

Simple Real Poles

If $X(s)$ is a proper rational function

$$X(s) = \frac{N(s)}{(s + p_1)(s + p_2)} \qquad (3.29)$$

where $\{s_k = -p_k\}$, $k = 1, 2$, are simple real poles of $X(s)$, its partial fraction expansion and its inverse are given by

$$X(s) = \frac{A_1}{s + p_1} + \frac{A_2}{s + p_2} \quad \Leftrightarrow \quad x(t) = [A_1 e^{-p_1 t} + A_2 e^{-p_2 t}]u(t) \qquad (3.30)$$

where the expansion coefficients are computed as

$$A_k = X(s)(s + p_k)|_{s=-p_k} \quad k = 1, 2.$$

According to the Laplace transform tables the time function corresponding to $A_k/(s + p_k)$ is $A_k e^{-p_k t} u(t)$ thus the form of the inverse $x(t)$. To find the coefficients of the expansion, say A_1, we multiply both sides of the equation by its corresponding denominator $(s + p_1)$ so that

$$X(s)(s + p_1) = A_1 + \frac{A_2(s + p_1)}{s + p_2}.$$

If we let $s + p_1 = 0$, or $s = -p_1$, in the above expression the second term in the right will be zero and we find that

$$A_1 = X(s)(s + p_1)|_{s=-p_1} = \frac{N(-p_1)}{-p_1 + p_2}.$$

Likewise for A_2.

Example 3.17. Consider the proper rational function

$$X(s) = \frac{3s + 5}{s^2 + 3s + 2} = \frac{3s + 5}{(s + 1)(s + 2)};$$

find its causal inverse.

Solution: The partial fraction expansion is

$$X(s) = \frac{A_1}{s + 1} + \frac{A_2}{s + 2}$$

Given that the two poles are real, the expected signal $x(t)$ will be a superposition of two decaying exponentials, with damping factors -1 and -2, or

$$x(t) = [A_1 e^{-t} + A_2 e^{-t}]u(t)$$

where, as indicated above,

$$A_1 = X(s)(s + 1)|_{s=-1} = \frac{3s + 5}{s + 2}\Big|_{s=-1} = 2 \quad \text{and} \quad A_2 = X(s)(s + 2)|_{s=-2} = \frac{3s + 5}{s + 1}\Big|_{s=-2} = 1.$$

Therefore

$$X(s) = \frac{2}{s+1} + \frac{1}{s+2},$$

and as such $x(t) = [2e^{-t} + e^{-2t}]u(t)$. To check that the solution is correct one could use the initial-value theorem, according to which $x(0) = 3$ should coincide with

$$\lim_{s \to \infty} \left[sX(s) = \frac{3s^2 + 5s}{s^2 + 3s + 2} \right] = \lim_{s \to \infty} \frac{3 + 5/s}{1 + 3/s + 2/s^2} = 3,$$

as it does.   □

Remarks
The coefficients A_1 and A_2 can be found using other methods:

- Since the following expression should be valid for any s, as long as we do not divide by zero

$$X(s) = \frac{3s + 5}{(s+1)(s+2)} = \frac{A_1}{s+1} + \frac{A_2}{s+2}, \tag{3.31}$$

  choosing two different values of s will permit us to find A_1 and A_2, for instance:

$$s = 0 \quad X(0) = \frac{5}{2} = A_1 + \frac{1}{2}A_2, \qquad s = 1 \quad X(1) = \frac{8}{6} = \frac{1}{2}A_1 + \frac{1}{3}A_2,$$

  which gives a set of two linear equations with two unknowns, and applying Cramer's rule we find that $A_1 = 2$ and $A_2 = 1$.
- Cross-multiplying the partial expansion in Equation (3.31) we obtain

$$X(s) = \frac{3s + 5}{s^2 + 3s + 2} = \frac{s(A_1 + A_2) + (2A_1 + A_2)}{s^2 + 3s + 2}.$$

  Comparing the numerators, we then see that $A_1 + A_2 = 3$ and $2A_1 + A_2 = 5$, two equations with two unknowns, which can be shown to have as unique solution $A_1 = 2$ and $A_2 = 1$, as before.

Simple Complex Conjugate Poles
Consider the Laplace transform

$$X(s) = \frac{N(s)}{(s+\alpha)^2 + \Omega_0^2} = \frac{N(s)}{(s+\alpha - j\Omega_0)(s+\alpha + j\Omega_0)}.$$

Because the numerator and the denominator polynomials of $X(s)$ have real coefficients, the zeros and poles whenever complex appear as complex conjugate pairs. One could thus think of the case of a pair of complex conjugate poles as similar to the case of two simple real poles presented above. Notice that the numerator $N(s)$ must be a first-order polynomial for $X(s)$ to be proper rational. The poles of $X(s)$, $s_{1,2} = -\alpha \pm j\Omega_0$, indicate that the signal $x(t)$ will have an exponential $e^{-\alpha t}$, given that the real part of the poles is $-\alpha$, multiplied by a sinusoid of frequency Ω_0 given that the imaginary parts of the poles are $\pm\Omega_0$. We have the following expansion:

$$X(s) = \frac{A}{s+\alpha - j\Omega_0} + \frac{A^*}{s+\alpha + j\Omega_0}$$

where the expansion coefficients are complex conjugate of each other. From the pole information, the general form of the inverse is

$$x(t) = Ke^{-\alpha t}\cos(\Omega_0 t + \Phi)u(t)$$

for some constants K and Φ. As before, we can find A as

$$A = X(s)(s + \alpha - j\Omega_0)|_{s=-\alpha+j\Omega_0} = |A|e^{j\theta}$$

and that $X(s)(s + \alpha + j\Omega_0)|_{s=-\alpha-j\Omega_0} = A^*$ can easily be verified. Then the inverse transform is given by

$$\begin{aligned} x(t) &= \left[Ae^{-(\alpha - j\Omega_0)t} + A^*e^{-(\alpha + j\Omega_0)t} \right]u(t) = |A|e^{-\alpha t}(e^{j(\Omega_0 t + \theta)} + e^{-j(\Omega_0 t + \theta)})u(t) \\ &= 2|A|e^{-\alpha t}\cos(\Omega_0 t + \theta)u(t). \end{aligned}$$

By appropriately choosing the numerator $N(s)$, the equivalent expansion separates $X(s)$ into two terms corresponding to sine and cosine multiplied by a real exponential. The last expression in (3.35) is obtained by adding the phasors corresponding to the sine and the cosine terms.

The partial fraction expansion of a proper rational function

$$X(s) = \frac{N(s)}{(s + \alpha)^2 + \Omega_0^2} = \frac{N(s)}{(s + \alpha - j\Omega_0)(s + \alpha + j\Omega_0)} \tag{3.32}$$

with complex conjugate poles $\{s_{1,2} = -\alpha \pm j\Omega_0\}$ is given by

$$X(s) = \frac{A}{s + \alpha - j\Omega_0} + \frac{A^*}{s + \alpha + j\Omega_0}$$

where

$$A = X(s)(s + \alpha - j\Omega_0)|_{s=-\alpha+j\Omega_0} = |A|e^{j\theta}$$

so that the inverse is the function

$$x(t) = 2|A|e^{-\alpha t}\cos(\Omega_0 t + \theta)u(t). \tag{3.33}$$

Equivalent partial fraction expansion: Let $N(s) = a + b(s + \alpha)$, for some constants a and b, so that

$$X(s) = \frac{a + b(s + \alpha)}{(s + \alpha)^2 + \Omega_0^2} = \frac{a}{\Omega_0}\frac{\Omega_0}{(s + \alpha)^2 + \Omega_0^2} + b\frac{s + \alpha}{(s + \alpha)^2 + \Omega_0^2}, \tag{3.34}$$

so that the inverse is

$$\begin{aligned} x(t) &= \left[\frac{a}{\Omega_0}e^{-\alpha t}\sin(\Omega_0 t) + be^{-\alpha t}\cos(\Omega_0 t) \right]u(t) \\ &= \sqrt{\frac{a^2}{\Omega_0^2} + b^2}\ e^{-\alpha t}\cos\left(\Omega_0 t - \tan^{-1}\left(\frac{a}{\Omega_0 b}\right)\right)u(t). \end{aligned} \tag{3.35}$$

Remarks

1. When $\alpha = 0$ Equation (3.34) indicates that the inverse Laplace transform of

$$X(s) = \frac{a+bs}{s^2 + \Omega_0^2}, \qquad \text{with poles } \pm j\Omega_0 \text{ is}$$

$$x(t) = \left[\frac{a}{\Omega_0}\sin(\Omega_0 t) + b\cos(\Omega_0 t)\right]u(t) = \sqrt{\frac{a^2}{\Omega_0^2} + b^2}\,\cos\left(\Omega_0 t - \tan^{-1}\left(\frac{a}{\Omega_0 b}\right)\right)u(t)$$

or the sum of a sine and a cosine functions.

2. When the frequency $\Omega_0 = 0$, we see that the inverse Laplace transform of

$$X(s) = \frac{a + b(s+\alpha)}{(s+\alpha)^2} = \frac{a}{(s+\alpha)^2} + \frac{b}{s+\alpha}$$

(corresponding to a double pole at $-\alpha$) is

$$x(t) = \lim_{\Omega_0 \to 0}\left[\frac{a}{\Omega_0}e^{-\alpha t}\sin(\Omega_0 t) + be^{-\alpha t}\cos(\Omega_0 t)\right]u(t) = [ate^{-\alpha t} + be^{-\alpha t}]u(t)$$

where the first limit is found by L'Hôpital's rule.

3. Notice above that when computing the partial fraction expansion of the double pole $s = -\alpha$ the expansion is composed of two terms, one with denominator $(s+\alpha)^2$ and the other with denominator $s+\alpha$ and constant numerators.

Example 3.18. Consider the Laplace function

$$X(s) = \frac{2s+3}{s^2 + 2s + 4} = \frac{2s+3}{(s+1)^2 + 3}.$$

Find the corresponding causal signal $x(t)$, then use MATLAB to validate your answer.

Solution: The poles are $-1 \pm j\sqrt{3}$, so that we expect that $x(t)$ be a decaying exponential with a damping factor of -1 (the real part of the poles) multiplied by a causal cosine of frequency $\sqrt{3}$. The partial fraction expansion is of the form

$$X(s) = \frac{2s+3}{s^2 + 2s + 4} = \frac{a + b(s+1)}{(s+1)^2 + 3}$$

so that $3 + 2s = (a+b) + bs$, or $b = 2$ and $a + b = 3$ or $a = 1$. Thus

$$X(s) = \frac{1}{\sqrt{3}}\frac{\sqrt{3}}{(s+1)^2 + 3} + 2\frac{s+1}{(s+1)^2 + 3},$$

which corresponds to

$$x(t) = \left[\frac{1}{\sqrt{3}}\sin(\sqrt{3}t) + 2\cos(\sqrt{3}t)\right]e^{-t}u(t).$$

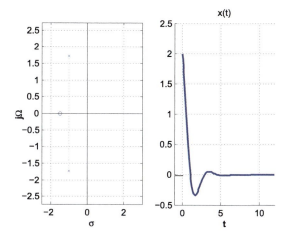

FIGURE 3.14

Inverse Laplace transform of $X(s) = (2s + 3)/(s^2 + 2s + 4)$: poles and zeros (left) and inverse $x(t) = 2e^{-t}(\cos(\sqrt{3}t) + \sqrt{3}\sin(\sqrt{3}t)/6)u(t)$.

The initial value $x(0) = 2$ and according to the initial value theorem the following limit should equal it:

$$\lim_{s \to \infty}\left[sX(s) = \frac{2s^2 + 3s}{s^2 + 2s + 4}\right] = \lim_{s \to \infty}\frac{2 + 3/s}{1 + 2/s + 4/s^2} = 2,$$

which is the case.

We use the MATLAB function *ilaplace* to compute symbolically the inverse Laplace transform and plot the response using *fplot* as shown in the following script.

```
%%
% Example 3.18---Inverse Laplace transform
%%
clear all; clf
syms s t w
num=[0 2 3]; den=[1 2 4];              % numerator and denominator
subplot(121)
splane(num,den)                        % plotting poles and zeros
disp('>> Inverse Laplace <<')
x=ilaplace((2*s+3)/(s^2+2*s+4));       % inverse Laplace transform
subplot(122)
fplot(x,[0,12]); title('x(t)')
axis([0 12 -0.5 2.5]); grid
```

The results are shown in Fig. 3.14.   □

Double Real Poles

If a proper rational function has double real poles

$$X(s) = \frac{N(s)}{(s+\alpha)^2} = \frac{a + b(s+\alpha)}{(s+\alpha)^2} = \frac{a}{(s+\alpha)^2} + \frac{b}{s+\alpha} \tag{3.36}$$

then its inverse is

$$x(t) = [ate^{-\alpha t} + be^{-\alpha t}]u(t) \tag{3.37}$$

where a can be computed as

$$a = X(s)(s+\alpha)^2|_{s=-\alpha},$$

and after replacing it, b is found by computing $X(s_0)$ for a value $s_0 \neq -\alpha$.

When we have double real poles we need to express the numerator $N(s)$ as a first-order polynomial, just as in the case of a pair of complex conjugate poles. The values of a and b can be computed in different ways, as we illustrate in the following examples.

Example 3.19. Typically the Laplace transforms appear as combinations of the different terms we have considered, for instance a combination of first- and second-order poles gives

$$X(s) = \frac{4}{s(s+2)^2},$$

which has a pole at $s = 0$ and a double pole at $s = -2$. Find the causal signal $x(t)$. Use MATLAB to plot the poles and zeros of $X(s)$ and to find the inverse Laplace transform $x(t)$.

Solution: The partial fraction expansion is

$$X(s) = \frac{A}{s} + \frac{a + b(s+2)}{(s+2)^2}$$

the value of $A = X(s)s|_{s=0} = 1$ and so

$$X(s) - \frac{1}{s} = \frac{4 - (s+2)^2}{s(s+2)^2} = \frac{-(s+4)}{(s+2)^2} = \frac{a + b(s+2)}{(s+2)^2}.$$

Comparing the numerators of $X(s) - 1/s$ and the one in the partial fraction expansion gives $b = -1$ and $a + 2b = -4$ or $a = -2$, then we have

$$X(s) = \frac{1}{s} + \frac{-2 - (s+2)}{(s+2)^2}$$

so that $x(t) = [1 - 2te^{-2t} - e^{-2t}]u(t)$.

Another way to do this type of problem is to express $X(s)$ as

$$X(s) = \frac{A}{s} + \frac{B}{(s+2)^2} + \frac{C}{s+2}.$$

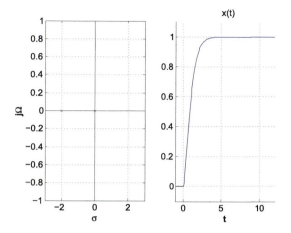

FIGURE 3.15

Inverse Laplace transform of $X(s) = 4/(s(s+2)^2)$: poles and zeros (left) and $x(t) = [1 - 2te^{-2t} - e^{-2t}]u(t)$.

We find the A as before, and then find B by multiplying both sides by $(s+2)^2$ and letting $s = -2$, which gives

$$X(s)(s+2)^2|_{s=-2} = \left[\frac{A(s+2)^2}{s} + B + C(s+2) \right]_{s=-2} = B \quad \Rightarrow \quad B = -2.$$

To find C we compute the partial fraction expansion for a value of s for which no division by zero is possible. For instance if we let $s = 1$ we can find the value of C, by solving

$$X(1) = \frac{4}{9} = \frac{A}{1} + \frac{B}{9} + \frac{C}{3} = 1 - \frac{2}{9} + \frac{C}{3} \quad \Rightarrow \quad C = -1$$

after which we can find the inverse.

The initial value $x(0) = 0$ coincides with

$$\lim_{s \to \infty} \left[sX(s) = \frac{4s}{s(s+2)^2} \right] = \lim_{s \to \infty} \frac{4/s^2}{(1 + 2/s)^2} = 0.$$

To find the inverse Laplace transform with MATLAB we use a similar script to the one used before, only the numerator and denominator description needs to be changed. The plots are shown in Fig. 3.15.   □

Example 3.20. Find the inverse Laplace transform of the function

$$X(s) = \frac{4}{s((s+1)^2 + 3)},$$

which has a simple real pole $s = 0$, and complex conjugate poles $s = -1 \pm j\sqrt{3}$.

Solution: The partial fraction expansion is

$$X(s) = \frac{A}{s+1-j\sqrt{3}} + \frac{A^*}{s+1+j\sqrt{3}} + \frac{B}{s};$$

we then have

$$B = sX(s)|_{s=0} = 1,$$

$$A = X(s)(s+1-j\sqrt{3})|_{s=-1+j\sqrt{3}} = 0.5(-1+\frac{j}{\sqrt{3}}) = \frac{1}{\sqrt{3}}\angle 150°,$$

so that

$$x(t) = \frac{2}{\sqrt{3}}e^{-t}\cos(\sqrt{3}t + 150°)u(t) + u(t) = -[\cos(\sqrt{3}t) + 0.577\sin(\sqrt{3}t)]e^{-t}u(t) + u(t). \quad \square$$

Remark

1. Following the above development, when the poles are complex conjugate and double the procedure for the double poles is repeated. For instance, consider the partial expansion

$$X(s) = \frac{N(s)}{(s+\alpha - j\Omega_0)^2(s+\alpha + j\Omega_0)^2} = \frac{a + b(s+\alpha - j\Omega_0)}{(s+\alpha - j\Omega_0)^2} + \frac{a^* + b^*(s+\alpha + j\Omega_0)}{(s+\alpha + j\Omega_0)^2}.$$

   After finding a and b we obtain the inverse.
2. The partial fraction expansion when complex conjugate poles exist of second or higher order should be done with MATLAB.

Example 3.21. In this example we use MATLAB to find the inverse Laplace transform of more complicated functions than the ones considered before. In particular we want to illustrate some of the additional information that our function *pfeLaplace* gives. Consider the Laplace transform

$$X(s) = \frac{3s^2 + 2s - 5}{s^3 + 6s^2 + 11s + 6}.$$

Find poles and zeros of $X(s)$, and obtain the coefficients of its partial fraction expansion (also called the residues). Use *ilaplace* to find its inverse and plot it using *fplot*.

Solution: The following is the function *pfeLaplace*

```
function pfeLaplace(num,den)
%
disp('>>>>> Zeros <<<<<')
z=roots(num)
[r,p,k]=residue(num,den);
disp('>>>>> Poles <<<<<')
p
disp('>>>>> Residues <<<<<')
```

```
r
splane(num,den)
>>>>> Zeros <<<<<
   z = -1.6667
        1.0000
   >>>>> Poles <<<<<
   p = -3.0000
       -2.0000
       -1.0000
   >>>>> Residues <<<<<
   r = 8.0000
      -3.0000
      -2.0000
   >>>>> Inverse Laplace <<<<<
   x =8*exp(-3*t)-3*exp(-2*t)-2*exp(-t)
```

The function *pfeLaplace* uses the MATLAB function *roots* to find the zeros of $X(s)$ defined by the coefficients of its numerator and denominator given in descending order of s. For the partial fraction expansion, *pfeLaplace* uses the MATLAB function *residue* which find coefficients of the expansion as well as the poles of $X(s)$ (the residue $r(i)$ in the vector r corresponds to the expansion term for the pole $p(i)$; for instance the residue $r(1) = 8$ corresponds to the expansion term corresponding to the pole $p(1) = -3$). The symbolic function *ilaplace* is then used to find the inverse $x(t)$, as input to *ilaplace* the function $X(s)$ is described in a symbolic way. The MATLAB function *fplot* is used for the plotting of the symbolic computations. The analytic results are shown below, and the plot of $x(t)$ is given in Fig. 3.16.   □

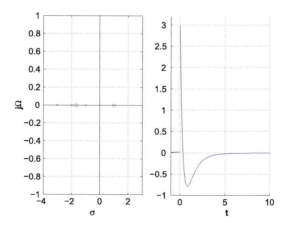

FIGURE 3.16

Inverse Laplace transform of $X(s) = (3s^2 + 2s - 5)/(s^3 + 6s^2 + 11s + 6)$. Poles and zeros (left) of $X(s)$ are given with the corresponding inverse $x(t) = [8e^{-3t} - 3e^{-2t} - 2e^{-t}]u(t)$.

3.5.2 INVERSE OF FUNCTIONS CONTAINING $e^{-\rho s}$ TERMS

Exponentials in the numerator—Given the Laplace transform

$$X(s) = \sum_k \frac{N_k(s)e^{-\rho_k s}}{D_k(s)},$$

the inverse $x(t)$ is found by determining first the inverse of each term $N_k(s)/D_k(s)$ and then using the time-shift property to obtain the inverse of each term $N_k(s)e^{-\rho_k s}/D_k(s)$.

Exponentials in the denominator—Two cases:

(a) Given the Laplace transform

$$X(s) = \frac{N(s)}{D(s)(1 - e^{-\alpha s})} = \frac{N(s)}{D(s)} + \frac{N(s)e^{-\alpha s}}{D(s)} + \frac{N(s)e^{-2\alpha s}}{D(s)} + \cdots$$

if $f(t)$ is the inverse of $N(s)/D(s)$ then

$$x(t) = f(t) + f(t - \alpha) + f(t - 2\alpha) + \cdots .$$

(b) Given the Laplace transform

$$X(s) = \frac{N(s)}{D(s)(1 + e^{-\alpha s})} = \frac{N(s)}{D(s)} - \frac{N(s)e^{-\alpha s}}{D(s)} + \frac{N(s)e^{-2\alpha s}}{D(s)} - \cdots$$

if $f(t)$ is the inverse of $N(s)/D(s)$ we then have

$$x(t) = f(t) - f(t - \alpha) + f(t - 2\alpha) - \cdots .$$

The presence of exponentials $e^{-\rho s}$ is connected with time-shifts of the signal and with the existence of infinite number of singularities. When the exponentials are in the numerator, we ignore the exponentials until after the inverse of each term is done and then consider them. When an infinite number of poles are present due to $1 - e^{-\alpha s}$ or $1 + e^{-\alpha s}$, we express the Laplace as in the previous case and find the inverse using the time-shift property. One should not attempt to do partial fraction expansion when there are an infinite number of poles.

The sums

$$\sum_{k=0}^{\infty} e^{-\alpha s k} = \frac{1}{1 - e^{-\alpha s}},$$

$$\sum_{k=0}^{\infty} (-1)^k e^{-\alpha s k} = \frac{1}{1 + e^{-\alpha s}},$$

can easily be verified by cross-multiplying. These terms have an infinite number of poles. Indeed, the poles of $1/(1 - e^{-\alpha s})$ are $s_k = \pm j2\pi k/\alpha$, for $k = 0, \pm 1, \pm 2, \cdots$, and the poles of $1/(1 + e^{-\alpha s})$ are

$s_k = \pm j(2k+1)\pi/\alpha$, for $k = 0, \pm 1, \pm 2, \cdots$. So when the function is

$$X_1(s) = \frac{N(s)}{D(s)(1 - e^{-\alpha s})} = \frac{N(s)}{D(s)} \sum_{k=0}^{\infty} e^{-\alpha s k} = \frac{N(s)}{D(s)} + \frac{N(s)e^{-\alpha s}}{D(s)} + \frac{N(s)e^{-2\alpha s}}{D(s)} + \cdots$$

and if $f(t)$ is the inverse of $N(s)/D(s)$ we then have

$$x_1(t) = f(t) + f(t - \alpha) + f(t - 2\alpha) + \cdots.$$

Likewise when

$$X_2(s) = \frac{N(s)}{D(s)(1 + e^{-\alpha s})} = \frac{N(s)}{D(s)} \sum_{k=0}^{\infty} (-1)^k e^{-\alpha s k} = \frac{N(s)}{D(s)} - \frac{N(s)e^{-\alpha s}}{D(s)} + \frac{N(s)e^{-2\alpha s}}{D(s)} - \cdots$$

if $f(t)$ is the inverse of $N(s)/D(s)$ we then have

$$x_2(t) = f(t) - f(t - \alpha) + f(t - 2\alpha) - \cdots.$$

Example 3.22. We wish to find the causal inverses of

$$(i) \;\; X_1(s) = \frac{1 - e^{-s}}{(s + 1)(1 + e^{-2s})}, \quad (ii) \;\; X_2(s) = \frac{2\pi(1 + e^{-s/2})}{(1 - e^{-s/2})(s^2 + 4\pi^2)}.$$

Solution: (i) We let

$$X_1(s) = F(s) \sum_{k=0}^{\infty} (-1)^k (e^{-2s})^k \quad \text{where} \;\; F(s) = \frac{1 - e^{-s}}{s + 1}.$$

The inverse of $F(s)$ is $f(t) = e^{-t}u(t) - e^{-(t-1)}u(t - 1)$, and the inverse of $X(s)$ is thus given by

$$x_1(t) = f(t) - f(t - 2) + f(t - 4) + \cdots.$$

(ii) $X_2(s)$ is the Laplace transform of a full-wave rectified signal (see Example 3.12). If we let $G(s) = 2\pi(1 + e^{-s/2})/(s^2 + 4\pi^2)$ then

$$X_2(s) = \frac{G(s)}{1 - e^{-s/2}} = G(s)(1 + e^{-s/2} + e^{-s} + e^{-3s/2} + \cdots \quad \text{so that}$$
$$x_2(t) = g(t) + g(t - 0.5) + g(t - 1) + g(t - 1.5) + \cdots \quad \text{where}$$
$$g(t) = \sin(2\pi t) + \sin(2\pi(t - 0.5)). \quad \square$$

3.6 THE TRANSFER FUNCTION OF LTI SYSTEMS

From the point of view of signal processing, the convolution property is the most important application of the Laplace transform to systems. The computation of the convolution integral is difficult even for simple signals. In the previous chapter we showed how to obtain the convolution integral analytically as well as graphically. As we will see in this section, it is not only that this property gives an efficient solution to the computation of the convolution integral, but that it introduces the **transfer function**, an important representation of a LTI system. Thus a system, like signals, is represented by the poles and zeros of the transfer function. But it is not only the pole/zero characterization of the system that can be obtained from the transfer function:

(i) the system's impulse response is uniquely obtained from the poles and zeros of the transfer function and the corresponding region of convergence;
(ii) the way the system responds to different frequencies is given by the transfer function;
(iii) stability and causality of the system can be equally related to the transfer function;
(iv) design of filters depends on the transfer function.

The Laplace transform of the convolution $y(t) = [x * h](t)$, where $x(t)$, $h(t)$ and $y(t)$ are the input, the impulse response and the output of a linear time-invariant system, is given by the product

$$Y(s) = X(s)H(s) \tag{3.38}$$

where $Y(s) = \mathcal{L}[y(t)]$, $X(s) = \mathcal{L}[x(t)]$ and $H(s)$ is the **transfer function of the LTI system** or

$$H(s) = \mathcal{L}[h(t)] = \frac{Y(s)}{X(s)}. \tag{3.39}$$

$H(s)$ transfers the Laplace transform $X(s)$ of the input into the Laplace transform of the output, $Y(s)$. Once $Y(s)$ is found, $y(t)$ is computed by means of the inverse Laplace transform.

Example 3.23. Use the Laplace transform to find the convolution $y(t) = [x * h](t)$ when

(i) the input is $x(t) = u(t)$, and the impulse response is a pulse $h(t) = u(t) - u(t - 1)$,
(ii) the input and the impulse response of the system are $x(t) = h(t) = u(t) - u(t - 1)$.

Solution:

(i) The Laplace transforms are $X(s) = \mathcal{L}[u(t)] = 1/s$ and $H(s) = \mathcal{L}[h(t)] = (1 - e^{-s})/s$, so that

$$Y(s) = H(s)X(s) = \frac{1 - e^{-s}}{s^2}.$$

Its inverse is $y(t) = r(t) - r(t - 1)$, where $r(t)$ is the ramp signal. This result coincides with the one obtained graphically in Chapter 2.

(ii) Now, $X(s) = H(s) = \mathcal{L}[u(t) - u(t - 1)] = (1 - e^{-s})/s$ so that

$$Y(s) = H(s)X(s) = \frac{(1 - e^{-s})^2}{s^2} = \frac{1 - 2e^{-s} + e^{-2s}}{s^2},$$

which corresponds to

$$y(t) = r(t) - 2r(t - 1) + r(t - 2)$$

or a triangular pulse as we obtained graphically in Chapter 2. □

Example 3.24. Consider an LTI system represented by the ordinary differential equation

$$\frac{d^2y(t)}{dt^2} + 2\frac{dy(t)}{dt} + 2y(t) = x(t) + \frac{dx(t)}{dt}$$

where $y(t)$ and $x(t)$ are the output and input of the system.

(i) Find the impulse response $h(t)$ of the system.
(ii) If the input of the system is $x(t) = (1 - t)e^{-t}u(t)$, find the corresponding output using the convolution property.

Solution:

(i) The transfer function of the system is obtained from the ordinary differential equation by letting the initial conditions be zero:

$$(s^2 + 2s + 2)Y(s) = (1 + s)X(s) \quad \Rightarrow \quad H(s) = \frac{Y(s)}{X(s)} = \frac{1+s}{(s+1)^2 + 1}$$

and the impulse response is then $h(t) = e^{-t}\cos(t)u(t)$.
(ii) The output of the system is the convolution integral, i.e., $y(t) = (h * x)(t)$, which is not easy to find. But finding the Laplace transform of $x(t)$,

$$X(s) = \frac{1}{s+1} - \frac{1}{(s+1)^2} = \frac{s}{(s+1)^2},$$

and using the transfer function we then have

$$Y(s) = H(s)X(s) = \frac{1+s}{(s+1)^2 + 1} \times \frac{s}{(s+1)^2} = \frac{s}{(s^2 + 2s + 2)(s+1)} = \frac{s}{s^3 + 3s^2 + 4s + 2}.$$

The poles of $Y(s)$ are $s_{1,2} = -1 \pm j1$, and $s_3 = -1$, while the zero is $s_1 = 0$. The following script computes and plots the poles and zeros of $Y(s)$ and then finds and plots the inverse

$$y(t) = \left[e^{-t}(\cos(t) + \sin(t) - 1)\right]u(t).$$

Fig. 3.17 displays the pole/zero distribution and the output $y(t)$.   □

```
%%
% Example 3.24---Inverse Laplace transform
%%
clear all; clf
syms s
num=[1 0];
den=[1 3 4 2];
figure(1)
subplot(211)
splane(num, den)
```

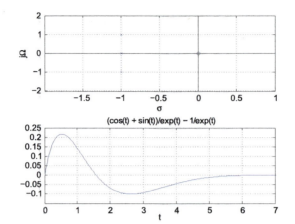

FIGURE 3.17

Poles and zeros of $Y(s)$ (top), response to input $x(t) = (1-t)e^{-t}u(t)$ of a system represented by an ordinary differential equation $d^2y(t)/dt^2 + 2dy(t)/dt + 2y(t) = x(t) + dx(t)/dt$ using the convolution property of the Laplace transform.

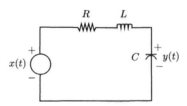

FIGURE 3.18

RLC circuit with input a voltage source $x(t)$ and output the voltage across the capacitor $y(t)$.

```
disp('>>>>> Inverse Laplace <<<<<')
x=ilaplace(s/(s^3+3*s^2+4*s+2))          % inverse Laplace transform
subplot(212)
fplot(x,[0,12]);
axis([0 7 -0.15 0.25]); grid
```

Example 3.25. Consider the RLC circuit shown in Fig. 3.18, where the input is a voltage source $x(t)$ and the output the voltage $y(t)$ across the capacitor. Let $LC = 1$ and $R/L = 2$.

1. Find the impulse response $h(t)$ of the circuit.
2. Use the convolution property to find the unit-step response $s(t)$.

Solution: The RLC circuit is represented by a second-order ordinary differential equation. Letting $i(t)$ be the current in the circuit, applying Kirchhoff's voltage law we have

$$x(t) = Ri(t) + L\frac{di(t)}{dt} + y(t)$$

where the voltage across the capacitor is given by

$$y(t) = \frac{1}{C}\int_0^t i(\sigma)d\sigma + y(0)$$

with $y(0)$ the initial voltage across the capacitor. The above two equations are called an *integro-differential* equation given that they are composed of an integral and an ordinary differential equation. To obtain an ordinary differential equation in terms of $x(t)$ and $y(t)$, we find the first and second derivative of $y(t)$, which gives

$$\frac{dy(t)}{dt} = \frac{1}{C}i(t) \quad \Rightarrow \quad i(t) = C\frac{dy(t)}{dt},$$

$$\frac{d^2y(t)}{dt^2} = \frac{1}{C}\frac{di(t)}{dt} \quad \Rightarrow \quad L\frac{di(t)}{dt} = LC\frac{d^2y(t)}{dt^2},$$

which when replaced in the KVL equation gives

$$x(t) = RC\frac{dy(t)}{dt} + LC\frac{d^2y(t)}{dt^2} + y(t), \tag{3.40}$$

a second-order ordinary differential equation with two initial conditions: $y(0)$, the initial voltage across the capacitor, and $i(0) = Cdy(t)/dt|_{t=0}$ or the initial current in the inductor.

1. To find the impulse response of this circuit, we let $x(t) = \delta(t)$ and the initial conditions be zero. The Laplace transform of Equation (3.40) gives

$$X(s) = [LCs^2 + RCs + 1]Y(s).$$

The impulse response of the system $h(t)$ is the inverse Laplace transform of the transfer function

$$H(s) = \frac{Y(s)}{X(s)} = \frac{1/LC}{s^2 + (R/L)s + 1/LC}.$$

If $LC = 1$ and $R/L = 2$, then the transfer function is

$$H(s) = \frac{1}{(s+1)^2},$$

which corresponds to the impulse response $h(t) = te^{-t}u(t)$.

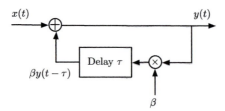

FIGURE 3.19

Positive feedback created by closeness of a microphone to a set of speakers.

2. To find the response to $x(t) = u(t)$, we could calculate the convolution integral:

$$
\begin{aligned}
y(t) &= \int_{-\infty}^{\infty} x(\tau)h(t-\tau)d\tau = \int_{-\infty}^{\infty} u(\tau)(t-\tau)e^{-(t-\tau)}u(t-\tau)d\tau \\
&= \int_{0}^{t} (t-\tau)e^{-(t-\tau)}d\tau = [1 - e^{-t}(1+t)]u(t).
\end{aligned}
$$

The unit-step response is then $s(t) = [1 - e^{-t}(1+t)]u(t)$.

The above can be more easily computed using the convolution property of the Laplace domain:

$$
Y(s) = H(s)X(s) = \frac{1}{(s+1)^2}\frac{1}{s}
$$

where we replaced the transfer function and the Laplace transform of $x(t) = u(t)$. The partial fraction expansion of $Y(s)$ is then

$$
Y(s) = \frac{A}{s} + \frac{B}{s+1} + \frac{C}{(s+1)^2}
$$

and after obtaining that $A = 1$, $C = -1$ and $B = -1$, we get

$$
y(t) = s(t) = u(t) - e^{-t}u(t) - te^{-t}u(t),
$$

which coincides with the solution of the convolution integral. It has been obtained, however, in a much easier way. □

Example 3.26. Consider the positive feedback system, from Chapter 2, created by a microphone close to a set of speakers that are putting out an amplified acoustic signal (see Fig. 3.19). For simplicity let $\beta = 1$, $\tau = 1$ and $x(t) = u(t)$.

- Find the impulse response $h(t)$ of the system using the Laplace transform, and use it to express the output in terms of a convolution.
- Determine the transfer function $H(s)$ and from it find the output $y(t)$.
- Show that the systems is not BIBO stable. Connect the location of the poles of the transfer function with the unstable behavior of the system.

Solution:
As indicated in Chapter 2, the impulse response of a feedback system cannot be explicitly obtained in the time domain, but it can now be done using the Laplace transform. The input–output equation for the positive feedback is $y(t) = x(t) + \beta y(t - \tau)$. If we let $x(t) = \delta(t)$, the output $y(t) = h(t)$ or $h(t) = \delta(t) + \beta h(t - \tau)$ from which we cannot find an expression for $h(t)$. But if $H(s) = \mathcal{L}[h(t)]$ then the Laplace transform of the above equation is $H(s) = 1 + \beta H(s)e^{-s\tau}$ or solving for $H(s)$:

$$H(s) = \frac{1}{1 - \beta e^{-s\tau}} = \frac{1}{1 - e^{-s}} = \sum_{k=0}^{\infty} e^{-sk} = 1 + e^{-s} + e^{-2s} + e^{-3s} + \cdots$$

after replacing the given values for β and τ. The impulse response $h(t)$ is the inverse Laplace transform of $H(s)$ or

$$h(t) = \delta(t) + \delta(t - 1) + \delta(t - 2) + \cdots = \sum_{k=0}^{\infty} \delta(t - k).$$

If $x(t)$ is the input, the output can now be written in terms of the convolution integral:

$$\begin{aligned} y(t) &= \int_{-\infty}^{\infty} x(t - \tau)h(\tau)d\tau = \int_{-\infty}^{\infty} \sum_{k=0}^{\infty} \delta(\tau - k)x(t - \tau)d\tau \\ &= \sum_{k=0}^{\infty} \int_{-\infty}^{\infty} \delta(\tau - k)x(t - \tau)d\tau = \sum_{k=0}^{\infty} x(t - k) \end{aligned}$$

and replacing $x(t) = u(t)$ we get

$$y(t) = \sum_{k=0}^{\infty} u(t - k),$$

which tends to infinity as t increases.

For this system to be BIBO stable, the impulse response $h(t)$ must be absolutely integrable, which is not the case for this system. Indeed,

$$\int_{-\infty}^{\infty} |h(t)|dt = \int_{-\infty}^{\infty} \sum_{k=0}^{\infty} \delta(t - k)dt = \sum_{k=0}^{\infty} \int_{-\infty}^{\infty} \delta(t - k)dt = \sum_{k=0}^{\infty} 1 \to \infty.$$

The poles of $H(s)$ are the roots of $1 - e^{-s} = 0$, which are the values of s such that $e^{-s_k} = 1 = e^{j2\pi k}$ or $s_k = \pm j2\pi k$. That is, there are an infinite number of poles on the $j\Omega$-axis, indicating that the system is not BIBO stable. □

Example 3.27. In Chapter 2 we considered the deconvolution problem in which the input $x(t)$ and the output $y(t)$ of a LTI system were

$$x(t) = e^{-2t}u(t), \qquad y(t) = (e^{-t} - e^{-2t})u(t)$$

when the initial conditions are zero. We wanted to determine first the impulse response $h(t)$ of the system, and then assuming $h(t)$ and $y(t)$ were given to find the input $x(t)$. We want to show in this example that these deconvolution problems which were difficult to solve with the convolution are very simple using the Laplace transform.

Solution: The Laplace transforms of the given $x(t)$ and $y(t)$ are

$$X(s) = \frac{1}{s+2}, \qquad Y(s) = \frac{1}{s+1} - \frac{1}{s+2} = \frac{1}{(s+1)(s+2)}.$$

The transfer function $H(s)$ is

$$H(s) = \mathcal{L}[h(t)] = \frac{Y(s)}{X(s)} = \frac{1}{s+1},$$

so that $h(t) = e^{-t}u(t)$, coinciding with the answer we obtained in Chapter 2.

Now, if $h(t) = e^{-t}u(t)$ and $y(t) = (e^{-t} - e^{-2t})u(t)$, then we see that the Laplace transform of the input is

$$X(s) = \frac{Y(s)}{H(s)} = \frac{1}{s+2}$$

and as expected, $x(t) = e^{-2t}u(t)$. Thus, whenever two of the three possible functions $x(t)$, $h(t)$ and $y(t)$ of an LTI system are given, an d the initial conditions are zero, the concept of the transfer function can be used to find the desired third function. ☐

3.7 ANALYSIS OF LTI SYSTEMS REPRESENTED BY DIFFERENTIAL EQUATIONS

Dynamic linear time-invariant (LTI) systems are typically represented by ordinary differential equations. Using the derivative property of the one-sided Laplace transform (allowing the inclusion of initial conditions) and the inverse transformation, ordinary differential equations are changed into easier-to-solve algebraic equations.

Two ways to characterize the response of a causal and stable LTI system are:

- **zero-state** and **zero-input** responses, which have to do with the effect of the input and the initial conditions of the system,
- **transient** and **steady-state** responses, which have to do with close and far away behavior of the response.

The **complete response** $y(t)$ of a system represented by an Nth-order linear ordinary differential equation with constant coefficients

$$y^{(N)}(t) + \sum_{k=0}^{N-1} a_k y^{(k)}(t) = \sum_{\ell=0}^{M} b_\ell x^{(\ell)}(t) \qquad N > M, \tag{3.41}$$

where $x(t)$ is the input and $y(t)$ the output of the system, and the initial conditions are

$$\{y^{(k)}(t), \ 0 \leq k \leq N-1\}, \tag{3.42}$$

is obtained by inverting the Laplace transform

$$Y(s) = \frac{B(s)}{A(s)} X(s) + \frac{1}{A(s)} I(s) \tag{3.43}$$

where $Y(s) = \mathcal{L}[y(t)]$, $X(s) = \mathcal{L}[x(t)]$ and

$$A(s) \quad = \quad \sum_{k=0}^{N} a_k s^k, \qquad a_N = 1,$$

$$B(s) \quad = \quad \sum_{\ell=0}^{M} b_\ell s^\ell,$$

$$I(s) \quad = \quad \sum_{k=1}^{N} a_k \left(\sum_{m=0}^{k-1} s^{k-m-1} y^{(m)}(0) \right), \qquad a_N = 1,$$

i.e., $I(s)$ depends on the initial conditions.

The notation $y^{(k)}(t)$ and $x^{(\ell)}(t)$ indicates the kth and the ℓth derivatives of $y(t)$ and of $x(t)$, respectively (it is to be understood that $y^{(0)}(t) = y(t)$ and likewise $x^{(0)}(t) = x(t)$ in this notation). The assumption $N > M$ avoids the presence of $\delta(t)$ and its derivatives in the solution—which are realistically not possible. To obtain the complete response $y(t)$ we compute the Laplace transform of Equation (3.41):

$$\underbrace{\left[\sum_{k=0}^{N} a_k s^k \right]}_{A(s)} Y(s) = \underbrace{\left[\sum_{\ell=0}^{M} b_\ell s^\ell \right]}_{B(s)} X(s) + \underbrace{\sum_{k=1}^{N} a_k \left(\sum_{m=0}^{k-1} s^{(k-1)-m} y^{(m)}(0) \right)}_{I(s)},$$

which can be written as

$$A(s)Y(s) = B(s)X(s) + I(s) \tag{3.44}$$

by defining $A(s)$, $B(s)$ and $I(s)$ as indicated above. Solving for $Y(s)$ in (3.44) we have

$$Y(s) = \frac{B(s)}{A(s)} X(s) + \frac{1}{A(s)} I(s)$$

and finding its inverse we obtain the complete response $y(t)$.

Letting

$$H(s) = \frac{B(s)}{A(s)} \quad \text{and} \quad H_1(s) = \frac{1}{A(s)}$$

the **complete response** $y(t) = \mathcal{L}^{-1}[Y(s)]$ of the system is obtained by the inverse Laplace transform of

$$Y(s) = H(s)X(s) + H_1(s)I(s),$$ (3.45)

which gives

$$y(t) = y_{zs}(t) + y_{zi}(t)$$ (3.46)

where

$$y_{zs}(t) = \mathcal{L}^{-1}[H(s)X(s)] \quad \text{is the system's \textbf{zero-state response},}$$
$$y_{zi}(t) = \mathcal{L}^{-1}[H_1(s)I(s)] \quad \text{is the system's \textbf{zero-input response}.}$$

In terms of convolution integrals

$$y(t) = \int_0^t x(\tau)h(t-\tau)d\tau + \int_0^t i(\tau)h_1(t-\tau)d\tau$$ (3.47)

where $h(t) = \mathcal{L}^{-1}[H(s)]$ and $h_1(t) = \mathcal{L}^{-1}[H_1(s)]$ and

$$i(t) = \mathcal{L}^{-1}[I(s)] = \sum_{k=1}^{N} a_k \left(\sum_{m=0}^{k-1} y^{(m)}(0)\delta^{(k-m-1)}(t) \right)$$

where $\{\delta^{(m)}(t)\}$ are mth derivatives of the impulse signal $\delta(t)$ (as indicated before $\delta^{(0)}(t) = \delta(t)$).

Zero-State and Zero-Input Responses

Despite the fact that linear ordinary differential equations, with constant coefficients, do not represent LTI systems unless the initial conditions are zero and the input is causal, linear system theory is based on these representations with nonzero initial conditions. Typically, the input is causal so it is the initial conditions not always being zero that causes problems. This can be remedied by a different way of thinking about the initial conditions. In fact, one can think of the input $x(t)$ and the initial conditions as two different inputs to the system, and then apply superposition to find the responses to these two different inputs. This defines two responses, one which is due completely to the input, with zero initial conditions, called the **zero-state solution**. The other component of the complete response is due exclusively to the initial conditions, assuming that the input is zero, and so it is called the **zero-input solution**. See Fig. 3.20.

Remarks

1. It is important to recognize that to compute the *transfer function of the system*

$$H(s) = \frac{Y(s)}{X(s)}$$

according to Equation (3.45) requires that the initial conditions be zero, or $I(s) = 0$.

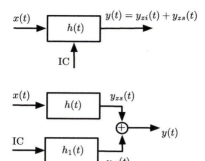

FIGURE 3.20

Zero-input $y_{zi}(t)$ and zero-state $y_{zs}(t)$ responses of LTI systems. The complete response is $y(t) = y_{zi}(t) + y_{zs}(t)$.

2. If there is no pole–zero cancellation, both $H(s)$ and $H_1(s)$ have the same poles, as both have $A(s)$ as denominator, and $h(t) = h_1(t)$ when $B(s) = 1$.

Transient and Steady-State Responses

Whenever the input of a causal and stable LTI system has a Laplace transform with poles in the closed left-hand s-plane—poles in the $j\Omega$-axis being simple—the complete response will be bounded. More-over, whether the response exists as $t \to \infty$, or the steady-state response, can be determined without finding the inverse Laplace transform.

The **complete response** $y(t)$ of a LTI system is made up of a **transient** component and a **steady-state** component. The transient response can be thought of as the inertia the system presents to the input, while the steady-state response is how the system reacts to the input away from the initial time.[6]

If the poles (simple or multiple, real or complex) of the Laplace transform of the output, $Y(s)$, of an LTI system are in the open left-hand s-plane (i.e., no poles on the $j\Omega$-axis) the steady-state response is

$$y_{ss}(t) = \lim_{t \to \infty} y(t) = 0.$$

In fact, for any real pole $s = -\alpha$, $\alpha > 0$, of multiplicity $m \geq 1$ we have

$$\mathcal{L}^{-1}\left[\frac{N(s)}{(s+\alpha)^m}\right] = \sum_{k=1}^{m} A_k t^{k-1} e^{-\alpha t} u(t)$$

where $N(s)$ is a polynomial of degree $m - 1$ or less. Clearly, for any value of $\alpha > 0$ and any order $m \geq 1$, the above inverse will tend to zero as t increases. The rate at which these terms go to zero

---

[6]Although typically the steady-state response is evaluated as $t \to \infty$, it can occur much earlier. Likewise, the steady-response can be obtained from either an input that is applied at a finite time, $t = 0$, and goes to infinity, $t \to \infty$, or an input that is applied at $-\infty$ and goes to a finite time.

depends on how close the pole(s) is (are) to the $j\Omega$-axis—the farther the faster the term goes to zero. Likewise, complex conjugate pairs of poles with negative real part also give terms that go to zero as $t \to \infty$, independent of their order. For complex conjugate pairs of poles, $s_{1,2} = -\alpha \pm j\Omega_0$ of order $m \geq 1$, we have

$$\mathcal{L}^{-1}\left[\frac{N(s)}{((s+\alpha)^2 + \Omega_0^2)^m}\right] = \sum_{k=1}^{m} 2|A_k|t^{k-1}e^{-\alpha t}\cos(\Omega_0 + \angle(A_k))u(t)$$

where $N(s)$ is a polynomial of degree $2m - 1$ or less. Due to the decaying exponentials this type of response will go to zero as t goes to infinity.

Simple complex conjugate poles, and a simple real pole at the origin of the s-plane cause a steady-state response. Indeed, if the pole of $Y(s)$ is $s = 0$ we know that its inverse transform is of the form $Au(t)$, and if the poles are complex conjugates $\pm j\Omega_0$ the corresponding inverse transform is a sinusoid—neither of which is transient. However, *multiple poles on the $j\Omega$-axis, or any poles in the right-hand s-plane will give inverses that grow as $t \to \infty$.* This statement is clear for the poles in the right-hand s-plane. For double or higher order poles in the $j\Omega$-axis their inverse transform is of the form

$$\mathcal{L}^{-1}\left[\frac{N(s)}{(s^2 + \Omega_0^2)^m}\right] = \sum_{k=1}^{m} 2|A_k|t^{m-1}\cos(\Omega_0 + \angle(A_k))u(t) \qquad m \geq 1,$$

which will continuously grow as t increases.

In summary, when solving ordinary differential equations—with or without initial conditions—using Laplace we have:

(i) The steady-state component of the complete solution is given by the inverse Laplace transforms of the partial fraction expansion terms of $Y(s)$ that have simple poles (real or complex conjugate pairs) in the $j\Omega$-axis.

(ii) The transient response is given by the inverse Laplace transform of the partial fraction expansion terms with poles in the left-hand s-plane, independent of whether the poles are simple or multiple, real or complex.

(iii) Multiple poles in the $j\Omega$-axis and poles in the right-hand s-plane give terms that will increase as t increases making the complete response unbounded.

Example 3.28. Consider a second-order ordinary differential equation

$$\frac{d^2y(t)}{dt^2} + 3\frac{dy(t)}{dt} + 2y(t) = x(t)$$

representing a LTI system with input $x(t)$ and output $y(t)$. Find the impulse response $h(t)$ and the unit-step response $s(t)$ of the system.

Solution: Letting the initial conditions be zero, computing the two- or one-sided Laplace transform of the two sides of this equation after letting $Y(s) = \mathcal{L}[y(t)]$ and $X(s) = \mathcal{L}[x(t)]$ and using the derivative property of Laplace we get

$$Y(s)[s^2 + 3s + 2] = X(s).$$

To find the impulse response of this system, i.e., the system response $y(t) = h(t)$, we let $x(t) = \delta(t)$ and the initial condition be zero. Since $X(s) = 1$, then $Y(s) = H(s) = \mathcal{L}[h(t)]$ is

$$H(s) = \frac{1}{s^2 + 3s + 2} = \frac{1}{(s+1)(s+2)} = \frac{A}{s+1} + \frac{B}{s+2}.$$

We obtain values $A = 1$ and $B = -1$, and the inverse Laplace transform is then

$$h(t) = \left[e^{-t} - e^{-2t} \right] u(t),$$

which is completely transient.

In a similar form we obtain the **unit-step response** $s(t)$, by letting $x(t) = u(t)$ and the initial conditions be zero. Calling $Y(s) = S(s) = \mathcal{L}[s(t)]$, since $X(s) = 1/s$ we obtain

$$S(s) = \frac{H(s)}{s} = \frac{1}{s(s^2 + 3s + 2)} = \frac{A}{s} + \frac{B}{s+1} + \frac{C}{s+2}.$$

It is found that $A = 1/2$, $B = -1$ and $C = 1/2$ so that

$$s(t) = 0.5u(t) - e^{-t}u(t) + 0.5e^{-2t}u(t).$$

The steady state of $s(t)$ is 0.5 as the two exponentials go to zero. The transient is $-e^{-t}u(t) + 0.5e^{-2t}u(t)$.

Interestingly, the relation $sS(s) = H(s)$ indicates that by computing the derivative of $s(t)$ we obtain $h(t)$. Indeed,

$$\begin{aligned}
\frac{ds(t)}{dt} &= 0.5\delta(t) + e^{-t}u(t) - e^{-t}\delta(t) - e^{-2t}u(t) + 0.5e^{-2t}\delta(t) \\
&= [0.5 - 1 + 0.5]\delta(t) + [e^{-t} - e^{-2t}]u(t) = [e^{-t} - e^{-2t}]u(t) = h(t). \quad \square
\end{aligned}$$

Remarks

1. Because the existence of the steady-state response depends on the poles of $Y(s)$ it is possible for an unstable causal system (recall that for such a system BIBO stability requires all the poles of the system transfer function be in the open left-hand s-plane) to have a steady-state response. It all depends on the input. Consider for instance an unstable system with $H(s) = 1/(s(s+1))$, being unstable due to the pole at $s = 0$, if the system input is $x_1(t) = u(t)$ so that $X_1(s) = 1/s$ then $Y_1(s) = 1/(s^2(s+1))$. There will be no steady state because of the double pole $s = 0$. On the other hand, if the input is $x_2(t) = (1 - 2t)e^{-2t}u(t)$, which has Laplace transform $X_2(s) = s/(s+2)^2$, then

$$Y_2(s) = H(s)X_2(s) = \frac{1}{s(s+1)} \frac{s}{(s+2)^2} = \frac{1}{(s+1)(s+2)^2},$$

which will give a zero steady state, even though the system is unstable. This is possible because of the zero–pole cancellation.

2. The steady-state response is the response of the system away from $t = 0$, and it can be found by letting $t \to \infty$ (even though the steady state can be reached at finite times, depending on how fast the transient goes to zero). In the above example, the steady-state response of $h(t) = (e^{-t} - e^{-2t})u(t)$ is 0 and it is obtained at $t = 0$, while for $s(t) = 0.5u(t) - e^{-t}u(t) + 0.5e^{-2t}u(t)$ is 0.5 and it will be obtained at a time $t_0 > 0$ for which $-e^{-t_0} + 0.5e^{-2t_0} = 0$, which is $t_0 = \infty$ since the exponentials are positive functions. The transient responses are then $h(t) - 0 = h(t)$ and $s(t) - 0.5u(t) = -e^{-t}u(t) + 0.5e^{-2t}u(t)$ and eventually disappear.

3. The relation found between the impulse response $h(t)$ and the unit-step response $s(t)$ can be extended to more cases by the definition of the transfer function, i.e., $H(s) = Y(s)/X(s)$ so that the response $Y(s)$ is connected with $H(s)$ by $Y(s) = H(s)X(s)$, giving the relation between $y(t)$ and $h(t)$. For instance, if $x(t) = \delta(t)$, then $Y(s) = H(s) \times 1$, with inverse the impulse response. If $x(t) = u(t)$ then $Y(s) = H(s)/s = S(s)$, the Laplace transform of the unit-step response, and so $s(t) = dh(t)/dt$. And if $x(t) = r(t)$ then $Y(s) = H(s)/s^2 = \rho(s)$, the Laplace transform of the ramp response, and so $\rho(t) = d^2h(t)/dt^2 = ds(t)/dt$.

Example 3.29. Consider the second-order ordinary differential equation in the previous example,

$$\frac{d^2y(t)}{dt^2} + 3\frac{dy(t)}{dt} + 2y(t) = x(t)$$

but now with initial conditions $y(0) = 1$ and $dy(t)/dt|_{t=0} = 0$, and $x(t) = u(t)$. Find the complete response $y(t)$. Could we find the impulse response $h(t)$ from this response? How could we do it?

Solution: The Laplace transform of the ordinary differential equation gives

$$\left[s^2Y(s) - sy(0) - \frac{dy(t)}{dt}\Big|_{t=0} \right] + 3[sY(s) - y(0)] + 2Y(s) = X(s),$$

$$Y(s)(s^2 + 3s + 2) - (s+3) = X(s),$$

so we have

$$
\begin{aligned}
Y(s) &= \frac{X(s)}{(s+1)(s+2)} + \frac{s+3}{(s+1)(s+2)} \\
&= \frac{1 + 3s + s^2}{s(s+1)(s+2)} = \frac{B_1}{s} + \frac{B_2}{s+1} + \frac{B_3}{s+2}
\end{aligned}
$$

after replacing $X(s) = 1/s$. We find that $B_1 = 1/2$, $B_2 = 1$ and $B_3 = -1/2$, so that the complete response is $y(t) = [0.5 + e^{-t} - 0.5e^{-2t}]u(t)$. The steady-state response is 0.5 and the transient $[e^{-t} - 0.5e^{-2t}]u(t)$.

If we are able to find the transfer function $H(s) = Y(s)/X(s)$ its inverse Laplace transform would be $h(t)$. However, that is not possible when the initial conditions are nonzero. As shown above, in the case of nonzero initial conditions we see that the Laplace transform

$$Y(s) = \frac{X(s)}{A(s)} + \frac{I(s)}{A(s)}$$

where in this case $A(s) = (s+1)(s+2)$ and $I(s) = s+3$, and thus we cannot find the ratio $Y(s)/X(s)$. If we make the second term zero, i.e., $I(s) = 0$, we then have $Y(s)/X(s) = H(s) = 1/A(s)$ and $h(t) = [e^{-t} - e^{-2t}]u(t)$.   ◻

Example 3.30. Consider an analog averager represented by

$$y(t) = \frac{1}{T} \int_{t-T}^{t} x(\tau)d\tau$$

where $x(t)$ is the input and $y(t)$ is the output. The derivative of $y(t)$ gives the first-order ordinary differential equation,

$$\frac{dy(t)}{dt} = \frac{1}{T}[x(t) - x(t-T)]$$

with a finite difference for the input. Find the impulse response of this analog averager.

Solution: The impulse response of the averager is found by letting $x(t) = \delta(t)$ and the initial condition be zero. Computing the Laplace transform of the two sides of the ordinary differential equation, we obtain

$$sY(s) = \frac{1}{T}[1 - e^{-sT}]X(s)$$

and substituting $X(s) = 1$, then

$$H(s) = Y(s) = \frac{1}{sT}[1 - e^{-sT}] \quad \text{and the impulse response is} \quad h(t) = \frac{1}{T}[u(t) - u(t-T)]. \quad ◻$$

3.8 INVERSE OF TWO-SIDED LAPLACE TRANSFORMS

When finding the inverse of a two-sided Laplace transform we need to pay close attention to the region of convergence and to the location of the poles with respect to the $j\Omega$-axis. Three regions of convergence are possible:

- a plane to the right of all the poles, corresponding to a causal signal,
- a plane to the left of all poles, corresponding to an anticausal signal, and
- a plane section in between poles on the right and poles on the left (no poles included in it) which corresponds to a two-sided signal.

When inverting a transfer function $H(s)$ if its ROC includes the $j\Omega$-axis it guarantees the bounded input–bounded output (BIBO) stability of the system or that the impulse response of the system is absolutely integrable. Furthermore, the system with that region of convergence would have a frequency response, and the signal a Fourier transform. The inverses of the causal and the anticausal components are obtained using the one-sided Laplace transform.

Example 3.31. Find the noncausal inverse Laplace transform of

$$X(s) = \frac{1}{(s+2)(s-2)} \qquad \text{ROC: } -2 < \mathcal{R}e(s) < 2.$$

Solution: The ROC $-2 < \mathcal{R}e(s) < 2$ is equivalent to $\{(\sigma, \Omega) : -2 < \sigma < 2, -\infty < \Omega < \infty\}$. The partial fraction expansion is

$$X(s) = \frac{1}{(s+2)(s-2)} = \frac{-0.25}{s+2} + \frac{0.25}{s-2} \qquad -2 < \mathcal{R}e(s) < 2$$

where the first term with the pole at $s = -2$ corresponds to a causal signal with a region of convergence $\mathcal{R}e(s) > -2$, and the second term corresponds to an anticausal signal with a region of convergence $\mathcal{R}e(s) < 2$. That this is so is confirmed by the intersection of these two regions of convergence,

$$[\mathcal{R}e(s) > -2] \cap [\mathcal{R}e(s) < 2] = -2 < \mathcal{R}e(s) < 2.$$

As such we have $x(t) = -0.25e^{-2t}u(t) - 0.25e^{2t}u(-t)$. $\qquad\square$

Example 3.32. Consider the transfer function

$$H(s) = \frac{s}{(s+2)(s-1)} = \frac{2/3}{s+2} + \frac{1/3}{s-1}$$

with a zero at $s = 0$, and poles at $s = -2$ and $s = 1$. Find out how many impulse responses can be connected with $H(s)$ by considering different possible regions of convergence and by determining in which cases the system with $H(s)$ as its transfer function is BIBO stable.

Solution: The following are the different possible impulse responses:

- If ROC: $\mathcal{R}e(s) > 1$, the impulse response

$$h_1(t) = (2/3)e^{-2t}u(t) + (1/3)e^t u(t)$$

corresponding to $H(s)$, with this region of convergence, is causal. The corresponding system is unstable—due to the pole in the right-hand s-plane which will make the impulse response grow.
- If ROC: $-2 < \mathcal{R}e(s) < 1$, the impulse response corresponding to $H(s)$ with this region of convergence is noncausal, but the system is stable. The impulse response is

$$h_2(t) = (2/3)e^{-2t}u(t) - (1/3)e^t u(-t).$$

Notice that the region of convergence includes the $j\Omega$-axis, and this guarantees the stability (verify that $h_2(t)$ is absolutely integrable), and as we will see later, also the existence of the Fourier transform of $h_2(t)$.
- If ROC: $\mathcal{R}e(s) < -2$, the impulse response in this case would be anticausal, and the system is unstable. The impulse response is

$$h_3(t) = (2/3)e^{-2t}u(-t) + (1/3)e^t u(-t),$$

for which the first term grows for $t < 0$ and as such $h_3(t)$ is not absolutely integrable, i.e., not BIBO stable. $\qquad\square$

Two very important generalizations of the results in the above example are:

1. A LTI with a transfer function $H(s)$ and region of convergence $\mathcal{R}$ is BIBO stable if the $j\Omega$-axis is contained in the region of convergence.
2. A casual LTI system with impulse response $h(t)$ or transfer function $H(s) = \mathcal{L}[h(t)]$ is <u>BIBO stable</u> if the following equivalent conditions are satisfied:

$$(i) \qquad H(s) = \mathcal{L}[h(t)] = \frac{N(s)}{D(s)}, \quad j\Omega\text{-axis in ROC of } H(s).$$

$$(ii) \qquad \int_{-\infty}^{\infty} |h(t)|dt < \infty, \quad h(t) \text{ is absolutely integrable.}$$

(iii) \qquad Poles of $H(s)$ are in the open left-hand s-plane (not including the $j\Omega$-axis).

If the transfer function $H(s)$ of a LTI system, causal or noncausal, has a region of convergence including the $j\Omega$-axis, then $H(j\Omega)$ exists, so that

$$|H(j\Omega)| = \left| \int_{-\infty}^{\infty} h(t)e^{-j\Omega t}dt \right| < \int_{-\infty}^{\infty} |h(t)|dt < \infty$$

or that the impulse response $h(t)$ is absolutely integrable. This implies the system is BIBO stable as indicated in Chapter 2.

If the LTI system is causal and BIBO stable, the region of convergence of its transfer function not only includes the $j\Omega$-axis, but the right-hand s-plane. Since no poles can be in the region of convergence then all the poles of the system should be in open left-hand s-plane (not including the $j\Omega$-axis).

Example 3.33. Consider causal LTI systems with impulse responses

(i) $h_1(t) = [4\cos(t) - e^{-t}]u(t)$, (ii) $h_2(t) = [te^{-t} + e^{-2t}]u(t)$, (iii) $h_3(t) = [te^{-t}\cos(2t)]u(t)$;

determine which of these systems is BIBO stable. For those that are not, indicate why they are not.

Solution: Testing if these impulse responses are absolutely integrable could be difficult. A better approach is to find the location of the poles of their transfer functions

$$(i) \qquad H_1(s) = \frac{4s}{s^2+1} - \frac{1}{s+1} = \frac{N_1(s)}{(s^2+1)(s+1)}, \quad \text{ROC: } \sigma > 0,$$

$$(ii) \qquad H_2(s) = \frac{1}{(s+1)^2} + \frac{1}{s+2} = \frac{N_2(s)}{(s+1)^2(s+2)}, \quad \text{ROC: } \sigma > -1,$$

$$(iii) \qquad H_3(s) = \frac{1}{[(s+1)^2+4]^2}, \quad \text{ROC: } \sigma > -1.$$

The causal system with transfer function $H_1(s)$ is not BIBO because it has poles in the $j\Omega$-axis: its poles are $s = \pm j$ (on the $j\Omega$-axis) and $s = -1$. Also its ROC does not include the $j\Omega$-axis. On the other hand, the causal systems with transfer functions $H_2(s)$ and $H_3(s)$ are BIBO stable given that they both have poles in the open left-hand s-plane: $H_2(s)$ has poles $s = -2$ and $s = -1$ (double), while $H_3(s)$ has double poles at $s = -1 \pm 2j$. Also, their ROCs include the $j\Omega$-axis.

3.9 WHAT HAVE WE ACCOMPLISHED? WHERE DO WE GO FROM HERE?

In this chapter you have learned the significance of the Laplace transform in the analysis of continuous-time signals as well as systems. The Laplace transform provides a complementary representation to the time representation of a signal, so that damping and frequency, poles and zeros, together with regions of convergence conform a new domain for signals. Moreover, you will see that these concepts will apply in the rest of this part of the book. When discussing the Fourier analysis of signal and systems we will come back to the Laplace transform. The solution of ordinary differential equations, and the different types of responses are obtained algebraically with the Laplace transform. Likewise, the Laplace transform provides a simple and yet very significant solution to the convolution integral. It also provides the concept of transfer function which will be fundamental in analysis and synthesis of linear time-invariant systems.

An important remark that will be clarified in the next two chapters is that the common thread of the Laplace and the Fourier transforms is the eigenfunction property of LTI systems. You will see that understanding this property provides you with the needed insight into the Fourier analysis.

3.10 PROBLEMS
3.10.1 BASIC PROBLEMS

3.1 Find the Laplace transform of the following:

(a) anticausal signals indicating their region of convergence:

$(i) \quad x(t) = e^t u(-t),$ $\qquad$ $(ii) \quad y(t) = e^t u(-t-1),$

$(iii) \quad z(t) = e^{t+1} u(-t-1),$ $\qquad$ $(iv) \quad w(t) = e^t(u(-t) - u(-t-1)),$

(b) noncausal signals indicating their regions of convergence:

$(i) \quad x_1(t) = u(t+1) - u(t-1),$ $\qquad$ $(ii) \quad y_1(t) = e^{-t} u(t+1),$

$(iii) \quad z_1(t) = e^t[u(t+1) - u(t-1)].$

Answers: (a) $Z(s) = e^s/(1-s)$, ROC $\sigma < 1$; (b) $Z_1(s) = 2\sinh(s-1)/(s-1)$, ROC: whole s-plane.

3.2 Consider the following cases involving sinusoids:

(a) Find the Laplace transform of $y(t) = \sin(2\pi t)[u(t) - u(t-1)]$ and its region of convergence. Carefully plot $y(t)$. Determine the region of convergence of $Y(s)$.

(b) A very smooth pulse, called the raised cosine, is $x(t) = 1 - \cos(2\pi t)$, $0 \le t \le 1$, and zero elsewhere. Find its Laplace transform and its corresponding region of convergence.

(c) Indicate three possible approaches to finding the Laplace transform of $\cos^2(t)u(t)$. Use two of these approaches to find the Laplace transform.

Answers: $Y(s) = 2\pi(1 - e^{-s})/(s^2 + 4\pi^2)$; $x(t)$ finite-support signal, ROC whole plane.

3.3 Find the Laplace transform of the following signals and in each case determine the corresponding region of convergence:

(a) the signal $x(t) = e^{-\alpha t}u(t) - e^{\alpha t}u(-t)$ when   (i) $\alpha > 0$,   (ii) $\alpha \to 0$,

(b) a sampled signal

$$x_1(t) = \sum_{n=0}^{N-1} e^{-2n}\delta(t-n),$$

(c) the "stairs to heaven" signal

$$s(t) = \sum_{n=0}^{\infty} u(t-n),$$

(d) the sinusoidal signal $v(t) = [\cos(2(t-1)) + \sin(2\pi t)]u(t-1)$,

(e) the signal $y(t) = t^2 e^{-2t}u(t)$ using that $x(t) = t^2 u(t)$ has $X(s) = 2/s^3$.

Answers: (a) As $\alpha \to 0$, $x(t) = u(t) - u(-t)$, has no Laplace transform; (b) the ROC is the whole s-plane; (c) $S(s) = 1/(s(1 - e^{-s}))$; (e) $Y(s) = 2/(s+2)^3$.

3.4 In the following problems properties of the Laplace transform are used.

(a) Show that the Laplace transform of $x(t)e^{-at}u(t)$ is $X(s+a)$, where $X(s) = \mathcal{L}[x(t)]$ and then use it to find the Laplace transform of $y(t) = \cos(t)e^{-2t}u(t)$.

(b) A signal $x_1(t)$ has as Laplace transform

$$X_1(s) = \frac{s+2}{(s+2)^2 + 1},$$

find poles and zeros of $X_1(s)$ and find $x_1(t)$ as $t \to \infty$ from the location of the poles.

(c) The signal $z(t) = de^{-t}u(t)/dt$.

    i. Compute the derivative $z(t)$ and then find its Laplace transform $Z(s)$.

    ii. Use the derivative property to find $Z(s)$. Compare your result with the one obtained above.

Answer: (a) $Y(s) = (s+2)/((s+2)^2 + 1)$; (b) $x_1(t) \to 0$ as $t \to \infty$; (c) $Z(s) = s/(s+1)$.

3.5 To find the Laplace transform of $x(t) = r(t) - 2r(t-1) + 2r(t-3) - r(t-4)$.

(a) Plot $x(t)$. Calculate $dx(t)/dt$, $d^2x(t)/dt^2$ and plot them.

(b) Use the Laplace transform of $d^2x(t)/dt^2$ to obtain $X(s)$.

Answer: $d^2x(t)/dt^2 = \delta(t) - 2\delta(t-1) + 2\delta(t-3) - \delta(t-4)$.

3.6 In the following problems we use the inverse Laplace transform and the relation between input and output of LTI systems.

(a) The Laplace transform of the output of a system is

$$Y_1(s) = \frac{e^{-2s}}{s^2 + 1} + \frac{(s+2)^2 + 2}{(s+2)^3};$$

find $y_1(t)$, assume it is causal.

(b) The Laplace transform of the output $y_2(t)$ of a second-order system is

$$Y_2(s) = \frac{-s^2 - s + 1}{s(s^2 + 3s + 2)}.$$

If the input of this system is $x_2(t) = u(t)$, find the ordinary differential equation that represents the system and the corresponding initial conditions $y_2(0)$ and $dy_2(0)/dt$.

(c) The Laplace transform of the output $y(t)$ of a system is

$$Y(s) = \frac{1}{s((s+1)^2 + 4)}.$$

Assume $y(t)$ to be causal. Find the steady-state response $y_{ss}(t)$, and the transient $y_t(t)$.
Answer: (a) $y_1(t) = \sin(t-2)u(t-2) + e^{-2t}u(t) + t^2 e^{-2t}u(t)$; (b) $y_2(0) = -1$ and $dy_2(0)/dt = 2$; (c) transient $y_t(t) = [-(1/5)e^{-t}\cos(2t) - (1/10)e^{-t}\sin(2t)]u(t)$.

3.7 A system is represented by the following ordinary differential equation:

$$\frac{d^2 y(t)}{dt^2} + 3\frac{dy(t)}{dt} + 2y(t) = x(t)$$

where $y(t)$ is the system output and $x(t)$ is the input.
(a) Find the transfer function $H(s) = Y(s)/X(s)$ of the system. From its poles and zeros determine if the system is BIBO stable or not.
(b) If $x(t) = u(t)$, and initial conditions are zero, determine the steady-state response $y_{ss}(t)$. What if the initial conditions were not zero, would you get the same steady state? Explain.
Answers: $H(s) = 1/(s^2 + 3s + 2)$; BIBO stable; $y_{ss}(t) = 0.5$.

3.8 Consider a LTI system with transfer function

$$H(s) = \frac{Y(s)}{X(s)} = \frac{s^2 + 4}{s((s+1)^2 + 1)}.$$

(a) Determine if the system is BIBO stable or not.
(b) Let the input be $x(t) = \cos(2t)u(t)$; find the response $y(t)$ and the corresponding steady-state response.
(c) Let the input be $x(t) = \sin(2t)u(t)$; find the response $y(t)$ and the corresponding steady-state response.
(d) Let the input be $x(t) = u(t)$; find the response $y(t)$ and the corresponding steady-state response.
(e) Explain why the results above seem to contradict the result about stability.
Answers: For $x(t) = \cos(2t)u(t)$, $\lim_{t \to \infty} y(t) = 0$; if $x(t) = u(t)$ there is no steady state.

3.9 Consider the following impulse responses:

$$h_1(t) = [(2/3)e^{-2t} + (1/3)e^t]u(t), \qquad h_2(t) = (2/3)e^{-2t}u(t) - (1/3)e^t u(-t),$$
$$h_3(t) = -(2/3)e^{-2t}u(-t) - (1/3)e^t u(-t).$$

(a) From the expression for $h_1(t)$ determine if the system is causal and BIBO stable. Find its Laplace transform $H_1(s)$ and its region of convergence.
(b) From the expression for $h_2(t)$ determine if the system is noncausal and BIBO stable. Find its Laplace transform $H_2(s)$ and its region of convergence.

(c) From the expression for $h_3(t)$ determine if the system is anticausal and BIBO stable. Find its Laplace transform $H_3(s)$ and its region of convergence.

(d) From the above, determine the general condition for a system to be BIBO stable.

Answers: $h_1(t)$ unbounded, so unstable causal system; $H_3(s) = (2/3)/(s+2) - (1/3)/(-s+1)$ ROC: $\sigma < -2$, unstable anticausal system.

3.10 The unit-step response of a system is $s(t) = [0.5 - e^{-t} + 0.5e^{-2t}]u(t)$.

(a) Find the transfer function $H(s)$ of the system.

(b) How could you use $s(t)$ to find the impulse response $h(t)$ and the ramp response $\rho(t)$ in the time and in the Laplace domains?

Answer: $H(s) = 1/(s+1)(s+2)$.

3.11 The transfer function of a causal LTI system is

$$H(s) = \frac{1}{s^2 + 4}.$$

(a) Find the ordinary differential equation that relates the system input $x(t)$ to the system output $y(t)$.

(b) Find the input $x(t)$ so that, for initial conditions $y(0) = 0$ and $dy(0)/dt = 1$, the corresponding output $y(t)$ is identically zero.

(c) Suppose we would like the output $y(t)$ to be identically zero, if we let $x(t) = \delta(t)$ what would be the initial conditions be equal to?

Answer: (b) $x(t) = -\delta(t)$; (c) $y(0) = 0$, $y'(0) = -1$.

3.12 The impulse response of an LTI is $h(t) = r(t) - 2r(t-1) + r(t-2)$ and the input is a sequence of impulses

$$x(t) = \sum_{k=0}^{\infty} \delta(t - kT).$$

(a) Find the system output $y(t)$ as the convolution integral of $x(t)$ and $h(t)$, and plot it for $T = 1$ and $T = 2$.

(b) For $T = 2$ obtain the Laplace transform $Y(s)$ of $y(t)$.

Answer: $Y(s) = (\cosh(s) - 1)/(s^2 \sinh(s))$.

3.13 A wireless channel is represented by $y(t) = \alpha x(t - T) + \alpha^3 x(t - 3T)$ where $0 < \alpha < 1$ is the attenuation and T the delay. The input is $x(t)$ and the output $y(t)$.

(a) Find the impulse response $h(t)$ of the channel.

(b) Find the transfer function $H(s)$, and its poles and zeros. Determine if the system is BIBO stable.

Answer: $h(t) = \alpha \delta(t - T) + \alpha^3 \delta(t - 3T)$; $H(s)$ has no poles, system is BIBO stable.

3.14 Consider the cascade of two LTI systems, see Fig. 3.21, where the input of the cascade is $z(t)$ and the output is $y(t)$, while $x(t)$ is the output of the first system and the input of the second system. The input to the cascaded systems is $z(t) = (1 - t)[u(t) - u(t - 1)]$.

(a) The input/output characterization of the first system is $x(t) = dz(t)/dt$. Find the corresponding output $x(t)$ of the first system.

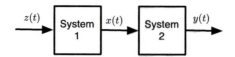

FIGURE 3.21

Problem 3.14.

(b) It is well known that for the second system when the input is $\delta(t)$ the corresponding output is $e^{-2t}u(t)$, and that when the input is $u(t)$ the output is $0.5(1 - e^{-2t})u(t)$. Use this information to calculate the output $y(t)$ of the second system when the input is $x(t)$ found above.

(c) Determine the ordinary differential equation corresponding to the cascaded system with input $z(t)$ and output $y(t)$.

Answer: $x(t) = \delta(t) - (u(t) - u(t-1))$; $dy(t)/dt + 2y(t) = dz(t)/dt$.

3.15 Consider the following cases where we want to determine different types of responses.

(a) The input to a LTI system is $x(t) = u(t) - 2u(t-1) + u(t-2)$ and the Laplace transform of the output is given by

$$Y(s) = \frac{(s+2)(1-e^{-s})^2}{s(s+1)^2},$$

determine the impulse response of the system.

(b) Without computing the inverse of the Laplace transform

$$X(s) = \frac{1}{s(s^2 + 2s + 10)},$$

corresponding to a causal signal $x(t)$, determine $\lim_{t \to \infty} x(t)$.

(c) The Laplace transform of the output of a LTI system is

$$Z(s) = \frac{1}{s((s+2)^2 + 1)},$$

what would be the steady-state response $z_{ss}(t)$?

(d) The Laplace transform of the output of a LTI system is

$$W(s) = \frac{e^{-s}}{s((s-2)^2 + 1)},$$

how would you determine if there is a steady state or not? Explain.

(e) The Laplace transform of the output of a LTI system is

$$V(s) = \frac{s+1}{s((s+1)^2 + 1)}.$$

Determine the steady state and the transient responses corresponding to $Y(s)$.

Answers: (a) $H(s) = (s + 2)/(s + 1)^2$, ROC: $\sigma > 0$; (b) $\lim_{t \to \infty} x(t) = 0.1$; (e) $v_t(t) = -0.5e^{-t}\cos(t)u(t) + 0.5e^{-t}\sin(t)u(t)$.

3.16 In convolution problems the impulse response $h(t)$ of the system and the input $x(t)$ are given and one is interested in finding the output of the system $y(t)$. The so-called "deconvolution" problem consists in giving two of $x(t)$, $h(t)$ and $y(t)$ to find the other. For instance, the output $y(t)$ and the impulse response $h(t)$ of the system and one wants to find the input. Consider the following cases:

(a) Suppose the impulse response of the system is $h(t) = e^{-t}\cos(t)u(t)$ and the output has a Laplace transform

$$Y(s) = \frac{4}{s((s+1)^2 + 1)},$$

what is the input $x(t)$?

(b) The output of a LTI system is $y_1(t) = r(t) - 2r(t - 1) + r(t - 2)$, where $r(t)$ is the ramp signal. Determine the impulse response $h_1(t)$ of the system if it is known that the input is $x_1(t) = u(t) - u(t - 1)$.

Answers: $x(t) = 4(1 - e^{-t})u(t)$; $h_1(t) = u(t) - u(t - 1)$.

3.17 The transfer function of a BIBO stable and causal system has poles only on the open left-hand s-plane (excluding the $j\Omega$ axis).

(a) Let the transfer function of a system be

$$H_1(s) = \frac{Y(s)}{X(s)} = \frac{1}{(s + 1)(s - 2)}$$

and let the poles of $X(s)$ be on the left-hand s-plane. Find $\lim_{t \to \infty} y(t)$. Determine if the system is BIBO stable. If not stable, determine what makes the system unstable.

(b) Let the transfer function be

$$H_2(s) = \frac{Y(s)}{X(s)} = \frac{1}{(s + 1)(s + 2)}$$

and $X(s)$ be as indicated above. Find $\lim_{t \to \infty} y(t)$. Can you use this limit to determine if the system is BIBO stable? If not, what would you do to check stability?

Answer: (a) Assuming no pole/zero cancellation $\lim_{t \to \infty} y(t) \to \infty$.

3.18 The steady-state solution of stable systems is due to simple poles in the $j\Omega$ axis of the s-plane coming from the input. Suppose the transfer function of the system is

$$H(s) = \frac{Y(s)}{X(s)} = \frac{1}{(s + 1)^2 + 4}.$$

(a) Find the poles and zeros of $H(s)$ and plot them in the s-plane. Find then the corresponding impulse response $h(t)$. Determine if the impulse response of this system is absolutely integrable so that the system is BIBO stable.

(b) Let the input $x(t) = u(t)$ and the initial conditions be zero, find $y(t)$ and from it determine the steady-state solution.

(c) Let the input $x(t) = tu(t)$ and the initial conditions be zero, find $y(t)$ and from it determine the steady-state response. What is the difference between this case and the previous one?

(d) To explain the behavior in the case above consider the following: Is the input $x(t) = tu(t)$ bounded? That is, is there some finite value M such that $|x(t)| < M$ for all times? So what would you expect the output to be, given that the system is stable?

Answers: System is BIBO stable; $y_{ss}(t) = 0.2$.

3.19 The input/output equation for an analog averager is given by the convolution integral

$$y(t) = \frac{1}{T} \int_{t-T}^{t} x(\tau)d\tau$$

where $x(t)$ is the input and $y(t)$ the output.

(a) Change the above equation to determine the impulse response $h(t)$.

(b) Graphically determine the output $y(t)$ corresponding to a pulse input $x(t) = u(t) - u(t-2)$ using the convolution integral (let $T = 1$) relating the input and the output. Carefully plot the input and the output. (The output can also be obtained intuitively from a good understanding of the averager.)

(c) Using the impulse response $h(t)$ found above, use now the Laplace transform to find the output corresponding to $x(t) = u(t) - u(t-2)$, let again $T = 1$ in the averager.

Answers: $h(t) = (1/T)[u(t) - u(t-T)]$; $y(t) = r(t) - r(t-1) - r(t-2) + r(t-3)$.

3.20 There are two types of feedback, negative and positive. In this problem we explore their difference.

(a) Consider negative feedback. Suppose you have a system with transfer function $H(s) = Y(s)/E(s)$ where $E(s) = C(s) - Y(s)$, and $C(s)$ and $Y(s)$ are the transforms of the feedback system's reference $c(t)$ and output $y(t)$. Find the transfer function of the overall system $G(s) = Y(s)/C(s)$.

(b) In positive feedback, the only equation that changes is $E(s) = C(s) + Y(s)$, the other equations remain the same. Find the overall feedback system transfer function $G(s) = Y(s)/C(s)$.

(c) Suppose that $C(s) = 1/s$, $H(s) = 1/(s+1)$ determine $G(s)$ for both negative and positive feedback. Find $y(t) = \mathcal{L}^{-1}[Y(s)]$ for both types of feedback and comment on the difference in these signals.

Answers: Positive feedback: $G(s) = H(s)/(1 - H(s))$.

3.21 The following problems consider approaches to stabilize an unstable system.

(a) An unstable system can be stabilized by using negative feedback with a gain K in the feedback loop. For instance consider an unstable system with transfer function

$$H(s) = \frac{2}{s-1},$$

which has a pole in the right-hand s-plane making the impulse response of the system, $h(t)$ grow at t increases. Use negative feedback with a gain $K > 0$ in the feedback loop, and put $H(s)$ in the forward loop. Draw a block diagram of the system. Obtain the transfer function $G(s)$ of the feedback system and determine the value of K that makes the overall system BIBO stable, i.e., its poles in the open left-hand s-plane.

(b) Another stabilization method consists in cascading an all-pass system with the unstable system to cancel the poles in the right-hand s-plane. Consider a system with a transfer function

$$H(s) = \frac{s+1}{(s-1)(s^2+2s+1)},$$

which has a pole in the right-hand s-plane, $s = 1$, so it is unstable.

(b)   The poles and zeros of an all-pass filter are such that if $p_{12} = -\sigma \pm j\Omega_0$ are complex conjugate poles of the filter then $z_{12} = \sigma \pm j\Omega_0$ are the corresponding zeros, and for real poles $p_0 = -\sigma$ there is a corresponding zero $z_0 = \sigma$. The orders of the numerator and the denominator of the all-pass filter are equal. Write the general transfer function of an all-pass filter $H_{ap}(s) = KN(s)/D(s)$.

(b)   Find an all-pass filter $H_{ap}(s)$ so that when cascaded with the given $H(s)$ the overall transfer function $G(s) = H(s)H_{ap}(s)$ has all its poles in the left-hand s-plane.

(b)   Find K of the all-pass filter so that when $s = 0$ the all-pass filter has a gain of unity. What is the relation between the magnitude of the overall system $|G(s)|$ and that of the unstable filter $|H(s)|$.

Answers: (a) $K > 0.5$; (b) $G(s) = -1/(s+1)^2$.

3.10.2 PROBLEM USING MATLAB

3.22 Inverse Laplace transform—Consider the following inverse Laplace problems:

(a)   Use MATLAB to compute the inverse Laplace transform of

$$X(s) = \frac{s^2+2s+1}{s(s+1)(s^2+10s+50)}.$$

Determine the value of $x(t)$ in the steady state. How would you be able to obtain this value without computing the inverse? Explain.

(b)   Find the inverse $x(t)$ of

$$X(s) = \frac{(1-se^{-s})}{s(s+2)}.$$

Use MATLAB to plot $x(t)$ and to verify your inverse.

Answer: (b) $x(t) = 0.5u(t) - 0.5e^{-2t}u(t) - e^{-2(t-1)}u(t-1)$.

3.23 Effect of poles and zeros—The poles corresponding to the Laplace transform $X(s)$ of a signal $x(t)$ are $p_{1,2} = -3 \pm j\pi/2$ and $p_3 = 0$.

(a)   Within some constants, give a general form of the signal $x(t)$.

(b)   Let

$$X(s) = \frac{1}{(s+3-j\pi/2)(s+3+j\pi/2)s}$$

from the location of the poles, obtain a general form for $x(t)$. Use MATLAB to find $x(t)$ and to plot it. How well did you guess the answer?

Answer: $x(t) = Ae^{-3t}\cos((\pi/2)t + \theta)u(t) + Bu(t)$.

3.24 **Differential equation, initial conditions and stability**—The following function $Y(s) = \mathcal{L}[y(t)]$ is obtained applying the Laplace transform to a differential equation representing a system with nonzero initial conditions and input $x(t)$, with Laplace transform $X(s)$

$$Y(s) = \frac{X(s)}{s^2 + 2s + 3} + \frac{s + 1}{s^2 + 2s + 3}.$$

(a) Find the differential equation in $y(t)$ and $x(t)$ representing the system.
(b) Find the initial conditions $y'(0)$ and $y(0)$.
(c) Use MATLAB to determine the impulse response $h(t)$ of this system and to plot it. Find the poles of the transfer function $H(s)$ and determine if the system is BIBO stable.
Answer: $d^2y(t)/dt^2 + 2dy(t)/dt + 3y(t) = x(t)$.

3.25 **Different responses**—Let $Y(s) = \mathcal{L}[y(t)]$ be the Laplace transform of the solution of a second-order differential equation representing a system with input $x(t)$ and some initial conditions,

$$Y(s) = \frac{X(s)}{s^2 + 2s + 1} + \frac{s + 1}{s^2 + 2s + 1}.$$

(a) Find the zero-state response (response due to the input only with zero initial conditions) for $x(t) = u(t)$.
(b) Find the zero-input response (response due to the initial conditions and zero input).
(c) Find the complete response when $x(t) = u(t)$.
(d) Find the transient and the steady-state response when $x(t) = u(t)$.
(e) Use MATLAB to verify the above responses.
Answer: For zero-state, $x(t) = u(t)$, $y_{zs}(t) = [1 - te^{-t} - e^{-t}]u(t)$.

3.26 **Transients for second-order systems**—The type of transient you get in a second-order system depends on the location of the poles of the system. The transfer function of the second-order system is

$$H(s) = \frac{Y(s)}{X(s)} = \frac{1}{s^2 + b_1 s + b_0}$$

and let the input be $x(t) = u(t)$.
(a) Let the coefficients of the denominator of $H(s)$ be $b_1 = 5$, $b_0 = 6$, find the response $y(t)$. Use MATLAB to verify the response and to plot it.
(b) Suppose then that the denominator coefficients of $H(s)$ are changed to $b_1 = 2$ and $b_0 = 6$, find the response $y(t)$. Use MATLAB to verify the response and to plot it.
(c) Explain your results above by relating your responses to the location of the poles of $H(s)$.
Answer: (a) $y(t) = [(1/6) - (1/2)e^{-2t} + (1/3)e^{-3t}]u(t)$.

3.27 Partial fraction expansion—Consider the following functions $Y_i(s) = \mathcal{L}[y_i(t)]$, $i = 1, 2$ and 3:

$$Y_1(s) = \frac{s+1}{s(s^2 + 2s + 4)}, \quad Y_2(s) = \frac{1}{(s+2)^2}, \quad Y_3(s) = \frac{s-1}{s^2((s+1)^2 + 9)}$$

where $\{y_i(t), i = 1, 2, 3\}$ are the complete responses of differential equations with zero initial conditions.

(a) For each of these functions determine the corresponding differential equation, if all of them have as input $x(t) = u(t)$.

(b) Find the general form of the complete response $\{y_i(t), i = 1, 2, 3\}$ for each of the $\{Y_i(s), i = 1, 2, 3\}$. Use MATLAB to plot the poles and zeros for each of the $\{Y_i(s)\}$, to find their partial fraction expansions and the complete responses.

Answer: $d^2y_1(t)/dt^2 + 2dy_1(t)/dt + 4y_1(t) = x(t) + dx(t)/dt$.

3.28 Iterative convolution integral—Consider the convolution of a pulse $x(t) = u(t + 0.5) - u(t - 0.5)$ with itself many times. Use MATLAB for the calculations and the plotting.

(a) Consider the result for $N = 2$ of these convolutions, i.e., $y_2(t) = (x * x)(t)$. Find $Y_2(s) = \mathcal{L}[y_2(t)]$ using the convolution property of the Laplace transform and find $y_2(t)$.

(b) Consider the result for $N = 3$ of these convolutions, i.e., $y_3(t) = (x * x * x)(t)$. Find $Y_3(s) = \mathcal{L}[y_3(t)]$ using the convolution property of the Laplace transform and find $y_3(t)$.

(c) The signal $x(t)$ can be considered the impulse response of an averager which "smooths" out a signal. Letting $y_1(t) = x(t)$, plot the three functions $y_i(t)$ for $i = 1, 2$ and 3. Compare these signals on their smoothness and indicate their supports in time (for $y_2(t)$ and $y_3(t)$ how do their supports relate to the supports of the signals convolved?).

Answers: $y_3(t) = \psi(t + 1.5) - 3\psi(t + 0.5) + 3\psi(t - 0.5) - \psi(t - 1.5)$, $\psi(t) = t^3 u(t)/6$.

3.29 Half-wave rectifier—In the generation of dc from ac voltage, the "half-wave" rectified signal is an important part. Suppose the ac voltage is $x(t) = \sin(2\pi t)u(t)$ and $y(t)$ is the half-wave rectified signal.

(a) Let $y_1(t)$ be the period of $y(t)$ between $0 \le t \le 1$. Show that $y_1(t)$ can be written equivalently as either

$$y_1(t) = \sin(2\pi t)u(t) + \sin(2\pi(t - 0.5))u(t - 0.5) \quad \text{or}$$
$$y_1(t) = \sin(2\pi t)[u(t) - u(t - 0.5)].$$

Use MATLAB to verify this. Find the Laplace transform $X_1(s)$ of $x_1(t)$.

(b) Express $y(t)$ in terms of $x_1(t)$ and find the Laplace transform $Y(s)$ of $y(t)$.

Answers: (a) Use $\sin(2\pi(t - 0.5)) = \sin(2\pi t - \pi) = -\sin(2\pi t)$; $Y(s) = Y_1(s)/(1 - e^{-s})$.

3.30 Feedback error—Consider a negative feedback system used to control a plant with transfer function

$$G(s) = 1/(s(s+1)(s+2)).$$

The output $y(t)$ of the feedback system is connected via a sensor with transfer function $H(s) = 1$ to a differentiator where the reference signal $x(t)$ is also connected. The output of the differentiator is the feedback error $e(t) = x(t) - v(t)$ where $v(t)$ is the output of the feedback sensor.

(a) Carefully draw the feedback system, and find an expression for $E(s)$, the Laplace transform of the feedback error $e(t)$.

(b) Two possible reference test signals for the given plant are $x(t) = u(t)$ and $x(t) = r(t)$, choose the one that would give a zero steady-state feedback error.

(c) Use numeric MATLAB do the partial fraction expansions for the two error functions $E_1(s)$, corresponding to when $x(t) = u(t)$ and $E_2(s)$ when $x(t) = r(t)$. Use these partial fraction expansions to find $e_1(t)$ and $e_2(t)$, and thus verify your results obtained before.

Answer: $E(s) = X(s)/(1 + H(s)G(s))$.

FREQUENCY ANALYSIS: THE FOURIER SERIES

4

CONTENTS

A Mathematician is a device for turning coffee into theorems.

Paul Erdos (1913–1996), Mathematician

4.1 INTRODUCTION

In this chapter and the next, we consider the frequency analysis of continuous-time signals and systems—the Fourier series for periodic signals in this chapter, and the Fourier transform for both periodic and aperiodic signals as well as for systems in the following chapter.

Signals and Systems Using MATLAB®. https://doi.org/10.1016/B978-0-12-814204-2.00014-4

The frequency representation of periodic and aperiodic signals indicates how their power or energy are distributed to different frequency components. Such a distribution over frequency is called the **spectrum of the signal**. For a periodic signal the spectrum is discrete, as its power is concentrated at frequencies multiples of the **fundamental frequency**, directly related to the fundamental period of the signal. On the other hand, the spectrum of an aperiodic signal is a continuous function of frequency. The concept of spectrum is similar to the one used in optics for light, or in material science for metals, each indicating the distribution of power or energy over frequency. The Fourier representation is also useful in finding the frequency response of linear time-invariant (LTI) systems, which is related to the transfer function obtained with the Laplace transform. The frequency response of a system indicates how a LTI system responds to sinusoids of different frequencies. Such a response characterizes the system, it permits easy computation of its steady-state response, and it will be equally important in the design or synthesis of systems.

It is important to understand the driving force behind the representation of signals in terms of basic signals when applied to LTI systems. The convolution integral which gives the output of a LTI system resulted from the representation of its input signal in terms of shifted impulses along with the concept of the impulse response of a LTI system. Likewise, the Laplace transform can be seen as the representation of signals in terms of complex exponentials or general eigenfunctions. In this chapter and the next, we will see that complex exponentials or sinusoids are used in the Fourier representation of periodic as well as aperiodic signals to take advantage of the eigenfunction property of LTI systems. The results of the Fourier series in this chapter will be extended to the Fourier transform in the next chapter.

Fourier analysis is in the steady state, while Laplace analysis considers both transient and steady state. Thus if one is interested in transients, as in control theory, Laplace is a meaningful transformation. On the other hand, if one is interested in the frequency analysis, or steady state, as in communications theory, the Fourier transform is the one to use. There will be cases, however, where in control and communications both Laplace and Fourier analysis are considered.

The frequency representation of signals and systems is extremely important in signal processing and communications. It explains filtering, modulation of messages in a communication system, the meaning of bandwidth, and how to design filters among other concepts. Likewise, the frequency representation turns out to be essential in the sampling of analog signals—the bridge between analog and digital signal processing.

4.2 EIGENFUNCTIONS REVISITED

The most important property of stable LTI systems is that when the system input is a complex exponential (or a combination of a cosine and a sine) of a certain frequency, the output of the system is the input times a complex constant connected with how the system responds to the frequency of the input.

Eigenfunction Property of LTI Systems

If $x(t) = e^{j\Omega_0 t}$, $-\infty < t < \infty$, for some frequency Ω_0, is the input to a causal and stable LTI system with impulse response $h(t)$, the output in the steady state is given by

$$y(t) = e^{j\Omega_0 t} H(j\Omega_0),$$ (4.1)

where

$$H(j\Omega_0) = \int_0^\infty h(\tau)e^{-j\Omega_0 \tau} d\tau = H(s)|_{s=j\Omega_0}$$ (4.2)

is the **frequency response of the system at** Ω_0 (or the system transfer function $H(s)$ at $s = j\Omega_0$). The signal $x(t) = e^{j\Omega_0 t}$ is said to be an **eigenfunction** of the LTI system as it appears at both input and output.

This can be seen by finding the output corresponding to $x(t) = e^{j\Omega_0 t}$ by means of the convolution integral:

$$y(t) = \int_0^\infty h(\tau)x(t - \tau)d\tau = e^{j\Omega_0 t} \underbrace{\int_0^\infty h(\tau)e^{-j\Omega_0 \tau} d\tau}_{H(j\Omega_0)}.$$

The input signal appears in the output modified by the frequency response of the system $H(j\Omega_0)$ at the frequency Ω_0 of the input. Notice that the convolution integral limits indicate that the input started at $-\infty$ (when $\tau = \infty$) and that we are considering the output due to inputs from $-\infty$ to a finite time t (when $\tau = 0$ in the integral) meaning that we are in a steady state.

The above result for one frequency can be easily extended to the case of several frequencies present at the input. If the input signal $x(t)$ is a linear combination of complex exponentials, with different amplitudes, frequencies and phases, or

$$x(t) = \sum_k X_k e^{j\Omega_k t}$$

where X_k are complex values, then since the output corresponding to $X_k e^{j\Omega_k t}$ is $X_k e^{j\Omega_k t} H(j\Omega_k)$ by superposition the response to $x(t)$ is

$$y(t) = \sum_k X_k e^{j\Omega_k t} H(j\Omega_k) = \sum_k X_k |H(j\Omega_k)| e^{j(\Omega_k t + \angle H(j\Omega_k))}.$$ (4.3)

The above is valid for any signal that is a combination of exponentials of arbitrary frequencies. As we will see in this chapter, when $x(t)$ is periodic it can be represented by the Fourier series which is a combination of complex exponentials harmonically related (i.e., the frequencies of the exponentials are multiples of the fundamental frequency of the periodic signal). Thus when a periodic signal is applied to a causal and stable LTI system its output is computed as in (4.3).

The significance of the eigenfunction property is also seen when the input signal is an integral (a sum, after all) of complex exponentials, with continuously varying frequency as the integrand. That is, if

$$x(t) = \int_{-\infty}^{\infty} X(\Omega) e^{j\Omega t} d\Omega$$

then using superposition and the eigenfunction property of a stable LTI system, with frequency response $H(j\Omega)$, the output is

$$
\begin{aligned}
y(t) &= \int_{-\infty}^{\infty} X(\Omega) e^{j\Omega t} H(j\Omega) d\Omega \\
&= \int_{-\infty}^{\infty} X(\Omega) |H(j\Omega)| e^{(j\Omega t + \angle H(j\Omega))} d\Omega .
\end{aligned}
\tag{4.4}
$$

The above representation of $x(t)$ relates to the Fourier representation of aperiodic signals, which will be covered in the next chapter. Here again the eigenfunction property of LTI systems provides an efficient way to compute the output. Furthermore, we also find that by letting $Y(\Omega) = X(\Omega) H(j\Omega)$ the above equation gives an expression to compute $y(t)$ from $Y(\Omega)$. The product $Y(\Omega) = X(\Omega) H(j\Omega)$ corresponds to the Fourier transform of the convolution integral $y(t) = [x * h](t)$, and is connected with the convolution property of the Laplace transform. It is important to start noticing these connections, to understand the link between the Laplace and the Fourier analyses.

Remarks

1. Notice the difference of notation for the frequency representation of signals and systems used above. If $x(t)$ is a periodic signal its frequency representation is given by $\{X_k\}$, and if aperiodic by $X(\Omega)$, while for a system with impulse response $h(t)$ its frequency response is given by $H(j\Omega)$ (although the j is not needed, it reminds us of the connection of $H(j\Omega)$ with a system rather than with a signal).
2. When considering the eigenfunction property, the stability of the LTI system is necessary to ensure that $H(j\Omega)$ exists for all frequencies.
3. The eigenfunction property applied to a linear circuit gives the same result as the one obtained from phasors in the sinusoidal steady state. That is, if

$$x(t) = A\cos(\Omega_0 t + \theta) = \frac{Ae^{j\theta}}{2} e^{j\Omega_0 t} + \frac{Ae^{-j\theta}}{2} e^{-j\Omega_0 t} \tag{4.5}$$

is the input of a circuit represented by the transfer function

$$H(s) = \frac{Y(s)}{X(s)} = \frac{\mathcal{L}[y(t)]}{\mathcal{L}[x(t)]},$$

then the corresponding steady-state output is given by

$$
\begin{aligned}
y_{ss}(t) &= \frac{Ae^{j\theta}}{2}e^{j\Omega_0 t}H(j\Omega_0) + \frac{Ae^{-j\theta}}{2}e^{-j\Omega_0 t}H(-j\Omega_0) \\
&= A|H(j\Omega_0)|\cos(\Omega_0 t + \theta + \angle H(j\Omega_0)) \qquad (4.6)
\end{aligned}
$$

where, very importantly, the frequency of the output coincides with that of the input, and the amplitude and phase of the input are changed by the magnitude and phase of the frequency response of the system at the frequency Ω_0. The frequency response is $H(j\Omega_0) = H(s)|_{s=j\Omega_0}$, and as we will see its magnitude is an even function of frequency, or $|H(j\Omega)| = |H(-j\Omega)|$, and its phase is an odd function of frequency or $\angle H(j\Omega_0) = -\angle H(-j\Omega_0)$. Using these two results we obtain Equation (4.6).

The phasor corresponding to the input

$$
x(t) = A\cos(\Omega_0 t + \theta)
$$

is defined as a vector

$$
X = A\angle\theta
$$

rotating in the polar plane at the frequency of Ω_0 rad/s. The phasor has a magnitude A and an angle of θ with respect to the positive real axis. The projection of the phasor onto the real axis, as it rotates at the given frequency, with time generates a cosine of the indicated frequency, amplitude and phase. The transfer function computed at $s = j\Omega_0$ or

$$
H(s)|_{s=j\Omega_0} = H(j\Omega_0) = \frac{Y}{X}
$$

(ratio of the phasors corresponding to the output, Y, and the input, X). The phasor for the output is thus

$$
Y = H(j\Omega_0)X = |Y|e^{j\angle Y}.
$$

Such a phasor is then converted into the sinusoid:

$$
\begin{aligned}
y_{ss}(t) &= Re\left[Ye^{j\Omega_0 t} = H(j\Omega_0)X\right] \\
&= A|H(j\Omega_0)|\cos(\Omega_0 t + \theta + \angle H(j\Omega_0)),
\end{aligned}
$$

which equals (4.6).

4. A very important application of LTI systems is **filtering**, where one is interested in preserving desired frequency components of a signal and getting rid of less desirable components. That a LTI system can be used for filtering is seen in Equations (4.3) and (4.4). In the case of a periodic signal, the magnitude $|H(j\Omega_k)|$ can be made ideally unity for those components we wish to keep and zero for those we wish to get rid of. Likewise, for an aperiodic signal, the magnitude $|H(j\Omega)|$ could be set ideally to one for those components we wish to keep, and zero for those components we wish to get rid of. Depending on the filtering application, a LTI system with the appropriate characteristics can be designed, obtaining the desired transfer function $H(s)$.

Phasor interpretation of the eigenfunction property of LTI systems
For a stable LTI system with transfer function $H(s)$ if the input is

$$x(t) = A\cos(\Omega_0 t + \theta) = \mathcal{Re}[Xe^{j\Omega_0 t}] \tag{4.7}$$

where $X = Ae^{j\theta}$ is the phasor corresponding to $x(t)$, the steady-state output of the system is

$$
\begin{aligned}
y(t) &= \mathcal{Re}[XH(j\Omega_0)e^{j\Omega_0 t}] = \mathcal{Re}[AH(j\Omega_0)e^{j(\Omega_0 t + \theta)}] \\
&= A|H(j\Omega_0)|\cos(\Omega_0 t + \theta + \angle H(j\Omega_0))
\end{aligned} \tag{4.8}
$$

where the frequency response of the system at Ω_0 is

$$H(j\Omega_0) = H(s)|_{s=j\Omega_0} = \frac{Y}{X} \tag{4.9}$$

and Y is the phasor corresponding to $y(t)$ in steady state.

Example 4.1. Consider the RC circuit shown in Fig. 4.1. Let the voltage source be $v_s(t) = 4\cos(t + \pi/4)$, the resistor be $R = 1\ \Omega$ and the capacitor $C = 1$ F. Find the steady-state voltage across the capacitor.

Solution: This problem can be approached in two equivalent ways.
Phasor approach: From the phasor circuit in Fig. 4.1, by voltage division and the impedances of the resistor and the capacitor we have the following phasor ratio, where V_s is the phasor corresponding to the source $v_s(t)$ and V_c the phasor corresponding to $v_c(t)$:

$$\frac{V_c}{V_s} = \frac{-j}{1-j} = \frac{-j(1+j)}{2} = \frac{\sqrt{2}}{2}\angle -\pi/4$$

and since $V_s = 4\angle\pi/4$

$$V_c = 2\sqrt{2}\angle 0,$$

so that in the steady state

$$v_c(t) = 2\sqrt{2}\cos(t).$$

Eigenfunction approach: Considering the voltage across the capacitor the output and the voltage source the input, the transfer function is obtained using voltage division and the impedances, as functions of s, of the resistor and the capacitor:

$$H(s) = \frac{V_c(s)}{V_s(s)} = \frac{1/s}{1 + 1/s} = \frac{1}{s+1}$$

so that the system frequency response at the frequency of the input, $\Omega_0 = 1$, is

$$H(j1) = \frac{\sqrt{2}}{2}\angle -\pi/4.$$

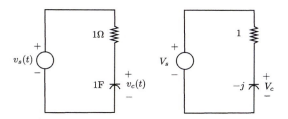

FIGURE 4.1

RC circuit and corresponding phasor circuit.

According to the eigenfunction property, the steady-state response of the capacitor is

$$v_c(t) = 4|H(j1)|\cos(t + \pi/4 + \angle H(j1)) = 2\sqrt{2}\cos(t)$$

which coincides with the solution found using phasors.   □

Example 4.2. An **ideal communication system** provides as output the input signal with only a possible delay in the transmission. Such an ideal system does not cause any distortion to the input signal beyond the delay. Find the frequency response of the ideal communication system, and use it to determine the steady-state response when the delay caused by the system is $\tau = 3$ s, and the input is $x(t) = 2\cos(4t - \pi/4)$.

Solution: Considering the ideal communication system LTI, its impulse response is $h(t) = \delta(t - \tau)$, where τ is the delay of the transmission. In fact, the output of the system according to the convolution integral is

$$y(t) = \int_0^\infty \underbrace{\delta(\rho - \tau)}_{h(\rho)} x(t - \rho)d\rho = x(t - \tau)$$

as expected. Let us then find the frequency response of the ideal communication system. According to the eigenvalue property, if the input $x(t) = e^{j\Omega_0 t}$ then the output is

$$y(t) = e^{j\Omega_0 t} H(j\Omega_0)$$

but also

$$y(t) = x(t - \tau) = e^{j\Omega_0(t - \tau)},$$

so that comparing these equations we have

$$H(j\Omega_0) = e^{-j\tau\Omega_0}.$$

For a generic frequency $0 \le \Omega < \infty$ we would get

$$H(j\Omega) = e^{-j\tau\Omega},$$

which is a complex function of Ω, with a unity magnitude $|H(j\Omega)| = 1$, and a linear phase $\angle H(j\Omega) = -\tau\Omega$. This system is called an **all-pass system**, since it allows all frequency components of the input to go through with a phase change only.

Consider the case when $\tau = 3$, and we input into this system $x(t) = 2\cos(4t - \pi/4)$, then $H(j\Omega) = e^{-j3\Omega}$, so that the output in the steady state is

$$y(t) = 2|H(j4)|\cos(4t - \pi/4 + \angle H(j4)) = 2\cos(4(t-3) - \pi/4) = x(t-3),$$

where we used $H(j4) = 1e^{-j12}$, i.e., $|H(j4)| = 1$ and $\angle H(j4) = 12$.   □

Example 4.3. Although there are better methods to compute the frequency response of a system represented by a differential equation, the eigenfunction property can be easily used for that. Consider the RC circuit shown in Fig. 4.1 where the input is

$$v_s(t) = 1 + \cos(10{,}000t)$$

with components of low frequency, $\Omega = 0$, and large frequency, $\Omega = 10{,}000$ rad/s. The output $v_c(t)$ is the voltage across the capacitor in the steady state. We wish to find the frequency response of this circuit to verify that it is a **low-pass filter** (it allows low-frequency components to go through, but filters out high-frequency components).

Solution: Using Kirchhoff's voltage law, this circuit is represented by a first-order differential equation

$$v_s(t) = v_c(t) + \frac{dv_c(t)}{dt}.$$

Now, if the input is $v_s(t) = e^{j\Omega t}$, for a generic frequency Ω, then the output is $v_c(t) = e^{j\Omega t} H(j\Omega)$, and replacing these in the differential equation we have

$$
\begin{aligned}
e^{j\Omega t} &= e^{j\Omega t} H(j\Omega) + \frac{de^{j\Omega t} H(j\Omega)}{dt} \\
&= e^{j\Omega t} H(j\Omega) + j\Omega e^{j\Omega t} H(j\Omega) \quad \text{so that} \\
H(j\Omega) &= \frac{1}{1 + j\Omega}
\end{aligned}
$$

or the frequency response of the filter for any frequency Ω. The magnitude of $H(j\Omega)$ is

$$|H(j\Omega)| = \frac{1}{\sqrt{1 + \Omega^2}}$$

which is close to 1 for small values of the frequency, and tends to zero when the frequency values are large—the characteristics of a low-pass filter. For the input

$$v_s(t) = 1 + \cos(10{,}000t) = \cos(0t) + \cos(10{,}000t)$$

(i.e., it has a zero frequency component and a 10,000 (rad/s) frequency component) the steady-state output of the circuit is

$$v_c(t) \approx 1 + \frac{1}{10{,}000}\cos(10{,}000t - \pi/2) \approx 1$$

since

$$H(j0) \quad = \quad 1,$$
$$H(j10{,}000) \quad \approx \quad \frac{1}{j\ 10^4} = \frac{-j}{10{,}000} = \frac{1}{10{,}000}\angle -\pi/2.$$

Thus this circuit acts like a low-pass filter by keeping the dc component (with the low frequency $\Omega = 0$) and essentially getting rid of the high frequency ($\Omega = 10{,}000$) component of the signal.

Notice that the frequency response can also be obtained by considering the phasor ratio for a generic frequency Ω, which by voltage division is

$$\frac{V_c}{V_s} = \frac{1/j\Omega}{1 + 1/j\Omega} = \frac{1}{1 + j\Omega},$$

which for $\Omega = 0$ is 1 and for $\Omega = 10{,}000$ is approximately $-j/10{,}000$, i.e., corresponding to $H(j0)$ and $H(j10{,}000)$. □

FOURIER AND LAPLACE

Jean-Baptiste-Joseph Fourier, French mathematician (1768–1830), was a contemporary of Laplace with whom he shared many scientific and political experiences [5,9]. Like Laplace, Fourier was from very humble origins but he was not as politically astute. Laplace and Fourier were affected by the political turmoil of the French Revolution and both came in close contact with Napoleon Bonaparte, French general and emperor. Named chair of the Mathematics Department of the Ecole Normale, Fourier led the most brilliant period of mathematics and science education in France in his time. His main work was "The Mathematical Theory of Heat Conduction" where he proposed the harmonic analysis of periodic signals. In 1807 he received the Grand Prize from the French Academy of Sciences for this work. This was despite the objections of Laplace, Lagrange and Legendre, who were the referees and who indicated that the mathematical treatment lacked rigor. Following the advice of "Never resent criticism, and never answer it," Fourier disregarded these criticisms and made no change to his treatise in heat-conduction of 1822. Although Fourier was an enthusiast for the revolution and followed Napoleon on some of his campaigns, in the Second Restoration he had to pawn his belongings to survive. Thanks to his friends, he became Secretary of the French Academy, the final position he held.

4.3 COMPLEX EXPONENTIAL FOURIER SERIES

The Fourier series is a representation of a periodic signal $x(t)$ in terms of complex exponentials or sinusoids of frequency multiples of the fundamental frequency of $x(t)$. The advantage of using the Fourier series to represent periodic signals is not only in obtaining their spectral characterization, but in finding the response of these signals when applied to LTI systems by means of the eigenfunction property.

Mathematically, the Fourier series is an expansion of periodic signals in terms of normalized orthogonal complex exponentials. The concept of orthogonality of functions is similar to the concept of perpendicularity of vectors: perpendicular vectors cannot be represented in terms of each other—orthogonal functions provide mutually exclusive information. The perpendicularity of two vectors can be established using the *dot or scalar* product of the vectors, the orthogonality of functions is established by the **inner product**, or the integration of the product of the function and its complex conjugate over the support where they are orthogonal. Consider a set of complex functions $\{\psi_k(t)\}$ defined in an interval $[a, b]$, and such that for any pair of these functions, let us say $\psi_\ell(t)$ and $\psi_m(t)$, their inner product is

$$\int_a^b \psi_\ell(t)\psi_m^*(t)dt = \begin{cases} 0 & \ell \neq m \\ 1 & \ell = m, \end{cases} \qquad (4.10)$$

where $*$ stands for complex conjugate. Such a set of functions is called **orthonormal** (i.e., orthogonal and normalized) in the interval $[a, b]$.

Using the functions $\{\psi_k(t)\}$, a finite energy signal $x(t)$ defined in $[a, b]$ can be approximated by a series

$$\hat{x}(t) = \sum_k \alpha_k \psi_k(t) \qquad (4.11)$$

according to a quadratic error criterion, i.e., we minimize the energy of the error function $\varepsilon(t) = x(t) - \hat{x}(t)$ or

$$\int_a^b |\varepsilon(t)|^2 dt = \int_a^b \left| x(t) - \sum_k \alpha_k \psi_k(t) \right|^2 dt \qquad (4.12)$$

by choosing appropriate coefficients $\{\alpha_k\}$. The expansion can be finite or infinite, and may or may not approximate the signal point by point.

Fourier proposed sinusoids as the functions $\{\psi_k(t)\}$ to represent periodic signals, and solved the quadratic minimization posed in (4.12) to obtain the coefficients of the representation. For most signals, the resulting Fourier series has an infinite number of terms and coincides point-wise with the signal. We will start with the general expansion that uses complex exponentials and from it obtain the sinusoidal form.

Recall that a periodic signal $x(t)$ is such that

- it is defined for $-\infty < t < \infty$, i.e., it has an infinite support,
- for any integer k, $x(t + kT_0) = x(t)$, where T_0 is the **fundamental period** of the signal or the smallest positive real number that makes this possible.

The **Fourier Series representation** of a periodic signal $x(t)$, of fundamental period T_0, is given by an infinite sum of weighted complex exponentials (cosines and sines) with frequencies multiples of the **fundamental frequency** $\Omega_0 = 2\pi / T_0$ (rad/s) of the signal:

$$x(t) = \sum_{k=-\infty}^{\infty} X_k e^{jk\Omega_0 t} \qquad (4.13)$$

where the Fourier coefficients $\{X_k\}$ are found according to

$$X_k = \frac{1}{T_0} \int_{t_0}^{t_0+T_0} x(t) e^{-jk\Omega_0 t} dt \qquad (4.14)$$

for $k = 0, \pm 1, \pm 2, \cdots$ and any t_0. The form of the last equation indicates that the information needed for the Fourier series can be obtained from any period of $x(t)$.

Remarks

1. The Fourier series uses the **Fourier basis** $\{e^{jk\Omega_0 t}, \; k \text{ integer}\}$ to represent the periodic signal $x(t)$ of fundamental period T_0. The Fourier functions are periodic of fundamental period T_0, i.e., for an integer m

$$e^{jk\Omega_0(t+mT_0)} = e^{jk\Omega_0 t} \underbrace{e^{jkm2\pi}}_{1} = e^{jk\Omega_0 t},$$

for harmonically related frequencies $\{k\Omega_0\}$.

2. The Fourier functions are **orthonormal** over a period, that is,

$$\frac{1}{T_0} \int_{t_0}^{t_0+T_0} e^{jk\Omega_0 t} [e^{j\ell\Omega_0 t}]^* dt = \begin{cases} 1 & k = \ell, \\ 0 & k \neq \ell, \end{cases} \tag{4.15}$$

that is, $e^{jk\Omega_0 t}$ and $e^{j\ell\Omega_0 t}$ are **orthogonal** when for $k \neq \ell$ the above integral is zero, and they are **normal** (or normalized) when for $k = \ell$ the above integral is unity. The functions $e^{jk\Omega_0 t}$ and $e^{j\ell\Omega_0 t}$ are orthogonal since

$$
\begin{aligned}
\frac{1}{T_0} \int_{t_0}^{t_0+T_0} e^{jk\Omega_0 t} [e^{j\ell\Omega_0 t}]^* dt &= \frac{1}{T_0} \int_{t_0}^{t_0+T_0} e^{j(k-\ell)\Omega_0 t} dt \\
&= \frac{1}{T_0} \int_{t_0}^{t_0+T_0} \left[\cos((k-\ell)\Omega_0 t) + j\sin((k-\ell)\Omega_0 t) \right] dt \\
&= 0 \quad k \neq \ell. \tag{4.16}
\end{aligned}
$$

The above integrals are zero given that the integrands are sinusoids and the limits of the integrals cover one or more periods of the integrand. Now when $k = \ell$ the above integral is

$$\frac{1}{T_0} \int_{t_0}^{t_0+T_0} e^{j0 t} dt = 1.$$

Notice that the orthonormality is not affected by the value t_0.

3. The Fourier coefficients $\{X_k\}$ are easily obtained using the orthonormality of the Fourier functions: first, we multiply the expression for $x(t)$ in Equation (4.13) by $e^{-j\ell\Omega_0 t}$ and then integrate over a period to get

$$\int_{T_0} x(t) e^{-j\ell\Omega_0 t} dt = \sum_k X_k \int_{T_0} e^{j(k-\ell)\Omega_0 t} dt = X_\ell T_0$$

given that when $k = \ell$ the integral in the summation is T_0, and zero otherwise as dictated by the orthonormality of the Fourier exponentials. This gives us then the expression for the Fourier coefficients $\{X_\ell\}$ in Equation (4.14) (you need to realize that the k and the ℓ are dummy variables in the Fourier series, and as such the expression for the coefficients is the same regardless of whether we use ℓ or k).

4. It is important to understand from the given Fourier series equation that for a periodic signal $x(t)$, of fundamental period T_0, any period

$$x(t), \quad t_0 \leq t \leq t_0 + T_0, \quad \text{for any } t_0$$

provides the necessary time-domain information characterizing $x(t)$. In an equivalent way, the coefficients and their corresponding harmonic frequencies $\{X_k, k\Omega_0\}$ provide all the necessary information about $x(t)$ in the frequency domain.

4.3.1 LINE SPECTRUM—POWER DISTRIBUTION OVER FREQUENCY

The Fourier series provides a way to determine the frequency components of a periodic signal and the significance of these frequency components. For a periodic signal its **power spectrum** provides information as to how the power of the signal is distributed over the different frequencies present in the signal. We thus learn not only what frequency components are present in the signal but also the strength of these frequency components. In practice, the power spectrum is computed and displayed using a **spectrum analyzer** which will be described in the next chapter.

Parseval's Power Relation

Although periodic signals are infinite-energy signals, they have finite power. The Fourier series provides a way to find how much of the signal power is in a certain band of frequencies. This can be done using **Parseval's power relation** for periodic signals:

The power P_x of a periodic signal $x(t)$, of fundamental period T_0, can be equivalently calculated in either the time or the frequency domain:

$$P_x = \frac{1}{T_0} \int_{t_0}^{t_0+T_0} |x(t)|^2 dt = \sum_{k=-\infty}^{\infty} |X_k|^2, \qquad \text{for any } t_0. \tag{4.17}$$

The power of a periodic signal $x(t)$ of fundamental period T_0 is given by

$$P_x = \frac{1}{T_0} \int_{t_0}^{t_0+T_0} |x(t)|^2 dt.$$

Replacing the Fourier series of $x(t)$ in the power equation we have

$$\frac{1}{T_0} \int_{t_0}^{t_0+T_0} |x(t)|^2 dt = \frac{1}{T_0} \int_{t_0}^{t_0+T_0} \sum_{k=-\infty}^{\infty} \sum_{m=-\infty}^{\infty} X_k X_m^* e^{j\Omega_0(k-m)t} dt$$

$$= \sum_{k=-\infty}^{\infty} \sum_{m=-\infty}^{\infty} X_k X_m^* \frac{1}{T_0} \int_{t_0}^{t_0+T_0} e^{j\Omega_0(k-m)t} dt = \sum_{k} |X_k|^2$$

after we apply the orthonormality of the Fourier exponentials. Even though $x(t)$ is real, we let $|x(t)|^2 = x(t)x^*(t)$ in the above equations permitting us to express them in terms of X_k and its conjugate. The above indicates that the power of $x(t)$ can be computed in either the time or the frequency domain giving exactly the same result.

Moreover, considering the signal to be the sum of harmonically related components or

$$x(t) = \sum_k \underbrace{X_k e^{jk\Omega_0 T}}_{x_k(t)},$$

the power of each of these components $x_k(t)$ is given by

$$\frac{1}{T_0} \int_{t_0}^{t_0+T_0} |x_k(t)|^2 dt = \frac{1}{T_0} \int_{t_0}^{t_0+T_0} |X_k e^{jk\Omega_0 t}|^2 dt = \frac{1}{T_0} \int_{t_0}^{t_0+T_0} |X_k|^2 dt = |X_k|^2,$$

and thus the power of $x(t)$ is the sum of the powers of the Fourier series components or the **superposition of power** that we referred to in Chapter 1.

A plot of $|X_k|^2$ versus the harmonic frequencies $k\Omega_0$, $k = 0, \pm 1, \pm 2, \cdots$, displays how the power of the signal is distributed over the harmonic frequencies. Given the discrete nature of the harmonic frequencies $\{k\Omega_0\}$ this plot consists of a line at each frequency and as such it is called the **power line spectrum** (that is, a periodic signal has no power at non-harmonic frequencies).

Since the Fourier series coefficients $\{X_k\}$ are complex, we define two additional spectra, one that displays the magnitude $|X_k|$ vs. $k\Omega_0$, the **magnitude line spectrum**, and the **phase line spectrum** $\angle X_k$ vs. $k\Omega_0$ showing the phase of the coefficients $\{X_k\}$ for $k\Omega_0$. The power line spectrum is simply the magnitude spectrum squared.

The above can be summarized as follows.

A periodic signal $x(t)$, of fundamental period T_0, is represented in the frequency by its

Magnitude line spectrum	$|X_k|$ vs. $k\Omega_0$	(4.18)
Phase line spectrum	$\angle X_k$ vs. $k\Omega_0$.	(4.19)

The **power line spectrum**, $|X_k|^2$ vs. $k\Omega_0$ of $x(t)$ displays the distribution of the power of the signal over frequency.

Symmetry of Line Spectra

To understand the spectrum displayed by a spectrum analyzer we need to consider its symmetry.

For a real-valued periodic signal $x(t)$, of fundamental period T_0, represented in the frequency-domain by the Fourier coefficients $\{X_k = |X_k| e^{j\angle X_k}\}$ at harmonic frequencies $\{k\Omega_0 = 2\pi k/T_0\}$, we have

$$X_k = X_{-k}^* \tag{4.20}$$

or equivalently

(i)	$|X_k| = |X_{-k}|$, i.e., magnitude $|X_k|$ is an even function of $k\Omega_0$.	
(ii)	$\angle X_k = -\angle X_{-k}$, i.e., phase $\angle X_k$ is an odd function of $k\Omega_0$.	(4.21)

Thus, for real-valued signals we only need to display for $k \geq 0$ the magnitude line spectrum or a plot of $|X_k|$ vs. $k\Omega_0$, and the phase line spectrum or a plot of $\angle X_k$ vs. $k\Omega_0$ and to remember the even and odd symmetries of these spectra.

For a real signal $x(t)$, the Fourier series of its complex conjugate $x^*(t)$ is

$$x^*(t) = \left[\sum_\ell X_\ell e^{j\ell\Omega_0 t} \right]^* = \sum_\ell X_\ell^* e^{-j\ell\Omega_0 t} = \sum_k X_{-k}^* e^{jk\Omega_0 t}.$$

Since $x(t) = x^*(t)$, the above equation is equal to

$$x(t) = \sum_k X_k e^{jk\Omega_0 t}.$$

Comparing the Fourier series coefficients in the two expressions, we have $X_{-k}^* = X_k$, which means that if $X_k = |X_k| e^{j\angle X_k}$ then

$$|X_k| = |X_{-k}|, \quad \angle X_k = -\angle X_{-k},$$

or that the magnitude is an even function of k, while the phase is an odd function of k. Thus the line spectra corresponding to real-valued signals is only given for positive harmonic frequencies, with the understanding that the magnitude line spectrum is even and the phase line spectrum odd.

4.3.2 TRIGONOMETRIC FOURIER SERIES

In this section we develop an equivalent expression for the Fourier series using sinusoids. First we need to show that sinusoids of harmonic frequencies are orthonormal. Orthogonality of the complex exponential Fourier basis indicates that an equivalent basis can be obtained from cosine and sine functions. Choosing $t_0 = -T_0/2$ in Equation (4.16) which displays the orthogonality of the exponentials, we have, for $k \neq \ell$:

$$
\begin{aligned}
0 &= \frac{1}{T_0} \int_{-T_0/2}^{T_0/2} e^{jk\Omega_0 t} [e^{j\ell\Omega_0 t}]^* dt \\
&= \underbrace{\frac{1}{T_0} \int_{-T_0/2}^{T_0/2} \cos((k-\ell)\Omega_0 t) dt}_{0} + j \underbrace{\frac{1}{T_0} \int_{-T_0/2}^{T_0/2} \sin((k-\ell)\Omega_0 t) dt}_{0}
\end{aligned}
\tag{4.22}
$$

or that the real and imaginary parts are zero. Expanding $\cos((k-\ell)\Omega_0 t)$ and $\sin((k-\ell)\Omega_0 t)$ in the integrals on the right in Equation (4.22) we obtain

$$0 = \frac{1}{T_0} \int_{-T_0/2}^{T_0/2} \cos(k\Omega_0 t) \cos(\ell\Omega_0 t) dt + \frac{1}{T_0} \int_{-T_0/2}^{T_0/2} \sin(k\Omega_0 t) \sin(\ell\Omega_0 t) dt,$$

$$0 = \frac{1}{T_0} \int_{-T_0/2}^{T_0/2} \sin(k\Omega_0 t) \cos(\ell\Omega_0 t) dt - \frac{1}{T_0} \int_{-T_0/2}^{T_0/2} \cos(k\Omega_0 t) \sin(\ell\Omega_0 t) dt.$$

Using the trigonometric identities

$$\sin(\alpha)\sin(\beta) = 0.5[\cos(\alpha - \beta) - \cos(\alpha - \beta)], \quad \cos(\alpha)\cos(\beta) = 0.5[\cos(\alpha - \beta) + \cos(\alpha - \beta)],$$

and the fact that the argument of the sinusoids is a nonzero integer times Ω_0 the top integrals are zero as the integrals are over one or more periods. Thus, $\cos(k\Omega t)$ is orthogonal to $\cos(\ell\Omega t)$ for any $k \neq \ell$, and $\sin(k\Omega t)$ is orthogonal to $\sin(\ell\Omega t)$ for $k \neq \ell$. Likewise, the second set of integrals are zero since the integrands are odd functions. Thus, $\cos(k\Omega t)$ is orthogonal to $\sin(\ell\Omega t)$ when $k \neq \ell$. Thus for different frequencies the cosine and the sine functions are orthogonal.

To normalize the sinusoidal basis, we find that

$$
\frac{1}{T_0} \int_{-T_0/2}^{T_0/2} \cos^2(k\Omega_0 t)dt = \frac{1}{T_0} \int_{-T_0/2}^{T_0/2} \sin^2(k\Omega_0 t)dt
$$

$$
= \frac{1}{T_0} \left[\int_{-T_0/2}^{T_0/2} 0.5dt \pm \int_{-T_0/2}^{T_0/2} 0.5\cos(2k\Omega_0 t)dt \right] = 0.5
$$

by using that $\cos^2(\theta) = 0.5(1 + \cos(2\theta))$ and $\sin^2(\theta) = 0.5(1 - \cos(2\theta))$. Therefore, if we choose as basis $\{\sqrt{2}\cos(k\Omega_0 t), \sqrt{2}\sin(k\Omega_0 t)\}$ we obtain the trigonometric Fourier series, equivalent to the exponential Fourier series presented before.

The **trigonometric Fourier Series** of a real-valued, periodic signal $x(t)$, of fundamental period T_0, is an equivalent representation that uses sinusoids rather than complex exponentials as the basis functions. It is given by

$$
x(t) = X_0 + 2\sum_{k=1}^{\infty} |X_k| \cos(k\Omega_0 t + \theta_k)
$$

$$
= c_0 + 2\sum_{k=1}^{\infty} [c_k \cos(k\Omega_0 t) + d_k \sin(k\Omega_0 t)] \qquad \Omega_0 = \frac{2\pi}{T_0} \qquad (4.23)
$$

where $X_0 = c_0$ is called the **dc component**, and $\{2|X_k|\cos(k\Omega_0 t + \theta_k)\}$ are the kth **harmonics** for $k = 1, 2 \cdots$. The coefficients $\{c_k, d_k\}$ are obtained from $x(t)$ as follows:

$$
c_k = \frac{1}{T_0} \int_{t_0}^{t_0+T_0} x(t)\cos(k\Omega_0 t) \, dt \qquad k = 0, 1, \cdots
$$

$$
d_k = \frac{1}{T_0} \int_{t_0}^{t_0+T_0} x(t)\sin(k\Omega_0 t) \, dt \qquad k = 1, 2, \cdots . \qquad (4.24)
$$

The coefficients $X_k = |X_k|e^{j\theta_k}$ are connected with the coefficients c_k and d_k by

$$
|X_k| = \sqrt{c_k^2 + d_k^2}
$$

$$
\theta_k = -\tan^{-1}\left[\frac{d_k}{c_k}\right].
$$

The sinusoidal basis functions $\{\sqrt{2}\cos(k\Omega_0 t), \sqrt{2}\sin(k\Omega_0 t)\}$, $k = 0, \pm 1, \cdots$, are orthonormal in $[0, T_0]$.

Using the relation $X_k = X^*_{-k}$, obtained in the previous section, we express the exponential Fourier series of a real-valued periodic signal $x(t)$ as

$$
\begin{aligned}
x(t) &= X_0 + \sum_{k=1}^{\infty}[X_k e^{jk\Omega_0 t} + X_{-k}e^{-jk\Omega_0 t}] = X_0 + \sum_{k=1}^{\infty}\left[|X_k|e^{j(k\Omega_0 t+\theta_k)} + |X_k|e^{-j(k\Omega_0 t+\theta_k)}\right] \\
&= X_0 + 2\sum_{k=1}^{\infty}|X_k|\cos(k\Omega_0 t + \theta_k),
\end{aligned}
$$

which is the top equation in (4.23).

Let us then show how the coefficients c_k and d_k can be obtained directly from the signal. Using the relation $X_k = X^*_{-k}$ and the fact that for a complex number $z = a + jb$, then $z + z^* = (a + jb) + (a - jb) = 2a = 2Re(z)$ we have

$$
\begin{aligned}
x(t) &= X_0 + \sum_{k=1}^{\infty}[X_k e^{jk\Omega_0 t} + X_{-k}e^{-jk\Omega_0 t}] = X_0 + \sum_{k=1}^{\infty}[X_k e^{jk\Omega_0 t} + X^*_k e^{-jk\Omega_0 t}] \\
&= X_0 + \sum_{k=1}^{\infty} 2Re[X_k e^{jk\Omega_0 t}].
\end{aligned}
$$

Since X_k is complex,

$$
\begin{aligned}
2Re[X_k e^{jk\Omega_0 t}] &= 2Re[(Re[X_k] + j\mathcal{I}m[X_k])(\cos(k\Omega_0 t) + j\sin(k\Omega_0 t))] \\
&= 2Re[X_k]\cos(k\Omega_0 t) - 2\mathcal{I}m[X_k]\sin(k\Omega_0 t).
\end{aligned}
$$

Now, if we let

$$
c_k = Re[X_k] = \frac{1}{T_0}\int_{t_0}^{t_0+T_0} x(t)\cos(k\Omega_0 t)\, dt \qquad k = 1, 2, \cdots,
$$

$$
d_k = -\mathcal{I}m[X_k] = \frac{1}{T_0}\int_{t_0}^{t_0+T_0} x(t)\sin(k\Omega_0 t)\, dt \qquad k = 1, 2, \cdots,
$$

we then have

$$
\begin{aligned}
x(t) &= X_0 + \sum_{k=1}^{\infty}(2Re[X_k]\cos(k\Omega_0 t) - 2\mathcal{I}m[X_k]\sin(k\Omega_0 t)) \\
&= X_0 + 2\sum_{k=1}^{\infty}(c_k\cos(k\Omega_0 t) + d_k\sin(k\Omega_0 t)),
\end{aligned}
$$

and since the average $X_0 = c_0$ we obtain the second form of the trigonometric Fourier series. Notice that $d_0 = 0$ and so it is not necessary to define it.

The coefficients $X_k = |X_k|e^{j\theta_k}$ are connected with the coefficients c_k and d_k by

$$|X_k| = \sqrt{c_k^2 + d_k^2}, \quad \theta_k = -\tan^{-1}\left[\frac{d_k}{c_k}\right].$$

This can be shown by adding the phasors corresponding to $c_k \cos(k\Omega_0 t)$ and $d_k \sin(k\Omega_0 t)$ and finding the magnitude and phase of the resulting phasor.

Example 4.4. Find the exponential Fourier series of a raised cosine signal ($B \geq A$),

$$x(t) = B + A\cos(\Omega_0 t + \theta),$$

which is periodic of fundamental period T_0 and fundamental frequency $\Omega_0 = 2\pi/T_0$. Let $y(t) = B + A\sin(\Omega_0 t)$, find its Fourier series coefficients and compare them to those for $x(t)$. Use symbolic MATLAB to compute the Fourier series of $y(t) = 1 + \sin(100t)$, find and plot its magnitude and phase line spectra.

Solution: We do not need to compute the Fourier coefficients in this case since $x(t)$ is in the Fourier series trigonometric form. According to Equation (4.23) the dc value of $x(t)$ is B, and A is the coefficient of the first harmonic in the trigonometric Fourier series. Thus $X_0 = B$ and $|X_1| = A/2$ and $\angle X_1 = \theta$. Likewise, using Euler's identity we obtain

$$x(t) = B + \frac{A}{2}\left[e^{j(\Omega_0 t + \theta)} + e^{-j(\Omega_0 t + \theta)}\right] = B + \frac{Ae^{j\theta}}{2}e^{j\Omega_0 t} + \frac{Ae^{-j\theta}}{2}e^{-j\Omega_0 t},$$

which gives

$$X_0 = B, \quad X_1 = \frac{Ae^{j\theta}}{2}, \quad X_{-1} = X_1^* = \frac{Ae^{-j\theta}}{2}.$$

If we let $\theta = -\pi/2$ in $x(t)$ we get

$$y(t) = B + A\sin(\Omega_0 t)$$

and from the above results its Fourier series coefficients are

$$Y_0 = B, \quad Y_1 = \frac{A}{2}e^{-j\pi/2} = Y_{-1}^* \quad \text{so that}$$

$$|Y_1| = |Y_{-1}| = \frac{A}{2} \quad \text{and} \quad \angle Y_1 = -\angle Y_{-1} = -\frac{\pi}{2}.$$

The line spectrum of the raised cosine ($\theta = 0$) and of the raised sine ($\theta = -\pi/2$) are shown in Fig. 4.2. For both $x(t)$ and $y(t)$ there are only two frequencies, the dc frequency and Ω_0, and as such the power of the signal is concentrated at those two frequencies. The difference between the line spectrum of the raised cosine and sine is in the phase.

We find the Fourier series coefficients with our symbolic MATLAB function *fourierseries*. The corresponding magnitude and phase are then plotted using *stem* to obtain the line spectra. The magnitude and phase line spectra corresponding to $y(t) = 1 + \sin(100t)$ are shown in Fig. 4.3.   □

```
%%
% Example 4.4---Computation of harmonics and harmonic frequencies
%%
clear all; syms t
% signal
y=1+sin(100*t); T0=2*pi/100; N=5;              % N harmonics
figure(1)
subplot(211)
fplot(y,[0,0.25]); grid; xlabel('t (s)');ylabel('y(t)')
% harmonics and harmonic frequencies
[Y1, w1]=fourierseries(y,T0,N);
Y=[conj(fliplr(Y1(2:N))) Y1];w=[-fliplr(w1(2:N)) w1];
subplot(223)
stem(w,abs(Y)); grid; axis([-400 400 -0.1 1.1])
xlabel('k\Omega_0 (rad/s)'); ylabel('|Y_k|')
subplot(224)
stem(w,angle(Y)); grid; axis([-400 400 -2 2])
xlabel('k\Omega_0 (rad/s)'); ylabel('\angle{Y_k}')

function [X, w]=fourierseries(x,T0,N)
% function fourierseries
% Computes harmonics of the Fourier series of a continuous-time signal
% symbolically
% input: periodic signal x(t), its period (T0), number of harmonics (N)
% output: harmonics X and corresponding harmonic frequency w
% use: [X, w]=fourier(x,T0,N)
syms t
for k=1:N,
    X1(k)=int(x*exp(-j*2*pi*(k-1)*t/T0),t,0,T0)/T0;
    X(k)=subs(X1(k));
    w(k)=(k-1)*2*pi/T0;
end
```

Remark

Just because a signal is a sum of sinusoids, each of which is periodic, is not enough for it to have a Fourier series. The whole signal should be periodic. The signal $x(t) = \cos(t) - \sin(\pi t)$ has components with fundamental periods $T_1 = 2\pi$ and $T_2 = 2$ so that the ratio $T_1/T_2 = \pi$ is not a rational number. Thus $x(t)$ is not periodic and no Fourier series for $x(t)$ is possible.

4.3.3 FOURIER SERIES AND LAPLACE TRANSFORM

The computation of the X_k coefficients (see Equation (4.14)) requires integration which for some signals can be rather complicated. The integration can be avoided whenever we know the Laplace transform of a period of the signal as we will show. The Laplace transform of a period of the signal

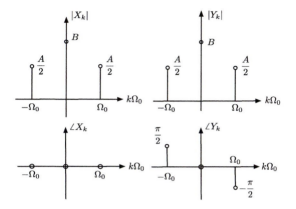

FIGURE 4.2

Line spectrum of raised cosine (left) and of raised sine (right).

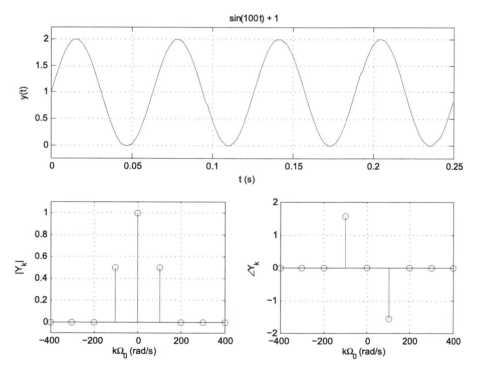

FIGURE 4.3

Line spectrum of Fourier series of $y(t) = 1 + \sin(100t)$. Notice the even and the odd symmetries of the magnitude and phase spectra. The phase is $-\pi/2$ at $\Omega = 100$ (rad/s).

exists over the whole s-plane, given that it is a finite support signal, and as such it can be computed at the harmonic frequencies permitting us to obtain an equation for $\{X_k\}$.

For a periodic signal $x(t)$, of fundamental period T_0, if we know or can easily compute the Laplace transform of a period of $x(t)$

$$x_1(t) = x(t)[u(t - t_0) - u(t - t_0 - T_0)] \qquad \text{for any } t_0$$

then the Fourier coefficients of $x(t)$ are given by

$$X_k = \frac{1}{T_0} \mathcal{L}\left[x_1(t)\right]_{s=jk\Omega_0} \quad \Omega_0 = \frac{2\pi}{T_0} \text{ (fundamental frequency)}, \ k = 0, \pm 1, \cdots. \qquad (4.25)$$

This can be seen by comparing the X_k coefficients with the Laplace transform of a period $x_1(t) = x(t)[u(t - t_0) - u(t - t_0 - T_0)]$ of $x(t)$. Indeed, we have

$$
\begin{aligned}
X_k &= \frac{1}{T_0} \int_{t_0}^{t_0+T_0} x(t)e^{-jk\Omega_0 t}\,dt = \frac{1}{T_0} \int_{t_0}^{t_0+T_0} x_1(t)e^{-st}\,dt \Big|_{s=jk\Omega_0} \\
&= \frac{1}{T_0} \mathcal{L}[x_1(t)]_{s=jk\Omega_0}.
\end{aligned}
$$

4.3.4 REFLECTION AND EVEN AND ODD PERIODIC SIGNALS

If the Fourier series of $x(t)$, periodic with fundamental frequency Ω_0, is

$$x(t) = \sum_k X_k e^{jk\Omega_0 t},$$

then the one for its reflected version $x(-t)$ is

$$x(-t) = \sum_m X_m e^{-jm\Omega_0 t} = \sum_k X_{-k} e^{jk\Omega_0 t}, \qquad (4.26)$$

so that the Fourier coefficients of $x(-t)$ are X_{-k} (remember that m and k are just dummy variables). This can be used to simplify the computation of Fourier series of even and odd signals.

For an even signal $x(t)$, we have $x(t) = x(-t)$ and as such $X_k = X_{-k}$ and therefore $x(t)$ is naturally represented in terms of cosines, and a dc term. Moreover, since in general $X_k = X^*_{-k}$ then $X_k = X^*_{-k} = X_{-k}$, as such these coefficients must be real-valued. Indeed, the Fourier series of $x(t)$ is

$$
\begin{aligned}
x(t) &= X_0 + \sum_{k=-\infty}^{-1} X_k e^{jk\Omega_0 t} + \sum_{k=1}^{\infty} X_k e^{jk\Omega_0 t} = X_0 + \sum_{k=1}^{\infty} X_k[e^{jk\Omega_0 t} + e^{-jk\Omega_0 t}] \\
&= X_0 + 2\sum_{k=1}^{\infty} X_k \cos(k\Omega_0 t). \qquad (4.27)
\end{aligned}
$$

That the Fourier series coefficients need to be real-valued when $x(t)$ is even can be shown directly:

$$X_k = \frac{1}{T_0} \int_{-T_0/2}^{T_0/2} x(t) e^{-jk\Omega_0 t} dt$$

$$= \frac{1}{T_0} \int_{-T_0/2}^{T_0/2} x(t) \cos(k\Omega_0 t) dt - j\frac{1}{T_0} \int_{-T_0/2}^{T_0/2} x(t) \sin(k\Omega_0) dt$$

$$= \frac{1}{T_0} \int_{-T_0/2}^{T_0/2} x(t) \cos(k\Omega_0 t) dt = X_{-k}$$

because $x(t)\sin(k\Omega_0 t)$ is odd and the corresponding integral is zero, and because the $\cos(k\Omega_0 t) = \cos(-k\Omega_0 t)$.

It will be similar for odd signals for which $x(t) = -x(-t)$, or $X_k = -X_{-k}$, in which case the Fourier series has a zero dc value and sine harmonics. The X_k are purely imaginary. Indeed, for $x(t)$ odd

$$X_k = \frac{1}{T_0} \int_{-T_0/2}^{T_0/2} x(t) e^{-jk\Omega_0 t} dt$$

$$= \frac{1}{T_0} \int_{-T_0/2}^{T_0/2} x(t) [\cos(k\Omega_0 t) - j\sin(k\Omega_0)] dt$$

$$= \frac{-j}{T_0} \int_{-T_0/2}^{T_0/2} x(t) \sin(k\Omega_0 t) dt$$

since $x(t)\cos(k\Omega_0 t)$ is odd and their integral is zero. The Fourier series of an odd function can thus be written as

$$x(t) = 2 \sum_{k=1}^{\infty} (jX_k) \sin(k\Omega_0 t). \tag{4.28}$$

According to the even and odd decomposition, any periodic signal $x(t)$ can be expressed as

$$x(t) = x_e(t) + x_o(t)$$

where $x_e(t)$ is the even and $x_o(t)$ is the odd component of $x(t)$. Finding the Fourier coefficients of $x_e(t)$, which will be real, and those of $x_o(t)$, which will be purely imaginary, we would then have $X_k = X_{ek} + X_{ok}$ since

$$x_e(t) = 0.5[x(t) + x(-t)] \quad \Rightarrow \quad X_{ek} = 0.5[X_k + X_{-k}],$$

$$x_o(t) = 0.5[x(t) - x(-t)] \quad \Rightarrow \quad X_{ok} = 0.5[X_k - X_{-k}].$$

Reflection: If the Fourier coefficients of a periodic signal $x(t)$ are $\{X_k\}$ then those of $x(-t)$, the time-reversed signal with the same period as $x(t)$, are $\{X_{-k}\}$.

Even periodic signal $x(t)$: Its Fourier coefficients X_k are real, and its trigonometric Fourier series is

$$x(t) = X_0 + 2\sum_{k=1}^{\infty} X_k \cos(k\Omega_0 t). \tag{4.29}$$

Odd periodic signal $x(t)$: Its Fourier coefficients X_k are imaginary, and its trigonometric Fourier series is

$$x(t) = 2\sum_{k=1}^{\infty} jX_k \sin(k\Omega_0 t). \tag{4.30}$$

Any periodic signal $x(t)$ can be written $x(t) = x_e(t) + x_o(t)$, where $x_e(t)$ and $x_o(t)$ are the even and the odd components of $x(t)$ then

$$X_k = X_{ek} + X_{ok} \tag{4.31}$$

where $\{X_{ek}\}$ are the Fourier coefficients of $x_e(t)$ and $\{X_{ok}\}$ are the Fourier coefficients of $x_o(t)$ or

$$X_{ek} = 0.5[X_k + X_{-k}],$$
$$X_{ok} = 0.5[X_k - X_{-k}]. \tag{4.32}$$

Example 4.5. Consider the periodic pulse train $x(t)$, of fundamental period $T_0 = 1$, shown in Fig. 4.4. Find its Fourier series.

Solution: Before finding the Fourier coefficients, we see that this signal has a dc component of 1, and that $x(t) - 1$ (zero-average signal) is well represented by cosines, given its even symmetry, and as such the Fourier coefficients X_k should be real. Doing this analysis before the computations is important so we know what to expect.

The Fourier coefficients are obtained either using their integral formula or from the Laplace transform of a period. Since $T_0 = 1$, the fundamental frequency of $x(t)$ is $\Omega_0 = 2\pi$ (rad/s). Using the integral

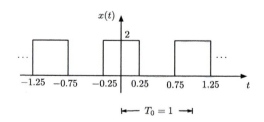

FIGURE 4.4

Train of rectangular pulses.

expression for the Fourier coefficients we have

$$
\begin{aligned}
X_k &= \frac{1}{T_0} \int_{-T_0/4}^{3T_0/4} x(t) e^{-j\Omega_0 kt} \, dt = \int_{-1/4}^{1/4} 2 e^{-j2\pi kt} \, dt \\
&= \frac{2}{\pi k} \left[\frac{e^{j\pi k/2} - e^{-j\pi k/2}}{2j} \right] = \frac{\sin(\pi k/2)}{(\pi k/2)}, \quad k \neq 0 \\
X_0 &= \frac{1}{T_0} \int_{-T_0/4}^{3T_0/4} x(t) \, dt = \int_{-1/4}^{1/4} 2 \, dt = 1,
\end{aligned}
$$

which are real as we predicted. The Fourier series is then

$$
x(t) = \sum_{k=-\infty}^{\infty} \frac{\sin(\pi k/2)}{(\pi k/2)} e^{jk2\pi t}.
$$

To find the Fourier coefficients with the Laplace transform, let the period be $x_1(t) = x(t)$ for $-0.5 \leq t \leq 0.5$. Delaying it by 0.25 we get $x_1(t - 0.25) = 2[u(t) - u(t - 0.5)]$ with a Laplace transform

$$
e^{-0.25s} X_1(s) = \frac{2}{s}(1 - e^{-0.5s})
$$

so that $X_1(s) = 2(e^{0.25s} - e^{-0.25s})/s$ and therefore

$$
X_k = \frac{1}{T_0} \mathcal{L}[x_1(t)] \Big|_{s=jk\Omega_0} = \frac{2}{jk\Omega_0 T_0} 2j \sin(k\Omega_0/4)
$$

and for $\Omega_0 = 2\pi$, $T_0 = 1$, we get

$$
X_k = \frac{\sin(\pi k/2)}{\pi k/2} \quad k \neq 0.
$$

To find X_0 (the above equation gives zero over zero when $k = 0$) we can use L'Hôpital's rule or use the integral formula as before. As expected, the Fourier coefficients coincide with the ones found before.

The following script is used to find the Fourier coefficients with our function *fourierseries* and to plot the magnitude and phase line spectra.

```
%%
% Example 4.5---Computation of harmonics and harmonic frequencies of
% train of pulses
%%
clear all;clf
syms t
T0=1;N=20;
% signal
m=heaviside(t)-heaviside(t-T0/4)+heaviside(t-3*T0/4);x=2*m;
% harmonics and harmonic frequencies
```

```
[X1,w1]=fourierseries(x,T0,N);
X=[conj(fliplr(X1(2:N))) X1];w=[-fliplr(w1(2:N)) w1];
figure(1)
subplot(221)
fplot(x,[0 T0]);grid; title('period of x(t)')
subplot(222)
stem(w,X); grid; axis([min(w) max(w) -0.5 1.1]); title('real X(k)')
xlabel('k\Omega_0 (rad/s)'); ylabel('X_k')
subplot(223)
stem(w,abs(X)); grid; title('magnitude line spectrum')
axis([min(w) max(w) -0.1 1.1])
xlabel('k\Omega_0 (rad/s)'); ylabel('|X_k|')
subplot(224)
stem(w,[-angle(X1(2:N)) angle(X1)]); grid; title('phase line spectrum')
axis([min(w) max(w) -3.5  3.5])
xlabel('k\Omega_0 (rad/s)'); ylabel('\angle{X_k}')
```

Notice that in this case:

1. The X_k Fourier coefficients of the train of pulses are given in terms of the $\sin(x)/x$ or the **sinc function**. This function was presented in Chapter 1. Recall the sinc

   - is even, i.e., $\sin(x)/x = \sin(-x)/(-x)$,
   - has a value at $x = 0$ that is found by means of L'Hôpital's rule because the numerator and the denominator of sinc are zero for $x = 0$, so

$$\lim_{x \to 0} \frac{\sin(x)}{x} = \lim_{x \to 0} \frac{d \sin(x)/dx}{dx/dx} = 1,$$

   - is bounded, indeed

$$\frac{-1}{x} \le \frac{\sin(x)}{x} \le \frac{1}{x}.$$

2. Since the dc component of $x(t)$ is 1, once it is subtracted it is clear that the rest of the series can be represented as a sum of cosines:

$$x(t) = 1 + \sum_{k=-\infty,k\neq 0}^{\infty} \frac{\sin(\pi k/2)}{(\pi k/2)} e^{jk2\pi t} = 1 + 2 \sum_{k=1}^{\infty} \frac{\sin(\pi k/2)}{(\pi k/2)} \cos(2\pi kt).$$

3. In general, the Fourier coefficients are complex and as such need to be represented by their magnitudes and phases. In this case, the X_k coefficients are real-valued, and in particular zero when $k\pi/2 = \pm m\pi$, m an integer (or when $k = \pm 2, \pm 4, \cdots$). Since the X_k values are real, the corresponding phase would be zero when $X_k \ge 0$, and $\pm\pi$ when $X_k < 0$. In Fig. 4.5 we show a period of the signal, and the magnitude and phase line spectra displayed only for positive values of frequency (with the understanding that the magnitude and the phase are even and odd functions of frequency).

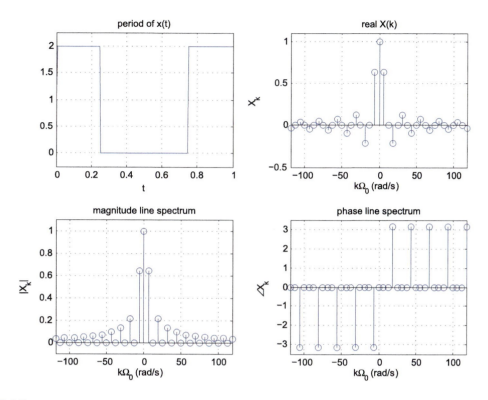

FIGURE 4.5

Top: period of $x(t)$ and real X_k vs. $k\Omega_0$; bottom: magnitude and phase line spectra.

4. The X_k coefficients and its squares (see Fig. 4.5), which are related to the power line spectrum, are:

k	$X_k = X_{-k}$	X_k^2
0	1	1
1	0.64	0.41
2	0	0
3	−0.21	0.041
4	0	0
5	0.13	0.016
6	0	0
7	−0.09	0.008.

We notice that the dc value and 5 harmonics, or 11 coefficients (including the zero values), provide a very good approximation of the pulse train, and would occupy a bandwidth of approximately $5\Omega_0 = 10\pi$ (rad/s). The power contribution, as indicated by X_k^2 after $k = \pm 6$ is about 3.3% of the

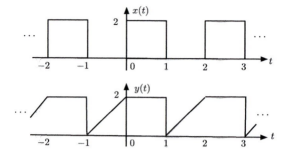

FIGURE 4.6

Non-symmetric periodic signals.

signal power. Indeed, the power of the signal is

$$P_x = \frac{1}{T_0} \int_{-T_0/4}^{3T_0/4} x^2(t)dt = \int_{-0.25}^{0.25} 4dt = 2$$

and the power of the approximation with 11 coefficients is

$$X_0^2 + 2\sum_{k=1}^{5} |X_k|^2 = 1.9340,$$

corresponding to $96.7\% \, P_x$. A very good approximation! $\qquad\square$

Example 4.6. Consider the periodic signals $x(t)$ and $y(t)$ shown in Fig. 4.6. Determine their Fourier coefficients by using the symmetry conditions and the even-and-odd decomposition. Verify your results for $y(t)$ by computing its Fourier coefficients using their integral or Laplace formulas.

Solution:
The given signal $x(t)$ is neither even nor odd, but the advanced signal $x(t + 0.5)$ is even with a fundamental period of $T_0 = 2$, $\Omega_0 = \pi$. Let $z(t) = x(t + 0.5)$ and call its period between -1 and 1

$$z_1(t) = 2[u(t + 0.5) - u(t - 0.5)],$$

so that its Laplace transform is

$$Z_1(s) = \frac{2}{s}[e^{0.5s} - e^{-0.5s}],$$

which gives the Fourier coefficients

$$Z_k = \frac{1}{2} \frac{2}{jk\pi}[e^{jk\pi/2} - e^{-jk\pi/2}] = \frac{\sin(0.5\pi k)}{0.5\pi k}$$

after replacing s by $jk\Omega_o = jk\pi$ and dividing by the fundamental period $T_0 = 2$. These coefficients are real-valued as corresponding to an even function. The dc coefficient is $Z_0 = 1$. Then

$$x(t) = z(t - 0.5) = \sum_k Z_k e^{jk\Omega_0(t-0.5)} = \sum_k \underbrace{\left[Z_k e^{-jk\pi/2} \right]}_{X_k} e^{jk\pi t}.$$

The X_k coefficients are complex since $x(t)$ is neither even nor odd.

The signal $y(t)$ is neither even nor odd, and cannot be made even or odd by shifting. A way to find its Fourier series is to decompose it into even and odd signals. The even and odd components of a period of $y(t)$ are shown in Fig. 4.7. The even and odd components of a period $y_1(t)$ between -1 and 1 are

$$y_{1e}(t) = \underbrace{[u(t+1) - u(t-1)]}_{\text{rectangular pulse}} + \underbrace{[r(t+1) - 2r(t) + r(t-1)]}_{\text{triangular pulse}},$$

$$y_{1o}(t) = \underbrace{[r(t+1) - r(t-1) - 2u(t-1)]}_{\text{triangular pulse}} - \underbrace{[u(t+1) - u(t-1)]}_{\text{rectangular pulse}}$$

$$= r(t+1) - r(t-1) - u(t+1) - u(t-1).$$

Thus the mean value of $y_e(t)$ is the area under $y_{1e}(t)$ divided by 2 or 1.5. For $k \neq 0$, $T_0 = 2$ and $\Omega_0 = \pi$,

$$\begin{aligned} Y_{ek} &= \frac{1}{T_0} Y_{1e}(s) \Big|_{s=jk\Omega_0} = \frac{1}{2} \left[\frac{1}{s}(e^s - e^{-s}) + \frac{1}{s^2}(e^s - 2 + e^{-s}) \right]_{s=jk\pi} \\ &= \frac{\sin(k\pi)}{\pi k} + \frac{1 - \cos(k\pi)}{(k\pi)^2} = 0 + \frac{1 - \cos(k\pi)}{(k\pi)^2} = \frac{1 - (-1)^k}{(k\pi)^2} \qquad k \neq 0, \end{aligned}$$

which are real as expected. The mean value of $y_o(t)$ is zero, and for $k \neq 0$, $T_0 = 2$ and $\Omega_0 = \pi$,

$$\begin{aligned} Y_{ok} &= \frac{1}{T_0} Y_{1o}(s) \Big|_{s=jk\Omega_0} = \frac{1}{2} \left[\frac{e^s - e^{-s}}{s^2} - \frac{e^s + e^{-s}}{s} \right]_{s=jk\pi} \\ &= -j \frac{\sin(k\pi)}{(k\pi)^2} + j \frac{\cos(k\pi)}{k\pi} = 0 + j \frac{\cos(k\pi)}{k\pi} = j \frac{(-1)^k}{k\pi} \qquad k \neq 0, \end{aligned}$$

which are purely imaginary as expected.

Finally, the Fourier series coefficients of $y(t)$ are

$$Y_k = \begin{cases} Y_{e0} + Y_{o0} = 1.5 + 0 = 1.5 & k = 0, \\ Y_{ek} + Y_{ok} = (1 - (-1)^k)/(k\pi)^2 + j(-1)^k/(k\pi) & k \neq 0. \end{cases} \qquad \square$$

Example 4.7. Find the Fourier series of the full-wave rectified signal $x(t) = |\cos(\pi t)|$ shown in Fig. 4.8. This signal is used in the design of dc sources; the rectification of an ac signal is the first step in this design.

Solution: Given that $T_0 = 1$, $\Omega_0 = 2\pi$, the Fourier coefficients are given by

$$X_k = \int_{-0.5}^{0.5} \cos(\pi t) e^{-j2\pi k t} dt,$$

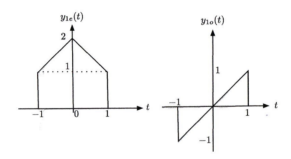

FIGURE 4.7

Even and odd components of the period of $y(t)$, $-1 \le t \le 1$.

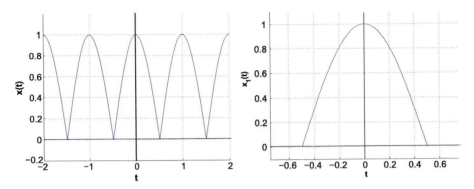

FIGURE 4.8

Full-wave rectified signal $x(t)$ and one of its periods $x_1(t)$.

which can be computed by using Euler's identity. We would like to show that it can easily be done using the Laplace transform.

A period $x_1(t)$ of $x(t)$ can be expressed as

$$x_1(t - 0.5) = \sin(\pi t)u(t) + \sin(\pi(t - 1))u(t - 1)$$

(show it graphically!) and using the Laplace transform we have

$$X_1(s)e^{-0.5s} = \frac{\pi}{s^2 + \pi^2}[1 + e^{-s}],$$

so that

$$X_1(s) = \frac{\pi}{s^2 + \pi^2}[e^{0.5s} + e^{-0.5s}].$$

For $T_0 = 1$ and $\Omega_0 = 2\pi$, the Fourier coefficients are then

$$X_k = \frac{1}{T_0}X_1(s)|_{s=j\Omega_0 k} = \frac{\pi}{(j2\pi k)^2 + \pi^2} \, 2\cos(\pi k) = \frac{2(-1)^k}{\pi(1 - 4k^2)}, \tag{4.33}$$

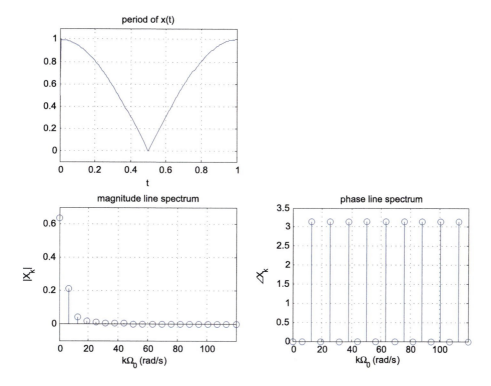

FIGURE 4.9

Period of full-wave rectified signal $x(t)$ and its magnitude and phase line spectra.

since $\cos(\pi k) = (-1)^k$. From Equation (4.33), the dc value of the full-wave rectified signal is $X_0 = 2/\pi$. Notice that the Fourier coefficients are real given the signal is even. The phase of X_k is zero when $X_k > 0$, and π (or $-\pi$) when $X_k < 0$.

A similar MATLAB script to the one in Example 4.5 can be written. In this case we only plot the spectra for positive frequencies—the rest are obtained by the symmetry of the spectra. The results are shown in Fig. 4.9.   □

```
%%
% Example 4.7---Fourier series of full-wave rectified signal
%%
% period generation
T0=1;
m=heaviside(t)-heaviside(t-T0);x=abs(cos(pi*t))*m
[X,w]=fourierseries(x,T0,N);
```

Example 4.8. Computing the derivative of a periodic signal enhances the higher harmonics of its Fourier series. To illustrate this consider the train of triangular pulses $y(t)$ (see left figure in Fig. 4.10)

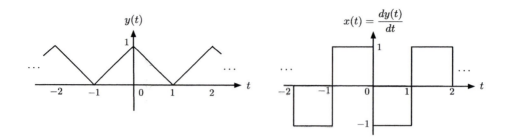

FIGURE 4.10

Train of triangular pulses $y(t)$ and its derivative $x(t)$. Notice that $y(t)$ is a continuous function while $x(t)$ is discontinuous.

with fundamental period $T_0 = 2$. Let $x(t) = dy(t)/dt$ (see right figure in Fig. 4.10), find its Fourier series and compare $|X_k|$ with $|Y_k|$ to determine which of these signals is smoother, i.e., which one has lower-frequency components.

Solution: A period of $y(t)$, $-1 \leq t \leq 1$, is given by $y_1(t) = r(t+1) - 2r(t) + r(t-1)$ with a Laplace transform

$$Y_1(s) = \frac{1}{s^2} \left[e^s - 2 + e^{-s} \right],$$

so that the Fourier coefficients are given by ($T_0 = 2$, $\Omega_0 = \pi$):

$$Y_k = \frac{1}{T_0} Y_1(s)|_{s=j\Omega_0 k} = \frac{1}{2(j\pi k)^2} [2\cos(\pi k) - 2] = \frac{1 - \cos(\pi k)}{\pi^2 k^2} \qquad k \neq 0,$$

which is equal to

$$Y_k = 0.5 \left[\frac{\sin(\pi k/2)}{(\pi k/2)} \right]^2 \qquad (4.34)$$

using the identity $1 - \cos(\pi k) = 2\sin^2(\pi k/2)$. By observing $y(t)$ we deduce its dc value is $Y_0 = 0.5$ (verify it!).

Let us then consider the periodic signal $x(t) = dy(t)/dt$ (shown on the right of Fig. 4.10) with a dc value $X_0 = 0$. For $-1 \leq t \leq 1$, its period is $x_1(t) = u(t+1) - 2u(t) + u(t-1)$ and

$$X_1(s) = \frac{1}{s} \left[e^s - 2 + e^{-s} \right],$$

which gives the Fourier series coefficients

$$X_k = j \frac{\sin^2(k\pi/2)}{k\pi/2}, \qquad (4.35)$$

since $X_k = \frac{1}{2} X_1(s)|_{s=j\pi k}$.

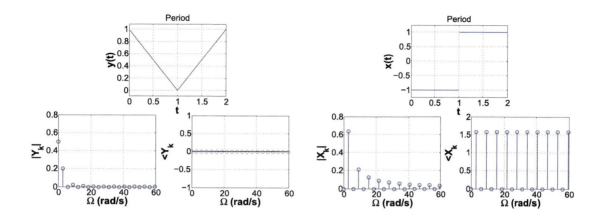

FIGURE 4.11

Magnitude and phase spectra of triangular signal $y(t)$ and its derivative $x(t)$. Ignoring the dc values, the magnitudes $\{|Y_k|\}$ decay faster to zero than the magnitudes $\{|X_k|\}$, thus $y(t)$ is smoother than $x(t)$.

For $k \neq 0$ we have $|Y_k| = |X_k|/(\pi k)$, so that as k increases the frequency components of $y(t)$ decrease in magnitude faster than the corresponding ones of $x(t)$: $y(t)$ is smoother than $x(t)$. The magnitude line spectrum $|Y_k|$ goes faster to zero than the magnitude line spectrum $|X_k|$ as $k \to \infty$ (see Fig. 4.11).

Notice that in this case $y(t)$ is even and its Fourier coefficients Y_k are real, while $x(t)$ is odd and its Fourier coefficients X_k are purely imaginary. If we subtract the average of $y(t)$, the signal $y(t)$ can be clearly approximated as a series of cosines, thus the need for real coefficients in its complex exponential Fourier series. The signal $x(t)$ is zero-average and as such it can be clearly approximated by a series of sines requiring its Fourier coefficients X_k to be imaginary. The following script does the computations and plotting. $\qquad \square$

```
%%
% Examples 4.8---Fourier series of derivative of signal
%%
clear all; syms t
T0=2;
m=heaviside(t)-heaviside(t-T0/2);
m1=heaviside(t-T0/2)-heaviside(t-T0);
for ind=1:2,
if ind==1
y=(1-t)*m+(t-1)*m1;
[Y,w]= fourierseries(y,T0,20);
figure(1)
subplot(221)
fplot(y,[0 T0]);grid;title('Period');xlabel('t');ylabel('y(t)')
```

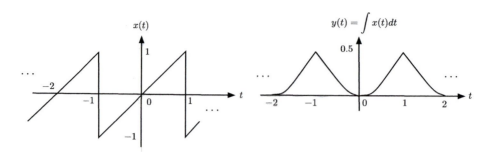

FIGURE 4.12

Sawtooth signal $x(t)$ and its integral $y(t)$. Notice that $x(t)$ is a discontinuous function while $y(t)$ is continuous.

```
subplot(223)
stem(w,abs(Y));grid;xlabel('\Omega (rad/s)');ylabel('|Y_k|')
subplot(224)
stem(w,angle(Y));grid;xlabel('\Omega (rad/s)');ylabel('\angle{Y_k}')
else
x=diff(y,t);
[X,w]=fourierseries(x,T0,20);
figure(2)
subplot(221)
fplot(x,[0 T0]);grid;title('Period');xlabel('t');ylabel('x(t)')
subplot(223)
stem(w,abs(X)); grid;xlabel('\Omega (rad/s)');ylabel('|X_k|')
subplot(224)
stem(w,angle(X));grid;xlabel('\Omega (rad/s)');ylabel('\angle{X_k}')
end; end
```

Example 4.9. Integration of a periodic signal, provided it has zero mean, gives a smoother signal. To see this, find and compare the magnitude line spectra of a sawtooth signal $x(t)$, of fundamental period $T_0 = 2$, and its integral

$$y(t) = \int_{-\infty}^{t} x(t)dt$$

shown in Fig. 4.12.

Solution: Before doing any calculations it is important to realize that the integral would not exist if the dc is not zero as it would accumulate as t grows. Indeed, if $x(t)$ had a Fourier series

$$x(t) = X_0 + \sum_k X_k e^{jk\Omega_0 t} \quad \text{the integral would be}$$

$$\int_{-\infty}^{t} x(\tau)d\tau = \int_{-\infty}^{t} X_0 d\tau + \sum_k X_k \int_{-\infty}^{t} e^{jk\Omega_0 \tau}d\tau$$

and the first integral would continuously increase.

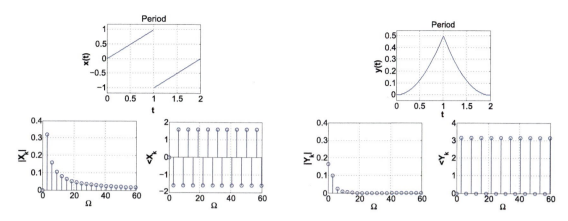

FIGURE 4.13

Periods of the sawtooth signal $x(t)$ and its integral $y(t)$ and their magnitude and phase line spectra.

Using the following script we can compute the Fourier series coefficients of $x(t)$ and $y(t)$. A period of $x(t)$ is

$$x_1(t) = tw(t) + (t-2)w(t-1) \quad 0 \le t \le 2$$

where $w(t) = u(t) - u(t-1)$ is a rectangular window.

The following script gives the basic code to find the Fourier series of the two signals.

```
%%
% Example 4.9---Saw-tooth signal and its integral
%%
syms t
T0=2;
m=heaviside(t)-heaviside(t-T0/2);
m1=heaviside(t-T0/2)-heaviside(t-T0);
x=t*m+(t-2)*m1;
y=int(x);
[X,w]=fourierseries(x,T0,20);
[Y,w]=fourierseries(y,T0,20);
```

Ignoring the dc components, the magnitudes $\{|Y_k|\}$ of $y(t)$ decay a lot faster to zero than the magnitudes $\{|X_k|\}$ of $x(t)$ as shown in Fig. 4.13. Thus the signal $y(t)$ is smoother than $x(t)$ as $x(t)$ has higher frequency components than $y(t)$. The discontinuities in $x(t)$ cause its higher frequencies.

As we will see in Section 4.5.3, computing the derivative of a periodic signal is equivalent to multiplying its Fourier series coefficients by $j\Omega_0 k$, which emphasizes the higher harmonics—differentiation makes the resulting signal rougher. If the periodic signal is zero mean, so that its integral exists, the Fourier coefficients of the integral can be found by dividing them by $j\Omega_0 k$ so that now the low harmonics are emphasized—integration makes the resulting signal smoother.   □

4.3.5 CONVERGENCE OF THE FOURIER SERIES

It can be said, without overstating it, that any periodic signal of practical interest has a Fourier series. Only very strange periodic signals would not have a converging Fourier series. Establishing convergence is necessary because the Fourier series has an infinite number of terms. To establish some general conditions under which the series converges we need to classify signals with respect to their smoothness.

A signal $x(t)$ is said to be **piecewise smooth** if it has a finite number of discontinuities, while a **smooth** signal has a derivative that changes continuously. Thus, smooth signals can be considered special cases of piecewise smooth signals.

The Fourier series of a piecewise smooth (continuous or discontinuous) periodic signal $x(t)$ converges for all values of t. The mathematician Dirichlet showed that for the Fourier series to converge to the periodic signal $x(t)$, the signal should satisfy the following sufficient (not necessary) conditions over a period:

1. be absolutely integrable,
2. have a finite number of maxima, minima and discontinuities.

The infinite series equals $x(t)$ at every continuity point and equals the average

$$0.5[x(t+0_+) + x(t+0_-)]$$

of the right-hand limit $x(t+0_+)$ and the left-hand limit $x(t+0_-)$ at every discontinuity point. If $x(t)$ is continuous everywhere, then the series converges absolutely and uniformly.

Although the Fourier series converges to the arithmetic average at discontinuities, it can be observed that there is some ringing before and after the discontinuity points. This is called the **Gibb's phenomenon.** To understand this phenomenon, it is necessary to explain how the Fourier series can be seen as an approximation to the actual signal, and how when a signal has discontinuities the convergence is not uniform around them. It will become clear that the smoother the signal $x(t)$ is, the easier it is to approximate it with a Fourier series with a finite number of terms.

When the signal is continuous everywhere, the convergence is such that at each point t the series approximates the actual value $x(t)$ as we increase the number of terms in the approximation. However, that is not the case when discontinuities occur in the signal. This is despite the fact that a minimum mean-square approximation seems to indicate that the approximation could give a zero error. Let

$$x_N(t) = \sum_{k=-N}^{N} X_k e^{jk\Omega_0 t} \tag{4.36}$$

be the Nth-order approximation of a periodic signal $x(t)$, of fundamental frequency Ω_0, which minimizes the average quadratic error over a period

$$E_N = \frac{1}{T_0} \int_{T_0} |x(t) - x_N(t)|^2 dt \tag{4.37}$$

with respect to the Fourier coefficients X_k. To minimize E_N with respect to the coefficients X_k we set its derivative with respect to X_k to zero. Let $\varepsilon(t) = x(t) - x_N(t)$, so that

$$\frac{dE_N}{dX_k} = \frac{1}{T_0}\int_{T_0} 2\varepsilon(t)\frac{d\varepsilon^*(t)}{dX_k}dt = -\frac{1}{T_0}\int_{T_0} 2[x(t) - x_N(t)]e^{-jk\Omega_0 t}dt = 0,$$

which after replacing $x_N(t)$ and using the orthogonality of the Fourier exponentials gives

$$X_k = \frac{1}{T_0}\int_{T_0} x(t)e^{-j\Omega_0 kt}dt \tag{4.38}$$

corresponding to the Fourier coefficients of $x(t)$ for $-N \leq k \leq N$. As $N \to \infty$ the average error $E_N \to 0$. The only issue left is how $x_N(t)$ converges to $x(t)$. As indicated before, if $x(t)$ is smooth $x_N(t)$ approximates $x(t)$ at every point, but if there are discontinuities the approximation is in an average fashion. The Gibbs phenomenon indicates that around discontinuities there will be ringing, regardless of the order N of the approximation. This phenomenon will be explained in Chapter 5 as the effect of using a rectangular window to obtain a finite frequency representation of a periodic signal.

Example 4.10. To illustrate the Gibbs phenomenon consider the approximation of a train of rectangular pulses $x(t)$ with zero mean and fundamental period $T_0 = 1$ (represented by the dashed signal in Fig. 4.14) with a Fourier series $x_N(t)$ approximation having a dc and $N = 20$ harmonics.

Solution: We compute analytically the Fourier coefficients of $x(t)$ and used them to obtain an approximation $x_N(t)$ of $x(t)$ having a zero dc component and 20 harmonics. The dashed figure in Fig. 4.14 is $x(t)$ and the continuous figure is $x_N(t)$ when $N = 20$. The discontinuities of the pulse train causes the Gibbs phenomenon. Even if we increase the number of harmonics there is an overshoot in the approximation around the discontinuities.

```
%%
% Example 4.10---Simulation of Gibb's phenomenon
%%
clf; clear all
w0=2*pi; DC=0; N=20;                  % parameters of periodic signal
% computation of Fourier series coefficients
for k=1:N,
X(k)=sin(k*pi/2)/(k*pi/2);
end
X=[DC X];                             % Fourier series coefficients
% computation of periodic signal
Ts=0.001; t=0:Ts:1-Ts;
L=length(t); x=[ones(1,L/4) zeros(1,L/2) ones(1,L/4)]; x=2*(x-0.5);
% computation of approximate
figure(1)
xN=X(1)*ones(1,length(t));
for k=2:N,
xN=xN+2*X(k)*cos(2*pi*(k-1).*t);     % approximate signal
```

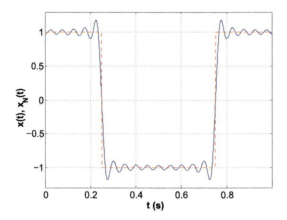

FIGURE 4.14

Approximate Fourier series $x_N(t)$ (continuous line) of the pulse train $x(t)$ (dashed line) using the dc component and 20 harmonics. The approximate $x_N(t)$ displays the Gibbs phenomenon around the discontinuities.

```
plot(t,xN); axis([0  max(t) 1.1*min(xN) 1.1*max(xN)])
hold on; plot(t,x,'r')
ylabel('x(t), x_N(t)'); xlabel('t (s)');grid
hold off
pause(0.1)
end
```

When you execute the above script it pauses to display the approximation for a number of harmonics increasing from 2 to N. At each of these values ringing around the discontinuities (the Gibbs phenomenon) is displayed. ☐

Example 4.11. Consider the mean-square error optimization to obtain an approximate $x_2(t) = \alpha + \beta \cos(\Omega_0 t)$ of the periodic signal $x(t)$ shown in Fig. 4.4, Example 4.5. Minimize the mean-square error

$$E_2 = \frac{1}{T_0} \int_{T_0} |x(t) - x_2(t)|^2 dt,$$

with respect to α and β, to find these values.

Solution: To minimize E_2 we set to zero its derivatives with respect to α and β, to get

$$\frac{d\,E_2}{d\alpha} = -\frac{1}{T_0} \int_{T_0} 2[x(t) - \alpha - \beta \cos(\Omega_0 t)]dt$$

$$= -\frac{1}{T_0} \int_{T_0} 2[x(t) - \alpha]dt = 0,$$

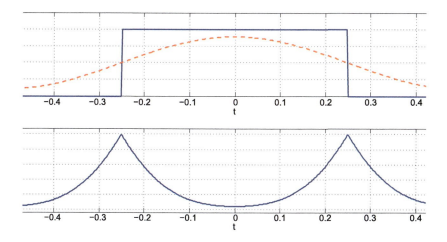

FIGURE 4.15

Top: a period of $x(t)$ and the approximating signal $x_2(t)$ (dashed line); bottom: plot of $\varepsilon(t) = |x(t) - x_2(t)|^2$ in a period with $E_2 = 0.3084$.

$$
\begin{aligned}
\frac{d E_2}{d\beta} &= -\frac{1}{T_0} \int_{T_0} 2[x(t) - \alpha - \beta \cos(\Omega_0 t)] \cos(\Omega_0 t) dt \\
&= -\frac{1}{T_0} \int_{T_0} 2[x(t) \cos(\Omega_0 t) - \beta \cos^2(\Omega_0 t)] dt = 0,
\end{aligned}
$$

which give after applying the orthonormality of the Fourier basis gives

$$
\alpha = \frac{1}{T_0} \int_{T_0} x(t) dt, \quad \beta = \frac{2}{T_0} \int_{T_0} x(t) \cos(\Omega_0 t) dt.
$$

For the signal in Fig. 4.4, Example 4.5, we obtain $\alpha = 1$, $\beta = \frac{4}{\pi}$ giving as approximation the signal

$$
x_2(t) = 1 + \frac{4}{\pi} \cos(2\pi t),
$$

corresponding to the dc and the first harmonic of the Fourier series found in Example 4.5. At $t = 0$, we have $x_2(0) = 2.27$ instead of the expected value of 2; and $x_2(0.25) = 1$ (because of the discontinuity at this point, this value is the average of 2 and 0 the values before and after the discontinuity) instead of 2 and $x_2(0.5) = -0.27$ instead of the expected 0. Fig. 4.15 displays $x(t)$, $x_2(t)$ and $\epsilon(t) = |x(t) - x_2(t)|^2$. The approximation error is $E_2 = 0.3084$.   □

4.3.6 TIME AND FREQUENCY SHIFTING

The time- and frequency-shifting properties are duals of each other.

Time shifting: A periodic signal $x(t)$, of fundamental period T_0, remains periodic of the same fundamental period when shifted in time. If X_k are the Fourier coefficients of $x(t)$, then for $x(t - t_0)$, $x(t)$ delayed t_0 seconds, its Fourier series coefficients are

$$X_k e^{-jk\Omega_0 t_0} = |X_k| e^{j(\angle X_k - k\Omega_0 t_0)}. \tag{4.39}$$

Likewise, $x(t)$ advanced t_0 seconds, or $x(t + t_0)$, has Fourier series coefficients

$$X_k e^{jk\Omega_0 t_0} = |X_k| e^{j(\angle X_k + k\Omega_0 t_0)}. \tag{4.40}$$

That is, only a change in phase is caused by the time shift; the magnitude spectrum remains the same.

Frequency shifting: When a periodic signal $x(t)$, of fundamental period T_0, modulates a complex exponential $e^{j\Omega_1 t}$,

- the modulated signal $x(t)e^{j\Omega_1 t}$ is periodic of fundamental period T_0 if $\Omega_1 = M\Omega_0$, for an integer $M \geq 1$,
- for $\Omega_1 = M\Omega_0$, $M \geq 1$, the Fourier coefficients X_k are shifted to frequencies $k\Omega_0 + \Omega_1 = (k + M)\Omega_0$,
- the modulated signal is real-valued by multiplying $x(t)$ by $\cos(\Omega_1 t)$.

If we delay or advance a periodic signal, the resulting signal is periodic of the same fundamental period. Only a change in the phase of the Fourier coefficients occurs to accommodate for the shift. Indeed, if

$$x(t) = \sum_k X_k e^{jk\Omega_0 t}$$

we then have

$$x(t - t_0) = \sum_k X_k e^{jk\Omega_0(t-t_0)} = \sum_k \left[X_k e^{-jk\Omega_0 t_0} \right] e^{jk\Omega_0 t},$$

$$x(t + t_0) = \sum_k X_k e^{jk\Omega_0(t+t_0)} = \sum_k \left[X_k e^{jk\Omega_0 t_0} \right] e^{jk\Omega_0 t},$$

so that the Fourier coefficients $\{X_k\}$ corresponding to $x(t)$ are changed to $\{X_k e^{\mp jk\Omega_0 t_0}\}$ for $x(t \mp t_0)$. In both cases, having the same magnitude $|X_k|$ but different phases.

In a dual way, if we multiply the above periodic signal $x(t)$ by a complex exponential of frequency Ω_1, $e^{j\Omega_1 t}$ we obtain a **modulated signal**

$$y(t) = x(t)e^{j\Omega_1 t} = \sum_k X_k e^{j(\Omega_0 k + \Omega_1)t},$$

indicating that the harmonic frequencies are shifted by Ω_1. The signal $y(t)$ is not necessarily periodic. Since T_0 is the fundamental period of $x(t)$

$$y(t + T_0) = x(t + T_0)e^{j\Omega_1(t+T_0)} = \underbrace{x(t)e^{j\Omega_1 t}}_{y(t)} e^{j\Omega_1 T_0}$$

and for it to be equal to $y(t)$ w need $\Omega_1 T_0 = 2\pi M$, for an integer $M > 0$ or

$$\Omega_1 = M\Omega_0 \qquad M \gg 1,$$

which goes along with the amplitude modulation condition that the modulating frequency Ω_1 is chosen much larger than Ω_0. The modulated signal is then given by

$$y(t) = \sum_k X_k e^{j(\Omega_0 k + \Omega_1)t} = \sum_k X_k e^{j\Omega_0(k+M)t} = \sum_\ell X_{\ell-M} e^{j\Omega_0 \ell t},$$

so that the Fourier coefficients are shifted to new frequencies $\{\Omega_0(k + M)\}$.

To keep the modulated signal real-valued, one multiplies the periodic signal $x(t)$ by a cosine of frequency $\Omega_1 = M\Omega_0$ for $M \gg 1$ to obtain a modulated signal

$$y_1(t) = x(t)\cos(\Omega_1 t) = \sum_k 0.5 X_k [e^{j(k\Omega_0 + \Omega_1)t} + e^{j(k\Omega_0 - \Omega_1)t}]$$

so that the harmonic components are now centered around $\pm\Omega_1$.

Example 4.12. To illustrate the modulation property using MATLAB consider modulating a sinusoid $\cos(20\pi t)$ with a periodic train of pulses given by

$$
\begin{aligned}
x_1(t) &= 0.5[1 + \text{sign}(\sin(\pi t))]\sin(\pi t) \\
&= \begin{cases} \sin(\pi t) & \sin(\pi t) \geq 0, \\ 0 & \sin(\pi t) < 0, \end{cases}
\end{aligned}
$$

i.e., the function sign applied to sine gives 1 when the sine is positive and -1 when the sine is negative. Use our function *fourierseries* to find the Fourier series of $x_1(t)$ and the modulated signal $x_2(t) = x_1(t)\cos(20\pi t)$ and to plot their magnitude line spectra.

Solution: A period of $x_1(t)$ is

$$
\begin{aligned}
x_{11}(t) &= [u(t) - u(t - 1)]\sin(\pi t) \\
&= \begin{cases} \sin(\pi t) & 0 \leq t \leq 1, \\ 0 & 1 < t \leq 2, \end{cases}
\end{aligned}
$$

which corresponds to the period of a train of pulses of fundamental period $T_0 = 2$. The following script allows us to compute the Fourier coefficients of the two signals.

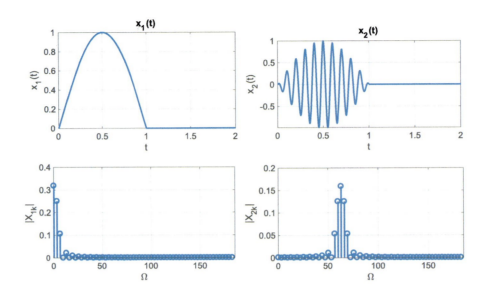

FIGURE 4.16

Modulated square-wave $x_1(t)$ and its spectrum $|X_1(k)|$ (left) and modulated cosine $x_2(t) = x_1(t)\cos(20\pi t)$ and its spectrum $|X_2(k)|$ (right).

```
%%
% Example 4.12---Modulation
%%
clear all;clf
syms t
T0=2;
m1=heaviside(t)-heaviside(t-T0/2);
x1=m1*sin(pi*t);
x2=x1*cos(20*pi*t);
[X1,w]=fourierseries(x1,T0,60);
[X2,w1]=fourierseries(x2,T0,60);
figure(1)
subplot(221)
fplot(x1,[0 T0]);grid;xlabel('t');ylabel('x_1(t)');title('x_1(t)')
subplot(223)
stem(w,abs(X1));grid;xlabel('\Omega');ylabel('|X_{1k}|')
subplot(222)
fplot(x2,[0 T0]);grid;xlabel('t');ylabel('x_2(t)');title('x_2(t)')
subplot(224)
stem(w1,abs(X2)); grid;xlabel('\Omega');ylabel('|X_{2k}|')
```

The modulated signals and their corresponding magnitude line spectra are shown in Fig. 4.16. The Fourier coefficients of the modulated signals are now clustered around the frequency 20π. □

4.4 RESPONSE OF LTI SYSTEMS TO PERIODIC SIGNALS

The following is the most important property of LTI systems.

Eigenfunction Property: In steady state, the response to a complex exponential (or a sinusoid) of a certain frequency is the same complex exponential (or sinusoid), but its amplitude and phase are affected by the frequency response of the system at that frequency.

Suppose the impulse response of a LTI system is $h(t)$ and that $H(s) = \mathcal{L}[h(t)]$ is the corresponding transfer function. If the input to this system is a periodic signal $x(t)$, of fundamental period T_0, with Fourier series

$$x(t) = \sum_{k=-\infty}^{\infty} X_k e^{jk\Omega_0 t} \qquad \Omega_0 = \frac{2\pi}{T_0} \tag{4.41}$$

then according to the eigenfunction property the output in the steady state (the input started at $-\infty$) is

$$y(t) = \sum_{k=-\infty}^{\infty} \left[X_k H(jk\Omega_0) \right] e^{jk\Omega_0 t} \tag{4.42}$$

and if we call $Y_k = X_k H(jk\Omega_0)$ we have a Fourier series representation of $y(t)$ with Y_k as its Fourier coefficients.

Since the input $x(t)$ is real, letting $X_k = |X_k| e^{j\angle X_k} = X_{-k}^*$ we have

$$x(t) = \sum_{k=-\infty}^{\infty} X_k e^{j(k\Omega_0 t)} = X_0 + \sum_{k=1}^{\infty} 2|X_k| \cos(k\Omega_0 t + \angle X_k)$$

where we use the symmetry conditions: $|X_k| = |X_{-k}|$, $\angle X_k = -\angle X_{-k}$, and $\angle X_0 = 0$. The obtained expression is the trigonometric Fourier series of $x(t)$. The steady-state output $y(t)$ is then according to the eigenfunction property:

$$\begin{aligned} y(t) &= \sum_{k=-\infty}^{\infty} \left[X_k H(jk\Omega_0) \right] e^{jk\Omega_0 t} \\ &= X_0 |H(j0)| + 2 \sum_{k=1}^{\infty} |X_k| |H(jk\Omega_0)| \cos(k\Omega_0 t + \angle X_k + \angle H(jk\Omega_0)) \end{aligned}$$

using that $H(jk\Omega_0) = |H(jk\Omega_0)| e^{j\angle H(jk\Omega_0)}$ and that its magnitude and phase are even and odd functions of frequency.

If the input $x(t)$ of a causal and stable LTI system, with impulse response $h(t)$, is periodic of fundamental period T_0 and has the Fourier series

$$x(t) = X_0 + 2\sum_{k=1}^{\infty} |X_k| \cos(k\Omega_0 t + \angle X_k) \qquad \Omega_0 = \frac{2\pi}{T_0} \qquad (4.43)$$

the steady-state response of the system is

$$y(t) = X_0 |H(j0)| + 2\sum_{k=1}^{\infty} |X_k||H(jk\Omega_0)| \cos(k\Omega_0 t + \angle X_k + \angle H(jk\Omega_0)) \qquad (4.44)$$

where

$$H(jk\Omega_0) = |H(jk\Omega_0)|e^{j\angle H(jk\Omega_0)} = \int_0^{\infty} h(\tau)e^{-jk\Omega_0\tau}d\tau = H(s)|_{s=jk\Omega_0} \qquad (4.45)$$

is the **frequency response of the system at** $k\Omega_0$.

Remarks

1. If the input signal $x(t)$ is a combination of sinusoids of frequencies which are not harmonically related, thus the signal is not periodic, the eigenfunction property still holds. For instance, if

$$x(t) = \sum_k A_k \cos(\Omega_k t + \theta_k)$$

and the frequency response of the LTI system is $H(j\Omega)$, the steady-state response is

$$y(t) = \sum_k A_k |H(j\Omega_k)| \cos(\Omega_k t + \theta_k + \angle H(j\Omega_k))$$

and it may not be periodic.

2. A relevant question is: for LTI systems, how does one know *a priori* that the steady state would be reached? The system must be BIBO stable for it to reach steady state. How is then the steady state reached? To see this consider a stable and causal LTI system, with impulse response $h(t)$. Let $x(t) = e^{j\Omega_0 t}$, $-\infty < t < \infty$, be the system input, the output is

$$y(t) = \int_0^{\infty} h(\tau)x(t-\tau)d\tau = e^{j\Omega_0 t} \underbrace{\int_0^{\infty} h(\tau)e^{-j\Omega_0\tau}d\tau}_{H(j\Omega_0)}.$$

The limits of the first integral indicate that the input $x(t-\tau)$ is applied starting at $-\infty$ (when $\tau = \infty$) to t (when $\tau = 0$), thus $y(t)$ is the steady-state response of the system. Suppose then that the input is $x(t) = e^{j\Omega_0 t}u(t)$, then the output of the causal and stable filter is

$$y(t) = \int_0^t h(\tau)x(t-\tau)d\tau = e^{j\Omega_0 t}\int_0^t h(\tau)e^{-j\Omega_0\tau}d\tau$$

where the integral limits indicate the input starts at 0 (when $\tau = t$) and ends at t (when $\tau = 0$); the lower limit of the integral is imposed by the causality of the system ($h(t) = 0$ for $t < 0$). Thus the output reaches steady state when $t \to \infty$, i.e.,

$$y_{ss}(t) = \lim_{t \to \infty} e^{j\Omega_0 t} \int_0^t h(\tau) e^{-j\Omega_0 \tau} d\tau = \lim_{t \to \infty} e^{j\Omega_0 t} H(j\Omega_0).$$

Thus, for a stable system steady state can be reached by starting the input either at $-\infty$ and considering the output at a finite time t, or by starting the input at $t = 0$ and considering the output at $t \to \infty$. Depending on the system, the steady state can be reached a lot faster than the implied infinite time. A transient occurs before attaining the steady state.

3. It is important to realize that if the LTI system is represented by a differential equation and the input is a sinusoid, or combination of sinusoids, it is not necessary to use the Laplace transform to obtain the complete response and then let $t \to \infty$ to find the sinusoidal steady-state response. The Laplace transform is only needed to find the transfer function of the system, which can then be used in Equation (4.44) to find the sinusoidal steady state.

Example 4.13. Use MATLAB to simulate the convolution of a sinusoid $x(t)$ of frequency $\Omega = 20\pi$, amplitude 10 and random phase with the impulse response $h(t) = 20e^{-10t} \cos(40\pi t)$ (a modulated decaying exponential) of a LTI system. Use the MATLAB function *conv* to approximate the convolution integral.

Solution: The following script simulates the convolution of $x(t)$ and $h(t)$. Notice how the convolution integral is approximated by multiplying the output of the function *conv* (which is the convolution sum) by the sampling period T_s used.

```
%%
% Example 4.13---Simulation of convolution
%%
clear all; clf
Ts=0.01; Tend=2; t=0:Ts:Tend;
x=10*cos(20*pi*t+pi*(rand(1,1)-0.5));        % input signal
h=20*exp(-10.^t).*cos(40*pi*t);              % impulse response
% approximate convolution integral
y=Ts*conv(x,h);
M=length(x);
figure(1)
x1=[zeros(1,5) x(1:M)];
z=y(1);y1=[zeros(1,5) z zeros(1,M-1)];
t0=-5*Ts:Ts:Tend;
for k=0:M-6,
   pause(0.05)
   h0=fliplr(h);
   h1=[h0(M-k-5:M) zeros(1,M-k-1)];
   subplot(211)
   plot(t0,h1,'r')
```

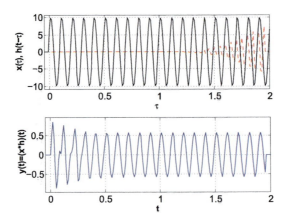

FIGURE 4.17

Convolution simulation. Top figure: input $x(t)$ (solid line) and $h(t - \tau)$ (dashed line); bottom figure: output $y(t)$ transient and steady-state response.

```
    hold on
    plot(t0,x1,'k')
    title('Convolution of x(t) and h(t)')
    ylabel('x(\tau), h(t-\tau)'); grid; axis([min(t0) max(t0) 1.1*min(x) 1.1*max(x)])
     hold off
    subplot(212)
    plot(t0,y1,'b')
    ylabel('y(t)=(x*h)(t)'); grid; axis([min(t0) max(t0) 0.1*min(x) 0.1*max(x)])
    z=[z y(k+2)];
    y1=[zeros(1,5) z zeros(1,M-length(z))];
  end
```

Fig. 4.17 displays the last step of the convolution integral simulation. The system response to the sinusoid starting at $t = 0$ displays a transient which eventually turns into a sinusoid of the same frequency as that of the input, with magnitude and phase depending on the response of the system at the frequency of the input. Notice also that the steady state is attained in a very short time (around $t = 0.5$ s). The transient changes every time that the script is executed due to the random phase.

4.4.1 FILTERING OF PERIODIC SIGNALS

A filter is a LTI system that allows us to retain, get rid of or attenuate frequency components of the input—i.e., to "filter" the input. According to (4.44) if we know the frequency response of the system at the harmonic frequencies of the periodic input $\{H(jk\Omega_0)\}$, the steady-state response of the output of the system is

$$y(t) = X_0|H(j0)| + 2\sum_{k=1}^{\infty}|X_k||H(jk\Omega_0)|\cos(k\Omega_0 t + \angle X_k + \angle H(jk\Omega_0)) \qquad (4.46)$$

where $\{X_k\}$ are the Fourier coefficients of the input $x(t)$, periodic of fundamental frequency Ω_0. Equation (4.46) indicates that depending on the frequency response of the filter at fundamental frequencies $\{k\Omega_0\}$, we can keep (letting $|H(j\ell\Omega_0)| = 1$ for those components of frequencies $\ell\Omega_0$ we want to keep) or get rid of frequency components of the input (letting $|H(j\ell\Omega_0)| = 0$ for those components of frequencies $\ell\Omega_0$ we want to get rid of). The filter output $y(t)$ is periodic of the same fundamental period as the input $x(t)$ with Fourier coefficients $Y_k = X_k H(jk\Omega_0)$.

Example 4.14. To illustrate using MATLAB the filtering of a periodic signal, consider a zero-mean train of pulses with Fourier series

$$x(t) = \sum_{k=-\infty, \neq 0}^{\infty} \frac{\sin(k\pi/2)}{k\pi/2} e^{j2k\pi t}$$

as the driving source of an RC circuit that realizes a low-pass filter (i.e., a system that tries to keep the low-frequency harmonics and to get rid of the high-frequency harmonics of the input). Let the transfer function of the RC low-pass be

$$H(s) = \frac{1}{1 + s/100}.$$

Solution: The following script computes the frequency response of the filter at the harmonic frequencies $H(jk\Omega_0)$ (see Fig. 4.18).

```
%%
% Example 4.14---Filtering of a periodic signal
%%
% Freq response of H(s)=1/(s/100+1)  -- low-pass filter
w0=2*pi;                                   % fundamental frequency
M=20; k=0:M-1; w1=k.*w0;                    % harmonic frequencies
H=1./(1+j*w1/100);  Hm=abs(H); Ha=angle(H);  % frequency response
subplot(211)
stem(w1,Hm,'filled'); grid; ylabel('|H(j\omega)|')
axis([0 max(w1) 0 1.3])
subplot(212)
stem(w1,Ha,'filled'); grid
axis([0 max(w1) -1 0])
ylabel('<H(j \omega)'); xlabel('w (rad/s)')
```

The response due to the train of pulses can be found by computing the response to each of its Fourier series components and adding them. In the simulation, we approximate $x(t)$ using $N = 20$ harmonics

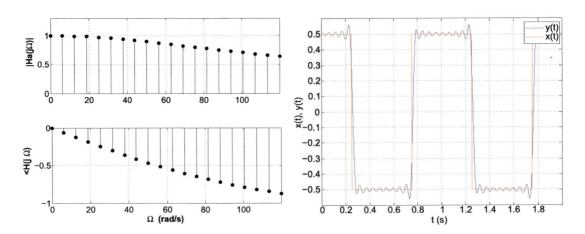

FIGURE 4.18

Left: magnitude and phase response of the low-pass RC filter at harmonic frequencies. Right: response due to the train of pulses $x(t)$. Actual signal values are given by the dashed line, and the filtered signal is indicated by the continuous line.

finding

$$x_N(t) = \sum_{k=-20,\neq 0}^{20} \frac{\sin(k\pi/2)}{k\pi/2} e^{j2k\pi t}$$

and the output voltage across the capacitor is given in the steady state by

$$y_{ss}(t) = \sum_{k=-20,\neq 0}^{20} H(j2k\pi)\frac{\sin(k\pi/2)}{k\pi/2} e^{j2k\pi t}.$$

Because of the small change of the magnitude response of the low-pass filter in the range of frequencies of the input, the output signal is very much like the input (see Fig. 4.18). The following script is used to find the response.

```
%%
% Low-pass filtering
%%
% FS coefficients of input
X(1)=0; % mean value
for k=2:M-1,
    X(k)=sin((k-1)*pi/2)/((k-1)*pi/2);
end
% periodic signal
Ts=0.001; t1=0:Ts:1-Ts;L=length(t1);
```

```
x1=[ones(1,L/4) zeros(1,L/2) ones(1,L/4)]; x1=x1-0.5; x=[x1 x1];
% output of filter
t=0:Ts:2-Ts;
y=X(1)*ones(1,length(t))*Ha(1);
plot(t,y); axis([0 max(t) -.6 .6])
for k=2:M-1,
    y=y+X(k)*Hm(k)*cos(w0*(k-1).*t+Ha(k));
    plot(t,y); axis([0 max(t) -.6 .6]); hold on
    plot(t,x,'r'); axis([0 max(t) -0.6 0.6]); grid
    ylabel('x(t), y(t)'); xlabel('t (s)'); hold off
    pause(0.1)
end
```

4.5 OPERATIONS USING FOURIER SERIES

In this section we show that it is possible to find the Fourier series of the sum, product, derivative and integral of periodic signals using the Fourier series of the signals involved in these operations.

4.5.1 SUM OF PERIODIC SIGNALS

<u>Same fundamental frequency:</u> If $x(t)$ and $y(t)$ are periodic signals with the same fundamental frequency Ω_0, then the Fourier series coefficients of $z(t) = \alpha x(t) + \beta y(t)$ for constants α and β are

$$Z_k = \alpha X_k + \beta Y_k \tag{4.47}$$

where X_k and Y_k are the Fourier coefficients of $x(t)$ and $y(t)$.

<u>Different fundamental frequencies:</u> If $x(t)$ is periodic of fundamental period T_1, and $y(t)$ is periodic of fundamental period T_2 such that $T_2/T_1 = N/M$, for non-divisible integers N and M, then $z(t) = \alpha x(t) + \beta y(t)$ is periodic of fundamental period $T_0 = MT_2 = NT_1$, and its Fourier coefficients are

$$Z_k = \alpha X_{k/N} + \beta Y_{k/M},$$
$$\text{for } k = 0, \pm 1, \cdots \text{ such that } k/N, \ k/M \text{ are integers,} \tag{4.48}$$

where X_k and Y_k are the Fourier coefficients of $x(t)$ and $y(t)$.

If $x(t)$ and $y(t)$ are periodic signals of the same fundamental period T_0 then $z(t) = \alpha x(t) + \beta y(t)$ is also periodic of fundamental period T_0. The Fourier coefficients of $z(t)$ are then

$$Z_k = \alpha X_k + \beta Y_k,$$

where X_k and Y_k are the Fourier coefficients of $x(t)$ and $y(t)$, respectively.

In general, if $x(t)$ is periodic of fundamental period T_1, and $y(t)$ is periodic of fundamental period T_2, their sum $z(t) = \alpha x(t) + \beta y(t)$ is periodic if the ratio T_2/T_1 is a rational number, i.e., $T_2/T_1 = N/M$ for non-divisible integers N and M. If so, the period of $z(t)$ is $T_0 = MT_2 = NT_1$. The fundamental frequency of $z(t)$ would then be $\Omega_0 = \Omega_1/N = \Omega_2/M$ for Ω_1 and Ω_2 the fundamental frequencies of $x(t)$ and $y(t)$. The Fourier series of $z(t)$ is then

$$
\begin{aligned}
z(t) &= \alpha x(t) + \beta y(t) = \alpha \sum_\ell X_\ell e^{j\Omega_1 \ell t} + \beta \sum_m Y_m e^{j\Omega_2 m t} \\
&= \alpha \sum_\ell X_\ell e^{jN\Omega_0 \ell t} + \beta \sum_m Y_m e^{jM\Omega_0 m t} \\
&= \alpha \sum_{k=0,\pm N,\pm 2N,\cdots} X_{k/N} e^{j\Omega_0 k t} + \beta \sum_{n=0,\pm M,\pm 2M,\cdots} Y_{n/M} e^{j\Omega_0 n t}
\end{aligned}
$$

where the last equation is obtained by letting $k = N\ell$ and $n = Mm$. Thus the coefficients are $Z_k = \alpha X_{k/N} + \beta Y_{k/M}$ for integers k such that k/N and k/M are integers.

Example 4.15. Consider the sum $z(t)$ of a periodic signal $x(t)$ of fundamental period $T_1 = 2$, with a periodic signal $y(t)$ with fundamental period $T_2 = 0.2$. Find the Fourier coefficients Z_k of $z(t)$ in terms of the Fourier coefficients X_k and Y_k of $x(t)$ and $y(t)$.

Solution: The ratio $T_2/T_1 = 1/10$ is rational, so $z(t)$ is periodic of fundamental period $T_0 = T_1 = 10T_2 = 2$. The fundamental frequency of $z(t)$ is $\Omega_0 = \Omega_1 = \pi$, and $\Omega_2 = 10\Omega_0 = 10\pi$ is the fundamental frequency of $y(t)$. Thus, the Fourier coefficients of $z(t)$ are

$$
Z_k = \begin{cases} X_k + Y_{k/10} & \text{when } k = 0, \pm 10, \pm 20, \cdots, \\ X_k & k = \pm 1, \cdots, \pm 9, \pm 11 \cdots \pm 19 \cdots. \end{cases} \qquad \square
$$

4.5.2 MULTIPLICATION OF PERIODIC SIGNALS

If $x(t)$ and $y(t)$ are periodic signals of the same fundamental period T_0, then their product

$$
z(t) = x(t)y(t) \tag{4.49}
$$

is also periodic of fundamental period T_0 and with Fourier coefficients which are the **convolution sum** of the Fourier coefficients of $x(t)$ and $y(t)$:

$$
Z_k = \sum_m X_m Y_{k-m}. \tag{4.50}
$$

If $x(t)$ and $y(t)$ are periodic with the same fundamental period T_0 then $z(t) = x(t)y(t)$ is also periodic of fundamental period T_0, since $z(t + kT_0) = x(t + kT_0)y(t + kT_0) = x(t)y(t) = z(t)$. Furthermore,

$$
\begin{aligned}
x(t)y(t) &= \sum_m X_m e^{jm\Omega_0 t} \sum_\ell Y_\ell e^{j\ell\Omega_0 t} = \sum_m \sum_\ell X_m Y_\ell e^{j(m+\ell)\Omega_0 t} \\
&= \sum_k \left[\sum_m X_m Y_{k-m} \right] e^{jk\Omega_0 t} = z(t)
\end{aligned}
$$

where we let $k = m + \ell$. The coefficients of the Fourier series of $z(t)$ are then

$$Z_k = \sum_m X_m Y_{k-m}$$

or the convolution sum of the sequences X_k and Y_k, to be formally defined in Chapter 9.

Example 4.16. Consider the train of rectangular pulses $x(t)$ shown in Fig. 4.4, let $z(t) = 0.25x^2(t)$. Use the Fourier series of $z(t)$ to show that

$$X_k = \alpha \sum_m X_m X_{k-m}$$

for some constant α. Determine α.

Solution:
The signal $0.5x(t)$ is a train of pulses of unit amplitude, so that $z(t) = (0.5x(t))^2 = 0.5x(t)$. Thus $Z_k = 0.5X_k$, but also as a product of $0.5x(t)$ with itself we have

$$Z_k = \sum_m [0.5X_m][0.5X_{k-m}]$$

and thus

$$\underbrace{0.5X_k}_{Z_k} = 0.25 \sum_m X_m X_{k-m} \quad \Rightarrow \quad X_k = \frac{1}{2} \sum_m X_m X_{k-m}, \tag{4.51}$$

so that $\alpha = 0.5$.
    The Fourier series of $z(t) = 0.5x(t)$ according to the results in Example 4.5 is

$$z(t) = 0.5x(t) = \sum_{k=-\infty}^{\infty} \frac{\sin(\pi k/2)}{\pi k} e^{jk2\pi t}.$$

If we define

$$S(k) = 0.5X_k = \frac{\sin(k\pi/2)}{k\pi} \quad \Rightarrow \quad X_k = 2S(k),$$

we have from (4.51) the interesting result

$$S(k) = \sum_{m=-\infty}^{\infty} S(m)S(k-m)$$

or the convolution sum of the discrete sinc function $S(k)$ with itself is $S(k)$. □

4.5.3 DERIVATIVES AND INTEGRALS OF PERIODIC SIGNALS

Derivative: The derivative $dx(t)/dt$ of a periodic signal $x(t)$, of fundamental period T_0, is periodic of the same fundamental period T_0. If $\{X_k\}$ are the coefficients of the Fourier series of $x(t)$, the Fourier coefficients of $dx(t)/dt$ are

$$jk\Omega_0 X_k \qquad (4.52)$$

where Ω_0 is the fundamental frequency of $x(t)$.

Integral: For a zero-mean, periodic signal $y(t)$, of fundamental period T_0, and Fourier coefficients $\{Y_k\}$, the integral

$$z(t) = \int_{-\infty}^{t} y(\tau)d\tau$$

is periodic of the same fundamental period as $y(t)$, with Fourier coefficients

$$Z_k = \frac{Y_k}{jk\Omega_0} \qquad k \text{ integer} \neq 0,$$

$$Z_0 = -\sum_{m \neq 0} Y_m \frac{1}{jm\Omega_0} \qquad \Omega_0 = \frac{2\pi}{T_0}. \qquad (4.53)$$

These properties come naturally from the Fourier series representation of the periodic signal. Once we find the Fourier series of a periodic signal, we can differentiate it or integrate it (only when the dc value is zero). The derivative of a periodic signal is obtained by computing the derivative of each of the terms of its Fourier series, i.e., if

$$x(t) = \sum_k X_k e^{jk\Omega_0 t} \quad \text{then}$$

$$\frac{dx(t)}{dt} = \sum_k X_k \frac{de^{jk\Omega_0 t}}{dt} = \sum_k \left[jk\Omega_0 X_k\right] e^{jk\Omega_0 t},$$

indicating that if the Fourier coefficients of $x(t)$ are X_k, the Fourier coefficients of $dx(t)/dt$ are $jk\Omega_0 X_k$.

To obtain the integral property we assume $y(t)$ is a zero-mean signal so that its integral $z(t)$ is finite. For some integer M, such that $MT_0 \leq t$, and using that the average in each period of $y(t)$ is zero we have

$$z(t) = \int_{-\infty}^{t} y(\tau)d\tau = \int_{-\infty}^{MT_0} y(\tau)d\tau + \int_{MT_0}^{t} y(\tau)d\tau = 0 + \int_{MT_0}^{t} y(\tau)d\tau.$$

Replacing $y(t)$ by its Fourier series gives

$$z(t) = \int_{MT_0}^{t} y(\tau)d\tau = \int_{MT_0}^{t} \sum_{k \neq 0} Y_k e^{jk\Omega_0 \tau} d\tau = \sum_{k \neq 0} Y_k \int_{MT_0}^{t} e^{jk\Omega_0 \tau} d\tau = \sum_{k \neq 0} Y_k \frac{1}{jk\Omega_0} \left[e^{jk\Omega_0 t} - 1\right]$$

$$= -\sum_{k \neq 0} Y_k \frac{1}{jk\Omega_0} + \sum_{k \neq 0} Y_k \frac{1}{jk\Omega_0} e^{jk\Omega_0 t}$$

where the first term corresponds to the average Z_0 and $Z_k = Y_k/(jk\Omega_0)$, $k \neq 0$, are the rest of the Fourier coefficients of $z(t)$.

Remark
It should be clear now why the derivative of a periodic signal $x(t)$ enhances its higher harmonics. Indeed, the Fourier coefficients of the derivative $dx(t)/dt$ are those of $x(t)$, X_k, multiplied by $j\Omega_0 k$, which increases with k. Likewise, the integration of a zero-mean periodic signal $x(t)$ does the opposite, i.e., it smooths out the signal, as we multiply X_k by decreasing terms $1/(jk\Omega_0)$ as k increases.

Example 4.17. Let $g(t)$ be the derivative of a triangular train of pulses $x(t)$, of fundamental period $T_0 = 1$. The period of $x(t)$, $0 \leq t \leq 1$, is

$$x_1(t) = 2r(t) - 4r(t - 0.5) + 2r(t - 1).$$

Use the Fourier series of $g(t)$ to find the Fourier series of $x(t)$.

Solution: According to the derivative property we have

$$X_k = \frac{G_k}{jk\Omega_0} \qquad k \neq 0$$

for the Fourier coefficients of $x(t)$. The signal $g(t) = dx(t)/dt$ has a corresponding period $g_1(t) = dx_1(t)/dt = 2u(t) - 4u(t - 0.5) + 2u(t - 1)$. Thus the Fourier series coefficients of $g(t)$ are

$$G_k = \frac{2e^{-0.5s}}{s}\left(e^{0.5s} - 2 + e^{-0.5s}\right)\Big|_{s=j2\pi k} = 2(-1)^k \frac{\cos(\pi k) - 1}{j\pi k} \qquad k \neq 0,$$

which are used to obtain the coefficients X_k for $k \neq 0$. The dc component of $x(t)$ is found to be 0.5 from its plot. □

Example 4.18. Consider the reverse of Example 4.17, that is given the periodic signal $g(t)$ of fundamental period $T_0 = 1$ and Fourier coefficients

$$G_k = 2(-1)^k \frac{\cos(\pi k) - 1}{j\pi k} \qquad k \neq 0$$

and $G_0 = 0$, find the integral

$$x(t) = \int_{-\infty}^{t} g(\tau)d\tau.$$

Solution:
The signal $x(t)$ is also periodic of the same fundamental period as $g(t)$, i.e., $T_0 = 1$ and $\Omega_0 = 2\pi$. The Fourier coefficients of $x(t)$ are

$$X_k = \frac{G_k}{j\Omega_0 k} = (-1)^k \frac{4(\cos(\pi k) - 1)}{(j2\pi k)^2} = (-1)^{(k+1)} \frac{\cos(\pi k) - 1}{\pi^2 k^2} \qquad k \neq 0$$

and the average term

$$
\begin{aligned}
X_0 &= -\sum_{m=-\infty, m\neq 0}^{\infty} G_m \frac{1}{j2m\pi} = \sum_{m=-\infty, m\neq 0}^{\infty} (-1)^m \frac{\cos(\pi m) - 1}{(\pi m)^2} \\
&= 0.5 \sum_{m=-\infty, m\neq 0}^{\infty} (-1)^{m+1} \left[\frac{\sin(\pi m/2)}{(\pi m/2)} \right]^2
\end{aligned}
$$

where we used $1 - \cos(\pi m) = 2\sin^2(\pi m/2)$. We used the following script to obtain the average, and to approximate the triangular signal using 100 harmonics. The mean is obtained as 0.498—the actual is 0.5. The approximate signal $x_N(t)$ is shown in Fig. 4.19.

```
%%
% Example 4.18---Approximation of triangular signal
%%
clf; clear all
w0=2*pi; N=100;
% computation of mean value
DC=0;
for m=1:N,
     DC=DC+2*(-1)^(m)*(cos(pi*m)-1)/(pi*m)^2;
end
% computation of Fourier series coefficients
Ts=0.001; t=0:Ts:2-Ts;
for k=1:N,
   X(k)=(-1)^(k+1)*(cos(pi*k)-1)/((pi*k)^2);
end
X=[DC X];                                       % FS coefficients
xa=X(1)*ones(1,length(t));
figure(1)
for k=2:N,
   xa=xa+2*abs(X(k))*cos(w0 *(k-1).*t+angle(X(k)));     % approximate signal
end
```

4.5.4 AMPLITUDE AND TIME SCALING OF PERIODIC SIGNALS

Scaling the amplitude and time of a periodic signal $x(t)$ a new periodic signal $y(t)$ can be generated. Indeed, if $x(t)$ is a periodic signal of fundamental period T_0, or fundamental frequency $\Omega_0 = 2\pi/T_0$, with a Fourier series representation

$$
x(t) = \sum_{k=-\infty}^{\infty} X_k e^{jk\Omega_0 t} \tag{4.54}
$$

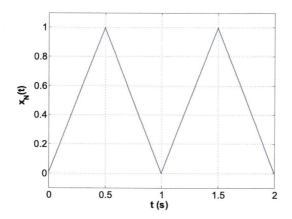

FIGURE 4.19

Two periods of the approximate triangular signal $x_N(t)$ using 100 harmonics.

then for a constant A and a scale value $\alpha \neq 0$, a new signal

$$y(t) = Ax(\alpha t) = \sum_{k=-\infty}^{\infty} (AX_k)e^{jk(\alpha\Omega_0)t} \tag{4.55}$$

is also periodic with Fourier coefficients $Y_k = AX_k$ and new fundamental frequencies $\Omega_1 = \alpha\Omega_0$. The fundamental period T_0 of the original signal becomes $T_1 = 2\pi/\Omega_1 = T_0/\alpha$. Thus, if $\alpha > 1$, then $x(\alpha t)$ is compressed (period $T_1 < T_0$) and the fundamental frequencies are expanded, and when $0 < \alpha < 1$, we have the opposite effect: time expansion and frequency compression. As observed before, time is inversely proportional to frequency. It is important to notice that if only time scaling is done, i.e., $A = 1$, for $\alpha > 0$ the Fourier coefficients of the original signal and the time-scaled signal are identical, it is the fundamental frequencies that are changed by the scaling factor.

For negative scaling values of reflection combined with expansion and contraction we see that the new signal

$$
\begin{aligned}
y(t) &= Ax(-|\alpha|t) = \sum_{k=-\infty}^{\infty} (AX_k)e^{jk(-|\alpha|\Omega_0)t} = \sum_{\ell=-\infty}^{\infty} (AX_{-\ell})e^{j|\ell|(\alpha\Omega_0)t} \\
&= \sum_{\ell=-\infty}^{\infty} (AX_{\ell}^*)e^{j\ell(\alpha\Omega_0)t} \tag{4.56}
\end{aligned}
$$

has Fourier coefficients $Y_\ell = AX_{-\ell} = AX_\ell^*$, the complex conjugates of those of $x(t)$ multiplied by A, and a new set of harmonic frequencies $\{\ell\Omega_1 = \ell|\alpha|\Omega_0\}$.

Example 4.19. Let $x(t)$ be a half-wave rectified signal. If we normalize $x(t)$ to have maximum amplitude of 1 and a fundamental period of $T_0 = 1$, obtain its Fourier series coefficients. Write a MATLAB function to generate $x(t)$ from the Fourier coefficients. Choose the value of A so that $Ax(10t)$ has a unity dc value. What would be a possible advantage of using this signal rather than the original $x(t)$ if we wish to use these signals for a dc source?

Solution:
A period $x_1(t)$, $0 \le t \le T_0$, of the half-wave rectified signal $x(t)$, normalized in amplitude and period $(T_0 = 1)$, is given by

$$x_1(t) = \begin{cases} \sin(2\pi t) & 0 \le t \le 0.5, \\ 0 & 0.5 < t \le 1, \end{cases}$$

and can be written as

$$x_1(t) = \sin(2\pi t)u(t) + \sin(2\pi(t - 0.5))u(t - 0.5)$$

with Laplace transform

$$X_1(s) = \frac{2\pi}{s^2 + 4\pi^2}(1 + e^{-0.5s})$$

and as such the Fourier coefficients of $x(t)$ are

$$X_k = \frac{1}{T_0}X_1(jk\Omega_0) = \frac{1}{2\pi(1 - k^2)}(1 + (-1)^k)$$

where we replaced $T_0 = 1$ and $\Omega_0 = 2\pi$. From this equation, we see that the dc coefficient is $X_0 = 1/\pi$, but the coefficient X_1 cannot be calculated using the above expression as it becomes zero divided by zero when $k = 1$, so instead we find it by its integral expression

$$X_1 = \int_0^{0.5} \sin(2\pi t)e^{-j2\pi t} dt = \int_0^{0.5} \frac{1 - e^{-j4\pi t}}{2j} dt = \frac{1}{4j} - \frac{1}{8\pi} + \frac{1}{8\pi} = \frac{1}{4j}.$$

For $k \ge 2$, $X_k = 0$ when $k \ge 2$ is odd and $X_k = 1/(\pi - k^2)$ when $k \ge 2$ is even. The following script generates the signal from the trigonometric Fourier coefficients which are obtained from the complex coefficients as $c_k = \mathcal{R}e[X_k]$, $d_k = -\mathcal{I}m[X_k]$ giving $c_0 = X_0 = 1/\pi$; $c_1 = 0$, $d_1 = 0.25$; $c_k = 1/(\pi(1 - k^2))$ and $d_k = 0$ for $k \ge 2$.

```
%%
% Examples 4.19---Generation of half-wave rectified signals
%%
clear all; clf
T0=1;N=400;omega0=2*pi;T=2;
% trigonometric coefficients
c=zeros(1,N); d=c;
X0=1/pi; c(1)=0; d(1)=0.25;
for k=2:2:N
    d(k)=0;
    c(k)=1/(pi*(1-k^2));
end
```

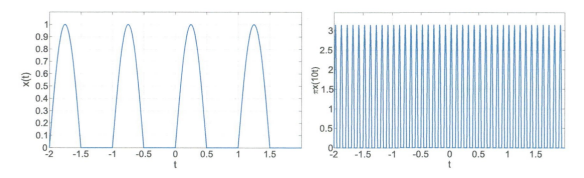

FIGURE 4.20

Half-wave rectified signals: normalized amplitude and period $T_0 = 1$, $x(t)$ (left); scaled amplitude and time $\pi x(10t)$ (right).

```
alpha=1;A=pi;
[x,t]=InvFSeries(T,X0,c,d,N,alpha,A);
figure(1)
plot(t,x); axis([min(t) max(t) 1.1*min(x) 1.1*max(x)]);grid
alpha=10;A=pi;
[x,t]=InvFSeries(T,X0,c,d,N,alpha,A);
figure(2)
plot(t,x); axis([min(t) max(t) 1.1*min(x) 1.1*max(x)]);grid
```

The function used to generate the periodic signals from the coefficients of the trigonometric Fourier series is given below.

```
function [x,t]=InvFseries(T,X0,c,d,N,alpha,A)
% generates periodic signals in [-T,T] from dc X0 and N Fourier
% coefficients c,d, amplitude scale A and time scale alpha
Ts=0.0001; t=-T:Ts:T-Ts;omega0=2*pi;
x=X0*ones(1,length(t));
for k=1:N,
        x=x+2*d(k)*sin(omega0*k*alpha*t)+2*+c(k)*cos(omega0*k*alpha*t);
end
x=A*x;
```

The resulting signals are shown in Fig. 4.20. Although these two signals look very different they have the same Fourier coefficients, thus we chose $A = \pi$ to get a unity dc value. A possible advantage of using $\pi x(10t)$ when obtaining a dc source is that its period is $T_1 = T_0/10 = 0.1$ and thus its harmonic frequencies are $\{k\Omega_1 = 10k\Omega_0, \ k = 0, 1, 2, \cdots\}$ providing ten times the frequency spacing between

Table 4.1 Basic properties of Fourier series

Signals and constants	$x(t)$, $y(t)$ periodic with period T_0, α, β	X_k, Y_k
Linearity	$\alpha x(t) + \beta y(t)$	$\alpha X_k + \beta Y_k$
Parseval's power relation	$P_x = \dfrac{1}{T_0}\displaystyle\int_{T_0} \lvert x(t)\rvert^2 dt$	$P_x = \displaystyle\sum_k \lvert X_k\rvert^2$
Differentiation	$\dfrac{dx(t)}{dt}$	$jk\Omega_0 X_k$
Integration	$\displaystyle\int_{-\infty}^{t} x(t')dt'$ only if $X_0 = 0$	$\dfrac{X_k}{jk\Omega_0} k \neq 0$
Time shifting	$x(t - \alpha)$	$e^{-j\alpha\Omega_0} X_k$
Frequency shifting	$e^{jM\Omega_0 t} x(t)$	X_{k-M}
Symmetry	$x(t)$ real	$\lvert X_k\rvert = \lvert X_{-k}\rvert$ even function of k $\angle X_k = -\angle X_{-k}$ odd function of k
Multiplication	$z(t) = x(t)y(t)$	$Z_k = \displaystyle\sum_m X_m Y_{k-m}$

the dc and the first harmonic compared to the spacing of Ω_0 given by the original signal. This means that a less sharp low-pass filter would be needed to get the dc source. □

4.6 WHAT HAVE WE ACCOMPLISHED? WHERE DO WE GO FROM HERE?

Periodic signals are not to be found in practice, so where did Fourier get the intuition to come up with a representation for them? As you will see, the fact that periodic signals are not found in practice does not mean that they are not useful. The Fourier representation of periodic signals will be fundamental in finding a representation for non-periodic signals. A very important concept you have learned in this chapter is that the inverse relation between time and frequency provides complementary information for the signal. The frequency domain constitutes the other side of the coin in representing signals. As mentioned before, it is the eigenfunction property of linear time invariant systems that holds the theory together. It will provide the fundamental principle for filtering. You should have started to experience deja vu in terms of the properties of the Fourier series (Table 4.1 summarizes them), some look like a version of the ones in the Laplace transform given the connection between these transforms. The Fourier series of some basic signals, normalized so they have unity fundamental period and amplitude, are shown in Table 4.2. You should have also noticed the usefulness of the Laplace transform in finding the Fourier coefficients, avoiding integration whenever possible. The next chapter will extend some of the results obtained in this chapter, thus unifying the treatment of periodic and non-periodic signals and the concept of spectrum. Also the frequency representation of systems will be introduced and exemplified by its application in filtering. Modulation is the basic tool in communications and can be easily explained in the frequency domain.

Table 4.2 Fourier series of normalized signals

Signal	Period	Fourier coefficients
Sinusoid	$x_1(t) = \cos(2\pi t + \theta)[u(t) - u(t-1)]$	$X_1 = 0.5e^{j\theta}$, $X_{-1} = X_1^*$, $X_k = 0, k \neq 1, -1$
Sawtooth	$x_1(t) = t[u(t) - u(t-1)] - u(t-1)$	$X_0 = 0.5$, $X_k = j\dfrac{1}{2\pi k}, k \neq 0$
Rectangular pulse	$x_1(t) = u(t) - u(t-d), 0 < d < 1$	$X_0 = d$, $X_k = d\dfrac{\sin(\pi kd)}{\pi kd}e^{-jk\pi d}, k \neq 0$
Square wave	$x_1(t) = u(t) - 2u(t-0.5) + u(t-1)$	$X_0 = 0, X_k = -j\dfrac{1-(-1)^k}{\pi k}$
Half-wave rectified	$x_1(t) = \sin(2\pi t)u(t) + \sin(2\pi(t-0.5))u(t-0.5)$	$X_0 = \dfrac{1}{\pi}, X_1 = \dfrac{-j}{4}$; $X_k = \dfrac{1+(-1)^k}{2\pi(1-k^2)}, k \neq 0, 1$
Full-wave rectified	$x_1(t) = \sin(\pi t)u(t) + \sin(\pi(t-1))u(t-1)$	$X_0 = \dfrac{2}{\pi}, X_k = \dfrac{1+(-1)^k}{\pi(1-k^2)}, k \neq 0$
Triangular	$x_1(t) = 2r(t) - 4r(t-0.5) + 2r(t-1)$	$X_0 = 0.5, X_k = -\dfrac{1-(-1)^k}{k^2\pi^2}, k \neq 0$
Impulse sequence	$x_1(t) = \delta(t)$	$X_k = 1$

4.7 PROBLEMS

4.7.1 BASIC PROBLEMS

4.1 The eigenfunction property is only valid for LTI systems. Consider the cases of non-linear and of time-varying systems.

   (a) A system represented by the input–output equation $y(t) = x^2(t)$ is non-linear. Let the input be $x(t) = e^{j\pi t/4}$ find the corresponding system output $y(t)$. Does the eigenfunction property hold? Explain.

   (b) Consider a time-varying system $y(t) = x(t)[u(t) - u(t-1)]$. Let $x(t) = e^{j\pi t/4}$ find the corresponding system output $y(t)$. Does the eigenfunction property hold? Explain.

Answers: (a) $y(t) = x^2(t) = e^{j\pi t/2}$ the eigenfunction property does not hold.

4.2 The input–output equation for an analog averager is

$$y(t) = \frac{1}{T}\int_{t-T}^{t} x(\tau)d\tau$$

let $x(t) = e^{j\Omega_0 t}$. Since the system is LTI then the output should be $y(t) = e^{j\Omega_0 t}H(j\Omega_0)$.

   (a) Find the integral for the given input and then compare it with the above equation to find $H(j\Omega_0)$, the response of the averager at frequency Ω_0.

   (b) Find the impulse response $h(t)$, from the input–output equation, and its Laplace transform $H(s)$ to verify that $H(j\Omega_0)$ obtained above is correct.

Answer: $H(j\Omega_0) = (1 - e^{-j\Omega_0 T})/(j\Omega_0 T)$.

4.3 A periodic signal $x(t)$ has a fundamental frequency $\Omega_0 = 1$ and a period of it is

$$x_1(t) = u(t) - 2u(t - \pi) + u(t - 2\pi).$$

(a) Find the Fourier series coefficients $\{X_k\}$ of $x(t)$ using their integral definition.
(b) Use the Laplace transform to find the Fourier series coefficients $\{X_k\}$ of $x(t)$.
Answers: $X_0 = 0$, $X_k = 0$ for k even.

4.4 Let a periodic signal $x(t)$ with a fundamental frequency $\Omega_0 = 2\pi$ have a period

$$x_1(t) = t[u(t) - u(t - 1)].$$

(a) Plot $x(t)$, and indicate its fundamental period T_0.
(b) Compute the Fourier series coefficients of $x(t)$ using their integral definition.
(c) Use the Laplace transform to compute the Fourier series coefficients X_k. Indicate how to compute the dc term.
(d) Suppose that $y(t) = dx(t)/dt$, find the Fourier transform of $x(t)$ by means of the Fourier series coefficients Y_k of $y(t)$. Explain.
Answers: $x_1(t)$ is periodic of fundamental period $T_0 = 1$; $X_k = j/(2\pi k)$, $k \neq 0$, $X_0 = 0.5$.

4.5 Consider the following problems related to the exponential Fourier series.
(a) The exponential Fourier series of a periodic signal $x(t)$ of fundamental period T_0 is

$$x(t) = \sum_{k=-\infty}^{\infty} \frac{3}{4 + (k\pi)^2} e^{jk\pi t}.$$

   i. Determine the value of the fundamental period T_0.
   ii. What is the average or dc value of $x(t)$?
   iii. Is $x(t)$ even, odd or neither even nor odd function of time?
   iv. One of the frequency components of $x(t)$ is expressed as $A\cos(3\pi t)$. What is A?
(b) A train of rectangular pulses $x(t)$, having a period $x_1(t) = u(t) - 2u(t - \pi) + u(t - 2\pi)$ has a Fourier series

$$x(t) = \frac{4}{\pi} \sum_{k=1}^{\infty} \frac{1}{2k - 1} \sin((2k - 1)t).$$

Figure out how to use these results to find an expression for π. Obtain an approximation of π using the first three terms of the Fourier series.
Answers: (a) $T_0 = 2$; $x(t)$ even; $A = 6/(4 + 9\pi^2)$; (b) use $x(1) = 1$ for instance.

4.6 Suppose you have the Fourier series of two periodic signals $x(t)$ and $y(t)$ of fundamental periods T_1 and T_2, respectively. Let X_k and Y_k be the Fourier series coefficients corresponding to $x(t)$ and $y(t)$.
(a) If $T_1 = T_2$ what would be the Fourier series coefficients of $z(t) = x(t) + y(t)$ in terms of X_k and Y_k?

(b) If $T_1 = 2T_2$ determine the Fourier series coefficients of $w(t) = x(t) + y(t)$ in terms of X_k and Y_k?

Answers: $Z_k = X_k + Y_k$; $W_k = X_k$ for k odd and $W_k = X_k + Y_k$ for k even.

4.7 Consider a periodic signal

$$x(t) = 0.5 + 4\cos(2\pi t) - 8\cos(4\pi t) \qquad -\infty < t < \infty.$$

(a) Determine the fundamental frequency Ω_0 of $x(t)$.

(b) Find the Fourier series coefficients $\{X_k\}$ and carefully plot their magnitude and phase. According to the line spectrum at what frequency is the power of $x(t)$ the largest?

(c) When $x(t)$ is passed through a filter with transfer function $H(s)$ the output of the filter is $y(t) = 2 - 2\sin(2\pi t)$. Determine the values of $H(j\Omega)$ at $\Omega = 0$, 2π, 4π rad/s.

Answers: $\Omega_0 = 2\pi$; $H(j0) = 4$, $H(j4\pi) = 0$.

4.8 Let $x(t)$ be a periodic signal $x(t)$ with fundamental period $T_0 = 1$ and a period $x_1(t) = -0.5t[u(t) - u(t-1)]$.

(a) Consider the derivative $g(t) = dx(t)/dt$. Indicate if $g(t)$ is periodic and if so give its fundamental period.

(b) Find the Fourier series of $g(t)$ and use it to find the Fourier series of $x(t)$.

(c) Suppose you consider the signal $y(t) = 0.5 + x(t)$. Find the Fourier series of $y(t)$.

Answers: $G_k = 0.5$, $X_k = 0.5/(jk2\pi)$, $k \neq 0$.

4.9 We are interested in designing a dc voltage source. To do so, we full-wave rectify an AC voltage to obtain the signal $x(t) = |\cos(\pi t)|$, $-\infty < t < \infty$. We propose then to obtain an ideal low-pass filter with frequency response $H(j\Omega)$ to which we input the signal $x(t)$. The output of this filter is $y(t)$, which is as well the output of the dc source.

(a) Specify the magnitude response $|H(j\Omega)|$ of the ideal low-pass filter so that only the dc component of $x(t)$ is passed through. Let the phase response be $\angle[H(j\Omega)] = 0$. Find the dc value of the full-wave rectified signal $x(t)$, and determine the dc gain of the filter so that the output $y(t) = 1$.

(b) If the input to the ideal filter obtained above is a half-wave rectified signal, what would be the corresponding output $y(t)$?

Answers: Full-wave rectified $x(t)$: $X_0 = 2/\pi$; $H(j0) = \pi/2$.

4.10 Consider the periodic signal $x(t)$ shown in Fig. 4.21.

(a) Use the Laplace transform to compute the Fourier series coefficients X_k, $k \neq 0$ of $x(t)$.

(b) Suppose that to find the Fourier series of $x(t)$ we consider finding its derivative, or $g(t) = dx(t)/dt$. Give an expression for $g(t)$ and use the Laplace transform to find its Fourier series

$$g(t) = \sum_{k=-\infty}^{\infty} G_k e^{jk\Omega_0 t}$$

and then use it to find the Fourier series of $x(t)$.

Answers: $X_k = -j/(2\pi k)$, $k \neq 0$; $G_k = 1$.

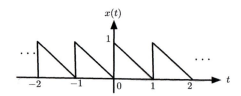

FIGURE 4.21

Problem 4.10.

4.11 The following problems are about the steady-state response of LTI systems due to periodic signals.

(a) The transfer function of a LTI system is

$$H(s) = \frac{Y(s)}{X(s)} = \frac{s+1}{s^2 + 3s + 2}.$$

If the input to this system is $x(t) = 1 + \cos(t + \pi/4)$ what is the output $y(t)$ in the steady state.

(b) The transfer function of a LTI system is given by

$$H(s) = \frac{Y(s)}{X(s)} = \frac{1}{s^2 + 3s + 2}$$

and its input is $x(t) = 4u(t)$.

    i. Use the eigenfunction property of LTI systems to find the steady-state response $y(t)$ of this system.

    ii. Verify your result above by means of the Laplace transform.

Answer: (a) $y_{ss}(t) = 0.5 + 0.447\cos(t + 18.4°)$; (b) $y_{ss}(t) = 2$.

4.12 We want to use the Fourier series of a square train of pulses (done in the chapter) to compute the Fourier series of the triangular signal $x(t)$ with a period

$$x_1(t) = r(t) - 2r(t-1) + r(t-2)$$

(a) Find the derivative of $x(t)$ or $y(t) = dx(t)/dt$ and carefully plot it. Plot also $z(t) = y(t) + 1$. Use the Fourier series of the square train of pulses to compute the Fourier series coefficients of $y(t)$ and $z(t)$.

(b) Obtain the sinusoidal forms of $y(t)$ and $z(t)$ and explain why they are represented by sines and why $z(t)$ has a nonzero mean.

(c) Obtain the Fourier series coefficients of $x(t)$ from those of $y(t)$.

(d) Obtain the sinusoidal form of $x(t)$ and explain why the cosine representation is more appropriate for this signal than a sine representation.

Answers: $y(t) = 4\sum_{k>0,\text{odd}} \sin(\pi kt)/(\pi k)$.

4.13 Consider the Fourier series of two periodic signals

$$x(t) = \sum_{k=-\infty}^{\infty} X_k e^{j\Omega_0 kt}, \quad y(t) = \sum_{k=-\infty}^{\infty} Y_k e^{j\Omega_1 kt}.$$

(a) Let $\Omega_1 = \Omega_0$, is $z(t) = x(t)y(t)$ periodic? If so, what is its fundamental period and its Fourier series coefficients?

(b) If $\Omega_1 = 2\Omega_0$, is $w(t) = x(t)y(t)$ periodic? If so, what is its fundamental period and its Fourier series coefficients?

Answers: If $\Omega_1 = 2\Omega_0$, $Z_n = \sum_k X_k Y_{(n-k)/2}$ for $(n-k)/2$ integer.

4.14 Let $x(t) = \sin^2(2\pi t)$, a periodic signal of fundamental period $T_0 = 1$, and $y(t) = |\sin(2\pi t)|$ also periodic of the same fundamental period.

(a) A trigonometric identity gives that $x(t) = 0.5[1 - \cos(4\pi t)]$. Use this result to find its complex exponential Fourier series.

(b) Use the Laplace transform to find the Fourier series of $y(t)$.

(c) Are $x(t)$ and $y(t)$ identical? Explain.

(d) Indicate how you would use an ideal low-pass filter to get a dc source from $x(t)$ and $y(t)$. Indicate the bandwidth and the magnitude of the filters. Which of these two signals is used for generating full-waved rectified dc source?

Answers: For $x(t)$: $X_0 = 0.5$, $X_2 = X_{-2} = -0.25$; dc value of $y(t)$ is $Y_0 = 2/\pi$.

4.15 The periodic impulse signal with period T_1 is

$$x(t) = \sum_{k=-\infty}^{\infty} \delta(t - kT_1)$$

and has Fourier coefficients $X_k = 1/T_1$. Suppose $y_1(t)$ is a pulse of support $0 \le t \le T_1$, determine the convolution $y(t) = (x * y_1)(t)$ using the above expression for $x(t)$ and the one given by its Fourier series. What do you get? Explain.

4.7.2 PROBLEMS USING MATLAB

4.16 Different ways to compute the Fourier coefficients—We would like to find the Fourier series of a sawtooth periodic signal $x(t)$ of period $T_0 = 1$. The period of $x(t)$ is

$$x_1(t) = r(t) - r(t - 1) - u(t - 1).$$

(a) Sketch $x(t)$ and compute the Fourier coefficients X_k using the integral definition.

(b) An easier way to do this is to use the Laplace transform of $x_1(t)$. Find X_k this way.

(c) Obtain a trigonometric Fourier series $\hat{x}(t)$ consisting of the dc term and 40 harmonics to approximate $x(t)$. Use it to find the values of $\hat{x}(t)$ for $t = 0$ to 10 in steps of 0.001. How does it compare with $x(t)$? Use MATLAB to plot the signal $\hat{x}(t)$ and its magnitude line spectrum.

Answers: $X_k = j/2\pi k$; $X_0 = 0.5$.

4.17 Half/full-wave rectifying and Fourier—Rectifying a sinusoid provides a way to create a dc source. In this problem we consider the Fourier series of the full and the half-wave rectified signals. The full-wave rectified signal $x_f(t)$ has a fundamental period $T_0 = 1$ and its period from 0 to 1 is

$$x_1(t) = \sin(\pi t) \qquad 0 \le t \le 1,$$

while the period for the half-wave rectifier signal $x_h(t)$ is

$$x_2(t) = \begin{cases} \sin(\pi t) & 0 \le t \le 1, \\ 0 & 1 < t \le 2, \end{cases}$$

with fundamental period $T_1 = 2$.

(a) Obtain the Fourier coefficients for both of these periodic signals.

(b) Use MATLAB and your analytic results obtained above to plot the magnitude line spectrum of the half-wave signal and use the dc and 40 harmonics to obtain an approximation of the half-wave signal. Plot them.

Answers: $x_1(t) = \sin(\pi t)u(t) + \sin(\pi(t-1))u(t-1)$; full-wave $X_k = 2/((1 - 4k^2)\pi)$.

4.18 Generation of signals from Fourier coefficients—Consider the entries in Table 4.2 corresponding to the sawtooth, the rectangular pulses of width d, and the triangular pulses. Verify analytically the Fourier coefficients, and then use our function *InvFSeries* to generate each of the three periodic signals. Let $d = 0.1$ for the rectangular pulse, and show the signals in time segment $[-2, 2]$.

4.19 Shifting and scaling of periodic signals—Consider the entries in Table 4.2 for the sawtooth and the triangular signals normalized in magnitude and period. Use their Fourier coefficients to obtain corresponding periodic signals that are zero-mean, maximum amplitude of one and that are odd.

4.20 Inverse sawtooth signal—An inverted sawtooth signal is given by the reflection $x(-t)$, where $x(t)$ is the sawtooth signal. Use the entry for the sawtooth signal in Table 4.2 to obtain a zero-mean inverted sawtooth signal $y(t)$ of period $T_1 = 2$ and maximum amplitude 2. Use our function *InvFSeries* to verify your result.

4.21 Smoothness and Fourier series—The smoothness of a period determines the way the magnitude line spectrum decays. Consider the following periodic signals $x(t)$ and $y(t)$ both of fundamental period $T_0 = 2$ s, and with a period from $0 \le t \le T_0$:

$$x_1(t) = u(t) - u(t-1), \quad y_1(t) = r(t) - 2r(t-1) + r(t-2).$$

Find the Fourier series coefficients of $x(t)$ and $y(t)$ and use MATLAB to plot their magnitude line spectrum for $k = 0, \pm 1, \pm 2, \cdots \pm 20$. Determine which of these spectra decays faster and how it relates to the smoothness of the period. (To see this relate $|X_k|$ to the corresponding $|Y_k|$.)
Answers: $y(t)$ is smoother than $x(t)$.

4.22 DC output from a full-wave rectified signal—Consider a full-wave rectifier that has as output a periodic signal $x(t)$ of fundamental period $T_0 = 1$ and a period of it is given as

$$x_1(t) = \begin{cases} \cos(\pi t) & -0.5 \le t \le 0.5, \\ 0 & \text{otherwise.} \end{cases}$$

(a) Obtain the Fourier coefficients X_k.

(b) Suppose we pass $x(t)$ through an ideal filter of transfer function $H(s)$, determine the values of this filter at harmonic frequencies $2\pi k, = 0, \pm 1, \pm 2, \cdots$, so that its output is a constant, i.e., we have a dc source. Use MATLAB to plot the magnitude line spectrum of $x(t)$.

Answers: $X_0 = 2/\pi$, $H(j0) = 1$, and $H(j2\pi k) = 0$ for $k \neq 0$.

4.23 Applying Parseval's result—We wish to approximate the triangular signal $\{x(t)\}$, with a period $x_1(t) = r(t) - 2r(t-1) + r(t-2)$, by a Fourier series with a finite number of terms, let us say $2N$. This approximation should have 99% of the power of the triangular signal; use MATLAB to find the value of N.

Answer: $P_x = (0.5)^2 + 2\sum_{k>0, \text{ odd}} 4/(\pi k)^4$.

4.24 Walsh functions—As seen in the previous problem, the Fourier series is one of a possible class of representations in terms of orthonormal functions. Consider the case of the *Walsh functions* which are a set of rectangular pulse signals that are orthonormal in a finite time interval $[0, 1]$. These functions are such that: (i) take only 1 and -1 values, (ii) $\phi_k(0) = 1$ for all k, and (iii) are ordered according to the number of sign changes.

(a) Consider obtaining the functions $\{\phi_k\}_{k=0,\cdots,5}$. The Walsh functions are clearly normal since when squared they are unity for $t \in [0, 1]$. Let $\phi_0(t) = 1$ for $t \in [0, 1]$ and zero elsewhere. Obtain $\phi_1(t)$ with one change of sign and that is orthogonal to $\phi_0(t)$. Find then $\phi_2(t)$, which has two changes of sign and is orthogonal to both $\phi_0(t)$ and $\phi_1(t)$. Continue this process. Carefully plot the $\{\phi_i(t)\}$, $i = 0, \cdots, 5$. Use the function *stairs* to plot these Walsh functions.

(b) Consider the Walsh functions obtained above as sequences of 1s and -1s of length 8, carefully write these 6 sequences. Observe the symmetry of the sequences corresponding to $\{\phi_i(t), i = 0, 1, 3, 5\}$, determine the circular shift needed to find the sequence corresponding to $\phi_2(t)$ from the sequence from $\phi_1(t)$, and $\phi_4(t)$ from $\phi_3(t)$. Write a MATLAB script that generates a matrix $\mathbf{\Phi}$ with entries the sequences, find the product $(1/8)\mathbf{\Phi}\mathbf{\Phi}^T$, explain how this result connects with the orthonormality of the Walsh functions.

(c) We wish to approximate a ramp function $x(t) = r(t)$, $0 \le t \le 1$, using $\{\phi_k\}_{k=0,\cdots,5}$. This could be written

$$\mathbf{r} = \mathbf{\Phi}\mathbf{a}$$

where $\mathbf{r}$ is a vector of $x(nT) = r(nT)$ where $T = 1/8$, $\mathbf{a}$ are the coefficients of the expansion, and $\mathbf{\Phi}$ the Walsh matrix found above. Determine the vector $\mathbf{a}$ and use it to obtain an approximation of $x(t)$. Plot $x(t)$ and the approximation $\hat{x}(t)$ (use *stairs* for this signal).

FREQUENCY ANALYSIS: THE FOURIER TRANSFORM

CONTENTS

Imagination is the beginning of creation. You imagine what you desire, you will what you imagine and at last you create what you will.

George Bernard Shaw, Irish dramatist, 1856–1950

5.1 INTRODUCTION

In this chapter we continue the frequency analysis of signals and systems. The frequency representation of signals and the frequency response of systems are tools of great significance in signal processing,

Signals and Systems Using MATLAB®. https://doi.org/10.1016/B978-0-12-814204-2.00015-6

communication and control theory. In this chapter we will complete the Fourier representation of signals by extending it to aperiodic signals. By a limiting process the harmonic representation of periodic signals is extended to the Fourier transform, a frequency-dense representation for non-periodic signals. The concept of spectrum introduced for periodic signals is generalized for both finite-power and finite-energy signals. Thus, the Fourier transform measures the frequency content of a signal, be it periodic or aperiodic.

In this chapter, the connection between the Laplace and the Fourier transforms will be highlighted computationally and analytically. The Fourier transform turns out to be a special case of the Laplace transform for signals with Laplace transforms with regions of convergence that include the $j\Omega$-axis. There are, however, signals where the Fourier transform cannot be obtained from the Laplace transform—for those cases properties of the Fourier transform will be used. The duality of the direct and the inverse transforms is of special interest in calculating the Fourier transform.

Filtering is an important application of the Fourier transform. The Fourier representation of signals and the eigenfunction property of LTI systems provide the tools to change the frequency content of a signal by processing it with a LTI system with the desired frequency response.

The idea of changing the frequency content of a signal via modulation is basic in analog communications. Modulation allows us to send signals over the airwaves using antennas of reasonable size. Voice and music are relatively low-frequency signals that cannot be easily radiated without the help of modulation. Continuous wave modulation can change the amplitude, the frequency or the phase of a sinusoidal carrier with frequency much greater than the frequencies present in the message we wish to transmit.

5.2 FROM THE FOURIER SERIES TO THE FOURIER TRANSFORM

In practice there are no periodic signals—such signals would have infinite supports and exact fundamental periods which are not possible. Since only finite-support signals can be processed numerically, signals in practice are treated as aperiodic. To obtain the Fourier representation of aperiodic signals, we use the Fourier series representation in a limiting process.

Any aperiodic signal is a periodic signal with infinite fundamental period. That is, an aperiodic signal $x(t)$ can be expressed as

$$x(t) = \lim_{T_0 \to \infty} \tilde{x}(t)$$

where $\tilde{x}(t)$ is a periodic signal of fundamental period T_0 with a Fourier series representation

$$\tilde{x}(t) = \sum_{n=-\infty}^{\infty} X_n e^{jn\Omega_0 t}, \qquad \Omega_0 = \frac{2\pi}{T_0} \quad \text{where}$$

$$X_n = \frac{1}{T_0} \int_{-T_0/2}^{T_0/2} \tilde{x}(t) e^{-jn\Omega_0 t} \, dt.$$

As $T_0 \to \infty$, X_n will tend to zero. To avoid this we define $X(\Omega_n) = T_0 X_n$ where $\{\Omega_n = n\Omega_0\}$ are the harmonic frequencies. Letting $\Delta\Omega = \Omega_0 = 2\pi/T_0$ be the frequency between harmonics, we can then write the above equations as

$$\tilde{x}(t) = \sum_{n=-\infty}^{\infty} \frac{X(\Omega_n)}{T_0} e^{j\Omega_n t} = \sum_{n=-\infty}^{\infty} X(\Omega_n) e^{j\Omega_n t} \frac{\Delta\Omega}{2\pi},$$

$$X(\Omega_n) = \int_{-T_0/2}^{T_0/2} \tilde{x}(t) e^{-j\Omega_n t} dt.$$

As $T_0 \to \infty$ then $\Delta\Omega \to d\Omega$, the line spectrum becomes denser, i.e. the lines in the line spectrum get closer, the sum becomes an integral, and $\Omega_n = n\Omega_0 = n\Delta\Omega \to \Omega$ so that in the limit we obtain

$$x(t) = \frac{1}{2\pi} \int_{-\infty}^{\infty} X(\Omega) e^{j\Omega t} d\Omega,$$

$$X(\Omega) = \int_{-\infty}^{\infty} x(t) e^{-j\Omega t} dt,$$

which are the **inverse** and the **direct Fourier transforms**, respectively. The first equation transforms a function in the frequency domain, $X(\Omega)$, into a signal in the time domain, $x(t)$, while the other equation does the opposite.

The Fourier transform measures the frequency content of a signal. As we will see, time and frequency representations are complementary, thus the characterization in one domain provides information that is not clearly available in the other.

An aperiodic, or non-periodic, signal $x(t)$ can be thought of as a periodic signal $\tilde{x}(t)$ with an infinite fundamental period. Using the Fourier series representation of this signal and a limiting process we obtain a Fourier transform pair

$$x(t) \quad \Leftrightarrow \quad X(\Omega)$$

where the signal $x(t)$ is transformed into a function $X(\Omega)$ in the frequency-domain by the

Fourier transform: $\quad X(\Omega) = \int_{-\infty}^{\infty} x(t) e^{-j\Omega t} dt,$ (5.1)

while $X(\Omega)$ is transformed into a signal $x(t)$ in the time-domain by the

Inverse Fourier transform: $\quad x(t) = \frac{1}{2\pi} \int_{-\infty}^{\infty} X(\Omega) e^{j\Omega t} d\Omega.$ (5.2)

Remark

1. Although we have obtained the Fourier transform from the Fourier series, the Fourier transform of a periodic signal cannot be obtained directly from the integral (5.1). Consider $x(t) = \cos(\Omega_0 t)$, $-\infty < t < \infty$, periodic of fundamental period $2\pi/\Omega_0$. If you attempt to compute its Fourier transform using the integral you have a not-well-defined problem (try computing the integral to convince yourself). But it is known from the line spectrum that the power of this signal is concentrated at the frequencies $\pm\Omega_0$, so somehow we should be able to find its Fourier transform. Sinusoids are basic functions.

2. On the other hand, if you consider a decaying exponential $x(t) = e^{-|a|t}$ signal which has a Laplace transform that is valid on the $j\Omega$-axis (i.e., the region of convergence of $X(s)$ includes this axis) then its Fourier transform is simply $X(s)$ computed at $s = j\Omega$ as we will see. There is no need for the integral formula in this case, although if you apply it your result coincides with the one from the Laplace transform.

3. Finally, consider finding the Fourier transform of a sinc function (which is the impulse response of a low-pass filter as we will see later). Neither the integral nor the Laplace transform can be used to find it. For this signal, we need to exploit the duality that exists between the direct and the inverse Fourier transforms (notice the duality, or similarity, that exists between Equations (5.1) and (5.2)).

5.3 EXISTENCE OF THE FOURIER TRANSFORM

For the Fourier transform to exist, $x(t)$ must be *absolutely integrable*, i.e.,

$$|X(\Omega)| \le \int_{-\infty}^{\infty} |x(t)e^{-j\Omega t}|dt = \int_{-\infty}^{\infty} |x(t)|dt < \infty.$$

Moreover, $x(t)$ must have only a finite number of discontinuities and a finite number of minima and maxima in any finite interval. (Given that the Fourier transform was obtained by a limiting procedure from the Fourier series, it is not surprising that the above conditions coincide with the existence conditions for the Fourier series.)

The Fourier transform

$$X(\Omega) = \int_{-\infty}^{\infty} x(t)e^{-j\Omega t}dt$$

of a signal $x(t)$ exists (i.e., we can calculate its Fourier transform via this integral) provided

- $x(t)$ is absolutely integrable or the area under $|x(t)|$ is finite,
- $x(t)$ has only a finite number of discontinuities and a finite number of minima and maxima in any finite interval.

From the definitions of the direct and the inverse Fourier transforms—both being integrals calculated over an infinite support—one wonders whether they exist in general, and if so, how to most efficiently compute them. Commenting on the existence conditions, Professor E. Craig wrote in [18]:

"It appears that almost nothing has a Fourier transform—nothing except practical communication signals. No signal amplitude goes to infinity and no signal lasts forever; therefore, no practical signal can have infinite area under it, and hence all have Fourier transforms."

Indeed, signals of practical interest have Fourier transforms and their spectra can be displayed using a **spectrum analyzer** (or better yet, any signal for which we can display its spectrum will have a Fourier transform). A spectrum analyzer is a device that displays the energy or the power of a signal distributed over frequencies.

5.4 FOURIER TRANSFORMS FROM THE LAPLACE TRANSFORM

The region of convergence (ROC) of the Laplace transform $X(s)$ indicates the region in the s-plane where $X(s)$ is defined.

If the region of convergence (ROC) of $X(s) = \mathcal{L}[x(t)]$ contains the $j\Omega$-axis, so that $X(s)$ is defined for $s = j\Omega$, then

$$
\begin{aligned}
\mathcal{F}[x(t)] \quad &= \quad \mathcal{L}[x(t)]|_{s=j\Omega} = \int_{-\infty}^{\infty} x(t)e^{-j\Omega t}\,dt \\
&= \quad X(s)|_{s=j\Omega}. \qquad\qquad\qquad\qquad\qquad (5.3)
\end{aligned}
$$

The above applies to causal, anticausal or noncausal signals.

The following rules of thumb will help you get a better understanding of the time–frequency relationship of a signal and its Fourier transform, and they indicate the best way to calculate it. On a first reading the use of these rules might not be obvious, but they will be helpful in understanding the discussions that follow and you might want to come back to these rules.

Rules of Thumb for Calculating the Fourier Transform of a Signal $x(t)$:

- If $x(t)$ has a finite time support and in that support $x(t)$ is bounded, its Fourier transform exists. To find it use the integral definition or the Laplace transform of $x(t)$.
- If $x(t)$ has infinite time support and a Laplace transform $X(s)$ with a region of convergence including the $j\Omega$-axis, its Fourier transform is $X(s)|_{s=j\Omega}$.
- If $x(t)$ is periodic, its Fourier transform is obtained using the signal Fourier series.
- If $x(t)$ is none of the above, if it has discontinuities (e.g., $x(t) = u(t)$), or it has discontinuities and it is not finite energy (e.g., $x(t) = \cos(\Omega_0 t)u(t)$) or it has possible discontinuities in the frequency domain even though it has finite energy (e.g., $x(t) = \text{sinc}(t)$) use properties of the Fourier transform.

Keep in mind to

- Consider the Laplace transform if the interest is in transients and steady state, and the Fourier transform if steady-state behavior is of interest.
- Represent periodic signals by their Fourier series before considering their Fourier transforms.
- Attempt other methods before performing integration to find the Fourier transform.

Example 5.1. Discuss whether it is possible to obtain the Fourier transform of the following signals:

$$(a)\ x_1(t) = u(t), \quad (b)\ x_2(t) = e^{-2t}u(t), \quad (c)\ x_3(t) = e^{-|t|},$$

using their Laplace transforms.

Solution: (a) The Laplace transform of $x_1(t)$ is $X_1(s) = 1/s$ with a region of convergence corresponding to the open right-hand s-plane, or ROC: $\{s = \sigma + j\Omega : \sigma > 0, -\infty < \Omega < \infty\}$, which does not include the $j\Omega$-axis so that the Laplace transform cannot be used to find the Fourier transform of $x_1(t)$.

(b) The signal $x_2(t)$ has as Laplace transform $X_2(s) = 1/(s+2)$ with a region of convergence ROC: $\{s = \sigma + j\Omega : \sigma > -2, -\infty < \Omega < \infty\}$ containing the $j\Omega$-axis. Then the Fourier transform of $x_2(t)$ is

$$X_2(\Omega) = \frac{1}{s+2} \Big|_{s=j\Omega} = \frac{1}{j\Omega + 2}.$$

(c) The Laplace transform of $x_3(t)$ is

$$X_3(s) = \frac{1}{s+1} + \frac{1}{-s+1} = \frac{2}{1-s^2},$$

with a region of convergence (ROC): $\{s = \sigma + j\Omega : -1 < \sigma < 1, -\infty < \Omega < \infty\}$, which contains the $j\Omega$-axis. Then the Fourier transform of $x_3(t)$ is

$$X_3(\Omega) = X_3(s)|_{s=j\Omega} = \frac{2}{1-(j\Omega)^2} = \frac{2}{1+\Omega^2}. \quad \square$$

5.5 LINEARITY, INVERSE PROPORTIONALITY AND DUALITY

Many of the properties of the Fourier transform are very similar to those of the Fourier series or of the Laplace transform, which is to be expected given the strong connection among these transformations. The linearity, the inverse time–frequency relationship, and the duality between the direct and the inverse Fourier transforms will help us determine the transform of signals that do not satisfy the Laplace transform condition.

5.5.1 LINEARITY

Just like the Laplace transform, the Fourier transform is linear.

If $\mathcal{F}[x(t)] = X(\Omega)$ and $\mathcal{F}[y(t)] = Y(\Omega)$, for constants α and β we have

$$
\begin{aligned}
\mathcal{F}[\alpha x(t) + \beta y(t)] &= \alpha\mathcal{F}[x(t)] + \beta\mathcal{F}[y(t)] \\
&= \alpha X(\Omega) + \beta Y(\Omega).
\end{aligned}
\tag{5.4}
$$

Example 5.2. Suppose you create a periodic sine

$$x(t) = \sin(\Omega_0 t) \qquad -\infty < t < \infty$$

by adding a causal sine $v(t) = \sin(\Omega_0 t)u(t)$ and an anticausal sine $y(t) = \sin(\Omega_0 t)u(-t)$ for each of which you can find Laplace transforms $V(s)$ and $Y(s)$. Discuss what would be wrong with this approach to find the Fourier transform of $x(t)$ by letting $s = j\Omega$.

Solution: Noticing that $y(t) = -v(-t)$, the Laplace transforms of $v(t)$ and $y(t)$ are

$$V(s) = \frac{\Omega_0}{s^2 + \Omega_0^2} \qquad \text{ROC}_1 : \mathcal{Re}\,[s] > 0,$$

$$Y(s) = \frac{-\Omega_0}{(-s)^2 + \Omega_0^2} \qquad \text{ROC}_2 : \mathcal{Re}\,[s] < 0,$$

giving $X(s) = V(s) + Y(s) = 0$. Moreover, the region of convergence of $X(s)$ is the intersection of the two given above ROCs, which is null, so it is not possible to obtain the Fourier transform of $x(t)$ this way. This is so even though the time signals add correctly to $x(t)$. The Fourier transform of the sine signal will be found using the periodicity of $x(t)$ or the duality property we consider later.   □

5.5.2 INVERSE PROPORTIONALITY OF TIME AND FREQUENCY

It is very important to realize that frequency is inversely proportional to time, and that as such time and frequency signal characterizations are complementary. Consider the following examples to illustrate this.

• The impulse signal $x_1(t) = \delta(t)$, although not a regular signal, has finite support (its support is only $t = 0$ as the signal is zero everywhere else), and is absolutely integrable so it has a Fourier transform

$$X_1(\Omega) = \mathcal{F}[\delta(t)] = \int_{-\infty}^{\infty} \delta(t)e^{-j\Omega t}\,dt = e^{-j0}\int_{-\infty}^{\infty} \delta(t)dt = 1 \qquad -\infty < \Omega < \infty,$$

displaying infinite support in frequency. (The Fourier transform could have also been obtained from the Laplace transform $\mathcal{L}[\delta(t)] = 1$ for all values of s. For $s = j\Omega$ we have $\mathcal{F}[\delta(t)] = 1$.) This result means that, since $\delta(t)$ changes so much in such a short time, its Fourier transform has all possible frequency components.

• Consider then the opposite case: a signal that is constant for all times, or a dc signal $x_2(t) = A$, $-\infty < t < \infty$. We know that the frequency of $\Omega = 0$ is assigned to it since the signal does not vary at all. The Fourier transform cannot be found by means of the integral because $x_2(t)$ is not absolutely integrable, but we can verify that it is given by $X_2(\Omega) = 2\pi A\delta(\Omega)$ (we will formally show this using the duality property). In fact, the inverse Fourier transform is

$$\frac{1}{2\pi}\int_{-\infty}^{\infty} X_2(\Omega)e^{j\Omega t}\,d\Omega = \frac{1}{2\pi}\int_{-\infty}^{\infty} 2\pi A \underbrace{\delta(\Omega)e^{j\Omega t}}_{\delta(\Omega)}\,d\Omega = A.$$

Notice the complementary nature of $x_1(t)$ and $x_2(t)$: $x_1(t) = \delta(t)$ has a one-point support while $x_2(t) = A$ has infinite support. Their corresponding Fourier transforms $X_1(\Omega) = 1$ and $X_2(\Omega) = 2\pi A\delta(\Omega)$ have infinite and one-point support in the frequency domain, respectively.

- To appreciate the transition from the dc signal to the impulse signal, consider a pulse signal

$$x_3(t) = A[u(t + \tau/2) - u(t - \tau/2)], \tag{5.5}$$

having finite energy, and whose Fourier transform can be found using its Laplace transform:

$$X_3(s) = \frac{A}{s}\left[e^{s\tau/2} - e^{-s\tau/2}\right]$$

with ROC the whole s-plane. Thus,

$$X_3(\Omega) = X_3(s)|_{s=j\Omega} = A\frac{(e^{j\Omega\tau/2} - e^{-j\Omega\tau/2})}{j\Omega} = A\tau\frac{\sin(\Omega\tau/2)}{\Omega\tau/2} \tag{5.6}$$

or a sinc function in frequency, where $A\tau$ corresponds to the area under $x_3(t)$. The Fourier transform $X_3(\Omega)$ is an even function of Ω; at $\Omega = 0$ using L'Hôpital's rule $X_3(0) = A\tau$; $X_3(\Omega)$ becomes zero when $\Omega = 2k\pi/\tau, k = \pm1, \pm2, \cdots$.

If we let $A = 1/\tau$ (so that the area of the pulse is unity), and let $\tau \to 0$ the pulse $x_3(t)$ becomes a delta function, $\delta(t)$, in the limit and the sinc function expands (for $\tau \to 0$, $X_3(\Omega)$ is not zero for any finite value) to become unity. On the other hand, if we let $\tau \to \infty$ the pulse becomes a constant signal A extending from $-\infty$ to ∞, and the Fourier transform gets closer and closer to $\delta(\Omega)$ (the sinc function becomes zero at values very close to zero and the amplitude at $\Omega = 0$ becomes larger and larger although the area under the curve remains $2\pi A$).

Example 5.3. To illustrate the change in the Fourier transform as the time support increases, consider two pulses $x_i(t) = [u(t + \tau_i/2) - u(t - \tau_i/2)]$, $i = 1, 2$, of unit amplitude, but different time supports $\tau_i = 1, 4, i = 1, 2$. Use MATLAB to find their Fourier transform and compare the results.

Solution: We use the following MATLAB script to compute the Fourier transform. The symbolic MATLAB function *fourier* computes the Fourier transform.

```
%%
% Example 5.3---Time vs frequency
%%
clear all; clf
syms t w
for i=1:2,
    tau=0.5+(i-1)*1.5;   x=heaviside(t+tau)-heaviside(t-tau); X=fourier(x);
    if(i==1)
        figure(1)
        subplot(211)
        fplot(x,[-3 3]);axis([-3 3 -0.1 1.1]);grid
        ylabel('x_1(t)');xlabel('t')
        subplot(212)
        fplot(X,[-50 50]);axis([-50 50 -1 5]);grid
        ylabel('X_1(\Omega)');xlabel('\Omega')
    else
```

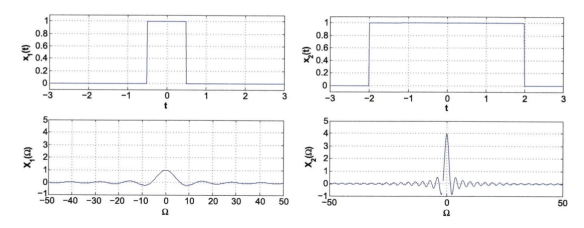

FIGURE 5.1

Fourier transform of pulses $x_1(t)$, with $A = 1$ and $\tau = 1$ (left), and $x_2(t)$ with $A = 1$ and $\tau = 4$ (right).

```
            figure(2)
            subplot(211)
            fplot(x,[-3 3]);axis([-3 3 -0.1 1.1]);grid
            ylabel('x_2(t)');xlabel('t')
            subplot(212)
            fplot(X,[-50 50]);axis([-50 50 -1 5]);grid
            ylabel('X_2(\Omega)');xlabel('\Omega')
    end
end
```

The results are shown in Fig. 5.1, and as shown there the wider the pulse, the more concentrated in frequency its Fourier transform. Notice that $X_i(0) = \tau$, $i = 1, 2$, the area under the pulses.   □

The time–frequency relation can be summarized as follows:

The support of $X(\Omega)$ is inversely proportional to the support of $x(t)$.
If $x(t)$ has a Fourier transform $X(\Omega)$ and $\alpha \neq 0$ is a real number, then $x(\alpha t)$

- is a contracted signal when $\alpha > 1$;
- is a contracted and reflected signal when $(\alpha < -1)$;
- is an expanded signal when $0 < \alpha < 1$;
- is a reflected and expanded signal when $-1 < \alpha < 0$; or
- is a reflected signal when $\alpha = -1$

and we have the pair

$$x(\alpha t) \quad \Leftrightarrow \quad \frac{1}{|\alpha|} X\left(\frac{\Omega}{\alpha}\right). \tag{5.7}$$

First let us mention that, like in the Laplace transform, the symbol $\Leftrightarrow$ indicates the relation between a signal $x(t)$ in the time domain to its corresponding Fourier transform $X(\Omega)$, in the frequency domain. This is *not* an equality, far from it!

This property is shown by a change of variable in the integration,

$$\mathcal{F}[x(\alpha t)] = \int_{-\infty}^{\infty} x(\alpha t) e^{-j\Omega t} dt = \begin{cases} \frac{1}{\alpha} \int_{-\infty}^{\infty} x(\rho) e^{-j\Omega \rho/\alpha} d\rho & \alpha > 0 \\ -\frac{1}{\alpha} \int_{-\infty}^{\infty} x(\rho) e^{-j\Omega \rho/\alpha} d\rho & \alpha < 0 \end{cases}$$

$$= \frac{1}{|\alpha|} X\left(\frac{\Omega}{\alpha}\right)$$

by the change of variable $\rho = \alpha t$. If $|\alpha| > 1$, the signal $x(\alpha t)$—when compared with $x(t)$—contracts while its corresponding Fourier transform expands. Likewise, when $0 < |\alpha| < 1$ the signal $x(\alpha t)$ expands, when compared with $x(t)$, and its Fourier transform contracts. If $\alpha < 0$ the corresponding contraction or expansion is accompanied by a reflection in time. In particular, if $\alpha = -1$ the reflected signal $x(-t)$ has $X(-\Omega)$ as its Fourier transform.

Example 5.4. Consider a pulse $x(t) = u(t) - u(t - 1)$. Find the Fourier transform of $x_1(t) = x(2t)$.

Solution: The Laplace transform of $x(t)$ is

$$X(s) = \frac{1 - e^{-s}}{s}$$

with the whole s-plane as its region of convergence. Thus its Fourier transform is

$$X(\Omega) = \frac{1 - e^{-j\Omega}}{j\Omega} = \frac{e^{-j\Omega/2}(e^{j\Omega/2} - e^{-j\Omega/2})}{2j\Omega/2} = \frac{\sin(\Omega/2)}{\Omega/2} e^{-j\Omega/2}.$$

To the finite-support signal $x(t)$ corresponds $X(\Omega)$ of infinite support. Then

$$x_1(t) = x(2t) = u(2t) - u(2t - 1) = u(t) - u(t - 0.5).$$

The Fourier transform of $x_1(t)$ is found, again using its Laplace transform, to be

$$X_1(\Omega) = \frac{1 - e^{-j\Omega/2}}{j\Omega} = \frac{e^{-j\Omega/4}(e^{j\Omega/4} - e^{-j\Omega/4})}{j\Omega} = \frac{1}{2} \frac{\sin(\Omega/4)}{\Omega/4} e^{-j\Omega/4} = \frac{1}{2} X(\Omega/2),$$

which is an expanded version of $X(\Omega)$ in the frequency domain and coincides with the result from the property. See Fig. 5.2.

The connection between the Fourier transforms $X_1(\Omega)$ and $X(\Omega)$ can be also found from the integral definitions:

$$X_1(\Omega) = \int_0^{0.5} 1 e^{j\Omega t} dt = \int_0^{0.5} e^{j(\Omega/2)2t} dt \qquad \text{letting } \rho = 2t$$

$$= \frac{1}{2} \int_0^1 e^{j(\Omega/2)\rho} d\rho = \frac{1}{2} X(\Omega/2)$$

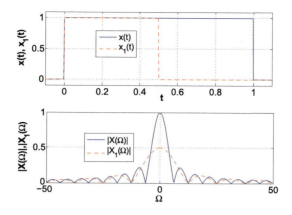

FIGURE 5.2

Pulse $x(t)$ and its compressed version $x_1(t) = x(2t)$, and the magnitude of their Fourier transforms. Notice that the signal contracts in time but expands in frequency.

where again using the integral definition

$$X(\Omega) = \int_0^1 1 e^{j\Omega t}\, dt = \frac{e^{-j\Omega t}}{-j\Omega}\Big|_0^1 = \frac{\sin(\Omega/2)}{\Omega/2} e^{-j\Omega/2}. \quad \square$$

Example 5.5. Apply the reflection property to find the Fourier transform of $x(t) = e^{-a|t|}$, $a > 0$. Let $a = 1$, use MATLAB to compute the Fourier transform and plot the signal and $|X(\Omega)|$ and $\angle X(\Omega)$.

Solution: The signal $x(t)$ can be expressed as $x(t) = e^{-at}u(t) + e^{at}u(-t) = x_1(t) + x_1(-t)$. The Fourier transform of $x_1(t)$ is

$$X_1(\Omega) = \frac{1}{s+a}\Big|_{s=j\Omega} = \frac{1}{j\Omega + a}$$

and according to the reflection property $x_1(-t)$ $(\alpha = -1)$ has

$$\mathcal{F}[x_1(-t)] = \frac{1}{-j\Omega + a},$$

so that

$$X(\Omega) = \frac{1}{j\Omega + a} + \frac{1}{-j\Omega + a} = \frac{2a}{a^2 + \Omega^2}.$$

If $a = 1$, using MATLAB the signal $x(t) = e^{-|t|}$ and the magnitude and phase of $X(\Omega)$ are computed and plotted as shown in Fig. 5.3. Since $X(\Omega)$ is real and positive, the corresponding phase is zero for all frequencies. This signal is called **low-pass** because its energy is concentrated in the low frequencies. $\quad \square$

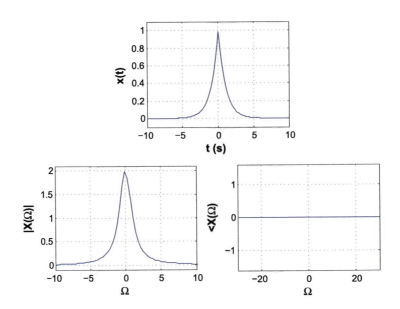

FIGURE 5.3

Magnitude (bottom left) and phase (bottom right) of the Fourier transform $X(\Omega)$ of the two-sided signal $x(t) = e^{-|t|}$ (top) for all frequencies. The magnitude of $X(\Omega)$ indicates $x(t)$ is a low-pass signal. Notice the phase is zero, because $X(\Omega)$ is real and positive.

```
%%
% Example 5.5---Fourier transform
%%
clear all;clf
syms t w
    x=exp(-abs(t))
    X=fourier(x)
    figure(1)
    subplot(221)
    ezplot(x,[-10,10]);grid;axis([-10 10 -0.2 1.2]); title(' ');
    ylabel('x(t)'); xlabel('t (s)')
    subplot(223)
    ezplot(sqrt((real(X))^2+(imag(X))^2),[-30,30]); grid; axis([-10 10 -0.2 2.2]);
    xlabel('\Omega'); ylabel('|X(\Omega)|'); title(' ');
    subplot(224)
    ezplot(imag(log(X)),[-30,30]); grid; title(' ');
    axis([-10 10 -1.8 1.8]); xlabel('\Omega'); ylabel('<X(\Omega)')
```

5.5.3 DUALITY

Besides the inverse relationship between time and frequency, by interchanging the frequency and the time variables in the definitions of the direct and the inverse Fourier transform (see Equations (5.1) and (5.2)) similar or dual equations are obtained.

To the Fourier transform pair

$$x(t) \quad \Leftrightarrow \quad X(\Omega) \tag{5.8}$$

corresponds the following dual Fourier transform pair:

$$X(t) \quad \Leftrightarrow \quad 2\pi x(-\Omega). \tag{5.9}$$

This can be shown by considering the inverse Fourier transform

$$x(t) = \frac{1}{2\pi} \int_{-\infty}^{\infty} X(\rho) e^{j\rho t} d\rho$$

and replacing in it t by $-\Omega$ and multiplying by 2π to get

$$2\pi x(-\Omega) = \int_{-\infty}^{\infty} X(\rho) e^{-j\rho\Omega} d\rho = \int_{-\infty}^{\infty} X(t) e^{-j\Omega t} dt = \mathcal{F}[X(t)]$$

after letting $\rho = t$. To understand the above equations you need to realize that ρ and t are dummy variables inside the top and the bottom integrals, and as such they are not reflected outside the integral.

Remark

1. This duality property allows us to obtain the Fourier transform of signals for which we already have a Fourier pair and that would be difficult to obtain directly. It is thus one more method to obtain the Fourier transform, besides the Laplace transform and the integral definition of the Fourier transform.
2. When computing the Fourier transform of a constant signal, $x(t) = A$, we indicated that it would be $X(\Omega) = 2\pi A\delta(\Omega)$. Indeed, we have the following dual pairs:

$$A\delta(t) \quad \Leftrightarrow \quad A$$
$$A \quad \Leftrightarrow \quad 2\pi A\delta(-\Omega) = 2\pi A\delta(\Omega) \tag{5.10}$$

where in the second equation we use that $\delta(\Omega)$ is an even function of Ω.

Example 5.6. Use the duality property to find the Fourier transform of the sinc signal

$$x(t) = A \frac{\sin(0.5t)}{0.5t} = A \operatorname{sinc}(0.5t) \qquad -\infty < t < \infty.$$

Use MATLAB to find the Fourier transforms for $A = 10$.

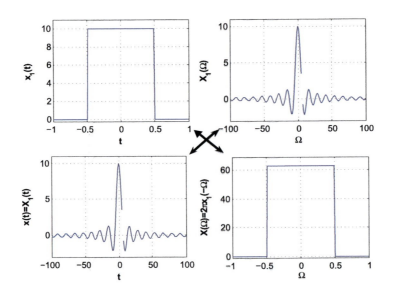

FIGURE 5.4

Application of duality to find Fourier transform of $x(t) = 10 \operatorname{sinc}(0.5t)$. Notice that $X(\Omega) = 2\pi x_1(\Omega) \approx$
$6.28x_1(\Omega) = 63.8[u(\Omega + 0.5) - u(\Omega - 0.5)]$.

Solution: The Fourier transform of the sinc signal cannot be found using the Laplace transform or the integral definition of the Fourier transform. The duality property provides a way to obtain it. We found before, see Equations (5.5) and (5.6) for $\tau = 0.5$, the following pair of Fourier transforms:

$$p(t) = A[u(t + 0.5) - u(t - 0.5)] \quad \Leftrightarrow \quad P(\Omega) = A\,\frac{\sin(0.5\Omega)}{0.5\Omega} = A \operatorname{sinc}(0.5\Omega).$$

Then, according to the duality property we have the Fourier transform pair

$$x(t) = P(t) = A \operatorname{sinc}(0.5t)$$
$$\Leftrightarrow \quad X(\Omega) = 2\pi p(-\Omega) = 2\pi p(\Omega) = 2\pi A[u(\Omega + 0.5) - u(\Omega - 0.5)] \qquad (5.11)$$

given that $p(.)$ is even. So the Fourier transform of the sinc is a rectangular pulse in frequency, in the same way that the Fourier transform of a pulse in time is a sinc function in frequency. Fig. 5.4 shows the dual pairs for $A = 10$. ☐

```
%%
% Example 5.6---Duality
%%
clear all; clf
syms t w
m=heaviside(t+0.5)-heaviside(t-0.5); A=10;
x=A*m; X=fourier(x); xx=ifourier(2*pi*x)
```

```
figure(1)
subplot(221)
ezplot(x,[-1,1]);grid;axis([-1 1 -0.2 11]);
 xlabel('t'); ylabel('x_1(t)')
subplot(222)
ezplot(X,[-100,100]); grid; axis([-100 100 -3 11]);
 xlabel('\Omega'); ylabel('X_1(\Omega)')
subplot(223)
ezplot(xx,[-100,100]);grid;axis([-100 100 -3 11]);
xlabel('t'); ylabel('x(t)=X_1(t)')
subplot(224)
ezplot(2*pi*x,[-1,1]);grid;axis([-1 1 -0.5 11*2*pi]);
 xlabel('\Omega'); ylabel('X(\Omega)=2\pi x_1(-\Omega)')
```

Example 5.7. Find the Fourier transform of $x(t) = A\cos(\Omega_0 t)$ using duality.

Solution: The Fourier transform of $x(t)$ cannot be computed using the integral definition, since this signal is not absolutely integrable, or the Laplace transform since $x(t)$ does not have a Laplace transform. As a periodic signal, $x(t)$ has a Fourier series representation and we will use it later to find its Fourier transform. For now, let us consider the following Fourier pair:

$$\delta(t - \rho_0) + \delta(t + \rho_0) \;\Leftrightarrow\; e^{-j\rho_0\Omega} + e^{j\rho_0\Omega} = 2\cos(\rho_0\Omega)$$

where we used the Laplace transform of $\delta(t - \rho_0) + \delta(t + \rho_0)$, which is $e^{-s\rho_0} + e^{s\rho_0}$ defined over the whole s-plane. At $s = j\Omega$ we get $2\cos(\rho_0\Omega)$. According to the duality property we thus have the following Fourier pair:

$$2\cos(\rho_0 t) \;\Leftrightarrow\; 2\pi[\delta(-\Omega - \rho_0) + \delta(-\Omega + \rho_0)] = 2\pi[\delta(\Omega + \rho_0) + \delta(\Omega - \rho_0)].$$

Replacing ρ_0 by Ω_0 and canceling the 2 in both sides we have

$$x(t) = \cos(\Omega_0 t) \;\Leftrightarrow\; X(\Omega) = \pi[\delta(\Omega + \Omega_0) + \delta(\Omega - \Omega_0)], \qquad (5.12)$$

indicating that $X(\Omega)$ only exists at $\pm\Omega_0$. □

5.6 SPECTRAL REPRESENTATION

In this section, we consider first the Fourier transform of periodic signals, using the modulation property, and then Parseval's relation for finite energy signals. It will be shown that these results unify the spectral representation of both periodic and aperiodic signals.

5.6.1 SIGNAL MODULATION

One of the most significant properties of the Fourier transform is modulation. Its application to signal transmission is fundamental in analog communications.

Frequency shift: If $X(\Omega)$ is the Fourier transform of $x(t)$, then we have the pair

$$x(t)e^{j\Omega_0 t} \quad \Leftrightarrow \quad X(\Omega - \Omega_0). \tag{5.13}$$

Modulation: The Fourier transform of the **modulated signal**

$$x(t)\cos(\Omega_0 t) \tag{5.14}$$

is given by

$$0.5\left[X(\Omega - \Omega_0) + X(\Omega + \Omega_0)\right], \tag{5.15}$$

i.e., $X(\Omega)$ is shifted to frequencies Ω_0 and $-\Omega_0$, added and multiplied by 0.5.

The frequency-shifting property is easily shown:

$$\mathcal{F}[x(t)e^{j\Omega_0 t}] = \int_{-\infty}^{\infty} [x(t)e^{j\Omega_0 t}]e^{-j\Omega t}\, dt = \int_{-\infty}^{\infty} x(t)e^{-j(\Omega - \Omega_0)t}\, dt = X(\Omega - \Omega_0).$$

Applying the frequency shifting to

$$x(t)\cos(\Omega_0 t) = 0.5x(t)e^{j\Omega_0 t} + 0.5x(t)e^{-j\Omega_0 t},$$

we obtain the Fourier transform of the modulated signal (5.15).

In communications, the **message** $x(t)$ (typically of lower frequency content than the frequency of the cosine) modulates the **carrier** $\cos(\Omega_0 t)$ to obtain the **modulated signal** $x(t)\cos(\Omega_0 t)$. Modulation is an important application of the Fourier transform as it allows one to change the original frequencies of a message to much higher frequencies making it possible to transmit the signal over the airwaves.

Remarks

1. As indicated in Chapter 2, amplitude modulation (AM) consists in multiplying a signal $x(t)$, or message, by a sinusoid of frequency Ω_0—higher than the maximum frequency of the message. Modulation thus shifts the frequencies of $x(t)$ to frequencies around $\pm\Omega_0$.
2. **Modulation using a sine**, instead of a cosine, changes the phase of the Fourier transform of the modulating signal besides performing the frequency shift. Indeed,

$$\begin{aligned}
\mathcal{F}[x(t)\sin(\Omega_0 t)] &= \mathcal{F}\left[\frac{x(t)e^{j\Omega_0 t} - x(t)e^{-j\Omega_0 t}}{2j}\right] = \frac{1}{2j}X(\Omega - \Omega_0) - \frac{1}{2j}X(\Omega + \Omega_0) \\
&= \frac{-j}{2}X(\Omega - \Omega_0) + \frac{j}{2}X(\Omega + \Omega_0)
\end{aligned} \tag{5.16}$$

where the $-j$ and the j terms add $-\pi/2$ and $\pi/2$ radians to the phase of the signal.
3. According to the eigenfunction property of LTI systems, **AM modulation systems are not LTI**: modulation shifts the frequencies at the input into new frequencies at the output. Non-linear or time-varying systems are typically used as amplitude-modulation transmitters.

Example 5.8. Consider modulating a carrier $\cos(10t)$ with the following signals:

1. $x_1(t) = e^{-|t|}$, $-\infty < t < \infty$. Use MATLAB to find the Fourier transform of $x_1(t)$, and to plot $x_1(t)$ and its magnitude and phase spectra.
2. $x_2(t) = 0.2[r(t+5) - 2r(t) + r(t+5)]$, where $r(t)$ is the ramp signal. Use MATLAB to plot $x_2(t)$, and $x_2(t)\cos(10t)$ and to compute and plot the magnitude of their Fourier transforms.

Solution: The modulated signals are

$$(i) \qquad y_1(t) = x_1(t)\cos(10t) = e^{-|t|}\cos(10t),$$
$$(ii) \qquad y_2(t) = x_2(t)\cos(10t) = 0.2[r(t+5) - 2r(t) + r(t+5)]\cos(10t).$$

The signal $x_1(t)$ is smooth and of infinite support, and thus most of its frequency components are low frequency. The signal $x_2(t)$ is smooth but has a finite support, so that its frequency components are mostly low-pass but its spectrum also displays higher frequencies.

   The following script indicates how to generate $y_1(t)$ and how to find the magnitude and phase of its Fourier transform $Y_1(\Omega)$. Notice the way the phase is computed. Symbolic MATLAB cannot use the function *angle* to compute the phase, thus the phase is computed using the inverse tangent function *atan* of the ratio of the imaginary over the real part of the Fourier transform.

```
%%
% Example 5.8---Modulation
%%
clear all;clf; syms t w
for ind=1:2,
if ind==1.
    x=exp(-abs(t)); y=x*0.5*exp(-j*10*t)+x*0.5*exp(j*10*t)
    X=fourier(x); Y=fourier(y);
    Xm=abs(X);Xa=atan(imag(X)/real(X));
    Ym=abs(Y); Ya=atan(imag(Y)/real(Y));
    figure(1)
else
    m=heaviside(t+5)-heaviside(t); m1=heaviside(t)-heaviside(t-5);
    x=(t+5)*m+m1*(-t+5);x=x/5; y=x*exp(-j*10*t)/2+x*exp(+j*10*t)/2;
    X=int(x*exp(-j*w*t), t,-5,5);Xm=abs(X);
    Y=int(y*exp(-j*w*t), t,-5,5);Ym=abs(Y);
    figure(2)
end
    subplot(221)
    fplot(x,[-6,6]);grid;axis([-6 6 -0.2 1.2]); xlabel('t'); ylabel('x(t)')
    subplot(222)
    fplot(y,[-6,6]);grid;axis([-6 6 -1.2 1.2]); xlabel('t');
    ylabel('y(t)=x(t)cos(10t)')
    subplot(223)
    fplot(Xm,[-8,8]); grid; axis([-8 8 -0.1 5.5]);
```

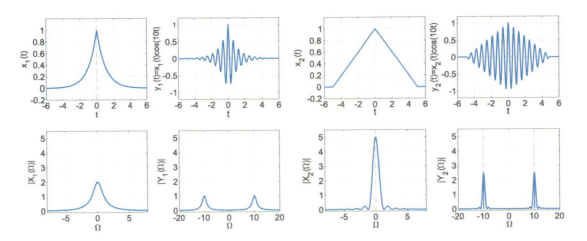

FIGURE 5.5

Left figure: the decaying exponential $x_1(t)$ and the modulated signal $y_1(t)$ and their magnitude spectra below.
Right figure: the triangular signal, $x_2(t)$, its modulated version, $y_2(t)$ and their corresponding magnitude spectra
underneath.

```
        xlabel('\Omega'); ylabel('|X(\Omega)|')
        subplot(224)
        fplot(Ym,[-20,20]); grid; axis([-20 20 -0.1 5.5]); title(' ')
        xlabel('\Omega'); ; ylabel('|Y(\Omega)|')
end
```

The results are shown in Fig. 5.5. Notice that in this example we used the function *fourier* but also the
integral definition to find the Fourier transform of $x_2(t)$ and $y_2(t)$. □

Why Amplitude Modulation?

The use of amplitude modulation to change the frequency content of a message from its baseband
frequencies to higher frequencies makes the transmission of the message over the airwaves possible.
Let us explore why it is necessary to use AM to transmit a music or a speech signal. Typically, music
signals are audible up to frequencies of about 22 kHz, while speech signals typically have frequencies
from about 100 Hz to about 5 kHz. Thus music and speech signals are relatively low-frequency signals. When **radiating a signal with an antenna**, the length of the antenna is about a quarter of the
wave-length

$$\lambda = \frac{3 \times 10^8}{f} \text{ m}$$

where f is the frequency in Hz (or 1/s) of the signal being radiated and 3×10^8 m/s is the speed of
light. Thus if we assume that frequencies up to $f = 30$ kHz are present in the signal (this would include
music and speech in the signal) the wavelength is 10 km and the size of the antenna is 2.5 km—a mile-
and-a-half long antenna! Thus, for music or a speech signal to be transmitted with a reasonable-size

antenna requires one to increase the frequencies present in the signal. Amplitude modulation provides an efficient way to shift an acoustic or speech signal to a desirable frequency.

5.6.2 FOURIER TRANSFORM OF PERIODIC SIGNALS

By applying the frequency-shifting property to compute the Fourier transform of periodic signals, we are able to unify the Fourier representation of aperiodic and periodic signals.

Representing a periodic signal $x(t)$, of period T_0, by its Fourier series we have the following Fourier pair:

$$x(t) = \sum_k X_k e^{jk\Omega_0 t} \quad \Leftrightarrow \quad X(\Omega) = \sum_k 2\pi X_k \delta(\Omega - k\Omega_0). \qquad (5.17)$$

Since a periodic signal $x(t)$ is not absolutely integrable, its Fourier transform cannot be computed using the integral formula. But we can use its Fourier series

$$x(t) = \sum_k X_k e^{jk\Omega_0 t}$$

where the $\{X_k\}$ are the Fourier coefficients, and $\Omega_0 = 2\pi/T_0$ is the fundamental frequency of the periodic signal $x(t)$ of fundamental period T_0. As such, according to the linearity and the frequency-shifting properties of the Fourier transform we obtain

$$X(\Omega) = \sum_k \mathcal{F}[X_k e^{jk\Omega_0 t}] = \sum_k 2\pi X_k \delta(\Omega - k\Omega_0)$$

where we used the fact that X_k is a constant that has a Fourier transform $2\pi X_k \delta(\Omega)$.

Remarks

1. When plotting $|X(\Omega)|$ vs. Ω, which we call the Fourier magnitude spectrum, for a periodic signal $x(t)$, we notice it being analogous to its line spectrum discussed before. Both indicate that the signal power is concentrated in multiples of the fundamental frequency—the only difference being in how the information is provided at each of the frequencies. The line spectrum displays the Fourier series coefficients at their corresponding frequencies, while the spectrum from the Fourier transform displays the concentration of the power at the harmonic frequencies by means of delta functions with amplitudes of 2π times the Fourier series coefficients. Thus, there is a clear relation between these two spectra, showing exactly the same information in slightly different form.

2. The Fourier transform of a cosine signal can now be computed directly as

$$\mathcal{F}[\cos(\Omega_0 t)] = \mathcal{F}[0.5e^{j\Omega_0 t} + 0.5e^{-j\Omega_0 t}] = \pi\delta(\Omega - \Omega_0) + \pi\delta(\Omega + \Omega_0)$$

and for a sine (compare this result with the one obtained before)

$$
\mathcal{F}[\sin(\Omega_0 t)] = \mathcal{F}\left[\frac{0.5}{j}e^{j\Omega_0 t} - \frac{0.5}{j}e^{-j\Omega_0 t}\right]
$$

$$
= \frac{\pi}{j}\delta(\Omega - \Omega_0) - \frac{\pi}{j}\delta(\Omega + \Omega_0) = \pi e^{-j\pi/2}\delta(\Omega - \Omega_0) + \pi e^{j\pi/2}\delta(\Omega + \Omega_0).
$$

The magnitude spectra of the two signals coincide, but the cosine has a zero phase spectrum while the phase spectrum for the sine displays a phase of $\mp\pi/2$ at frequencies $\pm\Omega_0$.

Example 5.9. Consider a periodic signal $x(t)$ with a period $x_1(t) = r(t) - 2r(t - 0.5) + r(t - 1)$. If the fundamental frequency of $x(t)$ is $\Omega_0 = 2\pi$, determine the Fourier transform $X(\Omega)$ analytically and using MATLAB. Plot several periods of the signal, and its Fourier transform.

Solution: The given period $x_1(t)$ corresponds to a triangular signal; its Laplace transform is

$$
X_1(s) = \frac{1}{s^2}\left(1 - 2e^{-0.5s} + e^{-s}\right) = \frac{e^{-0.5s}}{s^2}\left(e^{0.5s} - 2 + e^{-0.5s}\right),
$$

so that the Fourier coefficients of $x(t)$ are ($T_0 = 1$)

$$
X_k = \frac{1}{T_0}X_1(s)|_{s=j2\pi k} = \frac{1}{(j2\pi k)^2}2(\cos(\pi k) - 1)e^{-j\pi k}
$$

$$
= (-1)^{(k+1)}\frac{\cos(\pi k) - 1}{2\pi^2 k^2} = (-1)^k\frac{\sin^2(\pi k/2)}{\pi^2 k^2},
$$

after using the identity $\cos(2\theta) - 1 = -2\sin^2(\theta)$. The dc term is $X_0 = 0.5$. The Fourier transform of $x(t)$ is then

$$
X(\Omega) = 2\pi X_0\delta(\Omega) + \sum_{k=-\infty,\neq 0}^{\infty} 2\pi X_k\delta(\Omega - 2k\pi).
$$

To compute the Fourier transform using symbolic MATLAB, we approximate $x(t)$ by its Fourier series by means of its average and $N = 10$ harmonics (the Fourier coefficients are found using the *fourierseries* function from Chapter 4), and then create a sequence $\{2\pi X_k\}$ and the corresponding harmonic frequencies $\{\Omega_k = k\Omega_0\}$ and plot them as the spectrum $X(\Omega)$ (see Fig. 5.6). The following code gives some of the necessary steps (the plotting is excluded) to generate the periodic signal and to find its Fourier transform. The MATLAB function *fliplr* is used to reflect the Fourier coefficients.

```
%%
% Example 5.9---Fourier series
%%
syms t w  x x1
T0=1;N=10;w0=2*pi/T0;
m=heaviside(t)-heaviside(t-T0/2);
m1=heaviside(t-T0/2)-heaviside(t-T0);
```

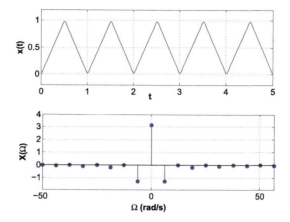

FIGURE 5.6

Triangular periodic signal $x(t)$, and its Fourier transform $X(\Omega)$, which is zero except at harmonic frequencies where it is $2\pi X_k \delta(\Omega - 2\pi k)$ with X_k the Fourier series coefficients of $x(t)$.

```
x=t*m+m1*(-t+T0);x=2*x;        % periodic signal
[Xk, w]=fourierseries(x,T0,N);  % Fourier coefficients,
                                % harmonic frequencies
% Fourier series approximation
for k=1:N,
    if k==1;
        x1=abs(Xk(k));
    else
        x1=x1+2*abs(Xk(k))*cos(w0*(k-1)*t+angle(Xk(k)));
    end
end
% Fourier coefficients and harmonic frequencies
 k=0:N-1; Xk1=2*pi*Xk;
wk=[-fliplr(k(2:N-1)) k]*w0; Xk=[fliplr(Xk1(2:N-1)) Xk1];
```

The Laplace transform simplifies the computation of the X_k values. Indeed, otherwise the Fourier series coefficients are given by

$$X_k = \int_0^{0.5} 2t e^{-j2\pi kt} dt + \int_{0.5}^1 2(1-t)e^{-j2\pi kt} dt,$$

which requires integration by parts.   □

5.6.3 PARSEVAL'S ENERGY RELATION

We saw in Chapter 4 that, for periodic signals, having finite power but infinite energy, Parseval's power relation indicates the power of the signal can be computed equally in either the time- or the frequency-domain, and how the power of the signal is distributed among the harmonic components. Likewise, for aperiodic signals of finite energy, an energy version of Parseval's relation indicates how the signal energy is distributed over frequencies.

For an aperiodic, finite-energy signal $x(t)$, with Fourier transform $X(\Omega)$, its energy E_x is conserved by the transformation:

$$E_x = \int_{-\infty}^{\infty} |x(t)|^2 dt = \frac{1}{2\pi} \int_{-\infty}^{\infty} |X(\Omega)|^2 d\Omega. \tag{5.18}$$

Thus $|X(\Omega)|^2$ is an energy density—indicating the amount of energy at each of the frequencies Ω.
The plot $|X(\Omega)|^2$ vs. Ω is called the **energy spectrum** of $x(t)$, and displays how the energy of the signal is distributed over frequency.

This energy conservation property is shown using the inverse Fourier transform. The finite energy of $x(t)$ is obtained in the frequency domain as follows:

$$
\begin{aligned}
\int_{-\infty}^{\infty} x(t)x^*(t)dt &= \int_{-\infty}^{\infty} x(t)\left[\frac{1}{2\pi}\int_{-\infty}^{\infty} X^*(\Omega)e^{-j\Omega t}d\Omega\right]dt \\
&= \frac{1}{2\pi}\int_{-\infty}^{\infty} X^*(\Omega)\left[\int_{-\infty}^{\infty} x(t)e^{-j\Omega t}dt\right]d\Omega = \frac{1}{2\pi}\int_{-\infty}^{\infty}|X(\Omega)|^2 d\Omega.
\end{aligned}
$$

Example 5.10. Parseval's energy relation helps us understand the nature of the impulse $\delta(t)$. It is clear from its definition that the area under an impulse is unity, which means $\delta(t)$ is absolutely integrable; but does it have finite energy? Show how Parseval's energy relation can help resolve this issue.

Solution: From the frequency point of view, using Parseval's energy relation, the Fourier transform of $\delta(t)$ is unity for all values of frequency and as such its energy is infinite. Such a result seems puzzling, because $\delta(t)$ was defined as the limit of a pulse of finite duration and unit area. This is what happens, if

$$p_\Delta(t) = \frac{1}{\Delta}[u(t + \Delta/2) - u(t - \Delta/2)]$$

is a pulse of unity area, and the signal

$$p_\Delta^2(t) = \frac{1}{\Delta^2}[u(t + \Delta/2) - u(t - \Delta/2)]$$

is a pulse of area $1/\Delta$. If we then let $\Delta \to 0$, we have

$$\lim_{\Delta \to 0} p_\Delta^2(t) = \lim_{\Delta \to 0}\left(\frac{1}{\Delta}\right)\lim_{\Delta \to 0}\frac{1}{\Delta}[u(t + \Delta/2) - u(t - \Delta/2)] = \left(\lim_{\Delta \to 0}\frac{1}{\Delta}\right)\delta(t),$$

that is, the squared pulse $p_\Delta^2(t)$ will tend to an impulse with infinite area under it. Thus $\delta(t)$ is not a finite energy signal.   □

Example 5.11. Consider a pulse $p(t) = u(t+1) - u(t-1)$; use its Fourier transform $P(\Omega)$ and Parseval's energy relation to show that

$$\int_{-\infty}^{\infty} \left(\frac{\sin(\Omega)}{\Omega}\right)^2 d\Omega = \pi.$$

Solution: The energy of the pulse is $E_p = 2$ (the area under the pulse). But according to Parseval's energy relation the energy computed in the frequency domain is given by

$$\frac{1}{2\pi} \int_{-\infty}^{\infty} \left(\frac{2\sin(\Omega)}{\Omega}\right)^2 d\Omega = E_p \quad \text{since}$$

$$P(\Omega) = \mathcal{F}(p(t)) = \frac{e^s - e^{-s}}{s}\Big|_{s=j\Omega} = \frac{2\sin(\Omega)}{\Omega}.$$

Replacing $E_p = 2$ we obtain the interesting and not obvious result

$$\int_{-\infty}^{\infty} \left(\frac{\sin(\Omega)}{\Omega}\right)^2 d\Omega = \pi.$$

One more way to compute π! ◻

5.6.4 SYMMETRY OF SPECTRAL REPRESENTATIONS

Now that the Fourier representation of aperiodic and periodic signals is unified, we can think of just one spectrum that accommodates both finite and infinite energy signals. The word "spectrum" is loosely used to mean different aspects of the frequency representation. The following provides definitions and the symmetry characteristic of the spectrum of real-valued signals.

If $X(\Omega)$ is the Fourier transform of a real-valued signal $x(t)$, periodic or aperiodic, the magnitude $|X(\Omega)|$ and the real part $\mathcal{Re}[X(\Omega)]$ are even functions of Ω:

$$|X(\Omega)| = |X(-\Omega)|,$$
$$\mathcal{Re}[X(\Omega)] = \mathcal{Re}[X(-\Omega)], \tag{5.19}$$

and the phase $\angle X(\Omega)$ and the imaginary part $\mathcal{Im}[X(\Omega)]$ are odd functions of Ω:

$$\angle X(\Omega) = -\angle X(-\Omega),$$
$$\mathcal{Im}[X(\Omega)] = -\mathcal{Im}[X(-\Omega)]. \tag{5.20}$$

We then call the plots

$|X(\Omega)|$ vs. Ω	**magnitude spectrum**
$\angle X(\Omega)$ vs. Ω	**phase spectrum**
$|X(\Omega)|^2$ vs. Ω	**energy/power spectrum.**

To show this consider the inverse Fourier transform of a real-valued signal $x(t)$,

$$x(t) = \int_{-\infty}^{\infty} X(\Omega)e^{j\Omega t} d\Omega,$$

which is identical, because of being real, to

$$x^*(t) = \int_{-\infty}^{\infty} X^*(\Omega)e^{-j\Omega t} d\Omega = \int_{-\infty}^{\infty} X^*(-\Omega')e^{j\Omega' t} d\Omega',$$

since the integral can be thought of as an infinite sum of complex values and by letting $\Omega' = -\Omega$. Comparing the two integrals, we have the following identities:

$$X(\Omega) = X^*(-\Omega),$$
$$|X(\Omega)|e^{j\angle X(\Omega)} = |X(-\Omega)|e^{-j\angle X(-\Omega)},$$
$$\mathcal{Re}[X(\Omega)] + j\mathcal{Im}[X(\Omega)] = \mathcal{Re}[X(-\Omega)] - j\mathcal{Im}[X(-\Omega)],$$

or that the magnitude is an even function of Ω and the phase an odd function of Ω. And that the real part of the Fourier transform is an even function and the imaginary part of the Fourier transform is an odd function of Ω.

Remarks

1. Clearly if the signal is complex, the above symmetry will not hold. For instance, if $x(t) = e^{j\Omega_0 t} = \cos(\Omega_0 t) + j\sin(\Omega_0 t)$, using the frequency shift property its Fourier transform is

$$X(\Omega) = 2\pi\delta(\Omega - \Omega_o),$$

   which occurs at $\Omega = \Omega_0$ only, so the symmetry in the magnitude and phase does not exist.
2. It is important to recognize the meaning of "negative" frequencies. In reality, only positive frequencies exist and can be measured, but as shown the spectrum, magnitude or phase, of a real-valued signal requires negative frequencies. It is only under this context that negative frequencies should be understood—as necessary to generate "real-valued" signals.

Example 5.12. Use MATLAB to compute the Fourier transform of the following signals:

$$(a) \ x_1(t) = u(t) - u(t-1), \quad (b) \ x_2(t) = e^{-t}u(t).$$

Plot their magnitude and phase spectra.

Solution: Three possible ways to compute the Fourier transforms of these signals using MATLAB are: (i) find their Laplace transforms, as in Chapter 3, using the symbolic function *laplace* and compute the magnitude and phase functions by letting $s = j\Omega$, (ii) use the symbolic function *fourier*, and (iii) sample the signals and approximate their Fourier transforms (this requires sampling theory to be discussed in Chapter 8, and the Fourier representation of sampled signals to be considered in Chapter 11).

The Fourier transform of $x_1(t) = u(t) - u(t-1)$ can be found by considering the advanced signal $z(t) = x_1(t+0.5) = u(t+0.5) - u(t-0.5)$ with Fourier transform

$$Z(\Omega) = \int_{-0.5}^{0.5} e^{-j\Omega t} dt = \frac{\sin(\Omega/2)}{\Omega/2}.$$

Since $z(t) = x_1(t+0.5)$ we have $Z(\Omega) = X_1(\Omega)e^{j0.5\Omega}$ so that

$$X_1(\Omega) = e^{-j0.5\Omega} Z(\Omega) \text{ and } |X_1(\Omega)| = \left| \frac{\sin(\Omega/2)}{\Omega/2} \right|.$$

Given that $Z(\Omega)$ is real, its phase is either zero when $Z(\Omega) \geq 0$ or $\pm\pi$ when $Z(\Omega) < 0$ (using these values so that the phase is an odd function of Ω) and as such the phase of $X_1(\Omega)$ is

$$\angle X_1(\Omega) = \angle Z(\Omega) - 0.5\Omega = \begin{cases} -0.5\Omega & Z(\Omega) \geq 0, \\ \pm\pi - 0.5\Omega & Z(\Omega) < 0. \end{cases}$$

The Fourier transform of $x_2(t) = e^{-t}u(t)$ is

$$X_2(\Omega) = \frac{1}{1 + j\Omega}.$$

The magnitude and phase are given by

$$|X_2(\Omega)| = \frac{1}{\sqrt{1 + \Omega^2}}, \quad \angle(X_2(\Omega)) = -\tan^{-1}\Omega.$$

Computing for different values of Ω we have

Ω	$|X_2(\Omega)|$	$\angle(X_2(\Omega))$
0	1	0
1	$\frac{1}{\sqrt{2}} = 0.707$	$-\pi/4$
∞	0	$-\pi/2,$

i.e., the magnitude spectrum decays as Ω increases. The following script gives the necessary instructions to compute and plot the signal $x_1(t)$ and the magnitude and phase of its Fourier transform using symbolic MATLAB. Similar calculations are done for $x_2(t)$.

```
%%
% Example 5.12---Magnitude and phase spectra
%%
clear all;clf
syms t w
x1=heaviside(t)-heaviside(t-1);
X1=fourier(x1)
X1m=sqrt((real(X1))^2+(imag(X1))^2);    % magnitude
X1a=imag(log(X1));                       % phase
```

Notice the way the magnitude and the phase are computed. The computation of the phase is complicated by the lack of the function *atan2* in symbolic MATLAB (*atan2* extends the principal values of the inverse tangent to $(-\pi, \pi]$ by considering the sign of the real part of the complex function). The phase computation can be done by using the *log* function:

$$\log(X_1(\Omega)) = \log[|X_1(\Omega)|e^{j\angle X_1(\Omega)}] = \log(|X_1(\Omega)|) + j\angle X_1(\Omega),$$

so that

$$\angle X_1(\Omega) = \mathcal{I}m[\log(X_1(\Omega))].$$

Changing the above script we can find the magnitude and phase of $X_2(\Omega)$. See Fig. 5.7 for results. ☐

Example 5.13. It is not always the case that the Fourier transform is a complex-valued function. Consider the signals

$$(a) \ \ x(t) = 0.5e^{-|t|}, \quad (b) \ \ y(t) = e^{-|t|}\cos(\Omega_0 t).$$

Find their Fourier transforms. Let $\Omega_0 = 1$; use the magnitudes $|X(\Omega)|$ and $|Y(\Omega)|$ to discuss the smoothness of the signals $x(t)$ and $y(t)$.

Solution: (a) The Fourier transform of $x(t)$ is

$$X(\Omega) = \frac{0.5}{s+1} + \frac{0.5}{-s+1}\Big|_{s=j\Omega} = \frac{1}{\Omega^2+1},$$

a positive and real-valued function of Ω. Thus $\angle(X(\Omega)) = 0$; $x(t)$ is a "low-pass" signal like $e^{-t}u(t)$ as its magnitude spectrum decreases with frequency:

Ω	$|X(\Omega)| = X(\Omega)$
0	1
1	0.5
∞	0

but $0.5e^{-|t|}$ is "smoother" than $e^{-t}u(t)$ because the magnitude response is more concentrated in the low frequencies. Compare the values of the magnitude responses at $\Omega = 0$ and 1 to verify this.
(b) The signal $y(t) = x(t)\cos(\Omega_0 t)$ is a "band-pass" signal. It is not as smooth as $x(t)$ given that the energy concentration of

$$Y(\Omega) = 0.5[X(\Omega - \Omega_0) + X(\Omega + \Omega_0)] = \frac{0.5}{(\Omega - \Omega_0)^2 + 1} + \frac{0.5}{(\Omega + \Omega_0)^2 + 1}$$

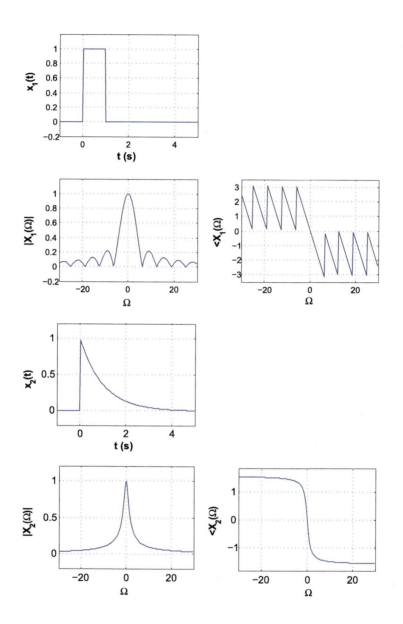

FIGURE 5.7

Top three figures: pulse $x_1(t) = u(t) - u(t-1)$ and its magnitude and phase spectra. Bottom three figures: decaying exponential $x_2(t) = e^{-t}u(t)$ and its magnitude and phase spectra.

is around the frequency Ω_0 and not the zero frequency as for $x(t)$. Since $Y(\Omega)$ is real-valued and positive the corresponding phase is zero. The magnitude of $Y(\Omega)$ for $\Omega_0 = 1$ is

Ω	$|Y(\Omega)| = Y(\Omega)$
0	0.5
1	1
2	0.5
∞	0

The higher the frequency Ω_0 the more variation is displayed by the signal. $\square$

The **bandwidth** of a signal $x(t)$ is the support—on the positive frequencies—of its Fourier transform $X(\Omega)$. There are different definitions of the bandwidth of a signal depending on how the support of its Fourier transform is measured. We will discuss some of the bandwidth measures used in filtering and in communications in Chapter 7.

The concept of bandwidth of a filter that you learned in circuit theory is one of its possible definitions, other possible definitions will be introduced later. The bandwidth together with the information about whether the energy or power of the signal is concentrated in the low, middle, high frequencies or a combination of them provides a good characterization of the signal. The spectrum analyzer, a device used to measure the spectral characteristics of a signal, will be presented after the section on filtering.

5.7 CONVOLUTION AND FILTERING

The modulation and the convolution properties are the most important properties of the Fourier transform. Modulation is essential in communications, and the convolution property is basic in the analysis and design of filters.

If the input $x(t)$ (periodic or aperiodic) of a stable LTI system has Fourier transform $X(\Omega)$, and the system has a frequency response $H(j\Omega) = \mathcal{F}[h(t)]$, where $h(t)$ is the impulse response of the system, the output of the LTI system is the convolution integral $y(t) = (x * h)(t)$, with Fourier transform

$$Y(\Omega) = X(\Omega)\, H(j\Omega). \tag{5.21}$$

In particular, if the input signal $x(t)$ is periodic the output is also periodic of the same fundamental period, and with Fourier transform

$$Y(\Omega) = \sum_{k=-\infty}^{\infty} 2\pi\, X_k\, H(jk\,\Omega_0)\delta(\Omega - k\Omega_0) \tag{5.22}$$

where $\{X_k\}$ are the Fourier series coefficients of $x(t)$ and Ω_0 its fundamental frequency.

This can be shown by considering the eigenfunction property of LTI systems. The Fourier representation of $x(t)$, if aperiodic, is an infinite summation of complex exponentials $e^{j\Omega t}$ multiplied by complex constants $X(\Omega)$, or

$$x(t) = \frac{1}{2\pi} \int_{-\infty}^{\infty} X(\Omega) e^{j\Omega t} d\Omega.$$

According to the eigenfunction property, the response of a LTI system to each term $X(\Omega)e^{j\Omega t}$ is $X(\Omega)e^{j\Omega t} H(j\Omega)$, where $H(j\Omega)$ is the frequency response of the system, and thus by superposition the response $y(t)$ is

$$y(t) = \frac{1}{2\pi} \int_{-\infty}^{\infty} \left[X(\Omega) H(j\Omega) \right] e^{j\Omega t} d\Omega = \frac{1}{2\pi} \int_{-\infty}^{\infty} Y(\Omega) e^{j\Omega t} d\Omega,$$

so that $Y(\Omega) = X(\Omega) H(j\Omega)$.

If $x(t)$ is periodic of fundamental period T_0 (or fundamental frequency $\Omega_0 = 2\pi/T_0$) then

$$X(\Omega) = \sum_{k=-\infty}^{\infty} 2\pi X_k \delta(\Omega - k\Omega_0),$$

so that the output $y(t)$ has as Fourier transform

$$Y(\Omega) = X(\Omega) H(j\Omega) = \sum_{k=-\infty}^{\infty} 2\pi X_k H(j\Omega)\delta(\Omega - k\Omega_0) = \sum_{k=-\infty}^{\infty} 2\pi X_k H(jk\Omega_0)\delta(\Omega - k\Omega_0).$$

Therefore, the output $y(t)$ is periodic with the same fundamental period as $x(t)$, i.e., its Fourier series is

$$y(t) = \sum_{k=-\infty}^{\infty} Y_k e^{jk\Omega_0 t}$$

where $Y_k = X_k H(jk\Omega_0)$.

An important consequence of the convolution property, just like in the Laplace transform, is that the ratio of the Fourier transforms of the input and the output gives the **frequency response of the system** or

$$H(j\Omega) = \frac{Y(\Omega)}{X(\Omega)}. \tag{5.23}$$

The magnitude and the phase of $H(j\Omega)$ are the **magnitude and phase frequency responses of the system**; they tell how the system responds to each particular frequency.

Remarks

1. It is important to keep in mind the following connection between the impulse response, $h(t)$, the transfer function $H(s)$ and the frequency response $H(j\Omega)$ characterizing a LTI system:

$$H(j\Omega) = \mathcal{L}[h(t)]|_{s=j\Omega} = H(s)|_{s=j\Omega} = \frac{Y(s)}{X(s)}\Big|_{s=j\Omega}.$$

2. As the Fourier transform of the impulse response $h(t)$, a real-valued function, $H(j\Omega)$ has a magnitude $|H(j\Omega)|$ and a phase $\angle H(j\Omega)$, which are even and odd functions of the frequency Ω, respectively.

3. The convolution property relates to the processing of an input signal by a LTI system. But it is possible, in general, to consider the case of convolving two signals $x(t)$ and $y(t)$ to get $z(t) = [x * y](t)$ in which case we have that $Z(\Omega) = X(\Omega)Y(\Omega)$, where $X(\Omega)$ and $Y(\Omega)$ are the Fourier transforms of $x(t)$ and $y(t)$.

5.7.1 BASICS OF FILTERING

The most important application of LTI systems is filtering. Filtering consists in getting rid of undesirable components of a signal. A typical example is when noise $\eta(t)$ has been added to a desired signal $x(t)$, i.e., $y(t) = x(t) + \eta(t)$, and the spectral characteristics of $x(t)$ and the noise $\eta(t)$ are known. The problem then is to design a filter, or a LTI system, that will get rid of the noise as much as possible. The filter design consists in finding a transfer function $H(s) = B(s)/A(s)$ that satisfies certain specifications that will allow getting rid of the noise. Such specifications are typically given in the frequency domain. This is a **rational approximation** problem, as we look for the coefficients of the numerator and denominator of $H(s)$ which make $H(j\Omega)$ in magnitude and phase approximate the filter specifications. The designed filter $H(s)$ should be implementable (i.e., its coefficients should be real, and the filter should be stable). In this section we discuss the basics of filtering and in Chapter 7 we introduce filter design.

Frequency discriminating filters keep the frequency components of a signal in a certain frequency band and attenuate the rest. Filtering an aperiodic signal $x(t)$ represented by its Fourier transform $X(\Omega)$, with magnitude $|X(\Omega)|$ and phase $\angle X(\Omega)$, using a filter with frequency response $H(j\Omega)$ gives an output $y(t)$, with a Fourier transform

$$Y(\Omega) = H(j\Omega)X(\Omega).$$

Thus, the output $y(t)$ is composed of only those frequency components of the input that are not filtered out by the filter. When designing the filter, we assign appropriate values to the magnitude in the desirable frequency band or bands, and let it be zero or close to zero in those frequencies in which we would not want components of the input signal.

If the input signal $x(t)$ is periodic of fundamental period T_0, or fundamental frequency $\Omega_0 = 2\pi/T_0$, the Fourier transform of the output is

$$Y(\Omega) = X(\Omega)H(j\Omega) = 2\pi \sum_k X_k H(jk\Omega_0)\delta(\Omega - k\Omega_0) \tag{5.24}$$

where the magnitude and the phase of each of the Fourier series coefficients of $x(t)$ is changed by the frequency response of the filter at the harmonic frequencies. Indeed, X_k corresponding to the frequency $k\Omega_0$ is changed into

$$X_k H(jk\Omega_0) = |X_k||H(jk\Omega_0)|e^{j(\angle X_k + \angle H(jk\Omega_0))}.$$

The filter output $y(t)$ is also periodic of fundamental period T_0 but is missing the harmonics of the input that have been filtered out.

The above shows that—independent of whether the input signal $x(t)$ is periodic or aperiodic—the output signal $y(t)$ has the frequency components of the input $x(t)$ allowed through by the filter.

Example 5.14. Consider how to obtain a dc source using a full-wave rectifier and a low-pass filter (it keeps only the low-frequency components). Let the full-wave rectified signal $x(t) = |\cos(\pi t)|$, $-\infty < t < \infty$, be the input of the filter and the output of the filter be $y(t)$, which we want to have a unit voltage. The rectifier and the low-pass filter constitute a system that converts alternating into direct voltage.

Solution: We found in Chapter 4 that the Fourier series coefficients of a full-wave rectified signal $x(t) = |\cos(\pi t)|$, $-\infty < t < \infty$, are given by

$$X_k = \frac{2(-1)^k}{\pi(1 - 4k^2)},$$

so that the average of $x(t)$ is $X_0 = 2/\pi$. To filter out all the harmonics and leave only the average component, we need an ideal low-pass filter with a magnitude A and a cut-off frequency $0 < \Omega_c < \Omega_0$, where $\Omega_0 = 2\pi/T_0 = 2\pi$ is the fundamental frequency of $x(t)$. Thus the filter is given by

$$H(j\Omega) = \begin{cases} A & -\Omega_c \le \Omega \le \Omega_c, \quad \text{where } 0 < \Omega_c < \Omega_0, \\ 0 & \text{otherwise.} \end{cases}$$

According to the convolution property then

$$Y(\Omega) = H(j\Omega)X(\Omega) = H(j\Omega) \left[2\pi X_0 \delta(\Omega) + \sum_{k \ne 0} 2\pi X_k \delta(\Omega - k\Omega_0) \right] = 2\pi A X_0 \delta(\Omega).$$

To get the output to have a unit amplitude we let $AX_0 = 1$, or $A = 1/X_0 = \pi/2$. Although the proposed filter is not realizable, the above shows what needs to be done to obtain a dc source from a full-wave rectified signal. □

Example 5.15. **Windowing** is a time-domain process by which we select part of a signal, this is done by multiplying the signal by a "window" signal $w(t)$. Consider the **rectangular window**

$$w(t) = u(t + \Delta) - u(t - \Delta) \qquad \Delta > 0.$$

For a given signal $x(t)$, the **windowed signal** is $y(t) = x(t)w(t)$. Discuss how windowing relates to the convolution property.

Solution: Windowing is the dual of filtering. In this case, the signal $y(t)$ has the support determined by the window, or $-\Delta \le t \le \Delta$, and as such it is zero outside this interval. The rectangular window gets rid of parts of the signal outside its support. The signal $y(t)$ can be written as

$$y(t) = w(t)x(t) = w(t)\frac{1}{2\pi} \int_{-\infty}^{\infty} X(\rho)e^{j\rho t}d\rho = \frac{1}{2\pi} \int_{-\infty}^{\infty} X(\rho)w(t)e^{j\rho t}d\rho$$

considering the integral an infinite summation, the Fourier transform of $y(t)$ is

$$Y(\Omega) = \frac{1}{2\pi} \int_{-\infty}^{\infty} X(\rho)\mathcal{F}[w(t)e^{j\rho t}]d\rho = \frac{1}{2\pi} \int_{-\infty}^{\infty} X(\rho)W(\Omega - \rho)d\rho$$

using the frequency-shifting property. Thus we have that rectangular windowing (or a multiplication in the time-domain of two signals $w(t)$ and $x(t)$) gives $Y(\Omega)$ as the convolution of $X(\Omega) = \mathcal{F}[x(t)]$ and

$$W(\Omega) = \mathcal{F}[w(t)] = \frac{1}{s}\left[e^{\Delta s} - e^{-\Delta s}\right]_{s=j\Omega} = \frac{2\sin(\Omega\Delta)}{\Omega}$$

multiplied by $1/(2\pi)$. This is one more example of the inverse relationship between time and frequency. In this case, the support of the result of the windowing is finite while the convolution in the frequency domain gives an infinite support for $Y(\Omega)$ since $W(\Omega)$ has an infinite support. □

5.7.2 IDEAL FILTERS

Frequency discriminating filters that keep low, middle and high frequency components, or a combination of these, are called **low-pass, band-pass, high-pass and multi-band** filters, respectively. A **band-eliminating or notch** filter gets rid of middle frequency components. It is also possible to have an **all-pass** filter that although does not filter out any of the input frequency components it changes the phase of the input signal.

The magnitude frequency response of an **ideal low-pass filter** is given by

$$|H_{lp}(j\Omega)| = \begin{cases} 1 & -\Omega_1 \leq \Omega \leq \Omega_1, \\ 0 & \text{otherwise,} \end{cases}$$

and the phase frequency response of this filter is

$$\angle H_{lp}(j\Omega) = -\alpha\Omega,$$

which as a function of Ω is a straight line, with slope $-\alpha$, through the origin of the frequency plane and thus the term **linear phase** for it. The frequency Ω_1 is called the **cut-off frequency** of the low-pass filter. (See Fig. 5.8.) The above magnitude and phase responses only need to be given for positive frequencies, given that the magnitude and the phase responses are even and odd function of Ω. The rest of the frequency response, if desired, is obtained by symmetry.

An **ideal band-pass** filter has a magnitude response

$$|H_{bp}(j\Omega)| = \begin{cases} 1 & \Omega_1 \leq \Omega \leq \Omega_2 \text{ and } -\Omega_2 \leq \Omega \leq -\Omega_1, \\ 0 & \text{otherwise,} \end{cases}$$

with cut-off frequencies Ω_1 and Ω_2. The magnitude response of an **ideal high-pass** filter is

$$|H_{hp}(j\Omega)| = \begin{cases} 1 & \Omega \geq \Omega_2 \text{ and } \Omega \leq -\Omega_2, \\ 0 & \text{otherwise,} \end{cases}$$

with a cut-off frequency of Ω_2. The phase for these filters is linear in the **pass-band** (where the magnitude is unity).

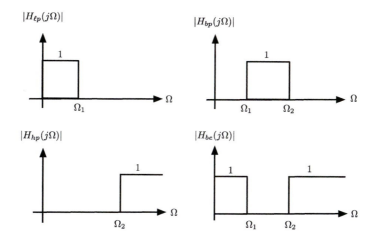

FIGURE 5.8

Ideal filters: (top-left clockwise) low-pass, band-pass, band-eliminating and high-pass.

From these definitions, we have the **ideal band-stop** filter has as magnitude response

$$|H_{bs}(j\Omega)| = 1 - |H_{bp}(j\Omega)|.$$

Noticing the intentional relation among the cut-off frequencies of the given low-, band-, and high-pass filters, the magnitude response of their sum (i.e., the filters are connected in parallel) gives the magnitude response of an **ideal all-pass** filter

$$|H_{ap}(j\Omega)| = |H_{lp}(j\Omega)| + |H_{bp}(j\Omega)| + |H_{hp}(j\Omega)| = 1$$

for all frequencies, since we chose the frequencies Ω_1 and Ω_2 so that the response of these filters do not overlap. An ideal all-pass filter can also be obtained as the parallel connection of the band-pass and the stop-band filters given above.

An **ideal multi-band** filter can be obtained as a combination of two of the low-, the band-, and the high-pass filters.

Remarks

1. If $h_{lp}(t)$ is the impulse response of a low-pass filter, applying the modulation property we get $2h_{lp}(t)\cos(\Omega_0 t)$, where $\Omega_0 >> \Omega_1$ and Ω_1 is the cut-off frequency of the ideal low-pass filter, corresponds to the impulse response of a band-pass filter centered around Ω_0. Indeed, its Fourier transform is given by

$$\mathcal{F}[2h_{lp}(t)\cos(\Omega_0 t)] = H_{lp}(j(\Omega - \Omega_0)) + H_{lp}(j(\Omega + \Omega_0)),$$

which is the frequency response of the low-pass filter shifted to a new center frequency Ω_0 and $-\Omega_0$, making it a band-pass filter.

2. A zero-phase ideal low-pass filter $H_{lp}(j\Omega) = u(\Omega + \Omega_1) - u(\Omega - \Omega_1)$, has as impulse response a sinc function with a support from $-\infty$ to ∞. This ideal low-pass filter is clearly noncausal as its impulse response is not zero for negative values of time. To make it causal, we first approximate its impulse response by a function $h_1(t) = h_{lp}(t)w(t)$ where $w(t) = u(t + \tau) - u(t - \tau)$ is a rectangular window where the value of τ is chosen so that outside the window the values of the impulse response $h_{lp}(t)$ are very close to zero. Although the magnitude of the Fourier transform of $h_1(t)$ is a very good approximation of the desired magnitude response $|H_{lp}(j\Omega)|$ for most frequencies, it displays ringing around the cut-off frequency Ω_1 because of the rectangular window. Finally, if we delay $h_1(t)$ by τ we get a causal filter with linear phase. That is, $h_1(t - \tau)$ has as magnitude response $|H_1(j\Omega)| \approx |H_{lp}(j\Omega)|$ and as phase response $\angle H_1(j\Omega) = -\tau\Omega$, assuming the phase of $H_{lp}(j\Omega)$ is zero. Although the above procedure is a valid way to obtain approximate low-pass filters with linear-phase, these filters are not guaranteed to be rational and would be difficult to implement. Thus other methods are used to design filters.

3. Since ideal filters are not causal they cannot be used in real-time applications, that is, when the input signal needs to be processed as it comes to the filter. Imposing causality on the filter restricts the frequency response of the filter in significant ways. According to the **Paley–Wiener integral condition**, a causal and stable filter with frequency response $H(j\Omega)$ should satisfy the following condition:

$$\int_{-\infty}^{\infty} \frac{|\log(H(j\Omega))|}{1 + \Omega^2} d\Omega < \infty. \tag{5.25}$$

To satisfy this condition, $H(j\Omega)$ cannot be zero in any band of frequencies (although it can be zero at a finite number of frequencies) because in such a case the numerator of the integrand would be infinite—the Paley–Wiener integral condition is clearly not satisfied by ideal filters. So ideal filters cannot be implemented or used in actual situations, but they can be used as models for designing filters.

4. That ideal filters are not realizable can be understood also by considering what it means to make the magnitude response of a filter zero in some frequency band. A measure of attenuation in filtering is given by the **loss function**, in decibels, defined as

$$\alpha(\Omega) = -10\log_{10}|H(j\Omega)|^2 = -20\log_{10}|H(j\Omega)| \quad \text{dB}.$$

Thus, when $|H(j\Omega)| = 1$ there is no attenuation and the loss is 0 dB, but when $|H(j\Omega)| = 10^{-5}$ there is a large attenuation and the loss is 100 dB. You quickly convince yourself that if a filter achieves a magnitude response of 0 at any frequency this would mean a loss or attenuation at that frequency of ∞ dB! Values of 60 to 100 dB attenuation are considered extremely good, and to obtain them the signal needs to be attenuated by a factor of 10^{-3} to 10^{-5}. Considering the gain, rather than the loss, in dB a curious term *JND or "just noticeable difference"* is used by experts in human hearing to characterize the smallest sound intensity that can be judged by a human as different. Such a value varies from 0.25 to 1 dB. Moreover, to illustrate what is loud in the dB scale consider that a sound pressure level higher than 130 dB causes pain, and that 110 dB is typically generated by an amplified rock band performance (now you see why young people are going deaf!) [37].

Example 5.16. The **Gibbs phenomenon**, which we mentioned in Chapter 4 when discussing the Fourier series of periodic signals with discontinuities, consists in ringing around these discontinuities. To explain this ringing, consider a periodic train of square pulses $x(t)$ of period T_0 displaying discontinuities at $kT_0/2$, for $k = \pm 1, \pm 2, \cdots$. Show how the Gibbs phenomenon is due to ideal low-pass filtering.

Solution: Choosing $2N + 1$ of the Fourier series coefficients to approximate the periodic signal $x(t)$, of fundamental frequency Ω_0, is equivalent to passing $x(t)$ through an ideal low-pass filter that keeps the dc and the first N harmonics or

$$H(j\Omega) = \begin{cases} 1 & -\Omega_c \leq \Omega \leq \Omega_c, \quad N\Omega_0 < \Omega_c < (N+1)\Omega_0, \\ 0 & \text{otherwise,} \end{cases}$$

having as impulse response a sinc function $h(t)$. If the Fourier transform of the periodic signal $x(t)$, of fundamental frequency $\Omega_0 = 2\pi/T_0$, is

$$X(\Omega) = \sum_{k=-\infty}^{\infty} 2\pi X_k \delta(\Omega - k\Omega_0).$$

The output of the filter is the signal

$$x_N(t) = \mathcal{F}^{-1}[X(\Omega)H(j\Omega)] = \mathcal{F}^{-1}\left[\sum_{k=-N}^{N} 2\pi X_k \delta(\Omega - k\Omega_0)\right]$$

or the inverse Fourier transform of $X(\Omega)$ multiplied by the frequency response $H(j\Omega)$ of the ideal low-pass filter. The filtered signal $x_N(t)$ is also the convolution

$$x_N(t) = [x * h](t)$$

where $h(t)$ is the inverse Fourier of $H(j\Omega)$, or a sinc function of infinite support. The convolution around the discontinuities of $x(t)$ causes ringing before and after them, and this ringing appears independent of the value of N. □

Example 5.17. Let the input of an RLC circuit (Fig. 5.9) be a voltage source with Laplace transform $V_i(s)$. Obtain different filters by choosing different outputs. For simplicity let $R = 1\ \Omega$, $L = 1$ H, and $C = 1$ F, and assume the initial conditions to be zero.

Solution: Low-pass filter: Let the output be the voltage across the capacitor. By voltage division we have

$$V_C(s) = \frac{V_i(s)/s}{1 + s + 1/s} = \frac{V_i(s)}{s^2 + s + 1},$$

so that the transfer function

$$H_{lp}(s) = \frac{V_C(s)}{V_i(s)} = \frac{1}{s^2 + s + 1}$$

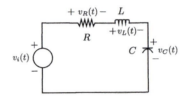

FIGURE 5.9

RLC circuit for implementing different filters.

corresponds to the transfer function of a second-order low-pass filter. Indeed, if the input is a dc source, so that its frequency is $\Omega = 0$, the inductor is a short circuit (its impedance would be 0) and the capacitor is an open circuit (its impedance would be infinite), so that the voltage across the capacitor is equal to the voltage in the source. On the other hand, if the frequency of the input source is very high, then the inductor is an open circuit and the capacitor a short circuit (its impedance is zero) so that the voltage across the capacitor is zero. This is a low-pass filter. Notice that this filter has no finite zeros.

High-pass filter: Suppose then that we let the output be the voltage across the inductor. Again by voltage division, the transfer function

$$H_{hp}(s) = \frac{V_L(s)}{V_i(s)} = \frac{s^2}{s^2 + s + 1}$$

corresponds to that of a high-pass filter. Indeed, for a dc input (frequency zero) the impedance in the inductor is zero, so that the inductor voltage is zero, and for very high frequency the impedance of the inductor is very large so that it can be considered open circuit and the voltage across the inductor equals that of the source. This filter has the same poles of the low-pass (this is determined by the overall impedance of the circuit which has not changed) and two zeros at zero. It is these zeros that make the frequency response close to zero for low frequencies.

Band-pass filter: Letting the output be the voltage across the resistor, its transfer function is

$$H_{bp}(s) = \frac{V_R(s)}{V_i(s)} = \frac{s}{s^2 + s + 1},$$

which corresponds to the transfer function of a band-pass filter. For zero frequency, the capacitor is an open circuit so the current is zero and the voltage across the resistor is zero. Similarly, for very high frequency the impedance of the inductor is very large, or an open circuit, making the voltage in the resistor zero because again the current is zero. For some middle frequency the serial combination of the inductor and the capacitor resonates and will have zero impedance. At the resonance frequency, the current achieves its largest value and the voltage across the resistor does too. This behavior is that of a band-pass filter. This filter again has the same poles as the other two, but only one zero at zero.

Band-stop filter: Finally, suppose we consider as output the voltage across the connection of the inductor and the capacitor. At low and high frequencies, the impedance of the LC connection is very high, or open circuit, and so the output voltage is the input voltage. At the resonance frequency $\Omega_r = \sqrt{LC} = 1$

the impedance of the LC connection is zero, so the output voltage is zero. The resulting filter is a band-stop filter with transfer function

$$H_{bs}(s) = \frac{s^2 + 1}{s^2 + s + 1}.$$

Second order filters can thus be easily identified by the numerator of their transfer functions. Second-order low-pass filters have no zeros, its numerator is $N(s) = 1$; band-pass filters have a zero at $s = 0$ so $N(s) = s$, and so on. We will see next that such a behavior can be easily seen from a geometric approach. $\square$

5.7.3 FREQUENCY RESPONSE FROM POLES AND ZEROS

Given a rational transfer function, $H(s) = B(s)/A(s)$, to calculate its frequency response we let $s = j\Omega$ and find the magnitude and phase for a discrete set of frequencies. This can be done using symbolic MATLAB. It could also be done numerically using MATLAB by discretizing the frequency and calculating the magnitude and phase for each frequency. A geometric way to obtain approximate magnitude and phase frequency responses is using the effects of zeros and poles on the frequency response of an LTI system.

Consider a function

$$G(s) = K\frac{s - z}{s - p}$$

with a zero z, a pole p and a gain $K \neq 0$. The frequency response corresponding to $G(s)$ at some frequency Ω_0 is found by letting $s = j\Omega_0$ or

$$G(s)|_{s=j\Omega_0} = K\frac{j\Omega_0 - z}{j\Omega_0 - p} = K\frac{\vec{Z}(\Omega_0)}{\vec{P}(\Omega_0)}.$$

Representing $j\Omega_0$, z and p, which are complex numbers, as vectors coming from the origin, when we add the vector corresponding to z, $\vec{z}$, to the vector $\vec{Z}(\Omega_0)$ shown in Fig. 5.10 we obtain the vector corresponding to $j\Omega_0$. Thus the vector $\vec{Z}(\Omega_0)$ corresponds to the numerator $j\Omega_0 - z$, and goes from the zero z to $j\Omega_0$. Likewise, the vector $\vec{P}(\Omega_0)$ corresponds to the denominator $j\Omega_0 - p$ and goes from the pole p to $j\Omega_0$. The argument Ω_0 in these vectors indicates that the magnitude and phase of these vectors depend on the frequency at which we are finding the frequency response. As we change the frequency the lengths and the phases of these vectors change. Thus in terms of vectors we have

$$G(j\Omega_0) = K\frac{\vec{Z}(\Omega_0)}{\vec{P}(\Omega_0)} = |K|e^{j\angle K}\frac{|\vec{Z}(\Omega_0)|}{|\vec{P}(\Omega_0)|}e^{j(\angle\vec{Z}(\Omega_0) - \angle\vec{P}(\Omega_0))}.$$

The magnitude response is then given by the ratio of the lengths of the vectors corresponding to the numerator and the denominator multiplied by $|K|$, or

$$|G(j\Omega_0)| = |K|\frac{|\vec{Z}(\Omega_0)|}{|\vec{P}(\Omega_0)|} \tag{5.26}$$

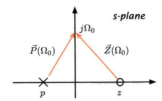

FIGURE 5.10

Geometric interpretation of computing frequency response from poles and zeros.

and the phase response is

$$\angle G(j\Omega_0) = \angle K + \angle \vec{Z}(\Omega_0) - \angle \vec{P}(\Omega_0) \tag{5.27}$$

where $\angle K$ is zero if $K > 0$ and $\pm\pi$ if $K < 0$. So that for $0 \leq \Omega_0 < \infty$ if we compute the length and the angle of $\vec{Z}(\Omega_0)$ and $\vec{P}(\Omega_0)$ at each frequency, the ratio of their lengths gives the magnitude response and the difference of their angles gives the phase response.

For a filter with transfer function

$$H(s) = K \frac{\prod_i (s - z_i)}{\prod_k (s - p_k)} \quad \text{where} \quad z_i, \ p_k \ \text{are zeros and poles of } H(s), \text{ and } K \text{ is a gain,}$$

vectors $\vec{Z}_i(\Omega)$ and $\vec{P}_k(\Omega)$, going from each of the zeros and poles to the frequency at which we are computing the magnitude and phase response in the $j\Omega$-axis, we have

$$
\begin{aligned}
H(j\Omega) = H(s)|_{s=j\Omega} \quad &= \quad K \frac{\prod_i \vec{Z}_i(\Omega)}{\prod_k \vec{P}_k(\Omega)} \\
&= \quad \underbrace{|K| \frac{\prod_i |\vec{Z}_i(\Omega)|}{\prod_k |\vec{P}_k(\Omega)|}}_{|H(j\Omega)|} \underbrace{e^{j\left[\angle K + \sum_i \angle(\vec{Z}_i(\Omega)) - \sum_k \angle(\vec{P}_k(\Omega))\right]}}_{e^{j\angle H(j\Omega)}}
\end{aligned} \tag{5.28}
$$

for $0 \leq \Omega < \infty$. It is understood the magnitude and phase are even and odd functions of frequency.

Example 5.18. Consider an RC circuit in series with a voltage source $v_i(t)$. Choose the output to obtain low-pass and high-pass filters and use the poles and zeros of the transfer functions to determine their frequency responses. Let $R = 1 \ \Omega$, $C = 1$ F and the initial condition be zero.

Solution: Low-pass filter: Let the output be the voltage in the capacitor. By voltage division, we obtain the transfer function of the filter:

$$H(s) = \frac{V_C(s)}{V_i(s)} = \frac{1/Cs}{R + 1/Cs}.$$

For $R = 1\,\Omega$, $C = 1\,$F we get

$$H(j\Omega) = \frac{1}{1 + j\Omega} = \frac{1}{\vec{P}(\Omega)}.$$

Drawing a vector from the pole $s = -1$ to any point on the $j\Omega$ axis gives $\vec{P}(\Omega)$ and for different frequencies we get

Ω	$\vec{P}(\Omega)$	$H(j\Omega)$
0	$1e^{j0}$	$1e^{j0}$
1	$\sqrt{2}e^{j\pi/4}$	$0.707e^{-j\pi/4}$
∞	$\infty\, e^{j\pi/2}$	$0e^{-j\pi/2}$

Since there are no zeros, the frequency response of this filter depends inversely on the behavior of the pole vector $\vec{P}(\Omega)$. Thus, the magnitude response is unity at $\Omega = 0$ and it decays as frequency increases. The phase is zero at $\Omega = 0$, $-\pi/4$ at $\Omega = 1$ and $-\pi/2$ at $\Omega \to \infty$. The magnitude response is even and the phase response is odd.

High-pass filter: Consider then the output being the voltage in the resistor. Again by voltage division we obtain the transfer function of this circuit as

$$H(s) = \frac{V_r(s)}{V_s(s)} = \frac{CRs}{CRs + 1}.$$

For $C = 1\,$F and $R = 1\,\Omega$, the frequency response is

$$H(j\Omega) = \frac{j\Omega}{1 + j\Omega} = \frac{\vec{Z}(\Omega)}{\vec{P}(\Omega)}.$$

The vector $\vec{Z}(\Omega)$ goes from the zero at the origin, $s = 0$, to $j\Omega$ on the $j\Omega$ axis, and the vector $\vec{P}(\Omega)$ goes from the pole $s = -1$ to $j\Omega$ in the $j\Omega$ axis. The vectors and the frequency response, at three different frequencies, are given by

Ω	$\vec{Z}(\Omega)$	$\vec{P}(\Omega)$	$H(j\Omega) = \vec{Z}(\Omega)/\vec{P}(\Omega)$
0	$0e^{j\pi/2}$	$1e^{j0}$	$0e^{j\pi/2}$
1	$1e^{j\pi/2}$	$\sqrt{2}e^{j\pi/4}$	$0.707e^{j\pi/4}$
∞	$\infty\, e^{j\pi/2}$	$\infty\, e^{j\pi/2}$	$1e^{j0}$

Thus, the magnitude response is zero at $\Omega = 0$ (this is due to the zero at $s = 0$, making $\vec{Z}(0) = 0$ as it is right on top of the zero), and it grows to unity as the frequency increases (at very high frequency, the lengths of the pole and the zero vectors are alike and so the magnitude response is unity and the phase response zero). The phase response starts at $\pi/2$ and it decreases to 0 as the frequency tends to infinity. $\qquad\square$

Remarks

1. Poles create "hills" at frequencies in the $j\Omega$-axis in front of the imaginary parts of the poles. The closer the pole is to the $j\Omega$-axis, the narrower and higher the hill. If, for instance, the poles are on the $j\Omega$-axis (this would correspond to an unstable and useless filter) the magnitude response at the frequencies of the poles will be infinity.
2. Zeros create "valleys" at the frequencies in the $j\Omega$-axis in front of the imaginary parts of the zeros. The closer a zero is to the $j\Omega$-axis approaching it from the left or the right (as the zeros are not restricted by stability to be in the open left-hand s-plane) the closer the magnitude response is to zero. If the zeros are on the $j\Omega$-axis, the magnitude response at the frequencies of the zeros is zero. Thus poles produce a magnitude response that looks like hills (or like the main pole of a circus) around the frequencies of the poles, and zeros make the magnitude response go to zero in the form of valleys around the frequencies of the zeros.

Example 5.19. Use MATLAB to find and plot the poles and zeros and the corresponding magnitude and phase frequency responses of:
(a) A band-pass filter with a transfer function

$$H_{bp}(s) = \frac{s}{s^2 + s + 1}.$$

(b) A high-pass filter with a transfer function

$$H_{hp}(s) = \frac{s^2}{s^2 + s + 1}.$$

(c) An all-pass filter with a transfer function

$$H(s) = \frac{s^2 - 2.5s + 1}{s^2 + 2.5s + 1}.$$

Solution: Our function *freqresp_s* computes and plots the poles and zeros of the filter transfer function, and the corresponding frequency response. This function needs the coefficients of the numerator and denominator of the transfer function in decreasing order of powers of s. It uses the MATLAB functions *freqs*, *abs* and *angle* to compute $H(j\Omega)$ for a set of discretized frequencies and then to find the magnitude and phase at each frequency. Our function *splane* computes the poles and zeros of the transfer function and plots them in the s-plane.

```
function [w,Hm,Ha]=freqresp_s(b,a,wmax)
w=0:0.01:wmax;
H=freqs(b,a,w);
Hm=abs(H);                          % magnitude
Ha=angle(H)*180/pi;                 % phase in degrees

function splane(num,den)
disp('>>>>>  Zeros <<<<<'); z=roots(num)
disp('>>>>>  Poles <<<<<'); p=roots(den)
```

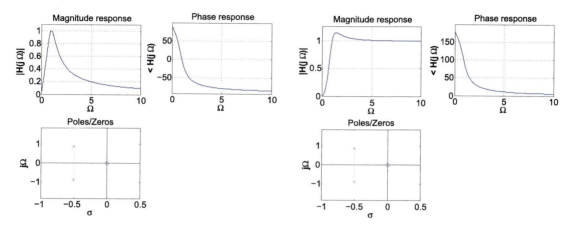

FIGURE 5.11

Frequency response and pole/zero location of band-pass (left) and high-pass RLC filters. High-pass filter has double zero at zero.

```
A1=[min(imag(z)) min(imag(p))];A1=min(A1)-1;
B1=[max(imag(z)) max(imag(p))];B1=max(B1)+1;
N=20;
D=(abs(A1)+abs(B1))/N;
im=A1:D:B1;
Nq=length(im);
re=zeros(1,Nq);
A=[min(real(z)) min(real(p))];A=min(A)-1;
B=[max(real(z)) max(real(p))];B=max(B)+1;
stem(real(z),imag(z),'o:'); hold on
stem(real(p),imag(p),'x:'); hold on
plot(re,im,'k');xlabel('\sigma');ylabel('j\Omega'); grid;
axis([A 3 min(im) max(im)]); hold off
```

(a)–(b) **Band-pass and high-pass filters:** For these filters we use the following script. The results are shown in Fig. 5.11. To get the coefficients of the numerator of the high-pass filter get rid of the comment % in front of the vector $n = [1\ 0\ 0]$.

```
%%
% Example 5.19---Frequency response
%%
n=[0 1 0];                      % numerator coefficients -- band-pass
% n=[1 0  0];                   % numerator coefficients -- high-pass
d=[1 1 1];                      % denominator coefficients
wmax=10;                            % maximum frequency
[w,Hm,Ha]=freqresp_s(n,d,wmax);     % frequency response
splane(n,d)                     % plotting of poles and zeros
```

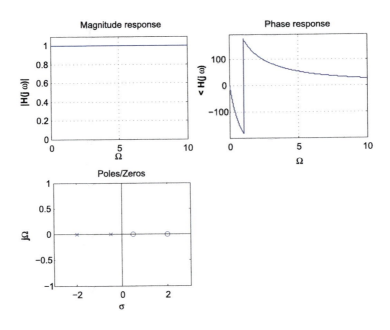

FIGURE 5.12

Frequency response and pole/zero location of the all-pass filter.

(c) **All-pass filter:** The poles and the zeros of an all-pass filter have the same imaginary parts, but the negative of its real part.

```
clear all
clf
n=[1 -2.5 1];
d=[1 2.5 1];
wmax=10;
freq_resp_s(n,d,wmax)
```

The results for the all-pass filter are shown in Fig. 5.12.   □

5.7.4 THE SPECTRUM ANALYZER

A **spectrum analyzer** is a device that measures and displays the spectrum of a signal. It can be implemented as a bank of narrow-band band-pass filters, with fixed bandwidths, covering the desired frequencies (see Fig. 5.13). The power of the output of each filter is computed and displayed at the corresponding center frequency. Another possible implementation is using a band-pass filter with an adjustable center frequency, with the power in its bandwidth being computed and displayed.

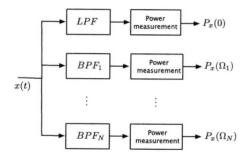

FIGURE 5.13

Bank-of-filter spectrum analyzer. LPF stands for low-pass filter, while BPF_i correspond to band-pass filters, $i = 1, \cdots, N$. The frequency response of the bank-of-filters is that of an all-pass filter covering the desired range of frequencies.

If the input of the spectrum analyzer is $x(t)$, the output $y(t)$ of either the fixed or the adjustable band-pass filter $H(j\omega)$ with unit magnitude response, center frequency Ω_0 and a very narrow bandwidth $\Delta\Omega$ would be according to the inverse Fourier transform:

$$
\begin{aligned}
y(t) &= \frac{1}{2\pi}\left[\int_{-\Omega_0-0.5\Delta\Omega}^{-\Omega_0+0.5\Delta\Omega} X(\Omega)\ 1 e^{j\angle H(j\Omega)} e^{j\Omega t}\,d\Omega + \int_{\Omega_0-0.5\Delta\Omega}^{\Omega_0+0.5\Delta\Omega} X(\Omega)\ 1 e^{j\angle H(j\Omega)} e^{j\Omega t}\,d\Omega\right] \\
&\approx \frac{\Delta\Omega}{2\pi}\left[X(-\Omega_0)e^{-j(\Omega_0 t + \angle H(j\Omega_0))} + X(\Omega_0)e^{j(\Omega_0 t + \angle H(j\Omega_0))}\right] \\
&= \frac{\Delta\Omega}{\pi}|X(\Omega_0)|\cos(\Omega_0 t + \angle H(j\Omega_0) + \angle X(\Omega_0)),
\end{aligned}
$$

i.e., approximately a periodic signal. Computing the power of this signal:

$$
\lim_{T\to\infty} \frac{1}{T}\int_T |y(t)|^2 dt = 0.5\left(\frac{\Delta\Omega}{\pi}\right)^2 |X(\Omega_0)|^2,
$$

the result is proportional to the power of the signal in $\Omega_0 \pm 0.5\Delta\Omega$ and $-\Omega_0 \pm 0.5\Delta\Omega$. A similar computation is done at each of the frequencies of the input signal. The spectrum analyzer puts all these measurements together and displays them.

Remarks

1. The bank of filter spectrum analyzer is used for the audio range of the spectrum.
2. Radio frequency spectrum analyzers resemble an AM demodulator. It usually consists of a single narrow-band intermediate frequency (IF) band-pass filter fed by a mixer. The local oscillator sweeps across the desired band, and the power at the output of the filter is computed and displayed on a monitor.

5.8 ADDITIONAL PROPERTIES

We consider now additional properties of the Fourier transform. Some look like those of the Laplace transform when $s = j\Omega$ and some are different.

5.8.1 TIME SHIFTING

If $x(t)$ has a Fourier transform $X(\Omega)$, then

$$
\begin{aligned}
x(t - t_0) &\Leftrightarrow X(\Omega)e^{-j\Omega t_0} \\
x(t + t_0) &\Leftrightarrow X(\Omega)e^{j\Omega t_0}.
\end{aligned}
\qquad (5.29)
$$

The Fourier transform of $x(t - t_0)$ is

$$
\begin{aligned}
\mathcal{F}[x(t - t_0)] &= \int_{-\infty}^{\infty} x(t - t_0)e^{-j\Omega t}\, dt \\
&= \int_{-\infty}^{\infty} x(\tau)e^{-j\Omega(\tau + t_0)}\, d\tau = e^{-j\Omega t_0} X(\Omega)
\end{aligned}
$$

where we changed the variable to $\tau = t - t_0$. Likewise for $x(t + t_0)$.

It is important to realize that shifting in time does not change the frequency content of a signal. That is, the signal does not change when delayed or advanced. This is clear when we see that the magnitude of the two transforms, corresponding to the original and the shifted signals, is the same,

$$
|X(\Omega)| = |X(\Omega)e^{\pm j\Omega t_0}|,
$$

and the effect of the time shift is only in the phase spectrum.

Example 5.20. Find the Fourier transform of $y(t) = \sin(\Omega_0 t)$ using the time-shift property and the Fourier transform of the cosine signal $x(t) = \cos(\Omega_0 t)$.

Solution: Since $y(t) = \cos(\Omega_0 t - \pi/2) = \cos(\Omega_0[t - \pi/(2\Omega_0)]) = x(t - \pi/(2\Omega_0))$ applying the time-shifting property we then get

$$
\begin{aligned}
\mathcal{F}[\sin(\Omega_0 t)] &= \mathcal{F}[x(t - \pi/2\Omega_0)] \\
&= \pi[\delta(\Omega - \Omega_0) + \delta(\Omega + \Omega_0)]e^{-j\Omega\pi/(2\Omega_0)} \\
&= \pi\delta(\Omega - \Omega_0)e^{-j\pi/2} + \pi\delta(\Omega + \Omega_0)e^{j\pi/2} \\
&= -j\pi\delta(\Omega - \Omega_0) + j\pi\delta(\Omega + \Omega_0)
\end{aligned}
$$

after applying the sifting property of $\delta(\Omega)$. The above shows that this Fourier transform is different from the one for the cosine in the phase only. $\qquad\square$

5.8.2 DIFFERENTIATION AND INTEGRATION

If $x(t)$, $-\infty < t < \infty$, has a Fourier transform $X(\Omega)$, then

$$\frac{d^N x(t)}{dt^N} \quad \Leftrightarrow \quad (j\Omega)^N X(\Omega), \tag{5.30}$$

$$\int_{-\infty}^{t} x(\sigma)d\sigma \quad \Leftrightarrow \quad \frac{X(\Omega)}{j\Omega} + \pi X(0)\delta(\Omega), \tag{5.31}$$

where

$$X(0) = \int_{-\infty}^{\infty} x(t)dt.$$

From the inverse Fourier transform

$$x(t) = \frac{1}{2\pi} \int_{-\infty}^{\infty} X(\Omega)e^{j\Omega t} d\Omega$$

we have

$$\frac{dx(t)}{dt} = \frac{1}{2\pi} \int_{-\infty}^{\infty} X(\Omega)\frac{d\,e^{j\Omega t}}{dt}d\Omega = \frac{1}{2\pi} \int_{-\infty}^{\infty} [X(\Omega)j\Omega]\,e^{j\Omega t}d\Omega,$$

indicating that

$$\frac{dx(t)}{dt} \Leftrightarrow j\Omega X(\Omega),$$

and similarly for higher derivatives.

The proof of the integration property can be done in two parts:
(i) The convolution of $u(t)$ and $x(t)$ gives the integral, that is,

$$\int_{-\infty}^{t} x(\tau)d\tau = \int_{-\infty}^{\infty} x(\tau)u(t-\tau)d\tau = [x*u](t),$$

since $u(t-\tau)$ as a function of τ equals

$$u(t-\tau) = \begin{cases} 1 & \tau < t, \\ 0 & \tau > t. \end{cases}$$

We thus have

$$\mathcal{F}\left[\int_{-\infty}^{t} x(\tau)d\tau\right] = X(\Omega)\mathcal{F}[u(t)]. \tag{5.32}$$

(ii) Since the unit-step signal is not absolutely integrable its Fourier transform cannot be found from the integral definition. We cannot use its Laplace transform either because its ROC does not include the $j\Omega$-axis.[1] The odd–even decomposition of $u(t)$ is

$$u(t) = 0.5\,\text{sgn}(t) + 0.5 \quad \text{where} \quad \text{sgn}(t) = \begin{cases} 1 & t > 0, \\ -1 & t < 0. \end{cases}$$

The derivative of $\text{sgn}(t)$ is

$$\frac{d\,\text{sgn}(t)}{dt} = 2\delta(t) \quad \Rightarrow \quad S(\Omega) = \mathcal{F}[\text{sgn}(t)] = \frac{2}{j\Omega}$$

using the derivative property (notice that $\text{sgn}(t)$ is an odd function and as such its Fourier transform must be purely imaginary). Thus

$$\mathcal{F}[u(t)] = 0.5S(\Omega) + \mathcal{F}(0.5) = \frac{1}{j\Omega} + \pi\delta(\Omega), \tag{5.33}$$

which is a complex function as it corresponds to $u(t)$, which is neither an even nor an odd function of time.

Replacing the Fourier transform of $u(t)$ in (5.32) we get

$$\mathcal{F}\left[\int_{-\infty}^{t} x(\tau)d\tau\right] = X(\Omega)\left[\frac{1}{j\Omega} + \pi\delta(\Omega)\right] = \frac{X(\Omega)}{j\Omega} + \pi X(0)\delta(\Omega).$$

Remark

1. Just like in the Laplace transform where the operator s corresponds to the derivative operation in time, in the Fourier transform $j\Omega$ becomes the corresponding operator for the derivative operation in time.
2. If $X(0)$, i.e., the value of $X(\Omega)$ at the dc frequency, is zero then the operator $1/(j\Omega)$ corresponds to integration in time of $x(t)$, just like $1/s$ in the Laplace domain.
3. As seen in the Fourier series, differentiation enhances the higher frequencies of the signal being differentiated, while integration enhances the lower frequencies of the signal being integrated.

Example 5.21. Suppose that a system is represented by a second-order differential equation with constant coefficients

$$2y(t) + 3\frac{dy(t)}{dt} + \frac{d^2y(t)}{dt^2} = x(t)$$

and that the initial conditions are zero. Let $x(t) = \delta(t)$. Find $y(t)$.

Solution: Computing the Fourier transform of this equation we get

$$[2 + 3j\Omega + (j\Omega)^2]Y(\Omega) = X(\Omega).$$

---

[1] From $du(t)/dt = \delta(t)$ we would get $sU(s) = 1$ with ROC the whole plane, and thus $U(\Omega) = 1/(j\Omega)$, a purely imaginary result indicating that $u(t)$ is odd which is not the case. Something is missing!

Replacing $X(\Omega) = 1$ and solving for $Y(\Omega)$, we have

$$Y(\Omega) = \frac{1}{2 + 3j\Omega + (j\Omega)^2} = \frac{1}{(j\Omega + 1)(j\Omega + 2)} = \frac{1}{(j\Omega + 1)} + \frac{-1}{(j\Omega + 2)}$$

and the inverse Fourier transform of these terms gives

$$y(t) = [e^{-t} - e^{-2t}]u(t).$$

Thus, the Fourier transform can be used to solve differential equations provided the initial conditions are zero; if they are not zero the Laplace transform must be used. ☐

Example 5.22. Find the Fourier transform of the triangular pulse $x(t) = r(t) - 2r(t - 1) + r(t - 2)$, which is piecewise linear, using the derivative property.

Solution: The first derivative gives

$$\frac{dx(t)}{dt} = u(t) - 2u(t - 1) + u(t - 2)$$

and the second derivative gives

$$\frac{d^2x(t)}{dt^2} = \delta(t) - 2\delta(t - 1) + \delta(t - 2).$$

Using the derivative and the time-shift properties we get

$$(j\Omega)^2 X(\Omega) = 1 - 2e^{-j\Omega} + e^{-j2\Omega} = e^{-j\Omega}[e^{j\Omega} - 2 + e^{-j\Omega}],$$

so that

$$X(\Omega) = \frac{2e^{-j\Omega}}{\Omega^2}[1 - \cos(\Omega)] = e^{-j\Omega}\left(\frac{\sin(\Omega/2)}{\Omega/2}\right)^2$$

after using $1 - \cos(\theta) = 2\sin^2(\theta/2)$. ☐

Example 5.23. Consider the integral

$$y(t) = \int_{-\infty}^{t} x(\tau)d\tau$$

where $x(t) = u(t + 1) - u(t - 1)$. Find the Fourier transform $Y(\Omega)$ by finding the integral, and by using the integration property.

Solution: The integral is

$$y(t) = \begin{cases} 0 & t < -1, \\ t + 1 & -1 \le t < 1, \\ 2 & t \ge 1, \end{cases}$$

or

$$y(t) = \underbrace{[r(t+1) - r(t-1) - 2u(t-1)]}_{y_1(t)} + 2u(t-1).$$

The Fourier transform of $y_1(t)$, of finite support $[-1, 1]$, is

$$Y_1(\Omega) = \left[\frac{e^s - e^{-s}}{s^2} - \frac{2e^{-s}}{s}\right]_{s=j\Omega} = \frac{-2j\sin(\Omega)}{\Omega^2} + j\frac{2e^{-j\Omega}}{\Omega}.$$

The Fourier transform of $2u(t-1)$ is $-2je^{-j\Omega}/\Omega + 2\pi\delta(\Omega)$ so that

$$Y(\Omega) = \frac{-2j\sin(\Omega)}{\Omega^2} + j\frac{2e^{-j\Omega}}{\Omega} - j\frac{2e^{-j\Omega}}{\Omega} + 2\pi\delta(\Omega) = \frac{-2j\sin(\Omega)}{\Omega^2} + 2\pi\delta(\Omega).$$

To use the integration property we first need $X(\Omega)$ which is

$$X(\Omega) = \frac{e^s - e^{-s}}{s}\Big|_{s=j\Omega} = \frac{2\sin(\Omega)}{\Omega}$$

and according to the integration property

$$Y(\Omega) = \frac{X(\Omega)}{j\Omega} + \pi X(0)\delta(\Omega) = \frac{-2j\sin(\Omega)}{\Omega^2} + 2\pi\delta(\Omega),$$

since $X(0) = 2$. As expected, the two results coincide. □

5.9 WHAT HAVE WE ACCOMPLISHED? WHAT IS NEXT?

By the end of this chapter you should have a very good understanding of the frequency representation of signals and systems. In this chapter, we have unified the treatment of periodic and non-periodic signals and their spectra and consolidated the concept of the frequency response of a linear time-invariant system. See Tables 5.1 and 5.2 for a summary of Fourier transform properties and pairs. Two significant applications are in filtering and in modulation. We introduced in this chapter the fundamentals of modulation and its application in communications, as well as the basics of filtering. These topics will be extended in Chapter 7.

Certainly the next step is to find where Laplace and Fourier analysis apply. That will be done in the next two chapters. After that, we will go into discrete-time signals and systems. The concepts of sampling, quantization and coding will bridge the continuous-time and the discrete-time and digital signals. Then transformations similar to Laplace and Fourier will permit us to do processing of discrete-time signals and systems.

Some additional mathematical and numerical analysis issues related to the Fourier analysis can be found in [36,10,73].

Table 5.1 Basic properties of the Fourier transform

Expansion/contraction	$x(\alpha t),\ \alpha \neq 0$	$\dfrac{1}{\lvert\alpha\rvert} X\left(\dfrac{\Omega}{\alpha}\right)$
Reflection	$x(-t)$	$X(-\Omega)$
Parseval's	$E_x = \displaystyle\int_{-\infty}^{\infty} \lvert x(t)\rvert^2 dt$	$E_x = \dfrac{1}{2\pi} \displaystyle\int_{-\infty}^{\infty} \lvert X(\Omega)\rvert^2 d\Omega$
Duality	$X(t)$	$2\pi x(-\Omega)$
Differentiation	$\dfrac{d^n x(t)}{dt^n},\ n \geq 1$	$(j\Omega)^n X(\Omega)$
Integration	$\displaystyle\int_{-\infty}^{t} x(t')dt'$	$\dfrac{X(\Omega)}{j\Omega} + \pi X(0)\delta(\Omega)$
Shifting	$x(t-\alpha),\ e^{j\Omega_0 t} x(t)$	$e^{-j\alpha\Omega} X(\Omega),\ X(\Omega - \Omega_0)$
Modulation	$x(t)\cos(\Omega_c t)$	$0.5[X(\Omega - \Omega_c) + X(\Omega + \Omega_c)]$
Periodic	$x(t) = \displaystyle\sum_k X_k e^{jk\Omega_0 t}$	$X(\Omega) = \displaystyle\sum_k 2\pi X_k \delta(\Omega - k\Omega_0)$
Symmetry	$x(t)$ real	$\lvert X(\Omega)\rvert = \lvert X(-\Omega)\rvert,$
		$\angle X(\Omega) = -\angle X(-\Omega)$
Convolution	$z(t) = [x * y](t)$	$Z(\Omega) = X(\Omega)Y(\Omega)$

Table 5.2 Fourier transform pairs

$\delta(t),\ \delta(t-\tau)$	$1,\ e^{-j\Omega\tau}$
$u(t),\ u(-t)$	$\dfrac{1}{j\Omega} + \pi\delta(\Omega),\ \dfrac{-1}{j\Omega} + \pi\delta(\Omega)$
$\text{sgn}(t) = 2[u(t) - 0.5]$	$\dfrac{2}{j\Omega}$
$A,\ Ae^{-at}u(t),\ a > 0$	$2\pi A\delta(\Omega),\ \dfrac{A}{j\Omega + a}$
$Ate^{-at}u(t),\ a > 0$	$\dfrac{A}{(j\Omega + a)^2}$
$e^{-a\lvert t\rvert},\ a > 0$	$\dfrac{2a}{a^2 + \Omega^2}$
$\cos(\Omega_0 t),\ -\infty < t < \infty$	$\pi[\delta(\Omega - \Omega_0) + \delta(\Omega + \Omega_0)]$
$\sin(\Omega_0 t),\ -\infty < t < \infty$	$-j\pi[\delta(\Omega - \Omega_0) - \delta(\Omega + \Omega_0)]$
$p(t) = A[u(t+\tau) - u(t-\tau)]$	$2A\tau\dfrac{\sin(\Omega\tau)}{\Omega\tau}$

5.10 PROBLEMS

5.10.1 BASIC PROBLEMS

5.1 A causal signal $x(t)$ having a Laplace transform with poles in the open-left s-plane (i.e., not including the $j\Omega$ axis) has a Fourier transform that can be found from its Laplace transform. Consider the following signals:

$$x_1(t) = e^{-2t}u(t), \quad x_2(t) = r(t), \quad x_3(t) = x_1(t)x_2(t).$$

(a) Determine the Laplace transform of the above signals indicating their corresponding region of convergence.

(b) Determine for which of these signals you can find its Fourier transform from its Laplace transform. Explain.

(c) Give the Fourier transform of the signals that can be obtained from their Laplace transform.
Answers: (a) $X_2(s) = 1/s^2$, $\sigma > 0$; (b) $x_1(t)$ and $x_3(t)$.

5.2 There are signals whose Fourier transforms cannot be found directly by either the integral definition or the Laplace transform. For instance, the sinc signal

$$x(t) = \frac{\sin(t)}{t}$$

is one of them.

(a) Let $X(\Omega) = A[u(\Omega + \Omega_0) - u(\Omega - \Omega_0)]$ be a possible Fourier transform of $x(t)$. Find the inverse Fourier transform of $X(\Omega)$ using the integral equation to determine the values of A and Ω_0.

(b) How could you use the duality property of the Fourier transform to obtain $X(\Omega)$? Explain.
Answer: $X(\Omega) = \pi[u(\Omega + 1) - u(\Omega - 1)]$.

5.3 The Fourier transforms of even and odd functions are very important. Let $x(t) = e^{-|t|}$ and $y(t) = e^{-t}u(t) - e^{t}u(-t)$.

(a) Plot $x(t)$ and $y(t)$, and determine whether they are even or odd.

(b) Show that the Fourier transform of $x(t)$ is found from

$$X(\Omega) = \int_{-\infty}^{\infty} x(t)\cos(\Omega t)dt,$$

which is a real function of Ω, therefore its computational importance. Show that $X(\Omega)$ is an even function of Ω. Find $X(\Omega)$ from the above equation.

(c) Show that the Fourier transform of $y(t)$ is found from

$$Y(\Omega) = -j\int_{-\infty}^{\infty} y(t)\sin(\Omega t)dt,$$

which is an imaginary function of Ω, thus its computational importance. Show that $Y(\Omega)$ is and odd function of Ω. Find $Y(\Omega)$ from the above equation (called the sine transform). Verify that your results are correct by finding the Fourier transform of $z(t) = x(t) + y(t)$ directly and using the above results.

(d) What advantages do you see in using the cosine and sine transforms? How would you use the cosine and the sine transforms to compute the Fourier transform of any signal, not necessarily even or odd? Explain.

Answers: $x(t)$ is even, $y(t)$ odd; $X(\Omega) = 2/(1 + \Omega^2)$; $z(t) = 2e^{-t}u(t)$.

5.4 Starting with the Fourier transform pair

$$x(t) = u(t+1) - u(t-1) \quad \Leftrightarrow \quad X(\Omega) = \frac{2\sin(\Omega)}{\Omega}$$

and using no integration, indicate the properties of the Fourier transform that will allow you to compute the Fourier transform of the following signals (do not find the Fourier transforms):

(a) $x_1(t) = -u(t+2) + 2u(t) - u(t-2)$,   (b) $x_2(t) = 2\sin(t)/t$,
(c) $x_3(t) = 2[u(t+0.5) - u(t-0.5)]$,   (d) $x_4(t) = \cos(0.5\pi t)[u(t+1) - u(t-1)]$,
(e) $x_5(t) = X(t)$.

You may want to plot the different signals to help you get the answers.
Answers: $X_1(\Omega) = -2jX(\Omega)\sin(\Omega)$, time-shift property; duality to get $X_5(\Omega)$.

5.5 Consider a signal $x(t) = \cos(t)$, $0 \le t \le 1$, and zero otherwise.
  (a) Find its Fourier transform $X(\Omega)$.
  (b) Let $y(t) = x(2t)$, find $Y(\Omega)$; let $z(t) = x(t/2)$, find $Z(\Omega)$.
  (c) Compare $Y(\Omega)$ and $Z(\Omega)$ with $X(\Omega)$.
  Answers: $X(\Omega) = 0.5[P(\Omega+1) + P(\Omega-1)]$, $P(\Omega) = 2e^{-j\Omega/2}\sin(\Omega/2)/\Omega$.

5.6 The derivative property can be used to simplify the computation of some Fourier transforms. Let

$$x(t) = r(t) - 2r(t-1) + r(t-2).$$

  (a) Find and plot the second derivative with respect to t of $x(t)$, $y(t) = d^2x(t)/dt^2$.
  (b) Find $X(\Omega)$ from $Y(\Omega)$ using the derivative property.
  (c) Verify the above result by computing the Fourier transform $X(\Omega)$ directly from $x(t)$ using the Laplace transform.
  Answer: $\mathcal{F}(\ddot{x}(t)) = (j\Omega)^2 X(\Omega)$.

5.7 Use the properties of the Fourier transform in the following problems.
  (a) Use the Fourier transform of $\cos(k\Omega_0 t)$ to find the Fourier transform of a periodic signal $x(t)$ with a Fourier series

$$x(t) = 1 + \sum_{k=1}^{\infty} (0.5)^k \cos(k\Omega_0 t).$$

  (b) Find the Fourier transform of

$$x_1(t) = \frac{1}{t^2 + a^2} \qquad a > 0$$

by considering the properties of the transform. *Hint:* use the FT of $e^{-a|t|}$ for $a > 0$.
Answer: (b) $X_1(\Omega) = (\pi/a)e^{-a|\Omega|}$.

5.8 In the following problems we want to find the Fourier transform of the signals.

(a) For the signal

$$y(t) = 0.5\,\text{sign}(t) = \begin{cases} 0.5 & t \geq 0, \\ -0.5 & t < 0, \end{cases}$$

find its Fourier transform by using the Fourier transform of

$$x(t) = 0.5e^{-at}u(t) - 0.5e^{at}u(-t), \quad a > 0$$

as $a \to 0$.

Use then $Y(\Omega)$ to determine the Fourier transform of the unit-step signal which can be written

$$u(t) = 0.5 + 0.5\,\text{sign}(t) = 0.5 + y(t).$$

(b) Consider the constant A being the superposition of $Au(t)$ and $Au(-t)$, both of which has Laplace transforms. Find these Laplace transforms, including their regions of convergence, and then try to use this results to find the Fourier transform of A. Is it possible?

Answer: (a) $\mathcal{F}[u(t)] = \pi\delta(\Omega) + 1/(j\Omega)$; (b) no, ROC is empty.

5.9 Consider the following problems related to the modulation and power properties of the Fourier transform.

(a) The carrier of an AM system is $\cos(10t)$, consider the following message signals:

    i. $m(t) = \cos(t)$,

    ii. $m(t) = r(t) - 2r(t-1) + r(t-2)$, where $r(t) = tu(t)$.

Sketch the modulated signals $y(t) = m(t)\cos(10t)$ for these two messages and find their corresponding spectrum.

(b) Find the power P_x of a sinc signal $x(t) = \sin(0.5t)/(\pi t)$, i.e., the integral

$$P_x = \int_{-\infty}^{\infty} |x(t)|^2 dt.$$

Answer: (a) $Y(\Omega) = 0.5[M(\Omega + 10) + M(\Omega - 10)]$; (b) use Parseval's relation.

5.10 Consider the sign signal

$$s(t) = \text{sign}(t) = \begin{cases} 1 & t \geq 0, \\ -1 & t < 0. \end{cases}$$

(a) Find the derivative of $s(t)$ and use it to find $S(\Omega) = \mathcal{F}(s(t))$.

(b) Find the magnitude and phase of $S(\Omega)$.

(c) Use the equivalent expression $s(t) = 2[u(t) - 0.5]$ to get $S(\Omega)$.

Answers: $ds(t)/dt = 2\delta(t)$; $S(\Omega) = 2/(j\Omega)$.

5.11 The following problems relate to the modulation property of the Fourier transform:

(a) Consider the signal

$$x(t) = p(t) + p(t)\cos(\pi t) \quad \text{where } p(t) = u(t+1) - u(t-1).$$

i. Use the modulation property to find the Fourier transform $X(\Omega)$ in terms of $P(\Omega)$, the Fourier transform of $p(t)$.

ii. Let $g(t) = x(t-1)$. Use the Laplace transform of $g(t)$ to obtain $X(\Omega)$. Verify the expression obtained for $X(\Omega)$ coincides with the previous one.

(b) Consider the signal $z(t) = \cos(t)[u(t) - u(t-\pi/2)]$.

i. If $Z(s)$ is the Laplace transform of $z(t)$ under what conditions can you obtain the Fourier transform by letting $s = j\Omega$?

ii. Assess if it true that the Fourier transform of $z(t)$ is

$$Z(\Omega) = \frac{j\Omega + e^{-j\pi\Omega/2}}{1 - \Omega^2}.$$

Answers: (a) $X(\Omega) = P(\Omega) + 0.5P(\Omega - \pi) + 0.5P(\Omega + \pi)$, $P(\Omega) = 2\sin(\Omega)/\Omega$; (b) no.

5.12 Consider the raised cosine pulse

$$x(t) = [1 + \cos(\pi t)](u(t+1) - u(t-1)).$$

(a) Carefully plot $x(t)$.

(b) Find the Fourier transform of the pulse $p(t) = u(t+1) - u(t-1)$

(c) Use the definition of the pulse $p(t)$ and the modulation property to find the Fourier transform of $x(t)$ in terms of $P(\Omega) = \mathcal{F}[p(t)]$.

Answer: $X(\Omega) = P(\Omega) + 0.5[P(\Omega - 2\pi) + P(\Omega + 2\pi)]$.

5.13 The frequency response of an ideal low-pass filter is

$$|H(j\Omega)| = \begin{cases} 1 & -2 \le \Omega \le 2, \\ 0 & \text{otherwise,} \end{cases} \qquad \angle H(j\Omega) = \begin{cases} -\pi/2 & \Omega \ge 0, \\ \pi/2 & \Omega < 0. \end{cases}$$

(a) Calculate the impulse response $h(t)$ of the ideal low-pass filter.

(b) If the input of the filter is a periodic signal $x(t)$ having a Fourier series

$$x(t) = \sum_{k=1}^{\infty} \frac{2}{k^2} \cos(3kt/2),$$

determine the steady-state response $y_{ss}(t)$ of the system.

Answers: $h(t) = (1 - \cos(2t))/(\pi t)$; $y_{ss}(t) = 2\sin(1.5t)$.

5.14 The transfer function of a filter is

$$H(s) = \frac{\sqrt{5}s}{s^2 + 2s + 2}.$$

(a) Find the poles and zeros of $H(s)$ and use this information to sketch the magnitude response $|H(j\Omega)|$ of the filter. Indicate the magnitude response at frequencies 0, 1 and ∞. Indicate what type of filter it is.

(b) Find the impulse response $h(t)$ of the filter.

(c) The above filter is connected in series with a defective sinusoidal signal generator that gives biased sinusoids $x(t) = B + \cos(\Omega t)$. If the signal generator can give all possible frequencies, for what frequency (or frequencies) Ω_0 is the filter output $y(t) = \cos(\Omega_0 t + \theta)$, i.e., the dc bias is filtered out. Determine the phase (or phases) θ_0 for that frequency (or frequencies).

Answers: Poles $s_{1,2} = -1 \pm j1$, zero $s = 0$; band-pass filter.

5.15 Consider the cascade of two filters with frequency responses

$$H_1(j\Omega) = j\Omega, \text{ and } H_2(j\Omega) = 1e^{-j\Omega}.$$

(a) Indicate what each of the filters does.

(b) Suppose that the input to the cascade is

$$x(t) = p(t)\cos(\pi t/2), \text{ where } p(t) = u(t+1) - u(t-1),$$

and the output is $y(t)$. Find $y(t)$.

(c) Suppose the cascading of the filters is done by putting first the filter with frequency response $H_2(j\Omega)$ followed by the filter with frequency response $H_1(j\Omega)$. Does the response change? Explain.

Answers: $H_1(j\Omega)$ computes the derivative; $y(t) = (\pi/2)\cos(\pi t/2)[u(t) - u(t-2)]$.

5.16 A continuous-time LTI system is represented by the ordinary differential equation

$$\frac{dy(t)}{dt} = -y(t) + x(t)$$

where $x(t)$ is the input and $y(t)$ the output.

(a) Determine the frequency response $H(j\Omega)$ of this system by considering the steady-state output of the system to inputs of the form $x(t) = e^{j\Omega t}$, for $-\infty < \Omega < \infty$.

(b) Carefully sketch the magnitude, $|H(j\Omega)|$, and the phase, $\angle H(j\Omega)$, frequency responses of the system. Indicate the magnitude and phase at frequencies of 0, ± 1 and $\pm\infty$ rad/s.

(c) If the input to this LTI is $x(t) = \sin(t)/(\pi t)$, determine and carefully plot the magnitude response $|Y(\Omega)|$ of the output, indicating the values at frequencies 0, ± 1 and $\pm\infty$ rad/s.

Answers: $H(j\Omega) = 1/(1 + j\Omega)$; $|H(j1)| = 0.707$, $\angle H(j1) = -\pi/4$.

5.17 If the Fourier transform of the pulse $x(t)$ given in Fig. 5.14 is $X(\Omega)$ (you do not need to compute it), then:

(a) Using the properties of the Fourier transform (no integration needed) obtain the Fourier transforms of the signals $x_i(t)$, $i = 1, 2$, shown in Fig. 5.14 in terms of $X(\Omega)$.

(b) Consider the modulated signal $x_3(t) = x(t)\cos(\pi t)$. Plot $x_3(t)$ and express its Fourier transform $X_3(\Omega)$ in terms of $X(\Omega)$.

(c) If $y(t) = x(t-1)$ find $Y(0)$.

Answers: $X_1(\Omega) = X(\Omega)(e^{j\Omega} - e^{-j\Omega})$; $Y(0) = 1$.

5.18 As indicated by the derivative property, if we multiply a Fourier transform by $(j\Omega)^N$ it corresponds to computing an Nth derivative of its time signal. Consider the dual of this property. That is, if we compute the derivative of $X(\Omega)$ what would happen to the signal in the time domain?

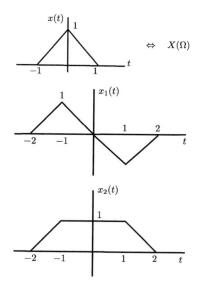

FIGURE 5.14

Problem 5.17: signals $x(t)$ and $x_i(t)$, $i = 1, 2$.

(a) Let $x(t) = \delta(t - 1) + \delta(t + 1)$; find its Fourier transform (using properties) $X(\Omega)$.
(b) Compute $dX(\Omega)/d\Omega$ and determine its inverse Fourier transform.
Answers: $X(\Omega) = 2\cos(\Omega)$; $Y(\Omega) = -2\sin(\Omega)$.

5.19 The sampling signal

$$\delta_{T_s}(t) = \sum_{n=-\infty}^{\infty} \delta(t - nT_s)$$

will be important in the sampling theory later on.
(a) As a periodic signal of fundamental period T_s, express $\delta_{T_s}(t)$ by its Fourier series.
(b) Determine then the Fourier transform $\Delta(\Omega) = \mathcal{F}[\delta_{T_s}(t)]$.
(c) Plot $\delta_{T_s}(t)$ and $\Delta(\Omega)$ and comment on the periodicity of these two functions.
Answer: $\mathcal{F}[\delta_{T_s}(t)] = (2\pi/T_s)\sum_k \delta(\Omega - k\Omega_s)$.

5.20 Suppose you want to design a dc source using a half-wave rectified signal $x(t)$ and an ideal filter. Let $x(t)$ be periodic, $T_0 = 2$, and with a period

$$x_1(t) = \begin{cases} \sin(\pi t) & 0 \le t \le 1, \\ 0 & 1 < t \le 2. \end{cases}$$

(a) Find the Fourier transform $X(\Omega)$ of $x(t)$, and plot the magnitude spectrum including the dc and the first three harmonics.

(b) Determine the magnitude and cut-off frequency of an ideal low-pass filter $H(j\Omega)$ such that when we have $x(t)$ as its input, the output is $y(t) = 1$. Plot the magnitude response of the ideal low-pass filter. (For simplicity assume the phase is zero.)

Answer: Low-pass filter has magnitude π and cut-off frequency $0 < \Omega_c < \pi$.

5.21 A pure tone $x(t) = 4\cos(1000t)$ is transmitted using an amplitude-modulation communication system with a carrier $\cos(10{,}000t)$. The output of the AM system is

$$y(t) = x(t)\cos(10{,}000t).$$

At the receiver, the sent signal $y(t)$ needs first to be separated from the thousands of other signals that are received. This is done with a band-pass filter with a center frequency equal to the carrier frequency. The output of this filter then needs to be demodulated.

(a) Consider an ideal band-pass filter $H(j\Omega)$. Let its phase be zero, determine its bandwidth, center frequency, and amplitude so we get as its output $10y(t)$. Plot the spectrum of $x(t)$, $10y(t)$ and the magnitude frequency response of $H(j\Omega)$.

(b) To demodulate $10y(t)$, we multiply it by $\cos(10{,}000t)$. You need then to pass the resulting signal through an ideal low-pass filter to recover the original signal $x(t)$. Plot the spectrum of $z(t) = 10y(t)\cos(10{,}000t)$, and from it determine the frequency response of the low-pass filter $G(j\Omega)$ needed to recover $x(t)$. Plot the magnitude response of $G(j\Omega)$.

Answer: $Z(\Omega) = 5X(\Omega) + 2.5[X(\Omega - 20{,}000) + X(\Omega + 20{,}000)]$, pass $z(t)$ through the low-pass filter of magnitude $1/5$ and take cut-off frequency slightly bigger than 1000 to get $x(t)$.

5.10.2 PROBLEMS USING MATLAB

5.22 **Fourier series vs. Fourier transform**—The connection between the Fourier series and the Fourier transform can be seen by considering what happens when the fundamental period of a periodic signal increases to a point at which the periodicity is not clear as only one period is seen. Consider a train of pulses $x(t)$ with fundamental period $T_0 = 2$, and a period of $x(t)$ is $x_1(t) = u(t + 0.5) - u(t - 0.5)$. Increase T_0 to 4, 8 and 16.

(a) Find the Fourier series coefficient X_0 for each of the values of T_0 and indicate how it changes for the different values of T_0.

(b) Find the Fourier series coefficients for $x(t)$ and carefully plot the line spectrum for each of the values of T_0. Explain what is happening in these spectra.

(c) If you were to let T_0 be very large what would you expect to happen to the Fourier coefficients? Explain.

(d) Write a MATLAB script that simulates the conversion from the Fourier Series to the Fourier transform of a sequence of rectangular pulses as the period is increased. The line spectrum needs to be multiplied by the period so that it does not become insignificant. Plot using *stem* the adjusted line spectrum for pulse sequences with periods from 4 to 62.

5.23 **Fourier transform from Laplace transform**—The Fourier transform of finite support signals, which are absolutely integrable or finite energy, can be obtained from their Laplace transform rather than doing the integral. Consider the following signals:

$$x_1(t) = u(t + 0.5) - u(t - 0.5), \qquad x_2(t) = \sin(2\pi t)[u(t) - u(t - 0.5)],$$
$$x_3(t) = r(t + 1) - 2r(t) + r(t - 1).$$

(a) Plot each of the above signals.

(b) Find the Fourier transforms $\{X_i(\Omega)\}$ for $i = 1, 2$ and 3 using the Laplace transform.

(c) Use MATLAB's symbolic integration function *int* to compute the Fourier transform of the given signals. Plot the magnitude spectrum corresponding to each of the signals.

Answers: $X_1(s) = (1/s)[e^{s/2} - e^{-s/2}]$; $X_3(\Omega) = 2(1 - \cos(\Omega))/\Omega^2$.

5.24 Time vs. frequency—The supports in time and in frequency of a signal $x(t)$ and its Fourier transform $X(\Omega)$ are inversely proportional. Consider a pulse

$$x(t) = \frac{1}{T_0}[u(t) - u(t - T_0)].$$

(a) Let $T_0 = 1$, and $T_0 = 10$ and find and compare the corresponding $|X(\Omega)|$.

(b) Use MATLAB to simulate the changes in the magnitude spectrum when $T_0 = 10^k$ for $k = 0, \cdots, 4$ for $x(t)$. Compute $X(\Omega)$ and plot its magnitude spectra for the increasing values of T_0 on the same plot. Explain the results. Use the symbolic function *fourier*.

Answers: For $T_0 = 10$, $X(\Omega) = \sin(5\Omega)e^{-j5\Omega}/(5\Omega)$.

5.25 Smoothness and frequency content—The smoothness of the signal determines the frequency content of its spectrum. Consider the signals

$$x(t) = u(t + 0.5) - u(t - 0.5), \qquad y(t) = (1 + \cos(\pi t))[u(t + 0.5) - u(t - 0.5)].$$

(a) Plot these signals. Can you tell which one is smoother?

(b) Find $X(\Omega)$ and carefully plot its magnitude $|X(\Omega)|$ versus frequency Ω.

(c) Find $Y(\Omega)$ (use the Fourier transform properties) and carefully plot its magnitude $|Y(\Omega)|$ versus frequency Ω.

(d) Which of these two signals has higher frequencies? Can you now tell which of the signals is smoother? Use MATLAB to decide. Make $x(t)$ and $y(t)$ have unit energy. Plot $20 \log_{10} |Y(\Omega)|$ and $20 \log_{10} |X(\Omega)|$ using MATLAB and see which of the spectra shows lower frequencies. Use *fourier* function to compute Fourier transforms.

Answers: $y(t)$ is smoother than $x(t)$.

5.26 Integration and smoothing—Consider the signal $x(t) = u(t + 1) - 2u(t) + u(t - 1)$ and let

$$y(t) = \int_{-\infty}^{t} x(\tau)d\tau.$$

(a) Plot $x(t)$ and $y(t)$.

(b) Find $X(\Omega)$ and carefully plot its magnitude spectrum. Is $X(\Omega)$ real? Explain. (Use MATLAB to do the plotting.)

(c) Find $Y(\Omega)$ and carefully plot its magnitude spectrum. Is $Y(\Omega)$ real? Explain. (Use MATLAB to do the plotting.)

(d) Determine from the above spectra which of these two signals is smoother. Use MATLAB to decide. Find the Fourier transform using the integration function *int*. Would you say that in general by integrating a signal you get rid of higher frequencies, or smooth out a signal?

Answers: $Y(\Omega) = \sin^2(\Omega/2)/(\Omega/2)^2$.

5.27 Passive RLC filters—Consider an RLC series circuit with a voltage source $v_s(t)$. Let the values of the resistor, capacitor and inductor be unity. Plot the poles and zeros and the corresponding frequency responses of the filters with output the voltage across the

- capacitor
- inductor
- resistor

Indicate the type of filter obtained in each case.

Use MATLAB to plot the poles and zeros, the magnitude and phase response of each of the filters obtained above.

Answers: Low-pass $V_C(s)/V_s(s)$; band-pass $V_R(s)/V_s(s)$.

5.28 Noncausal filter—Consider a filter with frequency response

$$H(j\Omega) = \frac{\sin(\pi\Omega)}{\pi\Omega}$$

or a sinc function in frequency.

(a) Find the impulse response $h(t)$ of this filter. Plot it and indicate whether this filter is a causal system or not.

(b) Suppose you wish to obtain a band-pass filter $G(j\Omega)$ from $H(j\Omega)$. If the desired center frequency of $|G(j\Omega)|$ is 5, and its desired magnitude is 1 at the center frequency, how would you process $h(t)$ to get the desired filter? Explain your procedure.

(c) Use symbolic MATLAB to find $h(t)$, $g(t)$ and $G(j\Omega)$. Plot $|H(j\Omega)|$, $h(t)$, $g(t)$, and $|G(j\Omega)|$. Compute the Fourier transform using the integration function.

Answer: $h(t) = 0.16[u(t+\pi) - u(t-\pi)]$.

5.29 Magnitude response from poles and zeros—Consider the following filters with the given poles and zeros, and dc constant:

$$
\begin{aligned}
H_1(s): \quad & K = 1 \ \text{poles } p_1 = -1, p_{2,3} = -1 \pm j\pi; \\
& \text{zeros } z_1 = 1, z_{2,3} = 1 \pm j\pi \\
H_2(s): \quad & K = 1 \ \text{poles } p_1 = -1, p_{2,3} = -1 \pm j\pi; \\
& \text{zeros } z_{1,3} = \pm j\pi \\
H_3(s): \quad & K = 1 \ \text{poles } p_1 = -1, p_{2,3} = -1 \pm j\pi; \\
& \text{zero } z_1 = 1.
\end{aligned}
$$

Use MATLAB to plot the magnitude responses of these filters and indicate the type of filters they are.

Answers: $H_1(s)$ is all-pass; $H_2(s)$ corresponds to a notch filter.

APPLICATION OF LAPLACE ANALYSIS TO CONTROL

CONTENTS

Who are you going to believe? Me or your own eyes.

Julius "Groucho" Marx (1890–1977), comedian and actor

6.1 INTRODUCTION

The Laplace transform finds application in many areas of engineering, in particular in control. In this chapter, we illustrate how the transform can be connected with the classical and modern theories of control.

In classical control, the objective is to change using frequency-domain methods the dynamics of a given system to achieve a desired response. This is typically done by connecting in feedback a **controller** to a **plant**. The plant is a system such as a motor, a chemical plant or an automobile we would like to control so that it responds in a certain way. The controller is a system we design to make the plant follow a prescribed reference input signal. By feeding back the response of the system to the input, it can be determined how the plant responds to the controller. The commonly used **negative feedback** generates an **error signal** that permits us to judge the performance of the controller. The concepts of transfer function, stability of systems and different types of responses obtained through the Laplace transform are very useful in the analysis and design of classical control systems.

Modern control, on the other hand, uses a time-domain approach to characterize and control systems. The state-variable representation is a more general representation than the transfer function, as it allows inclusion of initial conditions and it can easily be extended to the general case of multiple inputs and outputs. The theory is closely connected with linear algebra and differential equations. In this chapter we introduce the concept of state variables, its connection with the transfer function and the use of the Laplace transform to find the complete response and the transfer function from the state and output equations.

The aim of this chapter is to serve as an introduction to problems in classical and modern control and to link them with the Laplace analysis. More depth in these topics can be found in many excellent texts in control [60,22] and linear control systems [42,29].

6.2 SYSTEM CONNECTIONS AND BLOCK DIAGRAMS

Most systems, and in particular control and communication systems, consist of interconnection of sub-systems. As indicated in Chapter 2, three important connections of LTI systems are:

- cascade,
- parallel, and
- feedback.

The first two result from properties of the convolution integral, while the feedback connection is an ingenious connection[1] that relates the output of the overall system to its input. Using the background of the Laplace transform we present now a transform characterization of these connections. In Chapter 2, we obtained a complete time-domain characterizations for the cascade and the parallel connections, but not for the feedback connections.

The connection of two LTI continuous-time systems with transfer functions $H_1(s)$ and $H_2(s)$ (and corresponding impulse responses $h_1(t)$ and $h_2(t)$) can be done in

1. Cascade (Fig. 6.1)—provided that the two systems are isolated, the transfer function of the overall system is

$$H(s) = H_1(s)H_2(s). \qquad (6.1)$$

2. Parallel (Fig. 6.1)—the transfer function of the overall system is

$$H(s) = H_1(s) + H_2(s). \qquad (6.2)$$

3. Negative feedback (Fig. 6.4)—the transfer function of the overall system is

$$H(s) = \frac{H_1(s)}{1 + H_2(s)H_1(s)}. \qquad (6.3)$$

- **Open-loop** transfer function: $H_{o\ell}(s) = H_1(s)$.
- **Closed-loop** transfer function: $H_{c\ell}(s) = H(s)$.

---

[1]The invention of the negative-feedback amplifier by Black, an American electrical engineer, in 1927 not only revolutionized the field of electronics but the concept of feedback is considered one of the most important breakthroughs of the twentieth century [77].

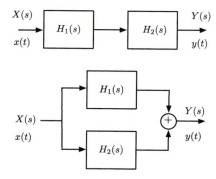

FIGURE 6.1

Cascade (top) and parallel (bottom) connections of systems with transfer function $H_1(s)$ and $H_2(s)$. The input/output are given in the time or frequency domains.

Cascading of LTI Systems. Given two LTI systems with transfer functions $H_1(s) = \mathcal{L}[h_1(t)]$, and $H_2(s) = \mathcal{L}[h_2(t)]$, where $h_1(t)$ and $h_2(t)$ are the corresponding impulse responses of the systems, the **cascading** of these systems gives a new system with transfer function

$$H(s) = H_1(s)H_2(s) = H_2(s)H_1(s),$$

provided that these systems are isolated from each other (i.e., they do not load each other). A graphical representation of the cascading of two systems is obtained by representing each of the systems with blocks with their corresponding transfer function (see Fig. 6.1). Although cascading of systems is a simple procedure, it has some disadvantages:

- it requires isolation of the systems,
- it causes delay as it processes the input signal, and it possibly compounds any errors in the processing.

Remarks

1. Loading, or lack of system isolation, needs to be considered when cascading two systems. Loading does not allow the overall transfer function to be the product of the transfer functions of the connected systems. Consider the cascade connection of two resistive voltage dividers (Fig. 6.2) each with a simple transfer function $H_i(s) = 1/2$, $i = 1, 2$. The cascade clearly will not have as transfer function $H(s) = H_1(s)H_2(s) = (1/2)(1/2)$ unless we include a buffer (such as an operational amplifier voltage follower) in between. The cascading of the two voltage dividers without the voltage follower gives a transfer function $H_1(s) = 1/5$ as can easily be shown by doing mesh analysis on the circuit.

2. The block diagrams of the cascade of two or more LTI systems can be interchanged with no effect on the overall transfer function—provided the connection is done with no loading. That is not true if the systems are not LTI. For instance, consider cascading a modulator (LTV system) and a differentiator

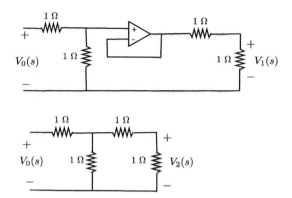

FIGURE 6.2

Cascading of two voltage dividers: using a voltage follower to avoid loading gives $V_1(s)/V_0(s) = (1/2)(1/2)$ (top); using no voltage follower $V_2(s)/V_0(s) = 1/5 \neq V_1(s)/V_0(s)$ due to loading.

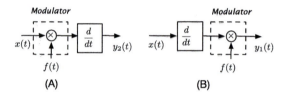

FIGURE 6.3

Cascading of a LTV system and a LTI system. The outputs are different, $y_1(t) \neq y_2(t)$.

(LTI) as shown in Fig. 6.3. If the modulator is first, the output of the overall system is

$$y_2(t) = \frac{dx(t)f(t)}{dt} = f(t)\frac{dx(t)}{dt} + x(t)\frac{df(t)}{dt},$$

while if we put the differentiator first, the output is

$$y_1(t) = f(t)\frac{dx(t)}{dt}.$$

It is obvious that, if $f(t)$ is not a constant, the two responses are very different.

Parallel Connection of LTI Systems. According to the distributive property of the convolution integral, the **parallel** connection of two LTI systems, with impulse responses $h_1(t)$ and $h_2(t)$, with the same input $x(t)$ applied to both gives as output the sum of the outputs of the systems being connected or

$$y(t) = (h_1 * x)(t) + (h_2 * x)(t) = [(h_1 + h_2) * x](t).$$

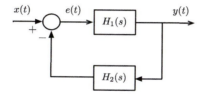

FIGURE 6.4

Negative-feedback connection of systems with transfer function $H_1(s)$ and $H_2(s)$. The input and the output are $x(t)$ and $y(t)$, $e(t)$ is the error signal.

The transfer function of the parallel system (see Fig. 6.1) is then

$$H(s) = \frac{Y(s)}{X(s)} = H_1(s) + H_2(s).$$

The parallel connection is better than the cascade connection as it does not require isolation between the systems, and it reduces the delay in processing an input signal.

Feedback Connection of LTI Systems. In control, **feedback** connections are more appropriate than cascade or parallel connections. In the negative-feedback connection of two LTI systems shown in Fig. 6.4, the output $y(t)$ of the first system is fed back through the second system into the input where it is subtracted from the input signal $x(t)$. In this case, like in the parallel connection, besides the blocks representing the systems we use **adders** to add/subtract two signals.

It is possible to have **positive-** or **negative-feedback** systems depending on whether we add or subtract the signal being fed back to the input. Typically negative feedback is found, as positive feedback can greatly increase the gain of the system. (Think of the screeching sound created by an open microphone near a loud-speaker: the microphone picks up the amplified sound from the loud-speaker continuously increasing the volume of the produced signal—this is caused by positive feedback.) For negative feedback, the connection of two systems is done by putting one in the feed-forward loop, $H_1(s)$, and the other in the feed-back loop, $H_2(s)$, (there are other possible connections). To find the overall transfer function we consider the Laplace transforms of the error signal $e(t)$, $E(s)$, and of the output $y(t)$, $Y(s)$, in terms of the Laplace transform of the input $x(t)$, $X(s)$, and of the transfer functions $H_1(s)$ and $H_2(s)$ of the systems:

$$E(s) = X(s) - H_2(s)Y(s), \quad Y(s) = H_1(s)E(s).$$

Replacing $E(s)$ in the second equation gives

$$Y(s)[1 + H_1(s)H_2(s)] = H_1(s)X(s)$$

and the transfer function of the feedback system is then

$$H(s) = \frac{Y(s)}{X(s)} = \frac{H_1(s)}{1 + H_1(s)H_2(s)}. \tag{6.4}$$

As you recall, in Chapter 2 we were not able to find an explicit expression for the impulse response of the overall system and now you can understand why.

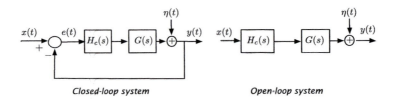

Closed-loop system   Open-loop system

FIGURE 6.5

Closed- and open-loop control of systems. The transfer function of the plant is $G(s)$ and the transfer function of the controller is $H_c(s)$.

6.3 APPLICATION TO CLASSICAL CONTROL

Because of different approaches, the theory of control systems can be divided into classical and modern. Classical control uses frequency-domain methods, while modern control uses time-domain methods. In classical **linear control**, the transfer function of the plant we wish to control is available, let us call it $G(s)$. The controller, with a transfer function $H_c(s)$, is designed to make the output of the overall system perform in a specified way. For instance, in a cruise control the plant is the car, and the desired performance is to automatically set the speed of the car to a desired value. There are two possible ways the controller and the plant are connected: either in **closed-loop** or in **open-loop** (see Fig. 6.5).

Open-Loop Control. In the **open-loop** approach the controller is cascaded with the plant (Fig. 6.5). To make the output $y(t)$ follow the reference signal at the input, $x(t)$, we minimize the error signal

$$e(t) = y(t) - x(t).$$

Typically, the output is affected by a disturbance $\eta(t)$, due to modeling or measurement errors. If we assume initially no disturbance, $\eta(t) = 0$, we find that the Laplace transform of the output of the overall system is

$$Y(s) = \mathcal{L}[y(t)] = H_c(s)G(s)X(s)$$

and that of the error is

$$E(s) = Y(s) - X(s) = [H_c(s)G(s) - 1]X(s).$$

To make the error zero, so that $y(t) = x(t)$, would require that $H_c(s) = 1/G(s)$ or the inverse of the plant, making the overall transfer function of the system $H_c(s)G(s)$ unity.

Remarks

Although open-loop control systems are simple to implement, they have several disadvantages:

- The controller $H_c(s)$ must cancel the poles and the zeros of $G(s)$ exactly, which is not very practical. In actual systems, the exact location of poles and zeros is not known due to measurement errors.
- If the plant $G(s)$ has zeros on the right-hand s-plane, then the controller $H_c(s)$ will be unstable, as its poles are the zeros of the plant.

- Due to ambiguity in the modeling of the plant, measurement errors, or simply the presence of noise, the output $y(t)$ is affected by a disturbance signal $\eta(t)$ mentioned above ($\eta(t)$ is typically random—we are going to assume for simplicity that it is deterministic so we can compute its Laplace transform). The Laplace transform of the overall system output is

$$Y(s) = H_c(s)G(s)X(s) + \eta(s)$$

where $\eta(s) = \mathcal{L}[\eta(t)]$. In this case, $E(s)$ is given by

$$E(s) = [H_c(s)G(s) - 1]X(s) + \eta(s).$$

Although we can minimize this error by choosing $H_c(s) = 1/G(s)$ as above, in this case $e(t)$ cannot be made zero, it remains equal to the disturbance $\eta(t)$—we have no control over this!

Closed-Loop Control. Assuming $y(t)$ and $x(t)$ in the open-loop control are the same type of signals, e.g., both are voltages, or temperatures, etc., if we feed back $y(t)$ and compare it with the input $x(t)$ we obtain a closed-loop control system. Considering the case of negative feedback (see Fig. 6.5) system, and assuming no disturbance ($\eta(t) = 0$) we have

$$E(s) = X(s) - Y(s), \quad Y(s) = H_c(s)G(s)E(s)$$

and replacing $Y(s)$ gives

$$E(s) = \frac{X(s)}{1 + G(s)H_c(s)}.$$

If we wish the error to go to zero in the steady-state, so that $y(t)$ **tracks** the input, the poles of $E(s)$ should be in the open left-hand s-plane.

If a disturbance signal $\eta(t)$ (consider it for simplicity deterministic and with Laplace transform $\eta(s)$) is present at the output the above analysis becomes

$$E(s) = X(s) - Y(s), \quad Y(s) = H_c(s)G(s)E(s) + \eta(s)$$

so that

$$E(s) = X(s) - H_c(s)G(s)E(s) - \eta(s)$$

or solving for $E(s)$

$$E(s) = \frac{X(s)}{1 + G(s)H_c(s)} - \frac{\eta(s)}{1 + G(s)H_c(s)} = E_1(s) + E_2(s).$$

If we wish $e(t)$ to go to zero in the steady state, then the poles of $E_1(s)$ and $E_2(s)$ should be in the open left-hand s-plane. Different from the open-loop control, the closed-loop control offers more flexibility in minimizing the effects of the disturbance.

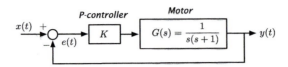

FIGURE 6.6

Proportional control of a motor.

Remarks

A control system includes two very important components:

- **Transducer**—Since it is possible that the output signal $y(t)$ and the reference signal $x(t)$ might not be of the same type, a transducer is used to change $y(t)$ so it is compatible with the reference input $x(t)$. Simple examples of transducer are light bulbs which convert voltage into light, a thermocouple which converts temperature into voltage, etc.
- **Actuator**—A device that makes possible the execution of the control action on the plant, so that the output of the plant follows the reference input.

Example 6.1 (Controlling an Unstable Plant). Consider a DC motor modeled as a LTI system with a transfer function

$$G(s) = \frac{1}{s(s+1)}.$$

The motor, modeled as such, is not a BIBO stable system given that its impulse response $g(t) = (1 - e^{-t})u(t)$ is not absolutely integrable. We wish the output of the motor $y(t)$ to track a given reference input $x(t)$, and propose using a so-called **proportional controller** with transfer $H_c(s) = K > 0$ to control the motor (see Fig. 6.6). The transfer function of the overall negative-feedback system is

$$H(s) = \frac{Y(s)}{X(s)} = \frac{KG(s)}{1 + KG(s)}.$$

Suppose that $X(s) = 1/s$, or that the reference signal is $x(t) = u(t)$. The question is: what should be the value of K so that in steady state the output of the system $y(t)$ coincides with $x(t)$? or, equivalently, what value of K makes the error signal $e(t) = x(t) - y(t)$ in the steady state zero?

We see that the Laplace transform of the error signal $e(t)$ is

$$E(s) = X(s)[1 - H(s)] = \frac{1}{s(1 + KG(s))} = \frac{s+1}{s(s+1) + K}.$$

The poles of $E(s)$ are the roots of the polynomial $s(s+1) + K = s^2 + s + K$, or

$$s_{1,2} = -0.5 \pm 0.5\sqrt{1 - 4K}.$$

For $0 < K \leq 0.25$ the roots are real, and complex for $K > 0.25$, and in both case in the left-hand s-plane. The partial fraction expansion corresponding to $E(s)$ would be

$$E(s) = \frac{B_1}{s - s_1} + \frac{B_2}{s - s_2}$$

for some values of B_1 and B_2. Given that the real parts of s_1 and s_2 are negative their corresponding inverse Laplace terms will have a zero steady-state response, thus

$$\lim_{t \to \infty} e(t) \to 0.$$

This can be found also with the final value theorem

$$sE(s)|_{s=0} = 0.$$

So for any $K > 0$ the output will tend to the input, $y(t) \to x(t)$, in the steady state.

Suppose then that $X(s) = 1/s^2$ or that $x(t) = tu(t)$, a ramp signal. Intuitively this a much harder situation to control, as the output needs to be continuously growing to try to follow the input. In this case the Laplace transform of the error signal is

$$E(s) = \frac{1}{s^2(1 + G(s)K)} = \frac{s+1}{s(s(s+1) + K)}.$$

In this case, even if we choose K so that the roots s_1 and s_2 of $s(s+1) + K$ are in the left-hand s-plane, we have a pole at $s = 0$. Thus, in the steady state, the partial fraction expansion terms corresponding to poles s_1 and s_2 will give a zero steady-state response, but the pole $s = 0$ will give a constant steady-state response A where

$$A = E(s)s|_{s=0} = \frac{1}{K}.$$

In the case of a ramp as input, it is not possible to make the output follow the input command signal, although by choosing a very large gain K, so that $A \to 0$, we can get them very close.

Example 6.2 (A Cruise Control). Consider controlling the speed of a car by means of a cruise control. How to choose the appropriate controller is not clear. We initially use a proportional plus integral (PI) controller $H_c(s) = 1 + 1/s$ (see Fig. 6.7) and ask you to use the proportional controller as an exercise.

Suppose we want to keep the speed of the car at V_0 miles/hour and as such we let the reference input be $x(t) = V_0 u(t)$. Assume for simplicity that the model for the car in motion is a system with transfer function

$$H_p(s) = \frac{\beta}{s + \alpha}$$

with $\beta > 0$ and $\alpha > 0$ related to the mass of the car and the friction coefficient. Let $\alpha = \beta = 1$. The Laplace transform of the output speed $v(t)$ of the car is

$$V(s) = \frac{H_c(s)H_p(s)}{1 + H_c(s)H_p(s)} X(s) = \frac{V_0/s}{s(1 + 1/s)} = \frac{V_0}{s(s+1)}.$$

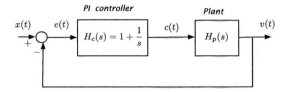

FIGURE 6.7

Cruise control system, reference speed $x(t) = V_0 u(t)$, output speed of car $v(t)$.

The poles of $V(s)$ are $s = 0$ and $s = -1$. We can then write $V(s)$ as

$$V(s) = \frac{B}{s+1} + \frac{A}{s},$$

where

$$A = sV(s)|_{s=0} = V_0.$$

Thus the steady-state response is

$$\lim_{t \to \infty} v(t) = V_0$$

since the inverse Laplace transform of the first term goes to zero due to its poles being in the left-hand s-plane. The error signal $e(t) = x(t) - v(t)$ in the steady state is zero. This value can be obtained using the final-value theorem of the Laplace transform applied to

$$E(s) = X(s) - V(s) = \frac{V_0}{s}\left[1 - \frac{1}{s+1}\right],$$

which gives

$$\lim_{t \to \infty} e(t) = \lim_{s \to 0} sE(s) = \lim_{s \to 0} V_0\left[1 - \frac{1}{s+1}\right] = 0,$$

coinciding with our previous result.

The controlling signal $c(t)$, see Fig. 6.7, that changes the speed of the car is

$$c(t) = e(t) + \int_0^t e(\tau)d\tau,$$

so that even if the error signal becomes zero at some point—indicating the desired speed had been reached—the value of $c(t)$ is not necessarily zero.

The PI controller used here is one of various possible controllers. Consider a simpler and cheaper controller such as a proportional controller with $H_c(s) = K$, would you be able to obtain the same results? Try it.   □

6.3.1 STABILITY AND STABILIZATION

A very important question related to the performance of systems is: how do we know that a given causal system has finite zero-input, zero-state or steady-state responses? This is the stability problem of great interest in control. Thus, if the system is represented by a finite-order, linear differential equation with constant coefficients the stability of the system determines that the zero-input, the zero-state and the steady-state response may exist. The stability of the system is also required when considering the frequency response in the Fourier analysis. It is important to understand that only the Laplace transform allows us to characterize stable as well as unstable systems, the Fourier transform does not.

Two possible ways to look at the stability of a causal LTI system are:

- when there is no input—the response of the system depends on initial energy in the system due to initial conditions—related to the zero-input response of the system, and
- when there is a bounded input and no initial condition—related to the zero-state response of the system.

Relating the zero-input response of a causal LTI system to stability, leads to **asymptotic stability**. A LTI system is said to be asymptotically stable, if its zero-input response (due only to initial conditions in the system) goes to zero as t increases, i.e.,

$$y_{zi}(t) \to 0 \qquad t \to \infty \tag{6.5}$$

for all possible initial conditions.

The second interpretation leads to the **bounded input–bounded output (BIBO) stability** which we considered in Chapter 2. A causal LTI system is BIBO stable if its response to a bounded input is also bounded. The condition we found in Chapter 2 for a causal LTI system to be BIBO stable was that the impulse response of the system be absolutely integrable, i.e.,

$$\int_0^\infty |h(t)| dt < \infty. \tag{6.6}$$

Such a condition is difficult to test, and we will see in this section that it is equivalent to the poles of the transfer function being in the open left-hand s-plane, a condition which can be more easily visualized and for which algebraic tests exist.

Consider a system being represented by the following differential equation:

$$y(t) + \sum_{k=1}^{N} a_k \frac{d^k y(t)}{dt^k} = b_0 x(t) + \sum_{\ell=1}^{M} b_\ell \frac{d^\ell x(t)}{dt^\ell} \qquad M < N, \ t > 0.$$

For some initial conditions and input $x(t)$, with Laplace transform $X(s)$, we see that the Laplace transform of the output is

$$Y(s) = Y_{zi}(s) + Y_{zs}(s) = \mathcal{L}[y(t)] = \frac{I(s)}{A(s)} + \frac{X(s)B(s)}{A(s)},$$

$$A(s) = 1 + \sum_{k=1}^{N} a_k s^k, \quad B(s) = b_0 + \sum_{m=1}^{M} b_m s^m,$$

where $I(s)$ is due to the initial conditions. To find the poles of $H_1(s) = 1/A(s)$, we set $A(s) = 0$, which corresponds to the **characteristic equation** of the system and its roots (either real, complex conjugate, simple or multiple) are the **natural modes or eigenvalues of the system**.

A causal LTI system with transfer function $H(s) = B(s)/A(s)$ exhibiting no pole–zero cancellation is said to be

- **Asymptotically stable** if the all-pole transfer function $H_1(s) = 1/A(s)$, used to determine the zero-input response, has all its poles in the open left-hand s-plane (the $j\Omega$-axis excluded), or equivalently

$$A(s) \neq 0 \quad \text{for} \quad \mathcal{R}e[s] \geq 0. \tag{6.7}$$

- **BIBO stable** if all the poles of $H(s)$ are in the open left-hand s-plane (the $j\Omega$-axis excluded), or equivalently

$$A(s) \neq 0 \quad \text{for} \quad \mathcal{R}e[s] \geq 0. \tag{6.8}$$

- If $H(s)$ exhibits pole–zero cancellations, the system can be BIBO stable but not asymptotically stable.

Testing the stability of a causal LTI system thus requires one to find the location of the roots of $A(s)$, or the poles of the system. This can be done for low-order polynomials $A(s)$ for which there are formulas to find the roots of a polynomial exactly. But as shown by Abel,[2] there are no equations to find the roots of higher than fourth-order polynomials. Numerical methods to find roots of these polynomials only provide approximate results which might not be good enough for cases where the poles are close to the $j\Omega$-axis. The Routh stability criterion is an algebraic test capable of determining whether the roots of $A(s)$ are on the left-hand s-plane or not, thus determining the stability of the system.

Example 6.3 (Stabilization of a Plant). Consider a plant with a transfer function $G(s) = 1/(s - 2)$, which has a pole in the right-hand s-plane, and therefore is unstable. Let us consider stabilizing it by cascading it with an all-pass filter and by using a negative-feedback system (Fig. 6.8).

Cascading the system with an all-pass system makes the overall system not only stable but also keeps the magnitude response of the original system. To get rid of the pole at $s = 2$ and to replace it with a new pole at $s = -2$ we let the all-pass filter be

$$H_a(s) = \frac{s - 2}{s + 2}.$$

To see that this filter has a constant magnitude response consider

$$H_a(s)H_a(-s) = \frac{(s - 2)(-s - 2)}{(s + 2)(-s + 2)} = \frac{(s - 2)(s + 2)}{(s + 2)(s - 2)} = 1.$$

---

[2]Niels H. Abel (1802–1829) Norwegian mathematician who accomplished brilliant work in his short lifetime. At *age 19*, he showed there is no general algebraic solution for the roots of equations of degree greater than four, in terms of explicit algebraic operations.

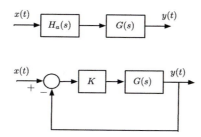

FIGURE 6.8

Stabilization of an unstable plant $G(s)$ using (top) an all-pass filter, and (bottom) a proportional controller of gain K.

If we let $s = j\Omega$ the above gives the magnitude squared function

$$H_a(j\Omega)H_a(-j\Omega) = H_a(j\Omega)H_a^*(j\Omega) = |H_a(j\Omega)|^2 = 1,$$

which is unity for all values of frequency. The cascade of the unstable system with the all-pass system gives a stable system

$$H(s) = G(s)H_a(s) = \frac{1}{s+2}$$

with the same magnitude response as $G(s)$, as $|H(j\Omega)| = |G(j\Omega)||H_a(j\Omega)| = |G(j\Omega)|$. This is an open-loop stabilization and it depends on the all-pass system having a zero exactly at 2 so that it cancels the pole causing the instability. Any small change on the zero and the overall system would not be stabilized. Another problem with the cascading of an all-pass filter to stabilize a filter is that it does not work when the pole causing the instability is at the origin, as we cannot obtain an all-pass filter able to cancel that pole.

Consider then a negative-feedback system (Fig. 6.8). Suppose we use a proportional controller with a gain K, the overall system transfer function is

$$H(s) = \frac{KG(s)}{1 + KG(s)} = \frac{K}{s + (K-2)}$$

and if the gain K is chosen so that $K - 2 > 0$ or $K > 2$ the feedback system will be stable. $\square$

6.3.2 TRANSIENT ANALYSIS OF FIRST- AND SECOND-ORDER CONTROL SYSTEMS

Although the input to a control system is not known *a priori*, there are many applications where the system is frequently subjected to a certain type of input and thus one can select a test signal. For instance, if a system is subjected to intense and sudden inputs, then an impulse signal might be the appropriate test input for the system; if the input applied to a system is constant or continuously increasing then a unit step or a ramp signal would be appropriate. Using test signals such as an impulse, a unit step, a ramp or a sinusoid, mathematical and experimental analyses of systems can be done.

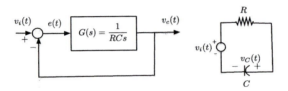

FIGURE 6.9

Feedback modeling of an RC circuit in series.

When designing a control system its stability becomes its most important attribute, but there are other system characteristics that need to be considered. The transient behavior of the system, for instance, needs to be stressed in the design. Typically, as we drive the system to reach a desired response, the system's response goes through a transient before reaching the desired response. Thus how fast the system responds, and what steady-state error it reaches, need to be part of the design considerations.

First-Order System. As an example of a first-order system consider an RC serial circuit with a voltage source $v_i(t) = u(t)$ as input (Fig. 6.9), and as output the voltage across the capacitor, $v_c(t)$. By voltage division, the transfer function of the circuit is

$$H(s) = \frac{V_c(s)}{V_i(s)} = \frac{1}{1 + RCs}.$$

Considering the RC circuit a feedback system with input $v_i(t)$ and output $v_c(t)$, the feedforward transfer function $G(s)$ in Fig. 6.9 is $1/RCs$. Indeed, from the feedback system we have

$$E(s) = V_i(s) - V_c(s), \quad V_c(s) = E(s)G(s).$$

Replacing $E(s)$ in the second of the above equations we have

$$\frac{V_c(s)}{V_i(s)} = \frac{G(s)}{1 + G(s)} = \frac{1}{1 + 1/G(s)},$$

so that the open-loop transfer function, when we compare the above equation to $H(s)$, is

$$G(s) = \frac{1}{RCs}.$$

The RC circuit can be seen as a feedback system: the voltage across the capacitor is constantly compared with the input voltage and if found smaller, the capacitor continues charging until its voltage coincides with the input voltage. How fast it charges depends on the RC value.

For $v_i(t) = u(t)$, so that $V_i(s) = 1/s$, then the Laplace transform of the output is

$$V_c(s) = \frac{1}{s(sRC + 1)} = \frac{1/RC}{s(s + 1/RC)} = \frac{1}{s} - \frac{1}{s + 1/RC},$$

so that

$$v_c(t) = (1 - e^{-t/RC})u(t).$$

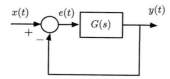

FIGURE 6.10

Second-order feedback system.

Second-Order System. A series RLC circuit with input a voltage source $v_s(t)$ and output the voltage across the capacitor, $v_c(t)$ has a transfer function

$$\frac{V_c(s)}{V_s(s)} = \frac{1/Cs}{R + Ls + 1/Cs} = \frac{1/LC}{s^2 + (R/L)s + 1/LC}.$$

If we define

$$\text{Natural frequency: } \Omega_n = \frac{1}{\sqrt{CL}}, \tag{6.9}$$

$$\text{Damping ratio: } \psi = 0.5R\sqrt{\frac{C}{L}}, \tag{6.10}$$

we can write

$$\frac{V_c(s)}{V_s(s)} = \frac{\Omega_n^2}{s^2 + 2\psi\Omega_n s + \Omega_n^2}. \tag{6.11}$$

A feedback system with this transfer function is given in Fig. 6.10 where the feedforward transfer function is

$$G(s) = \frac{\Omega_n^2}{s(s + 2\psi\Omega_n)}.$$

Indeed, the transfer function of the feedback system is given by

$$H(s) = \frac{Y(s)}{X(s)} = \frac{G(s)}{1 + G(s)} = \frac{\Omega_n^2}{s^2 + 2\psi\Omega_n s + \Omega_n^2}.$$

The dynamics of a second order system can be described in terms of the two parameters Ω_n and ψ, as these two parameters determine the location of the poles of the system and thus its response. We adapted the previously given script to plot the cluster of poles and the time response of the second-order system.

Example 6.4. Consider the simulation of a first and second-order systems. The following MATLAB script for ind=1, plots the poles $V_c(s)/V_i(s)$ and simulates the transients of $v_c(t)$ for $1 \le RC \le 10$ and a unit-step input. The results are shown in Fig. 6.11. Thus if we wish the system to respond fast to the unit-step input we locate the system pole far from the origin.

```
%
% Example 6.4---Transient analysis of first and second order systems
%
clf; clear all
syms s t
ind=input('first (1) or second (2) order system?    ')
if ind==1, % first-order system
num=[0 1];
figure(1)
for RC=1:2:10,
    den=[RC 1];
    subplot(211)
    splane(num,den);grid
    hold on
    vc=ilaplace(1/(RC*s^2+s))
    subplot(212)
    ezplot(vc,[0,50]); axis([0 50 0 1.2]); grid; xlabel('t'); ylabel('v_c(t)')
    hold on
    RC
    pause
end
  else % second-order system
  num=[0 0 1];figure(2)
  for phi=0:0.1:1,
    den=[1 2*phi 1]
    subplot(221)
    splane(num,den)
    hold on
    vc=ilaplace(1/(s^3+2*phi*s^2+s))
    subplot(222)
    ezplot(vc,[0,50]); axis([0 50 -0.1 2.5]); grid; xlabel('t'); ylabel('v_c(t)');
    hold on
    if phi>=0.707
      subplot(223)
      ezplot(vc,[0,50]); axis([0 50 -0.1 1.2]); grid; xlabel('t'); ylabel('v_c(t)');
      hold on
    else
    end
  pause
 end
 hold off
 end
```

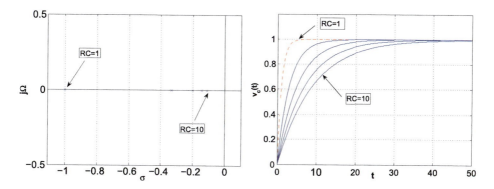

FIGURE 6.11

Clustering of poles (left) and time responses of first-order feedback system for $1 \leq RC \leq 10$.

For a second-order system simulation (ind=2), we assume $\Omega_n = 1$ (rad/s) and let $0 \leq \psi \leq 1$ (so that the poles of $H(s)$ are complex conjugate for $0 \leq \psi < 1$ and double real for $\psi = 1$). Let the input be unit-step signal so that $X(s) = 1/s$. We then have:

- If we plot the poles of $H(s)$ as ψ changes from 0 (poles in $j\Omega$-axis) to 1 (double real poles) the response $y(t)$ in the steady state changes from a sinusoid shifted up by 1 to a damped signal. The locus of the poles is a semicircle of radius $\Omega_n = 1$. Fig. 6.12 shows this behavior of the poles and the response.
- As in the first-order system, the location of the poles determines the response of the system. The system is useless if the poles are on the $j\Omega$-axis as the response is completely oscillatory and the input will never be followed. On the other extreme, the response of the system is slow when the poles become real. The designer would have to choose a value in between these two for ψ.
- For values of ψ between $\sqrt{2}/2$ to 1 the oscillation is minimal and the response is relatively fast. See Fig. 6.12 (top right). For values of ψ from 0 to $\sqrt{2}/2$ the response oscillates more and more, giving a large steady-state error. See Fig. 6.12 (bottom right). □

Example 6.5. In this example we find the response of a LTI system to different inputs by using functions in the Control Toolbox of MATLAB. You can learn more about the capabilities of this toolbox, or set of specialized functions for control, by running the demo *respdemo* and then using *help* to learn more about the functions *tf, impulse, step*, and *pzmap* which we will use here.

We want to create a MATLAB function that has as inputs the coefficients of the numerator $N(s)$ and of the denominator $D(s)$ of the transfer function $H(s) = N(s)/D(s)$ of the system (the coefficients are ordered from the highest order to the lowest order or constant term) and computes the impulse response, the unit-step response, and the response to a ramp. The function should show the transfer function, the poles and zeros and plot the corresponding responses. We need to figure out how to compute the ramp response using the *step* function. Consider the transfer functions

$$(i) \ \ H_1(s) = \frac{s+1}{s^2+s+1}, \quad (ii) \ \ H_2(s) = \frac{s}{s^3+s^2+s+1}$$

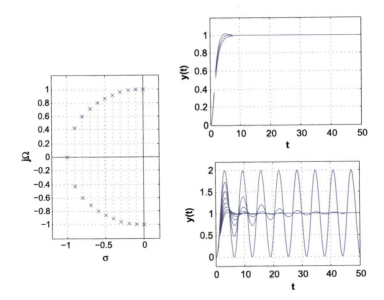

FIGURE 6.12

Clustering of poles (left) and time responses of second-order feedback system for $\sqrt{2}/2 \leq \psi \leq 1$ (top right) and $0 \leq \psi \leq \sqrt{2}/2$ (bottom right).

to test our function. To understand the results it is useful to consider the BIBO stability of the systems with these transfer functions.

The following script is used to look at the desired responses of the two systems and the location of their poles and zeros. You can choose either the first of the second transfer function. Our function *response* computes the impulse, step, and ramp responses.

```
%%
% Example 6.5---Control toolbox
%%
clear all; clf
ind=input('system (1) or (2)?    ')
if ind==1,
% H_1(s)
nu=[1 1];de=[1 1 1];    % stable
response(nu,de)
else
% H_2(s)
nu=[1 0];de=[1 1 1 1];  % unstable
response(nu,de)
end
```

```
function response(N,D)
sys=tf(N,D)
poles=roots(D)
zeros=roots(N)

figure(1)
pzmap(sys);grid
figure(2)
T=0:0.25:20;T1=0:0.25:40;
for t=1:3,
    if t==3,
        D1=[D 0];                    % for ramp response
    end
    if t==1,
        subplot(311)
        y=impulse(sys,T);
        plot(y);title(' Impulse response');ylabel('h(t)');xlabel('t'); grid
    elseif t==2,
        subplot(312)
        y=step(sys,T);
        plot(y);title(' Unit-step response');ylabel('s(t)');xlabel('t');grid
    else
        subplot(313)
        sys=tf(N,D1);                % ramp response
        y=step(sys,T1);
        plot(y); title(' Ramp response'); ylabel('q(t)'); xlabel('t');grid
    end
end
```

The transfer function $H_2(s)$ has

```
Transfer function:
         s
---------------
s^3 + s^2 + s + 1
poles =
  -1.0000
  -0.0000 + 1.0000i
  -0.0000 - 1.0000i
zeros =
     0
```

As you can see, two of the poles are on the $j\Omega$-axis, and so the system corresponding to $H_2(s)$ is unstable. The other system is stable. Results for the two systems are shown in Fig. 6.13.   □

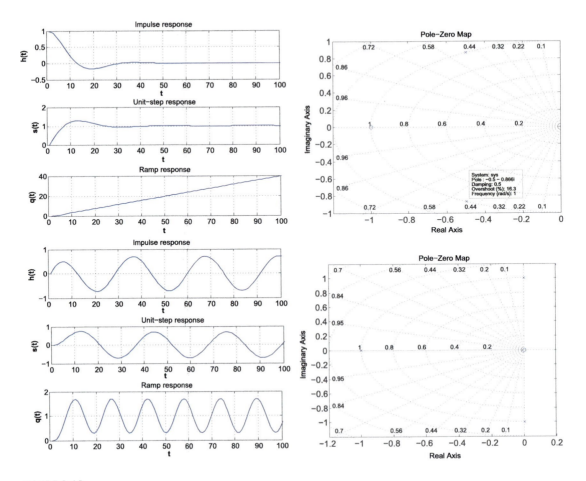

FIGURE 6.13

Impulse, unit-step and ramp responses (left) and poles and zeros (right) for system with transfer function $H_1(s)$ (top) and $H_2(s)$ (bottom).

6.4 STATE-VARIABLE REPRESENTATION OF LTI SYSTEMS

The modern approach to control uses the state-variable, or the internal, representation of the system which can be derived from the ordinary differential equation characterizing the system [42,29,28]. As a ratio of the Laplace transform of the output to that of the input, the transfer function—or the external representation of the system—does not include the effect of initial conditions and could omit some of the modes of the system by pole/zero cancellations. On the other hand, the state-variable representation by being directly connected with the system's ordinary differential equation, having no obvious cancellations, provides a more complete representation of the dynamics of the system. The states are the memory of the system, and the state-variable representation allows by simple extension, multiple inputs as well as multiple outputs. Modern control theory deals with the theory and application of concepts related to the state-variable representation. Although such a representation is not unique, it is amenable to computer computation.

To illustrate the concept of state, the state-variable representation and its advantages over the transfer function consider a series RC circuit. In this circuit the capacitor is the element capable of storing energy, i.e., the memory of the circuit. Assume that the voltage across the capacitor is the output of interest, and for simplicity let $C = 1$ F and $R = 1$ Ω. If $v_i(t)$ and $v_C(t)$ denote the input voltage and the voltage across the capacitor, with corresponding Laplace transforms $V_i(s)$ and $V_C(s)$, we have the following representations of the circuit.

- Transfer function representation. Assuming zero initial voltage across the capacitor, by voltage division we have

$$V_C(s) = \frac{V_i(s)/s}{1 + 1/s} = \frac{V_i(s)}{s+1} \quad \text{or} \quad H(s) = \frac{V_C(s)}{V_i(s)} = \frac{1}{s+1}, \tag{6.12}$$

the transfer function of the circuit. The impulse response is $h(t) = \mathcal{L}^{-1}[H(s)] = e^{-t}u(t)$. For an input $v_i(t)$, and zero initial conditions, the output $v_C(t)$ is the inverse Laplace transform of

$$V_C(s) = H(s)V_i(s)$$

or the convolution integral of the impulse response $h(t)$ and the input $v_i(t)$:

$$v_C(t) = \int_0^t h(t-\tau)v_i(\tau)d\tau. \tag{6.13}$$

- State-variable representation. The first-order ordinary differential equation for the circuit is

$$v_C(t) + \frac{dv_C(t)}{dt} = v_i(s) \qquad t \geq 0$$

with initial condition $v_C(0)$—the voltage across the capacitor at $t = 0$. Using the Laplace transform to find $v_C(t)$ we have

$$V_C(s) = \frac{v_C(0)}{s+1} + \frac{V_i(s)}{s+1}$$

with inverse Laplace transform

$$v_C(t) = e^{-t}v_C(0) + \int_0^t e^{-(t-\tau)}v_i(\tau)d\tau, \quad t \geq 0. \tag{6.14}$$

If $v_C(0) = 0$, the above equation gives the same result as the transfer function. But if the initial voltage is not zero, the transfer function cannot provide the above solution which includes the zero-input and the zero-state responses.

Considering that the voltage across the capacitor, $v_C(t)$, relates to the energy stored in the capacitor then $v_C(0)$ provides the voltage accumulated in the capacitor in the past, from $-\infty$ to $t = 0$, that according to (6.14) permits us to find future values of $v_C(t)$, for any value $t > 0$, for a given input

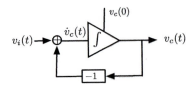

FIGURE 6.14

Block diagram of state equation for series RC circuit.

$v_i(t)$. Thus, for a given $v_i(t)$ knowing $v_C(0)$ is sufficient to compute the output of the system at *any* future time $t = t_0 > 0$, independent of what might have happened in the past in the circuit. Thus the voltage across the capacitor, the element in this circuit capable of storing energy or of having memory, is a state variable.

The state equation for this RC circuit can be written as

$$\left[\frac{dv_C(t)}{dt}\right] = [-1][v_C(t)] + [1][v_i(t)], \tag{6.15}$$

which is a fancy way of rewriting the ordinary differential equation obtained for the circuit. A block diagram to realize this equation requires an integrator, an adder and a constant multiplier as shown in Fig. 6.14. Notice that the output of the integrator—the integrator having an initial voltage $v_C(0)$—is the state variable.

Example 6.6. Consider the solution for the output $v_C(t)$ for the RC circuit given in (6.14). Show that for $t_1 > t_0 > 0$ we have

$$v_C(t_1) = e^{-(t_1 - t_0)} v_C(t_0) + \int_{t_0}^{t_1} e^{-(t_1 - \tau)} v_i(\tau) d\tau$$

or that given the state at $t_0 > 0$ and the input $v_i(t)$, we can compute a future value of $v_C(t_1)$ for $t_1 > t_0$ independent of how $v_C(t_0)$ is attained.

As given by (6.14), $v_C(t_0)$, $t_0 > 0$ and $v_C(t_1)$, $t_1 > t_0 > 0$ can be obtained by

$$
\begin{aligned}
v_C(t_0) &= e^{-t_0} v_C(0) + \int_0^{t_0} e^{-(t_0 - \tau)} v_i(\tau) d\tau \\
&= e^{-t_0}\left[v_C(0) + \int_0^{t_0} e^{\tau} v_i(\tau) d\tau \right], \quad t_0 > 0, \tag{6.16} \\
v_C(t_1) &= e^{-t_1}\left[v_C(0) + \int_0^{t_1} e^{\tau} v_i(\tau) d\tau \right], \quad t_1 > t_0. \tag{6.17}
\end{aligned}
$$

Solving for $v_C(0)$ in the top equation (6.16)

$$v_C(0) = e^{t_0} v_C(t_0) - \int_0^{t_0} e^{\tau} v_i(\tau) d\tau$$

and replacing it in the lower equation (6.17) we get

$$v_C(t_1) = e^{-(t_1-t_0)}v_C(t_0) - \int_0^{t_0} e^{-(t_1-\tau)}v_i(\tau)d\tau + \int_0^{t_1} e^{-(t_1-\tau)}v_i(\tau)d\tau$$

$$= e^{-(t_1-t_0)}v_C(t_0) + \int_{t_0}^{t_1} e^{-(t_1-\tau)}v_i(\tau)d\tau. \quad \Box$$

The <u>state</u> $\{x_k(t)\}$, $k = 1, \cdots, N$, of a LTI system is the smallest set of variables that if known at a certain time t_0 allows us to compute the response of the system at times $t > t_0$ for specified inputs $\{w_i(t)\}$, $i = 1, \cdots, M$. In the multiple-input multiple-output case a non-unique <u>state equation</u> for the system, in matrix form, is

$$\dot{\mathbf{x}}(t) = \mathbf{A}\mathbf{x}(t) + \mathbf{B}\mathbf{w}(t), \qquad (6.18)$$

$$\mathbf{x}^T(t) = [x_1(t) \ x_2(t) \cdots x_N(t)] \qquad \text{state vector,}$$

$$\dot{\mathbf{x}}^T(t) = [\dot{x}_1(t) \ \dot{x}_2(t) \cdots \dot{x}_N(t)],$$

$$\mathbf{A} = [a_{ij}] \ N \times N \text{ matrix,}$$

$$\mathbf{B} = [b_{ij}] \ N \times M \text{ matrix,}$$

$$\mathbf{w}^T(t) = [w_1(t) \ w_2(t) \cdots w_M(t)] \qquad \text{input vector,}$$

where $\dot{x}_i(t)$ stands for the derivative of $x_i(t)$.
The <u>output equation</u> of the system is a combination of the state variables and the inputs, i.e.,

$$\mathbf{y}(t) = \mathbf{C}\mathbf{x}(t) + \mathbf{D}\mathbf{w}(t), \qquad (6.19)$$

$$\mathbf{y}(t) = [y_1(t) \ y_2(t) \cdots y_L(t)] \qquad \text{output vector,}$$

$$\mathbf{C} = [c_{ij}] \ L \times N \text{ matrix,}$$

$$\mathbf{D} = [d_{ij}] \ L \times M \text{ matrix.}$$

Notice the notation used to write matrices and vectors.[3]

Example 6.7. Consider the series RLC circuit in Fig. 6.15 for which the output is the voltage across the capacitor is $v_C(t)$ and $v_s(t)$ is the input. Let $L = 1$ H, $R = 6 \ \Omega$ and $C = 1/8$ F. We want to:

1. Choose as states the voltage across the capacitor and the current in the inductor and obtain the state and the output equations for this circuit, and indicate how to find the initial conditions for the state variables in terms of the initial conditions for the circuit.
2. Do a partial fraction expansion for the transfer function $V_C(s)/V_s(s)$ and define a new set of state variables.
3. Find the output $v_C(t)$ using the Laplace transform from the second set of state variables.

---

[3] In the above $(.)^T$ stands for the transpose of the matrix or vector inside the parentheses. Also vectors are being considered as column vectors, so $\mathbf{z}$ is a column vector of some dimension $N \times 1$ (indicating there are N rows and 1 column.) Thus its transpose $\mathbf{z}^T$ is a row vector of dimension $1 \times N$, i.e., it has 1 row and N columns.

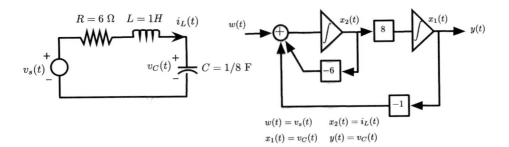

FIGURE 6.15

RLC circuit and state-variable realization.

In this case we have:

1. The equations for the voltage across the capacitor, and Kirchoff's voltage law for the circuit give the state equations:

$$\frac{dv_C(t)}{dt} = \frac{1}{C}i_L(t),$$

$$\frac{di_L(t)}{dt} = -\frac{1}{L}v_C(t) - \frac{R}{L}i_L(t) + \frac{1}{L}v_s(t).$$

If we let the state variables be $\{x_1(t) = v_C(t),\ x_2(t) = i_L(t)\}$, the input be $w(t) = v_s(t)$, and the output be $y(t) = v_C(t)$, we obtain after replacing the values of $C = 1/8$, $L = 1$, and $R = 6$ the state and the output equations for this circuit as

$$\begin{bmatrix} \dot{x}_1(t) \\ \dot{x}_2(t) \end{bmatrix} = \begin{bmatrix} 0 & 8 \\ -1 & -6 \end{bmatrix} \begin{bmatrix} x_1(t) \\ x_2(t) \end{bmatrix} + \begin{bmatrix} 0 \\ 1 \end{bmatrix} w(t),$$

$$y(t) = \begin{bmatrix} 1 & 0 \end{bmatrix} \begin{bmatrix} x_1(t) \\ x_2(t) \end{bmatrix} + [0]\, w(t). \tag{6.20}$$

For the initial conditions, we have

$$y(t) = x_1(t),$$
$$\dot{y}(t) = \dot{x}_1(t) = 8x_2(t),$$

and since $y(t) = v_C(t)$ and $\dot{y}(t) = i_L(t)/C$, the state-variable initial conditions are $x_1(0) = v_C(0)$ and $x_2(0) = \dot{y}(0)/8 = i_L(0)/(8C) = i_L(0)$.

2. For zero initial conditions, $C = 1/8$, $L = 1$ and $R = 6$ we have by voltage division that the transfer function and its partial fraction expansion are:

$$\frac{V_C(s)}{V_s(s)} = \frac{1}{LC(s^2 + (R/L)s + 1/(LC))} = \frac{8}{s^2 + 6s + 8}$$

$$= \frac{8}{(s+2)(s+4)} = \underbrace{\frac{4}{s+2}}_{N_1(s)/D_1(s)} + \underbrace{\frac{(-4)}{s+4}}_{N_2(s)/D_2(s)}$$

with the above definitions we then write

$$V_C(s) = \underbrace{\frac{N_1(s)V_s(s)}{D_1(s)}}_{\hat{X}_1(s)} + \underbrace{\frac{N_2(s)V_s(s)}{D_2(s)}}_{\hat{X}_2(s)}$$

or the output in the time domain

$$v_C(t) = \mathcal{L}^{-1}[\hat{X}_1(s) + \hat{X}_2(s)] = \hat{x}_1(t) + \hat{x}_2(t)$$

for new state variables $\hat{x}_i(t)$, $i = 1, 2$. After replacing $N_1(s)$, $N_2(s)$, $D_1(s)$, $D_2(s)$, letting the input be $w(t) = v_s(t)$ and output be $y(t) = v_C(t)$ we obtain from the above:

$$\frac{d\hat{x}_1(t)}{dt} = -2\hat{x}_1(t) + 4w(t),$$

$$\frac{d\hat{x}_2(t)}{dt} = -4\hat{x}_2(t) - 4w(t).$$

Thus another set of state and output equations in matrix form for the same circuit is

$$\begin{bmatrix} \dot{\hat{x}}_1(t) \\ \dot{\hat{x}}_2(t) \end{bmatrix} = \begin{bmatrix} -2 & 0 \\ 0 & -4 \end{bmatrix} \begin{bmatrix} \hat{x}_1(t) \\ \hat{x}_2(t) \end{bmatrix} + \begin{bmatrix} 4 \\ -4 \end{bmatrix} w(t),$$

$$y(t) = \begin{bmatrix} 1 & 1 \end{bmatrix} \begin{bmatrix} \hat{x}_1(t) \\ \hat{x}_2(t) \end{bmatrix}. \tag{6.21}$$

3. Assuming the initial conditions $\hat{x}_1(0)$ and $\hat{x}_2(0)$ are known, the state equations in (6.21) can be solved using the Laplace transform as follows:

$$s\hat{X}_1(s) - \hat{x}_1(0) = -2\hat{X}_1(s) + 4W(s),$$
$$s\hat{X}_2(s) - \hat{x}_2(0) = -4\hat{X}_2(s) - 4W(s),$$

which can be written in matrix form as

$$\begin{bmatrix} s+2 & 0 \\ 0 & s+4 \end{bmatrix} \begin{bmatrix} \hat{X}_1(s) \\ \hat{X}_2(s) \end{bmatrix} = \begin{bmatrix} \hat{x}_1(0) \\ \hat{x}_2(0) \end{bmatrix} + \begin{bmatrix} 4 \\ -4 \end{bmatrix} W(s);$$

solving for $\hat{X}_1(s)$ and $\hat{X}_2(s)$ we obtain

$$\begin{bmatrix} \hat{X}_1(s) \\ \hat{X}_2(s) \end{bmatrix} = \begin{bmatrix} s+2 & 0 \\ 0 & s+4 \end{bmatrix}^{-1} \begin{bmatrix} \hat{x}_1(0) \\ \hat{x}_2(0) \end{bmatrix} + \begin{bmatrix} s+2 & 0 \\ 0 & s+4 \end{bmatrix}^{-1} \begin{bmatrix} 4 \\ -4 \end{bmatrix} W(s)$$

$$= \begin{bmatrix} \hat{x}_1(0)/(s+2) \\ \hat{x}_2(0)/(s+4) \end{bmatrix} + \begin{bmatrix} 4W(s)/(s+2) \\ -4W(s)/(s+4) \end{bmatrix}.$$

Its inverse Laplace transform is thus

$$\begin{bmatrix} \hat{x}_1(t) \\ \hat{x}_2(t) \end{bmatrix} = \begin{bmatrix} \hat{x}_1(0)e^{-2t}u(t) + 4\int_0^t e^{-2(t-\tau)}w(\tau)d\tau \\ \hat{x}_2(0)e^{-4t}u(t) - 4\int_0^t e^{-4(t-\tau)}w(\tau)d\tau \end{bmatrix},$$

so that the output is

$$\begin{aligned} y(t) &= \hat{x}_1(t) + \hat{x}_2(t) \\ &= [\hat{x}_1(0)e^{-2t} + \hat{x}_2(0)e^{-4t}]u(t) + 4\int_0^t [e^{-2(t-\tau)} - e^{-4(t-\tau)}]w(\tau)d\tau. \quad \square \end{aligned}$$

Remarks

1. Although the transfer function is unique in the above example, there are two sets of state variables. The first set $\{x_1(t) = v_C(t), \; x_2(t) = i_L(t)\}$ clearly has physical significance as they relate to the voltage across the capacitor and the current in the inductor, while the second $\{\hat{x}_1(t), \hat{x}_2(t)\}$ does not.
2. To find the relation between the two set of state variables, consider the zero-state response $v_C(t)$. We see that the impulse response $h(t)$ of the system is the inverse Laplace transform of the transfer function

$$H(s) = \frac{V_C(s)}{V_s(s)} = \frac{8}{(s+2)(s+4)} = \frac{4}{s+2} + \frac{-4}{s+4}$$

or $h(t) = [4e^{-2t} - 4e^{-4t}]u(t)$. For a general input $v_s(t)$ then $V_C(s) = H(s)V_s(s)$ so that $v_C(t)$ is the convolution integral of $h(t)$ and $v_s(t)$ or after replacing $h(t)$:

$$\begin{aligned} v_C(t) &= \int_0^t [4e^{-2(t-\tau)} - 4e^{-4(t-\tau)}]v_s(\tau)d\tau \\ &= \underbrace{\int_0^t 4e^{-2(t-\tau)}v_s(\tau)d\tau}_{\hat{x}_1(t)} + \underbrace{\int_0^t [-4e^{-4(t-\tau)}]v_s(\tau)d\tau}_{\hat{x}_2(t)}, \quad\quad (6.22) \end{aligned}$$

which gives the output equation $y(t) = v_C(t) = \hat{x}_1(t) + \hat{x}_2(t)$.
Then from (6.22) and the definition of the first set of state variables:

$$\underbrace{\begin{bmatrix} x_1(t) \\ x_2(t) \end{bmatrix}}_{\mathbf{x}(t)} = \begin{bmatrix} v_C(t) \\ i_L(t) = Cdv_C(t)/dt \end{bmatrix} = \underbrace{\begin{bmatrix} 1 & 1 \\ -0.25 & -0.5 \end{bmatrix} \begin{bmatrix} \hat{x}_1(t) \\ \hat{x}_2(t) \end{bmatrix}}_{\mathbf{F}\hat{\mathbf{x}}(t)}$$

where the lower equation results from calculating the derivative of $v_C(t)$ in (6.22) and multiplying it by $C = 1/8$.
3. In matrix form, the connection between the first and second set of state equations is obtained as follows. The state and output equations (6.20) can be written in matrix form using appropriate

matrix $\mathbf{A}$ and vectors $\mathbf{b}$ and $\mathbf{c}^T$:

$$\dot{\mathbf{x}}(t) = \mathbf{A}\mathbf{x}(t) + \mathbf{b}w(t), \quad y(t) = \mathbf{c}^T\mathbf{x}(t), \quad \text{where}$$

$$\mathbf{A} = \begin{bmatrix} 0 & 8 \\ -1 & -6 \end{bmatrix}, \quad \mathbf{b} = \begin{bmatrix} 0 \\ 1 \end{bmatrix}, \quad \mathbf{c}^T = \begin{bmatrix} 1 & 0 \end{bmatrix}.$$

Letting $\mathbf{x}(t) = \mathbf{F}\hat{\mathbf{x}}(t)$, for an invertible transformation matrix $\mathbf{F}$ we rewrite the above as

$$\mathbf{F}\dot{\hat{\mathbf{x}}}(t) = \mathbf{A}\mathbf{F}\hat{\mathbf{x}}(t) + \mathbf{b}w(t) \quad \text{or} \quad \dot{\hat{\mathbf{x}}}(t) = \mathbf{F}^{-1}\mathbf{A}\mathbf{F}\hat{\mathbf{x}}(t) + \mathbf{F}^{-1}\mathbf{b}\, w(t),$$
$$y(t) = \mathbf{c}^T\mathbf{F}\hat{\mathbf{x}}(t).$$

In this case we have

$$\mathbf{F} = \begin{bmatrix} 1 & 1 \\ -0.25 & -0.5 \end{bmatrix}, \quad \mathbf{F}^{-1} = \begin{bmatrix} 2 & 4 \\ -1 & -4 \end{bmatrix}, \quad \text{so that}$$

$$\mathbf{F}^{-1}\mathbf{A}\mathbf{F} = \begin{bmatrix} -2 & 0 \\ 0 & -4 \end{bmatrix}, \quad \mathbf{F}^{-1}\mathbf{b} = \begin{bmatrix} 4 \\ -4 \end{bmatrix}, \quad \mathbf{c}^T\mathbf{F} = \begin{bmatrix} 1 & 1 \end{bmatrix}$$

or we have the matrix and vectors in (6.21).

The state variables of a system are not unique. For a single-input $w(t)$ single-output $y(t)$ system, for given state and output equations with state variables $\{x_i(t)\}$:

$$\dot{\mathbf{x}}(t) = \mathbf{A}\mathbf{x}(t) + \mathbf{b}w(t),$$
$$y(t) = \mathbf{c}^T\mathbf{x}(t) + dw(t), \tag{6.23}$$

a new set of state variables $\{z_i(t)\}$ can be obtained using an invertible transformation matrix $\mathbf{F}$

$$\mathbf{x}(t) = \mathbf{F}\mathbf{z}(t).$$

The matrix and vectors for the new state-variable representation are given by

$$\mathbf{A}_1 = \mathbf{F}^{-1}\mathbf{A}\mathbf{F}, \quad \mathbf{b}_1 = \mathbf{F}^{-1}\mathbf{b}, \quad \mathbf{c}_1^T = \mathbf{c}^T\mathbf{F}, \quad d_1 = d. \tag{6.24}$$

6.4.1 CANONICAL REALIZATIONS

In this section we consider well-accepted, or canonical, state-variable realizations. A desired characteristic of these realizations is that the number of integrators must coincide with the order of the ordinary differential equation of the system. Such realizations are called **minimal realizations**. Different characteristics of the system can be obtained from these canonical realizations, as you will learn in modern control.

Direct Minimal Realizations. Suppose you have the following ordinary differential equation representing a second-order LTI system with input $w(t)$ and output $y(t)$:

$$\frac{d^2y(t)}{dt^2} + a_1\frac{dy(t)}{dt} + a_0y(t) = w(t).$$

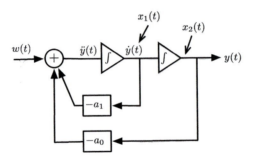

FIGURE 6.16

Direct realization of $d^2y(t)/dt^2 + a_1\,dy(t)/dt + a_0y(t) = w(t)$ with state variables $x_1(t) = \dot{y}(t)$ and $x_2(t) = y(t)$ as indicated.

A possible realization is shown in Fig. 6.16. Notice that in this case the input does not include derivatives of it, or the corresponding transfer function has only poles (an "all-pole" system). If we let the outputs of the two integrators be the state variables, we have the following state and output equations:

$$
\begin{aligned}
\dot{x}_1(t) &= \ddot{y}(t) = -a_1x_1(t) - a_0x_2(t) + w(t), \\
\dot{x}_2(t) &= x_1(t), \\
y(t) &= x_2(t),
\end{aligned}
$$

which in matrix form are

$$
\left[\begin{array}{c} \dot{x}_1(t) \\ \dot{x}_2(t) \end{array}\right] = \left[\begin{array}{cc} -a_1 & -a_0 \\ 1 & 0 \end{array}\right]\left[\begin{array}{c} x_1(t) \\ x_2(t) \end{array}\right] + \left[\begin{array}{c} 1 \\ 0 \end{array}\right]w(t),
$$

$$
y(t) = \left[\begin{array}{cc} 0 & 1 \end{array}\right]\left[\begin{array}{c} x_1(t) \\ x_2(t) \end{array}\right].
$$

Notice that the second order of the system is matched by the number of integrators. The realization in Fig. 6.16 is called a *minimal realization* (the number of integrators equals the order of the system) of the *all-pole* system with transfer function

$$
H(s) = \frac{Y(s)}{W(s)} = \frac{1}{s^2 + a_1s + a_0}.
$$

In general, the numerator of the transfer function is a polynomial is s, and then the input would consist of $w(t)$ and its derivatives. For such a system to be causal the order of the numerator should be less than that of the denominator, i.e., the transfer function be proper rational. To realize it directly would require differentiators, to obtain the derivatives of the input, which are undesirable components. It is possible, however, to attain minimal realizations by decomposing the transfer function as the product of an *all-pole* system and an *only-zeros* system as follows. Assume the transfer function of the LTI system is

$$H(s) = \frac{N(s)}{D(s)} = \frac{b_m s^m + b_{m-1}s^{m-1} + \cdots + b_0}{s^n + a_{n-1}s^{n-1} + \cdots + a_0} \qquad m \text{ (order of } N(s)) < n \text{ (order of } D(s)).$$

Since $H(s) = Y(s)/W(s)$ we write

$$Y(s) = \underbrace{\frac{W(s)}{D(s)}}_{Z(s)} N(s),$$

allowing us to define the all-pole and only-zeros transfer functions

$$(i) \quad \frac{Z(s)}{W(s)} = \frac{1}{D(s)}, \qquad (ii) \quad \frac{Y(s)}{Z(s)} = N(s),$$

from which we have

$$D(s)Z(s) = W(s) \quad \Rightarrow \quad \frac{d^n z(t)}{dt^n} + a_{n-1}\frac{d^{n-1}z(t)}{dt^{n-1}} + \cdots + a_0 z(t) = w(t), \qquad (6.25)$$

$$N(s)Z(s) = Y(s) \quad \Rightarrow \quad b_m \frac{d^m z(t)}{dt^m} + b_{m-1}\frac{d^{m-1}z(t)}{dt^{m-1}} + \cdots + b_0 z(t) = y(t). \qquad (6.26)$$

To realize (6.25) n integrators are needed, and the derivatives needed in (6.26) are already available in the first realization, so there is no need for differentiators. Thus the realization is minimal.

Example 6.8. Consider the ordinary differential equation

$$\frac{d^2 y(t)}{dt^2} + a_1 \frac{dy(t)}{dt} + a_0 y(t) = b_1 \frac{dw(t)}{dt} + b_0 w(t)$$

of a second-order LTI system with input $w(t)$ and output $y(t)$. Find a direct minimal realization.

The transfer function of the system is

$$H(s) = \left[\frac{Y(s)}{Z(s)}\right]\left[\frac{Z(s)}{W(s)}\right] = [b_0 + b_1 s]\left[\frac{1}{a_0 + a_1 s + a_2 s^2}\right].$$

Realizations of

$$w(t) = a_0 z(t) + a_1 \frac{dz(t)}{dt} + \frac{d^2 z(t)}{dt^2} \quad \text{and of}$$

$$y(t) = b_0 z(t) + b_1 \frac{dz(t)}{dt}$$

in that order give the realization (Fig. 6.17) of the system.

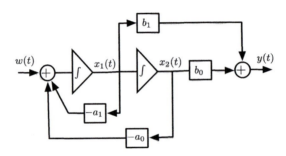

FIGURE 6.17

Direct minimal realization of $d^2 y(t)/dt^2 + a_1 \, dy(t)/dt + a_0 y(t) = b_1 \, dw(t)/dt + b_0 w(t)$ using state variables $x_1(t) = \dot{z}(t)$ and $x_2(t) = z(t)$ as shown.

Letting the state variables be the outputs of the integrators, we have the following state and output equations:

$$\begin{bmatrix} \dot{x}_1(t) \\ \dot{x}_2(t) \end{bmatrix} = \underbrace{\begin{bmatrix} -a_1 & -a_0 \\ 1 & 0 \end{bmatrix}}_{\mathbf{A}_c} \begin{bmatrix} x_1(t) \\ x_2(t) \end{bmatrix} + \underbrace{\begin{bmatrix} 1 \\ 0 \end{bmatrix}}_{\mathbf{b}_c} w(t),$$

$$y(t) = \underbrace{\begin{bmatrix} b_1 & b_0 \end{bmatrix}}_{\mathbf{c}_c^T} \begin{bmatrix} x_1(t) \\ x_2(t) \end{bmatrix}. \quad \square$$

This realization is called the **controller canonical form** with matrix and vectors $\{\mathbf{A}_c, \mathbf{b}_c, \mathbf{c}_c^T\}$ as indicated.

Remarks

- The factoring of the transfer function as given above makes it possible to obtain an all-pole realization so that an nth-order system is represented by an nth-order ordinary differential equation, with n initial conditions, and realized using n integrators.
- Another realization of interest is the **observer canonical form**, a dual of the controller form. The matrix and vectors of the observer realization are obtained from those of the controller as follows:

$$\mathbf{A}_o = \mathbf{A}_c^T, \quad \mathbf{b}_o^T = \mathbf{c}_c, \quad \mathbf{c}_o = \mathbf{b}_o^T. \tag{6.27}$$

Parallel and Cascade Realizations. Given a proper rational transfer function

$$H(s) = \frac{N(s)}{D(s)}$$

by partial fraction expansion and factorization of $N(s)$ and $D(s)$ we obtain parallel and cascade realizations.

The **parallel realization** is obtained from the partial expansion

$$H(s) = \sum_{i=1}^{N} H_i(s),$$

where each $H_i(s)$ is a proper rational function with real coefficients. The simplest case is when the poles of $H(s)$ are real and distinct, in which case we need to realize each of the first-order systems

$$H_i(s) = \frac{b_{0i}}{s + a_{0i}}, \quad i = 1, \cdots, N,$$

to obtain the parallel realization. The input of all of these components is $w(t)$ and the output is the sum of the state variables. When the poles are complex conjugate pairs, the corresponding second-order terms have real coefficients and can be realized. Thus for any transfer function with real and complex conjugate pairs, or equivalently with polynomials $N(s)$ and $D(s)$ with real coefficients, the realization is possible by combining the realization corresponding to the real poles (first-order systems) with the realization corresponding to the complex conjugate pairs of poles (second-order systems).

Example 6.9. Obtain a parallel realization of a system with transfer function

$$H(s) = \frac{1 + 2s}{2 + 3s + s^2} = \frac{1 + 2s}{(s+1)(s+2)} = \frac{-1}{s+1} + \frac{3}{s+2}.$$

Using the direct realizations of

$$H_1(s) = \frac{Y_1(s)}{W(s)} = \frac{-1}{s+1} \quad \text{and} \quad H_2(s) = \frac{Y_2(s)}{W(s)} = \frac{3}{s+2},$$

we obtain the realization for $H(s)$ shown in Fig. 6.18. Letting again the state variables be the outputs of the integrators we have the following state and output equations:

$$\begin{bmatrix} \dot{x}_1(t) \\ \dot{x}_2(t) \end{bmatrix} = \begin{bmatrix} -1 & 0 \\ 0 & -2 \end{bmatrix} \begin{bmatrix} x_1(t) \\ x_2(t) \end{bmatrix} + \begin{bmatrix} -1 \\ 3 \end{bmatrix} w(t),$$

$$y(t) = \begin{bmatrix} 1 & 1 \end{bmatrix} \begin{bmatrix} x_1(t) \\ x_2(t) \end{bmatrix}. \quad \square$$

Example 6.10. Consider the transfer function

$$H(s) = \frac{s^2 + 3s + 6}{[(s+1)^2 + 4](s+1)} = \frac{1}{s^2 + 2s + 5} + \frac{1}{s+1}$$

where the poles are $s_{1,2} = -1 \pm 2j$ and $s_3 = -1$. Obtain a realization in parallel and show the state and output equations for it.

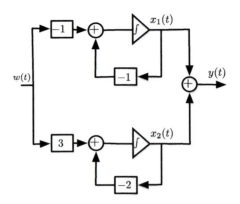

FIGURE 6.18

Parallel realization of $H(s) = (1 + 2s)/(2 + 3s + s^2)$.

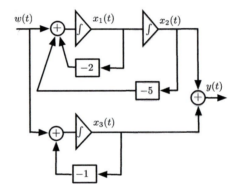

FIGURE 6.19

Parallel realization of $H(s) = (s^2 + 3s + 6)/[(s^2 + 2s + 5)(s + 1)]$.

We realize the second order system resulting from the complex poles in parallel with the first-order system resulting from the real pole. Fig. 6.19 shows the realization. The state and the output equations for the above realization is

$$
\begin{bmatrix} \dot{x}_1(t) \\ \dot{x}_2(t) \\ \dot{x}_3(t) \end{bmatrix} = \begin{bmatrix} -2 & -5 & 0 \\ 1 & 0 & 0 \\ 0 & 0 & -1 \end{bmatrix} \begin{bmatrix} x_1(t) \\ x_2(t) \\ x_3(t) \end{bmatrix} + \begin{bmatrix} 1 \\ 0 \\ 1 \end{bmatrix} w(t),
$$

$$
y(t) = \begin{bmatrix} 0 & 1 & 1 \end{bmatrix} \begin{bmatrix} x_1(t) \\ x_2(t) \\ x_3(t) \end{bmatrix}. \quad \square
$$

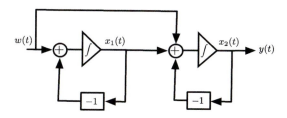

FIGURE 6.20

Cascade/parallel realization of $G(s) = (1/s + 1)[1 + 1/(s + 1)]$.

The **cascade realization** is another way to realize a LTI system by representing its transfer function as a product of first- and second-order transfer functions with real coefficients, i.e.,

$$H(s) = \prod_{i=1}^{M} H_i(s)$$

and to realize each $H_i(s)$ using direct first- and second-order minimal realizations.

Example 6.11. The following transfer function has repeated poles:

$$G(s) = \frac{s + 2}{(s + 1)^2}.$$

Find parallel and cascade/parallel realizations. Choose the one that is minimal and for it obtain a block diagram and determine the corresponding state and output equations.

Two possible realizations for $G(s)$ are:

- parallel realization of a first and a second-order filter by expressing

$$G(s) = \frac{1}{s + 1} + \frac{1}{(s + 1)^2},$$

- cascade and parallel realization by writing (factoring $1/(s + 1)$ in the partial fraction expansion)

$$G(s) = \frac{1}{s + 1}\left[1 + \frac{1}{s + 1}\right].$$

The parallel realization is not minimal as it uses three integrators for a second order system, but the cascade/parallel is. Fig. 6.20 displays the minimal realization.

The state and output equations for the minimal realization are

$$\begin{bmatrix} \dot{x}_1(t) \\ \dot{x}_2(t) \end{bmatrix} = \begin{bmatrix} -1 & 0 \\ 1 & -1 \end{bmatrix}\begin{bmatrix} x_1(t) \\ x_2(t) \end{bmatrix} + \begin{bmatrix} 1 \\ 1 \end{bmatrix} w(t),$$

$$y(t) = \begin{bmatrix} 0 & 1 \end{bmatrix}\begin{bmatrix} x_1(t) \\ x_2(t) \end{bmatrix}. \quad \square$$

6.4.2 COMPLETE SOLUTION FROM STATE AND OUTPUT EQUATIONS

Given the state representation the complete solution of the system, due to input and initial conditions, can be obtained in the time or in the Laplace domain. In the time domain, the solution can be expressed as an exponential matrix, while in the Laplace domain it can be obtained using Cramer's rule.

Exponential Matrix Solution

Suppose the following state and output equations are given:

$$\dot{\mathbf{x}}(t) = \mathbf{A}\mathbf{x}(t) + \mathbf{b}w(t),$$
$$y(t) = \mathbf{c}^T \mathbf{x}(t),$$

and we wish to find the complete solution $y(t)$, $t \geq 0$. Computing the Laplace transform of each of the state equations we have

$$s\mathbf{I}\,\mathbf{X}(s) - \mathbf{x}(0) = \mathbf{A}\mathbf{X}(s) + \mathbf{b}W(s) \quad \Rightarrow \quad (s\mathbf{I} - \mathbf{A})\mathbf{X}(s) = \mathbf{x}(0) + \mathbf{b}W(s),$$
$$Y(s) = \mathbf{c}^T \mathbf{X}(s),$$

where $\mathbf{X}(s) = [\mathcal{L}(x_i(t))]$ and $\mathbf{x}(0) = [x_i(0)]$ are vectors. Solving for $\mathbf{X}(s)$ in the top equation:

$$\mathbf{X}(s) = (s\mathbf{I} - \mathbf{A})^{-1}[\mathbf{x}(0) + \mathbf{b}W(s)]$$

and replacing it in the bottom equation we get

$$Y(s) = \mathbf{c}^T (s\mathbf{I} - \mathbf{A})^{-1}(\mathbf{x}(0) + \mathbf{b}\,W(s)), \tag{6.28}$$

for which we need to find its inverse Laplace transform to get $y(t)$.

An expression for the inverse $(s\mathbf{I} - \mathbf{A})^{-1}$ is

$$(s\mathbf{I} - \mathbf{A})^{-1} = \mathbf{I}s^{-1} + \mathbf{A}s^{-2} + \mathbf{A}^2 s^{-3} + \cdots .$$

Indeed, we have

$$(s\mathbf{I} - \mathbf{A})^{-1}(s\mathbf{I} - \mathbf{A}) \quad = \quad \mathbf{I} + \mathbf{A}s^{-1} + \mathbf{A}^2 s^{-2} + \mathbf{A}^3 s^{-3} + \cdots$$
$$- \mathbf{A}s^{-1} - \mathbf{A}s^{-2} - \mathbf{A}^2 s^{-3} - \cdots = \mathbf{I}.$$

The inverse Laplace transform of each of the terms of the expansion $(s\mathbf{I} - \mathbf{A})^{-1}$ gives

$$\mathcal{L}^{-1}[(s\mathbf{I} - \mathbf{A})^{-1}] = [\mathbf{I} + \mathbf{A}t + \mathbf{A}^2 \frac{t^2}{2!} + \mathbf{A}^3 \frac{t^3}{3!} \cdots]u(t) = e^{\mathbf{A}t} u(t) \tag{6.29}$$

or the **exponential matrix**. The inverse of Equation (6.28) can thus be found using the expression in (6.29) and the fact the inverse Laplace transform of the product of two functions of s is the convolution integral so that

$$y(t) = \mathbf{c}^T e^{\mathbf{A}(t)} \mathbf{x}(0) + \mathbf{c}^T \int_0^t e^{\mathbf{A}(t-\tau)} \mathbf{b} w(\tau) d\tau \quad t \geq 0.$$

In particular, the impulse response $h(t)$ of the system would be the response to $w(t) = \delta(t)$ and zero initial conditions or from the above

$$h(t) = \mathbf{c}^T \int_0^t e^{\mathbf{A}(t-\tau)} \mathbf{b} \delta(\tau) d\tau = \mathbf{c}^T e^{\mathbf{A}t} \mathbf{b}.$$

The complete solution of

$$\dot{\mathbf{x}}(t) = \mathbf{A}\mathbf{x}(t) + \mathbf{b}w(t),$$
$$y(t) = \mathbf{c}^T \mathbf{x}(t), \quad t > 0 \tag{6.30}$$

is given by

$$y(t) = \mathbf{c}^T e^{\mathbf{A}(t)} \mathbf{x}(0) + \mathbf{c}^T \int_0^t e^{\mathbf{A}(t-\tau)} \mathbf{b}w(\tau) d\tau, \quad t \geq 0 \tag{6.31}$$

where the exponential matrix is defined as

$$e^{\mathbf{A}t} = [\mathbf{I} + \mathbf{A}t + \mathbf{A}^2 \frac{t^2}{2!} + \mathbf{A}^3 \frac{t^3}{3!} \cdots]. \tag{6.32}$$

In particular, the impulse response of the system is

$$h(t) = \mathbf{c}^T e^{\mathbf{A}t} \mathbf{b}. \tag{6.33}$$

Cramer's Rule Solution

The exponential matrix solution depends on the inverse $(s\mathbf{I} - \mathbf{A})^{-1}$, which in the time domain relates to the exponential matrix. The result requires functions of matrices. An easier solution is obtained using Cramer's rule.

Applying Cramer's rule one can obtain the inverse $(s\mathbf{I} - \mathbf{A})^{-1}$ in closed form. Assuming nonzero initial conditions, the Laplace transform of the state equation is

$$s\mathbf{I}\mathbf{X}(s) - \mathbf{x}(0) = \mathbf{A}\mathbf{X}(s) + \mathbf{b}W(s) \quad \Rightarrow \quad [s\mathbf{I} - \mathbf{A}]\mathbf{X}(s) = [\mathbf{x}(0) + \mathbf{b}W(s)]$$

or in the form of a set of linear equations:

$$
\begin{bmatrix}
s - a_{11} & -a_{12} & \cdots & -a_{1N} \\
-a_{21} & s - a_{22} & \cdots & -a_{2N} \\
\vdots & \vdots & \cdots & \vdots \\
-a_{N1} & -a_{N2} & \cdots & s - a_{NN}
\end{bmatrix}
\begin{bmatrix}
X_1(s) \\
X_2(s) \\
\vdots \\
X_N(s)
\end{bmatrix}
=
\begin{bmatrix}
x_1(0) + b_1 W(s) \\
x_2(0) + b_2 W(s) \\
\vdots \\
x_N(0) + b_N W(s)
\end{bmatrix}
\tag{6.34}
$$

where Cramer's rule gives for $X_i(s)$, $i = 1, \cdots, N$:

$$X_i(s) = \frac{\det \begin{bmatrix} s - a_{11} & \cdots & x_1(0) + b_1 W(s) & \cdots & -a_{1N} \\ -a_{21} & \cdots & x_2(0) + b_2 W(s) & \cdots & -a_{2N} \\ \vdots & \vdots & \cdots & & \vdots \\ -a_{N1} & \cdots & x_N(0) + b_N W(s) & \cdots & s - a_{NN} \end{bmatrix}}{\det(s\mathbf{I} - \mathbf{A})}$$

where the vector on the right of Equation (6.34) replaces the ith column in the matrix in the left side of that equation. Repeating this for each $i = 1, \cdots, N - 1$, we obtain the $\{X_i(s)\}$ to be replaced in the output equation

$$Y(s) = \mathbf{c}^T \mathbf{X}(s) = c_1 X_1(s) + c_2 X_2(s) + \cdots + c_N X_N(s).$$

The inverse Laplace transform of this expression gives the complete response

$$y(t) = c_1 x_1(t) + c_2 x_2(t) + \cdots + c_N x_N(t).$$

Example 6.12. Consider a LTI system represented by the state and output equations

$$\begin{bmatrix} \dot{x}_1(t) \\ \dot{x}_2(t) \end{bmatrix} = \begin{bmatrix} 0 & 1 \\ -8 & -6 \end{bmatrix} \begin{bmatrix} x_1(t) \\ x_2(t) \end{bmatrix} + \begin{bmatrix} 0 \\ 8 \end{bmatrix} w(t),$$

$$y(t) = \begin{bmatrix} 1 & 0 \end{bmatrix} \begin{bmatrix} x_1(t) \\ x_2(t) \end{bmatrix}.$$

Use Cramer's rule to:

- Find the zero-input response if the initial conditions are $x_1(0) = 1$ and $x_2(0) = 0$.
- Find the zero-state response corresponding to $w(t) = u(t)$, unit-step signal, as input.
- Determine the impulse response $h(t)$ of the system.

The Laplace transform of the state equation, assuming nonzero initial conditions, is

$$\begin{bmatrix} s & -1 \\ 8 & s+6 \end{bmatrix} \begin{bmatrix} X_1(s) \\ X_2(s) \end{bmatrix} = \begin{bmatrix} x_1(0) \\ x_2(0) + 8W(s) \end{bmatrix},$$

$$Y(s) = X_1(s).$$

The Laplace transform of the output indicates that we need to obtain $X_1(s)$ from the state equations, so by Cramer's rule

$$Y(s) \quad = \quad X_1(s) = \frac{\det \begin{bmatrix} x_1(0) & -1 \\ x_2(0) + 8W(s) & s+6 \end{bmatrix}}{s(s+6) + 8} = \frac{x_1(0)(s+6) + x_2(0) + 8W(s)}{s^2 + 6s + 8}$$

$$= \quad \underbrace{\frac{s+6}{s^2 + 6s + 8}}_{Y_{zi}(s)} + \underbrace{\frac{8}{s(s^2 + 6s + 8)}}_{Y_{zs}(s)},$$

after replacing $W(s) = 1/s$ and the initial conditions $x_1(0) = 1$ and $x_2(0) = 0$.

For the zero-input response:

$$Y_{zi}(s) = \frac{s+6}{s^2+6s+8} = \frac{s+6}{(s+2)(s+4)} = \frac{2}{s+2} - \frac{1}{s+4} \quad \Rightarrow \quad y_{zi}(t) = [2e^{-2t} - e^{-4t}]u(t)$$

and for the zero-state equation we have

$$Y_{zs}(s) = \frac{8}{s(s^2+6s+8)} = \frac{8}{s(s+2)(s+4)} = \frac{1}{s} - \frac{2}{s+2} + \frac{1}{s+4}$$
$$\Rightarrow \quad y_{zs}(t) = [1 - 2e^{-2t} + e^{-4t}]u(t).$$

Thus the complete response is

$$y(t) = y_{zi}(t) + y_{zs}(t) = u(t).$$

To find the impulse response we let $W(s) = 1$, $Y(s) = H(s)$ and set the initial conditions to zero to get

$$H(s) = \frac{8}{s^2+6s+8} = \frac{8}{(s+2)(s+4)} = \frac{4}{s+2} - \frac{4}{s+4} \quad \Rightarrow \quad h(t) = [4e^{-2t} - 4e^{-4t}]u(t). \quad \square$$

6.4.3 EXTERNAL AND INTERNAL REPRESENTATION OF SYSTEMS

The transfer function provides the external representation of a LTI system, while the state-variable representation gives its internal representation. In most cases these representations are equivalent, but the non-uniqueness and the additional information that is provided by the state-variable representation, compared to that from the transfer function, makes the internal representation more general. For instance, the external representation only deals with the case of zero initial conditions, and unstable modes present in the system may disappear when pole–zero cancellations occur. Different canonical forms of the state variable provide more comprehensive consideration of the dynamics of the system. The following example illustrates the differences between the external and the internal representations of a system.

Example 6.13. A system is represented by the second-order ordinary differential equation

$$\frac{d^2y(t)}{dt^2} - \frac{dy(t)}{dt} - 2y(t) = \frac{dw(t)}{dt} - 2w(t), \quad t > 0$$

where $w(t)$ is the input and $y(t)$ is the output.

- Obtain the transfer function of this system and indicate the difference between it and the ordinary differential equation representation.
- Use the MATLAB function *tf2ss* to find the controller state representation and use the Laplace transform to obtain the complete response. Find a connection between the initial conditions of the system with those from the state variables.
- Change the state and output equations to obtain the observer form and determine the complete response using the Laplace transform. Find a connection between the initial conditions of the system with those from the state variables.
- Comment on the complete solutions of the two forms.

Solution: We have:

- The transfer function of the system is given by

$$H(s) = \frac{Y(s)}{W(s)} = \frac{s-2}{s^2-s-2} = \frac{s-2}{(s+1)(s-2)} = \frac{1}{s+1}.$$

Because of the pole–zero cancellation that occurs when determining the transfer function, the ordinary differential equation indicates the system is second-order while the transfer function shows it is first-order. Also the pole in the right-hand side of the s-plane is canceled in the transfer function but remains as an unstable mode in the ordinary differential equation and its effect will be felt in the state-variable representation. For an input $w(t)$, the response of the system for zero initial conditions is $Y(s) = H(s)W(s)$ or the convolution

$$y(t) = \int_0^t h(t-\tau)w(\tau)d\tau$$

where $h(t) = \mathcal{L}^{-1}[H(s)] = e^{-t}u(t)$ is the impulse response of the system. The transfer function requires zero initial conditions. The behavior due to nonzero initial conditions are displayed by state-variable representations differently.

- Using the function *tf2ss* we obtain the controller form:

$$\begin{bmatrix} \dot{x}_1(t) \\ \dot{x}_2(t) \end{bmatrix} = \begin{bmatrix} 1 & 2 \\ 1 & 0 \end{bmatrix} \begin{bmatrix} x_1(t) \\ x_2(t) \end{bmatrix} + \begin{bmatrix} 1 \\ 0 \end{bmatrix} w(t),$$

$$y(t) = \begin{bmatrix} 1 & -2 \end{bmatrix} \begin{bmatrix} x_1(t) \\ x_2(t) \end{bmatrix}.$$

Assuming initial conditions $x_1(0)$, $x_2(0)$ and a generic input $w(t)$, using the Laplace transform of the state and the output equations we have

$$\begin{bmatrix} s-1 & -2 \\ -1 & s \end{bmatrix} \begin{bmatrix} X_1(s) \\ X_2(s) \end{bmatrix} = \begin{bmatrix} x_1(0) + W(s) \\ x_2(0) \end{bmatrix},$$

$$Y(s) = X_1(s) - 2X_2(s).$$

Using Cramer's rule we find

$$X_1(s) = \frac{x_1(0)s + 2x_2(0) + sW(s)}{(s+1)(s-2)},$$

$$X_2(s) = \frac{x_1(0) + x_2(0)(s-1) + W(s)}{(s+1)(s-2)},$$

$$Y(s) = X_1(s) - 2X_2(s) = \frac{x_1(0)}{s+1} - \frac{2x_2(0)}{s+1} + \frac{W(s)}{s+1},$$

$$y(t) = [x_1(0) - 2x_2(0)]e^{-t}u(t) + \int_0^t h(t-\tau)w(\tau)d\tau, \quad \text{where } h(t) = e^{-t}u(t).$$

As the effect of the initial conditions disappears with time, the output $y(t)$ becomes the convolution of the input $w(t)$ with the impulse response $h(t) = e^{-t}u(t)$ thus coinciding with the inverse Laplace transform of the output $Y(s) = H(s)W(s) = W(s)/(s+1)$ or $y(t) = (h * w)(t)$. Therefore, for zero initial conditions, or for any initial conditions given the decaying exponential associated with them, the two responses will coincide.

The initial conditions $y(0)$ and $\dot{y}(0)$ are related to the state-variable initial conditions using the output equation:

$$y(t) = x_1(t) - 2x_2(t) \quad \to \quad y(0) = x_1(0) - 2x_2(0), \tag{6.35}$$

$$\dot{y}(t) = \begin{bmatrix} 1 & -2 \end{bmatrix} \begin{bmatrix} \dot{x}_1(t) \\ \dot{x}_2(t) \end{bmatrix} = \begin{bmatrix} 1 & -2 \end{bmatrix} \left(\begin{bmatrix} 1 & 2 \\ 1 & 0 \end{bmatrix} \begin{bmatrix} x_1(t) \\ x_2(t) \end{bmatrix} + \begin{bmatrix} 1 \\ 0 \end{bmatrix} w(t) \right) \quad \to$$

$$\dot{y}(0) = -x_1(0) + 2x_2(0), \tag{6.36}$$

assuming that the input is zero at $t = 0$. The determinant of Equations (6.35) and (6.36) is zero, thus the only solution of these equations is $y(0) = \dot{y}(0) = 0$, in which case $x_1(0) = 2x_2(0)$, and their effect disappears as $t \to \infty$.

• The MATLAB function *tf2ss* only gives the controller form of the state-variable representation. The observer form is obtained from the above controller form by using the following matrices:

$$\mathbf{A}_o = \mathbf{A}_c^T = \begin{bmatrix} 1 & 1 \\ 2 & 0 \end{bmatrix}, \quad \mathbf{b}_o^T = \mathbf{c}_c^T = \begin{bmatrix} 1 & -2 \end{bmatrix}, \quad \mathbf{c}_o^T = \mathbf{b}_c^T = \begin{bmatrix} 1 & 0 \end{bmatrix}$$

Calling the state variables $\hat{x}_1(t)$ and $\hat{x}_2(t)$ the Laplace transform of the state and output equations are

$$\begin{bmatrix} s-1 & -1 \\ -2 & s \end{bmatrix} \begin{bmatrix} \hat{X}_1(s) \\ \hat{X}_2(s) \end{bmatrix} = \begin{bmatrix} \hat{x}_1(0) + W(s) \\ \hat{x}_2(0) - 2W(s) \end{bmatrix}, \quad Y(s) = \hat{X}_1(s).$$

Then, using Cramer's rule

$$Y(s) = \hat{X}_1(s) = \frac{\hat{x}_1(0)s + \hat{x}_2(0) + W(s)(s-2)}{(s+1)(s-2)}$$

$$= \left[\frac{1}{3(s+1)} + \frac{2}{3(s-2)} \right] \hat{x}_1(0) + \left[\frac{-1}{3(s+1)} + \frac{1}{3(s-2)} \right] \hat{x}_2(0) + \frac{1}{s+1} W(s)$$

and the complete response is then

$$y(t) = \hat{x}_1(0) \left[\frac{e^{-t} + 2e^{2t}}{3} \right] u(t) + \hat{x}_2(0) \left[\frac{-e^{-t} + e^{2t}}{3} \right] u(t) + \int_0^t e^{-(t-\tau)} w(\tau) d\tau.$$

For this response to coincide with the zero-state response given by the transfer function it is necessary to guarantee that the initial conditions be zero, any small variation will be amplified by the unstable mode generated by the pole on the right-hand s-plane. This insight is not available in the transfer function because the initial conditions are assumed zero. Thus the internal representation provides information that is not available in the external representation given by the transfer function. If the initial conditions are not zero, the complete response would grow as t increases, i.e., the unstable mode is not controlled.

Using a similar approach as in the controller form, the relation between the initial conditions $y(0)$, $\dot{y}(0)$ in this case can be shown to be

$$\hat{x}_1(0) = y(0),$$
$$\hat{x}_2(0) = -y(0) + \dot{y}(0),$$

indicating that to have $\hat{x}_1(0) = \hat{x}_2(0) = 0$ we would need $y(0) = \dot{y}(0) = 0$, if these initial conditions vary from zero, the unstable mode will make the output become unbounded.

The following script can be used to verify the above results.   □

```
%%
% Example 6.13---Transfer function to state-variable representation of
% y^(2)(t)-y^(1)(t)-2 y(t)=w^(1)(t)- 2w(t)
%%
Num=[0 1 -2]; Den=[1 -1 -2];
% controller
disp('Controller form')
[A,b,c,d]=tf2ss(Num,Den)
[N,D]=ss2tf(A,b,c,d) % transfer function from state-variable equations
% observer
disp('Observer form')
Ao=A'
bo=c'
co=b'
do=d
[N,D]=ss2tf(Ao,bo,co,do)% transfer function from state-variable equations
% Application of Cramer's rule
syms x10 x20 w s
% controller
X1= det([x10+w -2;x20 s])/det([s-1 -2;-1 s])
X2= det([s-1 x10+w;-1  x20])/det([s-1 -2;-1 s])
pause
% observer
X1= det([x10+w -1;x20-2*w s])/det([s-1 -1;-2 s])
```

6.5 WHAT HAVE WE ACCOMPLISHED? WHAT IS NEXT?

In this chapter we have illustrated the application of the Laplace analysis to the theory of control, classical and modern. As you have seen, in classical control the Laplace transform is very appropriate for problems where transients as well as steady-state responses are of interest. Moreover, it is important to realize that stability can only be characterized in the Laplace domain, and that it is necessary when considering steady-state responses. Block diagrams help to visualize the interconnection of the different systems. The control examples show the importance of the transfer function, and the transient and the steady-state computations. In modern control although the emphasis is in the time domain, via the state approach, the Laplace transform becomes very useful in obtaining the complete solution and in connecting the internal and the external representations. The application of MATLAB in both classical and modern control was illustrated.

Although the material in this chapter does not have sufficient depth, this being reserved for texts in control and in linear system theory, it serves to connect the theory of continuous-time signals and systems with applications. In the next part of the book, we will consider how to process signals using computers and how to apply the resulting theory again in classical and modern control theory.

6.6 PROBLEMS
6.6.1 BASIC PROBLEMS

6.1 The transfer function $H(s) = 1/(s+1)^2$ of a filter is to be implemented by cascading two first order filters $H_i(s) = 1/(s+1)$, $i = 1, 2$.

(a) Implement $H_i(s)$ as a series RC circuit with input $v_i(t)$ and output $v_{i+1}(t)$, $i = 1, 2$. Cascade two of these circuits and find the overall transfer function $V_3(s)/V_1(s)$. Carefully draw the circuit.

(b) Use a voltage follower to connect the two circuits when cascaded and find the overall transfer function $V_3(s)/V_1(s)$. Carefully draw the circuit.

(c) Using the voltage follower circuit implement a new transfer function

$$G(s) = \frac{1}{(s+1000)(s+1)}.$$

Carefully draw your circuit.

Answer: Without voltage follower $V_3(s)/V_1(s) = 1/(s^2 + 3s + 1)$.

6.2 An ideal operational amplifier circuit can be shown to be equivalent to a negative-feedback system.

(a) Consider the inverting amplifier circuit and its two-port network equivalent shown in Fig. 6.21. Obtain a feedback system with input $V_i(s)$ and output $V_0(s)$.

(b) What is the effect of $A \to \infty$ on the above circuit?

Answers: $V_i(s) = R_1(V_i(s) - V_o(s))/(R_1 + R_2) - V_o(s)/A$; as $A \to \infty$ then $V_o(s)/V_i(s) = -R_2/R_1$.

6.3 A resistor $R = 1\Omega$, a capacitor $C = 1$ F and an inductor $L = 1$ H are connected in series with a source $v_i(t)$. Consider the output the voltage across the capacitor $v_o(t)$.

(a) Use integrators and adders to implement the differential equation that relates the input $v_i(t)$ and the output $v_o(t)$ of the circuit.

(b) Obtain a negative-feedback system block diagram with input $V_i(s)$ and output $V_0(s)$. Determine the feedforward transfer function $G(s)$ and the feedback transfer function $H(s)$ of the feedback system.

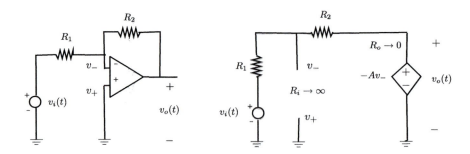

FIGURE 6.21

Problem 6.2.

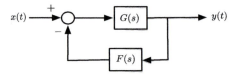

FIGURE 6.22

Problem 6.5.

(c) Find an equation for the error $E(s)$ and determine the steady-state error when the input is a unit-step signal, i.e., $V_i(s) = 1/s$.

Answers: $G(s) = 1/s$ in feedforward loop and $H(s) = s^2 + 1$ in feedback loop.

6.4 Suppose you would like to obtain a feedback implementation of an all-pass system with transfer function

$$T(s) = \frac{s^2 - 2\sqrt{2}s + 1}{s^2 + 2\sqrt{2}s + 1}.$$

(a) Determine the feedforward transfer function $G(s)$ and the feedback transfer function $H(s)$ of a negative-feedback system that has $T(s)$ as its overall transfer function.

(b) Would it be possible to implement $T(s)$ using a positive-feedback system? If so, indicate the feedforward transfer function $G(s)$ and the feedback transfer function $H(s)$ for it.

Answer: Negative feedback: let $H(s) = 1$ and $G(s) = (s^2 - \sqrt{2}s + 1)/(2\sqrt{2}s)$.

6.5 Consider the following problems connected with the feedback system shown in Fig. 6.22.

(a) The transfer function of the plant in Fig. 6.22 is $G(s) = 1/(s(s+1))$. If we want the impulse response of the feedback system to be

$$h(t) = e^{-t} \sin(\sqrt{3}t)u(t),$$

find the value α in the feedback transfer function $F(s) = s + \alpha$ that would give such an impulse response.

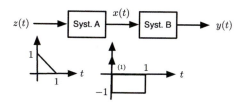

FIGURE 6.23

Problem 6.6.

(b) In the feedback system shown in Fig. 6.22, we know that when there is no feedback (i.e., $F(s) = 0$), for input $x(t) = e^{-t}u(t)$ the output is $y(t) = \cos(t)u(t)$.

    i. Find $G(s)$ and determine if it corresponds to a BIBO stable plant or not.

    ii. Find $F(s)$ so that the output is $y(t) = e^{-t}u(t)$ when $x(t) = \cos(t)u(t)$, i.e., we reverse the open-loop transfer function.

    iii. Is the system with transfer function $F(s)$ BIBO stable?

    iv. Find the impulse response $h(t)$ of the overall system. Is the overall system BIBO stable?

Answer: (a) $\alpha = 4$; (b) $G(s) = s(s+1)/(s^2+1)$; the system corresponding to $F(s)$ is unstable.

6.6 Consider the cascade of two continuous-time systems shown in Fig. 6.23. The input–output characterization of system A is $x(t) = dz(t)/dt$. It is known that system B is linear and time-invariant, and that when $x(t) = \delta(t)$ its output is $y(t) = e^{-2t}u(t)$, and that when the input is $x(t) = u(t)$ the output is $y(t) = 0.5(1 - e^{-2t})u(t)$.
If $z(t)$ is

$$z(t) = \begin{cases} 1 - t & 0 \le t \le 1, \\ 0 & \text{otherwise,} \end{cases}$$

(a) use the given information as regards system B to calculate the output $y(t)$ corresponding to $x(t)$ found above;

(b) find the differential equation characterizing the overall system with input $z(t)$ and output $y(t)$.

Answers: (a) A: $x(t) = \delta(t) - [u(t) - u(t-1)]$; (b) $x(t) = \delta(t)$ then $H(s) = Y(s)/X(s) = 1/(s+2)$.

6.7 The feedback system shown in Fig. 6.24 has two inputs: the conventional input $x(t) = e^{-t}u(t)$ and a disturbance input $v(t) = (1 - e^{-t})u(t)$.

(a) Find the transfer functions $Y(s)/X(s)$ and $Y(s)/V(s)$.

(b) Determine the output $y(t)$.

Answer: When $x(t) = 0$ then $Y(s)/V(s) = s(s+1)/(s^2+s+1)$.

6.8 To explore the performance of a proportional-plus-derivative controller on a second-order system, let $G_p(s) = 1/(s(s+1))$ be the transfer function of the plant and $G_c(s) = K_1 + K_2 s$ be the controller.

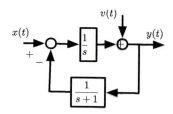

FIGURE 6.24

Problem 6.7.

(a) Find the transfer function $H(s) = Y(s)/X(s)$ of a negative-feedback system with $G_c(s)$ and $G_p(s)$ in the feedforward path and unity in the feedback. The input and the output of the feedback system are $x(t)$ and $y(t)$ with Laplace transforms $X(s)$ and $Y(s)$.

(b) Suppose (i) $K_1 = K_2 = 1$; (ii) $K_2 = 0$, $K_1 = 1$, and (iii) $K_1 = 1$, $K_2 = 0$. For each of these cases indicate the new location of the poles and zeros, and the corresponding steady-state responses due to $x(t) = u(t)$. Comparing the absolute value of the steady-state error $|\varepsilon(t)| = |y_{ss}(t) - 1|$ for the three cases, which of (i)–(iii) gives the largest error?

Answers: $H(s) = (K_1 + K_2 s)/(s^2 + s(1 + K_2) + K_1)$; (ii) gives the largest error.

6.9 Let the transfer function of a system be

$$H(s) = \frac{Y(s)}{X(s)} = \frac{b_0 + b_1 s}{a_0 + a_1 s + s^2}.$$

Show that by defining the state variables as

$$v_1(t) = y(t), \quad v_2(t) = \dot{y}(t) + a_1 y(t) - b_1 x(t)$$

we obtain a minimal state-variable and output realization (i.e., only two integrators are needed). Show the matrices/vectors for the state and the output equations. Draw a block diagram for the realization.

Answers: The matrices for the realization are

$$\mathbf{A} = \begin{bmatrix} -a_1 & 1 \\ -a_0 & 0 \end{bmatrix}, \quad \mathbf{b} = \begin{bmatrix} b_1 \\ b_0 \end{bmatrix}, \quad \mathbf{c}^T = \begin{bmatrix} 1 & 0 \end{bmatrix}.$$

6.10 You are given a state-variable realization of a second-order system having the following matrix and vectors:

$$\mathbf{A}_c = \begin{bmatrix} -a_1 & -a_0 \\ 1 & 0 \end{bmatrix}, \quad \mathbf{b}_c = \begin{bmatrix} 1 \\ 0 \end{bmatrix}, \quad \mathbf{c}_c^T = \begin{bmatrix} b_1 & b_0 \end{bmatrix}.$$

(a) Find an invertible matrix $\mathbf{F}$ that can be used to transform the given state and output equations into state and output equations with matrix and vectors:

$$\mathbf{A}_o = \mathbf{A}_c^T, \quad \mathbf{b}_o = \mathbf{c}_c, \quad \mathbf{c}_o^T = \mathbf{b}_c^T.$$

(b) Assume that

$$\mathbf{A}_c = \begin{bmatrix} 1 & 2 \\ 1 & 0 \end{bmatrix}, \quad \mathbf{b}_c = \begin{bmatrix} 1 \\ 0 \end{bmatrix}, \quad \mathbf{c}_c^T = \begin{bmatrix} 1 & -2 \end{bmatrix};$$

is there an invertible matrix $\mathbf{F}$ that would permit one to change the given realization into a realization with matrices $[\mathbf{A}_o, \mathbf{b}_o, \mathbf{c}_o]$? Explain.
Answers: $f_{11} = b_1$, $f_{12} = f_{21} = b_0$.

6.6.2 PROBLEMS USING MATLAB

6.11 Feedback error—Control systems attempt to follow the reference signal at the input, but in many cases they cannot follow particular types of inputs. Let the system we are trying to control have a transfer function $G(s)$, and the feedback transfer function be $H(s)$. If $X(s)$ is the Laplace transform of the reference input signal, $Y(s)$ the Laplace transform of the output, and

$$G(s) = \frac{1}{s(s+1)(s+2)} \quad \text{and} \quad H(s) = 1.$$

(a) Find an expression for $E(s)$ in terms of $X(s)$, $G(s)$ and $H(s)$.
(b) Let $x(t) = u(t)$, and the Laplace transform of the corresponding error be $E_1(s)$. Use the final value property of the Laplace transform to obtain the steady-state error e_{1ss}.
(c) Let $x(t) = tu(t)$, i.e., a ramp signal, and $E_2(s)$ be the Laplace transform of the corresponding error signal. Use the final value property of the Laplace transform to obtain the steady-state error e_{2ss}. Is this error value larger than the one above? Which of the two inputs $u(t)$ and $r(t)$ is easier to follow?
(d) Use MATLAB to find the partial fraction expansions of $E_1(s)$ and $E_2(s)$ and use them to find $e_1(t)$ and $e_2(t)$. Plot them.
Answers: $E_1(s) = (s+1)(s+2)/(1+s(s+1)(s+2))$; $e_{2ss} = 2$.

6.12 State-variable representation of system with unstable mode—Part 1. A LTI system is represented by an ordinary differential equation

$$\frac{d^2 y(t)}{dt^2} + \frac{dy(t)}{dt} - 2y(t) = \frac{dx(t)}{dt} - x(t).$$

(a) Obtain the transfer function $H(s) = Y(s)/X(s) = B(s)/A(s)$ and find its poles and zeros. Is this system BIBO stable? Is there any pole–zero cancellation?
(b) Decompose $H(s)$ as $W(s)/X(s) = 1/A(s)$ and $Y(s)/W(s) = B(s)$ for an auxiliary variable $w(t)$ with $W(s)$ as its Laplace transform. Obtain a state/output realization that uses only two integrators. Call the state variable $v_1(t)$ the output of the first integrator and $v_2(t)$ the output of the second integrator. Give the matrix $\mathbf{A}_1$, and the vectors $\mathbf{b}_1$, and $\mathbf{c}_1^T$ corresponding to this realization.
(c) Draw a block diagram for this realization.
(d) Use the MATLAB's function *tf2ss* to obtain state and output realization from $H(s)$. Give the matrix $\mathbf{A}_c$, and the vectors $\mathbf{b}_c$, and $\mathbf{c}_c^T$. How do these compare to the ones obtained before?

Answers: $H(s) = 1/(s+2)$; $d^2w(t)/dt^2 + dw(t)/dt - 2w(t) = x(t)$; $y(t) = dw(t)/dt - w(t)$.

6.13 State-variable representation of system with unstable mode—Part 2. For the system in Part 1, consider the state variables

$$v_1(t) = y(t) \quad v_2(t) = \dot{y}(t) + y(t) - x(t).$$

(a) Obtain the matrix $\mathbf{A}_2$ and the vectors $\mathbf{b}_2$ and $\mathbf{c}_2^T$ for the state and the output equations that realize the ordinary differential equation in Part 1.
(b) Draw a block diagram for the state variables and output realization.
(c) How do $\mathbf{A}_2$, $\mathbf{b}_2$ and $\mathbf{c}_2^T$ obtained above compare with the ones in Part 1?
(d) In the block diagram in Part 1, change the summers into nodes, and the nodes into summers, invert the direction of all the arrows, and interchange $x(t)$ and $y(t)$, i.e., make the input and output of the previous diagram into the output and the input of the diagram in this part. This is the dual of the block diagram in Part 1. How does it compare to your block diagram obtained in item (b)? How do you explain the duality?
Answers: $\dot{v}_1(t) = v_2(t) - v_1(t) + x(t)$; $\dot{v}_2(t) = 2v_1(t) - x(t)$.

6.14 Minimal realization—Suppose that a state realization has the following matrices:

$$\mathbf{A}_o = \begin{bmatrix} -2 & 1 \\ -1 & 0 \end{bmatrix}, \quad \mathbf{b}_o = \begin{bmatrix} 1 \\ -1 \end{bmatrix}, \quad \mathbf{c}_o^T = \begin{bmatrix} 1 & 0 \end{bmatrix}.$$

Find the corresponding transfer function and verify it using the function *ss2tf*. Obtain a minimal realization of this system. Draw a block diagram of the realization.
Answer: $H(s) = (s-1)/(s+1)^2$.

FOURIER ANALYSIS IN COMMUNICATIONS AND FILTERING

CONTENTS

Sometimes the questions are complicated and the answers are simple.

Theodor Seuss Geisel, (Dr. Seuss), (1904–1991), American poet, and cartoonist

7.1 INTRODUCTION

Communications is an area of electrical engineering where the Fourier analysis applies. In this chapter, we illustrate how the frequency representation of signals and systems, as well as the concepts of modulation, bandwidth, spectrum are used in communications. A communication system consists of three

Signals and Systems Using MATLAB®. https://doi.org/10.1016/B978-0-12-814204-2.00017-X

components: **a transmitter, a channel, and a receiver**. The objective of communications is to transmit a message over a channel to a receiver. The message is a signal, for instance a voice or a music signal, typically containing low frequencies. Transmission of the message can be done over the airwaves or through a line connecting the transmitter to the receiver, or a combination of the two—constituting channels with different characteristics. Telephone communication can be done with or without wires, radio and television are wireless. Frequency and the concepts of bandwidth, spectrum and modulation developed by means of the Fourier transform are fundamental in the analysis and design of communication systems. The aim of this chapter is to serve as an introduction to problems in communication and to link them with the Fourier analyses. More depth in these topics can be found in many excellent texts in communications [17,33,32,67,74,45].

The other topic covered in this chapter is analog filter design. Filtering is a very important application of LTI systems in communications, control and signal processing. The material in this chapter is introductory and will be complemented by the design of discrete filters in Chapter 12. Important issues related to signals and systems are illustrated in the design and implementation of filters.

7.2 APPLICATION TO COMMUNICATIONS

The application of the Fourier transform in communications is clear. The representation of signals in the frequency domain, and the concept of modulation and bandwidth are basic in communications. In this section we show examples of linear (amplitude modulation or AM) as well as non-linear (frequency and phase modulation or FM and PM) modulation methods. We also consider important extensions such as quadrature amplitude modulation (QAM) and frequency division multiplexing (FDM).

Given the low-pass nature of most message signals, it is necessary to shift in frequency the spectrum of the message to avoid using a very large antenna. This is attained by modulation, which is done by changing either the magnitude $A(t)$ or the phase $\theta(t)$ of a carrier

$$A(t)\cos(2\pi f_c + \theta(t)). \tag{7.1}$$

When $A(t)$ is proportional to the message, for constant phase, we have AM. On the other hand, if we let $\theta(t)$ change with the message, keeping the amplitude constant, we then have FM or PM, which are called **angle modulations**.

Remarks

1. A wireless communication system can be visualized as the cascading of three subsystems: the transmitter, the channel, and the receiver—none of which is LTI. The low-frequency nature of the message signals requires us to use as transmitter a system that can generate a signal with much higher frequencies, and that is not possible with LTI systems (recall the eigenfunction property). Transmitters are thus typically non-linear or linear time-varying. The receiver is also not LTI. A wireless channel is typically time-varying.

2. Some communication systems use parallel connections (see quadrature amplitude modulation or QAM later in this chapter). To make it possible for several users to communicate over the same channel, a combination of parallel and cascade connections are used—see frequency division

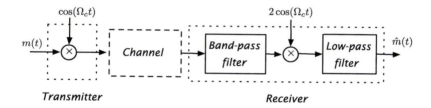

FIGURE 7.1

AM-SC transmitter, channel and receiver.

multiplexing (FDM) systems later in this chapter. But again, it should be emphasized that these subsystems are not LTI.

7.2.1 AM SUPPRESSED CARRIER (AM-SC)

Consider a message signal $m(t)$ (e.g., voice or music, or a combination of the two) modulating a cosine carrier $\cos(\Omega_c t)$ to give an amplitude modulated signal

$$s(t) = m(t)\cos(\Omega_c t). \qquad (7.2)$$

The carrier frequency $\Omega_c \gg 2\pi f_0$, where f_0 (Hz) is the maximum frequency in the message (for music f_0 is about 22 kHz and for voice about 5 kHz). The signal $s(t)$ is called **amplitude modulated with suppressed carrier or AM-SC** signal (the last part of this denomination will become clear soon). According to the modulation property of the Fourier transform, the transform of $s(t)$ is

$$S(\Omega) = \frac{1}{2}[M(\Omega - \Omega_c) + M(\Omega + \Omega_c)] \qquad (7.3)$$

where $M(\Omega)$ is the spectrum of the message. The frequency content of the message is now shifted to a much larger frequency Ω_c (rad/s) than that of the base-band signal $m(t)$. Accordingly, the antenna needed to transmit the amplitude modulated signal is of reasonable length. An AM-SC system is shown in Fig. 7.1. Notice that spectrum of the modulated signal does not include the information of the carrier, thus the name "suppressed carrier."

At the receiver, we need to first detect the desired signal among the many signals transmitted by several sources. This is possible with a tunable bandpass filter which selects the desired signal and rejects the others. Suppose that the signal obtained by the receiver, after the bandpass filtering, is exactly $s(t)$—we then need to demodulate this signal to get the original message signal $m(t)$. This is done by multiplying $s(t)$ by a cosine of exactly the same frequency of the carrier in the transmitter, i.e., Ω_c, which will give $r(t) = 2s(t)\cos(\Omega_c t)$, which again according to the modulation property has a Fourier transform

$$R(\Omega) = S(\Omega - \Omega_c) + S(\Omega + \Omega_c) = M(\Omega) + \frac{1}{2}[M(\Omega - 2\Omega_c) + M(\Omega + 2\Omega_c)]. \qquad (7.4)$$

The spectrum of the message, $M(\Omega)$, is obtained by passing the received signal $r(t)$ through a low-pass filter that rejects the other terms $M(\Omega \pm 2\Omega_c)$. In this ideal situation where we assumed that we received the modulated signal $s(t)$ and that no interference from other transmitters or channel noise are present, the obtained signal is the sent message $m(t)$.

The above is a simplification of the actual processing of the received signal. Besides the many other transmitted signals that the receiver encounters, there is channel noise caused from equipment in the transmission path as well as interference from other signals being transmitted around the carrier frequency. Thus the desired modulated message, signal interferences and the channel noise are picked up by the bandpass filter and a perfect recovery of $m(t)$ is not possible. Furthermore, the sent signal has no indication of the carrier frequency Ω_c—which is suppressed in the sent signal—and so the receiver needs to guess it and any deviation would give errors.

Remarks

1. The AM-SC transmitter is linear but time-varying. AM-SC is thus called a linear modulation. The fact that the modulated signal displays frequencies much higher than those in the message indicates the transmitter is not LTI—otherwise it would satisfy the eigenfunction property, giving a signal with a subset of the frequencies at the input.

2. A more general characterization than $\Omega_c >> 2\pi f_0$, where f_0 is the largest frequency in the message, is given by $\Omega_c >> BW$, where BW (rad/s) is the **bandwidth** of the message. You probably recall the definition of bandwidth of filters used in circuit theory. In communications there are several possible definitions for bandwidth. The **bandwidth** of a signal is the width of the range of positive frequencies for which some measure of the spectral content is satisfied. For instance, two possible definitions are:

   • **The half-power or 3-dB bandwidth** is the width of the range of positive frequencies where a peak value at zero or infinite frequency (low-pass and high-pass signals) or at a center frequency (bandpass signals) is attenuated to 0.707 the value at the peak. This corresponds to the frequencies for which the power at dc, infinity or at a center frequency reduces to half.
   • **The null-to-null bandwidth** determines the width of the range of positive frequencies of the spectrum of a signal that has a main lobe containing a significant part of the energy of the signal. If a low-pass signal has a clearly defined maximum frequency, then the bandwidth are frequencies from 0 to the maximum frequency, and if the signal is bandpass and has a minimum and a maximum frequency, its bandwidth is the maximum minus the minimum frequency.

3. In AM-SC demodulation it is important to know exactly the carrier frequency. Any small deviation would cause errors when recovering the message. Suppose, for instance, that there is a small error in the carrier frequency, i.e., instead of Ω_c the demodulator uses $\Omega_c + \Delta$, $\Delta > 0$, so that the demodulated received signal is

$$\tilde{r}(t) = s(t)\cos((\Omega_c + \Delta)t)$$

with Fourier transform

$$\begin{aligned}
\tilde{R}(\Omega) &= S(\Omega - \Omega_c - \Delta) + S(\Omega + \Omega_c + \Delta) \\
&= \frac{1}{2}[M(\Omega + \Delta) + M(\Omega - \Delta)] + \frac{1}{2}\left[M(\Omega - 2(\Omega_c + \Delta/2)) + M(\Omega + 2(\Omega_c + \Delta/2))\right].
\end{aligned}$$

Thus the low-pass filtered signal will not be the message even in the ideal case when no channel noise or interference is present.

7.2.2 COMMERCIAL AM

In commercial broadcasting, the carrier is added to the AM signal so that information on the carrier is available at the receiver helping in the identification of the radio station. For demodulation, however, such information is not important as commercial AM uses **envelope detectors** to obtain the message. By making the envelope of the modulated signal look like the message, detecting this envelope is all that is needed. Thus the commercial AM signal is of the form

$$s(t) = [K + m(t)]\cos(\Omega_c t)$$

where the AM modulation index K is chosen so that $K + m(t) > 0$ for all values of t so that the envelope of $s(t)$ is proportional to the message $m(t)$. The Fourier transform of the sent signal $s(t)$ is given by

$$S(\Omega) = K\pi\left[\delta(\Omega - \Omega_c) + \delta(\Omega + \Omega_c)\right] + \frac{1}{2}\left[M(\Omega - \Omega_c) + M(\Omega + \Omega_c)\right],$$

where the carrier with amplitude K appears at $\pm\Omega_c$ and the spectrum of the modulated message appears as in AM-SC. The **envelope detector receiver** determines the message by finding the envelope of the received signal.

Remarks

1. The advantage of adding the carrier to the message, which allows the use of a simple envelope detector, comes at the expense of increasing the power in the transmitted signal compared to the power of the suppressed-carrier AM.
2. The demodulation in commercial AM is called **non-coherent**. **Coherent demodulation** consists in multiplying the received signal with a sinusoid of the same frequency and phase of the carrier. A **local oscillator** generates this sinusoid.
3. A disadvantage of commercial as well as suppressed carrier AM is the doubling of the bandwidth of the transmitted signal compared to the bandwidth of the message. Given the symmetry of the spectrum, in magnitude as well as in phase, it is clear that it is not necessary to send the upper and the lower sidebands of the spectrum to get back the signal in the demodulation. It is thus possible to have **upper- and lower-sideband AM** modulations which are more efficient in spectrum utilization.
4. Most AM receivers use the **superheterodyne receiver technique** developed by Fessenden and Armstrong.[1]

Example 7.1 (Simulating Amplitude Modulation with MATLAB). For simulations, MATLAB provides different data files such as *train.mat* used here. Suppose the message $y(t)$ is the train signal and we wish to use it to modulate a cosine $\cos(\Omega_c t)$ to create an amplitude modulated signal $z(t)$. Because

---

[1] Reginald Fessenden was the first to suggest the heterodyne principle: mixing the radio-frequency signal using a local oscillator of different frequency, resulting in a signal that could drive the diaphragm of an earpiece at an audio frequency. Mixing a 101 kHz input with 100 kHz generated by the receiver gives a frequency of 1 kHz, in the audible range. Fessenden could not make a practical success of the heterodyne receiver, which was accomplished by Edwin H. Armstrong in the 1920s using electron tubes.

the *train* signal is given in a sampled form, the simulation requires discrete-time processing (you might want to come back to this when you have learned the discrete processing) but the results approximate well the analog AM.

```
%%
% Example 7.1---Discrete-time simulation of AM
%%
clear all; clf; load train
y1=y'; N=length(y1); t1=0:1/Fs:(N-1)/Fs;
figure(1)
plot(t1,y1); grid; xlabel('t'); ylabel('y(t)'); title('message')
fmax=Fs/2;              % bandwith of train signal
fc=5*fmax;              % carrier frequency (rad/s)
fs=2*(fc+fmax);         % sampling frequency of modulated signal
wc=2*pi*fc; L=ceil(fs/Fs);
y=interp(y1,L);
N1=1024*2; n=0:N1-1; t2=0:1/fs:(N1-1)/fs; K=1*abs(min(y(1:N1)));
% Modulation
z=(y(1:N1)+K).*cos(wc*t2);
figure(2)
subplot(211)
plot(t2,y(1:N1));grid;ylabel('y(t)')
axis([0 max(t2) -1 1])
subplot(212)
plot(t2,z,':r');hold on
plot(t2,y(1:N1)+K,'k');hold on
plot(t2,-y(1:N1)-K,'k');grid;hold off
axis([0 max(t2) -2 2])
xlabel('t');ylabel('z(t)')
figure(3)
w=0:N1-1;w=w*2*pi/N1;w=w-pi;W=w*fs;
subplot(211)
plot(W/(2*pi),abs(fftshift(fft(y(1:N1)))));grid;xlabel('f (Hz)');
ylabel('|Y(\Omega)|')
subplot(212)
plot(W/(2*pi),abs(fftshift(fft(z))));grid ;xlabel('f (Hz)');
ylabel('|Z(\Omega)|')
```

The carrier frequency is chosen to be $f_c = 20.48$ kHz. For the envelope detector to work at the transmitter we add a constant K to the message to make $y(t) + K > 0$. The envelope of the AM modulated signal should resemble the message. The AM signal is thus

$$z(t) = [K + y(t)]\cos(\Omega_c t) \qquad \Omega_c = 2\pi f_c.$$

In Fig. 7.2 we show the train signal, a segment of the signal and the corresponding modulated signal displaying the envelope, as well as the Fourier transform of the segment and of its modulated version.

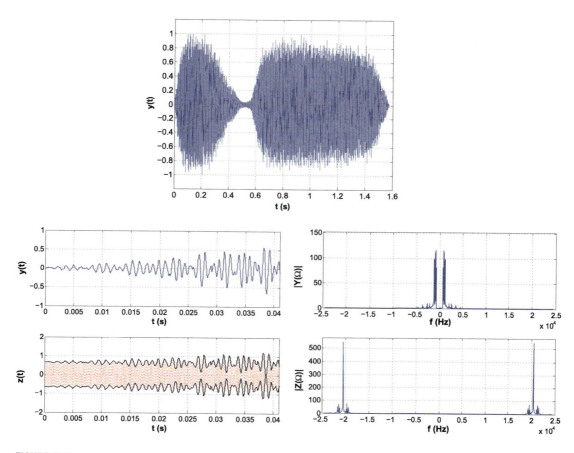

FIGURE 7.2

Commercial AM modulation: original signal (top), part of original signal and corresponding AM modulated signal (bottom left), spectrum of the original signal and of the modulated signal (top and bottom right).

Notice that the envelope resembles the original signal. Also from the spectrum of the train signal its bandwidth is about 5 kHz, while the spectrum of the modulated signal display the frequency-shifted spectrum of the signal with a bandwidth of about 10 kHz. Notice also the large peak at $\pm f_c$ corresponding to the carrier. $\square$

7.2.3 AM SINGLE SIDEBAND

The message $m(t)$ is typically a real-valued signal that, as indicated before, has a symmetric spectrum, i.e., the magnitude and the phase of the Fourier transform $M(\Omega)$ are even and odd functions of frequency. When using AM modulation the resulting spectrum has redundant information by providing the **upper and the lower sidebands** of $M(\Omega)$. To reduce the bandwidth of the transmitted signal, we could get rid of either the upper or the lower sideband of the AM signal by means of a bandpass filter

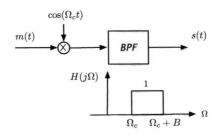

FIGURE 7.3

Upper sideband AM transmitter. Ω_c is the carrier frequency and B the bandwidth of the message.

(see Fig. 7.3). The resulting modulation is called **AM single sideband (AM-SSB)** (upper or lower sideband depending on which of the two sidebands is kept). This type of modulation is used whenever the quality of the received signal is not as important as the advantages of having a narrow band and less noise in the frequency band of the received signal. AM-SSB is used by amateur radio operators.

As shown in Fig. 7.3, the upper sideband modulated signal $s(t)$ is obtained by bandpass filtering $m(t)\cos(\Omega_c t)$. At the receiver, the received signal is filtered with a bandpass filter and the resulting output is then demodulated like in an AM-SC system. To obtain the sent message, $m(t)$, the output of the demodulator is low-pass filtered using as cutoff frequency for the low-pass filter the bandwidth of the message.

7.2.4 QUADRATURE AM AND FREQUENCY DIVISION MULTIPLEXING

Quadrature amplitude modulation (QAM) and frequency division multiplexing (FDM) are the precursors of many of the new communication systems. QAM and FDM are of great interest for their efficient use of the radio spectrum.

Quadrature Amplitude Modulation (QAM)

QAM enables two AM-SC signals to be transmitted over the same frequency band-conserving bandwidth. The messages can be separated at the receiver. This is accomplished by using two orthogonal carriers, such as a cosine and a sine. See Fig. 7.4. The QAM modulated signal is

$$s(t) = m_1(t)\cos(\Omega_c t) + m_2(t)\sin(\Omega_c t), \tag{7.5}$$

where $m_1(t)$ and $m_2(t)$ are the messages. You can think of $s(t)$ as having a phasor that is the sum of two phasors perpendicular to each other (the cosine leading the sine by $\pi/2$), indeed

$$s(t) = \mathcal{R}e[(m_1(t)e^{j0} + m_2(t)e^{-j\pi/2})e^{j\Omega_c t}]$$

has a phasor $m_1(t)e^{j0} + m_2(t)e^{-j\pi/2}$, which is the sum of two perpendicular phasors. Since

$$m_1(t)e^{j0} + m_2(t)e^{-j\pi/2} = m_1(t) - jm_2(t)$$

we could interpret the QAM signal as the result of amplitude modulating the real and the imaginary parts of a complex message $m(t) = m_1(t) - jm_2(t)$.

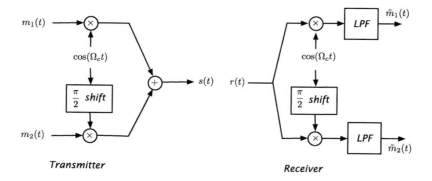

FIGURE 7.4

QAM transmitter and receiver: $s(t)$ is the transmitted signal and $r(t)$ the received signal.

To simplify the computation of the spectrum of $s(t)$, let us consider the message $m(t) = m_1(t) - jm_2(t)$, i.e., a complex message, with spectrum $M(\Omega) = M_1(\Omega) - jM_2(\Omega)$ so that

$$s(t) = \mathcal{R}e[m(t)e^{j\Omega_c t}] = 0.5[m(t)e^{j\Omega_c t} + m^*(t)e^{-j\Omega_c t}]$$

where $*$ stands for complex conjugate. The spectrum of $s(t)$ is then given by

$$
\begin{aligned}
S(\Omega) &= 0.5[M(\Omega - \Omega_c) + M^*(\Omega + \Omega_c)] \\
&= 0.5[M_1(\Omega - \Omega_c) - jM_2(\Omega - \Omega_c) + M_1(\Omega + \Omega_c) + jM_2(\Omega + \Omega_c)]
\end{aligned}
$$

where the superposition of the spectra of the two messages is clearly seen. At the receiver, if we multiply the received signal—for simplicity assume it to be $s(t)$—by $\cos(\Omega_c t)$ we get

$$
\begin{aligned}
r_1(t) &= s(t)\cos(\Omega_c t) = 0.25[m(t)e^{j\Omega_c t} + m^*(t)e^{-j\Omega_c t}][e^{j\Omega_c t} + e^{-j\Omega_c t}] \\
&= 0.25[m(t) + m^*(t)] + 0.25[m(t)e^{j2\Omega_c t} + m^*(t)e^{-j2\Omega_c t}],
\end{aligned}
$$

which, when passed through a low-pass filter, with the appropriate bandwidth, gives

$$
\begin{aligned}
0.25[m(t) + m^*(t)] &= 0.25[m_1(t) - jm_2(t) + m_1(t) + jm_2(t)] \\
&= 0.5m_1(t).
\end{aligned}
$$

Likewise, to get the second message we multiply $s(t)$ by $\sin(\Omega_c t)$ and pass the resulting signal through a low-pass filter.

Frequency Division Multiplexing (FDM)

FDM implements sharing of the radio spectrum by several users by allocating a specific frequency band to each. One could, for instance, think of the commercial AM or FM locally as an FDM system. In telephony, using a bank of filters it is possible to get several users in the same system—it is, however, necessary to have a similar system at the receiver to have a two way communication.

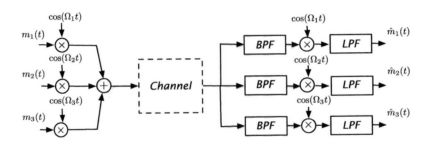

FIGURE 7.5

Frequency division multiplexing (FDM) system.

To illustrate an FDM system (Fig. 7.5), consider we have a set of messages of known finite bandwidth (we could low-pass filter the messages to satisfy this condition) we wish to transmit. Each of the messages modulates a different carrier so that the modulated signals are in different frequency bands without interfering with each other (if needed, a frequency guard could be used to insure this). These frequency-multiplexed modulated messages can now be transmitted. At the receiver, a bank of bandpass filters—centered at the carrier frequencies used at the transmitter—and followed by appropriate demodulators recover the different messages. Any of the AM modulation techniques could be used in the FDM system.

7.2.5 ANGLE MODULATION

Amplitude modulation is said to be linear modulation, because it behaves like a linear system. Frequency and phase- or angle-modulation systems on the other hand are non-linear. Interest in angle modulation is due to the decreasing effect of noise or interferences when compared with AM, although at the cost of a much wider bandwidth and complexity in implementation. The non-linear behavior of angle modulation systems makes their analyses more difficult than that for AM. Computing the spectrum of an FM or PM signal is much harder than that of an AM signal. In the following we consider the case of the so-called **narrow-band FM** where we are able to find its spectrum directly.

Professor Edwin H. Armstrong developed the first successful frequency modulation system— narrow-band FM.[2] If $m(t)$ is the message signal, and we modulate a carrier signal of frequency Ω_c (rad/s), the transmitted signal $s(t)$ in angle modulation is of the form

$$s(t) = A\cos(\Omega_c t + \theta(t)) \tag{7.6}$$

where the angle $\theta(t)$ depends on the message $m(t)$. In the case of **phase modulation**, the angle function is proportional to the message $m(t)$, i.e.,

$$\theta(t) = K_f\, m(t) \tag{7.7}$$

---

[2]Edwind H. Armstrong (1890–1954), Professor of electrical engineering at Columbia University, and inventor of some of the basic electronic circuits underlying all modern radio, radar, and television, was born in New York. His inventions and developments form the backbone of Radio Communications as we know it.

where $K_f > 0$ is called the **modulation index**. If the angle is such that

$$\frac{d\theta(t)}{dt} = \Delta\Omega \, m(t) \tag{7.8}$$

this relation defines **frequency modulation**. The **instantaneous frequency**, as a function of time, is the derivative of the argument of the cosine or

$$IF(t) = \frac{d[\Omega_c t + \theta(t)]}{dt} = \Omega_c + \frac{d\theta(t)}{dt} = \Omega_c + \Delta\Omega \, m(t), \tag{7.9}$$

indicating how the frequency is changing with time. For instance, if $\theta(t)$ is a constant—so that the carrier is just a sinusoid of frequency Ω_c and constant phase θ—the instantaneous frequency is simply Ω_c. The term $\Delta\Omega \, m(t)$ relates to the spreading of the frequency about Ω_c. Thus the **modulation paradox** Professor E. Craig proposed in his book [18]:

> *In* amplitude *modulation the bandwidth depends on the* frequency *of the message, while in* frequency *modulation the bandwidth depends on the* amplitude *of the message.*

Thus the modulated signals are

$$\text{PM:} \quad s_{PM}(t) = \cos(\Omega_c t + K_f m(t)), \tag{7.10}$$

$$\text{FM:} \quad s_{FM}(t) = \cos(\Omega_c t + \Delta\Omega \int_{-\infty}^{t} m(\tau)d\tau). \tag{7.11}$$

Narrow-band FM. In this case the angle $\theta(t)$ is small, so that $\cos(\theta(t)) \approx 1$ and $\sin(\theta(t)) \approx \theta(t)$, simplifying the spectrum of the transmitted signal:

$$\begin{aligned} S(\Omega) &= \mathcal{F}[\cos(\Omega_c t + \theta(t))] = \mathcal{F}[\cos(\Omega_c t)\cos(\theta(t)) - \sin(\Omega_c t)\sin(\theta(t))] \\ &\approx \mathcal{F}[\cos(\Omega_c t) - \sin(\Omega_c t)\theta(t)]. \end{aligned} \tag{7.12}$$

Using the spectrum of a cosine and the modulation theorem we get

$$S(\Omega) \approx \pi[\delta(\Omega - \Omega_c) + \delta(\Omega + \Omega_c)] + \frac{1}{2j}[\Theta(\Omega - \Omega_c) - \Theta(\Omega + \Omega_c)] \tag{7.13}$$

where the $\Theta(\Omega)$ is the spectrum of the angle which is found from (7.8) to be (using the derivative property of the Fourier transform):

$$\Theta(\Omega) = \frac{\Delta\Omega}{j\Omega} M(\Omega) \tag{7.14}$$

giving a spectrum similar to an AM system with carrier:

$$S(\Omega) \approx \pi[\delta(\Omega - \Omega_c) + \delta(\Omega + \Omega_c)] - \frac{\Delta\Omega}{2}\left[\frac{M(\Omega - \Omega_c)}{\Omega - \Omega_c} - \frac{M(\Omega + \Omega_c)}{\Omega + \Omega_c}\right]. \tag{7.15}$$

If the angle $\theta(t)$ is not small, we have **wide-band FM** and its spectrum is more difficult to obtain.

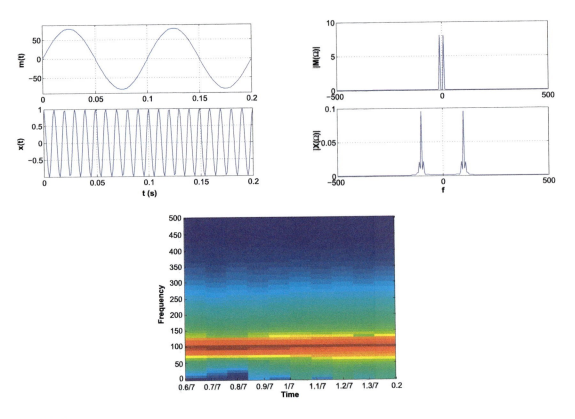

FIGURE 7.6

Narrow-band frequency modulation: top left—message $m(t)$ and narrow-band FM signal $x(t)$; top right—magnitude spectra of $m(t)$ and $x(t)$. Spectrogram of $x(t)$ displaying evolution of its Fourier transform with respect to time.

Example 7.2 (Simulation of FM Modulation with MATLAB). In these simulations we will concern ourselves with the results and leave the discussion of the code for the next chapter since the signals are approximated by discrete-time signals. For the narrow-band FM we consider the sinusoidal message

$$m(t) = 80\sin(20\pi t)u(t),$$

and a sinusoidal carrier of frequency $f_c = 100$ Hz, so that the FM signal is

$$x(t) = \cos(2\pi f_c t + 0.1\pi \int_{-\infty}^{t} m(\tau)d\tau).$$

Fig. 7.6 shows on the top left the message and the narrow-band FM signal $x(t)$, and on the top right their corresponding magnitude spectra $|M(\Omega)|$ and $|X(\Omega)|$. The narrow-band FM has only shifted the frequency of the message. The instantaneous frequency (the derivative of the argument of the cosine) is

$$IF(t) = 2\pi f_c + 0.1\pi m(t) = 200\pi + 8\pi \sin(20\pi t) \approx 200\pi,$$

i.e., it remains almost constant for all times. For the narrow-band FM, the spectrum of the modulated signal remains the same for all times. To illustrate this we compute the spectrogram of $x(t)$. Simply, the spectrogram can be thought of as the computation of the Fourier transform as the signal evolves with time. See bottom plot in Fig. 7.6.

To illustrate the wide-band FM, we consider two messages

$$m_1(t) = 80 \sin(20\pi t) u(t),$$
$$m_2(t) = 2000 t u(t),$$

giving FM signals

$$x_i(t) = \cos(2\pi f_{ci} t + 50\pi \int_{-\infty}^{t} m_i(\tau) d\tau) \qquad i = 1, 2$$

where $f_{c1} = 2500$ Hz and $f_{c2} = 25$ Hz. In this case the instantaneous frequency is

$$IF_i(t) = 2\pi f_{ci} + 50\pi m_i(t) \qquad i = 1, 2.$$

These instantaneous frequencies are not "almost constant" as before. The frequency of the carrier is now continuously changing with time. For instance, for the ramp message the instantaneous frequency is

$$IF_2(t) = 50\pi + 10^5 t\pi$$

so that for a small time interval $[0, 0.1]$ we get a chirp (sinusoid with time-varying frequency) as shown in Fig. 7.7. This figure displays the messages, the FM signals and their corresponding magnitude spectra and their spectrograms. These FM signals are broadband, occupying a band of frequencies much larger that the messages, and their spectrograms shows their spectra change with time—a joint time and frequency characterization. □

7.3 ANALOG FILTERING

The basic idea of filtering is to get rid of frequency components in a signal that are not desirable. Application of filtering can be found in control, in communications and in signal processing. In this section we provide a short introduction to the design of analog filters. Chapter 12 will be dedicated to the design of discrete filters and to some degree that chapter will be based on the material in this section.

According to the eigenfunction property of LTI systems (Fig. 7.8) the steady-state response of a LTI system to a sinusoidal input—with a certain magnitude, frequency and phase—is a sinusoid of the same frequency as the input, but with magnitude and phase affected by the response of the system at the input frequency. Since periodic as well as aperiodic signals have Fourier representations consisting of sinusoids of different frequencies, the frequency components of any signal can be modified by

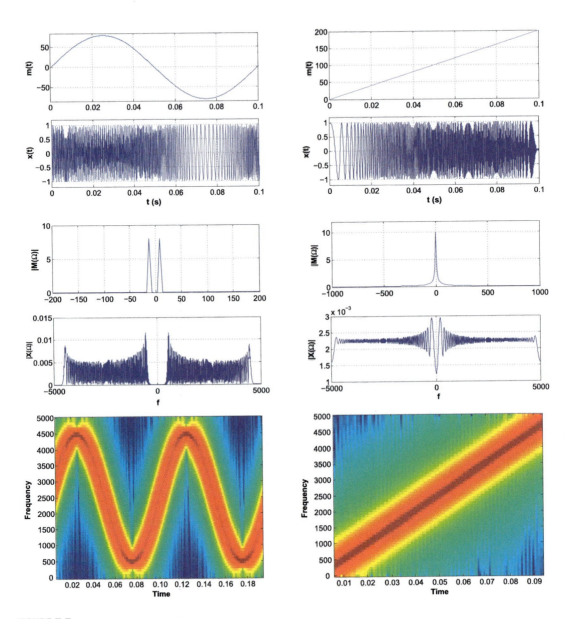

FIGURE 7.7

Wide-band frequency modulation, from top to bottom, on the left for the sinusoidal message and on the right for the ramp message: messages, FM modulated signals, spectra of messages, spectra of FM signals, and spectrograms of FM signals.

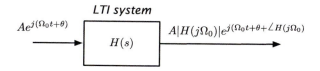

LTI system

$Ae^{j(\Omega_0 t+\theta)}$ → $H(s)$ → $A|H(j\Omega_0)|e^{j(\Omega_0 t+\theta+\angle H(j\Omega_0))}$

FIGURE 7.8

Eigenfunction property of continuous LTI systems.

appropriately choosing the frequency response of the LTI system, or filter. Filtering can thus be seen as a way of changing the frequency content of an input signal.

The appropriate filter for a certain application is specified using the spectral characterization of the input and the desired spectral characteristics of the output. Once the specifications of the filter are set, the problem becomes one of approximation as a ratio of polynomials in s. The classical approach in filter design is to consider low-pass prototypes, with normalized frequency and magnitude responses, which may be transformed into other filters with the desired frequency response. Thus many efforts are made for designing low-pass filters and developing frequency transformations to map low-pass filters into other types of filters. Using cascade and parallel connections of filters also provides a way to obtain different types of filters.

The resulting filter should be causal, stable and have real-valued coefficients so that it can be used in real-time applications and realized as a passive or an active filter. Resistors, capacitors, and inductors are used in the realization of passive filters, while resistors, capacitors, and operational amplifiers are used in active filter realizations.

7.3.1 FILTERING BASICS

A filter with transfer function $H(s) = B(s)/A(s)$ is an LTI system having a specific frequency response. The convolution property of the Fourier transform entails that the output of the filter $y(t)$ has Fourier transform

$$Y(\Omega) = X(\Omega)H(j\Omega) \tag{7.16}$$

where the frequency response of the system is obtained from the transfer function $H(s)$ as

$$H(j\Omega) = H(s)|_{s=j\Omega}$$

and $X(\Omega)$ is the Fourier transform of the input $x(t)$. Thus the frequency content of the input, represented by the Fourier transform $X(\Omega)$, is changed by the frequency response $H(j\Omega)$ of the filter so that the output signal with spectrum $Y(\Omega)$ only has desirable frequency components.

Magnitude Squared Function

The magnitude squared function of an analog low-pass filter has the general form

$$|H(j\Omega)|^2 = \frac{1}{1 + f(\Omega^2)} \tag{7.17}$$

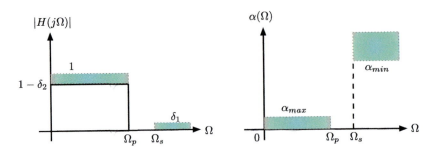

FIGURE 7.9

Magnitude specifications for a low-pass filter.

where for low frequencies $f(\Omega^2) << 1$ so that $|H(j\Omega)|^2 \approx 1$, and for high frequencies $f(\Omega^2) >> 1$ so that $|H(j\Omega)|^2 \to 0$. Accordingly, there are two important issues to consider:

1. selection of the appropriate function $f(.)$, and
2. the factorization needed to get $H(s)$ from the magnitude squared function.

Filter Specifications

Although an ideal low-pass filter is not realizable (recall the Paley–Wiener criteria) its magnitude response can be used as prototype for specifying low-pass filters. Thus the desired magnitude is specified as

$$1 - \delta_2 \leq |H(j\Omega)| \leq 1 \qquad 0 \leq \Omega \leq \Omega_p \quad \text{(passband)},$$
$$0 \leq |H(j\Omega)| \leq \delta_1 \qquad \Omega \geq \Omega_s \quad \text{(stopband)}, \tag{7.18}$$

for some small values δ_1 and δ_2. No specification is given in the transition region $\Omega_p < \Omega < \Omega_s$. Also the phase is not specified, although we wish it to be linear at least in the passband. See the figure on the left in Fig. 7.9.

To simplify the computation of the filter parameters, and to provide a scale that has more resolution and physiological significance than the specifications given above, the magnitude specifications are expressed in a logarithmic scale. Defining the loss function (in decibels, or dB) as

$$\alpha(\Omega) = -10\log_{10}|H(j\Omega)|^2 = -20\log_{10}|H(j\Omega)| \quad \text{dB} \tag{7.19}$$

an equivalent set of specifications to those in (7.18) is

$$0 \leq \alpha(\Omega) \leq \alpha_{max} \qquad 0 \leq \Omega \leq \Omega_p \quad \text{(passband)},$$
$$\alpha(\Omega) \geq \alpha_{min} \qquad \Omega \geq \Omega_s \quad \text{(stopband)}, \tag{7.20}$$

where $\alpha_{max} = -20\log_{10}(1 - \delta_2)$ and $\alpha_{min} = -20\log_{10}(\delta_1)$, both of which are positive given that $1 - \delta_2$ and δ_1 are less than one. See the figure on the right in Fig. 7.9.

In the above specifications, the dc loss was 0 dB corresponding to a normalized dc gain of 1. In more general cases, $\alpha(0) \neq 0$ and the loss specifications are given as $\alpha(0) = \alpha_1$, α_2 in the passband and α_3 in the stopband. To normalize these specifications we need to subtract α_1, so that the loss specifications are

$$\alpha(0) = \alpha_1 \quad \text{dc loss,}$$

$$\alpha_{max} = \alpha_2 - \alpha_1 \quad \text{maximum attenuation in passband,}$$

$$\alpha_{min} = \alpha_3 - \alpha_1 \quad \text{minimum attenuation in stopband.}$$

Thus, using $\{\alpha_{max}, \Omega_p, \alpha_{min}, \Omega_s\}$ we proceed to design a magnitude-normalized filter (having a unit dc gain) and then use α_1 to achieve the desired dc gain.

The design problem is then: Given the magnitude specifications in the passband ($\alpha(0)$, α_{max} and Ω_p) and in the stopband (α_{min} and Ω_s) we then:

1. Choose the rational approximation method (e.g., Butterworth, Chebyshev).
2. Solve for the parameters of the filter that satisfy the given specifications.
3. Factorize the magnitude squared function and choose the poles on the left-hand of a normalized S-plane (guaranteeing the filter stability) to obtain the transfer function $H_N(S)$ of the designed filter. Choose a gain K so that $K H_N(S)$ satisfies the dc gain constraint. Denormalize the designed filter by letting $S = s/\Omega_n$ ($\Omega_n = \Omega_{hp}$ in the Butterworth case, $\Omega_n = \Omega_p$ for the Chebyshev).
4. Verify that $K H_N(j\Omega)$ satisfies the given magnitude specifications and that its phase is approximately linear in the passband.

7.3.2 BUTTERWORTH LOW-PASS FILTER DESIGN

The magnitude squared approximation of an Nth-order low-pass Butterworth filter is

$$|H_N(j\Omega')|^2 = \frac{1}{1 + [\Omega']^{2N}} \qquad \Omega' = \frac{\Omega}{\Omega_{hp}} \tag{7.21}$$

where Ω_{hp} is the half-power or -3 dB frequency. This frequency response is normalized with respect to the half-power frequency (i.e., the normalized frequency is $\Omega' = \Omega/\Omega_{hp}$) and normalized in magnitude as the dc gain is $|H(j0)| = 1$. It is then said that the magnitude squared function is **normalized with respect to frequency** (giving a unity half-power frequency) and **normalized in magnitude** (giving a unity dc gain). As the order N increases and we choose Ω_p closer to Ω_s, i.e., the transition band is closer to zero, the rational approximation gets closer to an ideal low-pass filter.

Remarks

1. The half-power frequency is called the -3 dB frequency because in the case of the low-pass filter $G(s)$ with dc gain of 1, or $G(j0) = 1$, by definition at the half-power frequency Ω_{hp} the magnitude square function is

$$|G(j\Omega_{hp})|^2 = \frac{G|(j0)|^2}{2} = \frac{1}{2}. \tag{7.22}$$

In the logarithmic scale we have

$$10\log_{10}(|G(j\Omega_{hp})|^2) = -10\log_{10}(2) \approx -3 \text{ (dB)},\tag{7.23}$$

corresponding to a gain of -3 dB or a loss of 3 dB; the loss function goes from 0 dB at the dc frequency to 3 dB at the half-power frequency.

2. It is important to understand the advantages of the **frequency and magnitude normalizations** typical in filter design. Having a low-pass filter with normalized magnitude, its dc gain is 1, if one desires a filter with a dc gain $K \neq 1$ it can be easily obtained by multiplying the magnitude-normalized filter by the constant K. Likewise, a filter $H(S)$ designed with a normalized frequency, say $\Omega' = \Omega/\Omega_{hp}$ so that the normalized half-power frequency is 1, is easily converted into a denormalized filter $H(s)$ with a desired Ω_{hp} by replacing $S = s/\Omega_{hp}$ in $H(S)$.

Factorization

To obtain the transfer function $H_N(s)$ of a Butterworth low-pass filter that satisfies the specifications we need to factorize the magnitude squared function (7.21). By letting $S = s/\Omega_{hp}$ be the normalized Laplace variable then $S/j = \Omega' = \Omega/\Omega_{hp}$ and

$$H_N(S)H_N(-S) = \frac{1}{1 + (-S^2)^N}.$$

If the denominator can be factorized as

$$D(S)D(-S) = 1 + (-S^2)^N\tag{7.24}$$

and if there are no poles in the $j\Omega'$-axis we let $H_N(S) = 1/D(S)$, i.e., we assign to $H_N(S)$ the poles in the left-hand s-plane so that the resulting filter is stable. The $2N$ poles S_k of $H_N(S)H_N(-S)$ make $1 + (-S_k^2)^N = 0$, or $(-1)^N S_k^{2N} = -1$, and can be obtained as follows:

$$(-1)^N S_k^{2N} = e^{j(2k-1)\pi} \ \Rightarrow \ S_k^{2N} = \frac{e^{j(2k-1)\pi}}{e^{-j\pi N}} = e^{j(2k-1+N)\pi} \text{ for integers } k = 1, \cdots, 2N$$

where we let $-1 = e^{j(2k-1)\pi}$ and $(-1)^N = e^{-j\pi N}$. The $2N$ roots are then

$$S_k = e^{j(2k-1+N)\pi/(2N)} \qquad k = 1, \cdots, 2N.\tag{7.25}$$

Remarks

1. Since $|S_k| = 1$, the poles of the Butterworth filter are on a circle of unit radius symmetrically distributed around it. Because the complex poles should be complex conjugate pairs, the poles are symmetrically distributed with respect to the $j\Omega'$ and the σ' axes. Letting $S = s/\Omega_{hp}$ be the normalized Laplace variable, then $s = S\Omega_{hp}$, the denormalized filter $H(s)$ has its poles in a circle of radius Ω_{hp}.

2. No poles are on the $j\Omega'$-axis, or equivalently the angle of the poles are not equal to $\pi/2$ or $3\pi/2$. In fact, for $1 \leq k \leq N$ the angle of the poles are bounded below and above by

$$\frac{\pi}{2}\left(1 + \frac{1}{N}\right) \leq \frac{(2k - 1 + N)\pi}{2N} \leq \frac{\pi}{2}\left(3 - \frac{1}{N}\right)$$

and for integers $N \geq 1$ the above indicates that the angle will not equal either $\pi/2$ or $3\pi/2$.

3. Consecutive poles are separated by π/N rad from each other. Indeed, subtracting the angles of two consecutive poles gives $\pm \pi/N$.

Using the above remarks and the fact that the poles must be in conjugate pairs, since the coefficients of the filter are real-valued, it is easy to determine the location of the poles geometrically.

Example 7.3. A second-order low-pass Butterworth filter, normalized in magnitude and in frequency, has a transfer function

$$H(S) = \frac{1}{S^2 + \sqrt{2}S + 1}.$$

We would like to obtain a new filter $H(s)$ with a dc gain of 10 and a half-power frequency $\Omega_{hp} = 100$ rad/s.

The dc gain of $H(S)$ is unity, in fact when $\Omega = 0$, $S = j0$ gives $H(j0) = 1$. The half-power frequency of $H(S)$ is unity, indeed letting $\Omega' = 1$ then $S = j1$ and

$$H(j1) = \left[\frac{1}{j^2 + j\sqrt{2} + 1}\right] = \frac{1}{j\sqrt{2}}$$

so that $|H(j1)|^2 = |H(j0)|^2/2 = 1/2$, or $\Omega' = 1$ is the half-power frequency.

Thus, the desired filter with a dc gain of 10 is obtained by multiplying $H(S)$ by 10. Furthermore, if we let $S = s/100$ be the normalized Laplace variable when $S = j\Omega'_{hp} = j1$ we get $s = j\Omega_{hp} = j100$, or $\Omega_{hp} = 100$, the desired half-power frequency. Thus the denormalized-in-frequency filter $H(s)$ is obtained by replacing $S = s/100$. The denormalized filter, in magnitude and frequency, is then

$$H(s) = \frac{10}{(s/100)^2 + \sqrt{2}(s/100) + 1} = \frac{10^5}{s^2 + 100\sqrt{2}s + 10^4}. \qquad \square$$

Filter Design

For the Butterworth low-pass filter, the design consists in finding the parameters N, the minimum order, and Ω_{hp}, the half-power frequency, of the filter from the constraints in the passband and in the stopband. The loss function for the low-pass Butterworth is

$$\alpha(\Omega) = -10\log_{10}|H_N(\Omega/\Omega_{hp})|^2 = 10\log_{10}(1 + (\Omega/\Omega_{hp})^{2N})$$

and the loss specifications are

$$0 \leq \alpha(\Omega) \leq \alpha_{max}, \qquad 0 \leq \Omega \leq \Omega_p,$$
$$\alpha_{min} \leq \alpha(\Omega) < \infty, \qquad \Omega \geq \Omega_s.$$

At $\Omega = \Omega_p$ we see that the loss is

$$\alpha(\Omega_p) = 10\log_{10}(1 + (\Omega_p/\Omega_{hp})^{2N}) \leq \alpha_{max} \quad \text{so that} \quad \frac{\Omega_p}{\Omega_{hp}} \leq \left(10^{0.1\alpha_{max}} - 1\right)^{1/2N}. \tag{7.26}$$

Similarly for $\Omega = \Omega_s$ we see that the loss is

$$\alpha(\Omega_s) = 10\log_{10}(1 + (\Omega_s/\Omega_{hp})^{2N}) \geq \alpha_{min} \quad \text{so that} \quad \frac{\Omega_s}{\Omega_{hp}} \geq (10^{0.1\alpha_{min}} - 1)^{1/2N}. \tag{7.27}$$

Then from Equations (7.26) and (7.27) the half-power frequency is in the range

$$\frac{\Omega_p}{(10^{0.1\alpha_{max}} - 1)^{1/2N}} \leq \Omega_{hp} \leq \frac{\Omega_s}{(10^{0.1\alpha_{min}} - 1)^{1/2N}} \tag{7.28}$$

and from the log of the two extremes of Equation (7.28) we have

$$N \geq \frac{\log_{10}[(10^{0.1\alpha_{min}} - 1)/(10^{0.1\alpha_{max}} - 1)]}{2\log_{10}(\Omega_s/\Omega_p)}. \tag{7.29}$$

Remarks

- According to Equation (7.29) when either
  - the transition band is narrowed, i.e., $\Omega_p \to \Omega_s$, or
  - the loss α_{min} is increased, or
  - the loss α_{max} is decreased

  the quality of the filter is improved at the cost of having to implement a filter with a higher order N.
- The minimum order N is an integer larger or equal to the right-hand side of Equation (7.29). Any integer larger than the chosen N also satisfies the specifications but increases the complexity of the filter.
- Notice that because of the log function equation (7.29) is equivalent to

$$N \geq \frac{\log_{10}[(10^{0.1\alpha_{max}} - 1)/(10^{0.1\alpha_{min}} - 1)]}{2\log_{10}(\Omega_p/\Omega_s)}. \tag{7.30}$$

  Notice also that it is the ratio Ω_p/Ω_s, which is important in determining the minimum order, rather than the actual values of these frequencies.
- Although there is a range of possible values for the half-power frequency, it is typical to make the frequency response coincide with either the passband or the stopband specifications giving a value for the half-power frequency in the range. Thus, we can have either

$$\Omega_{hp} = \frac{\Omega_p}{(10^{0.1\alpha_{max}} - 1)^{1/2N}} \quad \text{or} \quad \Omega_{hp} = \frac{\Omega_s}{(10^{0.1\alpha_{min}} - 1)^{1/2N}} \tag{7.31}$$

  as possible values for the half-power frequency. In the first case the loss plot is moved up—improving the losses in the stopband—while in the second case the loss plot is moved down—improving the losses in the passband.

- The design aspect is clearly seen in the flexibility given by the equations. We can select out of an infinite possible set of values for N and Ω_{hp}. However, the optimal order, in terms of complexity, is the minimal order N and the corresponding Ω_{hp} as given by (7.31).
- After the factorization, or after obtaining $D(S)$ from the poles, we need to denormalize the obtained transfer function $H_N(S) = 1/D(S)$, by letting $S = s/\Omega_{hp}$ to obtain $H_N(s) = 1/D(s/\Omega_{hp})$, the filter that satisfies the given specifications. If the desired dc gain is not unit, the filter needs to be denormalized in magnitude by multiplying it by an appropriate gain K to get $H_N(s) = K/D(s/\Omega_{hp})$.

7.3.3 CHEBYSHEV LOW-PASS FILTER DESIGN

The normalized magnitude squared function for the Chebyshev low-pass filter is given by

$$|H_N(j\Omega')|^2 = \frac{1}{1 + \varepsilon^2 C_N^2(\Omega')}, \qquad \Omega' = \frac{\Omega}{\Omega_p}, \tag{7.32}$$

where the frequency is normalized with respect to the passband frequency Ω_p, N stands for the order of the filter, ε is a ripple factor, and $C_N(.)$ are the orthogonal **Chebyshev polynomials of the first kind** defined as

$$C_N(\Omega') = \begin{cases} \cos(N \cos^{-1}(\Omega')) & |\Omega'| \leq 1, \\ \cosh(N \cosh^{-1}(\Omega')) & |\Omega'| > 1. \end{cases} \tag{7.33}$$

The definition of the Chebyshev polynomials depends on whether the value of $|\Omega'|$ is less or equal to 1, or bigger than 1. Indeed, whenever $|\Omega'| > 1$ the definition based in the cosine is not possible since the inverse would not exist, thus the cosh(.) definition is used. Likewise whenever $|\Omega'| \leq 1$ the definition based in the hyperbolic cosine would not be possible since the inverse of this function only exists for values of Ω' bigger or equal to 1 and so the cos(.) definition is used. However, from this definition it is not clear that $C_N(\Omega')$ is an N-order polynomial in Ω', but it is. Indeed, if we let $\theta = \cos^{-1}(\Omega')$ when $|\Omega'| \leq 1$ we have $C_N(\Omega') = \cos(N\theta)$ and

$$\begin{aligned} C_{N+1}(\Omega') &= \cos((N+1)\theta) = \cos(N\theta)\cos(\theta) - \sin(N\theta)\sin(\theta), \\ C_{N-1}(\Omega') &= \cos((N-1)\theta) = \cos(N\theta)\cos(\theta) + \sin(N\theta)\sin(\theta), \end{aligned}$$

so that adding them we get

$$C_{N+1}(\Omega') + C_{N-1}(\Omega') = 2\cos(\theta)\cos(N\theta) = 2\Omega' C_N(\Omega').$$

This gives a three term expression, or a difference equation, for computing $C_N(\Omega')$:

$$C_{N+1}(\Omega') + C_{N-1}(\Omega') = 2\Omega' C_N(\Omega') \qquad N \geq 0 \tag{7.34}$$

with initial conditions

$$\begin{aligned} C_0(\Omega') &= \cos(0) = 1, \\ C_1(\Omega') &= \cos(\cos^{-1}(\Omega')) = \Omega'. \end{aligned}$$

We can then see that

$$C_0(\Omega') = 1, \quad C_1(\Omega') = \Omega', \quad C_2(\Omega') = -1 + 2\Omega'^2, \quad C_3(\Omega') = -3\Omega' + 4\Omega'^3, \quad \cdots, \tag{7.35}$$

which are polynomials in Ω' of order $N = 0, 1, 2, 3, \cdots$. In Chapter 0 we gave a script to compute and plot these polynomials using symbolic MATLAB.

Remarks

- Two fundamental characteristics of the $C_N(\Omega')$ polynomials are: (i) they vary between -1 and 1 in the range $\Omega' \in [-1, 1]$, and (ii) they grow to infinity outside this range (according to their definition, the Chebyshev polynomials outside $\Omega' \in [-1, 1]$ are defined as cosh(.) functions which are functions always bigger than 1). The first characteristic generates ripples in the passband, while the second makes these filters have a magnitude response that goes to zero faster than the Butterworth filter.
- The Chebyshev polynomials are unity at $\Omega' = 1$, i.e., $C_N(1) = 1$ for all N. In fact, according to (7.35) we have $C_0(1) = 1$, $C_1(1) = 1$ and if we assume that $C_{N-1}(1) = C_N(1) = 1$ we then have $C_{N+1}(1) = 1$. This indicates that the magnitude square function $|H_N(j1)|^2 = 1/(1 + \varepsilon^2)$ for any N.
- Different from the Butterworth filter that has a unit dc gain, the dc gain of the Chebyshev filter depends on the order of the filter. Noticing that the Chebyshev polynomials of odd-order do not have a constant term, while those of even order have 1 or -1 as the constant term we have

$$|C_N(0)| = \begin{cases} 0 & N \text{ is odd,} \\ 1 & N \text{ is even.} \end{cases}$$

Thus the dc gain is

$$|H_N(j0)| = \begin{cases} 1 & N \text{ odd,} \\ 1/\sqrt{1 + \varepsilon^2} & N \text{ even.} \end{cases} \tag{7.36}$$

- Related to the ripples in the passband, because the polynomials $C_N(\Omega')$ have N real roots between -1 and 1 there are $N/2$ ripples between 1 and $\sqrt{1 + \varepsilon^2}$ for normalized frequencies between 0 and 1.

Filter Design

The loss function for the Chebyshev filter is

$$\alpha(\Omega') = 10 \log_{10}\left[1 + \varepsilon^2 C_N^2(\Omega')\right] \qquad \Omega' = \frac{\Omega}{\Omega_p}. \tag{7.37}$$

The design equations for the Chebyshev filter are obtained as follows:

- Ripple factor ε and ripple width (RW): From $C_N(1) = 1$, and letting the loss equal α_{max} at $\Omega' = 1$ we obtain

$$\varepsilon = \sqrt{10^{0.1\alpha_{max}} - 1}, \quad RW = 1 - \frac{1}{\sqrt{1 + \varepsilon^2}}. \tag{7.38}$$

- Minimum order: The loss function at Ω'_s is bigger than or equal to α_{min} so that solving for the Chebyshev polynomial we get after replacing ε

$$C_N(\Omega'_s) = \cosh(N\cosh^{-1}(\Omega'_s)) \geq \left(\frac{10^{.1\alpha_{min}} - 1}{10^{.1\alpha_{max}} - 1}\right)^{0.5}$$

where we used the cosh(.) definition of the Chebyshev polynomials since $\Omega'_s > 1$. Solving for N we get

$$N \geq \frac{\cosh^{-1}\left(\left[\frac{10^{0.1\alpha_{min}} - 1}{10^{0.1\alpha_{max}} - 1}\right]^{0.5}\right)}{\cosh^{-1}(\frac{\Omega_s}{\Omega_p})}. \tag{7.39}$$

- Half-power frequency: Letting the loss at the half-power frequency equal 3 dB, we have

$$\alpha(\Omega_{hp}) = 10\log_{10}(1 + \varepsilon^2 C_N^2(\Omega'_{hp})) = 3 \text{ dB}, \quad \text{then} \quad 1 + \varepsilon^2 C_N^2(\Omega'_{hp}) = 10^{0.3} \approx 2$$

or

$$C_N(\Omega'_{hp}) = \frac{1}{\varepsilon} = \cosh(N\cosh^{-1}(\Omega'_{hp}))$$

where the last term is the definition of the Chebyshev polynomial for $\Omega'_{hp} > 1$. Thus we get

$$\Omega_{hp} = \Omega_p \cosh\left[\frac{1}{N}\cosh^{-1}\left(\frac{1}{\varepsilon}\right)\right]. \tag{7.40}$$

Factorization

The factorization of the magnitude squared function is a lot more complicated for the Chebyshev filter than for the Butterworth filter. If we replace $\Omega' = S/j$, where the normalized variable $S = s/\Omega_p$, the magnitude squared function can be written as

$$H(S)H(-S) = \frac{1}{1 + \varepsilon^2 C_N^2(S/j)} = \frac{1}{D(S)D(-S)}.$$

As in the Butterworth case, the poles in the left-hand S-plane give $H(S) = 1/D(S)$, a stable filter.

The poles of $H(S)$ can be found to be in an ellipse. They can be connected with the poles of the corresponding order Butterworth filter by an algorithm due to Professor Ernst Guillemin. The poles of $H(S)$ are given by the following equations for $k = 1, \cdots, N$, with N the minimal order of the filter:

$$a = \frac{1}{N}\sinh^{-1}\left(\frac{1}{\varepsilon}\right),$$

$$\sigma_k = -\sinh(a)\cos(\psi_k) \qquad \text{real part of pole,}$$

$$\Omega'_k = \pm\cosh(a)\sin(\psi_k) \qquad \text{imaginary part of pole,} \tag{7.41}$$

where $0 \leq \psi_k < \pi/2$ are the angles corresponding to the Butterworth filters (measured with respect to the negative real axis of the S plane).

Remarks

- Although the dc gain of the Chebyshev filter is not unity as in the Butterworth filter (it depends on the order N) we can, however, set the desired dc value by choosing the appropriate value of a gain K so that $\hat{H}(S) = K/D(S)$ satisfies the dc gain specification.
- The poles of the Chebyshev filter depend now on the ripple factor ε and so there is no simple way to find them geometrically as it was in the case of the Butterworth.
- The final step is to replace the normalized variable $S = s/\Omega_p$ in $H(S)$ to get the desired filter $H(s)$.

Example 7.4. Consider the low-pass filtering of an analog signal $x(t) = [-2\cos(5t) + \cos(10t) + 4\sin(20t)]u(t)$ with MATLAB. The filter is a third-order low-pass Butterworth filter with a half-power frequency $\Omega_{hp} = 5$ rad/s, i.e., we wish to attenuate the frequency components of frequency 10 and 20 rad/s.

The design of the filter is done using the MATLAB function *butter* for which besides the specification of the desired order, $N = 3$, and half-power frequency $\Omega_{hp} = 5$ rad/s, we need to indicate that the filter is analog by including an 's' as one of the arguments. Once the coefficients of the filter are obtained, we could then either solve the ordinary differential equation from these coefficients, or use the Fourier transform, which we choose to do, to find the filter output. Symbolic MATLAB is thus used to compute the Fourier transform of the input, $X(\Omega)$, and after generating the frequency response function $H(j\Omega)$ from the filter coefficients, we multiply these two to get $Y(\Omega)$, which is inversely transformed to obtain $y(t)$. To obtain $H(j\Omega)$ symbolically we multiply the coefficients of the numerator and denominator obtained from *butter* by variables $(j\Omega)^n$ where n corresponds to the order of the coefficient in the numerator or the denominator, and then add them.

The poles of the designed filter and its magnitude response are shown in Fig. 7.10, as well as the input $x(t)$ and the output $y(t)$. The following script was used for the filter design and the filtering of the given signal. □

```
%%
% Example 7.4---Filtering with Butterworth filter
%%
clear all; clf
syms t w
x=cos(10*t)-2*cos(5*t)+4*sin(20*t);     % input signal
X=fourier(x);
N=3; Whp=5;                             % filter parameters
[b,a]=butter(N,Whp,'s');               % filter design
W=0:0.01:30; Hm=abs(freqs(b,a,W));     % magnitude response in W
% filter output
n=N:-1:0;  U=(j*w).^n
num=b*conj(U'); den=a*conj(U');
H=num/den;                             % frequency response
Y=X*H;                                 % convolution property
y=ifourier(Y,t);                       % inverse Fourier
```

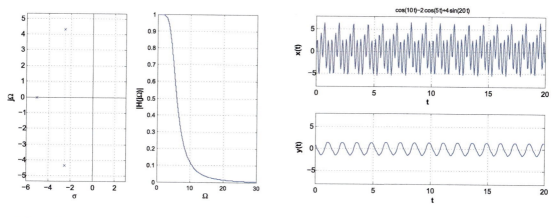

FIGURE 7.10

Filtering of an analog signal $x(t) = [-2\cos(5t) + \cos(10t) + 4\sin(20t)]u(t)$, shown on top right figure, using a low-pass Butterworth filter with poles and magnitude response shown on the left. The filtered signal, bottom right, is approximately the low-frequency component of $x(t)$.

Example 7.5. In this example we compare the performance of Butterworth and Chebyshev low-pass filters in the filtering of an analog signal $x(t) = [-2\cos(5t) + \cos(10t) + 4\sin(20t)]u(t)$ using MATLAB. We would like the two filters to have the same half-power frequency.

The magnitude specifications for the low-pass Butterworth filter are

$$\alpha_{max} = 0.1 \text{ dB}, \quad \Omega_p = 5 \text{ rad/s},$$
$$\alpha_{min} = 15 \text{ dB}, \quad \Omega_s = 10 \text{ rad/s},$$

and a dc loss of 0 dB. Once this filter is designed, we would like the Chebyshev filter to have the same half-power frequency as the Butterworth filter. To obtain this, we change the Ω_p specification for the Chebyshev filter by using the formulas for the half-power frequency of this type of filter to find the new value for Ω_p.

The Butterworth filter is designed by first determining the minimum order N and the half-power frequency Ω_{hp} that satisfy the specifications using the MATLAB function *buttord*, and then finding the filter coefficients by means of *butter*. Likewise, for the design of the Chebyshev filter we use the function *cheb1ord* to find the minimum order and the cutoff frequency (which coincides with the specified new Ω_p), and *cheby1*[3] to determine the filter coefficients. The filtering is implemented using the Fourier transform.

There are two significant differences between the designed Butterworth and Chebyshev filters. Although both of them have the same half-power frequency, the transition band of the Chebyshev filter is

---

[3]MATLAB provides two types of Chebyshev-based filter functions: type I (corresponding to the functions *cheb1ord*, *cheby1*) and type II (corresponding to *cheb2ord*, *cheby2*). Type I filters are equiripple in the passband and monotonic in the stopband, while type II filters are the opposite.

narrower, [6.88 10], than that of the Butterworth filter, [5 10], and the Chebyshev filter has a minimum order $N = 5$ compared to the sixth-order Butterworth. Fig. 7.11 displays the poles of the Butterworth and the Chebyshev filters, their magnitude responses, as well as the input signal $x(t)$ and the output $y(t)$ for the two filters (the two perform very similarly). ☐

```
%%
% Example 7.5---Filtering with Butterworth and Chebyshev filters
%%
clear all;clf
syms t w
x=cos(10*t)-2*cos(5*t)+4*sin(20*t); X=fourier(x);
wp=5;ws=10;alphamax=0.1;alphamin=15;        % filter parameters
% butterworth filter
[N,whp]=buttord(wp,ws,alphamax,alphamin,'s')
[b,a]=butter(N,whp,'s')
% cheby1 filter
epsi=sqrt(10^(alphamax/10) -1)
wp=whp/cosh(acosh(1/epsi)/N)                 % recomputing wp to get same whp
[N1,wn]=cheb1ord(wp,ws,alphamax,alphamin,'s');
[b1,a1]=cheby1(N1,alphamax,wn,'s');
% frequency responses
W=0:0.01:30;
Hm=abs(freqs(b,a,W));
Hm1=abs(freqs(b1,a1,W));
% generation of frequency response from coefficients
n=N:-1:0;  n1=N1:-1:0;
U=(j*w).^n;  U1=(j*w).^n1
num=b*conj(U'); den=a*conj(U');
num1=b1*conj(U1'); den1=a1*conj(U1')
H=num/den;                                   % Butterworth LPF
H1=num1/den1;                                % Chebyshev LPF
% output of filter
Y=X*H;
Y1=X*H1;
y=ifourier(Y,t)
y1=ifourier(Y1,t)
```

7.3.4 FREQUENCY TRANSFORMATIONS

As indicated before, the design of analog filters is typically done by transforming the frequency of a normalized prototype low-pass filter. The frequency transformations were developed by Professor Ronald Foster using the properties of reactance functions. The frequency transformations for the basic

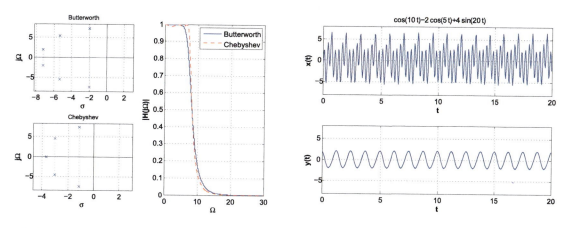

FIGURE 7.11

Comparison of filtering of an analog signal $x(t) = [-2\cos(5t) + \cos(10t) + 4\sin(20t)]u(t)$ using a low-pass Butterworth and Chebyshev filters with the same half-power frequency.

filters are given by

$$\text{Low-pass–Low-pass} \qquad S = \frac{s}{\Omega_0},$$

$$\text{Low-pass–High-pass} \qquad S = \frac{\Omega_0}{s},$$

$$\text{Low-pass–Bandpass} \qquad S = \frac{s^2 + \Omega_0}{s\,BW},$$

$$\text{Low-pass–Bandstop} \qquad S = \frac{s\,BW}{s^2 + \Omega_0}, \qquad (7.42)$$

where S is the normalized and s the final variables, while Ω_0 is a desired cutoff frequency and BW a desired bandwidth.

Remarks

1. The low-pass to low-pass (LP-LP) and low-pass to high-pass (LP-HP) transformations are linear in the numerator and denominator, thus the order of the numerator and the denominator of the transformed filter is that of the prototype. On the other hand, the low-pass to bandpass (LP-BP) and low-pass to bandstop (LP-BS) transformations are quadratic in either the numerator or the denominator, so that the order of the numerator and the denominator of the transformed filter is double that of the prototype. Thus to obtain a $2N$th-order bandpass or bandstop filter the prototype low-pass filter should be of order N. This is an important observation useful in the design of these filters with MATLAB.

2. It is important to realize that only frequencies are transformed, the magnitude of the prototype filter is preserved. Frequency transformations will be useful also in the design of discrete filters, where

these transformations are obtained in a completely different way as no reactance functions would be available in that domain.

Example 7.6. To illustrate how the above transformations can be used to convert a prototype low-pass filter we use the following script. First a low-pass prototype filter is designed using *butter*, and then to this filter we apply the different transformations with $\Omega_0 = 40$ and $BW = 10$. The magnitude responses are plotted with *ezplot*. Fig. 7.12 shows the results. □

```
%%
% Example 7.6---Frequency transformations
%%
   clear all; clf
   syms w
   N=5; [b,a]=butter(N,1,'s')              % low-pass prototype
   omega0=40;BW=10;                        % transformation parameters
   % low-pass prototype
   n=N:-1:0;
   U=(j*w).^n;  num=b*conj(U'); den=a*conj(U');
   H=num/den;
   % low-pass to high-pass
   U1=(omega0/(j*w)).^n;
   num1=b*conj(U1'); den1=a*conj(U1');
   H1=num1/den1;
   % low-pass to bandpass
   U2=((-w^2+omega0)/(BW*j*w)).^n
   num2=b*conj(U2'); den2=a*conj(U2');
   H2=num2/den2;
   % low-pass to band-eliminating
   U3=((BW*j*w)/(-w^2+omega0)).^n
   num3=b*conj(U3'); den3=a*conj(U3');
   H3=num3/den3
```

7.3.5 FILTER DESIGN WITH MATLAB

The design of filters, analog and discrete, is simplified by the functions that MATLAB provides. Functions to find the filter parameters from magnitude specifications, as well as functions to find the filter coefficients, its poles/zeros and to plot its magnitude and phase responses are available.

Low-Pass Filter Design

The design procedure is similar for all of the approximation methods (Butterworth, Chebyshev, elliptic) and consists of

* finding the filter parameters from loss specifications, and then
* obtaining the filter coefficients from these parameters.

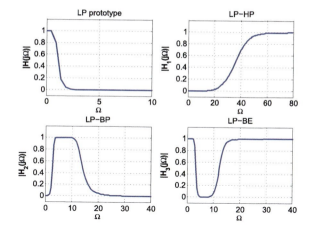

FIGURE 7.12

Frequency transformations: (top left) prototype low-pass filter, (top right) low-pass to high-pass transformation, (bottom left) low-pass to bandpass transformation, (bottom right) low-pass to band-eliminating transformation.

Thus to design an analog low-pass filter using the Butterworth approximation the loss specifications α_{max} and α_{min}, and the frequency specifications, Ω_p and Ω_s are first used by the function *buttord* to determine the minimum order N and the half-power frequency Ω_{hp} of the filter that satisfies the specifications. Then the function *butter* uses these two values to determine the coefficients of the numerator and the denominator of the designed filter. We can then use the function *freqs* to plot the designed filter magnitude and phase. Similarly, for the design of low-pass filters using the Chebyshev or the elliptic design methods. To include the design of low-pass filters using the Butterworth, Chebyshev (two types) and the elliptic methods we wrote the function *analogfil*.

```
function [b,a]=analogfil(Wp,Ws,alphamax,alphamin,Wmax,ind)
%%
% Analog filter design
% Parameters
% Input: loss specifications (alphamax, alphamin), corresponding
%             frequencies (Wp,Ws), frequency range [0,Wmax] and
%             indicator ind (1 for Butterworth, 2 for Chebyshev1, 3
%             for Chebyshev2 and 4 for elliptic).
% Output: coefficients of designed filter.
% Function plots magnitude, phase responses, poles and zeros of filter,
% and loss specifications
%%
if ind==1,           % Butterworth low-pass
    [N,Wn]=buttord(Wp,Ws,alphamax,alphamin, 's')
    [b,a]=butter(N,Wn,'s')
elseif ind==2,       % Chebyshev low-pass
```

```
        [N,Wn]=cheb1ord(Wp,Ws,alphamax, alphamin,'s')
        [b,a]=cheby1(N,alphamax,Wn,'s')
elseif ind==3,          % Chebyshev2 low-pass
        [N,Wn]=cheb2ord(Wp,Ws,alphamax, alphamin,'s')
        [b,a]=cheby2(N,alphamin,Wn,'s')
else                    % Elliptic low-pass
        [N,Wn]=ellipord(Wp,Ws,alphamax,alphamin,'s')
        [b,a]=ellip(N,alphamax,alphamin,Wn,'s')
end
W=0:0.001:Wmax;                             % frequency range
H=freqs(b,a,W); Hm=abs(H); Ha=unwrap(angle(H))    % magnitude and phase
N=length(W); alpha1=alphamax*ones(1,N);
alpha2=alphamin*ones(1,N);                  % loss specs
subplot(221)
plot(W,Hm); grid; axis([0 Wmax 0 1.1*max(Hm)])
subplot(222)
plot(W,Ha); grid; axis([0 Wmax 1.1*min(Ha) 1.1*max(Ha)])
subplot(223)
splane(b,a)
subplot(224)
plot(W,-20*log10(abs(H))); hold on
plot(W,alpha1,'r',W,alpha2,'r'); grid;  axis([0 max(W) -0.1 100])
hold off
```

Example 7.7. To illustrate the use of *analogfil* consider the design of low-pass filters using the Chebyshev2 and the Elliptic design methods. The specification for the designs are

$$\alpha(0) = 0, \quad \alpha_{max} = 0.1, \quad \alpha_{min} = 60 \text{ dB},$$
$$\Omega_p = 10, \quad \Omega_s = 15 \text{ rad/s}.$$

We wish to find the coefficients of the designed filters, plot their magnitude and phase, and plot the loss function for each of the filters and verify that the specifications have been met. The results are shown in Fig. 7.13.

```
%%
% Example 7.7----Filter design using analogfil
%%
clear all; clf
alphamax=0.1;
alphamin=60;
Wp=10;Ws=15;
Wmax=25;
ind=4                   % elliptic design
% ind=3                 % chebyshev2 design
[b,a]=analogfil(Wp,Ws,alphamax,alphamin,Wmax,ind)
```

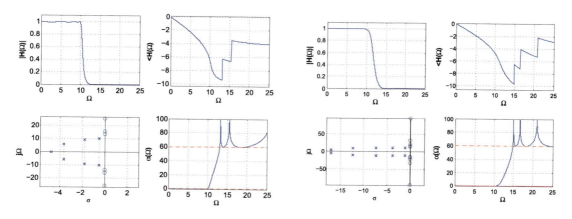

FIGURE 7.13

Elliptic (left) and Chebyshev2 (right) low-pass filter designs using *analogfil* function. Clockwise, magnitude, phase, loss function and poles and zeros are shown for each design.

The elliptic design is illustrated above. To obtain the Chebyshev2 design get rid of the comment symbol % in front of the corresponding indicator and put it in front of the one for the elliptic design. □

General comments on the design of low-pass filters using Butterworth, Chebyshev (1 and 2) and Elliptic methods are:

- The Butterworth and the Chebyshev2 designs are flat in the passband, while the others display ripples in that band.
- For identical specifications, the obtained order of the Butterworth filter is greater than the order of the other filters.
- The phase of all of these filters is approximately linear in the passband, but not outside it. Because of the rational transfer functions for these filters, it is not possible to have linear phase over all frequencies. However, the phase response is less significant in the stopband where the magnitude response is very small.
- The filter design functions provided by MATLAB can be used for analog or discrete filters. When designing an analog filter there is no constrain in the values of the frequency specifications and an 's' indicates that the filter being designed is analog.

General Filter Design

The filter design programs *butter, cheby1, cheby2, ellip* allow the design of other filters besides low-pass filters. Conceptually, a prototype low-pass filter is designed and then transformed into the desired filter by means of the frequency transformations. The filter is specified by the order and cutoff frequencies. In the case of a low-pass and a high-pass filter the specified cutoff frequency is a scalar, while for bandpass and stopband filters the specified cutoff frequencies are given as a vector. Also recall that the frequency transformations double the order of the low-pass prototype for the bandpass and band-eliminating filters, so when designing these filters half of the desired order should be given.

Example 7.8. To illustrate the general design consider

1. Using the *cheby2* function, design a bandpass filter with the following specifications:

$$\text{order } N = 20$$
$$\alpha(\Omega) = 60 \text{ dB in the stopband}$$
$$\text{passband frequencies } [10, \ 20] \text{ rad/s}$$
$$\text{unit gain in passband.}$$

2. Using the *ellip* function, design a bandstop filter with unit gain in the passbands and the following specifications:

$$\text{order } N = 10$$
$$\alpha(\Omega) = 0.1 \text{ dB in the passband}$$
$$\alpha(\Omega) = 40 \text{ dB in the stopband}$$
$$\text{passband frequencies } [10, \ 11] \text{ rad/s.}$$

The following script is used:

```
%%
% Example 7.8---General filter design
%%
clear all;clf
M=10;
%[b,a]=cheby2(M,60,[10 20],'s')         % cheby2 bandpass
[b,a]=ellip(M/2,0.1,40,[10 11],'stop','s')   % elliptic bandstop
W=0:0.01:30;
H=freqs(b,a,W);
```

Notice that the order given to *ellip* is 5 and 10 to *cheby2* since a quadratic transformation will be used to obtain the notch and the bandpass filters from a prototype low-pass filter. The magnitude and phase responses of the two designed filters are shown in Fig. 7.14. □

7.4 WHAT HAVE WE ACCOMPLISHED? WHAT IS NEXT?

In this chapter we have illustrated the application of the Fourier analysis to communications and filtering. Different from the application of the Laplace transform in control problems where transients as well as steady-state responses are of interest, in communications and filtering there is more interest in steady-state response and frequency characterization which are more appropriately treated with the Fourier transform. Different types of modulation systems are illustrated in the communication examples. Finally, this chapter provides an introduction to the design of analog filters. In all the examples, the application of MATLAB was illustrated.

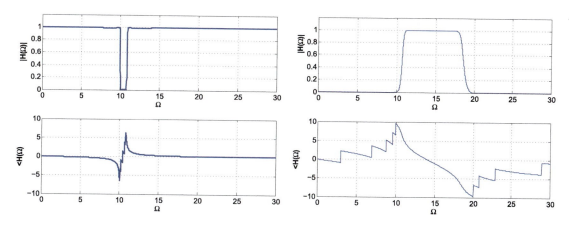

FIGURE 7.14

Design of a notch filter using *ellip* and of a bandpass filter using *cheby2*.

Although the material in this chapter does not have sufficient depth, reserved for texts in communications and filtering, it serves to connect the theory of continuous-time signals and systems with applications. In the next part of the book, we will consider how to process signals using computers and how to apply the resulting theory again in control, communication and signal processing problems.

7.5 PROBLEMS

7.5.1 BASIC PROBLEMS

7.1 A series RC circuit is connected to a voltage source $v_i(t)$, and the output is the voltage across the capacitor, $v_o(t)$.

   (a) Find the transfer function $H(s) = V_o(s)/V_i(s)$ of this filter when the resistor is $R = 10 \, k\Omega$, and the capacitor $C = 1 \, \mu F$.

   (b) To normalize this circuit we let $R_n = R/k_m$, $C_n = C(k_m k_f)$ where k_m is a magnitude scale and k_f is a frequency scale. Choose values of k_m and k_f so that the resistor and the capacitor are unity.

   (c) Find the normalized transfer function $T(S)$ of the filter with $R_n = 1$ and $C_n = 1$. Calculate the magnitude response of the normalized and of the not normalized filters, plot them together in the same figure and have two frequency scales—the actual and the normalized one. What do you notice? Explain.

   Answers: $H(s) = 1/(1 + s/100)$; $T(S) = 1/(1 + S)$ for $k_m = 10^4$ and $k_f = 10^2$.

7.2 Consider a second-order analog filter with transfer function

$$H(s) = \frac{1}{s^2 + \sqrt{2}s + 1}.$$

(a) Determine the dc gain of this filter. Plot the poles and zeros; determine the magnitude response $|H(j\Omega)|$ of this filter and carefully plot it.

(b) Find the frequency Ω_{max} at which the magnitude response of this filter is maximum.

(c) Is the half-power frequency of this filter $\Omega_{hp} = 1$?

Answers: $|H(j\Omega)|^2 = 1/((1 - \Omega^2)^2 + 2\Omega^2)$.

7.3 Consider a filter with a transfer function

$$H(s) = \frac{s/Q}{s^2 + (s/Q) + 1}.$$

(a) Determine the magnitude of this filter at $\Omega = 0$, 1 and ∞. What type of filter is it?

(b) Show that the bandwidth of this filter is $BW = \Omega_2 - \Omega_1 = 1/Q$ where the half-power frequencies Ω_1 and Ω_2 satisfy the condition $\Omega_1\Omega_2 = 1$.

Answer: Band-pass filter.

7.4 A second-order analog low-pass filter has a transfer function

$$H(s) = \frac{1}{s^2 + s/Q + 1},$$

where Q is called the quality factor of the filter.

(a) Show that the maximum of the magnitude

- for $Q < 1/\sqrt{2}$ occurs at $\Omega = 0$.
- for $Q \geq 1/\sqrt{2}$ occurs at a frequency different from zero.

(b) Show that, for $Q = 1/\sqrt{2} = 0.707$, the half-power frequency of this filter is $\Omega_{hp} = 1$ rad/s.

Answer: Maximum magnitude response occurs at frequency $\sqrt{1 - 1/(2Q^2)}$ when $Q > 0.707$.

7.5 A passive RLC filter is represented by the ordinary differential equation

$$\frac{d^2y(t)}{dt^2} + 2\frac{dy(t)}{dt} + y(t) = x(t) \qquad t \geq 0$$

where $x(t)$ is the input and $y(t)$ is the output.

(a) Find the transfer function $H(s)$ of the filter and indicate what type of filter it is.

(b) For an input $x(t) = 2u(t)$, find the corresponding output $y(t)$ and determine the steady-state response.

(c) If the input is $x_1(t) = 2[u(t) - u(t - 1)]$, express the corresponding output $y_1(t)$ in terms of the response $y(t)$ obtained above.

(d) For the above two cases find the difference between the outputs and the inputs of the filter and indicate if this difference signals tend to zero as t increases. If so, what does this mean?

Answers: Low-pass filter $H(s) = 1/(s^2 + 2s + 1)$; as $t \to \infty$, $y(t) - x(t) = -2e^{-t}(1 + t)u(t) \to 0$.

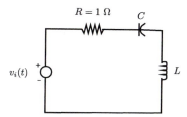

FIGURE 7.15

Problem 7.7.

7.6 The receiver of an AM system consists of a bandpass filter, a demodulator and a low-pass filter. The received signal is

$$r(t) = m(t)\cos(40{,}000\pi t) + q(t)$$

where $m(t)$ is a desired voice signal with bandwidth $BW = 5$ kHz that modulates the carrier $\cos(40{,}000\pi t)$ and $q(t)$ is the rest of the signals available at the receiver outside the band where the message is. The low-pass filter is ideal with magnitude 1 and bandwidth BW. Assume the bandpass filter is also ideal and that the demodulator is $\cos(\Omega_c t)$.

(a) What is the value of Ω_c in the demodulator?

(b) Suppose we input the received signal into the bandpass filter cascaded with the demodulator and the low-pass filter, determine the magnitude response of the bandpass filter that allows us to recover $m(t)$. Draw the overall system and indicate which of the components are LTI and which are LTV.

(c) By mistake we input the received signal into the demodulator, and the resulting signal into the cascade of the bandpass and the low-pass filters. If you use the bandpass filter obtained above, determine the recovered signal (i.e., the output of the low-pass filter). Would you get the same result regardless of what $m(t)$ is? Explain.

Answer: (c) $y(t) = 0$.

7.7 Consider the RLC circuit in Fig. 7.15 where $R = 1\ \Omega$.

(a) Determine the values of the inductor and the capacitor so that the transfer function of the circuit when the output is the voltage across the capacitor, $v_0(t)$, is

$$\frac{V_o(s)}{V_i(s)} = \frac{1}{s^2 + \sqrt{2}s + 1}.$$

(b) Find the transfer function of the circuit, with the values obtained above for the capacitor and the inductor, when the output is the voltage across the resistor. Carefully sketch the corresponding frequency response and determine the type of filter it is.

(c) What type of filter would we obtain by considering the output the voltage across the inductor and the capacitor obtained before? Find the corresponding transfer function.

Answers: $LC = 1,\ C = \sqrt{2}$.

7.5.2 PROBLEMS USING MATLAB

7.8 Wireless transmission—Consider the transmission of a sinusoid $x(t) = \cos(2\pi f_0 t)$ through a channel affected by multipath and Doppler. Let there be two paths, and assume the sinusoid is being sent from a moving object so that a Doppler frequency shift occurs. Let the received signal be

$$r(t) = \alpha_0 \cos(2\pi(f_0 - v)(t - L_0/c)) + \alpha_1 \cos(2\pi(f_0 - v)(t - L_1/c))$$

where $0 \leq \alpha_i \leq 1$, $i = 0, 1$, are attenuations, L_i are the distances from the transmitter to the receiver that the signal travels in the ith path, $c = 3 \times 10^8$ m/s and frequency shift v is caused by the Doppler effect.

(a) Let $f_0 = 2$ kHz, $v = 50$ Hz, $\alpha_0 = 1$ and $\alpha_1 = 0.9$. Let $L_0 = 10,000$ m, what would be L_1 if the two sinusoids have a phase difference of $\pi/2$?

(b) Is the received signal $r(t)$, with the parameters given above but $L_1 = 10,000$, periodic? If so what would be its period and how much does it differ from the period of the original sinusoid? If $x(t)$ is the input and $r(t)$ the output of the transmission channel, considered a system, is it linear and time invariant? Explain.

(c) Sample the signals $x(t)$ and $r(t)$ using a sampling frequency $F_s = 10$ kHz. Plot the sampled sent $x(nT_s)$ and received $r(nT_s)$ signals for $n = 0$ to 2000.

(d) Consider the situation where $f_0 = 2$ kHz, but the parameters of the paths are random trying to simulate real situations where these parameters are unpredictable—although somewhat related. Let

$$r(t) = \alpha_0 \cos(2\pi(f_0 - v)(t - L_0/c)) + \alpha_1 \cos(2\pi(f_0 - v)(t - L_1/c))$$

where $v = 50\eta$ Hz, $L_0 = 1000\eta$ and $L_1 = 10,000\eta$, $\alpha_0 = 1 - \eta$ and $\alpha_1 = \alpha_0/10$ and η is a random number between 0 and 1 with equal probability of being any of these values (this can be realized by using the *rand* MATLAB function). Generate the received signal for 10 different events, use $F_s = 10,000$ Hz as sampling rate, and plot them together to observe the effects of the multipath and Doppler. For those 10 different events find and plot the magnitude spectrum.

Answers: (a) $L_1 = 48.5 \times 10^3$ m; (b) $r(t) = 1.9\cos(2\pi \times 1950t - 2\pi \times 0.065)$.

7.9 Design of low-pass Butterworth and Chebyshev filters—The specifications for a low-pass filter are

$$\Omega_p = 1500 \text{ rad/s}, \quad \alpha_{max} = 0.5 \text{ dB},$$
$$\Omega_s = 3500 \text{ rad/s}, \quad \alpha_{min} = 30 \text{ dB}.$$

(a) Determine the minimum order of the low-pass Butterworth filter and compare it to the minimum order of the Chebyshev filter. Which is the smaller of the two? Use MATLAB functions *buttord* and *cheb1ord* to verify your results.

(b) Determine the half-power frequencies of the designed Butterworth and Chebyshev low-pass filters by letting $\alpha(\Omega_p) = \alpha_{max}$.

(c) For the Butterworth and the Chebyshev designed filters, find the loss function values at Ω_p and Ω_s. How are these values related to the α_{max} and α_{min} specifications? Explain.

(d) If new specifications for the passband and stopband frequencies are $\Omega_p = 750$ (rad/s) and $\Omega_s = 1750$ (rad/s), are the minimum orders of the Butterworth and the Chebyshev filters changed? Explain.

Answers: Minimum orders: Butterworth $N_b = 6$, Chebyshev $N_c = 4$; Chebyshev $\Omega_{hp} = 1639.7$.

7.10 Butterworth, Chebyshev and Elliptic filters—Design an analog low-pass filter satisfying the following magnitude specifications:

$$\alpha_{max} = 0.5 \text{ dB}, \qquad \alpha_{min} = 20 \text{ dB},$$
$$\Omega_p = 1000 \text{ rad/s}, \qquad \Omega_s = 2000 \text{ rad/s}.$$

(a) Use the Butterworth method. Plot the poles and zeros and the magnitude and phase of the designed filter. Verify by plotting the loss function that the specifications are satisfied.

(b) Use Chebyshev method (*cheby1*). Plot the poles and zeros and the magnitude and phase of the designed filter. Verify by plotting the loss function that the specifications are satisfied.

(c) Use the elliptic method. Plot the poles and zeros and the magnitude and phase of the designed filter. Verify by plotting the loss function that the specifications are satisfied.

(d) Compare the three filters and comment on their differences.

Answer: Minimum order of Butterworth is largest of the three filters.

7.11 Chebyshev filter design—Consider the following low-pass filter specifications:

$$\alpha_{max} = 0.1 \text{ dB}, \qquad \alpha_{min} = 60 \text{ dB},$$
$$\Omega_p = 1000 \text{ rad/s}, \qquad \Omega_s = 2000 \text{ rad/s}.$$

(a) Use MATLAB to design a Chebyshev low-pass filter that satisfies the above specifications. Plot the poles and zeros and the magnitude and phase of the designed filter. Verify by plotting the loss function that the specifications are satisfied.

(b) Compute the half-power frequency of the designed filter.

Answer: $\Omega_{hp} = 1051.9$ (rad/s).

7.12 Demodulation of AM—The signal at the input of an AM receiver is $u(t) = m_1(t)\cos(20t) + m_2(t)\cos(100t)$ where the messages $m_i(t)$, $i = 1, 2$ are the outputs of a low-pass Butterworth filter with inputs

$$x_1(t) = r(t) - 2r(t-1) + r(t-2) \quad \text{and} \quad x_2(t) = u(t) - u(t-4),$$

respectively. Suppose we are interested in recovering the message $m_1(t)$. The following is a discretized version of the required filtering and demodulation.

(a) Consider $x_1(t)$ and $x_2(t)$ pulses of duration 5 seconds and generate sampled signals $x_1(nT_s)$ and $x_2(nTs)$ with 512 samples. Determine T_s.

(b) Design a 10th order low-pass Butterworth filter with half-power frequency of 10 (rad/s). Find the frequency response $H(j\Omega)$ of this filter using the function *freqs* for the frequencies obtained for the discretized Fourier transforms obtained in the previous item. From the convolution property $M_i(\Omega) = H(j\Omega)X_i(\Omega)$, $i = 1, 2$.

(c) To recover the sampled version of the desired message $m_1(t)$, first use a bandpass filter to keep the desired signal containing $m_1(t)$ and to suppress the other. Design a bandpass Butterworth filter of order 20 with half-power frequencies $\Omega_\ell = 15$ and $\Omega_u = 25$ rad/s so that it will pass the signal $m_1(t)\cos(20t)$ and reject the other signal.

(d) Multiply the output of the bandpass filter by a sinusoid $\cos(20nT_s)$ (exactly the carrier in the transmitter used for the modulation of $m_1(t)$), and low-pass filter the output of the mixer (the system that multiplies by the carrier frequency cosine). Finally, use the previously designed low-pass Butterworth filter, of bandwidth 10 (rad/s) and order 10, to filter the output of the mixer. The output of the low-pass filter is found in the same way we did before.

(e) Plot the pulses $x_i(nT_s)$, $i = 1, 2$, their corresponding spectra and superpose the magnitude frequency response of the low-pass filter on the spectra. Plot the messages $m_i(nT_s)$, $i = 1, 2$, and their spectra. Plot together the spectrum of the received signal and of magnitude response of the bandpass filter. Plot the output of the bandpass filter and the recovered message signal. Use the function *plot* in these plots.

Use the following functions to compute the Fourier transform and its inverse.

```
function [X,W] = AFT(x,Ts)
N=length(x);
X=fft(x)*Ts; W=[0:N-1]*2*pi/N-pi;W=W/Ts;
X=fftshift(X);

function [x,t] = AIFT(X,Ts)
N=length(X);
X=fftshift(X);
x=real(ifft(X))/Ts;t=[0:N-1]*Ts;
```

7.13 Quadratic AM—Suppose we would like to send the two messages $m_i(t)$, $i = 1, 2$, created in Problem 7.12 using the same bandwidth and to recover them separately. To implement this consider the QAM approach where the transmitted signal is

$$s(t) = m_1(t)\cos(50t) + m_2(t)\sin(50t).$$

Suppose that at the receiver we receive $s(t)$ and that we demodulate it to obtain $m_i(t)$, $i = 1, 2$. This is a discretized version of the processing, similar to that in Problem 7.12. Use the functions *AFT* and *AIFT* given there to compute Fourier transform and inverse.

(a) Sample $s(t)$ using $T_s = 5/512$ and obtain its magnitude spectrum $|S(\Omega)|$ using the functions AFT. The sampled messages coincide with those in Problem 7.12.

(b) Multiply $s(t)$ by $\cos(50t)$, and filter the result using the low-pass filter designed in Problem 7.12.

(c) Multiply $s(t)$ by $\sin(50t)$, and filter the resulting signal using the low-pass filter designed in Problem 7.12.

(d) Use the function *plot* to plot the spectrum of $s(t)$ and the recovered messages. Comment on your results.

THEORY AND APPLICATIONS OF DISCRETE-TIME SIGNALS AND SYSTEMS

SAMPLING THEORY

CONTENTS

The pure and simple truth is rarely pure and never simple.

Oscar Wilde (1854–1900), Irish writer and poet

8.1 INTRODUCTION

Since many of the signals found in applications are continuous in time and amplitude, if we wish to process them with a computer it is necessary to sample, to quantize and to code them to obtain digital signals—discrete in both time and amplitude. Once the continuous-time signal is sampled in time, the amplitude of the obtained discrete-time signal is quantized and coded to give a binary sequence that can be either stored or processed with a computer.

As we will see, it is the inverse relation between time and frequency that provides the solution to the problem of how to preserve the information of a continuous-time signal when sampled. When sampling a continuous-time signal one could choose an extremely small value for the sampling period so that there is no significant difference between the continuous-time and the discrete-time signals—visually as well as from the information content point of view. Such a representation would, however, give redundant values that could be spared without losing the information provided by the continuous-time signal. On the other hand, if we choose a large value for the sampling period we achieve data compression but at the risk of losing some of the information provided by the continuous-time signal. So how should one choose an appropriate value for the sampling period? The answer is not clear in the time domain. It does become clear when considering the effects of sampling in the frequency domain: the sampling period depends on the maximum frequency present in the continuous-time signal. Furthermore, when using the correct sampling period the information in the continuous-time signal remains in the discrete-time signal after sampling, thus allowing the reconstruction of the original signal from the samples. These results, introduced by Nyquist and Shannon, constitute the bridge between continuous-time and discrete-time signals and systems and was the starting point for digital signal processing as a technical area.

The device that samples, quantizes and codes a continuous-time signal is called an **Analog-to-Digital Converter** or A/D converter, while the device that converts digital signals into continuous-time signal is called a **Digital-to-Analog Converter** or D/A converter. Besides the possibility of losing information by choosing a sampling period that is too large, the A/D converter also loses information in the quantization process. The quantization error is, however, made less significant by increasing the number of bits used to represent each sample. The D/A converter interpolates and smooths out the digital signal converting it back into a continuous-time signal. These two devices are essential in the processing of continuous-time signals with computers.

Over the years the principles of communications have remained the same, but their implementation has changed considerably—in great part due to the paradigm shift caused by the sampling and coding theory. Modern digital communications was initiated with the concept of Pulse Code Modulation (PCM) for the transmission of binary signals. PCM is a practical implementation of sampling, quantization and coding, or analog-to-digital conversion, of an analog message into a digital message. Efficient use of the radio spectrum has motivated the development of multiplexing techniques in time and in frequency. In this chapter, we highlight some of the communication techniques that relate to the sampling theory, leaving the technical details for texts in digital communications.

8.2 UNIFORM SAMPLING

The first step in converting a continuous-time signal $x(t)$ into a digital signal is to discretize the time variable, i.e., to consider samples of $x(t)$ at uniform times $t = nT_s$ or

$$x(nT_s) = x(t)|_{t=nT_s} \qquad n \text{ integer} \tag{8.1}$$

where T_s is the sampling period. The sampling process can be thought of as a modulation process, in particular connected with Pulse Amplitude Modulation (PAM), a basic approach in digital communications. A pulse amplitude modulated signal consists of a sequence of narrow pulses with amplitudes the values of the continuous-time signal at the sampling times. Assuming that the width of the pulses is much narrower than the sampling period T_s permits a simpler analysis based on impulse sampling.

8.2.1 PULSE AMPLITUDE MODULATION

A PAM system can be visualized as a switch that closes every T_s seconds for Δ seconds, and remains open otherwise. The PAM signal is thus the multiplication of the continuous-time signal $x(t)$ by a periodic signal $p(t)$ consisting of pulses of width Δ, amplitude $1/\Delta$ and period T_s. Thus, $x_{PAM}(t)$ consists of narrow pulses with the amplitudes of the signal within the pulse width (for narrow pulses, the pulse amplitude can be approximated by the signal amplitude at the sampling time—this is called flat-top PAM). For a small pulse width Δ, the PAM signal is

$$x_{PAM}(t) = x(t)p(t) = \frac{1}{\Delta}\sum_m x(mT_s)[u(t - mT_s) - u(t - mT_s - \Delta)]. \tag{8.2}$$

Now, as a periodic signal $p(t)$ is represented by its Fourier series

$$p(t) = \sum_{k=-\infty}^{\infty} P_k e^{jk\Omega_0 t} \qquad \Omega_0 = \frac{2\pi}{T_s}$$

where P_k are the Fourier series coefficients. Thus the PAM signal can be expressed as

$$x_{PAM}(t) = \sum_{k=-\infty}^{\infty} P_k\, x(t)\, e^{jk\Omega_0 t}$$

and its Fourier transform is according to the frequency-shifting property

$$X_{PAM}(\Omega) = \sum_{k=-\infty}^{\infty} P_k X(\Omega - k\Omega_0).$$

Thus, the above is a modulation of the train of pulses $p(t)$ by the signal $x(t)$. The spectrum of $x_{PAM}(t)$ is the spectrum of $x(t)$ shifted in frequency by $\{k\Omega_0\}$, weighted by P_k and added.

8.2.2 IDEAL IMPULSE SAMPLING

Given that the pulse width Δ in PAM is much smaller than T_s, $p(t)$ can be replaced by a periodic sequence of impulses of period T_s (see Fig. 8.1) or $\delta_{T_s}(t)$. This simplifies considerably the analysis and makes the results easier to grasp. Later in the chapter we will consider the effects of having pulses instead of impulses, a more realistic assumption.

The **sampling function** $\delta_{T_s}(t)$, or a periodic sequence of impulses of period T_s, is

$$\delta_{T_s}(t) = \sum_n \delta(t - nT_s) \tag{8.3}$$

where $\delta(t)$ is an approximation of the normalized pulse $p(t) = [u(t) - u(t - \Delta)]/\Delta$ when $\Delta << T_s$. The sampled signal is then given by

$$x_s(t) = x(t)\delta_{T_s}(t) \tag{8.4}$$

as shown in Fig. 8.1. There are two equivalent ways to view the sampled signal $x_s(t)$ in the frequency domain:

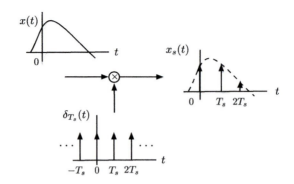

FIGURE 8.1

Ideal impulse sampling.

(i) Modulation: Since $\delta_{T_s}(t)$ is periodic, of fundamental frequency $\Omega_s = 2\pi/T_s$, its Fourier series is

$$\delta_{T_s}(t) = \sum_{k=-\infty}^{\infty} D_k e^{jk\Omega_s t}$$

where the Fourier series coefficients $\{D_k\}$ are

$$
\begin{aligned}
D_k &= \frac{1}{T_s} \int_{-T_s/2}^{T_s/2} \delta_{T_s}(t) e^{-jk\Omega_s t} dt = \frac{1}{T_s} \int_{-T_s/2}^{T_s/2} \delta(t) e^{-jk\Omega_s t} dt \\
&= \frac{1}{T_s} \int_{-T_s/2}^{T_s/2} \delta(t) e^{-j0} dt = \frac{1}{T_s}.
\end{aligned}
$$

The last equation is obtained using the sifting property of the impulse $\delta(t)$ and that its area is unity. Thus the Fourier series of the sampling signal is

$$\delta_{T_s}(t) = \sum_{k=-\infty}^{\infty} \frac{1}{T_s} e^{jk\Omega_s t}$$

and the sampled signal $x_s(t) = x(t)\delta_{T_s}(t)$ is then expressed as

$$x_s(t) = x(t)\delta_{T_s}(t) = \frac{1}{T_s} \sum_{k=-\infty}^{\infty} x(t) e^{jk\Omega_s t}$$

with Fourier transform

$$X_s(\Omega) = \frac{1}{T_s} \sum_{k=-\infty}^{\infty} X(\Omega - k\Omega_s) \tag{8.5}$$

where we used the frequency-shift property of the Fourier transform, and let $X(\Omega)$ and $X_s(\Omega)$ be the Fourier transforms of $x(t)$ and $x_s(t)$.

(ii) Discrete-Time Fourier Transform: The sampled signal can also be expressed as

$$x_s(t) = \sum_{n=-\infty}^{\infty} x(nT_s)\delta(t - nT_s)$$

with a Fourier transform

$$X_s(\Omega) = \sum_{n=-\infty}^{\infty} x(nT_s)e^{-j\Omega T_s n} \tag{8.6}$$

where we used the Fourier transform of a shifted impulse. This equation is equivalent to Equation (8.5) and will be used later in deriving the Fourier transform of discrete-time signals.

Remarks

1. The spectrum $X_s(\Omega)$ of the sampled signal, according to (8.5), is a superposition of shifted spectra $\{(1/T_s)X(\Omega - k\Omega_s)\}$ due to the modulation process in the sampling.
2. Considering that the output of the sampler displays frequencies which are not present in the input, according to the eigenfunction property the sampler is not LTI. It is a time-varying system. Indeed, if sampling $x(t)$ gives $x_s(t)$, sampling $x(t - \tau)$ where $\tau \neq kT_s$ for an integer k, will not be $x_s(t - \tau)$. The sampler is, however, a linear system.
3. Equation (8.6) provides the relation between the continuous frequency Ω (rad/s) of $x(t)$ and the discrete frequency ω (rad) of the discrete-time signal $x(nT_s)$ or $x[n]$[1]

$$\omega = \Omega T_s \quad [\text{rad/s}] \times [\text{s}] = [\text{rad}].$$

Sampling a continuous-time signal $x(t)$ at uniform times $\{nT_s\}$ gives a sampled signal

$$x_s(t) = x(t)\delta_{T_s}(t) = \sum_{n=-\infty}^{\infty} x(nT_s)\delta(t - nT_s) \tag{8.7}$$

or a sequence of samples $\{x(nT_s)\}$ at uniform times nT_s. Sampling is equivalent to modulating the sampling signal

$$\delta_{T_s}(t) = \sum_{n} \delta(t - nT_s) \tag{8.8}$$

periodic of fundamental period T_s (the sampling period) with $x(t)$.

---

[1] To help the reader visualize the difference between a continuous-time signal, which depends on a continuous variable t or a real number, and a discrete-time signal which depends on the integer variable n we will use square brackets for these. Thus $\eta(t)$ is a continuous-time signal, while $\rho[n]$ is discrete-time.

If $X(\Omega)$ is the Fourier transform of $x(t)$, the Fourier transform of the sampled signal $x_S(t)$ is given by the equivalent expressions

$$
\begin{aligned}
X_S(\Omega) &= \frac{1}{T_s} \sum_{k=-\infty}^{\infty} X(\Omega - k\Omega_s) \\
&= \sum_{n=-\infty}^{\infty} x(nT_s)e^{-j\Omega T_s n}, \qquad \Omega_s = \frac{2\pi}{T_s}.
\end{aligned} \qquad (8.9)
$$

Depending on the maximum frequency present in the spectrum of $x(t)$ and on the chosen sampling frequency Ω_s (or the sampling period T_s) it is possible to have overlaps when the spectrum of $x(t)$ is shifted and added to obtain the spectrum of the sampled signal $x_S(t)$. We have three possible situations:

1. If the signal $x(t)$ has a low-pass spectrum of finite support, i.e., $X(\Omega) = 0$ for $|\Omega| > \Omega_{max}$ (see Fig. 8.2A) where Ω_{max} is the maximum frequency present in the signal, such a signal is called **band-limited**. As shown in Fig. 8.2B, for band-limited signals it is possible to choose Ω_s so that the spectrum of the sampled signal consists of shifted non-overlapping versions of $(1/Ts)X(\Omega)$. Graphically, this is accomplished by letting $\Omega_s - \Omega_{max} \geq \Omega_{max}$ or

$$
\Omega_s \geq 2\Omega_{max},
$$

which is called the **Nyquist sampling rate condition**. As we will see, in this case we are able to recover $X(\Omega)$, or $x(t)$, from $X_S(\Omega)$ or from the sampled signal $x_S(t)$. Thus the information in $x(t)$ is preserved in the sampled signal $x_S(t)$.

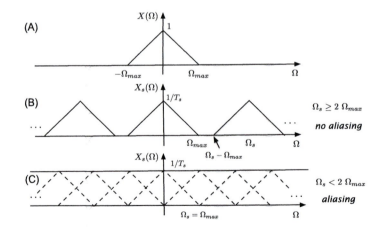

FIGURE 8.2

(A) Spectrum of band-limited signal, (B) spectrum of sampled signal when satisfying the Nyquist sampling rate condition, (C) spectrum of sampled signal with aliasing (superposition of spectra, shown in dashed lines, gives a constant shown by continuous line).

2. On the other hand, if the signal $x(t)$ is band-limited but we let $\Omega_s < 2\Omega_{max}$ then when creating $X_s(\Omega)$ the shifted spectra of $x(t)$ overlap (see Fig. 8.2C). In this case, due to the overlap it will not be possible to recover the original continuous-time signal from the sampled signal, and thus the sampled signal does not share the same information with the original continuous-time signal. This overlapping phenomenon is called **frequency aliasing** since due to the overlapping of the spectra some frequency components of the original continuous-time signal acquire a different frequency value or an "alias."

3. When the spectrum of $x(t)$ does not have a finite support, i.e., the signal is not **band-limited**, sampling using any sampling period T_s generates a spectrum for the sampled signal consisting of overlapped spectra of $x(t)$. Thus when sampling non-band-limited signals frequency aliasing is always present. The only way to sample a non-band-limited signal $x(t)$ without aliasing—at the cost of losing information provided by the high-frequency components of $x(t)$—is by obtaining an approximate signal $x_a(t)$ lacking the high-frequency components of $x(t)$, and thus permitting us to determine a maximum frequency for it. This is accomplished by **antialiasing filtering**, commonly used in samplers.

A band-limited signal $x(t)$, i.e., its low-pass spectrum $X(\Omega)$ is such that

$$|X(\Omega)| = 0 \quad \text{for } |\Omega| > \Omega_{max} \tag{8.10}$$

where Ω_{max} is the maximum frequency in $x(t)$, can be sampled uniformly and without frequency aliasing using a sampling frequency

$$\Omega_s = \frac{2\pi}{T_s} \geq 2\Omega_{max}, \tag{8.11}$$

called the Nyquist sampling rate condition.

BANDLIMITED OR NOT?

The following, taken from David Slepian's paper "On Bandwidth" [70], clearly describes the uncertainty about bandlimited signals:

> The Dilemma—Are signals really bandlimited? They seem to be, and yet they seem not to be.
> On the one hand, a pair of solid copper wires will not propagate electromagnetic waves at optical frequencies and so the signals I receive over such a pair must be bandlimited. In fact, it makes little physical sense to talk of energy received over wires at frequencies higher than some finite cutoff W, say 10^{20} Hz. It would seem, then, that signals must be bandlimited.
> On the other hand, however, signals of limited bandwidth W are finite Fourier transforms,

$$s(t) = \int_{-W}^{W} e^{2\pi i f t} S(f) df$$

> and irrefutable mathematical arguments show them to be extremely smooth. They possess derivatives of all orders. Indeed, such integrals are entire functions of t, completely predictable from any little piece, and they cannot vanish on any t interval unless they vanish everywhere. Such signals cannot start or stop, but must go on forever. Surely *real signals* start and stop, and they cannot be bandlimited!

> Thus we have a dilemma: to assume that real signals must go on forever in time (a consequence of bandlimitedness) seems just as unreasonable as to assume that real signals have energy at arbitrary high frequencies (no bandlimitation). Yet one of these alternatives must hold if we are to avoid mathematical contradiction, for either signals are bandlimited or they are not: there is no other choice. Which do you think they are?

Example 8.1. Consider the signal $x(t) = 2\cos(2\pi t + \pi/4)$, $-\infty < t < \infty$, determine if it is band-limited. Use $T_s = 0.4, 0.5$ and 1 s/sample as sampling periods and for each of these find out whether the Nyquist sampling rate condition is satisfied and if the sampled signal looks like the original signal or not.

Solution: Since $x(t)$ only has the frequency 2π, it is band-limited with $\Omega_{max} = 2\pi$ (rad/s). For any T_s the sampled signal is given as

$$x_s(t) = \sum_{n=-\infty}^{\infty} 2\cos(2\pi n T_s + \pi/4)\delta(t - nT_s), \qquad T_s \text{ s/sample.} \qquad (8.12)$$

Using $T_s = 0.4$ s/sample the sampling frequency in rad/s is $\Omega_s = 2\pi/T_s = 5\pi > 2\Omega_{max} = 4\pi$, satisfying the Nyquist sampling rate condition. The samples in (8.12) are then

$$x(nT_s) = 2\cos(2\pi\ 0.4n + \pi/4) = 2\cos\left(\frac{4\pi}{5}n + \frac{\pi}{4}\right) \qquad -\infty < n < \infty$$

and the corresponding sampled signal $x_s(t)$ repeats periodically every five samples. Indeed,

$$x_s(t + 5T_s) = \sum_{n=-\infty}^{\infty} x(nT_s)\delta(t - (n - 5)T_s) \quad \text{letting } m = n - 5$$

$$= \sum_{m=-\infty}^{\infty} x((m + 5)T_s)\delta(t - mT_s) = x_s(t),$$

since $x((m + 5)T_s) = x(mT_s)$. Notice in Fig. 8.3B that only three samples are taken in each period of the continuous-time sinusoid, and it is not obvious that the information of the continuous-time signal is preserved by the sampled signal. We will show in the next section that it is actually possible to recover $x(t)$ from this sampled signal $x_s(t)$, which allows us to say that $x_s(t)$ has the same information as $x(t)$.

When $T_s = 0.5$ the sampling frequency is $\Omega_s = 2\pi/T_s = 4\pi = 2\Omega_{max}$, barely satisfying the Nyquist sampling rate condition. The samples in (8.12) are now

$$x(nT_s) = 2\cos(2\pi\ n\ 0.5 + \pi/4) = 2\cos\left(\frac{2\pi}{2}n + \frac{\pi}{4}\right) \qquad -\infty < n < \infty.$$

In this case the sampled signal repeats periodically every 2 samples; as can easily be checked $x((n + 2)T_s) = x(nT_s)$. According to the Nyquist sampling rate condition, this is the minimum number of samples per period allowed before we start having aliasing. In fact, if we let $\Omega_s = \Omega_{max} = 2\pi$ corresponding to the sampling period $T_s = 1$, the samples in (8.12) are

$$x(nT_s) = 2\cos(2\pi n + \pi/4) = 2\cos(\pi/4) = \sqrt{2}$$

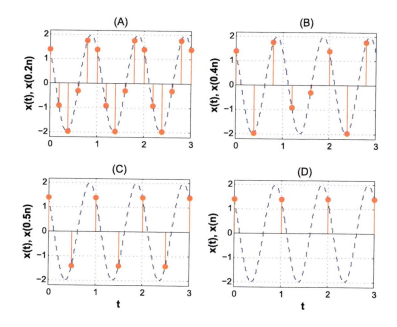

FIGURE 8.3

Sampling of $x(t) = 2\cos(2\pi t + \pi/4)$ with sampling periods (A) $T_s = 0.2$, (B) $T_s = 0.4$, (C) $T_s = 0.5$ and (D) $T_s = 1$ s/sample.

and the sampled signal is $\sqrt{2}\delta_{T_s}(t)$. In this case, the sampled signal cannot be possibly converted back into the continuous-time sinusoid. Thus we have lost the information provided by the sinusoid. **Under-sampling**, or acquiring too few samples per unit time, changes the nature of the reconstructed signal.

To illustrate the sampling process, we use MATLAB to plot the continuous signal and four sampled signals (see Fig. 8.3) for $T_s = 0.2$, 0.4, 0.5 and 1 s/sample. Clearly when $T_s = 1$ s/sample there is no similarity between the continuous-time and the discrete-time signals due to frequency aliasing, while for $T_s = 0.2$ s/sample the sampled signal looks very much like the original continuous-time signal. $\square$

Example 8.2. Consider the following signals:

$$(a)\ x_1(t) = u(t + 0.5) - u(t - 0.5), \quad (b)\ x_2(t) = e^{-t}u(t).$$

Determine if they are band-limited or not. If not, determine the frequency for which the energy of the non-band-limited signal corresponds to 99% of its total energy and use this result to approximate its maximum frequency.

Solution: (a) The signal $x_1(t) = u(t + 0.5) - u(t - 0.5)$ is a unit pulse signal. Clearly this signal can easily be sampled by choosing any value of $T_s \ll 1$. For instance, $T_s = 0.01$ s would be a good value, giving a discrete-time signal $x_1(nT_s) = 1$, for $0 \leq nT_s = 0.01n \leq 1$ or $0 \leq n \leq 100$. There seems to be no problem in sampling this signal; however, we see that the Fourier transform of $x_1(t)$,

$$X_1(\Omega) = \frac{e^{j0.5\Omega} - e^{-j0.5\Omega}}{j\Omega} = \frac{\sin(0.5\Omega)}{0.5\Omega}$$

does not have a maximum frequency and so $x_1(t)$ is not band-limited. Thus any chosen value of T_s will cause aliasing. Fortunately the values of the sinc function go fast to zero, so that one could compute an approximate maximum frequency that would include 99% of the energy of the signal.

Using Parseval's energy relation we see that the energy of $x_1(t)$ (the area under $x_1^2(t)$) is 1 and if we wish to find a value Ω_M, such that 99% of this energy is in the frequency band $[-\Omega_M, \Omega_M]$, we need to look for the limits of the following integral so it equals 0.99:

$$0.99 = \frac{1}{2\pi} \int_{-\Omega_M}^{\Omega_M} \left[\frac{\sin(0.5\Omega)}{0.5\Omega} \right]^2 d\Omega.$$

Since this integral is difficult to find analytically, we use the following script in MATLAB to approximate it.

```
%%
% Example 8.2---Parseval's relation and sampling
%%
clear all; clf; syms W
for k=16:24;
    E(k)=simplify(int((sin(0.5*W)/(0.5*W))^2,W,0,k*pi)/pi);
    e(k)=single(subs(E(k)));
end
figure(1);  stem(e);  hold on
EE=0.9900*ones(1,24); plot(EE,'r'); hold off
axis([1 24 0.98 1]); grid; xlabel('\Omega/\pi'); ylabel('Power')
```

By looking at Fig. 8.4, generated by the script, we find that for $\Omega_M = 20\pi$ (rad/s) 98.9% of the energy of the signal is included, and thus it could be used to approximate the actual maximum frequency and we could choose $T_s < \pi/\Omega_M = 0.05$ s/sample as the sampling period.

(b) For the causal exponential

$$x(t) = e^{-t}u(t)$$

its Fourier transform is

$$X(\Omega) = \frac{1}{1 + j\Omega} \quad \text{so that} \quad |X(\Omega)| = \frac{1}{\sqrt{1 + \Omega^2}},$$

which does not go to zero for any finite Ω, then $x(t)$ is not band-limited. To find a frequency Ω_M so that 99% of the energy is in $-\Omega_M \leq \Omega \leq \Omega_M$ we let

$$\frac{1}{2\pi} \int_{-\Omega_M}^{\Omega_M} |X(\Omega)|^2 d\Omega = \frac{0.99}{2\pi} \int_{-\infty}^{\infty} |X(\Omega)|^2 d\Omega,$$

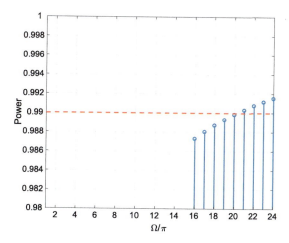

FIGURE 8.4

Determining the Ω_M that gives about 99% of the energy of the signal.

which gives

$$2\tan^{-1}(\Omega)|_0^{\Omega_M} = 2 \times 0.99\tan^{-1}(\Omega)|_0^\infty \quad \text{or} \quad \Omega_M = \tan\left(\frac{0.99\pi}{2}\right) = 63.66 \text{ rad/s.}$$

If we choose $\Omega_s = 2\pi/T_s = 5\Omega_M$ or $T_s = 2\pi/(5 \times 63.66) \approx 0.02$ s/sample, there will be hardly any aliasing or loss of information. $\qquad \square$

8.2.3 RECONSTRUCTION OF THE ORIGINAL CONTINUOUS-TIME SIGNAL

If the signal to be sampled, $x(t)$, has a Fourier transform $X(\Omega)$ and it is band-limited—so that its maximum frequency Ω_{max} is determined—by choosing the sampling frequency Ω_s satisfying the Nyquist sampling rate condition

$$\Omega_s > 2\Omega_{max}, \tag{8.13}$$

the spectrum of the sampled signal $x_s(t)$ displays a superposition of shifted versions of the spectrum $X(\Omega)$, multiplied by $1/T_s$ with no overlaps. In such a case, it is possible to recover the original continuous-time signal from the sampled signal by filtering. Indeed, if we consider an *ideal low-pass* filter with frequency response

$$H_{lp}(\Omega) = \begin{cases} T_s & -\Omega_s/2 \leq \Omega \leq \Omega_s/2, \\ 0 & \text{elsewhere,} \end{cases} \tag{8.14}$$

the Fourier transform of the output of the filter is

$$X_r(\Omega) = H_{lp}(\Omega)X_s(\Omega) = \begin{cases} X(\Omega) & -\Omega_s/2 \leq \Omega \leq \Omega_s/2, \\ 0 & \text{elsewhere,} \end{cases}$$

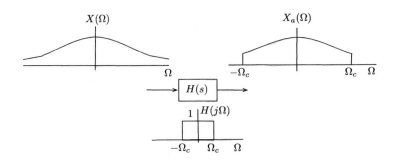

FIGURE 8.5

Antialiasing filtering of non-band-limited signal.

where $\Omega_s/2 = \Omega_{max}$, the maximum frequency of $x(t)$. The Fourier transforms of the recovered signal, $x_r(t)$, and of the original signal $x(t)$ coincide. Thus, when sampling a band-limited signal, using a sampling period T_s that satisfies the Nyquist sampling rate condition, the band-limited signal can be recovered exactly from the sampled signal by means of an ideal low-pass filter.

Remarks

1. In practice, the exact recovery of the original signal may not be possible for several reasons. One could be that the continuous-time signal is not exactly band-limited, so that it is not possible to obtain a maximum frequency—frequency aliasing occurs when sampling. Second, the sampling is not done exactly at uniform times, as random variation of the sampling times may occur. Third, the filter required for the exact recovery is an ideal low-pass filter which in practice cannot be realized, only an approximation is possible. Although this indicates the limitations of the sampling theory, in most cases where: (i) the signal is band-limited or approximately band-limited, (ii) the Nyquist sampling rate condition is satisfied, and (iii) the reconstruction filter approximates well the ideal low-pass reconstruction filter the recovered signal closely approximates the original signal.
2. For signals that do not satisfy the band-limitedness condition, one can obtain an approximate signal that satisfies that condition. This is done by passing the non-band-limited signal through an ideal low-pass filter. The filter output is guaranteed to have as maximum frequency the cutoff frequency of the filter (see Fig. 8.5 for an illustration). Because of the low-pass filtering, the filtered signal is a smoothed version of the original signal—high frequencies of the signal being sampled have been removed. The low-pass filter is called the **antialiasing filter**, since it makes the approximate signal band-limited, thus avoiding aliasing in the frequency domain.
3. In applications, the cutoff frequency of the antialiasing filter is set according to prior knowledge of the frequency content of the signal being sampled. For instance, when sampling speech, it is well known that speech has frequencies ranging from about 100 Hz to about 5 kHz (this range of frequencies provides understandable speech in phone conversations). Thus, when sampling speech an antialiasing filter with a cutoff frequency of 5 kHz is chosen and the sampling rate is then set to 10,000 samples/s or higher. Likewise, it is also known that music signals have a range of frequencies from 0 to approximately 22 kHz. Thus, when sampling music signals the antialiasing filter cutoff frequency is set to 22 kHz and the sampling rate is set to 44K samples/s or higher.

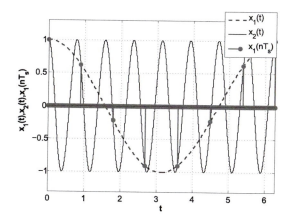

FIGURE 8.6

Sampling of two sinusoids of frequencies $\Omega_0 = 1$ and $\Omega_0 + \Omega_1 = 8$ with $T_s = 2\pi/\Omega_1$. The higher frequency signal is under-sampled, causing aliasing and making the two sampled signals coincide.

Example 8.3. Consider the two sinusoids

$$x_1(t) = \cos(\Omega_0 t), \quad x_2(t) = \cos((\Omega_0 + \Omega_1)t) \qquad -\infty \leq t \leq \infty$$

where $\Omega_1 > 2\Omega_0$. Show that if we sample these signals using $T_s = 2\pi/\Omega_1$, we cannot differentiate the sampled signals, i.e. $x_1(nT_s) = x_2(nT_s)$. Let $\Omega_0 = 1$ and $\Omega_1 = 7$ rad/s, use MATLAB to show the above graphically. Explain the significance of this.

Solution: Sampling the two signals using $T_s = 2\pi/\Omega_1$ we have

$$x_1(nT_s) = \cos(\Omega_0 nT_s), \quad x_2(nT_s) = \cos((\Omega_0 + \Omega_1)nT_s) \qquad -\infty \leq n \leq \infty,$$

but since $\Omega_1 T_s = 2\pi$ the sinusoid $x_2(nT_s)$ can be written

$$\begin{aligned} x_2(nT_s) &= \cos((\Omega_0 T_s + 2\pi)n) \\ &= \cos(\Omega_0 T_s n) = x_1(nT_s), \end{aligned}$$

making it impossible to differentiate the two sampled signals. Since the sampling frequency Ω_1 is smaller than twice the maximum frequency of $x_2(t)$ or $2(\Omega_0 + \Omega_1)$ the sampled version of this signal is aliased. See Fig. 8.6 for $\Omega_0 = 1$ and $\Omega_0 + \Omega_1 = 8$.

The following script shows the aliasing effect when $\Omega_0 = 1$ and $\Omega_1 = 7$ rad/s. Notice that $x_1(t)$ is sampled satisfying the Nyquits sampling rate condition ($\Omega_s = \Omega_1 = 7 > 2\Omega_0 = 2$ (rad/s)), while $x_2(t)$ is not ($\Omega_s = \Omega_1 = 7 < 2(\Omega_0 + \Omega_1) = 16$ rad/s).

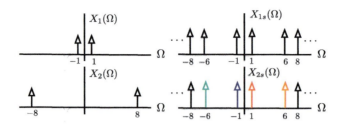

FIGURE 8.7

Spectra of sinusoids $x_1(t)$ and $x_2(t)$ (left). The spectra of the sampled signals $x_{1s}(t)$ and $x_{2s}(t)$ shown on the right look exactly the same due to the under-sampling of $x_2(t)$.

```
%%
% Example 8.3
%%
clear all; clf
omega_0=1;omega_s=7;
T=2*pi/omega_0; t=0:0.001:T;                      % a period of x_1
x1=cos(omega_0*t); x2=cos((omega_0+omega_s)*t);
N=length(t); Ts=2*pi/omega_s;                     % sampling period
M=fix(Ts/0.001); imp=zeros(1,N);
for k=1:M:N-1.
    imp(k)=1;
end
plot(t,x1,'b',t,x2,'k'); hold on
stem(t,imp.*x1,'r','filled');axis([0 max(t) -1.1 1.1]); xlabel('t'); grid
legend('x_1(t)','x_2(t)','x_1(nTs)'); xlabel('t');
ylabel('x_1(t), x_2(t), x_1(nTs)')
```

The result in the frequency domain is shown in Fig. 8.7: the spectra of the two sinusoids are different but the spectra of the sampled signals are identical. ☐

8.2.4 SIGNAL RECONSTRUCTION FROM SINC INTERPOLATION

The reconstruction of the continuous-time signal $x(t)$ from the discrete-time sampled signal $x_s(t)$ can be shown to be an interpolation using sinc functions. First, the ideal low-pass reconstruction filter $H_{lp}(s)$ in (8.14) has as impulse response

$$h_{lp}(t) = \frac{T_s}{2\pi} \int_{-\Omega_s/2}^{\Omega_s/2} e^{j\Omega t} d\Omega = \frac{\sin(\pi t/T_s)}{\pi t/T_s}, \qquad \Omega_s = \frac{2\pi}{T_s}, \qquad (8.15)$$

which is an infinite-time support sinc function that decays symmetrically with respect to $t = 0$. Thus, the reconstructed signal $x_r(t)$ is the convolution of the sampled signal $x_s(t)$ and $h_{lp}(t)$, which is found to be

$$x_r(t) = [x_s * h_{lp}](t) = \int_{-\infty}^{\infty} x_s(\tau)h_{lp}(t-\tau)d\tau = \int_{-\infty}^{\infty}\left[\sum_{n=-\infty}^{\infty} x(nT_s)\delta(\tau-nT_s)\right]h_{lp}(t-\tau)d\tau$$

$$= \sum_{n=-\infty}^{\infty} x(nT_s)\frac{\sin(\pi(t-nT_s)/T_s)}{\pi(t-nT_s)/T_s} \tag{8.16}$$

after replacing $x_s(t)$, $h_{lp}(t-\tau)$ and applying the sifting property of the delta function. The reconstructed signal is thus an interpolation in terms of time-shifted sinc functions with amplitudes the samples $\{x(nT_s)\}$. In fact, if we let $t = kT_s$, we can see that

$$x_r(kT_s) = \sum_{n} x(nT_s)\frac{\sin(\pi(k-n))}{\pi(k-n)} = x(kT_s),$$

since

$$\frac{\sin(\pi(k-n))}{\pi(k-n)} = \begin{cases} 1 & k-n = 0 \text{ or } n = k, \\ 0 & n \neq k, \end{cases}$$

according to L'Hopital's rule. Thus, the values at $t = kT_s$ are recovered exactly, and the rest are interpolated by a sum of sinc functions.

8.2.5 THE NYQUIST–SHANNON SAMPLING THEOREM

If a low-pass continuous-time signal $x(t)$ is <u>band-limited</u> (i.e., it has a spectrum $X(\Omega)$ such that $X(\Omega) = 0$ for $|\Omega| > \Omega_{max}$, where Ω_{max} is the maximum frequency in $x(t)$) we then have:

- The information in $x(t)$ is preserved by a sampled signal $x_s(t)$, with samples $x(nT_s) = x(t)|_{t=nT_s}$, $n = 0, \pm1, \pm2, \cdots$, provided that the sampling frequency $\Omega_s = 2\pi/T_s$ (rad/s) is such that

$$\Omega_s \geq 2\Omega_{max} \quad \underline{\text{Nyquist sampling rate condition}} \tag{8.17}$$

or equivalently if the sampling rate f_s (samples/s) or the sampling period T_s (s/sample) are

$$f_s = \frac{1}{T_s} \geq \frac{\Omega_{max}}{\pi}. \tag{8.18}$$

- When the Nyquist sampling rate condition is satisfied, the original signal $x(t)$ can be reconstructed by passing the sampled signal $x_s(t)$ through an ideal low-pass filter with frequency response:

$$H(j\Omega) = \begin{cases} T_s & -\Omega_s/2 < \Omega < \Omega_s/2, \\ 0 & \text{otherwise.} \end{cases}$$

The reconstructed signal is given by the following sinc interpolation from the samples:

$$x_r(t) = \sum_{n=-\infty}^{\infty} x(nT_s)\frac{\sin(\pi(t-nT_s)/T_s)}{\pi(t-nT_s)/T_s}. \tag{8.19}$$

Remarks

1. The value $2\Omega_{max}$ is called the **Nyquist sampling frequency**. The value $\Omega_s/2$ is called the **folding frequency**.

2. Considering the number of samples available every second, we can get a better understanding of the data storage requirement, the processing limitations imposed by real-time processing and the need for data compression algorithms. For instance, for a music signal sampled at 44,000 samples/s, with each sampled represented by 8 bits/sample, for every second of music we would need to store $44 \times 8 = 352$ kbits/s, and in an hour of sampling we would have $3600 \times 44 \times 8$ kbits. Likewise, if you were to process the raw signal you would have a new sample every $T_s = 1/44,000$ s/sample or 0.0227 ms, so that any real-time processing would have to be done very fast. If you would desire better quality you could increase the number of bits assigned to each sample, let us say to 16 bits/sample, which would double the above quantity, and if you wish more fidelity you could increase the sampling rate and have two channels for stereo sound but be ready to provide more storage or to come up with some data compression algorithm to be able to handle so much data.

ORIGINS OF THE SAMPLING THEORY—PART I

The sampling theory has been attributed to many engineers and mathematicians. It seems as if mathematicians and researchers in communications engineering came across these results from different perspectives. In the engineering community, the sampling theory has been attributed traditionally to Harry Nyquist and Claude Shannon, although other famous researchers such as V.A. Kotelnikov, E.T. Whittaker, and D. Gabor came out with similar results. Nyquist's work did not deal directly with sampling and reconstruction of sampled signals but it contributed to advances by Shannon in those areas.

Harry Nyquist was born in Sweden in 1889 and died in 1976 in the United States. He attended the University of North Dakota at Grand Forks and received his Ph.D. from Yale University in 1917. He worked for the American Telephone and Telegraph (AT&T) Company and the Bell Telephone Laboratories, Inc. He received 138 patents and published 12 technical articles. Nyquist's contributions range from thermal noise, stability of feedback amplifiers, telegraphy, television, to other important communications problems. His theoretical work on determining the bandwidth requirements for transmitting information provided the foundations for Claude Shannon's work on sampling theory [25,75].

As Hans D. Luke concludes in his paper "The Origins of the Sampling Theorem," [50], regarding the attribution of the sampling theorem to many authors:

... this history also reveals a process which is often apparent in theoretical problem in technology or physics: first the practicians put forward a rule of thumb, then theoreticians develop the general solution, and finally someone discovers that the mathematicians have long since solved the mathematical problem which it contains, but in 'splendid isolation'.

8.2.6 SAMPLING SIMULATIONS WITH MATLAB

The simulation of sampling with MATLAB is complicated by the representation of continuous-time signals and the computation of the continuous Fourier transform. Two sampling rates are needed: one being the sampling rate under study, f_s, and the other the one used to simulate the continuous-time signal, $f_{sim} >> f_s$. The computation of the continuous Fourier transform of $x(t)$ can be done approximately using the Fast Fourier Transform (FFT) multiplied by the sampling period. We will discuss the FFT in a latter chapter; for now just think of the FFT as an algorithm to compute the Fourier transform of a discretized signal.

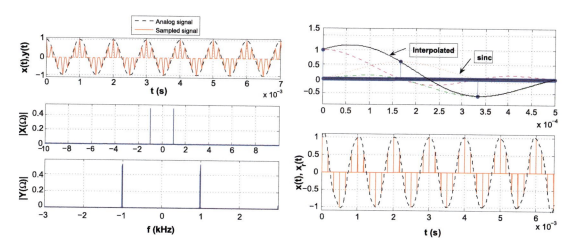

FIGURE 8.8

No aliasing: sampling simulation of $x(t) = \cos(2000\pi t)$ using $f_s = 6000$ samples/s. Plots on the left are of the signal $x(t)$ and the sampled signal $y(t)$, and their magnitude spectra $|X(\Omega)|$ and $|Y(\Omega)|$ (which is periodic and one period is shown). On the right, the top plot illustrates the sinc interpolation of 3 samples, and below the sinc–interpolated signal $x_r(t)$ (dashed lines) and the sampled signal $y(t)$. In this case, $x_r(t)$ is very close to the original signal $x(t)$.

To illustrate the sampling procedure consider sampling a sinusoid $x(t) = \cos(2\pi f_0 t)$ where $f_0 = 1$ kHz. To simulate this as a continuous-time signal we choose a sampling period $T_{sim} = 0.5 \times 10^{-4}$ s/sample or a sampling frequency $f_{sim} = 20{,}000$ samples/s.

No aliasing sampling—If we sample $x(t)$ with a sampling frequency $f_s = 6000 > 2f_0 = 2000$ Hz, the sampled signal $y(t)$ will not display aliasing in its frequency representation, as we are satisfying the Nyquist sampling rate condition. The figure on the left in Fig. 8.8 displays the signal $x(t)$ and its sampled version $y(t)$, as well as their approximate Fourier transforms. The magnitude spectrum $|X(\Omega)|$ corresponds to the sinusoid $x(t)$, while $|Y(\Omega)|$ is the first period of the spectrum of the sampled signal (recall the spectrum of the sampled signal is periodic of period $\Omega_s = 2\pi f_s$). In this case, when no aliasing occurs, the first period of the spectrum $|Y(\Omega)|$ coincides with the spectrum $|X(\Omega)|$ of $x(t)$ (notice that, as a sinusoid, the magnitude spectrum $|X(\Omega)|$ is zero except at the frequency of the sinusoid or ± 1 kHz, likewise $|Y(\Omega)|$ is zero except at ± 1 kHz and the range of frequencies is $[-f_s/2, f_s/2] = [-3, 3]$ kHz). On the right in Fig. 8.8 we show the sinc interpolation of 3 samples of $y(t)$, the solid line is the interpolated values or the sum of sincs centered at the three samples. At the bottom of that figure we show the sinc interpolation, for all the samples, obtained using our function *sincinterp*. The sampling is implemented using our function *sampling*.

Sampling with aliasing—In Fig. 8.9 we show the case when the sampling frequency is $f_s = 800 < 2f_0 = 2000$, so that in this case we have aliasing when sampling $x(t)$. This can be seen in the sampled signal $y(t)$ in the top-left of Fig. 8.9, which appears as if we were sampling a sinusoid of lower frequency. It can also be seen in the magnitude spectra of $x(t)$ and $y(t)$: $|X(\Omega)|$ is the same as in the pre-

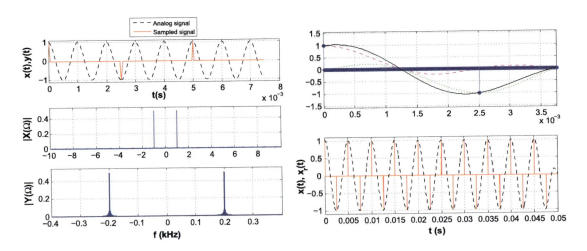

FIGURE 8.9

Aliasing: sampling simulation of $x(t) = \cos(2000\pi t)$ using $f_s = 800$ samples/s. Plots on the left display the original signal $x(t)$ and the sampled signal $y(t)$ (looks like a lower frequency signal being sampled). The bottom-left
figure displays the magnitude spectra of $x(t)$ and $y(t)$, or $|X(\Omega)|$ and $|Y(\Omega)|$ (periodic, the period shown displays
a lower frequency than $|X(\Omega)|$). Sinc interpolation for three samples and for the whole signal are displayed on
the right of the figure. The reconstructed signal $x_r(t)$ is a sinusoid of fundamental period 0.5×10^{-2} or of 200 Hz
frequency due to the aliasing.

vious case, but now $|Y(\Omega)|$, which is a period of the spectrum of the sampled signal $y(t)$, displays a fre-
quency of 200 Hz, lower than that of $x(t)$, within the frequency range $[-f_s/2, f_s/2] = [-400, 400]$ Hz.
Aliasing has occurred. Finally, the sinc interpolation gives a sinusoid of frequency 0.2 kHz, different
from $x(t)$, as shown on the bottom-right of Fig. 8.9.

Similar situations occur when a more complex signal is sampled. If the signal to be sampled is
$x(t) = 2 - \cos(\pi f_0 t) - \sin(2\pi f_0 t)$ where $f_0 = 500$ Hz, if we use a sampling frequency $f_s = 6000 >$
$2 f_{max} = 2 f_0 = 1000$ Hz, there will be no aliasing. On the other hand, if the sampling frequency is $f_s =$
$800 < 2 f_{max} = 2 f_0 = 1000$, frequency aliasing will occur. In the no aliasing sampling, the spectrum
$|Y(\Omega)|$ (in a frequency range $[-3000, 3000] = [-f_s/2, f_s/2]$) corresponding to a period of the Fourier
transform of the sampled signal $y(t)$, shows the same frequencies as $|X(\Omega)|$. The reconstructed signal
equals the original signal. See left figure in Fig. 8.10. When we use $f_s = 800$ Hz, the given signal $x(t)$
is under-sampled and aliasing occurs. The magnitude spectrum $|Y(\Omega)|$, corresponding to a period of
the Fourier transform of the under-sampled signal $y(t)$, does not show the same frequencies as $|X(\Omega)|$.
The reconstructed signal shown in the bottom-right figure in Fig. 8.10 does not resemble the original
signal.

The following function implements the sampling and computes the Fourier transform of the
continuous-time signal and of the sampled signal using the fast Fourier transform. It gives the range of
frequencies for each of the spectra.

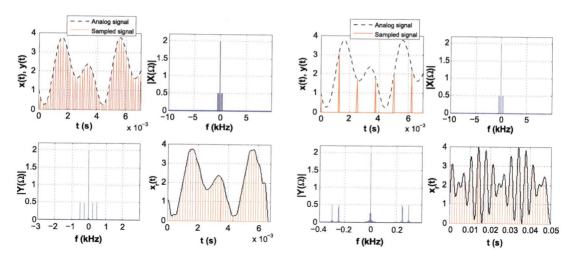

FIGURE 8.10

Sampling of $x(t) = 2 - \cos(500\pi t) - \sin(1000\pi t)$ with no aliasing ($f_s = 6000$ samples/s) on the left and with aliasing ($f_s = 800$ samples/s) on the right. When there is no aliasing the reconstructed signal coincides with the original, but that is not the case when aliasing occurs (see bottom-right figure).

```
function [y,y1,X,fx,Y,fy]=sampling(x,L,fs)
%
% Sampling
%   x   analog signal
%   L   length of simulated x
%   fs  sampling rate
%   y   sampled signal
%   X,Y magnitude spectra of x,y
%   fx, fy frequency ranges for X,Y
%
fsim=20000;                         % simulation sampling frequency
% sampling with rate fsim/fs
delta=floor(fsim/fs);
y1=zeros(1,L);
y1(1:delta:L)=x(1:delta:L);
y=x(1:delta:L);
% approximate analog FT of signals
dtx=1/fsim;
X=fftshift(abs(fft(x)))*dtx;
N=length(X); k=0:(N-1); fx=1/N.*k; fx=fx*fsim/1000-fsim/2000;
dty=1/fs;
Y=fftshift(abs(fft(y)))*dty;
N=length(Y); k=0:(N-1); fy=1/N.*k; fy=fy*fs/1000-fs/2000;
```

The following function computes the sinc interpolation of the samples.

```
function [t,xx,xr]= sincinterp(x,Ts)
%
% Sinc interpolation
% x sampled signal
% Ts sampling period of x
% xx,xr original samples and reconstructed in range t
%
dT=1/100; N=length(x)
t=0:dT:N;
xr=zeros(1,N*100+1);
for k=1:N,
   xr=xr+x(k)*sinc(t-(k-1));
end
xx(1:100:N*100)=x(1:N);
xx=[xx zeros(1,99)];
NN=length(xx); t=0:NN-1;t=t*Ts/100;
```

8.2.7 SAMPLING MODULATED SIGNALS

The given Nyquist sampling rate condition applies to low-pass or baseband signals. Sampling of band-pass signals is used in simulations of communication systems and in the implementation of modulation systems in software defined radio. For modulated signals it can be shown that the sampling rate depends on the bandwidth of the message or modulating signal, rather than on the maximum frequency of the modulated signal. This result provides significant savings in the number of samples given its independence from the carrier's frequency. Thus a voice message transmitted via a satellite communication system with a carrier of 6 GHz, for instance, would only need to be sampled at about a 10 kHz rate, rather than at 12 GHz as determined by the Nyquist sampling rate condition considering the maximum frequency of the modulated signal.

Consider a modulated signal $x(t) = m(t)\cos(\Omega_c t)$ where $m(t)$ is the message and $\cos(\Omega_c t)$ is the carrier with carrier frequency

$$\Omega_c >> \Omega_{max}$$

Ω_{max} being the maximum frequency present in the message. The sampling of $x(t)$ with a sampling period T_s generates in the frequency domain a superposition of the spectrum of $x(t)$ shifted in frequency by Ω_s and multiplied by $1/T_s$. Intuitively, to avoid aliasing the shifting in frequency should be such that there is no overlapping of the shifted spectra, which would require that

$$(\Omega_c + \Omega_{max}) - \Omega_s < (\Omega_c - \Omega_{max}) \Rightarrow \Omega_s > 2\Omega_{max} \text{ or } T_s < \frac{\pi}{\Omega_{max}}.$$

Thus, the sampling period depends on the bandwidth Ω_{max} of the message $m(t)$ rather than on the maximum frequency present in the modulated signal $x(t)$. A formal proof of this result requires the quadrature representation of bandpass signals typically considered in communication theory.

If the message $m(t)$ of a modulated signal $x(t) = m(t)\cos(\Omega_c)$ has a bandwidth B Hz, $x(t)$ can be reconstructed from samples taken at a sampling rate

$$f_s \geq 2B$$

independent of the frequency Ω_c of the carrier $\cos(\Omega_c)$.

Example 8.4. Consider the development of an AM transmitter that uses a computer to generate the modulated signal and is capable of transmitting music and speech signals. Indicate how to implement the transmitter.

Solution: Let the message be $m(t) = x(t) + y(t)$, where $x(t)$ is a speech signal and $y(t)$ is a music signal. Since music signals display larger frequencies than speech signals, the maximum frequency of $m(t)$ is that of the music signals or $f_{max} = 22$ kHz. To transmit $m(t)$ using AM, we modulate it with a sinusoid of frequency $f_c > f_{max}$, say $f_c = 3f_{max} = 66$ kHz.

To satisfy the Nyquist sampling rate condition: the maximum frequency of the modulated signal is $f_c + f_{max} = (66 + 22)$ kHz $= 88$ kHz, and so we choose $T_s = 10^{-3}/176$ s/sample as the sampling period. However, according to the above results we can also choose $T_s = 1/(2B)$ where B is the bandwidth of $m(t)$ in Hz or $B = f_{max} = 22$ kHz, which gives $T_{s1} = 10^{-3}/44$ s/sample—four times larger than the previous sampling period T_s, so we choose T_{s1} as the sampling period.

The continuous-time signal $m(t)$ to be transmitted is inputted into an A/D converter in the computer capable of sampling at $1/T_{s1} = 44,000$ samples/s. The output of the converter is then multiplied by a computer generated sinusoid

$$\cos(2\pi f_c n T_{s1}) = \cos(2\pi \times 66 \times 10^3 \times (10^{-3}/44)n) = \cos(3\pi n) = (-1)^n$$

to obtain the AM signal. The AM digital signal is then inputted into a D/A converter and its output sent to an antenna for broadcasting.   □

ORIGINS OF THE SAMPLING THEORY—PART II

As mentioned in Chapter 0, the theoretical foundations of digital communication theory were given in the paper "A Mathematical Theory of Communication" by Claude E. Shannon in 1948. His results on sampling theory made possible the new areas of digital communications and digital signal processing.

Shannon was born in 1916, in Petoskey, Michigan. He studied electrical engineering and mathematics at the University of Michigan, pursued graduate studies in electrical engineering and mathematics at MIT, and then joined Bell Telephone Laboratories. In 1956, he returned to MIT to teach. Besides being a celebrated researcher, Shannon was an avid chess player. He developed a juggling machine, rocket-powered frisbees, motorized Pogo sticks, a mind-reading machine, a mechanical mouse that could navigate a maze and a device that could solve the Rubik cube puzzle. At Bell Labs, he was remembered for riding the halls on a unicycle while juggling three balls [59,40,72].

8.3 PRACTICAL ASPECTS OF SAMPLING

To process continuous-time signals with computers it is necessary to convert them into digital signals, and back into continuous-time signals. The analog-to-digital and digital-to-analog conversions

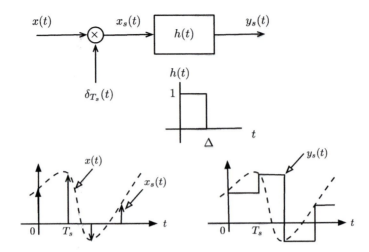

FIGURE 8.11

Sample-and-hold sampling system for $\Delta = T_s$. The sampled-and-hold signal $y_s(t)$ is a multi-level signal in this case.

are done by A/D and D/A converters. In practice these converters differ from the ideal versions we have discussed so far where the sampling is done with impulses, the discrete-time samples are assumed representable with infinite precision and the reconstruction is performed by an ideal low-pass filter. Pulses rather than impulses are needed, the discrete-time signals need to be discretized also in amplitude and the reconstruction filter needs to be reconsidered. These are the issues we consider in this section.

8.3.1 SAMPLE-AND-HOLD SAMPLING

In an actual A/D converter the time required to do the sampling, the quantization and the coding needs to be considered. Therefore, the width Δ of the sampling pulses cannot be zero as assumed. A **sample-and-hold sampling system** acquires a sample and holds it long enough for quantization and coding to be done before the next sample is acquired. The question is then how does this affect the sampling process and how does it differ from the ideal results obtained before? We hinted at the effects when we considered the PAM before.

The system shown in Fig. 8.11 generates the desired signal. Basically, we are modulating the ideal sampling signal $\delta_{T_s}(t)$ with the continuous-time input $x(t)$, giving an ideally sampled signal $x_s(t)$. This signal is then passed through a **zero-order-hold** filter, a LTI system having as impulse response $h(t)$ a pulse of the desired width $\Delta \leq T_s$. The output of the sample-and-hold system is a weighted sequence of shifted versions of the impulse response. Indeed, the output of the ideal sampler is $x_s(t) = x(t)\delta_{T_s}(t)$ and using the linearity and time invariance of the zero-order-hold system its output is

$$y_s(t) = (x_s * h)(t) \tag{8.20}$$

with a Fourier transform

$$Y_s(\Omega) = X_s(\Omega)H(j\Omega) = \left[\frac{1}{T_s}\sum_k X(\Omega - k\Omega_s)\right]H(j\Omega)$$

where the term in the brackets is the spectrum of the ideally sampled signal, and

$$H(j\Omega) = \frac{e^{-\Delta s/2}}{s}(e^{\Delta s/2} - e^{-\Delta s/2})|_{s=j\Omega} = \frac{\sin(\Delta\Omega/2)}{\Omega/2}e^{-j\Omega\Delta/2}$$

is the frequency response of the zero-order-hold system. Thus the spectrum of the zero-order-hold sampled signal is

$$Y_s(\Omega) = \left[\frac{1}{T_s}\sum_k X(\Omega - k\Omega_s)\right]\frac{\sin(\Delta\Omega/2)}{\Omega/2}e^{-j\Omega\Delta/2}. \qquad (8.21)$$

Remarks

1. Equation (8.20) can be written

$$y_s(t) = \sum_n x(nT_s)h(t - nT_s) \qquad (8.22)$$

i.e., $y_s(t)$ is a train of pulses $h(t) = u(t) - u(t - \Delta)$ shifted by nT_s, and weighted by the sample values $x(nT_s)$—a more realistic representation of the sampled signal.
2. Two significant changes due to considering pulses of width $\Delta > 0$ in the sampling are:

   - As indicated in equation (8.21) the spectrum of the ideal sampled signal $x_s(t)$ is now weighted by the sinc function of the frequency response $H(j\Omega)$ of the zero-order hold filter. Thus, the spectrum of the sampled signal using the sample-and-hold system will not be periodic and it will decay as Ω increases.
   - The reconstruction of the original signal $x(t)$ requires a more complex filter than the one used in the ideal sampling. Indeed, in this case the concatenation of the zero-order-hold filter with the reconstruction filter should be such that $H(s)H_r(s) = 1$, or the required filter be $H_r(s) = 1/H(s)$.

3. A circuit used for implementing the sample-and-hold is shown in Fig. 8.12. In this circuit the switch closes every T_s seconds and remains closed for a short time Δ. If the time constant $rC << \Delta$ the capacitor charges very fast to the value of the sample attained when the switch closes at some nT_s, and by setting the time constant $RC >> T_s$ when the switch opens Δ seconds later the capacitor slowly discharges. This cycle repeats providing a signal that approximates the output of the ideal sample-and-hold system explained before.
4. The digital-to-analog converter also uses a zero- or higher-order-holder to generate a continuous-time signal from the discrete signal coming out of the decoder into the D/A converter.

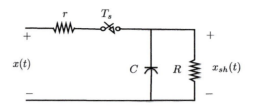

FIGURE 8.12

Sample-and-hold RC circuit. The input is $x(t)$ and the output is the sample-and-hold $x_{sh}(t)$.

8.3.2 QUANTIZATION AND CODING

Amplitude discretization of the sampled signal $x_s(t)$ is accomplished by a quantizer consisting of a number of fixed amplitude levels against which the sample amplitudes $\{x(nT_s)\}$ are compared. The output of the quantizer is one of the fixed amplitude levels that best represents $x(nT_s)$ according to some approximation scheme. The quantizer is a non-linear system.

Independent of how many levels or, equivalently, of how many bits are allocated to represent each level of the quantizer, in general there is a possible error in the representation of each sample. This is called the **quantization error**. To illustrate this, consider the 2-bit or four-level quantizer shown in Fig. 8.13. The input of the quantizer are the samples $\{x(nT_s)\}$, which are compared with the values in the bins $[-2\Delta, -\Delta]$, $[-\Delta, 0]$, $[0, \Delta]$ and $[\Delta, 2\Delta]$. Depending on which of these bins the sample falls in it is replaced by the corresponding levels -2Δ, $-\Delta$, 0 or Δ, respectively. The value of the quantization step Δ for the four-level quantizer is

$$\Delta = \frac{\text{dynamic range of signal}}{2^b} = \frac{2\max|x(t)|}{2^2} \tag{8.23}$$

where $b = 2$ is number of bits of the code assigned to each level. The bits assigned to each of the levels uniquely represents the different levels $[-2\Delta, -\Delta, 0, \Delta]$. As to how to approximate the given sample to one of these levels, it can be done by **rounding** or by **truncating**. The quantizer shown in Fig. 8.13 approximates by truncation, i.e., if the sample $k\Delta \leq x(nT_s) < (k+1)\Delta$, for $k = -2, -1, 0, 1$, then it is approximated by the level $k\Delta$.

To see how quantization and coding are done, and how to obtain the quantization error, let the sampled signal be

$$x(nT_s) = x(t)|_{t=nT_s}.$$

The given four-level quantizer is such that if the sample $x(nT_s)$ is such that

$$k\Delta \leq x(nT_s) < (k+1)\Delta \quad \Rightarrow \quad \hat{x}(nT_s) = k\Delta \qquad k = -2, -1, 0, 1. \tag{8.24}$$

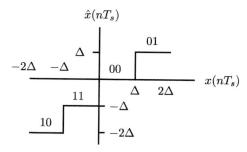

FIGURE 8.13

Four-level quantizer and coder.

The sampled signal $x(nT_s)$ is the input of the quantizer and the quantized signal $\hat{x}(nT_s)$ is its output. Thus whenever

$$
\begin{aligned}
-2\Delta \leq x(nT_s) < -\Delta &\Rightarrow \hat{x}(nT_s) = -2\Delta, \\
-\Delta \leq x(nT_s) < 0 &\Rightarrow \hat{x}(nT_s) = -\Delta, \\
0 \leq x(nT_s) < \Delta &\Rightarrow \hat{x}(nT_s) = 0, \\
\Delta \leq x(nT_s) < 2\Delta &\Rightarrow \hat{x}(nT_s) = \Delta.
\end{aligned}
$$

To transform the quantized values into unique binary 2-bit values, one could use a code such as

$$
\begin{array}{ccc}
\hat{x}(nT_s) & \Rightarrow & \text{binary code} \\
-2\Delta & & 10, \\
-\Delta & & 11, \\
0\Delta & & 00, \\
\Delta & & 01,
\end{array}
$$

which assigns a unique 2-bit binary number to each of the 4 quantization levels. Notice that the first bit of this code can be considered a sign bit, "1" for negative levels and "0" for positive levels.

If we define the quantization error as

$$
\varepsilon(nT_s) = x(nT_s) - \hat{x}(nT_s),
$$

and use the characterization of the quantizer given by equation (8.24) as

$$
\hat{x}(nT_s) \leq x(nT_s) \leq \hat{x}(nT_s) + \Delta
$$

by subtracting $\hat{x}(nT_s)$ from each of the terms shows that the quantization error is bounded as follows:

$$
0 \leq \varepsilon(nT_s) \leq \Delta \tag{8.25}
$$

i.e., the quantization error for the 4-level quantizer being considered is between 0 and Δ. This expression for the quantization error indicates that one way to decrease the quantization error is to make the quantization step Δ smaller. Increasing the number of bits of the A/D converter makes Δ smaller (see

Equation (8.23) where the denominator is 2 raised to the number of bits) which in turn makes smaller the quantization error, and improves the quality of the A/D converter.

In practice, the quantization error is random and so it needs to be characterized probabilistically. This characterization becomes meaningful when the number of bits is large, and when the input signal is not a deterministic signal. Otherwise, the error is predictable and thus not random. Comparing the energy of the input signal to the energy of the error, by means of the so-called signal to noise ratio (SNR), it is possible to determine the number of bits that are needed in a quantizer to get a reasonable quantization error.

Example 8.5. Suppose we are trying to decide between an 8- and a 9-bit A/D converter for a certain application where the signals in this application are known to have frequencies that do not exceed 5 kHz. The dynamic range of the signals is 10 V, so that the signal is bounded as $-5 \leq x(t) \leq 5$. Determine an appropriate sampling period and compare the percentage of error for the two A/Ds of interest.

Solution: The first consideration in choosing the A/D converter is the sampling period, so we need to get an A/D converter capable of sampling at $f_s = 1/T_s > 2 f_{max}$ samples/s. Choosing $f_s = 4 f_{max} = 20K$ samples/s then $T_s = 1/20$ ms/sample or 50 μs/sample. Suppose then we look at the 8-bit A/D converter, the quantizer has $2^8 = 256$ levels so that the quantization step is $\Delta = 10/256$ volts and if we use a truncation quantizer the quantization error would be

$$0 \leq \varepsilon(nTs) \leq 10/256.$$

If we find that objectionable we can then consider the 9-bit A/D converter, with a quantizer of $2^9 = 512$ levels and the quantization step $\Delta = 10/512$ or half that of the 8-bit A/D converter, and

$$0 \leq \varepsilon(nT_s) \leq 10/512.$$

Thus, by increasing one bit we cut the quantization error in half from the previous quantizer. Inputting a signal of constant amplitude 5 into the 9-bit A/D gives a quantization error of $[(10/512)/5] \times 100\% = (100/256)\% \approx 0.4\%$ in representing the input signal. For the 8-bit A/D it would correspond to 0.8% error.   □

8.3.3 SAMPLING, QUANTIZING AND CODING WITH MATLAB

The conversion of a continuous-time signal into a digital signal consists of three steps: sampling, quantizing and coding. These are the three operations an A/D converter does. To illustrate them consider a sinusoid $x(t) = 4\cos(2\pi t)$. Its sampling period, according to the Nyquist sampling rate condition, is

$$T_s \leq \pi/\Omega_{max} = 0.5 \text{ s/sample}$$

as the maximum frequency of $x(t)$ is $\Omega_{max} = 2\pi$. We let $T_s = 0.01$ (s/sample) to obtain a sampled signal $x_s(nT_s) = 4\cos(2\pi nT_s) = 4\cos(2\pi n/100)$, a discrete sinusoid of period 100. The following script is used to get the sampled $x[n]$ and the quantized $x_q[n]$ signals and the quantization error $\varepsilon[n]$ (see Fig. 8.14).

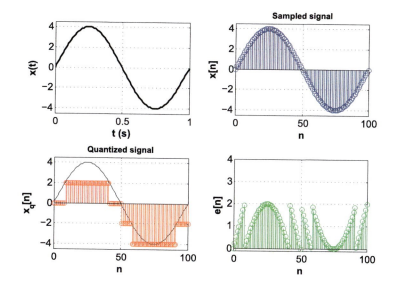

FIGURE 8.14

A period of sinusoid $x(t) = 4\cos(2\pi t)$ (left-top), sampled sinusoid using $T_s = 0.01$ (right-top), quantized sinusoid using 4 levels (left-bottom), quantization error (right-bottom) $0 \leq \varepsilon \leq \Delta = 2$.

```
%%
% Sampling, quantization and coding
%%
clear all; clf
% continuous-time signal
t=0:0.01:1; x=4*sin(2*pi*t);
% sampled signal
Ts=0.01; N=length(t); n=0:N-1;
xs=4*sin(2*pi*n*Ts);
% quantized signal
Q=2;          % quantization levels is 2Q
[d,y,e]=quantizer(x,Q);
% binary signal
z=coder(y,d)
```

The quantization of the sampled signal is implemented with our function *quantizer*, which compares each of the samples $x_s(nT_s)$ with 4 levels and assigns to each the corresponding level. Notice the approximation of the values given by the quantized signal to the actual values of the signal. The difference between the original and the quantized signal, or the quantization error, $\varepsilon(nT_s)$, is also computed and shown in Fig. 8.14.

```
function [d,y,e]=quantizer(x,Q)
%  Input: x, signal to be quantized at 2Q levels
%  Outputs: y, quantized signal
%              e, quantization error
%  USE [y,e]=midriser(x,Q)
%
N=length(x); d=max(abs(x))/Q;
for k=1:N,
  if x(k)>=0,
     y(k)=floor(x(k)/d)*d;
  else
    if x(k)==min(x),
       y(k)=(x(k)/abs(x(k)))*(floor(abs(x(k))/d)*d);
    else
        y(k)=(x(k)/abs(x(k)))*(floor(abs(x(k))/d)*d+d);
    end
  end
  if y(k)==2*d,
     y(k)=d;
  end
end
e=x-y
```

The binary signal corresponding to the quantized signal is computed using our function *coder* which assigns the binary codes '10', '11', '00' and '01' to the 4 possible levels of the quantizer. The result is a sequence of 0s and 1s, each pair of digits sequentially corresponding to each of the samples of the quantized signal. The following is the function used to effect this coding.

```
function z1=coder(y,delta)
%  Coder for 4-level quantizer
%  input: y quantized signal
%  output: z1 binary sequence
%  USE z1=coder(y)
%
z1='00';  % starting code
N=length(y);
for n=1:N,
 y(n)
    if y(n)== delta
       z='01';
    elseif y(n)==0
       z='00';
    elseif y(n)== -delta
       z='11';
    else
```

```
        z='10';
    end
    z1=[z1 z];
end
M=length(z1);
z1=z1(3:M) % get rid of starting code
```

8.4 APPLICATION TO DIGITAL COMMUNICATIONS

The concepts of sampling and binary signal representation introduced by Shannon in 1948 changed the implementation of communications. Analog communications transitioned into digital communications, while telephony and radio have coalesced into wireless communications. The scarcity of radio spectrum changed the original focus on bandwidth and energy efficiency into more efficient utilization of the available radio spectrum by sharing it, and by transmitting different types of data together. Wireless communications has allowed the growth of cellular telephony, personal communication systems and wireless local-area networks.

Modern digital communications was initiated with the concept of pulse code modulation (PCM) which allowed the transmission of binary signals. PCM is a practical implementation of sampling, quantization and coding, or analog-to-digital conversion, of an analog message into a digital message. Using the sample representation of a message, the idea of mixing several messages—possibly of different types—developed into time-division multiplexing (TDM) which is the dual of frequency-division multiplexing (FDM). In TDM, samples from different messages are interspersed into one message that can be quantized, coded and transmitted and then separated into the different messages at the receiver.

As multiplexing techniques, FDM and TDM became the basis for similar techniques used in wireless communications. Typically, three forms of sharing the available radio spectrum are: frequency-division multiple access (FDMA) where each user is assigned a band of frequencies all of the time; time-division multiple access (TDMA) where a user is given access to the available bandwidth for a limited time; and code division multiplexing (CDMA) where users share the available spectrum without interfering with each other thanks to a unique code given to each user.

To illustrate the application of the sampling theory we will introduce some of these techniques avoiding technical details, which we leave to texts in communications, telecommunications and wireless communications. As you will learn, digital communication has a number of advantages over analog communication:

- The cost of digital circuits decreases as digital technology becomes more prevalent.
- Data from voice and video can be merged with computer data and transmitted over a common transmission system.
- Digital signals can be denoised easier than analog signals, and errors in digital communications can be dealt with special coding.

However, digital communication systems require a much larger bandwidth than analog communication systems, and quantization noise is introduced when converting an analog signal into a digital signal.

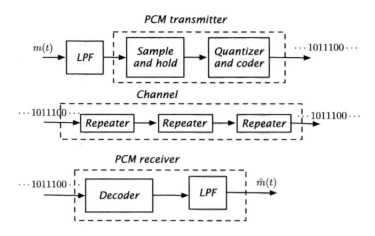

FIGURE 8.15

PCM system: transmitter, channel and receiver.

8.4.1 PULSE CODE MODULATION

Pulse code modulation (PCM) can be seen as an implementation of analog-to-digital conversion of analog signals to generate a serial bit stream. This means that sampling applied to a continuous-time message gives a pulse amplitude modulated (PAM) signal which is then quantized and assigned a binary code to uniquely represent the different quantization levels. If each of the digital words has b binary digits, there are 2^b levels and to each a different code word is assigned. An example of a code commonly used is the Gray code where only one bit changes from one quantization level to another. The most significant bit can be thought to correspond to the sign of the signal (1 for positive values, and 0 for negative values) and the others differentiate each level.

PCM is widely used in digital communications given that it is realized with inexpensive digital circuitry, and allows merging and transmission of data from different sources (audio, video, computers, etc.) by means of time-division multiplexing which we will see next. Moreover, PCM signals can easily be regenerated by repeaters in long-distance digital telephony. Despite all of these advantages, we need to remember that because of the sampling and the quantization the process used to obtain PCM signals is neither linear nor time-invariant, and as such its analysis is complicated. Fig. 8.15 shows the transmitter, channel and receiver of a PCM system.

The main disadvantage of PCM is that it has a wider bandwidth than that of the analog message it represents. This is due to the shape of the pulses in the transmitted signal. Representing the PCM signal $s(t)$ as a sum of rectangular pulses $\varphi(t)$, of width τ,

$$s(t) = \sum_{n=0}^{N-1} b_n \varphi(t - n\tau_s)$$

where b_n are the corresponding binary bits, its spectrum is

$$S(\Omega) = \sum_{n=0}^{N-1} b_n \mathcal{F}[\varphi(t - n\tau_s)] = \sum_{n=0}^{N} b_n \phi(\Omega) e^{-jn\tau_s \Omega}$$

with no finite bandwidth given that $\phi(\Omega)$ are sinc functions. On the other hand, if we could let the pulse be a sinc function, having an infinite time support, $s(t)$ would also have an infinite time support but since sincs have finite bandwidth, the spectrum $S(\Omega)$ would have finite bandwidth.

Example 8.6. Suppose you have a binary signal 01001001, with a duration of 8 s, and wish to represent it using rectangular and sinc pulses. Consider the bandwidth of each of the representations.

Solution: Using pulses $\varphi(t)$ the digital signal can be expressed as

$$s(t) = \sum_{n=0}^{7} b_n \varphi(t - n\tau_s)$$

where b_n are the binary digits of the digital signal, i.e., $b_0 = 0$, $b_1 = 1$, $b_2 = 0$, $b_3 = 0$, $b_4 = 1$, $b_5 = 0$, $b_6 = 0$, and $b_7 = 1$, and $\tau_s = 1$. Thus

$$s(t) = \varphi(t - 1) + \varphi(t - 4) + \varphi(t - 7)$$

and the spectrum of $s(t)$ is

$$
\begin{aligned}
S(\Omega) &= \phi(\Omega)(e^{-j\Omega} + e^{-j4\Omega} + e^{-j7\Omega}) = \phi(\Omega)e^{-j4\Omega}(e^{j3\Omega} + 1 + e^{-j3\Omega}) \\
&= \phi(\Omega)e^{-j4\Omega}(1 + 2\cos(3\Omega)) \qquad \text{so that} \\
|S(\Omega)| &= |\phi(\Omega)||(1 + 2\cos(3\Omega))|.
\end{aligned}
$$

If the pulses are rectangular

$$\varphi(t) = u(t) - u(t - 1)$$

the PCM signal will have an infinite support spectrum, because the pulse is of finite support. On the other hand, if we use sinc functions

$$\varphi(t) = \frac{\sin(\pi t/\tau_s)}{\pi t}$$

as pulses its time support is infinite but its frequency support is finite, i.e., the sinc function is band-limited. In which case, the spectrum of the PCM signal is also of finite support.

 If this digital signal is transmitted and received without any distortion, at the receiver we can use the orthogonality of the $\varphi(t)$ signals or sample the received signal at nT_s to obtain the b_n. Clearly each of these pulses has disadvantages, the advantage of having a finite support in the time or in the frequency becomes a disadvantage in the other domain.   □

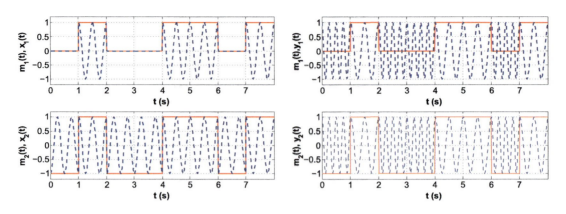

FIGURE 8.16

Pulse signals (continuous line) $m_1(t)$, and $m_2(t)$ corresponding to binary sequence 01001101. On the left, ASK signal $x_1(t) = m_1(t) \cos(4\pi t)$ (top) and PSK signal $x_2(t) = m_2(t) \cos(4\pi t)$ (bottom) in dashed lines. On the right, in dashed lines the FSK signals $y_1(t)$ (top) and $y_2(t)$ (bottom) equal $\cos(4\pi t)$ when $m_1(t)$, $m_2(t)$ are 1 and $\cos(8\pi t)$ when $m_1(t)$, $m_2(t)$ are 0 or -1.

Baseband and Band-Pass Communication Systems

A baseband signal can be transmitted over a pair of wires (like in a telephone), coaxial cables or optical fibers. But a baseband signal cannot be transmitted over a radio link or a satellite because this would require a large antenna to radiate the low-frequency spectrum of the signal. Hence the baseband signal spectrum must be shifted to a higher frequency by modulating a carrier with the baseband signal. This can be done by amplitude and by angle modulation (frequency and phase). Several forms are possible.

Example 8.7. Suppose the binary signal 01001101 is to be transmitted over a radio link using AM and FM modulation. Discuss the different band-pass signals obtained.

Solution: The binary message can be represented as a sequence of pulses with different amplitudes. For instance, we could represent the binary digit 1 by a pulse of constant amplitude, and the binary 0 is represented by switching off the pulse (see the corresponding modulating signal $m_1(t)$ in Fig. 8.16). Another possible representation would be to represent the binary digit 1 with a positive pulse of constant amplitude, and 0 with the negative of the pulse used for 1 (see the corresponding modulating signal $m_2(t)$ in Fig. 8.16).

In AM modulation, if we use $m_1(t)$ to modulate a sinusoidal carrier $\cos(\Omega_0 t)$ we obtain the amplitude-shift keying (ASK) signal shown in Fig. 8.16 (top-left). On the other hand, when using $m_2(t)$ to modulate the same carrier we obtain a phase-shift keying (PSK) signal shown in Fig. 8.16 (bottom-left). In this case, the phase of the carrier is shifted 180° as the pulse changes from positive to negative.

Using FM modulation, the symbol 0 is transmitted using a windowed cosine of some frequency Ω_{c0} and the symbol 1 is transmitted with a windowed cosine of frequency Ω_{c1} resulting in frequency-shift

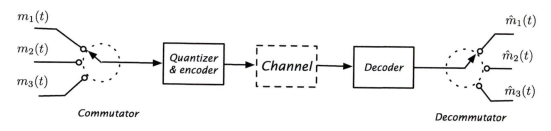

FIGURE 8.17

TDM system: transmitter, channel and receiver.

keying (FSK). The data is transmitted by varying the frequency. In this case it is possible to get the same modulated signal for both $m_1(t)$ and $m_2(t)$. The modulated signals are shown on the right in Fig. 8.16. The ASK, PSK, and FSK are also known as BASK, BPSK, and BFSK where the word binary (B) has been added to the corresponding amplitude-, phase-, and frequency-shift keying.   □

8.4.2 TIME-DIVISION MULTIPLEXING

In a telephone system, multiplexing enables multiple conversations to be carried across a single shared circuit. The first multiplexing system used was frequency-division multiplexing (FDM) which we covered in an earlier chapter. In FDM an available bandwidth is divided among different users. In the case of voice communications, each user is allocated a bandwidth of 4 kHz which provides good fidelity. In FDM, a user could use the allocated frequencies all of the time, but the user could not go outside the allocated band of frequencies.

Pulse modulated signals have large bandwidths and as such when transmitted together they overlap in frequency, interfering with each other. However, these signals only provide information at each of the sampling times, thus one could insert in between these times samples from other signals generating a combined signal with a smaller sampling time. The samples corresponding to each of the multiplexed signals are separated at the receiver. This is the principle of **time-division multiplexing**, or TDM, where pulses from different signals are interspersed into one signal and converted into a PCM signal and transmitted. See Fig. 8.17. At the receiver, the signal is changed back into the pulse modulated signal and separated into the number of signals interspersed at the input. Repeaters placed between the transmitter and the receiver regenerate noisy binary signal along the way. The noisy signal coming into the repeater is thresholded to known binary levels and resent.

TDM allows the transmission of different types of data, and mixture of analog and digital using different multiplexing techniques. Not to lose information, the switch at the receiver needs to be synchronized with the switch at the transmitter. **Frame synchronization** consists in sending a synchronizing signal for each frame. An additional channel is allocated for this purpose. To accommodate more users, the width of the pulses used for each user needs to become narrower, which increases the overall bandwidth of the multiplexed signal.

8.5 WHAT HAVE WE ACCOMPLISHED? WHERE DO WE GO FROM HERE?

The material in this chapter is the bridge between continuous-time and digital signal processing. The sampling theory provides the necessary information to convert a continuous-time signal into a discrete-time signal and then into a digital signal. It is the frequency representation of a continuous-time signal that determines the way in which it can be sampled and reconstructed. Analog-to-digital and digital-to-analog converters are the devices that in practice convert a continuous-time signal into a digital signal and back. Two parameters characterizing these devices are the sampling rate and the number of bits used to code each sample. The rate of change of a signal determines the sampling rate, while the precision in representing the samples determines the number of levels of the quantizer and the number of bits assigned to each sample. In the latter part of this chapter we illustrated the application of the sampling theory in communications. As seen, the motivation for an efficient use of the radio spectrum has generated very interesting communication techniques.

In the following chapters we will consider the analysis of discrete-time signals, as well as the analysis and synthesis of discrete systems. The effect of quantization in the processing and design of systems is an important problem that is left for texts in digital signal processing.

8.6 PROBLEMS
8.6.1 BASIC PROBLEMS

8.1 Consider the sampling of real signals.

(a) Typically, a speech signal that can be understood over a telephone line shows frequencies from about 100 Hz to about 5 kHz. What would be the sampling frequency f_s (samples/s) that would be used to sample speech without aliasing? How many samples would you need to save when storing an hour of speech? If each sample is represented by 8 bits, how many bits would you have to save for the hour of speech?

(b) A music signal typically displays frequencies from 0 up to 22 kHz, what would be the sampling frequency f_s that would be used in a CD player?

(c) If you have a signal that combines voice and musical instruments, what sampling frequency would you use to sample this signal? How would the signal sound if played at a frequency lower than the Nyquist sampling frequency?

Answers: (a) $f_s \geq 2f_{max} = 10^4$ samples/s; (c) $f_s \geq 2f_{max} = 44$ kHz.

8.2 Consider the signal $x(t) = 2\sin(0.5t)/t$.

(a) Is $x(t)$ band-limited? If so, indicate its maximum frequency Ω_{max}.

(b) Suppose that $T_s = 2\pi$, how does Ω_s relate to the Nyquist frequency $2\Omega_{max}$? Explain. What is the sampled signal $x(nT_s)$ equal to?

(c) Determine the spectrum of the sampled signal $X_s(\Omega)$ when $T_s = 2\pi$ and indicate how to reconstruct the original signal from the sampled signal.

Answer: $x(nT_s) = \sin(\pi n)/\pi n$.

8.3 The signal $x(t)$ has a Fourier transform $X(\Omega) = u(\Omega + 1) - u(\Omega - 1)$ thus it is band-limited, suppose we generate a new signal $y(t) = (x * x)(t)$, i.e., it is the convolution of $x(t)$ with itself.

(a) Find $x(t)$ and indicate its support.
(b) Is it true that $y(t) = x(t)$? To determine if this is so calculate $Y(\Omega)$.
(c) What would be the largest value of the sampling period T_s that would not cause aliasing when sampling $y(t)$?
(d) What happens if we choose $T_s = \pi$? How does this relate to the Nyquist condition?
Answers: $y(t) = x(t)$; $T_s \leq \pi$.

8.4 A message $m(t)$ with a bandwidth of $B = 2$ kHz modulates a cosine carrier of frequency 10 kHz to obtain a modulated signal $s(t) = m(t)\cos(20 \times 10^3 \pi t)$.
(a) What is the maximum frequency of $s(t)$? What would be the values of the sampling frequency f_s in Hz, according to the Nyquist sampling condition, that could be used to sample $s(t)$?
(b) Assume the spectrum of $m(t)$ is a triangle, with maximum amplitude 1, carefully plot the spectrum of $s(t)$. Would it be possible to sample $s(t)$ with a sampling frequency $f_s = 10$ kHz and recover the original message? Obtain the spectrum of the sampled signal and show how you would recover $m(t)$, if possible.
Answers: Nyquist: $f_s \geq 2f_{max} = 24 \times 10^3$ Hz; yes, it is possible.

8.5 You wish to recover the original analog signal $x(t)$ from its sampled form $x(nT_s)$.
(a) If the sampling period is chosen to be $T_s = 1$ so that the Nyquist sampling rate condition is satisfied, determine the magnitude and cutoff frequency of an ideal low-pass filter $H(j\Omega)$ to recover the original signal, plot them.
(b) What would be a possible maximum frequency of the signal? Consider an ideal and a non-ideal low-pass filters. Explain.
Answer: Ideal low-pass filter with cutoff frequency $\Omega_{max} < \Omega_c < 2\pi - \Omega_{max}$, magnitude 1.

8.6 A periodic signal has the following Fourier series representation:

$$x(t) = \sum_{k=-\infty}^{\infty} \sqrt{0.5^{|k|}} e^{jkt}.$$

(a) Is $x(t)$ band-limited?
(b) Calculate the power P_x of $x(t)$.
(c) If we approximate $x(t)$ as

$$\hat{x}(t) = \sum_{k=-N}^{N} \sqrt{0.5^{|k|}} e^{jkt}$$

by using $2N + 1$ terms, find N so that $\hat{x}(t)$ has 90% of P_x.
(d) If we use N found above, determine the sampling period that can be used to sample $\hat{x}(t)$ without aliasing.
Answers: $x(t)$ is not band-limited; $P_x = 3$.

8.7 Consider the periodic signals

$$x_1(t) = \cos(2\pi t), \quad x_2(t) = \cos((2\pi + \phi)t).$$

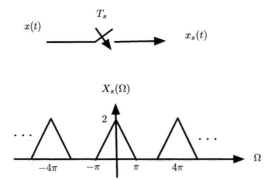

FIGURE 8.18

Problem 8.8.

    (a) Let $\phi = 4\pi$, show that if we sample these signals using $T_s = 0.5$ we get the same sample values from the two signals. Is any of these signals aliased? Explain.

    (b) What would be possible values of ϕ so that when using $T_s = 0.5$ we get the same sample values from the two signals? How does ϕ relate to the sampling frequency Ω_s?

Answer: $x_1(0.5n) = x_2(0.5n) = \cos(\pi n)$.

8.8 A signal $x(t)$ is sampled with no aliasing using an ideal sampler. The spectrum of the sampled signal is shown in Fig. 8.18.

    (a) Determine the sampling period T_s used.

    (b) Determine the signal $x(t)$.

    (c) Sketch and label the magnitude response of an ideal low-pass filter that can be used to recover the original signal $x(t)$ from the sampled signal $x_s(t)$ without distortion.

Answers: $\Omega_s = 4\pi$; ideal low-pass filter of dc gain $T_s = 1/2$ and bandwidth $\pi < B < 3\pi$.

8.6.2 PROBLEMS USING MATLAB

8.9 Sampling of time-limited signals—Consider the signals $x(t) = u(t) - u(t-1)$, and $y(t) = r(t) - 2r(t-1) + r(t-2)$.

    (a) Is any of these signals band-limited? Explain.

    (b) Use Parseval's energy result to determine the maximum frequency for the signals keeping 90% of the energy of the signals. Use the function *fourier* to find the Fourier transform $X(\Omega)$ and find $Y(\Omega)$ in terms of $X(\Omega)$. To find the integrals values use *int* and *subs*.

    (c) If we use the sampling period corresponding to $y(t)$ to sample $x(t)$, would aliasing occur? Explain.

    (d) Determine a sampling period that can be used to sample both $x(t)$ and $y(t)$ without causing aliasing in either signal.

Answers: $E_x = 1$, $E_y = 2/3$; $|Y(\Omega)| = |X(\Omega)|^2$.

8.10 Antialiasing—Suppose you want to find a reasonable sampling period T_s for the noncausal exponential $x(t) = e^{-|t|}$.

(a) Find the Fourier transform of $x(t)$, and plot $|X(\Omega)|$. Is $x(t)$ band-limited? Use the symbolic function *fourier* to find the Fourier transform.

(b) Find a frequency Ω_0 so that 99% of the energy of the signal is in $-\Omega_o \leq \Omega \leq \Omega_o$.

(c) If we let $\Omega_s = 2\pi/T_s = 5\Omega_0$ what would be T_s?

(d) Determine the magnitude and bandwidth of an antialiasing filter that would change the original signal into the band-limited signal with 99% of the signal energy.

Answer: $X(\Omega) = 2/(1 + \Omega^2)$, $E_x = 1$.

8.11 Two-bit analog-to-digital converter—Let $x(t) = 0.8 \ \cos(2\pi t) + 0.15$, $0 \leq t \leq 1$, and zero otherwise, be the input to a 2-bit analog-to-digital converter.

(a) For a sampling period $T_s = 0.25$ s determine and plot using MATLAB the sampled signal

$$x(nT_s) = x(t)|_{t=nT_s}.$$

(b) The 4-level quantizer (see Fig. 1.2) corresponding to the 2-bit A/D is defined as

$$k\Delta \leq x(nT_s) < (k+1)\Delta \quad \rightarrow \quad \hat{x}(nT_s) = k\Delta, \ \ k = -2, -1, 0, 1,$$

where $x(nT_s)$, found above, is the input and $\hat{x}(nT_s)$ is the output of the quantizer. Let the quantization step be $\Delta = 0.5$. Plot the input/output characterization of the quantizer, and find the quantized output for each of the sample values of the sampled signal $x(nT_s)$.

(c) To transform the quantized values into unique binary 2-bit values, consider the following code:

$$\hat{x}(nT_s) = -2\Delta \quad \rightarrow \quad 10,$$
$$\hat{x}(nT_s) = -\Delta \quad \rightarrow \quad 11,$$
$$\hat{x}(nT_s) = 0\Delta \quad \rightarrow \quad 00,$$
$$\hat{x}(nT_s) = \Delta \quad \rightarrow \quad 01.$$

Obtain the digital signal corresponding to $x(t)$.

Answer: $x(nT_s) = x[n] = 0.8\cos(2\pi n \ 0.025) + 0.15 \qquad 0 \leq n \leq 40$.

DISCRETE-TIME SIGNALS AND SYSTEMS

CONTENTS

It's like deja-vu, all over again.

Lawrence "Yogi" Berra, Yankees baseball player (1925)

9.1 INTRODUCTION

As you will see in this chapter, the basic theory of discrete-time signals and systems is very much like that of continuous-time signals and systems. However, there are significant differences that need to be understood.

Discrete-time signals resulting from sampling of continuous-time signals are only available at uniform times determined by the sampling period—they are not defined in between samples. It is important to emphasize the significance of sampling according to the Nyquist sampling rate condition since the characteristics of discrete-time signals will depend on it. Given the knowledge of the sampling period, discrete-time signals are functions of an integer variable n which unifies the treatment of discrete-time signals obtained from analog signals by sampling and those that are naturally discrete. It will be seen also that the frequency in the discrete domain differs from the analog frequency. This radian discrete frequency cannot be measured and depends on the sampling period used whenever the discrete-time signals result from sampling.

Although the concept of periodicity of discrete-time signals coincides with that, for continuous-time signals, there are significant differences. As functions of an integer variable, discrete-time periodic signals must have integer periods. This imposes some restrictions that do not exist in continuous-time periodic signals. For instance, analog sinusoids are always periodic as their period can be a positive real number, however, that will not be the case for discrete sinusoids. It is possible to have discrete sinusoids that are not periodic, even if they resulted from the uniform sampling of periodic continuous-time sinusoids. Characteristics such as energy, power, and symmetry of continuous-time signals are conceptually the same for discrete-time signals. Likewise, one can define a set of basic signals just like those for continuous-time signals. However, some of these basic signals do not display the mathematical complications of their analog counterparts. For instance, the discrete impulse signal is defined at every integer value in contrast with the continuous impulse which is not defined at zero.

The discrete approximation of derivatives and integrals provides an approximation of ordinary differential equations, representing dynamic continuous-time systems, by difference equations. Extending the concept of linear time invariance to discrete-time systems, we obtain a convolution sum to represent them. Thus dynamic discrete-time systems can be represented by difference equations and convolution sums. A computationally significant difference with continuous-time systems is that the solution of difference equations can be recursively obtained, and that the convolution sum provides a class of systems that do not have a counterpart in the analog domain.

In this chapter, we also introduce the theory of two-dimensional signals and systems. Particular two-dimensional signals of interest are images that are processed using two-dimensional systems. We will see how the theory of one-dimensional signals and systems can be extended to two dimensions, and how this theory is much richer. In fact, although many of the one-dimensional properties are valid for two dimensions not all are; it is actually the one-dimensional signals and systems that are subsets of the two-dimensional signals and systems.

9.2 DISCRETE-TIME SIGNALS

A **discrete-time signal** $x[n]$ can be thought of as a real- or complex-valued function of the integer sample index n:

$$x[.] : \mathcal{I} \to \mathcal{R} \ (\mathcal{C})$$
$$n \quad x[n]. \tag{9.1}$$

The above means that for discrete-time signals the independent variable is an integer n, the sample index, and that the value of the signal at n, $x[n]$, is either a real- or a complex value. Thus, the signal is only defined at integer values n, no definition exists for values between the integers.

Remarks

1. It should be understood that a sampled signal $x(nT_s) = x(t)|_{t=nT_s}$ is a discrete-time signal $x[n]$ which is a function of n only. Once the value of T_s is known, the sampled signal only depends on n, the sample index. However, this should not prevent us in some situations from considering a discrete-time signal obtained through sampling as a function of time t where the signal values only exist at discrete times $\{nT_s\}$.

2. Although in many situations, discrete-time signals are obtained from continuous-time signals by sampling, that is not always the case. There are many signals which are **inherently discrete**. Think, for instance, of a signal consisting of the final values attained daily by the shares of a company in the stock market. Such a signal would consist of the values reached by the share in the days when the stock market opens and has no connection with a continuous-time signal. This signal is naturally discrete. A signal generated by a random number generator in a computer would be a sequence of real values and can be considered a discrete-time signal. Telemetry signals, consisting of measurements—e.g., voltages, temperatures, pressures—from a certain process, taken at certain times are also naturally discrete.

Example 9.1. Consider a sinusoidal signal $x(t) = 3\cos(2\pi t + \pi/4)$, $-\infty < t < \infty$. Determine an appropriate sampling period T_s according to the Nyquist sampling rate condition, and obtain the discrete-time signal $x[n]$ corresponding to the largest allowed sampling period.

Solution: To sample $x(t)$ so that no information is lost, the Nyquist sampling rate condition indicates that the sampling period should be

$$T_s \le \frac{\pi}{\Omega_{max}} = \frac{\pi}{2\pi} = 0.5 \text{ s/sample.}$$

For the largest allowed sampling period $T_s = 0.5$ s/sample we obtain

$$x[n] = 3\cos(2\pi t + \pi/4)|_{t=0.5n} = 3\cos(\pi n + \pi/4) \qquad -\infty < n < \infty,$$

which is a function of the integer n. ☐

Example 9.2. To generate the celebrated Fibonacci sequence of numbers, $\{x[n], \ n \ge 0\}$, we use the following recursive equation:

$$x[n] = x[n-1] + x[n-2], \quad n \ge 2, \text{ initial conditions: } x[0] = 0, \ x[1] = 1,$$

which is a difference equation with zero-input and two initial conditions. The Fibonacci sequence has been used to model different biological systems.[1] Find the Fibonacci sequence.

---

[1] Leonardo of Pisa (also known as Fibonacci) in his book Liber Abaci described how his sequence could be used to model the reproduction of rabbits over a number of months assuming bunnies begin breeding when they are a few months old.

Solution: The given equation allows us to compute the Fibonacci sequence recursively. For $n \geq 2$ we find

$$x[2] = 1 + 0 = 1, \quad x[3] = 1 + 1 = 2, \quad x[4] = 2 + 1 = 3, \quad x[5] = 3 + 2 = 5, \quad \cdots,$$

where we are simply adding the previous two numbers in the sequence. The sequence is purely discrete as it is not related to a continuous-time signal.   □

9.2.1 PERIODIC AND APERIODIC DISCRETE-TIME SIGNALS

A discrete-time signal $x[n]$ is periodic if

- it is defined for all possible values of n, $-\infty < n < \infty$, and
- there is a positive integer N such that

$$x[n + kN] = x[n] \tag{9.2}$$

for any integer k. The smallest value of N is called the **fundamental period** of $x[n]$.

An aperiodic signal does not satisfy one or both of the above conditions.
Periodic discrete-time sinusoids, of fundamental period N, are of the form

$$x[n] = A \cos\left(\frac{2\pi m}{N} n + \theta\right) \qquad -\infty < n < \infty \tag{9.3}$$

where the discrete frequency is $\omega_0 = 2\pi m/N$ (rad), for positive integers m and N which are not divisible by each other, and θ is the phase angle.

The definition of a discrete-time periodic signal is similar to that of continuous-time periodic signals, except for the fundamental period being an integer. That periodic discrete-time sinusoids are of the given form can easily be shown: shifting the sinusoid in (9.3) by a multiple k of the fundamental period N, we have

$$x[n + kN] = A \cos\left(\frac{2\pi m}{N}(n + kN) + \theta\right) = A \cos\left(\frac{2\pi m}{N} n + 2\pi mk + \theta\right) = x[n],$$

since we add to the original angle a multiple mk (an integer) of 2π, which does not change the angle. If the discrete frequency is not in the form $2\pi m/N$, the sinusoid is not periodic.

Remark

The unit of the discrete frequency ω is radians. Moreover, discrete frequencies repeat every 2π, i.e., $\omega = \omega + 2\pi k$ for any integer k, and as such we only need to consider the range $-\pi \leq \omega < \pi$. This is in contrast with the analog frequency Ω which has rad/s as unit, and its range is from $-\infty$ to ∞.

Example 9.3. Consider the sinusoids

$$x_1[n] = 2\cos(\pi n - \pi/3), \quad x_2[n] = 3\sin(3\pi n + \pi/2) \qquad -\infty < n < \infty.$$

From their frequencies determine if these signals are periodic, and if periodic their corresponding fundamental periods.

Solution: The frequency of $x_1[n]$ can be written

$$\omega_1 = \pi = \frac{2\pi}{2}$$

for integers $m = 1$ and $N = 2$, so that $x_1[n]$ is periodic of fundamental period $N_1 = 2$. Likewise the frequency of $x_2[n]$ can be written

$$\omega_2 = 3\pi = \frac{2\pi}{2}3$$

for integers $m = 3$ and $N = 2$, so that $x_2[n]$ is also periodic of fundamental period $N_2 = 2$, which can be verified as follows:

$$x_2[n + 2] = 3\sin(3\pi(n + 2) + \pi/2) = 3\sin(3\pi n + 6\pi + \pi/2) = x_2[n]. \quad \square$$

Example 9.4. What is true for analog sinusoids—that they are always periodic—is not true for discrete sinusoids. Discrete sinusoids can be non-periodic even if they result from uniformly sampling a periodic continuous-time sinusoid. Consider the discrete-time signal $x[n] = \cos(n + \pi/4)$, $-\infty < n < \infty$, which is obtained by sampling the analog sinusoid $x(t) = \cos(t + \pi/4)$, $-\infty < t < \infty$, with a sampling period $T_s = 1$ s/sample. Is $x[n]$ periodic? If so, indicate its fundamental period. Otherwise, determine values of the sampling period satisfying the Nyquist sampling rate condition that when used in sampling $x(t)$ result in periodic signals.

Solution: The sampled signal $x[n] = x(t)|_{t=nT_s} = \cos(n + \pi/4)$ has a discrete frequency $\omega = 1$ (rad) which cannot be expressed as $2\pi m/N$ for any integers m and N because π is an irrational number. So $x[n]$ is not periodic.

Since the frequency of the continuous-time signal $x(t) = \cos(t + \pi/4)$ is $\Omega_0 = 1$, the sampling period, according to the Nyquist sampling rate condition, should be

$$T_s \leq \frac{\pi}{\Omega_0} = \pi$$

and for the sampled signal $x(t)|_{t=nT_s} = \cos(nT_s + \pi/4)$ to be periodic of fundamental period N or

$$\cos((n + N)T_s + \pi/4) = \cos(nT_s + \pi/4) \quad \text{it is necessary that} \quad NT_s = 2k\pi$$

for an integer k, i.e., a multiple of 2π. Thus $T_s = 2k\pi/N \leq \pi$ satisfies the Nyquist sampling condition at the same time that ensures the periodicity of the sampled signal. For instance, if we wish to have a discrete sinusoid with fundamental period $N = 10$, then $T_s = k\pi/5$, for k chosen so that the Nyquist sampling rate condition is satisfied, i.e.,

$$0 < T_s = k\pi/5 \leq \pi \quad \text{so that} \quad 0 < k \leq 5.$$

From these possible values for k we choose $k = 1$ and 3 so that N and k are not divisible by each other and we get the desired fundamental period $N = 10$ (the values $k = 2$ and 4 would give 5 as the period, and $k = 5$ would give a period of 2 instead of 10). Indeed, if we let $k = 1$ then $T_s = 0.2\pi$ satisfies the Nyquist sampling rate condition, and we obtain the sampled signal

$$x[n] = x(t)|_{t=0.2\pi n} = \cos(0.2\pi n + \pi/4) = \cos\left(\frac{2\pi}{10}n + \frac{\pi}{4}\right),$$

which according to its frequency is periodic of fundamental period 10. For $k = 3$, we have $T_s = 0.6\pi < \pi$ and

$$x[n] = x(t)|_{t=0.6\pi n} = \cos(0.6\pi n + \pi/4) = \cos\left(\frac{2\pi \times 3}{10}n + \frac{\pi}{4}\right)$$

also of fundamental period $N = 10$. $\qquad\qquad\square$

When sampling an analog sinusoid

$$x(t) = A\cos(\Omega_0 t + \theta) \qquad -\infty < t < \infty \qquad\qquad (9.4)$$

of fundamental period $T_0 = 2\pi/\Omega_0$, $\Omega_0 > 0$, we obtain a **periodic discrete sinusoid**

$$x[n] = A\cos(\Omega_0 T_s n + \theta) = A\cos\left(\frac{2\pi T_s}{T_0}n + \theta\right) \qquad\qquad (9.5)$$

provided that

$$\frac{T_s}{T_0} = \frac{m}{N} \qquad\qquad (9.6)$$

for positive integers N and m which are not divisible by each other. To avoid frequency aliasing the sampling period should also satisfy the Nyquist sampling condition,

$$T_s \leq \frac{\pi}{\Omega_0} = \frac{T_0}{2}. \qquad\qquad (9.7)$$

Indeed, sampling a continuous-time signal $x(t)$ using as sampling period T_s we obtain

$$x[n] = A\cos(\Omega_0 T_s n + \theta) = A\cos\left(\frac{2\pi T_s}{T_0}n + \theta\right)$$

where the discrete frequency is $\omega_0 = 2\pi T_s/T_0$. For this signal to be periodic we should be able to express this frequency as $2\pi m/N$ for non-divisible positive integers m and N. This requires that

$$\frac{T_s}{T_0} = \frac{m}{N}$$

be rational with integers m and N not divisible by each other, or that

$$mT_0 = NT_s, \qquad\qquad (9.8)$$

which says that a period ($m = 1$) or several periods ($m > 1$) should be divided into $N > 0$ segments of duration T_s s. If this condition is not satisfied, then the discretized sinusoid is not periodic. To avoid frequency aliasing, the sampling period should be chosen so that

$$T_s \leq \frac{\pi}{\Omega_0} = \frac{T_0}{2}$$

as determined by the Nyquist sampling condition.

The sum $z[n] = x[n] + y[n]$ of periodic signals $x[n]$ with fundamental period N_1, and $y[n]$ with fundamental period N_2 is periodic if the ratio of periods of the summands is rational, i.e.,

$$\frac{N_2}{N_1} = \frac{p}{q}$$

and p and q are integers not divisible by each other. If so, the fundamental period of $z[n]$ is $q N_2 = p N_1$.

If $q N_2 = p N_1$, since $p N_1$ and $q N_2$ are multiples of the periods of $x[n]$ and $y[n]$, we have

$$z[n + p N_1] = x[n + p N_1] + y[n + p N_1] = x[n] + y[n + q N_2] = x[n] + y[n] = z[n]$$

or $z[n]$ is periodic of fundamental period $q N_2 = p N_1$.

Example 9.5. The signal $z[n] = v[n] + w[n] + y[n]$ is the sum of three periodic signals $v[n]$, $w[n]$ and $y[n]$ of fundamental periods $N_1 = 2$, $N_2 = 3$ and $N_3 = 4$, respectively. Determine if $z[n]$ is periodic, and if so give its fundamental period.

Solution: Let $x[n] = v[n] + w[n]$, so that $z[n] = x[n] + y[n]$. The signal $x[n]$ is periodic since $N_2/N_1 = 3/2$ is a rational number with non-divisible factors, and its fundamental period is $N_4 = 3N_1 = 2N_2 = 6$. The signal $z[n]$ is also periodic since

$$\frac{N_4}{N_3} = \frac{6}{4} = \frac{3}{2}$$

is rational with non-divisible factors. The fundamental period of $z[n]$ is $N = 2N_4 = 3N_3 = 12$. Thus $z[n]$ is periodic of fundamental period 12, indeed

$$z[n + 12] = v[n + 6N_1] + w[n + 4N_2] + y[n + 3N_3] = v[n] + w[n] + y[n] = z[n]. \quad \square$$

Example 9.6. Determine if the signal

$$x[n] = \sum_{m=0}^{\infty} X_m \cos(m\omega_0 n), \qquad \omega_0 = \frac{2\pi}{N_0}$$

is periodic, and if so determine its fundamental period.

Solution: The signal $x[n]$ consists of the sum of a constant X_0 and cosines of frequency

$$m\omega_0 = \frac{2\pi m}{N_0} \qquad m = 1, 2, \cdots .$$

The periodicity of $x[n]$ depends on the periodicity of the cosines. According to the frequency of the cosines, they are periodic of fundamental period N_0. Thus $x[n]$ is periodic of fundamental period N_0, indeed

$$x[n + N_0] = \sum_{m=0}^{\infty} X_m \cos(m\omega_0(n + N_0)) = \sum_{m=0}^{\infty} X_m \cos(m\omega_0 n + 2\pi m) = x[n]. \quad \square$$

9.2.2 FINITE-ENERGY AND FINITE-POWER DISCRETE-TIME SIGNALS

For discrete-time signals, we obtain definitions for energy and power similar to those for continuous-time signals by replacing integrals by summations.

For a discrete-time signal $x[n]$ we have the following definitions:

Energy: $\qquad \mathcal{E}_x = \sum_{n=-\infty}^{\infty} |x[n]|^2$ $\qquad\qquad$ (9.9)

Power: $\qquad P_x = \lim_{N\to\infty} \frac{1}{2N+1} \sum_{n=-N}^{N} |x[n]|^2.$ $\qquad$ (9.10)

- $x[n]$ is said to have **finite energy** or to be **square summable** if $\mathcal{E}_x < \infty$.
- $x[n]$ is called **absolutely summable** if

$$\sum_{n=-\infty}^{\infty} |x[n]| < \infty. \qquad\qquad (9.11)$$

- $x[n]$ is said to have **finite power** if $P_x < \infty$.

Example 9.7. A "causal" sinusoid, obtained from a signal generator after it is switched on, is

$$x(t) = \begin{cases} 2\cos(\Omega_0 t - \pi/4) & t \ge 0 \\ 0 & \text{otherwise} \end{cases}$$

The signal $x(t)$ is sampled using a sampling period of $T_s = 0.1$ s to obtain a discrete-time signal

$$x[n] = x(t)|_{t=0.1n} = 2\cos(0.1\Omega_0 n - \pi/4) \qquad n \ge 0$$

and zero otherwise. Determine if this discrete-time signal has finite energy, finite power and compare these characteristics with those of the continuous-time signal $x(t)$ when $\Omega_0 = \pi$ and when $\Omega_0 = 3.2$ rad/s (an upper approximation of π).

Solution: The continuous-time signal $x(t)$ has infinite energy, and so does the discrete-time signal $x[n]$, for both values of Ω_0. Indeed, the energy of $x[n]$ is

$$\varepsilon_x = \sum_{n=-\infty}^{\infty} |x[n]|^2 = \sum_{n=0}^{\infty} 4\cos^2(0.1\Omega_0 n - \pi/4) \to \infty$$

for both values of Ω_0. Although the continuous-time and the discrete-time signals have infinite energy, they have finite power. That the continuous-time signal $x(t)$ has finite power can be shown as indicated in Chapter 1. For the discrete-time signal, $x[n]$, we have for the two frequencies:
(i) For $\Omega_0 = \pi$, $x[n] = 2\cos(\pi n/10 - \pi/4) = 2\cos(2\pi n/20 - \pi/4)$ for $n \geq 0$ and zero otherwise. Thus $x[n]$ repeats every $N_0 = 20$ samples for $n \geq 0$, and its power is

$$
\begin{aligned}
P_x &= \lim_{N\to\infty} \frac{1}{2N+1} \sum_{n=-N}^{N} |x[n]|^2 = \lim_{N\to\infty} \frac{1}{2N+1} \sum_{n=0}^{N} |x[n]|^2 \\
&= \lim_{N\to\infty} \frac{1}{2N+1} N \underbrace{\left[\frac{1}{N_0} \sum_{n=0}^{N_0-1} |x[n]|^2 \right]}_{\text{power of a period}} \\
&= \lim_{N\to\infty} \frac{1}{2+1/N} \left[\frac{1}{N_0} \sum_{n=0}^{N_0-1} |x[n]|^2 \right] = \frac{1}{2N_0} \sum_{n=0}^{N_0-1} |x[n]|^2 < \infty
\end{aligned}
$$

where we used the causality of the signal ($x[n] = 0$ for $n < 0$), and considered N periods of $x[n]$ for $n \geq 0$ and for each calculated its power to get the final result. Thus for $\Omega_0 = \pi$ the discrete-time signal $x[n]$ has finite power and can be computed using a period for $n \geq 0$. To find the power we use the trigonometric identity (or Euler's equation) $\cos^2(\theta) = 0.5(1 + \cos(2\theta))$ and so replacing $x[n]$ we have recalling that $N_0 = 20$:

$$P_x = \frac{4}{40} 0.5 \left[\sum_{n=0}^{19} 1 + \sum_{n=0}^{19} \cos\left(\frac{2\pi n}{10} - \pi/2\right) \right] = \frac{2}{40}[20 + 0] = 1$$

where the sum of the cosine is zero, as we are adding the values of the periodic cosine over two periods.
(ii) For $\Omega_0 = 3.2$, $x[n] = 2\cos(3.2n/10 - \pi/4)$ for $n \geq 0$ and zero otherwise. The signal now does not repeat periodically after $n = 0$, as the frequency $3.2/10$ (which equals the rational $32/100$) cannot be expressed as $2\pi m/N$ (since π is an irrational value) for integers m and N. Thus in this case we do not have a closed form for the power, we can simply say that the power is

$$P_x = \lim_{N\to\infty} \frac{1}{2N+1} \sum_{n=-N}^{N} |x[n]|^2 = \lim_{N\to\infty} \frac{1}{2N+1} \sum_{n=0}^{N} |x[n]|^2$$

and conjecture that because the analog signal has finite power, so would $x[n]$. Thus we use MATLAB to compute the power for both cases.

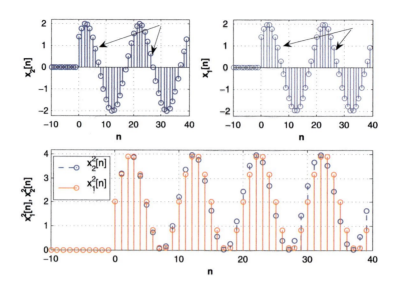

FIGURE 9.1

Top figures: signals $x_2[n]$ (non-periodic for $n \geq 0$) and $x_1[n]$ (periodic for $n \geq 0$). The arrows show that the values are not equal for $x_2[n]$ and equal for $x_1[n]$. Bottom figure: the squares of the signals differ slightly suggesting that if $x_1[n]$ has finite power so does $x_2[n]$.

```
%%
% Example 9.7---Power
%%
clear all;clf
n=0:100000;
x2=2*cos(0.1*n*3.2-pi/4);           % non-periodic for positive n
x1=2*cos(0.1*n*pi-pi/4);            % periodic for positive n
N=length(x1)
Px1=sum(x1.^2)/(2*N+1)              % power of x1
Px2=sum(x2.^2)/(2*N+1)              % power of x2
P1=sum(x1(1:20).^2)/(20);          % power in one period of x1
```

The signal $x_1[n]$ in the script has unit power and so does the signal $x_2[n]$ when we consider 100,001 samples. (See Fig. 9.1.) □

Example 9.8. Determine if a discrete-time exponential $x[n] = 2\,(0.5)^n$, $n \geq 0$, and zero otherwise, has finite energy, finite power or both.

Solution: The energy is given by

$$\varepsilon_x = \sum_{n=0}^{\infty} 4(0.5)^{2n} = 4 \sum_{n=0}^{\infty} (0.25)^n = \frac{4}{1 - 0.25} = \frac{16}{3}$$

thus $x[n]$ is a finite-energy signal. Just as with continuous-time signals, a finite-energy signal is a finite power (actually zero power) signal. Indeed,

$$P_x = \lim_{N \to \infty} \frac{1}{2N+1} \varepsilon_x = 0. \quad \square$$

9.2.3 EVEN AND ODD DISCRETE-TIME SIGNALS

Time shifting and scaling of discrete-time signals are very similar to the continuous-time cases, the only difference being that the operations are now done using integers.

A discrete-time signal $x[n]$ is said to be

- **delayed** by L (an integer) samples if $x[n - L]$ is $x[n]$ shifted to the right L samples,
- **advanced** by M (an integer) samples if $x[n + M]$ is $x[n]$ shifted to the left M samples,
- **reflected** if the variable n in $x[n]$ is negated, i.e., $x[-n]$.

The shifting to the right or the left can be readily seen by considering when $x[0]$ is attained. For $x[n - L]$, this is when $n = L$, i.e., L samples to the right of the origin or equivalently $x[n]$ is delayed by L samples. Likewise for $x[n + M]$ the $x[0]$ appears when $n = -M$, or advanced by M samples. Negating the variable n flips the signal over with respect to the origin.

Example 9.9. A triangular discrete pulse is defined as

$$x[n] = \begin{cases} n & 0 \le n \le 10, \\ 0 & \text{otherwise.} \end{cases}$$

Find an expression for $y[n] = x[n + 3] + x[n - 3]$ and $z[n] = x[-n] + x[n]$ in terms of n and carefully plot them.

Solution: Replacing n by $n + 3$ and by $n - 3$ in the definition of $x[n]$ we get the advanced and delayed signals

$$x[n + 3] = \begin{cases} n + 3 & -3 \le n \le 7, \\ 0 & \text{otherwise,} \end{cases}$$

and

$$x[n - 3] = \begin{cases} n - 3 & 3 \le n \le 13, \\ 0 & \text{otherwise,} \end{cases}$$

so that when added we get

$$y[n] = x[n + 3] + x[n - 3] = \begin{cases} n + 3 & -3 \le n \le 2, \\ 2n & 3 \le n \le 7, \\ n - 3 & 8 \le n \le 13, \\ 0 & \text{otherwise.} \end{cases}$$

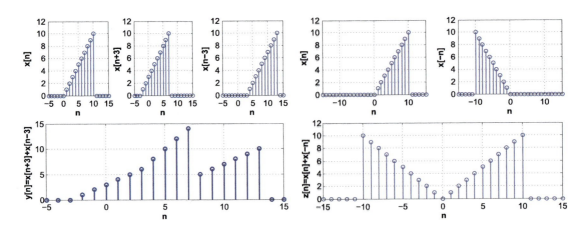

FIGURE 9.2

Generation of $y[n] = x[n+3] + x[n-3]$ (left) and $z[n] = x[n] + x[-n]$ (right).

Likewise, we have

$$z[n] = x[n] + x[-n] = \begin{cases} n & 1 \leq n \leq 10, \\ 0 & n = 0, \\ -n & -10 \leq n \leq -1, \\ 0 & \text{otherwise.} \end{cases}$$

The results are shown in Fig. 9.2. □

Example 9.10. We will see that in the convolution sum we need to figure out how a signal $x[n-k]$ behaves as a function of k for different values of n. Consider the signal

$$x[k] = \begin{cases} k & 0 \leq k \leq 3, \\ 0 & \text{otherwise.} \end{cases}$$

Obtain an expression for $x[n-k]$ for $-2 \leq n \leq 2$ and determine in which direction it shifts as n increases from -2 to 2.

Solution: Although $x[n-k]$, as a function of k, is reflected it is not clear if it is advanced or delayed as n increases from -2 to 2. If $n = 0$,

$$x[-k] = \begin{cases} -k & -3 \leq k \leq 0, \\ 0 & \text{otherwise.} \end{cases}$$

For $n \neq 0$, we have

$$x[n-k] = \begin{cases} n-k & n-3 \leq k \leq n, \\ 0 & \text{otherwise.} \end{cases}$$

As n increases from -2 to 2, $x[n-k]$ moves to the right. For $n=-2$ the support of $x[-2-k]$ is $-5 \le k \le -2$, while for $n=0$ the support of $x[-k]$ is $-3 \le k \le 0$, and for $n=2$ the support of $x[2-k]$ is $-1 \le k \le 2$, each shifted to the right.   □

We can thus use the above to define even and odd signals and obtain a general decomposition of any signal in terms of even and odd signals.

Even and odd discrete-time signals are defined as

$$x[n] \text{ is \textbf{even} if } x[n] = x[-n],$$ (9.12)

$$x[n] \text{ is \textbf{odd} if } x[n] = -x[-n].$$ (9.13)

Any discrete-time signal $x[n]$ can be represented as the sum of an even component and an odd component

$$x[n] = \underbrace{\frac{1}{2}(x[n] + x[-n])}_{x_e[n]} + \underbrace{\frac{1}{2}(x[n] - x[-n])}_{x_o[n]}$$

$$= x_e[n] + x_o[n].$$ (9.14)

The even and odd decomposition can easily be seen. The even component $x_e[n] = 0.5(x[n] + x[-n])$ is even since $x_e[-n] = 0.5(x[-n] + x[n])$ equals $x_e[n]$, and the odd component $x_o[n] = 0.5(x[n] - x[-n])$ is odd, since $x_o[-n] = 0.5(x[-n] - x[n]) = -x_o[n]$.

Example 9.11. Find the even and odd components of the following discrete-time signal:

$$x[n] = \begin{cases} 4-n & 0 \le n \le 4, \\ 0 & \text{otherwise.} \end{cases}$$

Solution: The even component of $x[n]$ is given by

$$x_e[n] = 0.5(x[n] + x[-n]).$$

When $n=0$ then $x_e[0] = 0.5 \times 2x[0] = 4$, when $n>0$ then $x_e[n] = 0.5x[n]$, and when $n<0$, then $x_e[n] = 0.5x[-n]$ giving

$$x_e[n] = \begin{cases} 2+0.5n & -4 \le n \le -1, \\ 4 & n=0, \\ 2-0.5n & 1 \le n \le 4, \\ 0 & \text{otherwise,} \end{cases}$$

while the odd component

$$x_o[n] = 0.5(x[n] - x[-n]) = \begin{cases} -2-0.5n & -4 \le n \le -1, \\ 0 & n=0, \\ 2-0.5n & 1 \le n \le 4, \\ 0 & \text{otherwise.} \end{cases}$$

The sum of these two components gives $x[n]$.   □

Remark

Expansion and compression of discrete-time signals are more complicated that in the continuous-time. In the discrete domain, expansion and compression can be related to the change of the sampling period in the sampling. Thus, if a continuous-time signal $x(t)$ is sampled using a sampling period T_s, by changing the sampling period to MT_s, for an integer $M > 1$, we obtain fewer samples, and by changing the sampling period to T_s/L, for an integer $L > 1$, we increase the number of samples.

For the corresponding discrete-time signal $x[n]$, increasing the sampling period would give $x[Mn]$ which is called the **down-sampling** of $x[n]$ by M. Unfortunately, because the arguments of the discrete-time signal must be integers, it is not clear what $x[n/L]$ is unless the values for n are multiples of L, i.e., $n = 0, \pm L, \pm 2L, \cdots$, with no clear definition when n takes other values. This leads to the definition of the **up-sampled** signal as

$$x_u[n] = \begin{cases} x[n/L] & n = 0, \pm L, \pm 2L, \cdots, \\ 0 & \text{otherwise.} \end{cases} \tag{9.15}$$

To replace the zero entries with the values obtained by decreasing the sampling period we need to low-pass filter the up-sampled signal—this is equivalent to interpolating the nonzero values to replace the zero values of the up-sampled signal. MATLAB provides the functions *decimate* and *interp* to implement decimation (related to down-sampling) and interpolation (related to up-sampling). In Chapter 10, we will continue the discussion of these operations including their frequency characterization.

9.2.4 BASIC DISCRETE-TIME SIGNALS

The representation of discrete-time signals via basic signals is simpler than in the continuous-time domain. This is due to the lack of ambiguity in the definition of the impulse and the unit-step discrete-time signals. The definitions of impulses and unit-step signals in the continuous-time domain are more abstract.

Discrete-Time Complex Exponential

Given complex numbers $A = |A|e^{j\theta}$ and $\alpha = |\alpha|e^{j\omega_0}$, a **discrete-time complex exponential** is a signal of the form

$$\begin{aligned} x[n] &= A\alpha^n = |A||\alpha|^n e^{j(\omega_0 n + \theta)} \\ &= |A||\alpha|^n \left[\cos(\omega_0 n + \theta) + j \sin(\omega_0 n + \theta) \right] \end{aligned} \tag{9.16}$$

where ω_0 is a discrete frequency in radians.

Remarks

1. The discrete-time complex exponential looks different from the continuous-time complex exponential. This can be explained by sampling the continuous-time complex exponential

$$x(t) = e^{(-a + j\Omega_0)t}$$

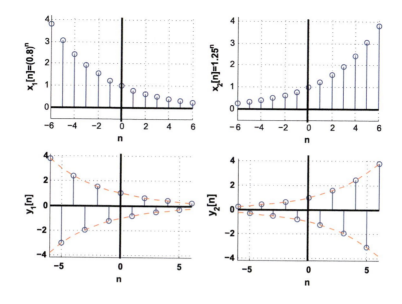

FIGURE 9.3

Real exponential $x_1[n] = 0.8^n$, $x_2[n] = 1.25^n$ (top) and modulated $y_1[n] = x_1[n] \cos(\pi n)$ and $y_2[n] = x_2[n] \cos(\pi n)$.

using as sampling period T_s. The sampled signal is

$$x[n] = x(nT_s) = e^{(-anT_s + j\Omega_0 nT_s)} = (e^{-aT_s})^n e^{j(\Omega_0 T_s)n} = A\alpha^n e^{j\omega_0 n}$$

where we let $\alpha = e^{-aT_s}$ and $\omega_0 = \Omega_0 T_s$.

2. Just as with the continuous-time complex exponential, we obtain different signals depending on the chosen parameters A and α (see Fig. 9.3 for examples). For instance, the real part of $x[n]$ in (9.16) is a real signal

$$g[n] = \mathcal{R}e(x[n]) = |A||\alpha|^n \cos(\omega_0 n + \theta)$$

when $|\alpha| < 1$ it is a damped sinusoid, and when $|\alpha| > 1$ it is a growing sinusoid (see Fig. 9.3). If $\alpha = 1$ then the above signal is a sinusoid.

3. It is important to realize that for a real $\alpha > 0$ the exponential

$$x[n] = (-\alpha)^n = (-1)^n \alpha^n = \alpha^n \cos(\pi n),$$

i.e., a modulated exponential.

Example 9.12. Given the analog signal $x(t) = e^{-at} \cos(\Omega_0 t) u(t)$, determine the values of $a > 0$, Ω_0, and T_s that permit us to obtain a discrete-time signal

$$y[n] = \alpha^n \cos(\omega_0 n) \qquad n \geq 0$$

and zero otherwise. Consider the case when $\alpha = 0.9$ and $\omega_0 = \pi/2$, find a, Ω_0, and T_s that will permit us to obtain $y[n]$ from $x(t)$ by sampling. Plot $x(t)$ and $y[n]$ using MATLAB.

Solution: Comparing the sampled continuous-time signal $x(nT_s) = (e^{-aT_s})^n \cos((\Omega_0 T_s)n)u[n]$ with $y[n]$ we obtain the following two equations:

$$(i)\ \alpha = e^{-aT_s}, \quad (ii)\ \omega_0 = \Omega_0 T_s$$

with three unknowns (a, Ω_0, and T_s) when α and ω_0 are given, so there is no unique solution. However, we know that according to the Nyquist condition

$$T_s \leq \frac{\pi}{\Omega_{max}}.$$

Assuming the maximum frequency is $\Omega_{max} = N\Omega_0$ for $N \geq 2$ (since the signal $x(t)$ is not band-limited the maximum frequency is not known, to estimate it we could use Parseval's result as indicated in the previous chapter; instead we are assuming that is a multiple of Ω_0) if we let $T_s = \pi/N\Omega_0$ after replacing it in the above equations we get

$$\alpha = e^{-a\pi/N\Omega_0}, \quad \omega_0 = \Omega_0 \pi/N\Omega_0 = \pi/N.$$

If we want $\alpha = 0.9$ and $\omega_0 = \pi/2$, we have $N = 2$ and

$$a = -\frac{2\Omega_0}{\pi}\log 0.9$$

for any frequency $\Omega_0 > 0$. For instance, if $\Omega_0 = 2\pi$ then $a = -4\log 0.9$, and $T_s = 0.25$. Fig. 9.4 displays the continuous and the discrete-time signals generated using the above parameters. The following script is used. The continuous-time and the discrete-time signals coincide at the sample times. □

```
%%
% Example 9.12
%%
a=-4*log(0.9);Ts=0.25;                              % parameters
alpha=exp(-a*Ts);
n=0:30; y=alpha.^n.*cos(pi*n/2);                    % discrete-time signal
t=0:0.001:max(n)*Ts; x=exp(-a*t).*cos(2*pi*t);     % analog signal
stem(n,y,'r'); hold on
plot(t/Ts,x); grid; legend('y[n]','x(t)'); hold off
```

Example 9.13. Show how to obtain the casual discrete-time exponential $x[n] = (-1)^n$ for $n \geq 0$ and zero otherwise, by sampling a continuous-time signal $x(t)$.

Solution: Because the values of $x[n]$ are 1 and -1, $x[n]$ cannot be generated by sampling an exponential $e^{-at}u(t)$, indeed $e^{-at} > 0$ for any values of a and t. The discrete signal can be written

$$x[n] = (-1)^n = \cos(\pi n)$$

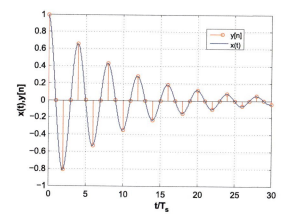

FIGURE 9.4

Determination of parameters for a continuous signal $x(t) = e^{-at} \cos(\Omega_0 t) u(t)$ that when sampled gives a desired discrete-time signal $y[n]$.

for $n \geq 0$ and zero otherwise. If we sample an analog signal $x(t) = \cos(\Omega_0 t) u(t)$ with a sampling period T_s and equate it to the desired discrete signal we get

$$x[n] = x(nT_s) = \cos(\Omega_0 n T_s) = \cos(\pi n)$$

for $n \geq 0$ and zero otherwise. Thus, $\Omega_0 T_s = \pi$ giving $T_s = \pi / \Omega_0$. For instance, for $\Omega_0 = 2\pi$, then $T_s = 0.5$.   □

Discrete-Time Sinusoids

A discrete-time sinusoid is a special case of the complex exponential, letting $\alpha = e^{j\omega_0}$ and $A = |A| e^{j\theta}$, we have according to Equation (9.16)

$$x[n] = A\alpha^n = |A| e^{j(\omega_0 n + \theta)} = |A| \cos(\omega_0 n + \theta) + j|A| \sin(\omega_0 n + \theta) \tag{9.17}$$

so the real part of $x[n]$ is a cosine, while the imaginary part is a sine. As indicated before, discrete sinusoids of amplitude A and phase shift θ are periodic if they can be expressed as

$$A \cos(\omega_0 n + \theta) = A \sin(\omega_0 n + \theta + \pi/2) \qquad -\infty < n < \infty \tag{9.18}$$

where $w_0 = 2\pi m/N$ (rad) is the discrete frequency, for integers m and $N > 0$ which are not divisible. Otherwise, discrete-time sinusoids are not periodic.

Because ω is given in radians, it repeats periodically with 2π as fundamental period, i.e., adding a multiple $2\pi k$ (k positive or negative) to a discrete frequency ω_0 gives back the frequency ω_0

$$\omega_0 + 2\pi k = \omega_0 \qquad k \text{ positive or negative integer.} \tag{9.19}$$

To avoid this ambiguity, we will let $-\pi < \omega \leq \pi$ as the possible range of discrete frequencies and convert any frequency ω_1 outside this range using the following modular representation. For a positive

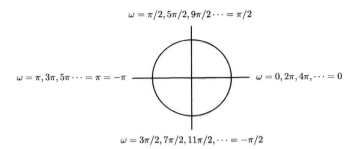

FIGURE 9.5

Discrete frequencies ω and their equivalent frequencies in mod 2π.

integer k two frequencies $\omega_1 = \omega + 2\pi k$, and ω where $-\pi \le \omega \le \pi$ are equal modulo 2π, which can be written

$$\omega_1 \equiv \omega \quad (\text{mod } 2\pi).$$

Thus as shown in Fig. 9.5, frequencies not in the range $[-\pi, \pi)$ can be converted into equivalent discrete frequencies in that range. For instance, $\omega_0 = 2\pi$ can be written as $\omega_0 = 0 + 2\pi$ so that an equivalent frequency is 0; $\omega_0 = 7\pi/2 = (8 - 1)\pi/2 = 2 \times 2\pi - \pi/2$ is equivalent to a frequency $-\pi/2$. According to this (mod 2π) representation of the frequencies, a signal $\sin(3\pi n)$ is identical to $\sin(\pi n)$, and a signal $\sin(1.5\pi n)$ is identical to $\sin(-0.5\pi n) = -\sin(0.5\pi n)$.

Example 9.14. Consider the following four sinusoids:

$$x_1[n] = \sin(0.1\pi n), \quad x_2[n] = \sin(0.2\pi n), \quad x_3[n] = \sin(0.6\pi n), \quad x_4[n] = \sin(0.7\pi n) \quad -\infty < n < \infty.$$

Find if they are periodic and if so find their fundamental periods. Are these signals harmonically related? Use MATLAB to plot these signals from $n = 0, \cdots, 40$. Comment on which of these signals resemble sampled analog sinusoids and indicate why some do not.

Solution: To find if they are periodic, we rewrite the given signals as

$$x_1[n] = \sin(0.1\pi n) = \sin\left(\frac{2\pi}{20}n\right), \quad x_2[n] = \sin(0.2\pi n) = \sin\left(\frac{2\pi}{20}2n\right),$$

$$x_3[n] = \sin(0.6\pi n) = \sin\left(\frac{2\pi}{20}6n\right), \quad x_4[n] = \sin(0.7\pi n) = \sin\left(\frac{2\pi}{20}7n\right)$$

indicating the signals are periodic of fundamental periods 20, with frequencies harmonically related. When plotting these signals using MATLAB the first two resemble analog sinusoids but not the other two. See Fig. 9.6.

One might think that $x_3[n]$ and $x_4[n]$ look like that because of aliasing, but that is not the case. To obtain $\cos(\omega_0 n)$ we could sample an analog sinusoid $\cos(\Omega_0 t)$ using a sampling period $T_s = 1$ so that according to the Nyquist condition

$$T_s = 1 \le \frac{\pi}{\Omega_0}$$

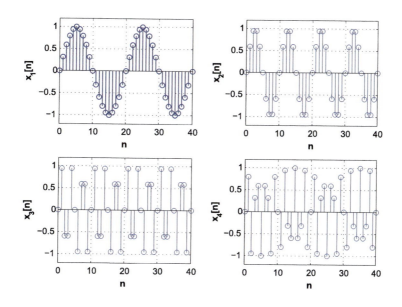

FIGURE 9.6

Periodic signals $x_i[n]$, $i = 1, 2, 3, 4$, given in Example 9.14.

where π/Ω_0 is the maximum value permitted for the sampling period for no aliasing. Thus $x_3[n] = \sin(0.6\pi n) = \sin(0.6\pi t)|_{t=nT_s=n}$ when $T_s = 1$ and in this case

$$T_s = 1 \leq \frac{\pi}{0.6\pi} \approx 1.66.$$

Comparing this with $x_2[n] = \sin(0.2\pi n) = \sin(0.2\pi t)|_{t=nT_s=n}$ we get

$$T_s = 1 \leq \frac{\pi}{0.2\pi} = 5.$$

Thus when obtaining $x_2[n]$ from $\sin(0.2\pi t)$ we are oversampling more than when we obtain $x_3[n]$ from $\sin(0.6\pi t)$ using the same sampling period, as such $x_2[n]$ resembles more an analog sinusoid than $x_3[n]$, but no aliasing is present. Similarly for $x_4[n]$. □

Remarks

1. The discrete-time sine and cosine signals, as in the continuous-time case, are out of phase $\pi/2$ radians.
2. The discrete frequency ω is given in radians since n, the sample index, does not have units. This can also be seen when we sample a sinusoid using a sampling period T_s so that

$$\cos(\Omega_0 t)|_{t=nT_s} = \cos(\Omega_0 T_s n) = \cos(\omega_0 n)$$

where we defined $\omega_0 = \Omega_0 T_s$, and since Ω_0 has as unit rad/s and T_s has s as unit, ω_0 has rad as unit.

3. The frequency Ω of analog sinusoids can vary from 0 (dc frequency) to ∞. Discrete frequencies ω as radian frequencies can only vary from 0 to π. Negative frequencies are needed in the analysis of real-valued signals, thus $-\infty < \Omega < \infty$ and $-\pi < \omega \leq \pi$. A discrete-time cosine of frequency 0 is constant for all n, and a discrete-time cosine of frequency π varies from -1 to 1 from sample to sample, giving the largest variation possible for the discrete-time signal.

Discrete-Time Unit-Step and Unit-Sample Signals

The unit-step $u[n]$ and the unit-sample $\delta[n]$ discrete-time signals are defined as

$$u[n] = \begin{cases} 1 & n \geq 0, \\ 0 & n < 0, \end{cases} \tag{9.20}$$

$$\delta[n] = \begin{cases} 1 & n = 0, \\ 0 & \text{otherwise.} \end{cases} \tag{9.21}$$

These two signals are related as follows:

$$\delta[n] = u[n] - u[n-1], \tag{9.22}$$

$$u[n] = \sum_{k=0}^{\infty} \delta[n-k] = \sum_{m=-\infty}^{n} \delta[m]. \tag{9.23}$$

It is easy to see the relation between the two signals $u[n]$ and $\delta[n]$:

$$\delta[n] = u[n] - u[n-1],$$

$$u[n] = \delta[n] + \delta[n-1] + \cdots = \sum_{k=0}^{\infty} \delta[n-k] = \sum_{m=-\infty}^{n} \delta[m]$$

where the last expression[2] is obtained by a change of variable, $m = n - k$. These two equations should be contrasted with the ones for $u(t)$ and $\delta(t)$. Instead of the derivative relation $\delta(t) = du(t)/dt$ we have a difference relation and instead of the integral connection

$$u(t) = \int_0^\infty \delta(t - \zeta)d\zeta = \int_{-\infty}^t \delta(\tau)d\tau$$

we now have a sum relation between $u[n]$ and $\delta[n]$.

---

[2]It may not be clear how the second sum gives $u[n]$. For $n < 0$, the sum is of unit-sample signals $\{\delta[m]\}$ for which $m \leq n < 0$ (i.e., the arguments m are all negative and so all are zero) gives $u[n] = 0, n < 0$. For $n = n_0 \geq 0$, the sum gives

$$u[n_0] = \cdots + \delta[n_0 - 3] + \delta[n_0 - 2] + \delta[n_0 - 1] + \delta[n_0]$$

so that only one of the unit-sample signals would have an argument of 0 and that one would be equal to 1, the other unit-sample signals would give zero. For instance, $u[2] = \cdots + \delta[-1] + \delta[0] + \delta[1] + \delta[2] = 1$. As such $u[n] = 1$ for $n \geq 0$.

Remarks

1. Notice that there is no ambiguity in the definition of $u[n]$ or of $\delta[n]$ as there is for their continuous-time counterparts $u(t)$ and $\delta(t)$. Moreover, the definitions of these signals do not depend on $u(t)$ or $\delta(t)$; $u[n]$ and $\delta[n]$ are not sampled versions of $u(t)$ and $\delta(t)$.
2. The discrete ramp function $r[n]$ is defined as

$$r[n] = nu[n]. \tag{9.24}$$

As such, it can be expressed in terms of delta and unit-step signals as

$$r[n] = \sum_{k=0}^{\infty} k\delta[n-k] = \sum_{k=0}^{\infty} u[n-k]. \tag{9.25}$$

Moreover, $u[n] = r[n] - r[n-1]$ which can easily be shown.[3] Recall that $dr(t)/dt = u(t)$, so that the derivative is replaced by the difference in the discrete domain.

Generic Representation of Discrete-Time Signals

Any discrete-time signal $x[n]$ is represented using unit-sample signals as

$$x[n] = \sum_{k=-\infty}^{\infty} x[k]\delta[n-k]. \tag{9.26}$$

The representation of any signal $x[n]$ in terms of $\delta[n]$ results from the **sifting property** of the unit-sample signal:

$$x[n]\delta[n-n_0] = x[n_0]\delta[n-n_0],$$

which is due to

$$\delta[n-n_0] = \begin{cases} 1 & n = n_0, \\ 0 & \text{otherwise.} \end{cases}$$

Thus considering $x[n]$ a sequence of samples

$$\cdots x[-1]\, x[0]\, x[1] \cdots,$$

---

[3] We have

$$r[n] - r[n-1] \quad = \quad \sum_{k=0}^{\infty} u[n-k] - \sum_{k=0}^{\infty} u[n-(k-1)]$$

$$= \quad u[n] + \sum_{k=1}^{\infty} u[n-k] - \sum_{m=1}^{\infty} u[n-m] = u[n]$$

according to the new variable $m = k + 1$.

at sample times $\cdots -1,\ 0,\ 1, \cdots$, we can write $x[n]$ as

$$x[n] = \cdots + x[-1]\delta[n+1] + x[0]\delta[n] + x[1]\delta[n-1] + \cdots = \sum_{k=-\infty}^{\infty} x[k]\delta[n-k].$$

The generic representation (9.26) of any signal $x[n]$ will be useful in finding the output of a discrete-time linear time-invariant system.

Example 9.15. Consider a discrete pulse

$$x[n] = \begin{cases} 1 & 0 \leq n \leq N-1, \\ 0 & \text{otherwise;} \end{cases}$$

obtain representations of $x[n]$ using unit-sample and unit-step signals.

Solution: The signal $x[n]$ can be represented as

$$x[n] = \sum_{k=0}^{N-1} \delta[n-k]$$

and using $\delta[n] = u[n] - u[n-1]$ we obtain a representation of the discrete pulse in terms of unit-step signals

$$\begin{aligned} x[n] &= \sum_{k=0}^{N-1} (u[n-k] - u[n-k-1]) \\ &= (u[n] - u[n-1]) + (u[n-1] - u[n-2]) + \cdots (u[n-N+1] - u[n-N]) \\ &= u[n] - u[n-N] \end{aligned}$$

because of the cancellation of consecutive terms. □

Example 9.16. Consider how to generate a periodic train of triangular, discrete-time pulses $t[n]$ of fundamental period $N = 11$. A period of $t[n]$ is

$$\tau[n] = t[n](u[n] - u[n-11]) = \begin{cases} n & 0 \leq n \leq 5, \\ -n+10 & 6 \leq n \leq 10, \\ 0 & \text{otherwise.} \end{cases}$$

Find then an expression for its finite difference $d[n] = t[n] - t[n-1]$.

Solution: The periodic signal can be generated by adding shifted versions of $\tau[n]$, or

$$t[n] = \cdots + \tau[n+11] + \tau[n] + \tau[n-11] + \cdots = \sum_{k=-\infty}^{\infty} \tau[n-11k].$$

The finite difference $d[n]$ is then

$$d[n] = t[n] - t[n-1] = \sum_{k=-\infty}^{\infty} (\tau[n-11k] - \tau[n-11k-1]).$$

The signal $d[n]$ is also periodic of the same fundamental period $N = 11$ as $t[n]$. If we let

$$s[n] = \tau[n] - \tau[n-1] = \begin{cases} 0 & n = 0, \\ 1 & 1 \le n \le 5, \\ -1 & 6 \le n \le 10, \\ 0 & \text{otherwise,} \end{cases}$$

then

$$d[n] = \sum_{k=-\infty}^{\infty} s[n-11k]. \quad \square$$

Example 9.17. Consider the discrete-time signal

$$y[n] = 3r(t+3) - 6r(t+1) + 3r(t) - 3u(t-3)|_{t=0.15n},$$

obtained by sampling with a sampling period $T_s = 0.15$ a continuous-time signal formed by ramp and unit-step signals. Write MATLAB functions to generate the ramp and the unit-step signals and obtain $y[n]$. Write then a MATLAB function that provides the even and odd decomposition of $y[n]$.

Solution: The signal $y(t)$ is obtained by sequentially adding the different signals as we go from $-\infty$ to ∞:

$$y(t) = \begin{cases} 0 & t < -3, \\ 3r(t+3) = 3t+9 & -3 \le t < -1, \\ 3t+9 - 6r(t+1) = 3t+9 - 6(t+1) = -3t+3 & -1 \le t < 0, \\ -3t+3+3r(t) = -3t+3+3t = 3 & 0 \le t < 3, \\ 3 - 3u(t-3) = 3-3 = 0 & t \ge 3. \end{cases}$$

The three functions *ramp, unitstep,* and *evenodd* for this example are shown below. The following script shows how they can be used to generate the ramp signals, with the appropriate slopes and time shifts, as well as the unit-step signal with the desired delay and then how to compute the even and odd decomposition of $y[n]$:

```
%%
% Example 9.17
%%
Ts=0.15;                              % sampling period
t=-5:Ts:5;                            % time support
y1=ramp(t,3,3); y2=ramp(t,-6,1);
y3=ramp(t,3,0);                       % ramp signals
y4=-3*unitstep(t,-3);                 % unit-step signal
y=y1+y2+y3+y4;
[ye,yo]=evenodd(y);
```

We choose as support $-5 \leq t \leq 5$ for the continuous-time signal $y(t)$ which translates into a support $-5 \leq 0.15n \leq 5$ or $-5/0.15 \leq n \leq 5/0.15$ for the discrete-time signal when $T_s = 0.15$. Since the limits are not integers, to make them integers (as required because n is an integer) we use the MATLAB function *floor* to find integers smaller than $-5/0.15$ and $5/0.15$ giving a range $[-34, 32]$. This is used when plotting $y[n]$.

The following function generates a ramp signal for a range of time values, for different slopes and time shifts:

```
function y=ramp(t,m,ad)
% ramp generation
% t: time support
% m: slope of ramp
% ad : advance (positive), delay (negative) factor
N=length(t);
y=zeros(1,N);
for i=1:N,
    if t(i)>=-ad,
    y(i)=m*(t(i)+ad);
    end
end
```

Likewise, the following function generates unit-step signals with different time-shifts (notice the similarities with the *ramp* function).

```
function y=unitstep(t,ad)
% generation of unit step
% t: time support
% ad : advance (positive), delay (negative)
N=length(t);
y=zeros(1,N);
for i=1:N,
    if t(i)>=-ad,
    y(i)=1;
    end
end
```

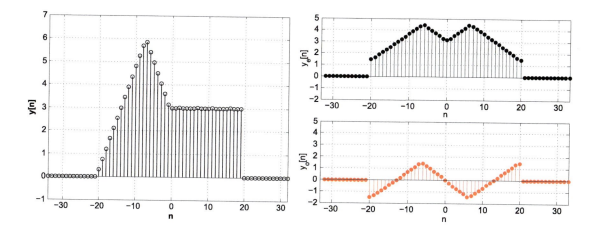

FIGURE 9.7

Discrete-time signal $y[n]$ (left), even $y_e[n]$ and odd $y_o[n]$ components (right).

Finally, the following function can be used to compute the even and the odd decomposition of a discrete-time signal. The MATLAB function *fliplr* reflects the signal as needed in the generation of the even and odd components.

```
function [ye,yo]=evenodd(y)
% even/odd decomposition
% NOTE: the support of the signal should be
%          symmetric about the origin
% y: analog signal
% ye, yo: even and odd components
yr=fliplr(y);      % reflect
ye=0.5*(y+yr);
yo=0.5*(y-yr);
```

The results are shown in Fig. 9.7.   □

9.3 DISCRETE-TIME SYSTEMS

Just as with continuous-time systems, a discrete-time system is a transformation of a discrete-time input signal $x[n]$ into a discrete-time output signal $y[n]$, i.e.,

$$y[n] = \mathcal{S}\{x[n]\}. \tag{9.27}$$

We are interested in dynamic systems $\mathcal{S}\{.\}$ having the following properties:

- Linearity
- Time invariance
- Stability
- Causality

just as we were when we studied the continuous-time systems.

A discrete-time system $\mathcal{S}$ is said to be

1. **Linear:** if for inputs $x[n]$ and $v[n]$, and constants a and b, it satisfies the following:

   - **Scaling:** $\mathcal{S}\{ax[n]\} = a\mathcal{S}\{x[n]\}$
   - **Additivity:** $\mathcal{S}\{x[n] + v[n]\} = \mathcal{S}\{x[n]\} + \mathcal{S}\{v[n]\}$

   or equivalently if **superposition** applies, i.e.,

   $$\mathcal{S}\{ax[n] + bv[n]\} = a\mathcal{S}\{x[n]\} + b\mathcal{S}\{v[n]\}. \tag{9.28}$$

2. **Time invariant:** if for an input $x[n]$ the corresponding output is $y[n] = \mathcal{S}\{x[n]\}$, the output corresponding to an advanced or a delayed version of $x[n]$, $x[n \pm M]$, for an integer M, is $y[n \pm M] = \mathcal{S}\{x[n \pm M]\}$, or the same as before but shifted as the input. In other words, the system is not changing with time.

Example 9.18 (A Square-Root Computation System). The input–output relation characterizing a discrete-time system is non-linear if there are non-linear terms that include the input, $x[n]$, the output, $y[n]$, or both (e.g., a square root of $x[n]$, products of $x[n]$ and $y[n]$, etc.). Consider the development of an iterative algorithm to compute the square root of a positive real number α. If the result of the algorithm is $y[n]$ as $n \to \infty$, then $y^2[n] = \alpha$ and likewise $y^2[n-1] = \alpha$, thus $y[n] = 0.5(y[n-1] + y[n-1])$ and replacing $y[n-1] = \alpha/y[n-1]$ in this equation, the following difference equation, with some initial condition $y[0]$, can be used to find the square root of α:

$$y[n] = 0.5\left[y[n-1] + \frac{\alpha}{y[n-1]}\right] \qquad n > 0.$$

Find recursively the solution of this difference equation. Use the results of finding the square roots of 4 and 2 to show the system is non-linear. Use MATLAB to solve the difference equation, and then plot the results for $\alpha = 4$, and 2.

Solution: The given difference equation is of first order, non-linear (expanding it you get the product of $y[n]$ with $y[n-1]$ and $y^2[n-1]$, which are non-linear terms) with constant coefficients. This equation can be solved recursively for $n > 0$ by replacing $y[0]$ to get $y[1]$, and we use this to get $y[2]$ and so on, i.e.,

$$y[1] = 0.5\left[y[0] + \frac{\alpha}{y[0]}\right], \quad y[2] = 0.5\left[y[1] + \frac{\alpha}{y[1]}\right], \quad y[3] = 0.5\left[y[2] + \frac{\alpha}{y[2]}\right], \quad \cdots.$$

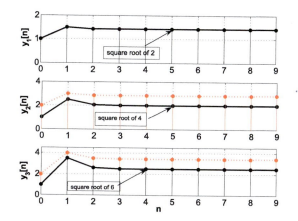

FIGURE 9.8

Square root of 2 (top); square root of 4 compared with twice the square root of 2 (middle); sum of previous responses compared with response of square root of 2 + 4 (bottom). Unity initial condition. Middle figure shows scaling property does not hold, and the bottom figure that additivity property does not hold, either. System is nonlinear.

For instance, let $y[0] = 1$, and $\alpha = 4$ (i.e., we wish to find the square root of 4),

$$y[0] = 1, \quad y[1] = 0.5\left[1 + \frac{4}{1}\right] = 2.5, \quad y[2] = 0.5\left[2.5 + \frac{4}{2.5}\right] = 2.05, \quad \cdots,$$

which is converging to 2, the square root of 4 (see Fig. 9.8). Thus, as indicated before, when $n \to \infty$ then $y[n] = y[n-1] = Y$, for some value Y which according to the difference equation satisfies the relation $Y = 0.5Y + 0.5(4/Y)$ or $Y = \sqrt{4} = 2$.

Suppose then that the input is $\alpha = 2$, half of what it was before. If the system is linear we should get half the previous output according to the scaling property. That is not the case, however. For the same initial condition $y[0] = 1$ we obtain recursively for $\alpha = 2$

$$y[0] = 1, \quad y[1] = 0.5[1 + 2] = 1.5, \quad y[2] = 0.5\left[1.5 + \frac{2}{1.5}\right] = 1.4167, \quad \cdots ;$$

this solution is clearly not half of the previous one. Moreover, as $n \to \infty$, we expect $y[n] = y[n-1] = Y$, for Y, that satisfies the relation $Y = 0.5Y + 0.5(2/Y)$ or $Y = \sqrt{2} = 1.4142$, so that the solution is tending to $\sqrt{2}$ and not to 2 as it should if the system were linear. Finally, if we add the signals in the above two cases and compare the resulting signal with the one we obtain when finding the square root of $2 + 4$, they do not coincide. The additive condition is not satisfied either, verifying once more that the system is not linear. See Fig. 9.8. □

9.3.1 RECURSIVE AND NON-RECURSIVE DISCRETE-TIME SYSTEMS

Depending on the relation between the input $x[n]$ and the output $y[n]$, two types of discrete-time systems of interest are:

- **Recursive system**

$$y[n] = -\sum_{k=1}^{N-1} a_k y[n-k] + \sum_{m=0}^{M-1} b_m x[n-m] \qquad n \geq 0,$$

initial conditions $y[-k], \ k = 1, \cdots, N-1.$ \hfill (9.29)

This system is also called an **infinite impulse response (IIR)** system.
- **Non-recursive system**

$$y[n] = \sum_{m=0}^{M-1} b_m x[n-m].$$ \hfill (9.30)

This system is also called a **finite impulse response (FIR)** system.

The recursive system is analogous to a continuous-time system represented by an ordinary differential equation. For this type of system the discrete-time input $x[n]$ and the discrete-time output $y[n]$ are related by a **difference equation** such as

$$y[n] = -\sum_{k=1}^{N-1} a_k y[n-k] + \sum_{m=0}^{M-1} b_m x[n-m] \qquad n \geq 0,$$

initial conditions $y[-k], \ k = 1, \cdots, N-1.$

As in the continuous-time case, if the difference equation is linear, with constant coefficients, zero initial conditions and the input is zero for $n < 0$, then it represents a linear and time-invariant system. For these systems, the output at a present time n, $y[n]$, depends or recurs on past values of the output $\{y[n-k], \ k = 1, \cdots, N-1\}$ and thus they are called recursive. We will see that these systems are also called infinite impulse response or IIR because their impulse responses are typically of infinite length.

On the other hand, if the output $y[n]$ does not depend on previous values of the output, but only on weighted and shifted inputs $\{b_m x[n-m], \ m = 0, \cdots, M-1\}$ the system with an input/output equation such as

$$y[n] = \sum_{m=0}^{M-1} b_m x[n-m]$$

is called non-recursive. We will see that the impulse response of non-recursive systems is of finite length, and as such these systems are also called finite impulse response or FIR.

Example 9.19 (Moving-Average Discrete System). A third-order moving-average system (also called a **smoother** as it smooths out the input signal) is an FIR system for which the input $x[n]$ and the output

$y[n]$ are related by

$$y[n] = \frac{1}{3}(x[n] + x[n-1] + x[n-2]).$$

Show that this system is linear and time invariant.

Solution: This is a non-recursive system that uses a present sample, $x[n]$, and two past values $x[n-1]$ and $x[n-2]$ to get an average $y[n]$ at every n. Thus its name of moving-average system.

Linearity—If we let the input be $ax_1[n] + bx_2[n]$ and assume that $\{y_i[n], \ i = 1, 2\}$ are the corresponding outputs to $\{x_i[n], \ i = 1, 2\}$, the system output is

$$\frac{1}{3}[(ax_1[n] + bx_2[n]) + (ax_1[n-1] + bx_2[n-1]) + (ax_1[n-2] + bx_2[n-2])] = ay_1[n] + by_2[n],$$

thus linear.

Time invariance—If the input is $x_1[n] = x[n-N]$ the corresponding output to it is

$$\frac{1}{3}(x_1[n] + x_1[n-1] + x_1[n-2]) = \frac{1}{3}(x[n-N] + x[n-N-1] + x[n-N-2]) = y[n-N],$$

i.e., the system is time invariant.   □

Example 9.20 (Autoregressive Discrete System). The recursive discrete-time system represented by the following first-order difference equation (with initial condition $y[-1]$):

$$y[n] = ay[n-1] + bx[n] \qquad n \geq 0$$

is called an **autoregressive (AR)** system. "Autoregressive" refers to the feedback in the output, i.e., the present value of the output $y[n]$ depends on its previous value $y[n-1]$. Find recursively the solution of the difference equation and determine under what conditions the system represented by this difference equation is linear and time invariant.

Solution: Let us first discuss why the initial condition is $y[-1]$. The initial condition is the value needed to compute $y[0]$. According to the difference equation to compute

$$y[0] = ay[-1] + bx[0]$$

we need the initial condition $y[-1]$ since $x[0]$ is known.

   Assume that the initial condition $y[-1] = 0$, and that the input $x[n] = 0$ for $n < 0$, i.e., the system is not energized for $n < 0$. The solution of the difference equation, when the input $x[n]$ is not defined can be found by a repetitive substitution of the input/output relationship. Thus replacing $y[n-1] = ay[n-2] + bx[n-1]$ in the difference equation, and then letting $y[n-2] = ay[n-3] + bx[n-2]$, replacing it and so on, we obtain

$$
\begin{aligned}
y[n] \quad &= \quad a(ay[n-2] + bx[n-1]) + bx[n] = a(a(ay[n-3] + bx[n-2])) + abx[n-1] + bx[n] \\
&\cdots = bx[n] + abx[n-1] + a^2bx[n-2] + a^3bx[n-3] + \cdots
\end{aligned}
$$

until we reach a term with $x[0]$. The solution can be written as

$$y[n] = \sum_{k=0}^{n} ba^k x[n-k] \qquad n \geq 0, \tag{9.31}$$

which we will see in the next section is the **convolution sum** of the **impulse response** of the system and the input.

To verify that (9.31) is actually the solution of the above difference equation, we need to show that when replacing the above expression for $y[n]$ in the right-hand term of the difference equation we obtain the left-hand term $y[n]$. Indeed, we have

$$
\begin{aligned}
ay[n-1] + bx[n] &= a\left[\sum_{k=0}^{n-1} ba^k x[n-1-k]\right] + bx[n] = \sum_{m=1}^{n} ba^m x[n-m] + bx[n] \\
&= \sum_{m=0}^{n} ba^m x[n-m] = y[n]
\end{aligned}
$$

where the dummy variable k in the first sum was changed to $m = k+1$, so that the limits of the summation became $m = 1$ when $k = 0$, and $m = n$ when $k = n-1$. The final equation is identical to $y[n]$.

To establish if the system represented by the difference equation is linear, we use the solution (9.31) with input $x[n] = \alpha x_1[n] + \beta x_2[n]$, where the outputs $\{y_i[n], \; i = 1, 2\}$ correspond to inputs $\{x_i[n], \; i = 1, 2\}$, and α, β are constants. The output for $x[n]$ is

$$
\begin{aligned}
\sum_{k=0}^{n} ba^k x[n-k] &= \sum_{k=0}^{n} ba^k (\alpha x_1[n-k] + \beta x_2[n-k]) \\
&= \alpha \sum_{k=0}^{n} ba^k x_1[n-k] + \beta \sum_{k=0}^{n} ba^k x_2[n-k] = \alpha y_1[n] + \beta y_2[n].
\end{aligned}
$$

So the system is linear.

The time invariance is shown by letting the input be $v[n] = x[n-N]$, $n \geq N$, and zero otherwise. The corresponding output according to (9.31) is

$$
\begin{aligned}
\sum_{k=0}^{n} ba^k v[n-k] &= \sum_{k=0}^{n} ba^k x[n-N-k] \\
&= \sum_{k=0}^{n-N} ba^k x[n-N-k] + \sum_{k=n-N+1}^{n} ba^k x[n-N-k] = y[n-N],
\end{aligned}
$$

since the summation

$$\sum_{k=n-N+1}^{n} ba^k x[n-N-k] = 0$$

given that $x[-N] = \cdots = x[-1] = 0$, as assumed. Thus the system represented by the above difference equation is linear and time invariant. As in the continuous-time case, however, if the initial condition $y[-1]$ is not zero, or if $x[n] \neq 0$ for $n < 0$ the system characterized by the difference equation is not LTI. □

Example 9.21 (Autoregressive Moving-Average System). A recursive system represented by the first-order difference equation

$$y[n] = 0.5y[n-1] + x[n] + x[n-1] \qquad n \geq 0, \; y[-1]$$

is called an **autoregressive moving-average** system given that it is the combination of the two systems discussed before. Consider two cases:

- Let the initial condition be $y[-1] = -2$, and the input $x[n] = u[n]$ first and then $x[n] = 2u[n]$. Find the corresponding outputs.
- Let the initial condition be $y[-1] = 0$, and the input $x[n] = u[n]$ first and then $x[n] = 2u[n]$. Find the corresponding outputs.

Use the above results to determine in each case if the system is linear. Find the steady-state response, i.e., $\lim_{n \to \infty} y[n]$.

Solution: For an initial condition $y[-1] = -2$ and $x[n] = u[n]$ we get recursively

$$y[0] = 0.5y[-1] + x[0] + x[-1] = 0, \quad y[1] = 0.5y[0] + x[1] + x[0] = 2,$$
$$y[2] = 0.5y[1] + x[2] + x[1] = 3, \quad \cdots .$$

Let us then double the input, i.e., $x[n] = 2u[n]$, and call the response $y_1[n]$. As the initial condition remains the same, i.e., $y_1[-1] = -2$ we get

$$y_1[0] = 0.5y_1[-1] + x[0] + x[-1] = 1, \quad y_1[1] = 0.5y_1[0] + x[1] + x[0] = 4.5,$$
$$y_1[2] = 0.5y_1[1] + x[2] + x[1] = 6.25, \quad \cdots .$$

It is clear that the response $y_1[n]$ is not $2y[n]$. Due to the initial condition not being zero, the system is non-linear.

If the initial condition is set to zero, and the input $x[n] = u[n]$, the response is

$$y[0] = 0.5y[-1] + x[0] + x[-1] = 1, \quad y[1] = 0.5y[0] + x[1] + x[0] = 2.5$$
$$y[2] = 0.5y[1] + x[2] + x[1] = 3.25, \quad \cdots$$

and if we double the input, i.e., $x[n] = 2u[n]$, and call the response $y_1[n]$, $y_1[-1] = 0$, we obtain

$$y_1[0] = 0.5y_1[-1] + x[0] + x[-1] = 2, \quad y_1[1] = 0.5y_1[0] + x[1] + x[0] = 5,$$
$$y_1[2] = 0.5y_1[1] + x[2] + x[1] = 6.5, \quad \cdots .$$

For the zero initial condition, it is clear that $y_1[n] = 2y[n]$ when we double the input. One can also show that superposition holds for this system. For instance if we let the input be the sum of the previous inputs, $x[n] = u[n] + 2u[n] = 3u[n]$ and let $y_{12}[n]$ be the response when the initial condition is zero, $y_{12}[0] = 0$, we obtain

$$y_{12}[0] = 0.5y_{12}[-1] + x[0] + x[-1] = 3, \quad y_{12}[1] = 0.5y_{12}[0] + x[1] + x[0] = 7.5$$
$$y_{12}[2] = 0.5y_{12}[1] + x[2] + x[1] = 9.75, \quad \cdots,$$

showing that $y_{12}[n]$ is the sum of the responses for inputs $u[n]$ and $2u[n]$. Thus, the system represented by the given difference equation with a zero initial condition is linear.

Although when the initial condition is zero and $x[n] = u[n]$ we cannot find a closed form for the response, we can see that the response is going toward a final value or a steady-state response. Assuming that as $n \to \infty$ we have $Y = y[n] = y[n-1]$ and since $x[n] = x[n-1] = 1$, according to the difference equation the steady-state value Y is found from

$$Y = 0.5Y + 2 \quad \text{or} \quad Y = 4.$$

For this system, the steady-state response is independent of the initial condition. Likewise, when $x[n] = 2u[n]$, the steady-state solution Y is obtained from $Y = 0.5Y + 4$ or $Y = 8$, and again independent of the initial condition. □

Remarks

1. Like in the continuous-time, to show that a discrete-time system is linear and time invariant an explicit expression relating the input and the output is needed, i.e., the output should be expressed as a function of the input only.
2. Although the solution of linear difference equations can be obtained in the time domain, just like with ordinary differential equations, we will see that their solution can also be obtained using the Z-transform, just like the Laplace transform is used to solve linear ordinary differential equations.

9.3.2 DYNAMIC DISCRETE-TIME SYSTEMS REPRESENTED BY DIFFERENCE EQUATIONS

A recursive discrete-time system is represented by a difference equation

$$y[n] = -\sum_{k=1}^{N-1} a_k y[n-k] + \sum_{m=0}^{M-1} b_m x[n-m] \qquad n \geq 0,$$
$$\text{initial conditions} \quad y[-k], \ k = 1, \cdots, N-1, \tag{9.32}$$

characterizing the dynamics of the discrete-time system. This difference equation could be the approximation of an ordinary differential equation representing a continuous-time system being processed discretely. For instance, to approximate a second-order ordinary differential equation by a difference equation, we could approximate the first derivative as

$$\frac{dv_c(t)}{dt} \approx \frac{v_c(t) - v_c(t - T_s)}{T_s}$$

and the second derivative as

$$\frac{d^2v_c(t)}{dt^2} = \frac{d(dv_c(t)/dt)}{dt} \approx \frac{d((v_c(t) - v_c(t - T_s))/T_s)}{dt} = \frac{v_c(t) - 2v_c(t - T_s) + v_c(t - 2T_s)}{T_s^2}$$

to obtain a second-order difference equation. Choosing a small value for T_s provides an accurate approximation to the ordinary differential equation. Other transformations can be used; in Chapter 0 we indicated that approximating integrals by the trapezoidal rule gives the **bilinear transformation** which can also be used to change differential into difference equations.

Just as in the continuous-time case, the system being represented by the difference equation is not LTI unless the initial conditions are zero and the input is causal. The complete response of a system represented by the difference equation can be shown to be composed of **zero-input** and **zero-state** responses, i.e., if $y[n]$ is the solution of the difference equation (9.32) with initial conditions not necessarily equal to zero, then

$$y[n] = y_{zi}[n] + y_{zs}[n]. \tag{9.33}$$

The component $y_{zi}[n]$ is the response when the input $x[n]$ is set to zero, thus it is completely due to the initial conditions. The response $y_{zs}[n]$ is due to the input only, as we set the initial conditions to zero. The complete response $y[n]$ is thus seen as the superposition of these two responses. The Z-transform provides an algebraic way to obtain the complete response, whether the initial conditions are zero or not. It is important, as in continuous time, to differentiate the zero-input and the zero-state responses from the **transient** and the **steady-state** responses. Examples illustrating how to obtain these responses using the Z-transform will be given in the next chapter.

9.3.3 THE CONVOLUTION SUM

Let $h[n]$ be the **impulse response** of a linear time-invariant (LTI) discrete-time system, or the output of the system corresponding to an impulse $\delta[n]$ as input, and initial conditions (if needed) equal to zero.

Using the **generic representation** of the input $x[n]$ of the LTI system

$$x[n] = \sum_{k=-\infty}^{\infty} x[k]\delta[n - k] \tag{9.34}$$

the output of the LTI system is given by either of the following equivalent forms of the **convolution sum**:

$$y[n] = \sum_{k=-\infty}^{\infty} x[k]h[n - k] = \sum_{m=-\infty}^{\infty} x[n - m]h[m]. \tag{9.35}$$

The impulse response $h[n]$ of a recursive LTI discrete-time system is due exclusively to an input $\delta[n]$, as such the initial conditions are set to zero. Clearly no initial conditions are needed when finding the impulse response of non-recursive systems, as no recursion exists, just the input being $\delta[n]$ is needed.

Now, if $h[n]$ is the response due to $\delta[n]$, by time invariance the response to $\delta[n-k]$ is $h[n-k]$. By superposition, the response due to $x[n]$ with the generic representation

$$x[n] = \sum_k x[k]\delta[n-k]$$

is the sum of responses due to $x[k]\delta[n-k]$ which is $x[k]h[n-k]$ ($x[k]$ is not a function of n) or

$$y[n] = \sum_k x[k]h[n-k]$$

or the convolution sum of the input $x[n]$ with the impulse response $h[n]$ of the system. The second expression of the convolution sum in Equation (9.35) is obtained by a change of variable $m = n - k$.

Remarks

1. The output of non-recursive or FIR systems is the convolution sum of the input and the impulse response of the system. Indeed, if the input/output expression of an FIR system is

$$y[n] = \sum_{k=0}^{N-1} b_k x[n-k] \tag{9.36}$$

its impulse response is found by letting $x[n] = \delta[n]$, which gives

$$h[n] = \sum_{k=0}^{N-1} b_k \delta[n-k] = b_0\delta[n] + b_1\delta[n-1] + \cdots + b_{N-1}\delta[n-(N-1)]$$

so that $h[n] = b_n$ for $n = 0, \cdots, N-1$ and zero otherwise. Now, replacing the b_k coefficients in (9.36) by $h[k]$ we find that the output can be written

$$y[n] = \sum_{k=0}^{N-1} h[k]x[n-k]$$

or the convolution sum of the input and the impulse response. This is a very important result, indicating that FIR systems are obtained by means of the convolution sums rather than difference equations, which gives great significance to the efficient computation of the convolution sum.

2. Considering the convolution sum as an operator, i.e.,

$$y[n] = [h * x][n] = \sum_{k=-\infty}^{\infty} x[k]h[n-k]$$

it is easily shown to be linear. Indeed, whenever the input is $ax_1[n] + bx_2[n]$, and $\{y_i[n]\}$ are the outputs corresponding to $\{x_i[n]\}$ for $i = 1, 2$, then we have

$$
\begin{aligned}
[h * (ax_1 + bx_2)][n] &= \sum_k (ax_1[k] + bx_2[k])h[n-k] \\
&= a\sum_k x_1[k]h[n-k] + b\sum_k x_2[k]h[n-k] \\
&= a[h * x_1][n] + b[h * x_2][n] = ay_1[n] + by_2[n]
\end{aligned}
$$

as expected since the system was assumed to be linear when the expression for the convolution sum was obtained.

We also see that if the output corresponding to $x[n]$ is $y[n]$, given by the convolution sum, then the output corresponding to a shifted version of the input, $x[n - N]$ should be $y[n - N]$. In fact, if we let $x_1[n] = x[n - N]$ the corresponding output is

$$
\begin{aligned}
[h * x_1][n] &= \sum_k x_1[n-k]h[k] = \sum_k x[n - N - k]h[k] \\
&= [h * x][n - N] = y[n - N];
\end{aligned}
$$

again this result is expected given that the system was considered time invariant when the convolution sum was obtained.

3. From the equivalent representations for the convolution sum we have

$$
[h * x][n] = \sum_k x[k]h[n-k] = \sum_k x[n-k]h[k] = [x * h][n]
$$

indicating that the convolution commutes with respect to the input $x[n]$ and the impulse response $h[n]$.

4. Just as with continuous-time systems, when connecting two LTI discrete-time systems (with impulse responses $h_1[n]$ and $h_2[n]$) in cascade or in parallel, their respective impulse responses are given by $[h_1 * h_2][n]$ and $h_1[n] + h_2[n]$, respectively. See Fig. 9.9 for the block diagrams. In particular notice that in the cascade connection we can interchange the order of the systems with not change on the output. This is due to the systems being LTI, such an interchange is not valid for non-linear or time-varying systems.

5. There are situations when instead of giving the input and the impulse response to compute the output, the information available is, for instance, the input and the output and we wish to find the impulse response of the system, or we have the output and the impulse response and wish to find the input. This type of problem is called **deconvolution**. We will consider this problem later in this chapter after considering causality, and in the next chapter where we will show that the deconvolution problem can easily be solved using the Z-transform.

6. The computation of the convolution sum is typically difficult. It is made easier when the Z-transform is used as we will see. MATLAB provides the function *conv* to compute the convolution sum.

Example 9.22. Consider a moving-average system

$$
y[n] = \frac{1}{3}(x[n] + x[n-1] + x[n-2])
$$

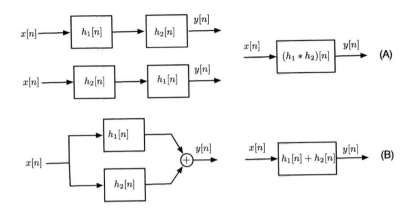

FIGURE 9.9

Cascade (A) and parallel (B) connections of LTI systems with impulse responses $h_1[n]$ and $h_2[n]$. Equivalent systems on the right. Notice the interchange of systems in the cascade connection.

where the input is $x[n]$ and the output is $y[n]$. Find its impulse response $h[n]$. Then:

1. Let $x[n] = u[n]$, find the output $y[n]$ using the input–output relation and the convolution sum.
2. If the input is $x[n] = A\cos(2\pi n/N)u[n]$, determine the values of A, and N, so that the steady-state response of the system is zero. Explain. Use MATLAB to verify your results.

Solution: If the input is $x[n] = \delta[n]$, the output is $y[n] = h[n]$ or the impulse response of the system. No initial conditions are needed. We thus have

$$h[n] = \frac{1}{3}(\delta[n] + \delta[n-1] + \delta[n-2])$$

so that $h[0] = h[1] = h[2] = 1/3$ and $h[n] = 0$ for $n \neq 0,\ 1,\ 2$. Notice that the coefficients of the filter equal the impulse response at $n = 0, 1$, and 2.

For an input $x[n]$ such that $x[n] = 0$ for $n < 0$, let us find a few values of the convolution sum to see what happens as n grows. If $n < 0$, the arguments of $x[n]$, $x[n-1]$ and $x[n-2]$ are negative giving zero values, and so the output is also zero, i.e., $y[n] = 0$, $n < 0$. For $n \geq 0$ we have

$$y[0] = \frac{1}{3}(x[0] + x[-1] + x[-2]) = \frac{1}{3}x[0],$$

$$y[1] = \frac{1}{3}(x[1] + x[0] + x[-1]) = \frac{1}{3}(x[0] + x[1]),$$

$$y[2] = \frac{1}{3}(x[2] + x[1] + x[0]) = \frac{1}{3}(x[0] + x[1] + x[2]),$$

$$y[3] = \frac{1}{3}(x[3] + x[2] + x[1]) = \frac{1}{3}(x[1] + x[2] + x[3]),$$

$$\cdots.$$

Thus if $x[n] = u[n]$ then we see that $y[0] = 1/3$, $y[1] = 2/3$, and $y[n] = 1$ for $n \geq 2$.

Noticing that, for $n \geq 2$, the output is the average of the present and past two values of the input, when the input is $x[n] = A\cos(2\pi n/N)$ if we let $N = 3$, and A be any real value the input repeats every 3 samples and the local average of 3 of its values is zero, giving $y[n] = 0$ for $n \geq 2$. Thus, the steady-state response will be zero.

The following MATLAB script uses the function *conv* to compute the convolution sum when the input is either $x[n] = u[n]$ or $x[n] = \cos(2\pi n/3)u[n]$.

```
%%
% Example 9.22---Convolution sum
%%
clear all; clf
x1=[0 0 ones(1,20)]                    % unit-step input
n=-2:19; n1=0:19;
x2=[0 0 cos(2*pi*n1/3)];               % cosine input
h=(1/3)*ones(1,3);                     % impulse response
y=conv(x1,h); y1=y(1:length(n));       % convolution sums
y=conv(x2,h); y2=y(1:length(n));
```

Notice that each of the input sequences has two zeros at the beginning so that the response can be found at $n \geq -2$. Also, when the input is of infinite support, we can only approximate it as a finite sequence in MATLAB and as such the final values of the convolution obtained from *conv* are not correct and should not be considered—in this case, the final two values of the convolution results are not correct and are not considered. The results are shown in Fig. 9.10.   □

Example 9.23. Consider an autoregressive system represented by a first-order difference equation

$$y[n] = 0.5y[n-1] + x[n] \qquad n \geq 0.$$

Find the impulse response $h[n]$ of the system and then compute the response of the system to $x[n] = u[n] - u[n-3]$ using the convolution sum. Verify the results with MATLAB.

Solution: The impulse response $h[n]$ can be found recursively. Letting $x[n] = \delta[n]$, $y[n] = h[n]$ and initial condition $y[-1] = h[-1] = 0$ we have

$$h[0] = 0.5h[-1] + \delta[0] = 1, \quad h[1] = 0.5h[0] + \delta[1] = 0.5,$$
$$h[2] = 0.5h[1] + \delta[2] = 0.5^2, \quad h[3] = 0.5h[2] + \delta[3] = 0.5^3, \quad \cdots,$$

from which the general expression for the impulse response is $h[n] = 0.5^n u[n]$.

The response to $x[n] = u[n] - u[n-3]$ using the convolution sum is then given by

$$y[n] = \sum_{k=-\infty}^{\infty} x[k]h[n-k] = \sum_{k=-\infty}^{\infty} (u[k] - u[k-3])0.5^{n-k}u[n-k].$$

Since as functions of k, $u[k]u[n-k] = 1$ for $0 \leq k \leq n$, zero otherwise, and $u[k-3]u[n-k] = 1$ for $3 \leq k \leq n$, zero otherwise (in the two cases, draw the two signals as functions of k and verify this

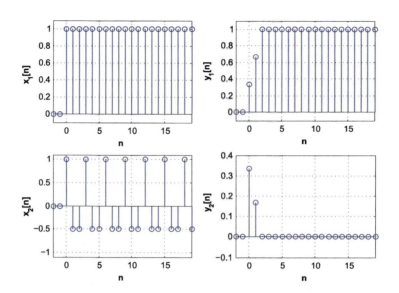

FIGURE 9.10

Convolution sums for a moving-averaging system $y[n] = (x[n] + x[n-1] + x[n-2])/3$ with inputs $x_1[n] = u[n]$ (top) and $x_2[n] = \cos(2\pi n/3)u[n]$ (bottom).

is true), $y[n]$ can be expressed as

$$y[n] = 0.5^n \left[\sum_{k=0}^{n} 0.5^{-k} - \sum_{k=3}^{n} 0.5^{-k} \right] u[n] = \begin{cases} 0 & n < 0, \\ 0.5^n \sum_{k=0}^{n} 0.5^{-k} = 0.5^n(2^{n+1} - 1) & n = 0, 1, 2, \\ 0.5^n \sum_{k=0}^{2} 0.5^{-k} = 7(0.5^n) & n \geq 3. \end{cases}$$

Another way to solve this problem is to notice that the input can be rewritten as

$$x[n] = \delta[n] + \delta[n-1] + \delta[n-2]$$

and since the system is LTI, the output can be written

$$y[n] = h[n] + h[n-1] + h[n-2] = 0.5^n u[n] + 0.5^{n-1} u[n-1] + 0.5^{n-2} u[n-2],$$

which gives zero for $n < 0$ and

$$y[0] = 0.5^0 = 1, \quad y[1] = 0.5^1 + 0.5^0 = \frac{3}{2},$$

$$y[2] = 0.5^2 + 0.5^1 + 0.5^0 = \frac{7}{4}, \quad y[3] = 0.5^3 + 0.5^2 + 0.5 = \frac{7}{8}, \quad \cdots,$$

which coincides with the above more general solution. It should be noticed that even in a simple example like this the computation required by the convolution sum is quite high. We will see that the

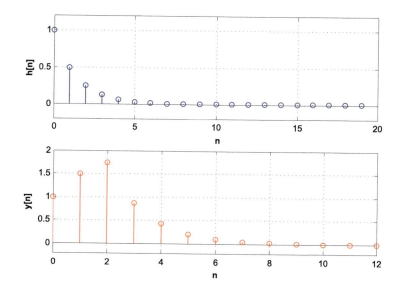

FIGURE 9.11

Impulse response $h[n]$ of first-order autoregressive system $y[n] = 0.5y[n-1] + x[n]$, $n \geq 0$, (top), and response $y[n]$ due to $x[n] = u[n] - u[n-3]$ (bottom).

Z-transform reduces the computational load, just like the Laplace transform does in the computation of the convolution integral.

The following MATLAB script is used to verify the above results. The MATLAB function *filter* is used to compute the impulse response and the response of the filter to the pulse. The output obtained with *filter* coincided with the output found using *conv*, as it should. Fig. 9.11 displays the results.   □

```
%%
% Example 9.23
%%
a=[1 -0.5];b=1;              % coefficients
d=[1 zeros(1,99)];           % approximate delta function
h=filter(b,a,d);             % impulse response
x=[ones(1,3) zeros(1,10)];     % input
y=filter(b,a,x);             % output from filter function
y1=conv(h,x); y1=y1(1:length(y))   % output from conv
```

9.3.4 LINEAR AND NON-LINEAR FILTERING WITH MATLAB

A recursive or a non-recursive discrete-time system can be used to get rid of undesirable components in a signal. These systems are called linear filters. In this section, we illustrate the use and possible advantages of non-linear filters.

Linear Filtering

To illustrate the way a linear filter works, consider getting rid of a random disturbance $\eta[n]$, which we model as Gaussian noise (this is one of the possible noise signals MATLAB provides) that has been added to a sinusoid $x[n] = \cos(\pi n/16)$. Let $y[n] = x[n] + \eta[n]$. We will use an averaging filter having an input/output equation

$$z[n] = \frac{1}{M} \sum_{k=0}^{M-1} y[n-k].$$

This Mth-order filter averages M past input values $\{y[n-k], k = 0, \cdots, M-1\}$ and assigns this average to the output $z[n]$. The effect is to smooth out the input signal by attenuating the high-frequency components of the signal due to the noise. The larger the value of M the better the results, but at the expense of more complexity and a larger delay in the output signal (this is due to the linear phase frequency response of the filter, as we will see later).

We use 3rd- and 15th-order filters, implemented by our function *averager* given below. The denoising is done by means of the following script:

```
%%
% Linear filtering
%%
N=200;n=0:N-1;
x=cos(pi*n/16);          % input signal
noise=0.2*randn(1,N);    % noise
y=x+noise;               % noisy signal
z=averager(3,y);         % averaging linear filter with M=3
z1=averager(15,y);       % averaging linear filter with M=15
```

Our function *averager* defines the coefficients of the averaging filter and then uses the MATLAB function *filter* to compute the filter response. The inputs of *filter* are the vector $\mathbf{b} = (1/M)[1 \cdots 1]$, or the coefficients connected of the numerator, the coefficient of the denominator (1), and $\mathbf{x}$ a vector with entries the signal samples we wish to filter. The results of filtering using these two filters are shown in Fig. 9.12. As expected the performance of the filter with $M = 15$ is a lot better, but a delay of 8 samples (or the integer larger than $M/2$) is shown in the filter output.

```
function y=averager(M,x)
% Moving average of signal x
%   M: order of averager
%   x: input signal
%
b=(1/M)*ones(1,M);
y=filter(b,1,x);
```

Non-linear Filtering

Is linear filtering always capable of getting rid of noise? The answer is: It depends on the type of noise. In the previous example we showed that a high-order averaging filter, which is linear, performs well

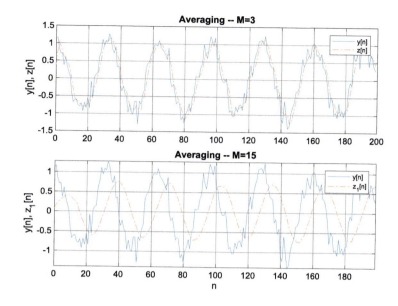

FIGURE 9.12

Averaging filtering with filters of order $M = 3$ (top figure), and of order $M = 15$ result (bottom figure) used to get rid of Gaussian noise added to a sinusoid $x[n] = \cos(\pi n/16)$. Solid line corresponds to the noisy signal, while the dashed line is for the filtered signal. The filtered signal is very much like the noisy signal (see top figure) when $M = 3$ is the order of the filter, while the filtered signal looks like the sinusoid, but shifted, when $M = 15$.

for Gaussian noise. Let us now consider an **impulsive** noise that is either zero or a certain value at random. This is the type of noise occurring in communications whenever cracking sounds are heard in the transmission, or the "salt-and-pepper" noise that appears in images.

It will be shown that even the 15th-order averager—which did well before—is not capable of denoising the signal with impulsive noise. A **median filter** considers a certain number of samples (the example shows the case of a 5th-order median filter), orders them according to their values and chooses the one in the middle (i.e., the median) as the output of the filter. Such a filter is non-linear as it does not satisfy superposition. The following script is used to filter the noisy signal using a linear and a non-linear filter. The results shown in Fig. 9.13. In this case the non-linear filter is able to denoise the signal much better than the linear filter.

```
%%
% Linear and non-linear filtering
%%
clear all; clf
N=200;n=0:N-1;
% impulsive noise
for m=1:N,
    d=rand(1,1);
```

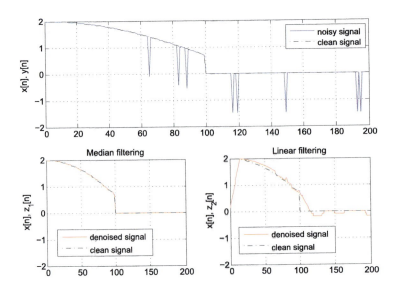

FIGURE 9.13

Non-linear 5th-order median filtering (bottom left) versus linear 15th-order averager (bottom right) corresponding to the noisy signal (dash line) and clean signal (solid line) on top plots. Clean signal (solid line) is superposed on de-noised signal (dashed line) in the bottom figures. Linear filter does not perform as well as the non-linear filter.

```
      if d>=0.95,
          noise(m)=-1.5;
        else
          noise(m)=0 ;
        end
   end
 x=[2*cos(pi*n(1:100)/256) zeros(1,100)];
 y1=x+noise;
 % linear filtering
 z2=averager(15,y1);
 % non-linear filtering -- median filtering
 z1(1)=median([0 0 y1(1) y1(2) y1(3)]);
 z1(2)=median([0 y1(1) y1(2) y1(3) y1(4)]);
 z1(N-1)=median([y1(N-3) y1(N-2) y1(N-1) y1(N) 0]);
 z1(N)=median([y1(N-2) y1(N-1) y1(N) 0 0]);
 for k=3:N-2,
     z1(k)=median([y1(k-2) y1(k-1) y1(k) y1(k+1) y1(k+2)]);
 end
```

Although the theory of non-linear filtering is beyond the scope of this book, it is good to remember that in cases like this when linear filters do not seem to do well, there are other methods to use.

9.3.5 CAUSALITY AND STABILITY OF DISCRETE-TIME SYSTEMS

As with continuous-time systems, two additional independent properties of discrete-time systems are causality and stability. Causality relates to the conditions under which computation can be performed in real-time, while stability relates to the usefulness of the system.

Causality

In many situations signals need to be processed in **real-time**, i.e., the processing must be done as the signal comes into the system. In those situations, the system must be causal. Whenever the data can be stored, not a real-time situation, is not necessary to use a causal system.

A discrete-time system S is **causal** if:

- whenever the input $x[n] = 0$, and there are no initial conditions, the output is $y[n] = 0$,
- the present output $y[n]$ does not depend on future inputs.

Causality is independent of the linearity and time-invariance properties of a system. For instance, the system represented by the input/output equation

$$y[n] = x^2[n],$$

where $x[n]$ is the input and $y[n]$ the output, is non-linear, time invariant, and according to the above definition causal: the output is zero whenever the input is zero, and the output depends on the present value of the input. Likewise, an LTI system can be noncausal, such is the case of the following LTI system that computes the moving average of the input:

$$y[n] = \frac{1}{3}(x[n+1] + x[n] + x[n-1]).$$

The input/output equation indicates that at the present time n to compute $y[n]$ we need a present value $x[n]$, a past value $x[n-1]$, and a future value $x[n+1]$ of the input. Thus, the system is LTI but noncausal since it requires future values of the input.

- An LTI discrete-time system is **causal** if the impulse response of the system is such that

$$h[n] = 0 \qquad n < 0. \tag{9.37}$$

- A signal $x[n]$ is said to be **causal** if

$$x[n] = 0 \qquad n < 0. \tag{9.38}$$

- For a causal LTI discrete-time system with a causal input $x[n]$ its output $y[n]$ is given by

$$y[n] = \sum_{k=0}^{n} x[k]h[n-k] \qquad n \geq 0 \tag{9.39}$$

where the lower limit of the sum depends on the input causality, $x[k] = 0$ for $k < 0$, and the upper limit on the causality of the system, $h[n-k] = 0$ for $n - k < 0$ or $k > n$.

That $h[n] = 0$ for $n < 0$ is the condition for an LTI discrete-time system to be causal is understood by considering that when computing the impulse response, the input $\delta[n]$ only occurs at $n = 0$ and there are no initial conditions so the response for $n < 0$ should be zero. Extending the notion of causality to signals we can then see that the output of a causal LTI discrete-time system can be written in terms of the convolution sum as

$$y[n] = \sum_{k=-\infty}^{\infty} x[k]h[n-k] = \sum_{k=0}^{\infty} x[k]h[n-k] = \sum_{k=0}^{n} x[k]h[n-k]$$

where we first used the causality of the input ($x[k] = 0$ for $k < 0$) and then that of the system, i.e., $h[n-k] = 0$ whenever $n - k < 0$ or $k > n$. According to this equation the output depends on inputs $\{x[0], \cdots, x[n]\}$ which are past and present values of the input.

Example 9.24. So far we have considered the convolution sum as a way of computing the output $y[n]$ of a LTI system with impulse response $h[n]$ for a given input $x[n]$. But it actually can be used to find either of these three variables given the other two. The problem is then called **deconvolution**. Assume the input $x[n]$ and the output $y[n]$ of a causal LTI system are given, find equations to compute recursively the impulse response $h[n]$ of the system. Consider finding the impulse response $h[n]$ of a causal LTI system with input $x[n] = u[n]$ and output $y[n] = \delta[n]$. Use the MATLAB function *deconv* to find $h[n]$.

Solution: If the system is causal and LTI, the input $x[n]$ and the output $y[n]$ are connected by the convolution sum

$$y[n] = \sum_{m=0}^{n} h[n-m]x[m] = h[n]x[0] + \sum_{m=1}^{n} h[n-m]x[m].$$

To find $h[n]$ from given input and output values, under the condition that $x[0] \neq 0$, the above equation can be rewritten as

$$h[n] = \frac{1}{x[0]} \left[y[n] - \sum_{m=1}^{n} h[n-m]x[m] \right]$$

so that the impulse response of the causal LTI can be found recursively as follows:

$$h[0] = \frac{1}{x[0]} y[0], \quad h[1] = \frac{1}{x[0]} (y[1] - h[0]x[1]),$$

$$h[2] = \frac{1}{x[0]} (y[2] - h[0]x[2] - h[1]x[1]) \quad \cdots.$$

For the given case where $y[n] = \delta[n]$ and $x[n] = u[n]$ we get according to the above

$$h[0] = \frac{1}{x[0]} y[0] = 1,$$

$$h[1] = \frac{1}{x[0]} (y[1] - h[0]x[1]) = 0 - 1 = -1,$$

$$h[2] = \frac{1}{x[0]} (y[2] - h[0]x[2] - h[1]x[1]) = 0 - 1 + 1 = 0,$$

$$h[3] = \frac{1}{x[0]} (y[3] - h[0]x[3] - h[1]x[2] - h[2]x[3]) = 0 - 1 + 1 - 0 = 0, \cdots$$

and in general $h[n] = \delta[n] - \delta[n-1]$.

The length of the convolution $y[n]$ is the sum of the lengths of the input $x[n]$ and of the impulse response $h[n]$ minus one. Thus,

$$\text{length of } h[n] = \text{length of } y[n] - \text{length of } x[n] + 1.$$

When using the MATLAB function *deconv* we need to make sure that the length of $y[n]$ is always larger than that of $x[n]$. If $x[n]$ is of infinite length, like when $x[n] = u[n]$, this would require an even longer $y[n]$, which is not possible. However, MATLAB can only provide a finite support input, so we make the support of $y[n]$ larger. In this example we have found analytically that the impulse response $h[n]$ is of length 2. Thus, if the length of $y[n]$ is chosen larger than the length of $x[n]$ by one we get the correct answer (case (a) in the script below). Otherwise we do not (case (b)). Run the two cases to verify this (get rid of the symbol % to run case (b)).

```
%%
% Example 9.24---Deconvolution
%%
clear all
x=ones(1,100);
y=[1 zeros(1,100)];        % case (a), correct h
% y=[1 zeros(1,99)];       % case (b), wrong h
[h,r]=deconv(y,x)
```

Bounded Input–Bounded Output (BIBO) Stability

Stability characterizes useful systems. A stable system provides well-behaved outputs for well-behaved inputs. Bounded input–bounded output (BIBO) stability establishes that for a bounded (which is what is meant by 'well-behaved') input $x[n]$ the output of a BIBO stable system $y[n]$ is also bounded. This means that if there is a finite bound $M < \infty$ such that $|x[n]| < M$ for all n (you can think of it as an envelope $[-M, M]$ inside which the input $x[n]$ is) the output is also bounded, i.e., $|y[n]| < L$ for $L < \infty$ and all n.

An LTI discrete-time system is said to be BIBO stable if its impulse response $h[n]$ is absolutely summable

$$\sum_k |h[k]| < \infty \quad \text{(absolutely summable).} \tag{9.40}$$

Assuming that the input $x[n]$ of the system is bounded, or that there is a value $M < \infty$ such that $|x[n]| < M$ for all n, the output $y[n]$ of the system represented by a convolution sum is also bounded or

$$|y[n]| \leq \left| \sum_{k=-\infty}^{\infty} x[n-k]h[k] \right| \leq \sum_{k=-\infty}^{\infty} |x[n-k]||h[k]|$$

$$\leq M \sum_{k=-\infty}^{\infty} |h[k]| \leq \underbrace{MN}_{L} < \infty$$

provided that $\sum_{k=-\infty}^{\infty} |h[k]| < N < \infty$, or that the impulse response be absolutely summable.

Remarks

1. Non-recursive or FIR systems are BIBO stable. Indeed, the impulse response of such a system is of finite length and therefore absolutely summable.
2. For a recursive or IIR system represented by a difference equation, to established stability according to the above result we need to find the system's impulse response $h[n]$ and determine whether it is absolutely summable or not. A much simpler way to test the stability of an IIR system will be based on the location of the poles of the Z-transform of $h[n]$, as we will see in the next chapter.

Example 9.25. Consider an autoregressive system

$$y[n] = 0.5y[n-1] + x[n].$$

Determine if the system is BIBO stable.

Solution: As shown in Example 9.23, the impulse response of the system is $h[n] = 0.5^n u[n]$, checking the BIBO stability condition we have

$$\sum_{n=-\infty}^{\infty} |h[n]| = \sum_{n=0}^{\infty} 0.5^n = \frac{1}{1-0.5} = 2;$$

thus the system is BIBO stable. □

9.4 TWO-DIMENSIONAL DISCRETE SIGNALS AND SYSTEMS

In this section we consider two-dimensional signals (with special interests in images) and systems. The theory of two-dimensional signals and systems has a great deal of similarities, but also significant differences, with the theory of one-dimensional signals and systems; it implies the characteristics of one-dimensional signals and systems, but not the other way around. For a more detailed presentation refer to [24,49,81].

9.4.1 TWO-DIMENSIONAL DISCRETE SIGNALS

A discrete two-dimensional signal $x[m,n]$ is a mapping of integers $[m,n]$ into real values that is not defined for non-integer values. Two-dimensional signals have a greater variety of supports than one-dimensional signals: they can have finite or infinite support in either of the quadrants of the two-dimensional space, or in a combination of these.

To represent any signal $x[m, n]$ consider a **two-dimensional impulse** $\delta[m, n]$ defined as

$$\delta[m, n] = \begin{cases} 1 & [m, n] = [0, 0], \\ 0 & [m, n] \neq [0, 0], \end{cases} \tag{9.41}$$

so that a signal $x[m, n]$ defined in a support $[M_1, N_1] \times [M_2, N_2]$, $M_1 < M_2$, $N_1 < N_2$ can be written

$$x[m, n] = \sum_{k=M_1}^{M_2} \sum_{\ell=N_1}^{N_2} x[k, \ell] \delta[m - k, n - \ell]. \tag{9.42}$$

Thus simple signals such as the **two-dimensional unit-step** signal $u_1[m, n]$, with support in the first quadrant,[4] is given by

$$u_1[m, n] = \begin{cases} 1 & m \geq 0, \, n \geq 0 \\ 0 & \text{otherwise} \end{cases} = \sum_{k=0}^{\infty} \sum_{\ell=0}^{\infty} \delta[m - k, n - \ell] \tag{9.43}$$

and a **two-dimensional unit-ramp** signal $r_1[m, n]$, with support in the first quadrant, is given by

$$r_1[m, n] = \begin{cases} mn & m \geq 0, \, n \geq 0 \\ 0 & \text{otherwise} \end{cases} = \sum_{k=0}^{\infty} \sum_{\ell=0}^{\infty} k\ell \, \delta[m - k, n - \ell]. \tag{9.44}$$

A class of two-dimensional signals of interest are **separable signals** $y[m, n]$ that are the product of two one-dimensional signals, one being a function of m and the other of n:

$$y[m, n] = y_1[m] y_2[n]. \tag{9.45}$$

Given that $\delta[m, n] = \delta[m]\delta[n]$ is separable, one can easily see that $u_1[m, n]$ and $r_1[m, n]$ are also separable. There are, however, simple two-dimensional signals that are not separable. For instance, consider

$$\delta_1[m, n] = \begin{cases} 1 & m = n, \, m \geq 0, n \geq 0, \\ 0 & \text{otherwise}, \end{cases} \tag{9.46}$$

or a sequence of delta functions supported on the diagonal of the first quadrant. This signal cannot be expressed as a product of two one-dimensional sequences. It can, however, be represented by a sum of separable delta functions, or

$$\delta_1[m, n] = \sum_{k=0}^{\infty} \delta[m - k, n - k] = \sum_{k=0}^{\infty} \delta[m - k]\delta[n - k]. \tag{9.47}$$

---

[4]Whenever it is necessary to indicate the quadrant or quadrants of support the subscript number or numbers will indicate that support. Thus, as $u_1[m, n]$ has support in the first quadrant, $u_3[m, n]$ has support in the third quadrant and $u_{23}[m, n]$ has support in the second and third quadrants.

The best example of two-dimensional discrete signals are discretized images. A discretized image with MN picture elements or pixels, is a positive array $i[m, n]$, $0 \le m \le M - 1$, $0 \le n \le N - 1$ obtained by sampling an analog image.[5] Such signals are not separable in most cases, but they have a finite support, and they have positive values given that each pixel represents the average illumination of a small area in the analog image.

Example 9.26. In two dimensions (either space and space, or time and space, or any other two variables) the sampling of a continuous two-variable signal $x(\xi, \zeta)$ is much more general than the sampling of one-variable signal. The signal $x(\xi, \zeta)$ can be sampled in general by expressing ξ, ζ for integers m and n and real values $\{M_{i,j}, \ 1 \le i, j \le 2\}$ as

$$\begin{bmatrix} \xi \\ \zeta \end{bmatrix} = \begin{bmatrix} M_{11} & M_{12} \\ M_{21} & M_{22} \end{bmatrix} \begin{bmatrix} m \\ n \end{bmatrix} \tag{9.48}$$

so that $x(M_{11}m + M_{12}n, M_{21}m + M_{22}n)$ is a discrete signal. Suppose $M_{12} = M_{21} = 0$, how are the values M_{11} and M_{22} chosen? Suppose $M_{11} = M_{12} = 1$ and $M_{21} = -1$ and $M_{22} = 1$; what would be the geometry of the sampling? Is it possible to change the variables ξ and ζ to have the matrix $\mathbf{M}$ with entries $\{M_{i,j}\}$ as the identity matrix?

Solution: When $M_{12} = M_{21} = 0$ we have a rectangular sampling, where the values of the discrete signal are obtained by sampling in a rectangular way with two sampling spacings M_{11} and M_{22}. This is an extension of the sampling in one-dimension, where the samples are at the intersection of $M_{11}m$ and $M_{22}n$, and in that case for band-limited two-dimensional signals $x(\xi, \zeta)$ with maximum frequencies Ω_{1max} and Ω_{2max}, we would let

$$M_{11} \le \frac{\pi}{\Omega_{1max}}, \qquad M_{22} \le \frac{\pi}{\Omega_{2max}},$$

a simple extension of the Nyquist criterion.

When $M_{11} = M_{12} = 1$ and $M_{21} = -1$ and $M_{22} = 1$, we will then have $\xi = m + n$ and $\zeta = -m + n$, and instead of a rectangle in this case we get a hexagonal shape as shown in Fig. 9.14.

The matrix $\mathbf{M}$ formed by the $\{M_{i,j}, 1 \le i, j \le 2\}$ should be invertible. If $\mathbf{M}^{-1}$ is the inverse, then premultiplying Equation (9.48) by it gives the new variables ξ' and ζ' such that

$$\begin{bmatrix} \xi' \\ \zeta' \end{bmatrix} = \begin{bmatrix} 1 & 0 \\ 0 & 1 \end{bmatrix} \begin{bmatrix} m \\ n \end{bmatrix}$$

so that we can have a rectangular sampling when we transform the original variables ξ and ζ using the transformation $\mathbf{M}^{-1}$. $\square$

If the two-dimensional signal is of infinite support it is possible for it to be periodic. A periodic two-dimensional signal $\tilde{x}[m, n]$, $-\infty < m < \infty$, $-\infty < n < \infty$, is such that for positive integers M and N

$$\tilde{x}[m + kM, n + \ell N] = \tilde{x}[m, n], \qquad \text{for any integers } k, \ell. \tag{9.49}$$

---

[5]When quantized and coded discrete images become digital images.

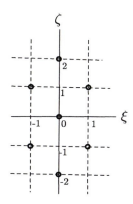

FIGURE 9.14

Sampling pattern when $M_{11} = M_{12} = 1$ and $M_{21} = -1$, $M_{22} = 1$ repeated with a period of 2 in the horizontal direction and 4 in the vertical direction. Notice that there are 7 samples (denoted by dark circles) instead of 11 (given by a rectangular sampling with $M_{11} = M_{22} = 1$ and $M_{12} = M_{21} = 0$) in the area bounded by the hexagon.

Because of the required infinite dimension, periodic signals are not common in practical applications but conceptually they are of great significance. The integer nature of the periods causes that the sampling of periodic analog signals does not always result in periodic discrete signals.

Example 9.27. The periodic analog sinusoid

$$x(t_1, t_2) = \sin(2\pi t_1 + \pi t_2), \qquad -\infty < t_1 < \infty, \ -\infty < t_2 < \infty,$$

has continuous frequencies $f_1 = 1$ Hz and $f_2 = 0.5$ Hz so that $T_1 = 1$ s and $T_2 = 2$ s are the corresponding periods. Suppose $x(t_1, t_2)$ is sampled at times $t_1 = 0.1T_1 m$ and $t_2 = T_2 n/\pi$, for integers m and n. Is the resulting discrete signal still periodic? How would one come up with a two-dimensional periodic signal using approximate sampling times?

Solution: Sampling $x(t_1, t_2)$ by taking samples at $t_1 = 0.1T_1 m$ and $t_2 = T_2 n/\pi$, for integers m, and n, gives the discrete signal

$$x(0.1T_1 m, T_2 n/\pi) = \sin(0.2\pi T_1 m + T_2 n) = \sin\left(\frac{2\pi}{10}m + 2n\right) = x_1[m, n],$$

which is periodic in m with period $M = 10$, but is not periodic in n as there is no positive integer value N such that the discrete frequency $\omega_2 = 2\pi/N$ be equal to 2. The reason for the non-periodicity of the discrete signal is that samples in t_2 are taken at irrational times $T_2 n/\pi$, which in practice cannot be done. To remedy that, suppose the analog sinusoid is sampled at $t_1 = 0.1T_1 m$ as before but at $t_2 = 0.32T_2 n$ (where $0.32 \approx 1/\pi$). We then have

$$x(0.1T_1 m, 0.32T_2 n) = \sin(0.2\pi T_1 m + 0.32\pi T_2 n) = \sin\left(\frac{2\pi}{10}m + \frac{2\pi \times 32}{100}n\right) = x_1[m, n]$$

with discrete frequencies $\omega_1 = 2\pi/10$ and $\omega_2 = 2\pi \times 32/100$ and thus being periodic with periods $M = 10$ and $N = 100$. The expectation that sampling an analog sinusoid, periodic by nature, results in a periodic signal is not always true in two dimensions, just as it was not always true in one dimension. $\quad\square$

A difference between one- and two-dimensional periodicity is that the period of a one-dimensional periodic signal can be chosen so as to give the smallest possible period, i.e., the fundamental period, but that is not always the case in two dimensions. A general way to represent the periodicity in two dimensions is by assigning a periodicity matrix

$$\mathbf{N} = \begin{bmatrix} N_{11} & N_{12} \\ N_{21} & N_{22} \end{bmatrix}$$

such that

$$
\begin{aligned}
\tilde{x}[m, n] &= \tilde{x}[m + N_{11}, n + N_{12}] \\
&= \tilde{x}[m + N_{21}, n + N_{22}].
\end{aligned} \tag{9.50}
$$

If $N_{11} = M$, $N_{22} = N$, where M and N are the periodicities in Equation (9.49), and $N_{12} = N_{21} = 0$ this is called **rectangular periodicity**, otherwise it is a **block periodicity**. The number of terms in a rectangular period is $N_{11}N_{22} = MN$ which is the absolute value of the determinant of the matrix $\mathbf{N}$ for that case. There are cases when N_{12} and N_{21} are different from zero, and in such cases there is a smaller array that can be taken as the period. Such an array has $|N_{11}N_{22} - N_{12}N_{21}|$ elements, a value that could be smaller than the one corresponding to the rectangular periodicity. The rectangular periodicity does not necessarily have the smallest period but it is a valid periodicity. Because of the complexity in finding the general or block periodicity and a valid, smaller sized period we will only consider the rectangular periodicity.

Example 9.28. To illustrate the connection between the rectangular and the block periodicities, consider the two-dimensional sinusoid

$$\tilde{x}[m, n] = \cos(2\pi \ m/8 + 2\pi \ n/16),$$

which is rectangular periodic in m with a period $M = 8$, and in n with a period $N = 16$. The number of entries of the corresponding period is $M \times N = 8 \times 16 = 128$ which equals the absolute value of the determinant of the periodicity matrix

$$\mathbf{N} = \begin{bmatrix} 8 & 0 \\ 0 & 16 \end{bmatrix}.$$

If the periodicity matrix is chosen to be

$$\mathbf{N}_1 = \begin{bmatrix} 4 & 8 \\ 1 & -2 \end{bmatrix}$$

show the signal is still periodic. What would be an advantage of choosing this matrix over the diagonal matrix?

Solution: Using $\mathbf{N}_1$ we have

$$\tilde{x}[m, n] = \tilde{x}[m + 4, n + 8] = \tilde{x}[m + 1, n - 2]$$

as can easily be verified:

$$\tilde{x}[m + 4, n + 8] = \cos(2\pi m/8 + \pi + 2\pi n/16 + \pi) = \tilde{x}[m, n],$$
$$\tilde{x}[m + 1, n - 2] = \cos(2\pi m/8 + 2\pi/8 + 2\pi n/16 - 2\pi/8) = \tilde{x}[m, n].$$

This general periodicity provides a period with fewer terms than the previous rectangular periodicity. The absolute value of the determinant of the matrix $\mathbf{N}_1$ is 16. $\qquad\square$

9.4.2 TWO-DIMENSIONAL DISCRETE SYSTEMS

A two-dimensional system is an operator S that maps an input $x[m, n]$ into a unique output

$$y[m, n] = S(x[m, n]). \tag{9.51}$$

The desirable characteristics of two-dimensional systems are like those of one-dimensional systems. Thus for given inputs $\{x_i[m, n]\}$ having as output $\{y_i[m, n] = S(x_i[m, n])\}$ and real-valued constants $\{a_i\}$, for $i = 1, 2, \cdots, I$, we see that if

$$S\left(\sum_{i=1}^{I} a_i x_i[m, n]\right) = \sum_{i=1}^{I} a_i S(x_i[m, n]) = \sum_{i=1}^{I} a_i y_i[m, n], \tag{9.52}$$

then the system S is **linear**, and if the shifting of the input does not change the output, i.e., for some integers M and N if

$$S(x_i[m - M, n - N]) = y_i[m - M, n - N], \tag{9.53}$$

the system S is **shift-invariant**. A system satisfying these two conditions is called **linear shift-invariant** or **LSI**.

Suppose then the input $x[m, n]$ of an LSI system is represented as in Equation (9.42) and that the response of the system to $\delta[m, n]$ is $h[m, n]$ or the **impulse response** of the system. According to the linearity and shift-invariance characteristics of the system the response to $x[m, n]$ is

$$
\begin{aligned}
y[m, n] &= \sum_k \sum_\ell x[k, \ell] S(\delta[m - k, n - \ell]) \\
&= \sum_k \sum_\ell x[k, \ell] h[m - k, n - \ell] = (x * h)[m, n]
\end{aligned} \tag{9.54}
$$

or the **two-dimensional convolution sum**. The impulse response is the response of the system exclusively to $\delta[m, n]$ and zero-boundary conditions, or the zero-boundary conditions response.

An LSI system is **separable** if its impulse response $h[m, n]$ is a separable sequence. The convolution sum when the system is separable, i.e., its impulse response is $h[m, n] = h_1[m]h_2[n]$, and both the input $x[m, n]$ and the impulse response $h[m, n]$ have finite first quadrant support is

$$y[m, n] = \sum_{k=-\infty}^{\infty} \sum_{\ell=-\infty}^{\infty} x[k, \ell] h_1[m - k] h_2[n - \ell]$$

$$= \sum_{k=-\infty}^{\infty} h_1[m - k] \left[\sum_{\ell=-\infty}^{\infty} x[k, \ell] h_2[n - \ell] \right].$$

Noticing that the term in the brackets is the convolution sum of $h_2[n]$ and the input for fixed values of k, if we let

$$y_1[k, n] = \sum_{\ell=-\infty}^{\infty} x[k, \ell] h_2[n - \ell] = \sum_{\ell=0}^{\infty} x[k, \ell] h_2[n - \ell], \tag{9.55}$$

we then have

$$y[m, n] = \sum_{k=-\infty}^{\infty} y_1[k, n] h_1[m - k] = \sum_{k=0}^{\infty} y_1[k, n] h_1[m - k], \tag{9.56}$$

which is a one-dimensional convolution sum of $h_1[m]$ and $y_1[k, n]$ for fixed values of n. Thus, for a system with separable impulse response the two-dimensional convolution results from performing a one-dimensional convolution of columns (or rows) and then rows (or columns).[6]

Example 9.29. To illustrate the above, consider a separable impulse response

$$h[m, n] = \begin{cases} 1 & 0 \le m \le 1, 0 \le n \le 1, \\ 0 & \text{otherwise,} \end{cases}$$

$$= (u[m] - u[m - 2])(u[n] - u[n - 2]) = h_1[m]h_2[n].$$

For an input

$$x[m, n] = \begin{cases} 1 & 0 \le m \le 1, 0 \le n \le 1, \\ 0 & \text{otherwise,} \end{cases}$$

find the output of the system $y[m, n]$.

Solution: For $k = 0, 1, 2$ we have

$$y_1[0, n] = \sum_{\ell=0}^{1} x[0, \ell] h_2[n - \ell] = x[0, 0] h_2[n] + x[0, 1] h_2[n - 1] = h_2[n] + h_2[n - 1],$$

---

[6]The final forms of Equations (9.55) and (9.56) are obtained using that both input and impulse response are supported in the first quadrant.

$$y_1[1, n] = \sum_{\ell=0}^{1} x[1, \ell] h_2[n - \ell] = x[1, 0] h_2[n] + x[1, 1] h_2[n - 1] = h_2[n] + h_2[n - 1],$$

$$y_1[2, n] = \sum_{\ell=0}^{1} x[2, \ell] h_2[n - \ell] = x[2, 0] h_2[n] + x[2, 1] h_2[n - 1] = 0 + 0,$$

and zero for any value of k larger than 2. For values of $k < 0$ we have $y_1[k, n] = 0$ because the input will be zero. We thus have

$$y_1[0, 0] = 1, \quad y_1[0, 1] = 2, \quad y_1[0, 2] = 1,$$
$$y_1[1, 0] = 1, \quad y_1[1, 1] = 2, \quad y_1[1, 2] = 1,$$

and the rest of the values are zero. Notice that the support of $y_1[m, n]$ is 2×3 which is the support of $x[m, n]$ (2×2) plus the support of $h_2[n]$ (1×2) minus 1 for both row and column. The final result is then the one-dimensional convolution

$$y[0, 0] = \sum_{k=0}^{1} y_1[k, 0] h_1[-k] = y_1[0, 0] h_1[0],$$

$$y[0, 1] = \sum_{k=0}^{1} y_1[k, 1] h_1[-k] = y_1[0, 1] h_1[0],$$

$$y[0, 2] = \sum_{k=0}^{1} y_1[k, 2] h_1[-k] = y_1[0, 2] h_1[0],$$

$$y[1, 0] = \sum_{k=0}^{1} y_1[k, 0] h_1[1 - k] = y_1[0, 0] h_1[1] + y_1[1, 0] h_1[0],$$

$$y[1, 1] = \sum_{k=0}^{1} y_1[k, 1] h_1[1 - k] = y_1[0, 1] h_1[1] + y_1[1, 1] h_1[0],$$

$$y[1, 2] = \sum_{k=0}^{1} y_1[k, 2] h_1[1 - k] = y_1[0, 2] h_1[1] + y_1[1, 2] h_1[0],$$

$$y[2, 0] = \sum_{k=0}^{1} y_1[k, 0] h_1[2 - k] = y_1[1, 0] h_1[1],$$

$$y[2, 1] = \sum_{k=0}^{1} y_1[k, 1] h_1[2 - k] = y_1[1, 1] h_1[1],$$

$$y[2, 2] = \sum_{k=0}^{1} y_1[k, 2] h_1[2 - k] = y_1[1, 2] h_1[1],$$

which gives after replacing the values of $h_1[m]$

$$y[0,0]=1, \quad y[0,1]=2, \quad y[0,2]=1,$$
$$y[1,0]=2, \quad y[1,1]=4, \quad y[1,2]=2,$$
$$y[2,0]=1, \quad y[2,1]=2, \quad y[2,2]=1,$$

which is the result of convolving by columns and then by rows. The size of $y[m,n]$ is $(2+2-1) \times (2+2-1)$ or 3×3. To verify this result use the MATLAB function *conv2*. The advantage of using 2D systems that are separable is that only one-dimensional processing is needed. □

A **bounded input–bounded output (BIBO) stable** two-dimensional LSI system is such that if its input $x[m,n]$ is bounded (i.e., there is a positive finite value L such that $|x[m,n]| < L$) the corresponding output $y[m,n]$ is also bounded. Thus, from the convolution sum we have

$$|y[m,n]| \leq \sum_{k=-\infty}^{\infty} \sum_{\ell=-\infty}^{\infty} |x[k,\ell]| \, |h[m,n]| \leq L \sum_{k=-\infty}^{\infty} \sum_{\ell=-\infty}^{\infty} |h[m,n]| < \infty \qquad (9.57)$$

or that the impulse response $h[m,n]$ be absolutely summable in order for the system to be BIBO stable.

As in the one-dimensional case, two-dimensional systems are **recursive** and **non-recursive**. For instance, a non-recursive two-dimensional system is represented by an input–output equation:

$$y[m,n] = b_{00}x[m,n] + b_{01}x[m,n-1] + b_{10}x[m-1,n] + b_{11}x[m-1,n-1] \qquad (9.58)$$

where $x[m,n]$ is the input and $y[m,n]$ is the output, and the $\{b_{i,j}, i=0,1; \; j=0,1\}$ are real-valued coefficients. Such a system is LSI, and it is also called a **Finite Impulse Response (FIR)** system given that it has an impulse response $h[m,n]$ of finite support. Indeed, letting $x[m,n] = \delta[m,n]$ the output is $y[m,n] = h[m,n]$ or the impulse response of the system computed according to the above input/output relation:

$$h[m,n] = b_{00}\delta[m,n] + b_{01}\delta[m,n-1] + b_{10}\delta[m-1,n] + b_{11}\delta[m-1,n-1] \qquad (9.59)$$

having only four values $h[0,0] = b_{00}$, $h[0,1] = b_{01}$, $h[1,0] = b_{10}$, $h[1,1] = b_{11}$ in its support, the rest are zero. Thus such a system is BIBO if the coefficients are bounded. This result is generalized: any FIR filter is BIBO stable if its coefficients are bounded.

On the other hand, if the two-dimensional system is recursive, or is said to be an **Infinite Impulse Response (IIR)** system, the BIBO stability is more complicated as these systems typically have an impulse response with an infinite size support and as such the absolute summability might be hard to ascertain.

Example 9.30. A recursive system is represented by the difference equation

$$y[m,n] = x[m,n] + y[m-1,n] + y[m,n-1], \qquad m \geq 0, n \geq 0$$

where $x[m,n]$ is the input and $y[m,n]$ is the output. Determine the impulse response $h[m,n]$ of this recursive system and from it determine if the system is BIBO stable.

Solution: The impulse response is the response of the system when $x[m, n] = \delta[m, n]$ and zero-boundary conditions. Since for $m < 0$ and/or $n < 0$ the input as well as the boundary condition is zero, $h[m, n] = 0$ for $m < 0$ and/or $n < 0$. For other values, $h[m, n]$ is computed recursively (this is the reason for it being called a recursive system) in some order, we thus have

$$h[0, 0] = x[0, 0] + h[-1, 0] + h[0, -1] = 1 + 0 + 0 = 1,$$
$$h[0, 1] = x[0, 1] + h[-1, 1] + h[0, 0] = 0 + 0 + 1 = 1,$$
$$h[0, 2] = x[0, 2] + h[-1, 2] + h[0, 1] = 0 + 0 + 1 = 1,$$
$$\cdots$$
$$h[1, 0] = x[1, 0] + h[0, 0] + h[1, -1] = 0 + 1 + 0 = 1,$$
$$h[1, 1] = x[1, 1] + h[0, 1] + h[1, 0] = 0 + 1 + 1 = 2,$$
$$h[1, 2] = x[1, 2] + h[0, 2] + h[1, 1] = 0 + 1 + 2 = 3,$$
$$\cdots$$
$$h[2, 0] = x[2, 0] + h[1, 0] + h[2, -1] = 0 + 1 + 0 = 1,$$
$$h[2, 1] = x[2, 1] + h[1, 1] + h[2, 0] = 0 + 2 + 1 = 3,$$
$$h[2, 2] = x[2, 2] + h[1, 2] + h[2, 1] = 0 + 3 + 3 = 6,$$
$$\cdots .$$

Unless we were able to compute the rest of the values of the impulse response, or to obtain a closed form for, it the BIBO stability would be hard to determine from the impulse response. As it will be shown later, a closed form for this impulse response is

$$h[m, n] = \frac{(m + n)!}{n! \, m!} u_1[m, n]$$

where $n! = 1 \times 2 \cdots (n - 1) \times n$ is the n factorial.[7]

Using the MATLAB function *nchoosek* to compute the values of the impulse response (see Fig. 9.15) according to this formula, it is seen that as m and n increase the value of $h[m, n]$ also increases, so the system is not BIBO stable. □

Example 9.31. Consider a recursive system is represented by the difference equation

$$y[m, n] = by[m, n - 1] + ay[m - 1, n] - aby[m - 1, n - 1] + x[m, n], \quad |a| < 1, \; |b| < 1. \quad (9.60)$$

Is this system BIBO stable according to its impulse response?

---

[7]The impulse response can also be expressed as

$$h[m, n] = \binom{m + n}{n} u_1[m, n]$$

using the binomial coefficient notation.

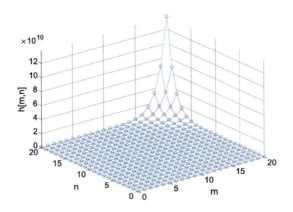

FIGURE 9.15

Impulse response of recursive system $y[m, n] = x[m, n] + y[m - 1, n] + y[m, n - 1]$, $m \geq 0$, $n \geq 0$.

Solution: The impulse response is obtained by letting $x[m, n] = \delta[m, n]$, and zero-boundary conditions. Then $h[m, n] = y[m, n]$ satisfies the difference equation

$$h[m, n] = bh[m, n - 1] + ah[m - 1, n] - abh[m - 1, n - 1] + \delta[m, n].$$

Letting $h_1[m, n] = h[m, n] - ah[m - 1, n]$, when replaced in the above difference equation gives

$$\begin{aligned} h_1[m, n] &= b(h[m, n - 1] - ah[m - 1, n - 1]) + \delta[m, n] \\ &= bh_1[m, n - 1] + \delta[m, n], \end{aligned}$$

which for a fixed m is a first-order difference equation in n with solution $h_1[m, n] = b^n u[n]$. From the definition $h_1[m, n] = h[m, n] - ah[m - 1, n]$ replacing $h_1[m, n] = b^n u[n]$ we get a first-order difference equation in m with input a function in n, i.e.,

$$b^n u[n] = h[m, n] - ah[m - 1, n]$$

with solution $h[m, n] = (b^n u[n])(a^m u[m]) = h_1[n]h_2[m]$, i.e., separable. This system is BIBO stable as $h[m, n]$ is absolutely summable given that $h_1[n]$ as well as $h_2[m]$ are absolutely summable. □

Causality is imposed on one-dimensional systems to allow computations in real-time, i.e., the output of the system at a particular instance is obtainable from present and past values of the input. Lack of causality requires adding delays to the processing units which is not desirable. Causality ensures that the necessary data is available when computing the output. Causality, as such, is the same characteristic for two-dimensional systems, but in practice it is not as necessary as in one-dimension. For instance, processing images can be executed in a causal manner by processing the pixels as they come from the raster but they could also be processed by considering the whole image. In most cases, images are available in frames and their processing does not require causality.

9.5 WHAT HAVE WE ACCOMPLISHED? WHERE DO WE GO FROM HERE?

As you saw in this chapter, the theory of discrete-time signals and systems is very similar to the theory of continuous-time signals and systems. Many of the results in the continuous-time theory are changed by replacing integrals by sums, derivatives by differences and ordinary differential equations by difference equations. However, there are significant differences imposed by the way the discrete-time signals and systems are generated. For instance, the discrete frequency is finite but circular, and it depends on the sampling time. Discrete sinusoids, as another example, are not necessarily periodic. Thus, despite the similarities there are also significant differences between the continuous-time and the discrete-time signals and systems.

Now that we have a basic structure for discrete-time signals and systems, we will start developing the theory of linear time-invariant discrete-time systems. Again you will find a great deal of similarity with the linear time-invariant continuous-time systems but also some very significant differences. Also, notice the relation that exists between the Z-transform and the Fourier representations of discrete-time signals and systems, not only with each other but with the Laplace and Fourier transforms. There is a great deal of connection among all of these transforms, and a clear understanding of this would help you with the analysis and synthesis of discrete-time signals and systems.

A large number of books have been written covering various aspects of digital signal processing [39,55,62,19,30,47,68,61,13,51]. Some of these are classical books in the area that the reader should look into for a different perspective of the material presented in this part of the book.

Extending the theory of one-dimensional signals and systems to two dimensions is very doable, but it is important to recognize that the theory of two-dimensional signals is more general than the one-dimensional. As such, it is clear that the two-dimensional theory is richer conceptually, and that many of the one-dimensional properties are no valid in two dimensions, given that the theory of one-dimensional signals and systems is a subset of the two-dimensional theory. In the next chapters we will see how to obtain a two-dimensional Z-transform and a discrete Fourier transform, and how they can be used to process two-dimensional signals and in particular images.

9.6 PROBLEMS

9.6.1 BASIC PROBLEMS

9.1 For the discrete-time signal

$$x[n] = \begin{cases} 1 & n = -1, \ 0, \ 1, \\ 0.5 & n = 2, \\ 0 & \text{otherwise,} \end{cases}$$

sketch and label carefully the following signals:
- **(a)** $x[n-1]$, $x[-n]$, and $x[2-n]$.
- **(b)** The even component $x_e[n]$ of $x[n]$.
- **(c)** The odd component $x_o[n]$ of $x[n]$.

Answers: $x[2-n] = 0.5\delta[n] + \delta[n-1] + \delta[n-2] + \delta[n-3]$; $x_o[n] = 0.25\delta[n-2] - 0.25\delta[n+2]$.

9.2 For the discrete-time periodic signal $x[n] = \cos(0.7\pi n)$, $-\infty < n < \infty$,
   (a) Determine its fundamental period N_0.
   (b) Suppose we sample the continuous-time signal $x(t) = \cos(\pi t)$ with a sampling period $T_s = 0.7$. Is the Nyquist sampling condition satisfied? How does the sampled signal compare to the given $x[n]$?
   (c) Under what conditions would sampling a continuous-time signal $x(t) = \cos(\pi t)$ give a discrete-time sinusoid $x[n]$ that resembles $x(t)$? Explain and give an example.
   Answers: $N_0 = 20$; yes, $T_s = 0.7$ satisfies the Nyquist condition; let $T_s = 2/N$ with $N \gg 2$.

9.3 Consider the following problems related to the periodicity of discrete-time signals.
   (a) Determine whether the following discrete-time signals defined in $-\infty < n < \infty$ are periodic or not. If periodic, determine its fundamental period N_0.

$$(i)\ x[n] = 2\cos(\pi n - \pi/2), \quad (ii)\ y[n] = \sin(n - \pi/2),$$
$$(iii)\ z[n] = x[n] + y[n], \quad (iv)\ v[n] = \sin(3\pi n/2).$$

   (b) Consider two periodic signals $x_1[n]$, of fundamental periods $N_1 = 4$, and $y_1[n]$, of period $N_2 = 6$. Determine what would be the fundamental period of the following:

$$(i)\ z_1[n] = x_1[n] + y_1[n], \quad (ii)\ v_1[n] = x_1[n]y_1[n], \quad (iii)\ w_1[n] = x_1[2n].$$

   Answers: (a) $y[n]$ is not periodic; (b) $v_1[n]$ is periodic of fundamental period $N_0 = 12$.

9.4 The following problems relate to periodicity and power of discrete-time signals.
   (a) Is the signal $x[n] = e^{j(n-8)/8}$ periodic? If so determine its fundamental period N_0. What if $x_1[n] = e^{j((n-8)\pi/8)}$ (notice the difference with $x[n]$) would this new signal be periodic? If so what would the fundamental period N_1 be?
   (b) Given the discrete-time signal $x[n] = \cos(\pi n/5) + \sin(\pi n/10)$, $-\infty < n < \infty$.
      i. Is $x[n]$ periodic? If so determine its fundamental frequency ω_0.
      ii. Is the power of $x[n]$ the sum of the powers of $x_1[n] = \cos(\pi n/5)$ and $x_2[n] = \sin(\pi n/10)$ defined for $-\infty < n < \infty$? If so, show it.
   Answers: (a) $x_1[n]$ periodic, $N_1 = 16$; (b) $\omega_0 = 0.1\pi$; (c) yes.

9.5 The following problems relate to linearity, time invariance, causality, and stability of discrete-time systems.
   (a) The output $y[n]$ of a system is related to its input $x[n]$ by $y[n] = x[n]x[n-1]$. Is this system
      i. linear? time invariant?
      ii. causal? bounded input–bounded output stable?
    You may consider $x[n] = u[n]$ as the input to verify your results.
   (b) Given the discrete-time system in Fig. 9.16.
      i. Is this system time invariant?
      ii. Suppose that the input is $x[n] = \cos(\pi n/4)$, $-\infty < n < \infty$, so that the output is $y[n] = \cos(\pi n/4)\cos(n/4)$, $-\infty < n < \infty$. Determine the fundamental period N_0 of $x[n]$. Is $y[n]$ periodic? If so, determine its fundamental period N_1.
   Answers: (a) non-linear, time-invariant, BIBO stable system; (b) no; $N_0 = 8$, $y[n]$ not periodic.

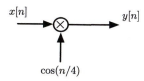

FIGURE 9.16

Problem 9.5.

9.6 Consider a discrete-time system with output $y[n]$ given by $y[n] = x[n]f[n]$ and $x[n]$ is the input and $f[n]$ is a function.

(a) Let the input be $x[n] = 4\cos(\pi n/2)$ and $f[n] = \cos(6\pi n/7)$, $-\infty < n < \infty$. Is $x[n]$ periodic? If so, indicate its fundamental period N_0. Is the output of the system $y[n]$ periodic? If so, indicate its fundamental period N_1.

(b) Suppose now that $f[n] = u[n] - u[n-2]$ and $x[n] = u[n]$. Determine if the system with the above input–output equation is time invariant.

Answers: The system is time varying.

9.7 Consider a system represented by

$$y[n] = \sum_{k=n-2}^{n+4} x[k]$$

where the input is $x[n]$ and the output $y[n]$. Is the system

(a) linear? time invariant?

(b) causal? bounded input–bounded output stable?

Answers: The system is linear, time invariant, and noncausal.

9.8 Determine the impulse response $h[n]$ of a LTI system represented by the difference equation

$$y[n] = -0.5y[n-1] + x[n],$$

where $x[n]$ is the input, $y[n]$ is the output and the initial conditions are zero. Find two different ways to compute the output $y[n]$ when the input is

$$x[n] = \begin{cases} 1 & 0 \le n \le 2, \\ 0 & \text{otherwise.} \end{cases}$$

Answers: $h[n] = (-0.5)^n u[n]$; let $x[n] = \delta[n] + \delta[n-1] + \delta[n-2]$ and use $h[n]$ to get $y[n]$.

9.9 The input of an LTI continuous-time system is $x(t) = u(t) - u(t-3.5)$. The system's impulse response is $h(t) = u(t) - u(t-2.5)$.

(a) Find the system's output $y(t)$ by graphically computing the convolution integral of $x(t)$ and $h(t)$. Sketch $x(t)$, $h(t)$ and the found $y(t)$.

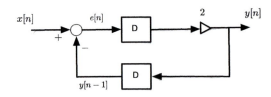

FIGURE 9.17

Problem 9.11.

(b) Suppose we sample $x(t)$, $h(t)$ and $y(t)$ using a sampling period $T_s = 0.5$ s/sample. Sketch these sampled signals as functions of n.

(c) Graphically calculate the convolution sum of $x[n]$ and $h[n]$, and compare it to $y(t)$. Explain your results.

Answers: $y(t) = r(t) - r(t - 2.5) - r(t - 3.5) + r(t - 6)$; $y[n]/6$ approximates $y(t)$.

9.10 You are testing a 1 V dc source and have the following measurements obtained from the source every minute starting at time 0:

n	$x[n]$	n	$x[n]$
0	1.0	3	0.7
1	1.2	4	1.2
2	0.9	5	1.0

To find the average voltages for the first 5 minutes, i.e. to get rid of some of the noise in the data, you use the following averager:

$$y[n] = \frac{x[n] + x[n-1] + x[n-2]}{3}.$$

(a) Use the input/output equation of the filter to compute and then plot the moving averages $y[n]$ for $2 \le n \le 5$.

(b) Find the impulse response $h[n]$ of the averager and compute the output $y[n]$ using the convolution sum. Do you obtain the same results as before?

(c) Suppose you use a median filter of length 3, so that at some sample n we consider the samples values at n, $n - 1$ and $n - 2$, order these values and the one in the middle is the median output $y_m[n]$. Moving one sample we consider the next 3 values and find the new median. Compute and plot $y_m[n]$ for $2 \le n \le 5$.

Answers: Identical results using difference equation and convolution.

9.11 A causal, LTI discrete-time system is represented by the block diagram shown in Fig. 9.17 where D stands for a one-sample delay.

(a) Find the difference equation relating the input $x[n]$ and the output $y[n]$.

(b) Find the impulse response $h[n]$ of the system and use it to determine if the system is BIBO stable? Explain.

Answer: $h[n] = 2(-2)^{(n-1)/2}$, n odd, 0 otherwise; system is not BIBO stable.

9.12 The input and the output of an LTI discrete-time system are

$$\text{Input: } x[n] = u[n] - u[n-3], \quad \text{Output: } y[n] = u[n-1] - u[n-4].$$

(a) What should be the length of the impulse response $h[n]$ of the system?

(b) Find the impulse response of the system, $h[n]$, and verify the previous answer.

Answers: Impulse response: $h[0] = 0$, $h[1] = 1$ and the rest of the values are zero.

9.13 The following problems relate to the response of LTI discrete-time systems.

(a) The unit-step response of a LTI discrete-time system is found to be

$$s[n] = (3 - 3(0.5)^{n+1})u[n].$$

Use $s[n]$ to find the impulse response $h[n]$ of the system.

(b) The output $y[n]$ of a discrete-time system is the even component of the input $x[n]$, i.e.,

$$y[n] = 0.5(x[n] + x[-n]).$$

  i. Consider an input $x[n] = u[n] - u[n-3]$, find the corresponding output $y[n]$. You might want to carefully sketch the input and the output. Is the system causal? Explain.

  ii. Use the same input as before, $x[n] = u[n] - u[n-3]$, with the obtained output $y[n]$. If we consider as input $x_1[n] = x[n-1]$, find the corresponding output $y_1[n]$, sketch it and from these results determine if the system is time invariant.

  iii. Suppose then that the input is $x[n] = \cos(2\pi n/5)u[n]$, find the corresponding output $y[n]$ and carefully sketch and label the input and output. Is the output periodic?

Answers: (a) $h[n] = (3 - 3(0.5)^{n+1})u[n] - (3 - 3(0.5)^n)u[n-1]$; (b) noncausal.

9.14 An LTI discrete-time system has an impulse response $h[n] = u[n] - u[n-4]$, and as input the signal $x[n] = u[n] - u[n-(N+1)]$ for a positive integer N. The output of the system $y[n]$ is calculated using the convolution sum.

(a) If $N = 4$ what is the length of the output $y[n]$? Explain. For $N = 4$, carefully sketch and label the output $y[n]$ resulting from the convolution sum.

(b) Determine the value of $N \leq 5$ for $x[n]$ so that $y[3] = 3$ and $y[6] = 0$.

Answers: $y[0] = y[7] = 1$, $y[1] = y[6] = 2$, $y[2] = y[5] = 3$, $y[3] = y[4] = 4$, 0 otherwise.

9.15 Consider a discrete-time system represented by the difference equation $y[n] = 0.5y[n-1] + x[n]$ where $x[n]$ is the input and $y[n]$ the output.

(a) An equivalent representation of the system is given by the difference equation

$$y[n] = 0.25y[n-2] + 0.5x[n-1] + x[n].$$

Is it true? Let $x[n] = \delta[n]$ and zero initial conditions and solve the two difference equations to verify this. Determine how to obtain the second difference equation from the first?

(b) Using the first initial difference equation show that the output is

$$y[n] = \sum_{k=0}^{\infty} (0.5)^k x[n-k].$$

What is this expression? determine the impulse response $h[n]$ of the system from this equation? Explain.

(c) If the output is computed using the convolution sum, and the input is

$$x[n] = u[n] - u[n - 11],$$

find $y[n]$. Determine the steady-state value of the output, i.e., $y[n]$ as $n \to \infty$.

(d) What is the maximum value achieved by the output $y[n]$? when is it attained?

Answers: Both equations give $y[n] = 0.5^n u[n]$; (d) maximum $y[10] = 2(1 - 0.5^{11})$.

9.16 The following difference equation is used to obtain recursively the ratio α/β:

$$c[n + 1] = (1 - \beta)c[n] + \alpha \qquad n \geq 0$$

with $c[0]$ as an initial condition. Solve the difference equation, and find under what condition(s) the solution $c[n]$ will converge to the desired answer of α/β as n tends to infinity.

Answers: $0 < \beta < 2$, independent of $c[0]$.

9.17 An LTI causal discrete-time system has the input/output relationship

$$y[n] = \sum_{k=-\infty}^{n} (n - k + 2)x[k]$$

where $x[n]$ is the input of the system, $y[n]$ is the response of the system. There is zero initial energy in the system prior to applying $x[n]$.

(a) Find the impulse response $h[n]$.

(b) Find the unit-step response of the given system.

Answers: $h[n] = (n + 2)u[n]$.

9.18 A discrete-time averager is characterized by the following equation relating the input $x(nT_s)$ with the output $y(nT_s)$:

$$y(nT_s) = \frac{1}{2N + 1} \sum_{k=-N}^{N} x(nT_s - kT_s).$$

(a) Is this system causal? Explain.

(b) Let $N = 2$ in the above equation. Find and plot the impulse response $h(nT_s)$ of the averager.

(c) For $N = 2$, if the input to the averager is

$$x(nT_s) = \begin{cases} 5 & n = 0, 1, 2 \\ 0 & \text{otherwise} \end{cases}$$

calculate the output $y(nT_s)$.

Answers: (a) This averager is noncausal; (c) $y(nT_s)$ is noncausal of length 7.

9.19 Consider a causal LTI system with impulse response $h[n]$, and input

$$x[n] = x_1[n] - x_1[n-2] + x_1[n-4]$$

where $x_1[n] = u[n] - u[n-2]$. The impulse response of the system is $h[n] = u[n] - u[n-2]$.
(a) Use the convolution sum to find the output of the system, $y[n]$.
(b) If the given system is cascaded with another causal LTI system with impulse response $g[n]$, we know the overall impulse response of the two cascaded systems is

$$h_T[n] = \begin{cases} 1 & n = 0, 3, \\ 0 & \text{otherwise.} \end{cases}$$

Find the impulse response $g[n]$.
Answers: If $y_1[n] = (x_1 * h)[n]$ then $y[n] = y_1[n] - y_1[n-2] + y_1[n-4]$.

9.20 A discrete-time averager is represented by the input/output equation

$$y[n] = (1/3)(x[n+1] + x[n] + x[n-1]),$$

where $x[n]$ is the input and $y[n]$ the output.
(a) Determine whether this system is causal or not. Explain.
(b) Determine whether this system is BIBO stable or not. Explain.
(c) The input of the system is generated by sampling an analog signal $x(t) = 2\cos(10t)$ using sampling periods $T_{s1} = 1$ or $T_{s2} = \pi$ s/sample. If we want the discrete-time signal $x[n] = x(t)|_{t=nT_s}$ to be periodic, which of the two sampling periods $\{T_{si}, i = 1, 2\}$ would you use? For the chosen sampling period what would be the period of $x[n]$?
(d) If $x[n]$ is periodic, would the output of the averager be also periodic? If so, what would be the period of the output? Explain.
Answers: The system is noncausal, but BIBO stable.

9.21 A finite impulse response (FIR) filter has an input/output relation $y[n] = x[n] - x[n-5]$ where $x[n]$ is the input and $y[n]$ the output.
(a) Find the impulse response $h[n]$ of this filter. Plot $h[n]$ as a function of n, and indicate if the filter is causal and BIBO stable or not.
(b) Suppose the input is $x[n] = u[n]$, find the corresponding output $y[n]$ and carefully plot it. Are $x[n]$ and $y[n]$ finite-energy signals?
(c) If $x[n] = \sin(2\pi n/5)u[n]$, find its corresponding output $y[n]$. Determine the energies of the input $x[n]$ and of $y[n]$. Are both finite energy?
(d) Determine the frequencies $\{\omega_0\}$ of the input $x[n] = \sin(\omega_0 n)u[n]$ for which the corresponding output $y[n]$ is finite energy. If you choose a frequency different from these frequencies, is the output finite energy?
Answers: $h[n] = \delta[n] - \delta[n-5]$; for frequencies $\{\omega_0 = 2\pi m/5\}$ for $m = 0, \pm 1, \pm 2, \cdots$, the energy of the output is finite.

9.22 Consider the following problems related to properties of filters.
(a) Filters that operate under real-time conditions need to be causal. When no real-time processing is needed the filter can be noncausal.

i. Consider the case of averaging an input signal $x[n]$ under real-time conditions. Suppose you are given two different filters

$$\text{(A)} \quad y[n] = \frac{1}{N}\sum_{k=0}^{N-1} x[n-k], \quad \text{(B)} \quad y[n] = \frac{1}{N}\sum_{k=-N+1}^{N-1} x[n-k],$$

which of these would you use and why.

ii. If you are given a tape with the data which of the two filters would you use? Why? Would you use either? Explain.

(b) A significant difference between IIR and FIR discrete-time systems is stability. Consider an IIR filter represented by the difference equation

$$y_1[n] = x[n] - 0.5y_1[n-1]$$

where $x[n]$ is the input and $y_1[n]$ is the output. Then consider an FIR filter

$$y_2[n] = x[n] + 0.5x[n-1] + 3x[n-2] + x[n-5]$$

where $x[n]$ is the input and $y_2[n]$ is the output.

i. Since to check the stability of these filters we need their impulse responses, find the impulse responses $h_1[n]$ corresponding to the IIR filter by recursion, and $h_2[n]$ corresponding to the FIR filter.

ii. Use the impulse responses $h_1[n]$ and $h_2[n]$ to check the stability of the IIR filter and of the FIR filter, respectively.

iii. Since the impulse response of an FIR filter has a finite number of nonzero terms, would it be correct to say that FIR filters are always stable? Explain.

Answers: (a) Use (A), it is causal; yes, use either; (b) $h_2[n] = \delta[n] + 0.5\delta[n-1] + 3\delta[n-2] + \delta[n-5]$.

9.23 For the two-dimensional signals where $|\alpha| < 1$

$$x[m, n] = \alpha^{m+n} u_{12}[m, n]$$
$$y[m, n] = \alpha^{m+n} u_{14}[m, n].$$

(a) Draw their supports, and express these domains in terms of $u_1[m, n]$.
(b) Let $z[m, n] = x[m, n] - y[m, n]$, and draw its support.
Answer: (b) The support of $z[m, n]$ is in the second and fourth quadrant.

9.24 Consider the line impulses

$$x[m, n] = \sum_{k=-\infty}^{\infty} \delta[0, n-k], \quad y[m, n] = \delta[0] \sum_{\ell=-\infty}^{\infty} \delta[m-\ell]$$

where $-\infty < m < \infty$ and $-\infty < n < \infty$.

(a) Draw the line impulses $x[m, n]$ and $y[m, n]$. Determine if they are separable.

(b) Consider the product $z[m, n] = x[m, n]y[m, n]$. Express $z[m, n]$ in terms of unit-step functions $u_{14}[m, n]$, with support in the first and fourth quadrants, and $u_{23}[m, n]$, with support in the second and third quadrants.

Answer: (b) $z[m, n] = u_{14}[m, n] + u_{23}[m + 1, n]$.

9.25 The impulse response

$$h[m, n] = \binom{m + n}{m}$$

satisfies the difference equation

$$h[m, n] = h[m - 1, n] + h[m, n - 1] + \delta[m, n]$$

with zero-boundary conditions. By definition $\binom{0}{0} = 1$.

(a) Use the difference equation to show that $\binom{-1}{-1} = \binom{-1}{0} = 0$.

(b) Show that for $m \geq 0, n > 0$

$$h[m, n] = \frac{m + n}{n} h[m, n - 1] + \delta[m, n]$$

determine the values of $h[m, 0]$ for $m \geq 0$, and use these equations to compute the impulse response $h[m, n]$. Plot the resulting $h[m, n]$. Is the system with this impulse response BIBO stable? Explain.

9.6.2 PROBLEMS USING MATLAB

9.26 Finite-energy signals—Given the discrete signal $x[n] = 0.5^n u[n]$.

(a) Use the function *stem* to plot $x[n]$ for $n = -5$ to 20.

(b) Is this a finite-energy discrete-time signal? i.e., compute the infinite sum

$$\varepsilon_x = \sum_{n=-\infty}^{\infty} |x[n]|^2.$$

(c) Verify your results by using symbolic MATLAB to find an expression for the above sum.
Answers: $\varepsilon_x = 4/3$.

9.27 Periodicity of sampled signals—Consider an analog periodic sinusoid $x(t) = \cos(3\pi t + \pi/4)$ being sampled using a sampling period T_s to obtain the discrete-time signal $x[n] = x(t)|_{t=nT_s} = \cos(3\pi T_s n + \pi/4)$.

(a) Determine the discrete frequency of $x[n]$.

(b) For what values of T_s is the discrete-time signal $x[n]$ periodic? Let $T_s = 1/3$, use function *stem* to plot $x[n]$, $0 \leq n \leq 100$ and determine if it is periodic. Let $T_s = 1$, use function *stem* to plot $x[n]$, $0 \leq n \leq 100$ and determine if it is periodic.
Answers: $\omega_0 = 3\pi T_s$ (rad), for $T_s = 1$ the signal $x[n]$ is not periodic.

9.28 Even and odd decomposition and energy—Suppose you sample the analog signal

$$x(t) = \begin{cases} 1-t & 0 \le t \le 1 \\ 0 & \text{otherwise} \end{cases}$$

with a sampling period $T_s = 0.25$ to generate $x[n] = x(t)|_{t=nT_s}$.

(a) Use function *stem* to plot $x[n]$ and $x[-n]$ for an appropriate interval.

(b) Find the even, $x_e[n]$, and the odd, $x_o[n]$, components of $x[n]$. Plot them using *stem*. Verify that $x_e[n] + x_o[n] = x[n]$ graphically.

(c) Compute the energy ε_x of $x[n]$ and compare it to the sum of the energies ε_{x_e} and ε_{x_o} of $x_e[n]$ and $x_o[n]$.

Answer: $\varepsilon_x = \varepsilon_{x_e} + \varepsilon_{x_o}$.

9.29 Expansion and compression of discrete-time signals—Consider the discrete-time signal $x[n] = \cos(2\pi n/7)$.

(a) The discrete-time signal can be compressed by getting rid of some of its samples (**down-sampling**). Consider the down-sampling by 2. Write a script to obtain and plot $z[n] = x[2n]$. Plot also $x[n]$ and compare it with $z[n]$, what happened? Explain.

(b) The expansion for discrete-time signals requires interpolation. However, a first step of this process is the so called *up-sampling*. Up-sampling by 2, consists in defining a new signal $y[n]$ such that $y[n] = x[n/2]$ for n even, and $y[n] = 0$ otherwise. Write a script to perform up-sampling on $x[n]$. Plot the resulting signal $y[n]$ and explain its relation with $x[n]$.

(c) If $x[n]$ resulted from sampling a continuous-time signal $x(t) = \cos(2\pi t)$ using a sampling period T_s and with no frequency aliasing, determine T_s. How would you sample the analog signal $x(t)$ to get the down-sampled signals $z[n]$? That is, choose values for the sampling period T_s to get $z[n]$ directly from $x(t)$. Can you choose T_s to get $y[n]$ from $x(t)$ directly? Explain.

Answers: $z[n] = \cos(4\pi n/7)$; $y[n] = \cos(\pi n/7)$.

9.30 Absolutely summable and finite-energy discrete-time signals—Suppose we sample the analog signal $x(t) = e^{-2t} u(t)$, using a sample period $T_s = 1$.

(a) Expressing the sampled signal as $x(nT_s) = x[n] = \alpha^n u[n]$, what is the corresponding value of α? Use *stem* to plot $x[n]$.

(b) Show that $x[n]$ is absolutely summable, i.e., show the following sum is finite:

$$\sum_{n=-\infty}^{\infty} |x[n]|.$$

(c) If you know that $x[n]$ is absolutely summable, could you say that $x[n]$ is a finite-energy signal? Use the function *stem* to plot $|x[n]|$ and $x^2[n]$ in the same plot to help you decide.

(d) In general, for what values of α are signals $y[n] = \alpha^n u[n]$ finite energy? Explain.

Answers: $\alpha = e^{-2} < 1$; $\sum_{n=0}^{\infty} (\alpha^2)^n = e^4/(e^4 - 1) < \infty$.

9.31 Periodicity of sum and product of periodic signals—If $x[n]$ is periodic of period $N_1 > 0$ and $y[n]$ is periodic of period $N_2 > 0$:

(a) What should be the condition for the sum $z[n]$ of $x[n]$ and $y[n]$ to be periodic?

(b) What would be the period of the product $v[n] = x[n]y[n]$?

(c) Would the formula

$$\frac{N_1 N_2}{gcd[N1, N2]}$$

(gcd[N_1, N_2] stands for the greatest common divisor of N_1 and N_2) give the period of the sum and the product of the two signals $x[n]$ and $y[n]$?

(d) If $x[n] = \cos(2\pi n/3)$, $y[n] = 1 + \sin(6\pi n/7)$, generate and plot using MATLAB their sum $z[n]$ and product $v[n]$, find their periods and verify your analytic results.

Answers: (d) For $N_1 = 3$ and $N_2 = 7$, $z[n]$ and $v[n]$ are periodic of fundamental period $N_0 = 21$.

9.32 Echoing of music—An effect similar to multi-path in acoustics is echoing or reverberation. To see the effects of an echo in an acoustic signal consider the simulation of echoes on the *handel.mat* signal $y[n]$. Pretend that this piece is being played in a round theater where the orchestra is in the middle of two concentric circles and the walls on one half side are at a radial distances of 17 m (corresponding to the inner circle), and 34 m (corresponding to the outer circle) on the other side (yes, a usual theater!) from the orchestra. The speed of sound is 345 m/s. Assume that the recorded signal is the sum of the original signal $y[n]$, and attenuated echoes from the two walls so that the recorded signal is given by

$$r[n] = y[n] + 0.8y[n - N_1] + 0.6y[n - N_2]$$

where N_1 is the delay caused by the closest wall and N_2 the delay caused by the farther wall. The recorder is at the center of the auditorium where the orchestra is and we record for 10 s.

(a) Find the values of the two delays N_1 and N_2 (remember these are integers). The sampling frequency F_s of *handel* is given when you load it.

(b) Simulate the echo signal. Plot $r[n]$. Use *sound* to listen to the original and the echoed signals.

Answers: $T_1 = 0.10$, $T_2 = 0.20$ s.

9.33 A/D converter—An A/D converter can be thought of composed of three subsystems: a sampler, a quantizer, and a coder.

(a) The sampler, as a system, has as input an analog signal $x(t)$ and as output a discrete-time signal $x(nT_s) = x(t)|_{t=nT_s}$, where T_s is the sampling period. Determine whether the sampler is a linear system or not.

(b) Sample $x(t) = \cos(0.5\pi t)u(t)$ and $x(t - 0.5)$ using $T_s = 1$ to get $y(nT_s)$ and $z(nT_s)$, respectively. Plot $x(t)$, $x(t - 0.5)$ and $y(nT_s)$ and $z(nT_s)$. Is $z(nT_s)$ a shifted version of $y(nT_s)$ so that you can say the sampler is time invariant? Explain.

Answers: Sampler is linear but time varying.

9.34 Rectangular windowing—A window is a signal w[n] that is used to highlight part of another signal. The windowing process consists in multiplying an input signal $x[n]$ by the window signal w[n], so that the output is $y[n] = x[n]$w[n]. There are different types of windows used in signal processing, one of them is the so called *rectangular window* which is given by

$$w[n] = u[n] - u[n - N].$$

(a) Determine whether the rectangular windowing system is linear. Explain.
(b) Suppose $x[n] = nu[n]$, use MATLAB to plot the output $y[n]$ of the windowing system (with $N = 6$).
(c) Let the input be $x[n-5]$, use MATLAB to plot the corresponding output of the rectangular windowing system, and indicate whether the rectangular windowing system is time invariant.

Answers: Windowing is linear, but time varying.

9.35 Impulse response of an IIR system—A discrete-time IIR system is represented by the following difference equation:

$$y[n] = 0.15y[n-2] + x[n], \qquad n \geq 0,$$

where $x[n]$ is the input and $y[n]$ is the output.

(a) To find the impulse response $h[n]$ of the system, let $x[n] = \delta[n]$, $y[n] = h[n]$, and the initial conditions be zero. Find recursively the values of $h[n]$ for values of $n \geq 0$.
(b) As a second way to find $h[n]$, replace the relation between the input and the output given by the difference equation to obtain a convolution sum representation which will give the impulse response $h[n]$. What is $h[n]$?
(c) Use the MATLAB function *filter* to get the impulse response $h[n]$ (use *help* to learn about the function *filter*).

Answers: $h[n] = 0.5(1 + (-1)^n)0.15^{n/2}u[n]$.

9.36 FIR filters—An FIR filter has a non-recursive input/output relation

$$y[n] = \sum_{k=0}^{5} kx[n-k].$$

(a) Find the impulse response $h[n]$ of this filter. Is this a causal and stable filter? Explain.
(b) Find the unit-step response $s[n]$ for this filter and plot it.
(c) If the input $x[n]$ for this filter is bounded, $|x[n]| < 3$, what would be a minimum bound M for the output, i.e., $|y[n]| \leq M$.
(d) Use the function *filter* to compute the impulse response $h[n]$ and the unit-step response $s[n]$ for the given filter and plot them.

Answers: $s[n] = u[n-1] + 2u[n-2] + 3u[n-3] + 4u[n-4] + 5u[n-5]$.

9.37 Steady state of IIR systems—Suppose an IIR system is represented by a difference equation

$$y[n] = ay[n-1] + x[n],$$

where $x[n]$ is the input and $y[n]$ is the output.

(a) If the input is $x[n] = u[n]$ and it is known that the steady-state response is $y[n] = 2$, what would be a for that to be possible (hint: in steady state $x[n] = 1$ and $y[n] = y[n-1] = 2$ since $n \to \infty$).
(b) Writing the system input as $x[n] = u[n] = \delta[n] + \delta[n-1] + \delta[n-2] + \cdots$ then according to the linearity and time invariance the output should be

$$y[n] = h[n] + h[n-1] + h[n-2] + \cdots .$$

Use the value for a found above, that the initial condition is zero, i.e., $y[-1] = 0$, and that the input is $x[n] = u[n]$, to find the values of the impulse response $h[n]$ for $n \geq 0$ using the above equation. The system is causal.

(c) Use the function *filter* to compute the impulse response $h[n]$ and compare it with the one obtained above.

Answers: $a = 0.5$; $h[n] = 0.5^n u[n]$.

9.38 Unit-step vs. impulse response—The unit-step response of a discrete-time LTI system is

$$s[n] = 2[(-0.5)^n - 1]u[n].$$

Use this information to find:

(a) The impulse response $h[n]$ of the discrete-time LTI system.

(b) The response of the LTI system to a ramp signal $x[n] = nu[n]$.

Answers: $h[n] = -2(0.5)^n u[n-1]$.

9.39 Convolution sum—A discrete-time system has a unit impulse response $h[n]$.

(a) Let the input to the discrete-time system be a pulse $x[n] = u[n] - [n-4]$ compute the output of the system in terms of the impulse response.

(b) Let $h[n] = 0.5^n u[n]$; what would be the response of the system $y[n]$ to $x[n] = u[n] - u[n-4]$? Plot the output $y[n]$.

(c) Use the convolution sum to verify your response $y[n]$.

(d) Use the function *conv* to compute the response $y[n]$ to $x[n] = u[n] - u[n-4]$. Plot both the input and the output.

Answers: Use the result that when the input is $x_1[n] = u[n]$ then the output is $y_1[n] = 0.5^n(2^{n+1} - 1)u[n]$.

9.40 Discrete envelope detector—Consider an *envelope detector* that would be used to detect the message sent in an AM system. Consider the envelope detector as a system composed of the cascading of two systems one which computes the absolute value of the input, and a second one that low-pass filters its input. A circuit that is used as an envelope detector consists of a diode circuit that does the absolute value operation, and an RC circuit that does the low-pass filtering. The following is an implementation of these operations in discrete time. Let the input to the envelope detector be a sampled signal $x(nT_s) = p(nT_s)\cos(2000\pi nT_s)$ where

$$p(nT_s) = u(nT_s) - u(nT_s - 20T_s) + u(nT_s - 40T_s) - u(nT_s - 60T_s)$$

two pulses of duration $20T_s$ and amplitude equal to one.

(a) Choose $T_s = 0.01$, and generate 100 samples of the input signal $x(nT_s)$ and plot it.

(b) Consider then the subsystem that computes the absolute value of the input $x(nT_s)$ and compute and plot 100 samples of $y(nT_s) = |x(nT_s)|$.

(c) Let the low-pass filtering be done by a moving averager of order 15, i.e., if $y(nT_s)$ is the input, then the output of the filter is

$$z(nT_s) = \frac{1}{15}\sum_{k=0}^{14} y(nT_s - kT_s).$$

Implement this filter using the function *filter*, and plot the result. Explain your results.

(d) Is this a linear system? Come up with an example using the script developed above to show that the system is linear or not.

9.41 Two-dimensional convolution—Let the input be

$$x[m, n] = \delta[m, n] + 2\delta[m - 1, n] + 3\delta[m - 2, n] + 4\delta[m, n - 1] + 5\delta[m - 1, n - 1]$$
$$+ 6\delta[m - 2, n - 1] + 7\delta[m, n - 2] + 8\delta[m - 1, n - 2] + 9\delta[m - 2, n - 2]$$

and the impulse response of an FIR filter be

$$h[m, n] = \delta[m, n] + \delta[m - 1, n] - \delta[m, n - 1] - \delta[m - 1, n - 1].$$

(a) Use the two-dimensional convolution function *conv2* to find the output $y[m, n]$.
(b) Is the impulse response separable, i.e., $h[m, n] = h_1[m]h_2[n]$? If so, determine the one-dimensional impulse responses $h_1[m]$, and $h_2[n]$ and use them to verify the above result.
Answer: Let $x = [1\ 2\ 3; 4\ 5\ 6; 7\ 8\ 9]$ be an array representing $x[m, n]$.

9.42 Filtering and binarization—Use the function *imread, rgb2gray* and *double* to read the color image *peppers.png*. Convert it into a gray-level image $I[m, n]$ with double precision. Add noise to it using the function *randn* (Gaussian number generator) multiplied by 40. Create a one-dimensional moving-average filter with impulse response

$$h_1[n] = 0.25(\delta[n] + \delta[n - 1] + \delta[n - 2] + \delta[n - 3]).$$

Letting a second filter have the same impulse response, i.e., $h_2[n] = h_1[n]$ create a separable two-dimensional filter $h[m, n] = h_1[m]h_2[n]$. Convolve the image with the two-dimensional filter to obtain an image $y[m, n]$.
(a) Use the function *histogram* to look at the distribution of the pixel values of the two images. Explain the effect of the moving-average filtering on the image $I[m, n]$.
(b) Binarizing an image consists in using a gray-level threshold to convert the given image into one that has two levels only. The threshold can be chosen from a histogram of the image to be binarized. We wish to binarize the $y[m, n]$ to obtain two images: $y_1[m, n]$ with gray level of 250 whenever the $y[m, n] > 100$ and 0 otherwise; $y_2[m, n]$ with gray level of 200 whenever $y[m, n] \leq 100$ and 0 otherwise. Use *imshow* to display these images, and *histogram* to display their gray-level distributions.
Answer: The noise is given by the array $40 * randn(M(1), M(2))$ where $M(1)$ and $M(2)$ are the dimensions of the image.

9.43 Edge detection—Detection of edges is a very important application in image processing. Taking the gradient of a two-dimensional function detects the changes the edges of an image. A filter than is commonly used in edge detection is Sobel's filter. Consider the generation of two related impulse responses

$$h_u[m, n] = \delta[m, n] - \delta[m - 2, n] + 2\delta[m, n - 1] - 2\delta[m - 2, n - 1]$$
$$+ \delta[m, n - 2] - \delta[m - 2, n - 2]$$
$$h_v[m, n] = \delta[m, n] + 2\delta[m - 1, n] + \delta[m - 2, n] - \delta[m, n - 2]$$
$$- 2\delta[m - 1, n - 2] - \delta[m - 2, n - 2].$$

For a given input $x[m, n]$ the output of the Sobel is

$$y_1[m, n] = (h_u * x)[m, n], \quad y_2[m, n] = (h_v * x)[m, n]$$
$$y[m, n] = \sqrt{(y_1[m, n])^2 + (y_2[m, n])^2}.$$

(a) Use *imread* to read in the image *peppers.png*, and convert it into a gray image with double precision by means of the functions *rgb2gray* and *double*. Letting this image be input $x[m, n]$ implement the Sobel filter to obtain an image $y[m, n]$. Determine a threshold T such that whenever $y[m, n] > T$ we let a final image $z[m, n] = 1$ and zero otherwise. Display the complement of the image $z[m, n]$.

(b) Repeat the above for the image *circuit.tif*, choose a threshold T so that you obtain a binarized image displaying the edges of the image.

Answer: The Sobel filters are given by the arrays $hu = [1, 0, -1; 2, 0, -2; 1, 0, -1]$ and $hv = hu'$, the transpose of hu.

THE Z-TRANSFORM

CONTENTS

I was born not knowing and have had only a little time to change that here and there.

Richard Feynman, Professor and Physicist (1918–1988)

10.1 INTRODUCTION

The Z-transform provides a way to represent and process discrete-time signals and systems. Although the Z-transform can be related to the Laplace transform, the relation is operationally not very useful.

Signals and Systems Using MATLAB®. https://doi.org/10.1016/B978-0-12-814204-2.00021-1

However, it shows that the complex Z-plane is in a polar form where the radius is a damping factor and the angle corresponds to the discrete frequency ω in radians. Thus, the unit circle in the Z-plane is analogous to the $j\Omega$-axis in the Laplace plane, and the inside of the unit circle is analogous to the left-hand s-plane. We will see that once the connection between the Laplace and the Z-plane is established, the significance of poles and zeros in the Z-plane can be obtained like in the Laplace plane.

The representation of discrete-time signals by the Z-transform is very intuitive: it converts a sequence of samples into a polynomial. The inverse Z-transform can be obtained by many more methods than the inverse Laplace transform, but the partial fraction expansion is the most commonly used method. Using the one-sided Z-transform, solving difference equations that could result from the discretization of ordinary differential equations, but not exclusively, is an important application of the Z-transform.

As it was the case with the Laplace transform and the convolution integral, the most important property of the Z-transform is the implementation of the convolution sum as a multiplication. This is not only important as a computational tool but also as a way to represent a discrete system by its transfer function. Filtering is again an important application as it was in the continuous-time domain, and as before the localization of poles and zeros determines the type of filter. However, in the discrete domain there is a greater variety of filters than in the analog domain given that they can be recursive (IIR) or non-recursive (FIR).

Modern control uses the state-variable representation and the Z-transform is useful in obtaining the system realization and in determining the complete solution of the system. Just like in the continuous-time, the state represents the memory of the system, and it generalizes the transfer function representation. The state-variable models for discrete-time systems are very similar to the ones discussed before for continuous-time systems.

The extension of the Z-transform to two dimensions is possible, although complicated by the four-dimensional domain imposed by the two complex variables. The two-dimensional Z-transform aids in the representation and processing of two-dimensional signals and systems. It is particularly useful in dealing with the convolution and in developing stability testing procedures for two-dimensional filters. The transfer function thus becomes the basic representation of two-dimensional linear shift-invariant systems.

10.2 LAPLACE TRANSFORM OF SAMPLED SIGNALS

The Laplace transform of a sampled signal

$$x(t) = \sum_n x(nT_s)\delta(t - nT_s) \tag{10.1}$$

is given by

$$X(s) = \sum_n x(nT_s)\mathcal{L}[\delta(t - nT_s)] = \sum_n x(nT_s)e^{-nsT_s}. \tag{10.2}$$

By letting $z = e^{sT_s}$, we can rewrite (10.2) as

$$\mathcal{Z}[x(nT_s)] = \mathcal{L}[x_s(t)]\big|_{z=e^{sT_s}} = \sum_n x(nT_s)z^{-n}, \tag{10.3}$$

which is called the Z-transform of the sampled signal.

Remarks

The function $X(s)$ in (10.2) is different from the Laplace transforms we considered before:

1. Letting $s = j\Omega$, $X(\Omega)$ is periodic of fundamental period $2\pi/T_s$, i.e., $X(\Omega + 2k\pi/T_s) = X(\Omega)$ for an integer k. Indeed,

$$X(\Omega + 2k\pi/T_s) = \sum_n x(nT_s)e^{-jn(\Omega+2k\pi/T_s)T_s} = \sum_n x(nT_s)e^{-jn(\Omega T_s+2k\pi)} = X(\Omega).$$

2. $X(s)$ may have an infinite number of poles or zeros—complicating the partial fraction expansion when finding its inverse. Fortunately, the presence of the $\{e^{-nsT_s}\}$ terms suggests that the inverse should be done using the time-shift property of the Laplace transform instead.

Example 10.1. To see the possibility of an infinite number of poles and zeros in the Laplace transform of a sampled signal, consider a pulse $x(t) = u(t) - u(t - T_0)$ sampled with a sampling period $T_s = T_0/N$, N a positive integer. Find the Laplace transform of the sampled signal and determine its poles and zeros.

Solution: The discrete-time signal obtained by sampling is

$$x(nT_s) = \begin{cases} 1 & 0 \leq nT_s \leq T_0 \text{ or } 0 \leq n \leq N, \\ 0 & \text{otherwise,} \end{cases}$$

with Laplace transform

$$X(s) = \sum_{n=0}^{N} e^{-nsT_s} = \frac{1 - e^{-(N+1)sT_s}}{1 - e^{-sT_s}}.$$

The poles are the s_k values that make the denominator zero, i.e.,

$$e^{-s_k T_s} = 1 = e^{j2\pi k} \qquad k \text{ integer}, \quad -\infty < k < \infty$$

or $s_k = -j2\pi k/T_s$ for any integer k—an infinite number of poles. Similarly, one can show that $X(s)$ has an infinite number of zeros by finding the values s_m that make the numerator zero, or

$$e^{-(N+1)s_m T_s} = 1 = e^{j2\pi m} \qquad m \text{ integer}, \quad -\infty < m < \infty$$

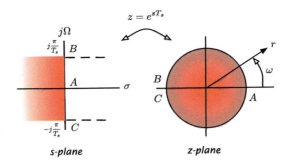

FIGURE 10.1

Mapping of the Laplace plane into the Z-plane. Strips of width $2\pi/T_s$ in the left-hand s-plane are mapped into the inside of a unit circle in the Z-plane. The right-hand side of the strips are mapped outside the unit circle. The whole s-plane as union of these strips is mapped into the same Z-plane.

or $s_m = -j2\pi m/((N+1)T_s)$ for any integer m—again, an infinite number of zeros. Such a behavior can be better understood when we consider the connection between the s- and the Z-plane. □

The relation $z = e^{sT_s}$ provides the connection between the s-plane and the Z-plane:

$$z = e^{sT_s} = e^{(\sigma + j\Omega)T_s} = e^{\sigma T_s} e^{j\Omega T_s}.$$

Letting $r = e^{\sigma T_s}$ and $\omega = \Omega T_s$, we have $z = re^{j\omega}$, which is a complex variable in polar form, with radius $0 \le r < \infty$ and angle ω in radians. The variable r is a damping factor, and ω is the discrete frequency in radians. So the Z-plane corresponds to circles of radius r and angles $-\pi \le \omega < \pi$.

Let us see how $z = e^{sT_s}$ maps the s-plane into the Z-plane. Consider the strip of width $2\pi/T_s$ across the s-plane shown in Fig. 10.1. The width of this strip is related to the Nyquist condition, establishing that the maximum frequency of the analog signals we are considering is $\Omega_M = \Omega_s/2 = \pi/T_s$, where Ω_s is the sampling frequency and T_s the sampling period. If $T_s \to 0$, we would be considering the class of signals with maximum frequency approaching ∞, i.e. all signals.

The relation $z = e^{sT_s}$ maps the real part of $s = \sigma + j\Omega$, $\mathcal{R}e(s) = \sigma$, into the radius $r = e^{\sigma T_s} \ge 0$, and the analog frequencies $-\pi/T_s \le \Omega \le \pi/T_s$ into $-\pi \le \omega < \pi$, according to the frequency connection $\omega = \Omega T_s$. Thus, the mapping of the $j\Omega$-axis in the s-plane, corresponding to $\sigma = 0$, gives a circle of radius $r = 1$ or the unit circle. The right-hand s-plane, $\sigma > 0$, maps into circles with radius $r > 1$, and the left-hand s-plane, $\sigma < 0$, maps into circles of radius $r < 1$. Points A, B, and C in the strip are mapped into corresponding points in the Z-plane as shown in Fig. 10.1. So the given strip in the s-plane maps into the whole Z-plane. Similarly for other strips of the same width. Thus the s-plane, as a union of these strips, is mapped onto the same Z-plane.

The mapping $z = e^{sT_s}$ can be used to illustrate the sampling process. Consider a band-limited signal $x(t)$ with maximum frequency π/T_s with a spectrum in the band $[-\pi/T_s \quad \pi/T_s]$. According to the relation $z = e^{sT_s}$ the spectrum of $x(t)$ in $[-\pi/T_s \ \pi/Ts]$ is mapped into the unit circle of the Z-plane from $[-\pi, \pi)$. Going around the unit circle in the Z-plane, the mapped frequency response repeats periodically just like the spectrum of the sampled signal.

THE HISTORY OF THE Z-TRANSFORM

The history of the Z-transform goes back to the work of the French mathematician De Moivre, who in 1730 introduced the characteristic function to represent the probability mass function of a discrete random variable. Also the Z-transform is a special case of the Laurent series, used to represent complex functions.

In the 1950s the Russian engineer and mathematician Yakov Tsypkin (1919–1997) proposed the discrete Laplace transform which he applied to the study of pulsed systems. Then Professor John Ragazzini and his students Eliahu Jury and Lofti Zadeh at Columbia University developed the Z-transform. Ragazzini (1912–1988) was chairman of the Department of Electrical Engineering at Columbia University. Three of his students are well recognized in electrical engineering for their accomplishments: Jury for the Z-transform, non-linear systems and the inners stability theory, Zadeh for the Z-transform and fuzzy set theory, and Rudolf Kalman for the Kalman filtering.

Jury was born in Iraq, and received his Doctor of Engineering Science degree from Columbia University in 1953. He was professor of Electrical Engineering at the University of California, Berkeley, and at the University of Miami. Among his publications, Professor Jury's "Theory and Application of the Z-transform" [41] is a seminal work on the theory and application of the Z-transform.

10.3 TWO-SIDED Z-TRANSFORM

Given a discrete-time signal $x[n]$, $-\infty < n < \infty$, its two-sided Z-transform is given by

$$X(z) = \sum_{n=-\infty}^{\infty} x[n]z^{-n} \tag{10.4}$$

defined in a region of convergence (ROC) in the Z-plane.

Considering the sampled signal $x(nT_s)$ a function of n in Equation (10.3), we obtain the two-sided Z-transform.

The Z-transform can be thought of as the transformation of the sequence $\{x[n]\}$ into a polynomial $X(z)$ (possibly of infinite degree) in positive and negative powers of z, where to a sample $x[n_0]$ we attach a monomial z^{-n_0}. Thus, given a sequence of samples $\{x[n]\}$ its Z-transform simply consists of a polynomial with coefficients $x[n]$ corresponding to z^{-n}. In the same way, given a Z-transform, as in (10.4), its inverse is easily obtained by looking at the coefficients attached to the z^{-n} monomials, for positive and negative values of the sample value n. Clearly this inverse is not in a closed form. We will see ways to compute these closed-form inverses later in this chapter.

The two-sided Z-transform is not useful in solving difference equations with nonzero initial conditions, just as the two-sided Laplace transform was not useful either in solving ordinary differential equations with nonzero initial conditions. To include initial conditions in the transformation it is necessary to define the **one-sided Z-transform**.

The **one-sided Z-transform** is defined for causal signals, $x[n] = 0$ for $n < 0$, or for signals that are made causal by multiplying them with the unit-step signal $u[n]$:

$$X_1(z) = \mathcal{Z}(x[n]u[n]) = \sum_{n=0}^{\infty} x[n]u[n]z^{-n} \qquad (10.5)$$

in a region of convergence $\mathcal{R}_1$. The two-sided Z-transform can thus be expressed in terms of the one-sided Z-transform as follows:

$$X(z) = \mathcal{Z}(x[n]u[n]) + \mathcal{Z}(x[-n]u[n])|_z - x[0]. \qquad (10.6)$$

The region of convergence of $X(z)$ is

$$\mathcal{R} = \mathcal{R}_1 \cap \mathcal{R}_2$$

where $\mathcal{R}_1$ is the region of convergence of $\mathcal{Z}(x[n]u[n])$ and $\mathcal{R}_2$ the region of convergence of $\mathcal{Z}(x[-n]u[n])|_z$.

The one-sided Z-transform coincides with the two-sided Z-transform whenever $x[n]$ is *causal*, i.e., $x[n] = 0$ for $n < 0$. If the signal is noncausal, multiplying it by $u[n]$ makes it causal. To express the two-sided Z-transform in terms of one-sided Z-transform we separate the sum into two and make each into a causal sum:

$$
\begin{aligned}
X(z) &= \sum_{n=-\infty}^{\infty} x[n]z^{-n} = \sum_{n=0}^{\infty} x[n]u[n]z^{-n} + \sum_{n=-\infty}^{0} x[n]u[-n]z^{-n} - x[0] \\
&= \mathcal{Z}(x[n]u[n]) + \sum_{m=0}^{\infty} x[-m]u[m]z^{m} - x[0] \\
&= \mathcal{Z}(x[n]u[n]) + \mathcal{Z}(x[-n]u[n])|_z - x[0]
\end{aligned}
$$

where the inclusion of the additional term $x[0]$ in the sum from $-\infty$ to 0 is compensated by subtracting it, and in the same sum a change of variable ($m = -n$) gives a one-sided Z-transform in terms of positive powers of z, as indicated by the notation $\mathcal{Z}(x[-n]u[n])|_z$.

10.3.1 REGION OF CONVERGENCE

The infinite summation of the two-sided Z-transform must converge for some values of z. For $X(z)$ to converge it is necessary that

$$|X(z)| = \left| \sum_{n} x[n]z^{-n} \right| \leq \sum_{n} |x[n]||r^{-n}e^{j\omega n}| = \sum_{n} |x[n]|r^{-n} < \infty.$$

Thus, the convergence of $X(z)$ depends on r. The region in the Z-plane where $X(z)$ converges or its **region of convergence** (ROC) connects the signal and its Z-transform uniquely. As with the Laplace transform, the poles of $X(z)$ are connected with its region of convergence.

The poles of a Z-transform $X(z)$ are complex values $\{p_k\}$ such that

$$X(p_k) \to \infty,$$

while the zeros of $X(z)$ are the complex values $\{z_k\}$ that make

$$X(z_k) = 0.$$

Example 10.2. Find the poles and zeros of the following Z-transforms:

$$(i) \ X_1(z) = 1 + 2z^{-1} + 3z^{-2} + 4z^{-3}, \quad (ii) \ X_2(z) = \frac{(z^{-1} - 1)(z^{-1} + 2)^2}{z^{-1}(z^{-2} + \sqrt{2}z^{-1} + 1)}.$$

Solution: To see the poles and zeros more clearly let us express $X_1(z)$ as a function of positive powers of z:

$$X_1(z) = \frac{z^3(1 + 2z^{-1} + 3z^{-2} + 4z^{-3})}{z^3} = \frac{z^3 + 2z^2 + 3z + 4}{z^3} = \frac{N_1(z)}{D_1(z)}.$$

There are three poles at $z = 0$, the roots of $D_1(z) = 0$. The zeros are the roots of $N_1(z) = z^3 + 2z^2 + 3z + 4 = 0$, which can be found using MATLAB's function *roots* as $z_1 = -1.65$, $z_2 = -0.175 + j1.547$, $z_3 = -0.175 - j1.547$. Notice that the coefficients of $N_1(z)$ are real and as such the complex roots are a complex conjugate pair.

Likewise, expressing $X_2(z)$ as a function of positive powers of z,

$$X_2(z) = \frac{z^3(z^{-1} - 1)(z^{-1} + 2)^2}{z^3(z^{-1}(z^{-2} + \sqrt{2}z^{-1} + 1))} = \frac{(1 - z)(1 + 2z)^2}{1 + \sqrt{2}z + z^2} = \frac{N_2(z)}{D_2(z)}.$$

The poles of $X_2(z)$ are the roots of $D_2(z) = 1 + \sqrt{2}z + z^2 = 0$, or $p_{1,2} = -0.707 \pm j0.707$, while the zeros of $X_2(z)$ are the roots of $N_2(z) = (1 - z)(1 + 2z)^2 = 0$, or $z_1 = 1$ and $z_{2,3} = -0.5$ (double). $\quad\square$

The region of convergence depends on the support of the signal. If it is finite, the ROC is very much the whole Z-plane, if it is infinite the ROC depends on whether the signal is causal, anticausal or noncausal. Something to remember is that in no case the ROC includes any poles of the Z-transform.

ROC of Finite-Support Signals

The region of convergence (ROC) of the Z-transform of a signal $x[n]$ of finite support $[N_0, N_1]$, where $-\infty < N_0 \leq n \leq N_1 < \infty$,

$$X(z) = \sum_{n=N_0}^{N_1} x[n]z^{-n} \tag{10.7}$$

is the whole Z-plane, excluding the origin $z = 0$ and/or $z = \pm\infty$ depending on N_0 and N_1.

Given the finite support of $x[n]$ its Z-transform has no convergence problem. Indeed, letting $|z| \neq 0$ and $|z| \neq \infty$ if negative or positive powers of z occur in (10.7), we have

$$|X(z)| \leq \sum_{n=N_0}^{N_1} |x[n]||z^{-n}| < \infty.$$

This sum is finite given that it has a finite number of terms, and that each term is finite (i.e., $|x[n]| < \infty$ and $|z^{-n}| = (1/r)^n$ cannot be infinite because r is not allowed to be zero, when $n > 0$, or infinity when $n < 0$). You can see, for instance, that when $N_0 \geq 0$ the poles of $X(z)$ are at the origin of the Z-plane, or that when $N_1 \leq 0$ there are no poles, only zeros, and as such $X(z)$ would only become infinite when $|z| = 0$ or $|z| = \infty$. Thus, the ROC of $X(z)$ is the whole z-plane excluding $|z| = 0$ or $|z| = \infty$ as the case may be.

Example 10.3. Find the Z-transform of a discrete-time pulse $x[n] = u[n] - u[n - 10]$. Determine the region of convergence of $X(z)$.

Solution: The Z-transform of $x[n]$ is

$$X(z) = \sum_{n=0}^{9} 1 \, z^{-n} = \frac{1 - z^{-10}}{1 - z^{-1}} = \frac{z^{10} - 1}{z^9(z - 1)}. \tag{10.8}$$

That this sum equals the term on the right can be verified by multiplying the left term by the denominator $1 - z^{-1}$ and showing the result is the same as the numerator. In fact,

$$\begin{aligned}
(1 - z^{-1}) \sum_{n=0}^{9} 1 \, z^{-n} &= \sum_{n=0}^{9} 1 \, z^{-n} - \sum_{n=0}^{9} 1 \, z^{-n-1} \\
&= (1 + z^{-1} + \cdots + z^{-9}) - (z^{-1} + \cdots + z^{-9} + z^{-10}) = 1 - z^{-10}.
\end{aligned}$$

Since $x[n]$ is a finite sequence there is no problem with the convergence of the sum, although $X(z)$ in (10.8) seems to indicate the need for $z \neq 1$ ($z = 1$ makes the numerator and denominator go to zero). However, from the sum if we let $z = 1$ then $X(1) = 10$, so there is no need to restrict z to be different from 1. This is caused by the pole at $z = 1$ being canceled by a zero at $z = 1$. Indeed, the zeros z_k of $X(z)$ (see Equation (10.8)) are the roots of $z^{10} - 1 = 0$, which are $z_k = e^{j2\pi k/10}$ for $k = 0 \cdots 9$. Therefore, the zero when $k = 0$, or $z_0 = 1$, cancels the pole at 1 so that

$$X(z) = \frac{\prod_{k=1}^{9}(z - e^{j\pi k/5})}{z^9},$$

i.e., $X(z)$ has nine poles at the origin of the z-plane, and nine zeros around the unit circle except at $z = 1$. Thus the whole z-plane excluding the origin is the region of convergence of $X(z)$.

Notice that $X(z)$ can also be written as a polynomial:

$$X(z) = 1 + z^{-1} + z^{-2} + z^{-3} + z^{-4} + z^{-5} + z^{-6} + z^{-7} + z^{-8} + z^{-9},$$

which only tends to infinity when $z = 0$, so the region of convergence of $X(z)$ is the whole z-plane excluding the origin. □

ROC of Infinite-Support Signals

Signals of infinite support are either causal, anticausal or a combination of these—noncausal. For the Z-transform of a causal signal $x_c[n]$, i.e., $x_c[n] = 0$, $n < 0$,

$$X_c(z) = \sum_{n=0}^{\infty} x_c[n] z^{-n} = \sum_{n=0}^{\infty} x_c[n] r^{-n} e^{-jn\omega}$$

to converge we need to determine appropriate values of r, the damping factor, as the frequency ω has no effect on the convergence. If R_1 is the radius of the farthest out pole of $X_c(z)$, then there is an exponential $R_1^n u[n]$ such that $|x_c[n]| < M R_1^n$ for $n \geq 0$ for some value $M > 0$. Then, for $X(z)$ to converge we need that

$$|X_c(z)| \leq \sum_{n=0}^{\infty} |x_c[n]| r^{-n} < M \sum_{n=0}^{\infty} \left(\frac{R_1}{r}\right)^n < \infty$$

or that $R_1/r < 1$, which is equivalent to $|z| = r > R_1$. The ROC is thus the outside of a circle containing all the poles of $X_c(z)$, i.e., it does not include any poles of $X_c(z)$.

Likewise, for an anticausal signal $x_a[n]$ (i.e., $x_a[n] = 0$, $n > 0$) if we choose a radius R_2 which is smaller than the radius of all the poles of $X_a(z)$, the region of convergence is $|z| = r < R_2$. This ROC is the inside of a circle that does not include any poles of $X_a(z)$.

If the signal $x[n]$ is noncausal, it can be expressed as

$$x[n] = x_c[n] + x_a[n]$$

where the supports of $x_a[n]$ and $x_c[n]$ can be finite or infinite or any possible combination of these two. The corresponding ROC of $X(z) = \mathcal{Z}\{x[n]\}$ would then be

$$0 < R_1 < |z| < R_2 < \infty.$$

This ROC is a torus surrounded on the inside by the poles of the causal component, and in the outside by the poles of the anticausal component. If the signal has finite support, then $R_1 = 0$ and $R_2 = \infty$, coinciding with the result for finite-support signals.

The Z-transform $X(z)$ of an infinite support:

1. **causal signal** $x[n]$ has a region of convergence $|z| > R_1$ where R_1 is the largest radius of the poles of $X(z)$, i.e., the region of convergence is the outside of a circle of radius R_1,
2. **anticausal signal** $x[n]$ has as region of convergence the inside of the circle defined by the smallest radius R_2 of the poles of $X(z)$, or $|z| < R_2$,
3. **noncausal signal** $x[n]$ has as region of convergence $R_1 < |z| < R_2$, or the inside of a torus of inside radius R_1 and outside radius R_2 corresponding to the maximum and minimum radii of the poles of $X_c(z)$ and $X_a(z)$, or the Z-transforms of the causal and anticausal components of $x[n]$.

Example 10.4. The poles of $X(z)$ are $z = 0.5$ and $z = 2$; find all the possible signals that can be associated with $X(z)$ according to different regions of convergence.

Solution: Possible regions of convergence are

- $\{\mathcal{R}_1 : |z| > 2\}$, the outside of a circle of radius 2, we associate with $X(z)$ a causal signal $x_1[n]$.
- $\{\mathcal{R}_2 : |z| < 0.5\}$, the inside of a circle of radius 0.5, an anticausal signal $x_2[n]$ can be associated with $X(z)$.
- $\{\mathcal{R}_3 : 0.5 < |z| < 2\}$, a torus of radii 0.5 and 2, a noncausal signal $x_3[n]$ can be associated with $X(z)$.

Three different signals can be connected with $X(z)$ by considering three different regions of convergence. $\qquad\square$

Example 10.5. Find the regions of convergence of the Z-transforms of the following signals:

$$(i)\ x_1[n] = \left(\frac{1}{2}\right)^n u[n], \quad (ii)\ x_2[n] = -\left(\frac{1}{2}\right)^n u[-n-1].$$

Find then the Z-transform of $x_1[n] + x_2[n]$.

Solution: The signal $x_1[n]$ is causal, while $x_2[n]$ is anticausal. The Z-transform of $x_1[n]$ is

$$X_1(z) = \sum_{n=0}^{\infty} \left(\frac{1}{2}\right)^n z^{-n} = \frac{1}{1 - 0.5z^{-1}} = \frac{z}{z - 0.5}$$

provided that $|0.5z^{-1}| < 1$ or that its region of convergence be $\mathcal{R}_1 : |z| > 0.5$. The region $\mathcal{R}_1$ is the outside of a circle of radius 0.5.

The signal $x_2[n]$ grows as n decreases from -1 to $-\infty$, and the rest of its values are zero. Its Z-transform is found to be

$$X_2(z) = -\sum_{n=-\infty}^{-1} \left(\frac{1}{2}\right)^n z^{-n} = -\sum_{m=0}^{\infty} \left(\frac{1}{2}\right)^{-m} z^m + 1 = -\sum_{m=0}^{\infty} 2^m z^m + 1 = \frac{-1}{1 - 2z} + 1 = \frac{z}{z - 0.5}$$

with a region of convergence $\mathcal{R}_2 : |z| < 0.5$.

It is very important to notice that although the signals are clearly different, their Z-transforms are identical. It is their regions of convergence that differentiate them. The Z-transform of $x_1[n] + x_2[n]$ does not exist given that the intersection of $\mathcal{R}_1$ and $\mathcal{R}_2$ is empty. $\qquad\square$

Remark

The uniqueness of the Z-transform requires that the Z-transform of a signal be accompanied by a region of convergence. It is possible to have identical Z-transforms with different regions of convergence, corresponding to different signals.

Example 10.6. Let $c[n] = \alpha^{|n|}$, $0 < \alpha < 1$, be a discrete-time signal (it is actually an autocorrelation function related to the power spectrum of a random signal). Determine its Z-transform.

Solution: To find its two-sided Z-transform $C(z)$ we consider its causal and anticausal components. First for the causal component we have

$$\mathcal{Z}(c[n]u[n]) = \sum_{n=0}^{\infty} \alpha^n z^{-n} = \frac{1}{1 - \alpha z^{-1}}$$

with the region of convergence $|\alpha z^{-1}| < 1$ or $|z| > \alpha$. For the anticausal component,

$$\mathcal{Z}(c[-n]u[n])_z = \sum_{n=0}^{\infty} \alpha^n z^n = \frac{1}{1 - \alpha z}$$

with a region of convergence $|\alpha z| < 1$ or $|z| < 1/\alpha$. Thus, the two-sided Z-transform of $c[n]$ is (notice that the term for $n = 0$ was used twice in the above calculations so we need to subtract it)

$$C(z) = \frac{1}{1 - \alpha z^{-1}} + \left(\frac{1}{1 - \alpha z} - 1 \right) = \frac{z}{z - \alpha} - \frac{z}{(z - 1/\alpha)} = \frac{(\alpha - 1/\alpha)z}{(z - \alpha)(z - 1/\alpha)}$$

with a region of convergence

$$\alpha < |z| < \frac{1}{\alpha}. \quad \square$$

10.4 ONE-SIDED Z-TRANSFORM

In many situations where the Z-transform is used the system is LTI, causal (its impulse response $h[n] = 0$ for $n < 0$) and the input signal is also causal ($x[n] = 0$ for $n < 0$). In such cases the one-sided Z-transform is very appropriate. Moreover, as we saw before, the two-sided Z-transform can be expressed in terms of one-sided Z-transforms. Another reason to study the one-sided Z-transform in more detail is its use in solving difference equations with initial conditions.

Recall that the one-sided Z-transform is defined as

$$X_1(z) = \mathcal{Z}(x[n]u[n]) = \sum_{n=0}^{\infty} x[n]u[n]z^{-n} \tag{10.9}$$

in a region of convergence $\mathcal{R}_1 : |z| > R_1$. Also recall that the computation of the two-sided Z-transform using the one-sided Z-transform is given in Equation (10.6).

10.4.1 SIGNAL BEHAVIOR AND POLES

The Z-transform is a **linear transformation** meaning that

$$\mathcal{Z}(ax[n] + by[n]) = a\mathcal{Z}(x[n]) + b\mathcal{Z}(y[n]) \tag{10.10}$$

for signals $x[n]$ and $y[n]$ and constants a and b.

The linearity of the transform can be easily verified using its definition as a summation. In this section, we use the linearity property of the Z-transform to connect the behavior of a signal with the poles of its Z-transform.

Consider the signal $x[n] = \alpha^n u[n]$ for real or complex values α. Its Z-transform will be used to compute the Z-transform of the following signals:

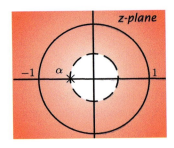

FIGURE 10.2

Region of convergence (shaded area) of $X(z)$ with a pole at $z = \alpha$, $\alpha < 0$ (same ROC if pole is at $z = -\alpha$).

- $x[n] = \cos(\omega_0 n + \theta)u[n]$ for frequency $0 \le \omega_0 \le \pi$, and phase θ,
- $x[n] = \alpha^n \cos(\omega_0 n + \theta)u[n]$ for frequency $0 \le \omega_0 \le \pi$, and phase θ,

and to show how the poles of the corresponding Z-transforms connect with the signals.

The Z-transform of the causal signal $x[n] = \alpha^n u[n]$ is

$$X(z) = \sum_{n=0}^{\infty} \alpha^n z^{-n} = \sum_{n=0}^{\infty} (\alpha z^{-1})^n = \frac{1}{1 - \alpha z^{-1}} = \frac{z}{z - \alpha} \qquad \text{ROC: } |z| > |\alpha|.$$

Using the last expression for $X(z)$ in the above equation, its zero is $z = 0$ and its pole is $z = \alpha$. For α real, be it positive or negative, the region of convergence is the same, but the poles are located in different places. See Fig. 10.2 for $\alpha < 0$.

If $\alpha = 1$ the signal $x[n] = u[n]$ is constant for $n \ge 0$ and the pole of $X(z)$ is at $z = 1e^{j0}$ (radius $r = 1$ and the lowest discrete frequency $\omega = 0$ rad). On the other hand, when $\alpha = -1$ the signal is $x[n] = (-1)^n u[n]$, which varies from sample to sample for $n \ge 0$, its Z-transform has a pole at $z = -1 = 1e^{j\pi}$ (a radius $r = 1$ and the highest discrete frequency $\omega = \pi$ rad). As we move the pole towards the center of the Z-plane, i.e., $|\alpha| \to 0$, the corresponding signal decays exponentially for $0 < \alpha < 1$, and is a modulated exponential $|\alpha|^n (-1)^n u[n] = |\alpha|^n \cos(\pi n) u[n]$ for $-1 < \alpha < 0$. When $|\alpha| > 1$ the signal becomes either a growing exponential ($\alpha > 1$) or a growing modulated exponential ($\alpha < -1$).

For a real value $\alpha = |\alpha| e^{j\omega_0}$, with $\omega_0 = 0$ or π,

$$x[n] = \alpha^n u[n] \quad \Leftrightarrow \quad X(z) = \frac{1}{1 - \alpha z^{-1}} = \frac{z}{z - \alpha} \qquad \text{ROC: } |z| > |\alpha|$$

and the location of the pole of $X(z)$ determines the behavior of the signal:

- when $\alpha > 0$ then $\omega_0 = 0$ and the signal is less and less damped as $\alpha \to \infty$, and
- when $\alpha < 0$ then $\omega_0 = \pi$ and the signal is a modulated exponential that grows as $\alpha \to -\infty$.

To compute the Z-transform of $x[n] = \cos(\omega_0 n + \theta)u[n]$ we use Euler's identity to write $x[n]$ as

$$x[n] = \left[\frac{e^{j(\omega_0 n + \theta)}}{2} + \frac{e^{-j(\omega_0 n + \theta)}}{2} \right] u[n].$$

Applying the linearity property and using the above Z-transform when $\alpha = e^{j\omega_0}$ and its conjugate $\alpha^* = e^{-j\omega_0}$ we get

$$
\begin{aligned}
X(z) &= \frac{1}{2} \left[\frac{e^{j\theta}}{1 - e^{j\omega_0} z^{-1}} + \frac{e^{-j\theta}}{1 - e^{-j\omega_0} z^{-1}} \right] = \frac{1}{2} \left[\frac{2\cos(\theta) - 2\cos(\omega_0 - \theta)z^{-1}}{1 - 2\cos(\omega_0)z^{-1} + z^{-2}} \right] \\
&= \frac{\cos(\theta) - \cos(\omega_0 - \theta)z^{-1}}{1 - 2\cos(\omega_0)z^{-1} + z^{-2}}.
\end{aligned}
\tag{10.11}
$$

Expressing $X(z)$ in terms of positive powers of z, we get

$$X(z) = \frac{z(z\cos(\theta) - \cos(\omega_0 - \theta))}{z^2 - 2\cos(\omega_0)z + 1} = \frac{z(z\cos(\theta) - \cos(\omega_0 - \theta))}{(z - e^{j\omega_0})(z - e^{-j\omega_0})} \tag{10.12}$$

valid for any value of θ. If $x[n] = \cos(\omega_0 n)u[n]$ then $\theta = 0$ and the poles of $X(z)$ are a complex conjugate pair on the unit circle at frequency ω_0 radians. The zeros are at $z = 0$ and $z = \cos(\omega_0)$. When $x[n] = \sin(\omega_0 n)u[n] = \cos(\omega_0 n - \pi/2)u[n]$, then $\theta = -\pi/2$ and the poles are at the same location as those for the cosine, but the zeros are at $z = 0$ and $z = \cos(\omega_0 + \pi/2)/\cos(\pi/2) \to \infty$, so there is only one finite zero at zero. For any other value of θ, the poles are located in the same place but there is a zero at $z = 0$ and another at $z = \cos(\omega_0 - \theta)/\cos(\theta)$.

When we let $\theta = 0$, and $\omega_0 = 0$, one of the double poles at $z = 1$ is canceled by one of the zeros at $z = 1$, resulting in the poles and zeros of $\mathcal{Z}(u[n])$ or $X(z) = z/(z - 1)$. Indeed, the signal when $\omega_0 = 0$ and $\theta = 0$ is $x[n] = \cos(0n)u[n] = u[n]$. When the frequency $\omega_0 > 0$ the poles move along the unit circle from the lowest ($\omega_0 = 0$ rad) to the highest ($\omega_0 = \pi$ rad) frequency.

The Z-transform pairs of a cosine and a sine are

$\cos(\omega_0 n)u[n]$	$\Leftrightarrow$	$\dfrac{z(z - \cos(\omega_0))}{(z - e^{j\omega_0})(z - e^{-j\omega_0})}$	ROC: $\lvert z \rvert > 1$,	(10.13)
$\sin(\omega_0 n)u[n]$	$\Leftrightarrow$	$\dfrac{z\sin(\omega_0)}{(z - e^{j\omega_0})(z - e^{-j\omega_0})}$	ROC: $\lvert z \rvert > 1$.	(10.14)

The Z-transforms for these sinusoids have identical poles $1e^{\pm j\omega_0}$, but different zeros. The frequency of the sinusoid increases from the lowest $\omega_0 = 0$ (rad) to the highest $\omega_0 = \pi$ (rad) as the poles move along the unit circle from 1 to -1.

Consider then the signal $x[n] = r^n \cos(\omega_0 n + \theta)u[n]$, which is a combination of the above cases. As before, the signal is expressed as a linear combination

$$x[n] = \left[\frac{e^{j\theta}(re^{j\omega_0})^n}{2} + \frac{e^{-j\theta}(re^{-j\omega_0})^n}{2} \right] u[n]$$

and it can be shown its Z-transform is

$$X(z) = \frac{z(z\cos(\theta) - r\cos(\omega_0 - \theta))}{(z - re^{j\omega_0})(z - re^{-j\omega_0})}. \tag{10.15}$$

The Z-transform of a sinusoid is a special case of the above, i.e., when $r = 1$. It also becomes clear that, as the value of r decreases towards zero, the exponential in the signal decays, and that whenever $r > 1$ the exponential in the signal grows making the signal unbound.

The Z-transform pair

$$r^n \cos(\omega_0 n + \theta)u[n] \quad \Leftrightarrow \quad \frac{z(z\cos(\theta) - r\cos(\omega_0 - \theta))}{(z - re^{j\omega_0})(z - re^{-j\omega_0})} = \frac{z(z\cos(\theta) - r\cos(\omega_0 - \theta))}{z^2 - 2r\cos(\omega_0)z + r^2} \tag{10.16}$$

shows how complex conjugate pairs of poles inside the unit circle represent the damping indicated by the radius r and the frequency given by ω_0 in radians.

Finally, double poles are related to the derivative of $X(z)$ or to the multiplication of the signal by n. If

$$X(z) = \sum_{n=0}^{\infty} x[n]z^{-n},$$

its derivative with respect to z is

$$\frac{dX(z)}{dz} = \sum_{n=0}^{\infty} x[n]\frac{dz^{-n}}{dz} = -z^{-1}\sum_{n=0}^{\infty} nx[n]z^{-n}.$$

Or the pair

$$nx[n]u[n] \quad \Leftrightarrow \quad -z\frac{dX(z)}{dz}. \tag{10.17}$$

For instance, if $X(z) = 1/(1 - \alpha z^{-1}) = z/(z - \alpha)$, the Z-transform of $x[n] = \alpha^n u[n]$, we then have

$$\frac{dX(z)}{dz} = -\frac{\alpha}{(z - \alpha)^2},$$

and as such

The pair

$$n\alpha^n u[n] \quad \Leftrightarrow \quad \frac{\alpha z}{(z - \alpha)^2}$$

indicates that double poles correspond to multiplication of $x[n]$ by n.

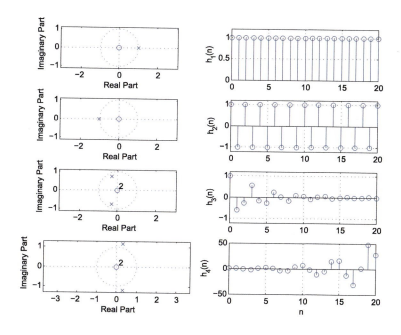

FIGURE 10.3

Effect of pole location on the inverse Z-transform (from top to bottom): if pole is at $z = 1$ the signal is $u(n)$, constant for $n \geq 0$; if pole is at $z = -1$ the signal is a cosine of frequency π continuously changing, constant amplitude; when poles are complex, if inside the unit circle the signal is a decaying modulated exponential, and if outside the unit circle the signal is a growing modulated exponential.

According to the above discussion, the location of the poles of $X(z)$ provides basic information about the signal $x[n]$. This is illustrated in Fig. 10.3, where we display the signal and its corresponding poles.

10.4.2 COMPUTING Z-TRANSFORMS WITH SYMBOLIC MATLAB

Similar to the computation of the Laplace transform, the computation of the Z-transform can be done using the symbolic toolbox of MATLAB.

Example 10.7. Write the necessary code for computing the Z-transform of

$$h_1[n] = 0.9u[n], \qquad h_2[n] = u[n] - u[n - 10],$$
$$h_3[n] = \cos(\omega_0 n)u[n], \quad h_4[n] = h_1[n]h_3[n].$$

Solution: The results are shown at the bottom. (As in the continuous case, in MATLAB the *heaviside* function is the same as the unit-step function.)

```
%%
% Example 10.7---Z-transform computation
%%
  syms n w0
  h1=0.9.^n; H1 = ztrans(h1)
  h2=heaviside(n)-heaviside(n-10); H2= ztrans(h2)
  h3=cos(w0*n)*heaviside(n); H3=ztrans(h3)
  H4=ztrans(h1*h3)
```

```
H1 = z/(z - 9/10)
H2 = 1/(z - 1) - (1/(z - 1) + 1/2)/z^10 + 1/2
H3 = (z*(z - cos(w0)))/(z^2 - 2*cos(w0)*z + 1)
H4 =(10*z*((10*z)/9 - cos(w0)))/(9*((100*z^2)/81 - (20*cos(w0)*z)/9 + 1))
```

The function *iztrans* computes the inverse Z-transform. We will illustrate its use later on. $\qquad\square$

10.4.3 CONVOLUTION SUM AND TRANSFER FUNCTION

The most important property of the Z-transform, as it was for the Laplace transform, is the convolution property.

The output $y[n]$ of a causal LTI system is calculated using the convolution sum

$$y[n] = [x * h][n] = \sum_{k=0}^{n} x[k]h[n-k] = \sum_{k=0}^{n} h[k]x[n-k] \tag{10.18}$$

where $x[n]$ is a causal input and $h[n]$ the impulse response of the system. The Z-transform of $y[n]$ is the product

$$Y(z) = \mathcal{Z}\{[x * h][n]\} = \mathcal{Z}\{x[n]\}\mathcal{Z}\{h[n]\} = X(z)H(z) \tag{10.19}$$

and the transfer function of the system is thus defined as

$$H(z) = \frac{Y(z)}{X(z)} = \frac{\mathcal{Z}[\text{ output } y[n]]}{\mathcal{Z}[\text{ input } x[n]]}, \tag{10.20}$$

i.e., $H(z)$ transfers the input $X(z)$ into the output $Y(z)$.

The convolution sum property can be seen as a way to obtain the coefficients of the product of two polynomials. Whenever we multiply two polynomials $X_1(z)$ and $X_2(z)$, of finite or infinite order, the coefficients of the resulting polynomial can be obtained by means of the convolution sum. For instance, consider

$$X_1(z) = 1 + a_1 z^{-1} + a_2 z^{-2} \quad \text{and} \quad X_2(z) = 1 + b_1 z^{-1};$$

their product is

$$
\begin{aligned}
X_1(z)X_2(z) &= 1 + b_1 z^{-1} + a_1 z^{-1} + a_1 b_1 z^{-2} + a_2 z^{-2} + a_2 b_1 z^{-3} \\
&= 1 + (b_1 + a_1)z^{-1} + (a_1 b_1 + a_2)z^{-2} + a_2 b_1 z^{-3}.
\end{aligned}
$$

The convolution sum of the two sequences $[1 \ a_1 \ a_2]$ and $[1 \ b_1]$, formed by the coefficients of $X_1(z)$ and $X_2(z)$, is $[1 \ (a_1 + b_1) \ (a_2 + b_1 a_1) \ a_2]$ corresponding to the coefficients of the product of the polynomials $X_1(z)X_2(z)$. Also notice that the sequence of length 3 (corresponding to the first-order polynomial $X_1(z)$) and the sequence of length 2 (corresponding to the second-order polynomial $X_2(z)$) when convolved give a sequence of length $3 + 2 - 1 = 4$ (corresponding to a third-order polynomial $X_1(z)X_2(z)$).

A finite impulse response or FIR filter is implemented by means of the convolution sum. Consider an Nth-order FIR with an input/output equation

$$
y[n] = \sum_{k=0}^{N-1} b_k x[n-k] \tag{10.21}
$$

where $x[n]$ is the input and $y[n]$ the output. The impulse response of this filter is (let $x[n] = \delta[n]$, set initial conditions to zero, so that $y[n] = h[n]$)

$$
h[n] = \sum_{k=0}^{N-1} b_k \delta[n-k]
$$

giving $h[n] = b_n$ for $n = 0, \cdots, N-1$ and zero otherwise. Accordingly, we can write Equation (10.21) as

$$
y[n] = \sum_{k=0}^{N-1} h[k]x[n-k],
$$

which is the convolution of the input $x[n]$ and the impulse response $h[n]$ of the FIR filter. Thus, if $X(z) = \mathcal{Z}(x[n])$ and $H(z) = \mathcal{Z}(h[n])$ then

$$
Y(z) = H(z)X(z) \quad \text{and} \quad y[n] = \mathcal{Z}^{-1}[Y(z)].
$$

- The length of the convolution of two sequences of lengths M and N is $M + N - 1$.
- If one of the sequences is of infinite length, the length of the convolution is infinite. Thus, for infinite impulse response IIR or recursive filters the output is always of infinite length for any input signal, given that the impulse response of these filters is of infinite length.

Example 10.8. Consider computing the output of an FIR filter

$$
y[n] = \frac{1}{2}(x[n] + x[n-1] + x[n-2])
$$

for an input $x[n] = u[n] - u[n-4]$ using the convolution sum, analytically and graphically, and the Z-transform.

Solution: The impulse response is $h[n] = 0.5(\delta[n] + \delta[n-1] + \delta[n-2])$, so that $h[0]$, $h[1]$, $h[2]$ are 0.5, 0.5, 0.5 and $h[n]$ is zero otherwise.

<u>Analytic convolution sum:</u> The equation

$$y[n] = \sum_{k=0}^{n} x[k]h[n-k] = x[0]h[n] + x[1]h[n-1] + \cdots + x[n]h[0] \quad n \geq 0$$

with the condition that in each entry the arguments of $x[.]$ and $h[.]$ add to $n \geq 0$, gives

$$
\begin{aligned}
y[0] &= x[0]h[0] = 0.5, \\
y[1] &= x[0]h[1] + x[1]h[0] = 1, \\
y[2] &= x[0]h[2] + x[1]h[1] + x[2]h[0] = 1.5, \\
y[3] &= x[0]h[3] + x[1]h[2] + x[2]h[1] + x[3]h[0] = x[1]h[2] + x[2]h[1] + x[3]h[0] = 1.5, \\
y[4] &= x[0]h[4] + x[1]h[3] + x[2]h[2] + x[3]h[1] + x[4]h[0] = x[2]h[2] + x[3]h[1] = 1, \\
y[5] &= x[0]h[5] + x[1]h[4] + x[2]h[3] + x[3]h[2] + x[4]h[1] + x[5]h[0] = x[3]h[2] = 0.5,
\end{aligned}
$$

and the rest are zero. In the above computation, we notice that the length of $y[n]$ is $4 + 3 - 1 = 6$ since the length of $x[n]$ is 4, and that of $h[n]$ is 3.

<u>Graphical approach:</u> See Fig. 10.4. The convolution sum is given by

$$y[n] = \sum_{k=0}^{n} x[k]h[n-k] = \sum_{k=0}^{n} h[k]x[n-k].$$

Choosing one of these equations, let us say the first one, we need $x[k]$ and $h[n-k]$, as functions of k, for different values of n, multiply them and then add the nonzero values. For instance, for $n = 0$ the sequence $h[-k]$ is the reflection of $h[k]$, multiplying $x[k]$ by $h[-k]$ gives only one value different from zero at $k = 0$, or $y[0] = 1/2$. For $n = 1$, the sequence $h[1-k]$, as a function of k, is $h[-k]$ shifted to the right one sample. Multiplying $x[k]$ by $h[1-k]$ gives two values different from zero, which when added gives $y[1] = 1$, and so on. For increasing values of n we shift to the right one sample to get $h[n-k]$, multiply it by $x[k]$ and then add the nonzero values to obtain the output $y[n]$.

<u>Convolution sum property:</u> We have

$$X(z) = 1 + z^{-1} + z^{-2} + z^{-3}, \quad H(z) = \frac{1}{2}[1 + z^{-1} + z^{-2}],$$

and according to the convolution sum property

$$Y(z) = X(z)H(z) = \frac{1}{2}(1 + 2z^{-1} + 3z^{-2} + 3z^{-3} + 2z^{-4} + z^{-5}).$$

Thus $y[0] = 0.5$, $y[1] = 1$, $y[2] = 1.5$, $y[3] = 1.5$, $y[4] = 1$, and $y[5] = 0.5$, just as before.

In MATLAB the function *conv* is used to compute the convolution sum giving the results shown in Fig. 10.5, which coincide with the ones obtained in the other approaches.   □

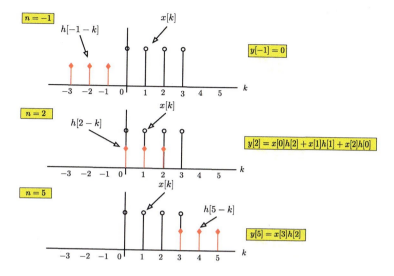

FIGURE 10.4

Graphical approach: convolution sum for $n = -1$, $n = 2$ and $n = 5$ with corresponding outputs $y[-1]$, $y[2]$ and $y[5]$. Both $x[k]$ and $h[n-k]$ are plotted as functions of k for a given value of n. The signal $x[k]$ remains stationary, while $h[n-k]$ moves linearly from left to right. Thus the convolution sum is also called a linear convolution.

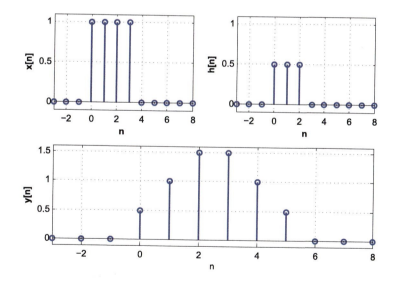

FIGURE 10.5

Convolution sum for an averager FIR: $x[n]$ (top left), $h[n]$ (top right) and $y[n]$ (bottom). The output $y[n]$ is of length 6 given that $x[n]$ is of length 4 and $h[n]$ is the impulse response of a second-order FIR filter of length 3.

Example 10.9. Consider an FIR filter with impulse response

$$h[n] = \delta[n] + \delta[n-1] + \delta[n-2].$$

Find the filter output for an input $x[n] = \cos(2\pi n/3)(u[n] - u[n-14])$. Use the convolution sum to find graphically the output, and the convolution property of the Z-transform.

Solution:

Graphical approach: Let us use the formula

$$y[n] = \sum_{k=0}^{n} x[k]h[n-k],$$

which keeps the more complicated signal $x[k]$ as the unchanged signal. The term $h[n-k]$ is $h[k]$ reversed for $n = 0$, and then shifted to the right for $n \geq 1$. The output is zero for negative values of n, and for $n \geq 0$ we have

$$\begin{aligned}
y[0] &= 1, & y[1] &= 0.5, \\
y[n] &= 0, & 2 &\leq n \leq 13, \\
y[14] &= 0.5, & y[15] &= -0.5,
\end{aligned} \tag{10.22}$$

and zero otherwise. The first value is obtained by reflecting the impulse response to get $h[-k]$ and when multiplied by $x[k]$ we only have the value at $k = 0$ different from zero, or 1. As we shift the impulse response to the right to get $h[1-k]$, for $n = 1$, and multiply it by $x[k]$ we get two values different from zero, and when added equal 0.5. The result for $2 \leq n \leq 13$ is zero because we add 3 values -0.5, 1, and -0.5 from the cosine.

Convolution property approach: In the following script we find the Z-transform $X(z)$ and $H(z)$ and multiply them to get $Y(z)$. Notice the two ways we compute the Z-transforms of $x(n)$ and $h(n)$. The *kroneckerDelta( )* is the discrete delta function $\delta[]$.

```
clear all; clf
syms z Y X H y n x
w0=2*pi/3;x1(1:14)=cos(w0*[0:13])
x=x1(1)*kroneckerDelta(n,0);
for m=2:14,
    x=x+x1(m)*kroneckerDelta(n-m+1,0);
end
X=ztrans(x);
h=[1 1 1];
H=h(1);
for m=1:2,
    H=H+h(m)/z^m;
end
Y=H*X; y=iztrans(Y);
```

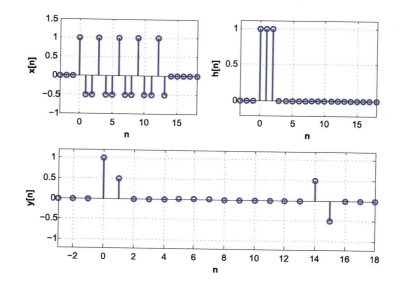

FIGURE 10.6

Convolution sum for FIR filter: $x[n]$ (top left), $h[n]$ (top right) and $y[n]$ (bottom).

which gives

$$Y(z) = \underbrace{1}_{y[0]} + \underbrace{0.5}_{y[1]} z^{-1} + \underbrace{0.5}_{y[14]} z^{-14} \underbrace{-0.5}_{y[15]} z^{-15},$$

where the coefficients coincide with the $y(n)$ values obtained by the graphical method (see (10.22)) and by the inverse Z-transform. To validate this result we use the function *conv* as shown in the continuation of the above script:

```
y1=conv(x1,h)
n=0:15;
figure(1)
subplot(221)
stem([0:13],x1);grid;axis([0 13 -0.6 1.1])
subplot(222)
stem([0:2],h);axis([0 2 -0.6 1.1]);grid
subplot(223)
stem(n,y1);grid;axis([0 15 -0.6 1.1])
```

These results are shown in Fig. 10.6 (the cosine does not look like a sampled cosine given that only three values are used per period).

Notice from this example that:

- the convolution sum is simply calculating the coefficients of the polynomial product $X(z)H(z)$,
- the length of the convolution sum $y[n] = $ length of $x[n] + $ length of $h[n] - 1 = 14 + 3 - 1 = 16$, i.e., $Y(z)$ is a polynomial of order 15.   □

Example 10.10. For noncausal input or impulse response, the convolution sum is more complicated graphically than when both input and impulse response are causal. Let

$$h_1[n] = \frac{1}{3}(\delta[n+1] + \delta[n] + \delta[n-1])$$

be the impulse response of a noncausal averager FIR filter, and $x[n] = u[n] - u[n-4]$ be the input. Indicate how to compute the filter output using the convolution property.

Solution: Graphically it is a bit confusing to plot $h_1[n-k]$, as a function of k, to do the convolution sum. Using the convolution and the time-shifting properties of the Z-transform we can view the computation more clearly. According to the convolution property, the Z-transform of the output of the noncausal filter is

$$Y_1(z) = X(z)H_1(z) = X(z)[zH(z)] \tag{10.23}$$

where we let

$$H_1(z) = \mathcal{Z}[h_1[n]] = \frac{1}{3}(z + 1 + z^{-1}) = z\left[\frac{1}{3}(1 + z^{-1} + z^{-2})\right] = zH(z)$$

where $H(z) = (1/3)\mathcal{Z}(\delta[n] + \delta[n-1] + \delta[n-2])$ is the transfer function of a causal filter. Letting $Y(z) = X(z)H(z)$ be the Z-transform of the convolution sum $y[n] = [x * h][n]$ of $x[n]$ and $h[n]$ both of which are causal and can be calculated as before. According to Equation (10.23), we then have $Y_1(z) = zY(z)$ or $y_1[n] = [x * h_1][n] = y[n+1]$. $\square$

Let $x_1[n]$ be the input to a noncausal LTI system, with an impulse response $h_1[n]$ such that $h_1[n] = 0$ for $n < N_1 < 0$. Assume $x_1[n]$ is also noncausal, i.e., $x_1[n] = 0$ for $n < N_0 < 0$. The output $y_1[n] = [x_1 * h_1][n]$ has a Z-transform

$$Y_1(z) = X_1(z)H_1(z) = [z^{N_0}X(z)][z^{N_1}H(z)]$$

where $X(z)$ and $H(z)$ are the Z-transforms of a causal signal $x[n]$ and of a causal impulse response $h[n]$. If we let

$$y[n] = [x * h][n] = \mathcal{Z}^{-1}[X(z)H(z)]$$

then

$$y_1[n] = [x_1 * h_1][n] = y[n + N_0 + N_1].$$

Remarks

1. For an IIR system represented by a difference equation, its impulse response $h[n]$ is found by setting the initial conditions to zero, and as such the transfer function $H(z)$ also requires a similar condition. If the initial conditions are not zero, the Z-transform of the total response $Y(z)$ is the sum of the Z-transforms of the zero-state and the zero-input responses, i.e., its Z-transform is of the form

$$Y(z) = \frac{X(z)B(z)}{A(z)} + \frac{I_0(z)}{A(z)} \tag{10.24}$$

and it does not permit us to compute the ratio $Y(z)/X(z)$ unless the component due to the initial conditions is zero, i.e., $I_0(z) = 0$.

2. It is important to remember the following relations:

$$H(z) = \mathcal{Z}[h[n]] = \frac{Y(z)}{X(z)} = \frac{\mathcal{Z}[y[n]]}{\mathcal{Z}[x[n]]}$$

where $H(z)$ is the transfer function and $h[n]$ the impulse response of the system, with $x[n]$ as input and $y[n]$ as output.

Example 10.11. Consider a discrete-time IIR system represented by the difference equation

$$y[n] = 0.5y[n-1] + x[n] \tag{10.25}$$

with $x[n]$ as input and $y[n]$ as output. Determine the transfer function of the system and from it find the impulse and the unit-step responses. Determine under what conditions the system is BIBO stable. If stable, determine the transient and steady state responses of the system.

Solution: The system transfer function is given by

$$H(z) = \frac{Y(z)}{X(z)} = \frac{1}{1 - 0.5z^{-1}}$$

and its impulse response is

$$h[n] = \mathcal{Z}^{-1}[H(z)] = 0.5^n u[n].$$

The response of the system to any input can be easily obtained by the transfer function. If the input is $x[n] = u[n]$, we have

$$Y(z) = H(z)X(z) = \frac{1}{1 - 0.5z^{-1}} \frac{1}{1 - z^{-1}} = \frac{1}{(1 - 0.5z^{-1})(1 - z^{-1})} = \frac{-1}{1 - 0.5z^{-1}} + \frac{2}{1 - z^{-1}}$$

so that the total solution is

$$y[n] = -0.5^n u[n] + 2u[n].$$

From the transfer function $H(z)$ of the LTI system, we can test the stability of the system by finding the location of its poles—very much like in the analog case:

$$H(z) = \frac{1}{1 - 0.5z^{-1}} = \frac{z}{z - 0.5}$$

has a pole at $z = 0.5$, inside the unit circle. Thus the system is BIBO stable. Equivalently, the system is BIBO stable if and only if the impulse response of the system is absolutely summable, i.e., $\sum_n |h[n]| \leq \infty$, which is the case here, indeed

$$\sum_{n=0}^{\infty} |0.5^n| = \sum_{n=0}^{\infty} 0.5^n = \frac{1}{1 - 0.5} = 2.$$

Since the system is stable, its transient and steady-state responses exist. As $n \to \infty$, $y[n] = 2$, which is the steady-state response, and $-0.5^n u[n]$ is the transient solution. □

Example 10.12. An FIR system has the following input/output equation:

$$y[n] = \frac{1}{3}(x[n] + x[n-1] + x[n-2])$$

where $x[n]$ is the input and $y[n]$ the output. Determine the transfer function and the impulse response of the system, and from them indicate whether the system is BIBO stable or not.

Solution: The transfer function is

$$H(z) = \frac{Y(z)}{X(z)} = \frac{1}{3}[1 + z^{-1} + z^{-2}] = \frac{z^2 + z + 1}{3z^2},$$

and the corresponding impulse response is

$$h[n] = \frac{1}{3}(\delta[n] + \delta[n-1] + \delta[n-2]).$$

The impulse response of this system only has three nonzero values $h[0] = h[1] = h[2] = 1/3$, the rest of the values are zero. As such, $h[n]$ is absolutely summable and the filter is BIBO stable. FIR filters are always BIBO stable given that their impulse responses will be absolutely summable, due to their final support, and equivalently because the poles of the transfer function of these systems are at the origin of the Z-plane, very much inside the unit circle. □

Non-recursive or FIR Systems. The impulse response $h[n]$ of an FIR or non-recursive system

$$y[n] = b_0 x[n] + b_1 x[n-1] + \cdots + b_M x[n-M]$$

has finite length and is given by

$$h[n] = b_0 \delta[n] + b_1 \delta[n-1] + \cdots + b_M \delta[n-M].$$

Its transfer function is

$$
\begin{aligned}
H(z) &= \frac{Y(z)}{X(z)} = b_0 + b_1 z^{-1} + \cdots + b_M z^{-M} \\
&= \frac{b_0 z^M + b_1 z^{M-1} + \cdots + b_M}{z^M}
\end{aligned}
$$

with all its poles at the origin $z = 0$ (multiplicity M) and as such the system is BIBO stable.

Recursive or IIR Systems. The impulse response $h[n]$ of an IIR or recursive system

$$y[n] = -\sum_{k=1}^{N} a_k y[n-k] + \sum_{m=0}^{M} b_m x[n-m], \qquad n \geq 0$$

has (possible) infinite length and is given by

$$h[n] \quad = \quad \mathcal{Z}^{-1}[H(z)] = \mathcal{Z}^{-1}\left[\frac{\sum_{m=0}^{M} b_m z^{-m}}{1 + \sum_{k=1}^{N} a_k z^{-k}}\right] = \mathcal{Z}^{-1}\left[\frac{B(z)}{A(z)}\right]$$

where $H(z)$ is the transfer function of the system. If the poles of $H(z)$ are inside the unit circle, or $A(z) \neq 0$ for $|z| \geq 1$, the system is BIBO stable.

10.4.4 INTERCONNECTION OF DISCRETE-TIME SYSTEMS

Just like with analog systems, two discrete-time LTI systems with transfer functions $H_1(z)$ and $H_2(z)$ (or with impulse responses $h_1[n]$ and $h_2[n]$) can be connected in cascade, parallel or in feedback. The first two forms result from properties of the convolution sum.

The transfer function of the cascading of the two LTI systems is

$$H(z) = H_1(z)H_2(z) = H_2(z)H_1(z), \tag{10.26}$$

showing that there is no effect on the overall system if we interchange the two systems (see Fig. 10.7A)—a property that is only valid for LTI systems.

In the parallel system, Fig. 10.7B, both systems have the same input and the output is the sum of the output of the subsystems. The overall transfer function is

$$H(z) = H_1(z) + H_2(z). \tag{10.27}$$

Finally, the negative feedback connection of the two systems shown in Fig. 10.7C gives in the feed-forward path

$$Y(z) = H_1(z)E(z) \tag{10.28}$$

where $Y(z) = \mathcal{Z}[y[n]]$ is the Z-transform of the output $y[n]$ and $E(z) = X(z) - W(z)$ is the Z-transform of the error function $e[n] = x[n] - w[n]$. The feedback path gives that

$$W(z) = \mathcal{Z}[w[n]] = H_2(z)Y(z).$$

Replacing $W(z)$ in $E(z)$, and then replacing $E(z)$ in (10.28) we obtain the overall transfer function

$$H(z) = \frac{Y(z)}{X(z)} = \frac{H_1(z)}{1 + H_1(z)H_2(z)}. \tag{10.29}$$

10.4.5 INITIAL- AND FINAL-VALUE PROPERTIES

In some control applications and to check the correctness of a partial fraction expansion, it is useful to find the initial or the final value of a discrete-time signal $x[n]$ from its Z-transform. These values can be found as follows:

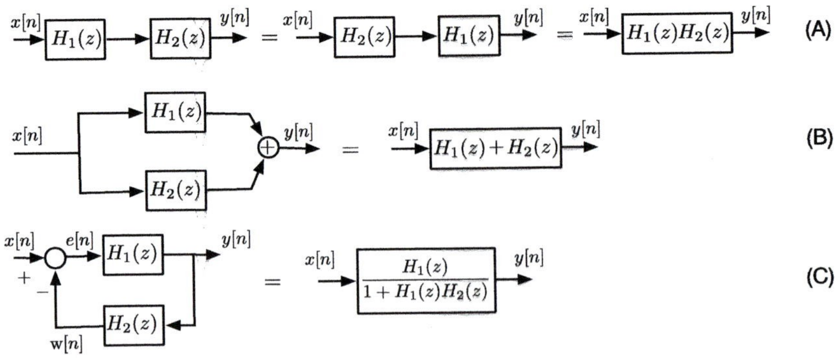

FIGURE 10.7

Connections of LTI systems: (A) cascade, (B) parallel, and (C) negative feedback.

If $X(z)$ is the Z-transform of a causal signal $x[n]$ with an initial value $x[0]$ and a final value $\lim_{n\to\infty} x[n]$ are obtained from $X(z)$ according to:

$$\text{Initial value:} \quad x[0] = \lim_{z\to\infty} X(z),$$

$$\text{Final value:} \quad \lim_{n\to\infty} x[n] = \lim_{z\to 1} (z-1)X(z). \tag{10.30}$$

The initial-value result results from the definition of the one-sided Z-transform, i.e.,

$$\lim_{z\to\infty} X(z) = \lim_{z\to\infty}\left(x[0] + \sum_{n\geq 1} \frac{x[n]}{z^n} \right) = x[0].$$

To show the final-value result, we have

$$(z-1)X(z) = \sum_{n=0}^{\infty} x[n]z^{-n+1} - \sum_{n=0}^{\infty} x[n]z^{-n} = x[0]z + \sum_{n=0}^{\infty} [x[n+1] - x[n]]z^{-n}$$

and thus the limit

$$\begin{aligned}
\lim_{z\to 1}(z-1)X(z) &= x[0] + \sum_{n=0}^{\infty}(x[n+1] - x[n]) \\
&= x[0] + (x[1] - x[0]) + (x[2] - x[1]) + (x[3] - x[2]) \cdots \\
&= \lim_{n\to\infty} x[n],
\end{aligned}$$

given that the entries in the sum cancel out as n increases, leaving $x[\infty]$.

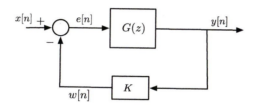

FIGURE 10.8

Negative feedback system with plant $G(z)$.

Example 10.13. Consider a negative feedback connection of a plant with a transfer function

$$G(z) = 1/(1 - 0.5z^{-1})$$

and a constant feedback gain K (see Fig. 10.8). If the reference signal is a unit-step, $x[n] = u[n]$, determine the behavior of the error signal $e[n]$. What is the effect of the feedback, from the error point of view, on an unstable plant $G(z) = 1/(1 - z^{-1})$?

Solution: For $G(z) = 1/(1 - 0.5z^{-1})$, the Z-transform of the error signal is

$$E(z) = X(z) - W(z) = X(z) - KG(z)E(z),$$

and for $X(z) = 1/(1 - z^{-1})$

$$E(z) = \frac{X(z)}{1 + KG(z)} = \frac{1}{(1 - z^{-1})(1 + KG(z))}.$$

The initial value of the error signal is then

$$e[0] = \lim_{z \to \infty} E(z) = \frac{1}{1 + K},$$

since $G(\infty) = 1$.

The steady state or final value of the error is

$$
\begin{aligned}
\lim_{n \to \infty} e[n] &= \lim_{z \to 1}(z - 1)E(z) = \lim_{z \to 1}\frac{(z - 1)X(z)}{1 + KG(z)} \\
&= \lim_{z \to 1}\frac{z(z - 1)}{(z - 1)(1 + KG(z))} = \frac{1}{1 + 2K},
\end{aligned}
$$

since $G(1) = 2$. If we want the steady-state error to go to zero, then K must be large. In that case, the initial error is also zero.

If $G(z) = 1/(1 - z^{-1})$, i.e., the plant is unstable, the initial value of the error function remains the same, $e[0] = 1/(1 + K)$ but the steady-state error goes to zero since $G(1) \to \infty$. $\square$

10.5 ONE-SIDED Z-TRANSFORM INVERSE

Different from the inverse Laplace transform, which was done mostly by the partial fraction expansion, the inverse Z-transform can be done in different ways. For instance, if the Z-transform is given as a finite-order polynomial, the inverse can be found by inspection. Indeed, if the given Z-transform is

$$X(z) = \sum_{n=0}^{N} x[n]z^{-n} = x[0] + x[1]z^{-1} + x[2]z^{-2} + \cdots + x[N]z^{-N} \qquad (10.31)$$

by the definition of the Z-transform, $x[k]$ is the coefficient of the monomial z^{-k}, for $k = 0, 1, \cdots, N$, thus the inverse Z-transform is given by the sequence $\{x[0], x[1], \cdots x[n]\}$. For instance, if we have a Z-transform

$$X(z) = 1 + 2z^{-10} + 3z^{-20},$$

the inverse is a sequence

$$x[n] = \delta[n] + 2\delta[n - 10] + 3\delta[n - 20]$$

so that $x[0] = 1$, $x[10] = 2$, $x[20] = 3$ and $x[n] = 0$ for $n \neq 0$, 10, 20. In this case it makes sense to do this because N is finite, but if $N \to \infty$, this way of finding the inverse Z-transform might not be very practical. In that case, the **long-division** method and the **partial fraction expansion** method, which we consider next, are more appropriate. In this section we will consider the inverse of one-sided Z-transforms, in the next section we consider the inverse of two-sided transforms.

10.5.1 LONG-DIVISION METHOD

When a rational function $X(z) = B(z)/A(z)$, having as ROC the outside of a circle of radius R (i.e., $x[n]$ is causal), is expressed as

$$X(z) = x[0] + x[1]z^{-1} + x[2]z^{-2} + \cdots$$

by dividing $B(z)$ by $A(z)$, then the inverse is the sequence $\{\cdots 0, 0, x[0], x[1], x[2], \cdots\}$ or

$$x[n] = x[0]\delta[n] + x[1]\delta[n - 1] + x[2]\delta[n - 2] + \cdots.$$

Thus to find the inverse we simply divide the polynomial $B(z)$ by $A(z)$ to obtain a possible infinite-order polynomial in negative powers of z^{-1}. The coefficients of this polynomial are the inverse values. The disadvantage of this method is that it does not provide a closed-form solution, unless there is a clear connection between the terms of the sequence. But this method is useful when we are interested in finding some of the initial values of the sequence $x[n]$.

Example 10.14. Find the inverse Z-transform of

$$X(z) = \frac{1}{1 + 2z^{-2}} \qquad |z| > \sqrt{2}.$$

Solution: We can perform the long division to find the $x[n]$ values, or equivalently let

$$X(z) = x[0] + x[1]z^{-1} + x[2]z^{-2} + \cdots,$$

and find the $\{x[n]\}$ samples so that the product $X(z)(1 + 2z^{-2}) = 1$. Thus

$$
\begin{aligned}
1 &= (1 + 2z^{-2})(x[0] + x[1]z^{-1} + x[2]z^{-2} + \cdots) \\
&= x[0] + x[1]z^{-1} + x[2]z^{-2} + x[3]z^{-3} + \cdots + 2x[0]z^{-2} + 2x[1]z^{-3} + \cdots \\
&= x[0] + x[1]z^{-1} + (x[2] + 2x[0])z^{-2} + (x[3] + 2x[1])z^{-3} + \cdots
\end{aligned}
$$

and comparing the terms on the two sides of the equality gives

$$x[0] = 1, \quad x[1] = 0, \quad x[2] = -2x[0] = -2, \quad x[3] = 2x[1] = 0, \quad x[4] = (-2)^2, \quad \cdots .$$

So the inverse Z-transform is $x[0] = 1$ and $x[n] = (-2)^{\log_2(n)}$ for $n > 0$ and even, and zero otherwise. Notice that this sequence grows as $n \to \infty$.

Another possible way to find the inverse is to use the geometric series equation,

$$\sum_{k=0}^{\infty} \alpha^n = \frac{1}{1 - \alpha} \qquad |\alpha| < 1$$

with $-\alpha = 2z^{-2}$ (notice that $|\alpha| = 2/|z|^2 < 1$ or $|z| > \sqrt{2}$, the given ROC). So that

$$X(z) = \frac{1}{1 + 2z^{-2}} = 1 + (-2z^{-2})^1 + (-2z^{-2})^2 + (-2z^{-2})^3 + \cdots,$$

but this method is not as general as the long-division method. $\qquad\square$

10.5.2 PARTIAL FRACTION EXPANSION

The basics of partial fraction expansion remain the same for the Z-transform as for the Laplace transform. A rational function is a ratio of polynomials $N(z)$ and $D(z)$ in z or z^{-1}

$$X(z) = \frac{N(z)}{D(z)}.$$

The poles of $X(z)$ are the roots of $D(z) = 0$ and the zeros of $X(z)$ are the roots of the equation $N(z) = 0$.

Remarks

1. The basic characteristic of the partial fraction expansion is that $X(z)$ must be a **proper rational function**, or that the degree of the numerator polynomial $N(z)$ be smaller than the degree of the denominator polynomial $D(z)$ (assuming both $N(z)$ and $D(z)$ are polynomials in either z^{-1} or z). If this condition is not satisfied, we perform long division until the residue polynomial is of degree less than that of the denominator.

2. It is more common in the Z-transform than in the Laplace transform to find that the numerator and the denominator are of the same degree—$\delta[n]$ is not as unusual as the analog impulse function $\delta(t)$.
3. The partial fraction expansion is generated, from the poles of the proper rational function, as a sum of terms whose inverse Z-transforms are easily found in a table of Z-transforms. By plotting the poles and zeros of a proper $X(z)$, the location of the poles provides a general form of the inverse within some constants that are found from the poles and the zeros.
4. Given that the numerator and the denominator polynomials of a proper rational function $X(z)$ can be expressed in terms of positive and negative powers of z, it is possible to do partial fraction expansions in either z or z^{-1}. We will see that the partial fraction expansion in negative powers is more like the partial fraction expansion in the Laplace transform, and as such we will prefer it. Partial fraction expansion in positive powers of z requires more care.

Example 10.15. Consider the non-proper rational function

$$X(z) = \frac{2 + z^{-2}}{1 + 2z^{-1} + z^{-2}}$$

(numerator and denominator of the same degree in powers of z^{-1}). Determine how to obtain an expansion of $X(z)$ containing a proper rational term to find $x[n]$.

Solution: By division we obtain

$$X(z) = 1 + \frac{1 - 2z^{-1}}{1 + 2z^{-1} + z^{-2}}$$

where the second term is proper rational as the denominator is of higher degree in powers of z^{-1} than the numerator. The inverse Z-transform of $X(z)$ will then be

$$x[n] = \delta[n] + \mathcal{Z}^{-1}\left[\frac{1 - 2z^{-1}}{1 + 2z^{-1} + z^{-2}}\right].$$

The inverse of the proper rational term is done as indicated in this section.   □

Example 10.16. Find the inverse Z-transform of

$$X(z) = \frac{1 + z^{-1}}{(1 + 0.5z^{-1})(1 - 0.5z^{-1})} = \frac{z(z + 1)}{(z + 0.5)(z - 0.5)} \qquad |z| > 0.5$$

by using the negative and the positive powers of z representations.

Solution: Clearly $X(z)$ is proper if it is considered a function of negative powers z^{-1} (in z^{-1}, the numerator is of degree 1 and the denominator of degree 2), but it is not if considered a function of positive powers z (numerator and denominator are both of degree 2). It is, however, unnecessary to perform long division to make $X(z)$ proper when considered as a function of z, one simple approach is to consider $X(z)/z$ as the function we wish to find its partial fraction expansion, i.e.,

$$\frac{X(z)}{z} = \frac{z + 1}{(z + 0.5)(z - 0.5)}, \tag{10.32}$$

Table 10.1 One-sided Z-transforms

$\delta[n]$	1, whole z-plane
$u[n]$	$\dfrac{1}{1 - z^{-1}}$, $\quad |z| > 1$
$nu[n]$	$\dfrac{z^{-1}}{(1 - z^{-1})^2}$, $\quad |z| > 1$
$n^2 u[n]$	$\dfrac{z^{-1}(1 + z^{-1})}{(1 - z^{-1})^3}$, $\quad |z| > 1$
$\alpha^n u[n]$, $\quad |\alpha| < 1$	$\dfrac{1}{1 - \alpha z^{-1}}$, $\quad |z| > |\alpha|$
$n\alpha^n u[n]$, $\quad |\alpha| < 1$	$\dfrac{\alpha z^{-1}}{(1 - \alpha z^{-1})^2}$, $\quad |z| > |\alpha|$
$\cos(\omega_0 n)u[n]$	$\dfrac{1 - \cos(\omega_0)z^{-1}}{1 - 2\cos(\omega_0)z^{-1} + z^{-2}}$, $\quad |z| > 1$
$\sin(\omega_0 n)u[n]$	$\dfrac{\sin(\omega_0)z^{-1}}{1 - 2\cos(\omega_0)z^{-1} + z^{-2}}$, $\quad |z| > 1$
$\alpha^n \cos(\omega_0 n)u[n]$, $\quad |\alpha| < 1$	$\dfrac{1 - \alpha\cos(\omega_0)z^{-1}}{1 - 2\alpha\cos(\omega_0)z^{-1} + \alpha^2 z^{-2}}$, $\quad |z| > |\alpha|$
$\alpha^n \sin(\omega_0 n)u[n]$, $\quad |\alpha| < 1$	$\dfrac{\alpha\sin(\omega_0)z^{-1}}{1 - 2\alpha\cos(\omega_0)z^{-1} + \alpha^2 z^{-2}}$, $\quad |z| > |\alpha|$

which is proper. Thus whenever $X(z)$, as a function of z terms, is not proper it is always possible to divide it by some power in z to make it proper. After obtaining the partial fraction expansion then the z term is put back.

Consider then the partial fraction expansion in z^{-1} terms,

$$X(z) = \frac{1 + z^{-1}}{(1 + 0.5z^{-1})(1 - 0.5z^{-1})} = \frac{A}{1 + 0.5z^{-1}} + \frac{B}{1 - 0.5z^{-1}}.$$

Given that the poles are real, one at $z = -0.5$ and the other at $z = 0.5$, from the Z-transform Table 10.1 we get that a general form of the inverse is

$$x[n] = (A\,(-0.5)^n + B\,0.5^n)\,u[n].$$

The A and B coefficients can be found (by analogy with the Laplace transform partial fraction expansion) as

$$\begin{aligned} A &= X(z)(1 + 0.5z^{-1})|_{z^{-1} = -2} = -0.5, \\ B &= X(z)(1 - 0.5z^{-1})|_{z^{-1} = 2} = 1.5, \end{aligned}$$

so that

$$x[n] = (-0.5(-0.5)^n + 1.5(0.5)^n)\,u[n].$$

Table 10.2 Basic properties of one-sided Z-transform

Causal signals	$\alpha x[n], \; \beta y[n]$	$\alpha X(z), \; \beta Y(z)$
Linearity	$\alpha x[n] + \beta y[n]$	$\alpha X(z) + \beta Y(z)$
Convolution sum	$\displaystyle\sum_k x[n]y[n-k]$	$X(z)Y(z)$
Time-shifting	$x[n-N], \; N > 0$	$z^{-N}X(z) + x[-1]z^{-N+1}$ $+ x[-2]z^{-N+2} + \cdots + x[-N]$
Time reversal	$x[-n]$	$X(z^{-1})$
Multiplication	$n\,x[n]$	$-z\dfrac{dX(z)}{dz}$
	$n^2\,x[n]$	$z^2\dfrac{d^2X(z)}{dz^2} + z\dfrac{dX(z)}{dz}$
Finite difference	$x[n] - x[n-1]$	$(1-z^{-1})X(z) - x[-1]$
Accumulation	$\displaystyle\sum_{k=0}^{n} x[k]$	$\dfrac{X(z)}{1-z^{-1}}$
Initial value	$x[0]$	$\displaystyle\lim_{z\to\infty} X(z)$
Final value	$\displaystyle\lim_{n\to\infty} x[n]$	$\displaystyle\lim_{z\to 1}(z-1)X(z)$

Consider then the partial fraction expansion in positive powers of z. From Equation (10.32) the proper rational function $X(z)/z$ can be expanded as

$$\frac{X(z)}{z} = \frac{z+1}{(z+0.5)(z-0.5)} = \frac{C}{z+0.5} + \frac{D}{z-0.5}.$$

The values of C and D are obtained as follows:

$$C = \frac{X(z)}{z}(z+0.5)|_{z=-0.5} = -0.5,$$

$$D = \frac{X(z)}{z}(z-0.5)|_{z=0.5} = 1.5.$$

We then have

$$X(z) = \frac{-0.5z}{z+0.5} + \frac{1.5z}{z-0.5},$$

which according to Table 10.1 (if entries are in negative powers of z convert them into positive powers of z) we get

$$x[n] = (-0.5(-0.5)^n + 1.5(0.5)^n)\, u[n],$$

which coincides with the above result. Also expressing $X(z)$ in negative powers of z we obtain the exact same partial fraction as before.

Noticing that the obtained $x[n]$ has finite initial and final values (see Table 10.2), two simple checks on our result are given by obtaining these values from $X(z)$ and comparing them with the ones from the $x[n]$ just obtained. For the initial value,

$$x[0] = 1 = \lim_{z \to \infty} X(z)$$

and

$$\lim_{n \to \infty} x[n] = \lim_{z \to 1} (z - 1) X(z) = 0.$$

Check both. It is important to recognize that these two checks do not guarantee that we did not make mistakes in computing the inverse, but if the initial or the final values were not to coincide with our results, our inverse would be wrong. ☐

10.5.3 INVERSE Z-TRANSFORM WITH MATLAB

Symbolic MATLAB can be used to compute the inverse one-side Z-transform. The function *iztrans* provides the sequence that corresponds to its argument.

Example 10.17. Find the inverse Z-transforms of

$$X_1(z) = \frac{z(z+1)}{(z+0.5)(z-0.5)},$$

$$X_2(z) = \frac{2-z}{2(z-0.5)},$$

$$X_3(z) = \frac{8 - 4z^{-1}}{z^{-2} + 6z^{-1} + 8}.$$

Solution: The following script shows how to find the inverse using MATLAB:

```
%%
% Example 10.17---Inverse Z-transform
%%
syms n z
         x1=iztrans((z*(z+1))/((z+0.5)*(z-0.5)))
         x2=iztrans((2-z)/(2*(z-0.5)))
         x3=iztrans((8-4*z^(-1))/(z^(-2)+6*z^(-1)+8))

x1 =(3*(1/2)^n)/2 - ((-1/2))^n/2
x2 =(3*(1/2)^n)/2 - 2*kroneckerDelta(n, 0)
x3 =4*(-1/2)^n - 3*(-1/4)^n
```

Notice that the Z-transform can be given in positive or negative powers of z, and that when it is non-proper the function *kroneckerDelta*$(n, 0)$ corresponds to $\delta[n]$. ☐

Partial Fraction Expansion

MATLAB provides also numerical methods to find the inverse. Several numerical functions are available in MATLAB to perform partial fraction expansion of a Z-transform, and to obtain the corresponding inverse as shown below.

A. Simple Poles

Consider finding the inverse Z-transform of

$$X(z) = \frac{z(z+1)}{(z-0.5)(z+0.5)} = \frac{(1+z^{-1})}{(1-0.5z^{-1})(1+0.5z^{-1})} \qquad |z| > 0.5.$$

The MATLAB function *residuez* provides the partial fraction expansion coefficients or residues, $r[k]$, the poles $p[k]$ and the gain k corresponding to $X(z)$ when the coefficients of its denominator and of the numerator are inputted. If the numerator or the denominator are given in a factored form (as is the case of the denominator above) we need to multiply the terms to obtain the denominator polynomial. Recall that multiplication of polynomials corresponds to convolution of the polynomials coefficients. Thus to perform the multiplication of the terms in the denominator, we use the MATLAB function *conv* to obtain the coefficients of the product. The convolution of the coefficients $[1 \; -0.5]$, corresponding to the pole $p_1(z) = 1 - 0.5z^{-1}$, and $[1 \; 0.5]$ corresponding to $p_2(z) = 1 + 0.5z^{-1}$ gives the denominator coefficients.

To find the poles and zeros of $X(z)$, given the coefficients $\{b[k]\}$ and $\{a[k]\}$ of the numerator and denominator, we use the MATLAB function *roots*. To get a plot of the poles and zeros of $X(z)$, the MATLAB function *zplane*, with inputs the coefficients of the numerator and the denominator of $X(z)$ is used (conventionally, an '$\times$' is used to denote poles and an 'o' for zeros).

Two possible approaches can now be used to compute the inverse Z-transform $x[n]$, we can compute the inverse (below we call it $x_1[n]$ to differentiate it from the other possible solution which we call $x[n]$) by using the information on the partial fraction expansion (the residues or coefficients of the expansion $r[k]$ and the corresponding poles). An alternative is to use the MATLAB function *filter* which considers $X(z)$ as a transfer function, with numerator and denominator defined by the b and a vectors of coefficients. If we assume the input is a delta function, of Z-transform unity, the function *filter* computes as output the inverse Z-transform $x[n]$ (i.e., we have tricked *filter* to give us the desired result).

Example 10.18.   The following script is used to implement the generation of the terms in the numerator and denominator, to obtain the corresponding coefficients, to plot them, and to find the inverse in the two different ways indicated above. Plotting code is omitted.

```
%%
% Example 10.18---Two methods for inverse Z-transform
%%
p1=poly(0.5); p2=poly(-0.5);    % generation of terms in denominator
a=conv(p1,p2)                   % denominator coefficients
z1=poly(0); z2=poly(-1);        % generation of terms in numerator
b=conv(z1,z2)                   % numerator coefficients
z=roots(b)                      % zeros of X(z)
[r,p,k]=residuez(b,a)           % residues, poles and gain
zplane(b,a)                     % plot of poles and zeros
d=[1 zeros(1,99)];              % impulse delta[n]
x=filter(b,a,d);                % x[n] computation from filter
n=0:99;
x1=r(1)*p(1).^n+r(2)*p(2).^n;   % x[n] computation from residues
```

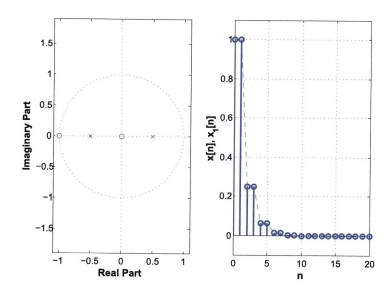

FIGURE 10.9

Poles and zeros of $X(z)$ (left) and inverse Z-transforms $x[n]$ and $x_1[n]$ found using *filter* and the residues.

```
a =1.0000          0   -0.2500
b = 1      1     0
z = 0
    -1
r =1.5000
   -0.5000
p =0.5000
   -0.5000
```

Fig. 10.9 displays the plot of the zeros and poles and the comparison between the inverses $x_1[n]$ and $x[n]$, for $0 \leq n \leq 99$, which coincide sample by sample.   □

B. Multiple Poles

Whenever multiple poles are present one has to be careful in interpreting the MATLAB results. First, use *help* to get more information on *residuez* and how the partial fraction expansion for multiple poles is done. Notice from the help file that the residues are ordered the same way the poles are. Furthermore, the residues corresponding to the multiple poles are ordered from the lowest to the highest order. Also notice the difference between the partial fraction expansion of MATLAB and ours. For instance, consider the Z-transform

$$X(z) = \frac{az^{-1}}{(1 - az^{-1})^2} \qquad |z| > a$$

with inverse $x[n] = na^n u[n]$. Writing the partial fraction expansion as MATLAB does gives

$$X(z) = \frac{r_1}{1 - az^{-1}} + \frac{r_2}{(1 - az^{-1})^2}, \qquad r_1 = -1, r_2 = 1 \qquad (10.33)$$

where the second term is not found in the Z-transform table (it is missing a first-order numerator). To write it so that each of the terms in the expansion are in the table of Z-transforms, we need to obtain values for A and B in the expansion

$$X(z) = \frac{A}{1 - az^{-1}} + \frac{Bz^{-1}}{(1 - az^{-1})^2} \qquad (10.34)$$

so that Equations (10.33) and (10.34) are equal. We find that $A = r_1 + r_2$, while $B - Aa = -r_1 a$ or $B = ar_2$. With these values we find the inverse to be

$$x[n] = [(r_1 + r_2)a^n + nr_2 a^n]u[n] = na^n u[n],$$

as expected.

Example 10.19. To illustrate the computation of the inverse Z-transform from the residues in the case of multiple poles, consider the transfer function

$$X(z) = \frac{0.5z^{-1}}{1 - 0.5z^{-1} - 0.25z^{-2} + 0.125z^{-3}}.$$

The following script is used:

```
%%
%  Example 10.19---Inverse Z-transform---multiple poles
%%
    clear all; clf
        b=[0 0.5 0 0 ];
        a=[1 -0.5 -0.25 0.125]
        [r,p,k]=residuez(b,a)
        zplane(b,a)                            % plot of poles and zeros
        n=0:99; xx=p(1).^n; yy=xx.*n;
        x1=(r(1)+r(2)).*xx+r(2).*yy+r(3)*p(3).^n;  % inverse
```

The poles and zeros, and the inverse Z-transform are shown in Fig. 10.10—there is a double pole at 0.5. The residues and the corresponding poles are

```
r =-0.2500
    0.5000
   -0.2500
p =0.5000
    0.5000
   -0.5000
```

Computationally, our method and MATLAB's are comparable but the inverse transform in our method is found directly from the table—while in the case of MATLAB's you need to change the expansion to get it into the forms found in tables.   □

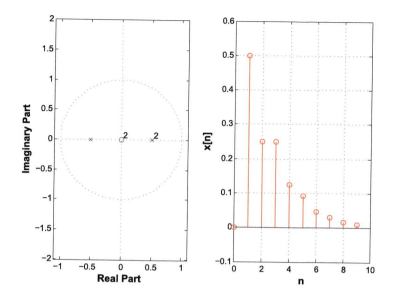

FIGURE 10.10

Poles and zeros of $X(z)$ (left) and inverse Z-transform $x[n]$ (right).

10.5.4 SOLUTION OF DIFFERENCE EQUATIONS

In this section we will use the shifting in time property of the Z-transform in the solution of difference equations with initial conditions. You will see that the partial fraction expansion used to find the inverse Z-transform is exactly like the one used in the inverse Laplace transform.

If $x[n]$ has a one-sided Z-transform $X(z)$, then $x[n - N]$ has the following one-sided Z-transform:

$$\mathcal{Z}[x[n - N]] = z^{-N} X(z) + x[-1]z^{-N+1} + x[-2]z^{-N+2} + \cdots + x[-N]. \qquad (10.35)$$

Indeed, we have

$$
\begin{aligned}
\mathcal{Z}(x[n - N]) &= \sum_{n=0}^{\infty} x[n - N]z^{-n} = \sum_{m=-N}^{\infty} x[m]z^{-(m+N)} \\
&= z^{-N} \sum_{m=0}^{\infty} x[m]z^{-m} + \sum_{m=-N}^{-1} x[m]z^{-(m+N)} \\
&= z^{-N} X(z) + x[-1]z^{-N+1} + x[-2]z^{-N+2} + \cdots + x[-N]
\end{aligned}
$$

where we first let $m = n - N$ and then separated the sum into two, one corresponding to the Z-transform of $x[n]$ multiplied by z^{-N} (the delay on the signal) and a second sum that corresponds to initial values $\{x[i], \ -N \leq i \leq -1\}$.

Remarks

1. If the signal is causal, so that $\{x[i], \ -N \leq i \leq -1\}$ are all zero, we then have $\mathcal{Z}(x[n-N]) = z^{-N}X(z)$, indicating that the operator z^{-1} is a delay operator. Thus $x[n-N]$ has been delayed N samples and its Z-transform is then simply $X(z)$ multiplied by z^{-N}.
2. The shifting in time property is useful in the solution of difference equations, especially when it has nonzero initial conditions as we will see next. On the other hand, if the initial conditions are zero either the one-sided or the two-sided Z-transforms could be used.

The analogs of ordinary differential equations are difference equations, which result directly from the modeling of a discrete system or from discretizing ordinary differential equations. The solution of ordinary differential equations requires that these equations be converted into difference equations since computers cannot perform integration. Many methods are used to solve ordinary differential equations with different degrees of accuracy and sophistication. This is a topic of numerical analysis, outside the scope of this text, and thus only simple methods are illustrated here.

Example 10.20. A discrete IIR system is represented by a first-order difference equation

$$y[n] = ay[n-1] + x[n] \qquad n \geq 0 \tag{10.36}$$

where $x[n]$ is the input of the system and $y[n]$ its output. Discuss how to solve it using recursive methods and the Z-transform. Obtain a general form for the complete solution $y[n]$ in terms of the impulse response $h[n]$ of the system. For input $x[n] = u[n] - u[n-11]$, zero initial conditions, and $a = 0.8$, use the MATLAB function *filter* to find $y[n]$. Plot the input and the output.

Solution: In the time domain a unique solution is obtained by using the recursion given by the difference equation. We would need an initial condition to compute $y[0]$, indeed

$$y[0] = ay[-1] + x[0],$$

and as $x[0]$ is given, we need $y[-1]$ as initial condition. Once we obtain $y[0]$, recursively we find the rest of the solution:

$$y[1] = ay[0] + x[1], \quad y[2] = ay[1] + x[2], \quad y[3] = ay[2] + x[3], \quad \cdots$$

where at each step the needed output values are given by the previous step of the recursion. However, this solution is not in closed form.

To obtain a closed-form solution, we use the Z-transform. Taking the one-sided Z-transform of the two sides of the equation we get

$$\begin{aligned}
\mathcal{Z}(y[n]) &= \mathcal{Z}(ay[n-1]) + \mathcal{Z}(x[n]), \\
Y(z) &= a(z^{-1}Y(z) + y[-1]) + X(z).
\end{aligned}$$

Solving for $Y(z)$ in the above equation we obtain

$$Y(z) = \frac{X(z)}{1 - az^{-1}} + \frac{ay[-1]}{1 - az^{-1}} \tag{10.37}$$

where the first term depends exclusively on the input and the second exclusively on the initial condition. If the input $x[n]$ and the initial condition $y[-1]$ are given, we can then find the inverse Z-transform to obtain the complete solution $y[n]$ of the form

$$y[n] = y_{zs}[n] + y_{zi}[n]$$

where the **zero-state response** $y_{zs}[n]$ is due exclusively to the input and zero initial conditions, and the **zero-input response** $y_{zi}[n]$ is the response due to the initial conditions with zero-input.

In this simple case we can obtain the complete solution for any input $x[n]$ and any initial condition $y[-1]$. Indeed, expressing $1/(1 - az^{-1})$ as its Z-transform sum, i.e.,

$$\frac{1}{1 - az^{-1}} = \sum_{k=0}^{\infty} a^k z^{-k},$$

Equation (10.37) becomes

$$
\begin{aligned}
Y(z) &= \sum_{k=0}^{\infty} X(z)a^k z^{-k} + ay[-1] \sum_{k=0}^{\infty} a^k z^{-k} \\
&= X(z) + aX(z)z^{-1} + a^2 X(z)z^{-2} + \cdots + ay[-1](1 + az^{-1} + a^2 z^{-2} + \cdots);
\end{aligned}
$$

using the time-shift property we then get the complete solution

$$
\begin{aligned}
y[n] &= x[n] + ax[n-1] + a^2 x[n-2] + \cdots + ay[-1](1 + a\delta[n-1] + a^2\delta[n-2] + \cdots) \\
&= \sum_{k=0}^{\infty} a^k x[n-k] + ay[-1] \sum_{k=0}^{\infty} a^k \delta[n-k]
\end{aligned}
\tag{10.38}
$$

for any input $x[n]$, initial condition $y[-1]$ and a.

To solve the difference equation with $a = 0.8$, $x[n] = u[n] - u[n-11]$ and zero initial condition $y[-1] = 0$, using MATLAB, we use the following script:

```
%%
% Example 10.20---Solution of difference equation
%%
N=100;n=0:N-1; x=[ones(1,10) zeros(1,N-10)];
den=[1 -0.8]; num=[1 0];
y=filter(num, den,x)
```

The function *filter* requires that the initial conditions be zero. The results are shown in Fig. 10.11.

Let us now find the impulse response $h[n]$ of the system. For that let $x[n] = \delta[n]$ and $y[-1] = 0$, then we have $y[n] = h[n]$ or $Y(z) = H(z)$. Thus we have

$$H(z) = \frac{1}{1 - az^{-1}} \quad \text{so that} \quad h[n] = a^n u[n].$$

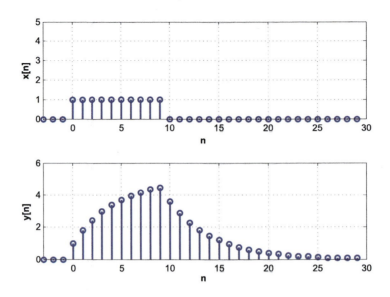

FIGURE 10.11

Solution of the first-order difference equation (bottom) with input (top).

You can then see that the first term in (10.38) is a convolution sum, and the second is the impulse response multiplied by $ay[-1]$, i.e.,

$$y[n] = y_{zs}[n] + y_{zi}[n]$$
$$= \sum_{k=0}^{\infty} h[k]x[n-k] + ay[-1]h[n]. \quad \square$$

Example 10.21. Consider a discrete-time system represented by a second-order difference equation with constant coefficients

$$y[n] - a_1 y[n-1] - a_2 y[n-2] = x[n] + b_1 x[n-1] + b_2 x[n-2] \quad n \geq 0$$

where $x[n]$ is the input, $y[n]$ is the output and the initial conditions are $y[-1]$, and $y[-2]$. Obtain an expression for $Y(z)$ that can be used to find the complete solution.

Solution: Applying the one-sided Z-transform to the two sides of the difference equation, we have

$$\mathcal{Z}(y[n] - a_1 y[n-1] - a_2 y[n-2]) = \mathcal{Z}(x[n] + b_1 x[n-1] + b_2 x[n-2]),$$
$$Y(z) - a_1(z^{-1}Y(z) + y[-1]) - a_2(z^{-2}Y(z) + y[-1]z^{-1} + y[-2]) = X(z)(1 + b_1 z^{-1} + b_2 z^{-2}),$$

where we used the linearity and the time-shift properties of the Z-transform. It was also assumed that the input is causal, $x[n] = 0$ for $n < 0$, so that the Z-transforms of $x[n-1]$ and $x[n-2]$ are simply

$z^{-1}X(z)$ and $z^{-2}X(z)$. Rearranging the above equation we have

$$Y(z)(1 - a_1 z^{-1} - a_2 z^{-2}) = (y[-1](a_1 + a_2 z^{-1}) + a_2 y[-2]) + X(z)(1 + b_1 z^{-1} + b_2 z^{-2})$$

and solving for $Y(z)$ we have

$$Y(z) = \frac{X(z)(1 + b_1 z^{-1} + b_2 z^{-2})}{1 - a_1 z^{-1} - a_2 z^{-2}} + \frac{y[-1](a_1 + a_2 z^{-1}) + a_2 y[-2]}{1 - a_1 z^{-1} - a_2 z^{-2}}$$

where again the first term is the Z-transform of the zero-state response, due to the input only, and the second term is the Z-transform of the zero-input response which is due to the initial conditions alone. The inverse Z-transform of $Y(z)$ will give us the complete response. ☐

Remarks
As we saw in Chapter 9, if either the initial conditions are not zero, or the input is not causal the system is not linear time-invariant (LTI). However, the time-shift property of the Z-transform allows us to find the complete response in that case. We can think of two inputs applied to the system: one due to the initial conditions and the other the regular input. By using superposition, we obtain the zero-state and the zero-input responses which add to the total response.

Just as with the Laplace transform, the steady-state response of a difference equation

$$y[n] + \sum_{k=1}^{N} a_k y[n-k] = \sum_{m=0}^{M} b_m x[n-m]$$

is due to simple poles of $Y(z)$ on the unit circle. Simple or multiple poles inside the unit circle give a transient, while multiple poles on the unit circle or poles outside the unit circle create an increasing response.

Example 10.22. Solve the difference equation

$$y[n] = y[n-1] - 0.25 y[n-2] + x[n] \qquad n \geq 0$$

with zero initial conditions and $x[n] = u[n]$.

Solution: The Z-transform of the terms of the difference equation gives

$$Y(z) = \frac{X(z)}{1 - z^{-1} + 0.25 z^{-2}} = \frac{1}{(1 - z^{-1})(1 - z^{-1} + 0.25 z^{-2})} = \frac{z^3}{(z-1)(z^2 - z + 0.25)} \qquad |z| > 1.$$

$Y(z)$ has three zeros at $z = 0$, a pole at $z = 1$, and a double pole at $z = 0.5$. The partial fraction expansion of $Y(z)$ is of the form

$$Y(z) = \frac{A}{1 - z^{-1}} + \frac{B(1 - 0.5 z^{-1}) + C z^{-1}}{(1 - 0.5 z^{-1})^2} \qquad (10.39)$$

where the terms of the expansion can be found in the table of Z-transforms. Within some constants, the complete response is

$$y[n] = Au[n] + [B(0.5)^n + Cn(0.5)^n]u[n].$$

The steady state is then $y_{ss}[n] = A$ (corresponding to the pole on the unit circle $z = 1$) as the other two terms, corresponding to the double pole $z = 0.5$ inside the unit circle, make up the transient response. The value of A is obtained:

$$A = Y(z)(1 - z^{-1})|_{z^{-1}=1} = 4.$$

Notice in Equation (10.39) that the expansion term corresponding to the double pole $z = 0.5$ has as numerator a first-order polynomial, with constants B and C to be determined, to ensure that term is proper rational. That term equals

$$\frac{B(1 - 0.5z^{-1}) + Cz^{-1}}{(1 - 0.5z^{-1})^2} = \frac{B}{1 - 0.5z^{-1}} + \frac{Cz^{-1}}{(1 - 0.5z^{-1})^2},$$

which is very similar to the expansion for multiple poles in the inverse Laplace transform. Once we find the values of B and C, the inverse Z-transforms are obtained from the table of Z-transforms. A simple method to obtain the coefficients B and C is to first obtain C by multiplying by $(1 - 0.5z^{-1})^2$ the two sides of (10.39) to get

$$Y(z)(1 - 0.5z^{-1})^2 = \frac{A(1 - 0.5z^{-1})^2}{1 - z^{-1}} + B(1 - 0.5z^{-1}) + Cz^{-1}$$

and then letting $z^{-1} = 2$ on both sides to find that

$$C = \frac{Y(z)(1 - 0.5z^{-1})^2}{z^{-1}}\bigg|_{z^{-1}=2} = -0.5.$$

The B value is then obtained by choosing a value for z^{-1} different from 1 or 0.5, to compute $Y(z)$. For instance, assume you choose $z^{-1} = 0$ and that you have found A and C then

$$Y(z)|_{z^{-1}=0} = A + B = 1$$

from which $B = -3$, so that

$$y[n] = 4u[n] + \left[-3(0.5)^n + n(0.5)^{n+1}\right]u[n]. \quad \square$$

Example 10.23. Find the complete response of the following difference equation:

$$y[n] + y[n-1] - 4y[n-2] - 4y[n-3] = 3x[n] \qquad n \geq 0,$$
$$y[-1] = 1, \ y[-2] = y[-3] = 0,$$

to an input $x[n] = u[n]$. Determine if the discrete-time system corresponding to this difference equation is BIBO stable or not, and the effect this has in the steady-state response.

Solution: Using the time-shifting and linearity properties of the Z-transform, and replacing the initial conditions we get

$$Y(z)[1 + z^{-1} - 4z^{-2} - 4z^{-3}] = 3X(z) + [-1 + 4z^{-1} + 4z^{-2}].$$

Letting

$$A(z) = 1 + z^{-1} - 4z^{-2} - 4z^{-3} = (1 + z^{-1})(1 + 2z^{-1})(1 - 2z^{-1})$$

we can write

$$Y(z) = 3\frac{X(z)}{A(z)} + \frac{-1 + 4z^{-1} + 4z^{-2}}{A(z)} \qquad |z| > 2. \qquad (10.40)$$

To determine whether the steady-state response exists or not let us first consider the stability of the system associated with the given difference equation. The transfer function $H(z)$ of the system is calculated by letting the initial conditions be zero (this makes the second term on the right of the above equation zero) so that we can get the ratio of the Z-transform of the output to the Z-transform of the input. If we do that then

$$H(z) = \frac{Y(z)}{X(z)} = \frac{3}{A(z)}.$$

Since the poles of $H(z)$ are the zeros of $A(z)$, which are $z = -1$, $z = -2$ and $z = 2$, the impulse response $h[n] = \mathcal{Z}^{-1}[H(z)]$ will not be absolutely summable, as required by the BIBO stability, because of the poles of $H(z)$ are on and outside the unit circle. Indeed, a general form of the impulse response is

$$h[n] = [C + D\,(2)^n + E\,(-2)^n]u[n]$$

where C, D, and E are constants that can be found by doing a partial fraction expansion of $H(z)$. Thus, $h[n]$ will grow as n increases and it would not be absolutely summable, i.e., the system is not BIBO stable.

Since the system is unstable, we expect the total response to grow as n increases. Let us see this. The partial fraction expansion of $Y(z)$, after replacing $X(z)$ in (10.40), is given by

$$Y(z) = \frac{2 + 5z^{-1} - 4z^{-3}}{(1 - z^{-1})(1 + z^{-1})(1 + 2z^{-1})(1 - 2z^{-1})} = \frac{B_1}{1 - z^{-1}} + \frac{B_2}{1 + z^{-1}} + \frac{B_3}{1 + 2z^{-1}} + \frac{B_4}{1 - 2z^{-1}},$$

$$B_1 = Y(z)(1 - z^{-1})|_{z^{-1}=1} = -\tfrac{1}{2}, \qquad B_2 = Y(z)(1 + z^{-1})|_{z^{-1}=-1} = -\tfrac{1}{6},$$

$$B_3 = Y(z)(1 + 2z^{-1})|_{z^{-1}=-1/2} = 0, \qquad B_4 = Y(z)(1 - 2z^{-1})|_{z^{-1}=1/2} = \tfrac{8}{3},$$

so that

$$y[n] = \left(-0.5 - \frac{1}{6}(-1)^n + \frac{8}{3}2^n\right)u[n],$$

which as expected will grow as n increases—there is no steady-state response.

In a problem like this the chance of making computational errors is large, so it is important to figure out a way to partially check your answer. In this case we can check the value of $y[0]$ using the difference equation which is $y[0] = -y[-1] + 4y(-2) + 4y(-3) + 3 = -1 + 3 = 2$ and compare it with the one obtained using our solution, which gives $y[0] = -3/6 - 1/6 + 16/6 = 2$. They coincide. Another way to partially check your answer is to use the initial and the final values properties shown in Table 10.2.   □

Approximate Solution of Ordinary Differential Equations

The solution of ordinary differential equations requires converting them into difference equations, which can then be solved in closed form by means of the Z-transform.

Example 10.24. Consider an RLC circuit represented by the second-order ordinary differential equation

$$\frac{d^2 v_c(t)}{dt^2} + \frac{d v_c(t)}{dt} + v_c(t) = v_s(t)$$

where the voltage across the capacitor $v_c(t)$ is the output and the source $v_s(t) = u(t)$ is the input. Let the initial conditions be zero. Approximate the derivatives by their definition, find and solve the resulting difference equation.

Solution: The Laplace transform of the output is found from the ordinary differential equation as

$$V_c(s) = \frac{V_s(s)}{1 + s + s^2} = \frac{1}{s(s^2 + s + 1)} = \frac{1}{s((s + 0.5)^2 + 3/4)}$$

where the final equation is obtained after replacing $V_s(s) = 1/s$. The solution of the ordinary differential equation is of the general form

$$v_c(t) = [A + Be^{-0.5t} \cos(\sqrt{3}t/2 + \theta)]u(t)$$

for constants A, B, and θ. To convert the ordinary differential equation into a difference equation we approximate the first derivative as

$$\frac{d v_c(t)}{dt} \approx \frac{v_c(t) - v_c(t - T_s)}{T_s}$$

and the second derivative as

$$\frac{d^2 v_c(t)}{dt^2} = \frac{d(d v_c(t)/dt)}{dt} \approx \frac{d(v_c(t) - v_c(t - T_s))/T_s}{dt} = \frac{v_c(t) - 2v_c(t - T_s) + v_c(t - 2T_s)}{T_s^2},$$

which when on replacing in the ordinary differential equation and computing the resulting equation for $t = nT_s$ gives

$$\left(\frac{1}{T_s^2} + \frac{1}{T_s} + 1\right) v_c(nT_s) - \left(\frac{2}{T_s^2} + \frac{1}{T_s}\right) v_c((n-1)T_s) + \left(\frac{1}{T_s^2}\right) v_c((n-2)T_s) = v_s(nT_s). \quad (10.41)$$

Although we know that we need to choose a very small value for T_s to get a good approximation to the exact result, for simplicity let us first set $T_s = 1$, so that the difference equation is

$$3v_c[n] - 3v_c[n-1] + v_c[n-2] = v_s[n] \qquad n > 0.$$

For zero initial conditions, we can recursively compute this equation to get

$$v_c[0] = 1/3, \quad \text{and} \quad v_c[n] = 1 \quad \text{as } n \to \infty.$$

A closed-form solution can be obtained using the Z-transform, giving (assuming zero initial conditions)

$$[3 - 3z^{-1} + z^{-2}]V_c(z) = \frac{1}{1 - z^{-1}}$$

so that

$$V_c(z) = \frac{z^3}{(z-1)(3z^2 - 3z + 1)},$$

from which we find that there is a triple zero $z = 0$, and poles at 1 and at $-0.5 \pm j\sqrt{3}/6$. The partial fraction expansion will be of the form

$$V_c(z) = \frac{A}{1 - z^{-1}} + \frac{B}{1 + (0.5 + j\sqrt{3}/6)z^{-1}} + \frac{B^*}{1 + (0.5 - j\sqrt{3}/6)z^{-1}}.$$

Since the complex poles are inside the unit circle, the steady-state response is due to the input which has a single pole at 1, i.e., the steady state $\lim_{n \to \infty} v_c[n] = A = 1$.

In solving the ordinary differential equation, we first use the symbolic MATLAB functions *ilaplace* and *ezplot* to find the exact solution. We then sample the input signal using a sampling period $T_s = 0.1$ s and use the approximations of the first and second derivatives to obtain the difference equation (10.41) which is computed using *filter*. The results are shown in Fig. 10.12. The exact solution of the ordinary differential equation is well approximated by the solution of the difference equation obtained by approximating the first and second derivatives.

```
%%
% Example 10.24---Approximate solution of RLC differential equation
%%
clear all; clf
 syms s
 vc=ilaplace(1/(s^3+s^2+s));            % exact solution
 ezplot(vc,[0,10]);grid; hold on
 Ts=0.1;                                % sampling period
 a1=1/Ts^2+1/Ts+1;a2=-2/Ts^2-1/Ts;a3=1/Ts^2;  % coefficients
 a=[1 a2/a1 a3/a1];b=1;
 t=0:Ts:10; N=length(t);
 vs=ones(1,N);                          % input
 vca=filter(b,a,vs);vca=vca/vca(N);     % solution
```

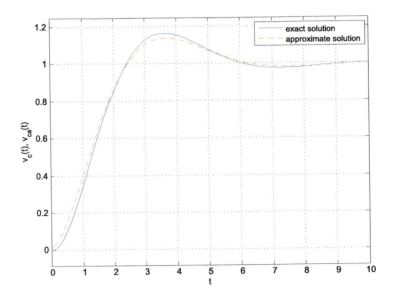

FIGURE 10.12

Solution of the ordinary differential equation $d^2 v_c(t)/dt^2 + dv_c(t)/dt + v_c(t) = v_s(t)$ (solid line). Solution of the difference equation approximating the ordinary differential equation for $T_s = 0.1$ (dotted line). Exact and approximate solutions are very close.

10.5.5 INVERSE OF TWO-SIDED Z-TRANSFORMS

When finding the inverse of a two-sided Z-transform, or a noncausal discrete-time signal, it is important to relate the poles to the causal and the anticausal components. The region of convergence plays a very important role in making this determination. Once this is done, the inverse is found by looking for the causal and the anticausal partial fraction expansion components in a Z-transform table. The coefficients of the partial fraction expansion are calculated like those in the case of causal signals.

Example 10.25. Consider finding the inverse Z-transform of

$$X(z) = \frac{2z^{-1}}{(1 - z^{-1})(1 - 2z^{-1})^2} = \frac{2z^2}{(z - 1)(z - 2)^2} \qquad 1 < |z| < 2,$$

which corresponds to a noncausal signal.

Solution: The function $X(z)$ has two zeros at $z = 0$, and a pole at $z = 1$ and a double pole at $z = 2$. For the region of convergence to be a torus of internal radius 1 and outer radius 2, we need to associate with the pole at $z = 1$ the region of convergence $\mathcal{R}_1 : |z| > 1$, corresponding to a causal signal, and with the pole at $z = 2$ the region of convergence $\mathcal{R}_2 : |z| < 2$ corresponding to an anticausal signal. Thus, we have

$$1 < |z| < 2 = \mathcal{R}_1 \cap \mathcal{R}_2.$$

The partial fraction expansion is then done so that

$$X(z) = \underbrace{\frac{A}{1 - z^{-1}}}_{\mathcal{R}_1 : |z| > 1} + \underbrace{\left[\frac{B}{1 - 2z^{-1}} + \frac{Cz^{-1}}{(1 - 2z^{-1})^2} \right]}_{\mathcal{R}_2 : |z| < 2},$$

i.e., the first term has $\mathcal{R}_1$ as the region of convergence and the terms in the square brackets have $\mathcal{R}_2$ as their region of convergence. The inverse of the first term will be a causal signal, and the inverse of the other two terms will be an anticausal signal.

The coefficients are found like in the case of causal signals. In this case, we have

$$A = X(z)(1 - z^{-1})|_{z^{-1}=1} = 2, \quad C = X(z)\frac{(1 - 2z^{-1})^2}{z^{-1}}|_{z^{-1}=0.5} = 4.$$

To calculate B we compute $X(z)$ and its expansion for a value of $z^{-1} \neq 1$ or 0.5. For instance, $z^{-1} = 0$ gives $X(0) = A + B = 0$ so that $B = -A = -2$.

Using the Z-transform property that if $x_i[n]$ has a transform $X_i(z)$ then $x_i[-n]$ has $X_i(z^{-1})$ ($i = 1, 2$) as its Z-transform, we have the following pairs:

$$x_1[n] = \alpha^{n-1}u[n-1] \quad \leftrightarrow \quad X_1(z) = \frac{z^{-1}}{1 - \alpha z^{-1}} \quad |z| > |\alpha|,$$

$$x_2[n] = n\alpha^n u[n] \quad \leftrightarrow \quad X_2(z) = \frac{\alpha z^{-1}}{(1 - \alpha z^{-1})^2} \quad |z| > |\alpha|,$$

so that the anticausal signals $x_1[-n]$ and $x_2[-n]$ have the following Z-transforms:

$$x_1[-n] = \alpha^{-n-1}u[-n-1] \quad \leftrightarrow \quad X_1(z^{-1}) = \frac{z}{1 - \alpha z} \quad |z| < \frac{1}{|\alpha|},$$

$$x_2[-n] = -n\alpha^{-n}u[-n] \quad \leftrightarrow \quad X_2(z^{-1}) = \frac{\alpha z}{(1 - \alpha z)^2} \quad |z| < \frac{1}{|\alpha|}.$$

Replacing the values for A, B, and C and expressing the Z-transforms corresponding to the anticausal components in positive powers of z we have

$$X(z) = \frac{2}{1 - z^{-1}} + \frac{-2z}{z - 2} + \frac{4z}{(z - 2)^2} = \frac{2}{1 - z^{-1}} + \frac{z}{1 - 0.5z} + \frac{z}{(1 - 0.5z)^2}$$

and using the above pairs, the inverse is found to be

$$x[n] = \underbrace{2u[n]}_{\text{causal}} + \underbrace{\left[2^{(n+1)}u[-n-1] - 2^{(n+1)}nu[-n] \right]}_{\text{anticausal}}. \quad \square$$

Example 10.26. Find all the possible impulse responses connected with the following transfer function of a discrete filter:

$$H(z) = \frac{1 + 2z^{-1} + z^{-2}}{(1 - 0.5z^{-1})(1 - z^{-1})},$$

having poles at $z = 1$ and $z = 0.5$.

Solution: As a function of z^{-1} this function is not proper since both its numerator and its denominator are of degree 2. After division we have the following partial fraction expansion:

$$H(z) = B_0 + \frac{B_1}{1 - 0.5z^{-1}} + \frac{B_2}{1 - z^{-1}}.$$

There are three possible regions of convergence that can be attached to $H(z)$:

- $\mathcal{R}_1 : |z| > 1$ so that the corresponding impulse response $h_1[n] = \mathcal{Z}^{-1}[H(z)]$ is causal with a general form

$$h_1[n] = B_0\delta[n] + (B_1(0.5)^n + B_2)u[n].$$

  For a nonzero B_2, the pole at $z = 1$ makes this filter unstable, as its impulse response is not absolutely summable and the region of convergence does not include the unit circle.

- $\mathcal{R}_2 : |z| < 0.5$ for which the corresponding impulse response $h_2[n] = \mathcal{Z}^{-1}[H(z)]$ is anticausal with a general form

$$h_2[n] = B_0\delta[n] - (B_1(0.5)^n + B_2)u[-n-1].$$

  The region of convergence $\mathcal{R}_2$ does not include the unit circle and so the impulse response is not absolutely summable and so the filter is not stable.

- $\mathcal{R}_3 : 0.5 < |z| < 1$, which gives a two-sided impulse response $h_2[n] = \mathcal{Z}^{-1}[H(z)]$ of general form

$$h_3[n] = B_0\delta[n] + \underbrace{B_1(0.5)^n u[n]}_{\text{causal}} \underbrace{- B_2 u[-n-1]}_{\text{anticausal}}.$$

  Again this filter is not stable as the unit circle is not included in the region of convergence.   □

10.6 STATE VARIABLE REPRESENTATION

Modern control theory uses the state-variable representation of systems, whether continuous- or discrete-time. In this section we introduce the discrete-time state-variable representation of systems. In many respects, it is very similar to the state-variable representation of continuous-time systems.

State variables are the memory of a system. In the discrete-time, just like in the continuous-time, knowing the state of a system at a present index n provides the necessary information from the past that together with present and future inputs allows us to calculate the present and future outputs of the system. The advantage of a state-variable representation over a transfer function is the inclusion of initial conditions in the analysis and the ability to use it in multiple-input and multiple-output systems.

Assume a discrete-time system is represented by a difference equation (which could have been obtained from an ordinary differential equation representing a continuous-time system) where $x[n]$ is the input and $y[n]$ the output:

$$y[n] + a_1 y[n-1] + \cdots + a_N y[n-N] = b_0 x[n] + b_1 x[n-1] + \cdots + b_M x[n-M], \quad n \geq 0,$$
$$(10.42)$$

where $M \leq N$ and the initial conditions of the system are $\{y[i], -N \leq i \leq -1\}$. As in the continuous-time, the state-variable representation of a discrete-time system is not unique. To obtain a state-variable representation from the difference equation in (10.42) let

$$v_1[n] = y[n-1],$$
$$v_2[n] = y[n-2],$$
$$\vdots$$
$$v_N[n] = y[n-N]. \tag{10.43}$$

We then obtain from the difference equation the state-variable equations:

$$v_1[n+1] = y[n] = -a_1 v_1[n] - \cdots - a_N v_N[n] + b_0 x[n] + \cdots + b_M x[n-M],$$
$$v_2[n+1] = v_1[n],$$
$$\vdots$$
$$v_N[n+1] = v_{N-1}[n], \tag{10.44}$$

and the output equation

$$y[n] = -a_1 v_1[n] - \cdots - a_N v_N[n] + b_0 x[n] + b_1 x[n-1] + \cdots + b_M x[n-M].$$

These state and output equations can then be written in a matrix form

$$\mathbf{v}[n+1] = \mathbf{A}\mathbf{v}[n] + \mathbf{B}\mathbf{x}[n],$$
$$y[n] = \mathbf{c}^T \mathbf{v}[n] + \mathbf{d}^T \mathbf{x}[n], \qquad n \geq 0, \tag{10.45}$$

by appropriately defining the matrices $\mathbf{A}$ and $\mathbf{B}$, the vectors $\mathbf{c}$ and $\mathbf{d}$ as well as the state vector $\mathbf{v}[n]$ and the input vector $\mathbf{x}[n]$.

Fig. 10.13 displays the block diagrams for a delay, a constant multiplier and an adder that are used in obtaining block diagrams for discrete-time system. Different from the continuous-time system representation, the representation of discrete-time systems does not require the equivalent of integrators, instead it uses delays.

Example 10.27. A continuous-time system is represented by the ordinary differential equation

$$\frac{d^2 y(t)}{dt^2} + \frac{dy(t)}{dt} + y(t) = x(t) \qquad t \geq 0.$$

To convert this ordinary differential equation into a difference equation we approximate the derivatives by

$$\frac{dy(t)}{dt} \approx \frac{y(t) - y(t - T_s)}{T_s},$$
$$\frac{d^2 y(t)}{dt^2} \approx \frac{y(t) - 2y(t - T_s) + y(t - 2T_s)}{T_s^2}.$$

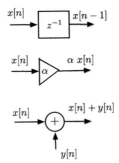

FIGURE 10.13

Block diagrams of different components used to represent discrete-time systems (top to bottom): delay, constant multiplier and adder.

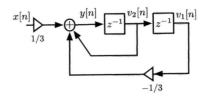

FIGURE 10.14

Block diagram of $y[n] - y[n-1] + (1/3)y[n-2] = (1/3)x[n]$ displaying the state variables $v_1[n] = y[n-2]$ and $v_2[n] = y[n-1]$.

Find the difference equation when $T_s = 1$ and $t = nT_s$, and find the corresponding state and output equations.

Solution: Replacing the approximations of the derivatives for $T_s = 1$ and letting $t = nT_s = n$ we obtain the difference equation

$$(y[n] - 2y[n-1] + y[n-2]) + (y[n] - y[n-1]) + y[n] = x[n] \quad \text{or}$$

$$y[n] - y[n-1] + \frac{1}{3}y[n-2] = \frac{1}{3}x[n],$$

which can be realized as indicated by the block diagram in Fig. 10.14. Because the second-order difference equation is realized using two unit delays, this realization is called minimal.

Letting the outputs of the delays be the state variables

$$v_1[n] = y[n-2], \quad v_2[n] = y[n-1],$$

we then have the following matrix equation for the state:

$$\begin{bmatrix} v_1[n+1] \\ v_2[n+1] \end{bmatrix} = \underbrace{\begin{bmatrix} 0 & 1 \\ -1/3 & 1 \end{bmatrix}}_{\mathbf{A}} \begin{bmatrix} v_1[n] \\ v_2[n] \end{bmatrix} + \underbrace{\begin{bmatrix} 0 \\ 1/3 \end{bmatrix}}_{\mathbf{b}} x[n].$$

The output equation is then

$$y[n] = -\frac{1}{3}v_1[n] + v_2[n] + \frac{1}{3}x[n] \quad \text{or in matrix form}$$

$$y[n] = \underbrace{\begin{bmatrix} -\frac{1}{3} & 1 \end{bmatrix}}_{\mathbf{c}^T} \begin{bmatrix} v_1[n] \\ v_2[n] \end{bmatrix} + \underbrace{\begin{bmatrix} \frac{1}{3} \end{bmatrix}}_{d} x[n].$$

The state variables are not unique. Indeed we can use an invertible transformation matrix $\mathbf{F}$ to define a new set of state variables

$$\mathbf{w}[n] = \mathbf{F}\mathbf{v}[n]$$

with a matrix representation for the new state variables and the output given by

$$\begin{aligned} \mathbf{w}[n+1] &= \mathbf{F}\mathbf{v}[n+1] = \mathbf{F}\mathbf{A}\mathbf{v}[n] + \mathbf{F}\mathbf{b}x[n] \\ &= \mathbf{F}\mathbf{A}\mathbf{F}^{-1}\mathbf{w}[n] + \mathbf{F}\mathbf{b}x[n], \\ y[n] &= \mathbf{c}^T\mathbf{v}[n] + \mathbf{d}x[n] = \mathbf{c}^T\mathbf{F}^{-1}\mathbf{w}[n] + \mathbf{d}x[n]. \quad \square \end{aligned}$$

Example 10.28. A discrete-time system is represented by the difference equation

$$y[n] - y[n-1] - y[n-2] = x[n] + x[n-1]$$

where $x[n]$ is the input and $y[n]$ the output. Obtain a state-variable representation for it.

Solution: Notice that this difference equation has $x[n-1]$ as well as $x[n]$ in the input. The transfer function corresponding to the difference equation is

$$H(z) = \frac{Y(z)}{X(z)} = \frac{1 + z^{-1}}{1 - z^{-1} - z^{-2}},$$

i.e., it is not a "constant-numerator" transfer function. A direct realization of the system in this case will not be minimal. Indeed, a block diagram obtained from the difference equation (see Fig. 10.15) shows that three delays are needed to represent this second-order system. It is important to realize that this representation despite being non-minimal is a valid representation. This is different from an analogous situation in the continuous-time representation where the input and its derivatives are present. Differentiators in the continuous-time representation are deemed not acceptable, while delays in the discrete-time representations are. $\square$

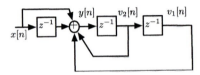

FIGURE 10.15

Block diagram of $y[n] - y[n-1] - y[n-2] = x[n] + x[n-1]$ displaying the state variables $v_1[n]$ and $v_2[n]$, and corresponding to a non-minimal realization.

Solution of the State and Output Equations

The solution of the state-variable equations

$$\mathbf{v}[n+1] = \mathbf{A}\mathbf{v}[n] + \mathbf{B}\mathbf{x}[n], \qquad n \geq 0,$$

can be obtained recursively:

$$\mathbf{v}[1] = \mathbf{A}\mathbf{v}[0] + \mathbf{B}\mathbf{x}[0],$$
$$\mathbf{v}[2] = \mathbf{A}\mathbf{v}[1] + \mathbf{B}\mathbf{x}[1] = \mathbf{A}^2\mathbf{v}[0] + \mathbf{A}\mathbf{B}\mathbf{x}[0] + \mathbf{B}\mathbf{x}[1],$$
$$\vdots$$
$$\mathbf{v}[n] = \mathbf{A}^n\mathbf{v}[0] + \sum_{k=0}^{n-1} \mathbf{A}^{n-1-k}\mathbf{B}\mathbf{x}[k],$$

where $\mathbf{A}^0 = \mathbf{I}$, the identity matrix. The complete solution is then obtained,

$$y[n] = \underbrace{\mathbf{c}^T\mathbf{A}^n\mathbf{v}[0]}_{y_{zi}[n] \text{ zero-input response}} + \underbrace{\sum_{k=0}^{n-1} \mathbf{c}^T\mathbf{A}^{n-1-k}\mathbf{B}\mathbf{x}[k] + \mathbf{d}\mathbf{x}[n]}_{y_{zs}[n] \text{ zero-state response}}. \qquad (10.46)$$

By definition of the state variables in Equation (10.43) the initial conditions of the state variables coincide with the initial conditions of the system,

$$v_1[0] = y[-1], \quad v_2[0] = y[-2], \quad \cdots \quad v_N[0] = y[-N]. \qquad (10.47)$$

Using the Z-transform we can obtain a close solution to the state and output equations. Indeed, if the state and output equations are in their matrix form

$$\mathbf{v}[n+1] = \mathbf{A}\mathbf{v}[n] + \mathbf{B}\mathbf{x}[n],$$
$$y[n] = \mathbf{c}^T\mathbf{v}[n] + \mathbf{d}^T\mathbf{x}[n], \qquad n \geq 0, \qquad (10.48)$$

calling the Z-transforms of the state variables $V_i(z) = \mathcal{Z}(v_i[n])$, for $i = 1, \cdots, N$; of the input $X_m(z) = \mathcal{Z}(x[n-m])$, for $m = 0, \cdots, M$, and of the output $Y(z) = \mathcal{Z}(y[n])$ we have the following matrix expression for the Z-transforms of the state equations in (10.48):

$$z\mathbf{V}(z) - z\mathbf{v}[0] = \mathbf{A}\mathbf{V}(z) + \mathbf{B}\mathbf{X}(z) \quad \text{or} \quad (z\mathbf{I} - \mathbf{A})\mathbf{V}(z) = z\mathbf{v}[0] + \mathbf{B}\mathbf{X}(z)$$

using the Z-transform of $v_i[n+1]$, and $\mathbf{v}[0]$ is the vector of initial conditions of the state variables and $\mathbf{I}$ the unit matrix. Assuming that the inverse of $(z\mathbf{I} - \mathbf{A})$ exists, i.e., $\det(z\mathbf{I} - \mathbf{A}) \neq 0$, we can solve for $\mathbf{V}(z)$:

$$\begin{aligned}
\mathbf{V}(z) &= (z\mathbf{I} - \mathbf{A})^{-1} z\mathbf{v}[0] + (z\mathbf{I} - \mathbf{A})^{-1}\mathbf{B}\mathbf{X}(z) \\
&= \frac{\mathrm{Adj}(z\mathbf{I} - \mathbf{A})}{\det(z\mathbf{I} - \mathbf{A})} z\mathbf{v}[0] + \frac{\mathrm{Adj}(z\mathbf{I} - \mathbf{A})}{\det(z\mathbf{I} - \mathbf{A})}\mathbf{B}\mathbf{X}(z)
\end{aligned} \tag{10.49}$$

by expressing the inverse of a matrix in terms of its adjoint and determinant. We can then obtain the Z-transform of the output as

$$Y(z) = \frac{\mathbf{c}^T \mathrm{Adj}(z\mathbf{I} - \mathbf{A})}{\det(z\mathbf{I} - \mathbf{A})} z\mathbf{v}[0] + \left[\frac{\mathbf{c}^T \mathrm{Adj}(z\mathbf{I} - \mathbf{A})}{\det(z\mathbf{I} - \mathbf{A})}\mathbf{B} + \mathbf{d} \right] X(z). \tag{10.50}$$

If the initial conditions $\mathbf{v}[0]$ are zero, and the input is $x[n]$, then we find that the transfer function is given by

$$H(z) = \frac{Y(z)}{X(z)} = \frac{\mathbf{c}^T \mathrm{Adj}(z\mathbf{I} - \mathbf{A})}{\det(z\mathbf{I} - \mathbf{A})} \mathbf{b} + d. \tag{10.51}$$

Example 10.29. Consider the state-variable representation of the system in Example 10.27 with matrices:

$$\mathbf{A} = \begin{bmatrix} 0 & 1 \\ -1/3 & 1 \end{bmatrix}, \qquad \mathbf{b} = \begin{bmatrix} 0 \\ 1/3 \end{bmatrix},$$

$$\mathbf{c}^T = \begin{bmatrix} -\frac{1}{3} & 1 \end{bmatrix}, \qquad d = \begin{bmatrix} \frac{1}{3} \end{bmatrix}.$$

Determine the transfer function of the system.

Solution: Instead of finding the inverse $(z\mathbf{I} - \mathbf{A})^{-1}$ we can use Cramer's rule to find $Y(z)$. Indeed, writing the state equations with zero initial conditions

$$\underbrace{\begin{bmatrix} z & -1 \\ 1/3 & z-1 \end{bmatrix}}_{(z\mathbf{I}-\mathbf{A})} \underbrace{\begin{bmatrix} V_1(z) \\ V_2(z) \end{bmatrix}}_{\mathbf{V}(z)} = \underbrace{\begin{bmatrix} 0 \\ X(z)/3 \end{bmatrix}}_{\mathbf{b}X(z)}$$

according to Cramer's rule we have

$$V_1(z) = \frac{\det \begin{bmatrix} 0 & -1 \\ X(z)/3 & z-1 \end{bmatrix}}{\Delta(z)} = \frac{X(z)/3}{\Delta(z)},$$

$$V_2(z) = \frac{\det\begin{bmatrix} z & 0 \\ 1/3 & X(z)/3 \end{bmatrix}}{\Delta(z)} = \frac{zX(z)/3}{\Delta(z)},$$

$$\Delta(z) = z^2 - z + 1/3,$$

and the Z-transform of the output is obtained after replacing the Z-transforms of the state variables:

$$
\begin{aligned}
Y(z) &= \begin{bmatrix} -1/3 & 1 \end{bmatrix}\begin{bmatrix} V_1(z) \\ V_2(z) \end{bmatrix} + \frac{X(z)}{3} \\
&= \frac{-1/9 + z/3 + (z^2/3 - z/3 + 1/9)}{z^2 - z + 1/3}X(z) = \frac{z^2/3}{z^2 - z + 1/3}X(z), \quad \text{so that} \\
H(z) &= \frac{Y(z)}{X(z)} = \frac{z^2/3}{z^2 - z + 1/3} = \frac{1/3}{1 - z^{-1} + z^{-2}/3}. \quad \square
\end{aligned}
$$

Example 10.30. The transfer function of a LTI discrete-time system is

$$H(z) = \frac{z^{-2}}{1 - 0.5z^{-1} + 0.5z^{-2}}.$$

Obtain a minimal state-variable realization (i.e., a realization that only uses two delays, corresponding to the second-order system). Determine the initial conditions of the state variables in terms of initial values of the output.

Solution: The given transfer function is not "constant-numerator" transfer function and the numerator indicates the system has delayed inputs. Letting $x[n]$ be the input and $y[n]$ the output with Z-transforms $X(z)$ and $Y(z)$, respectively, we factor $H(z)$ as follows:

$$H(z) = \frac{Y(z)}{X(z)} = \underbrace{z^{-2}}_{Y(z)/W(z)} \times \underbrace{\frac{1}{1 - 0.5z^{-1} + 0.5z^{-2}}}_{W(z)/X(z)}$$

so that we have the following equations:

$$w[n] = 0.5w[n-1] - 0.5w[n-2] + x[n],$$
$$y[n] = w[n-2].$$

The corresponding block diagram in Fig. 10.16 is a minimal realization as it only uses two delays. As shown in the figure, the state variables are defined as

$$v_1[n] = w[n-1], \quad v_2[n] = w[n-2].$$

The state and output equations are in matrix form

$$\mathbf{v}[n+1] = \begin{bmatrix} 1/2 & -1/2 \\ 1 & 0 \end{bmatrix}\mathbf{v}[n] + \begin{bmatrix} 1 \\ 0 \end{bmatrix}x[n],$$
$$y[n] = \begin{bmatrix} 0 & 1 \end{bmatrix}\mathbf{v}[n].$$

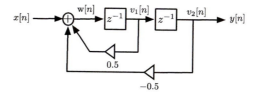

FIGURE 10.16

Minimal realization of a system with transfer function $H(z) = z^{-2}/(1 - 0.5z^{-1} + 0.5z^{-2})$.

Expressing the state variables in terms of the output:

$$v_1[n] = w[n-1] = y[n+1], \quad v_2[n] = y[n],$$

we then have $v_2[0] = y[0]$ and $v_1[0] = y[1]$. The reason for these initial conditions is that in the difference equation

$$y[n] - 0.5y[n-1] + 0.5y[n-2] = x[n-2]$$

the input is delayed two samples and so the initial conditions should be $y[0]$ and $y[1]$.   □

Example 10.31. Consider the state-variable representation of a system having the matrices

$$\mathbf{A} = \begin{bmatrix} 0 & 1 \\ 1/8 & 1/4 \end{bmatrix}, \quad \mathbf{b} = \begin{bmatrix} 0 \\ 1 \end{bmatrix},$$

$$\mathbf{c}^T = \begin{bmatrix} 1 & 0 \end{bmatrix}, \quad d = [0].$$

Determine the zero-input response of the system for arbitrary initial conditions $\mathbf{v}[0]$.

Solution: The zero-input response in the time-domain is

$$y[n] = \mathbf{c}^T \mathbf{A}^n \mathbf{v}[0]$$

requiring one to compute the matrix $\mathbf{A}^n$. To do so, consider

$$\mathcal{Z}[\mathbf{A}^n] = \sum_{n=0}^{\infty} \mathbf{A}^n z^{-n} = \left(\mathbf{I} - \mathbf{A}z^{-1} \right)^{-1}, \quad \det\left(\mathbf{I} - \mathbf{A}z^{-1} \right) \neq 0,$$

which can be verified by pre-multiplying by $\mathbf{I} - \mathbf{A}z^{-1}$ the above equation, giving

$$\left[\mathbf{I} - \mathbf{A}z^{-1} \right] \sum_{n=0}^{\infty} \mathbf{A}^n z^{-n} = \sum_{n=0}^{\infty} \mathbf{A}^n z^{-n} - \sum_{n=0}^{\infty} \mathbf{A}^{n+1} z^{-(n+1)} = \mathbf{I}$$

or that the infinite summation is the inverse of $(\mathbf{I} - \mathbf{A}z^{-1})$. In positive powers of z the Z-transform of $\mathbf{A}^n$ is

$$\mathcal{Z}[\mathbf{A}^n] = \left(z^{-1}[\mathbf{I}z - \mathbf{A}] \right)^{-1} = z(\mathbf{I}z - \mathbf{A})^{-1}.$$

Now, using the fact that for a 2×2 matrix

$$\mathbf{F} = \begin{bmatrix} a & b \\ c & d \end{bmatrix} \quad \Rightarrow \quad \mathbf{F}^{-1} = \frac{1}{ad - bc} \begin{bmatrix} d & -b \\ -c & a \end{bmatrix}$$

provided that the determinant of the matrix, $ad - bc \neq 0$, we then have

$$\mathbf{I} - \mathbf{A}z^{-1} = \begin{bmatrix} 1 & -z^{-1} \\ -z^{-1}/8 & 1 - z^{-1}/4 \end{bmatrix}$$

$$\Rightarrow \quad (\mathbf{I} - \mathbf{A}z^{-1})^{-1} = \frac{1}{1 - z^{-1}/4 - z^{-2}/8} \begin{bmatrix} 1 - z^{-1}/4 & z^{-1} \\ z^{-1}/8 & 1 \end{bmatrix}.$$

We then need to determine the inverse Z-transform of the four entries to find $\mathbf{A}^n$. If we let

$$P(z) = \frac{1}{1 - z^{-1}/4 - z^{-2}/8} = \frac{2/3}{1 - 0.5z^{-1}} + \frac{1/3}{1 + 0.25z^{-1}} \quad \text{with inverse Z-transform}$$

$$p[n] = \left(\frac{2}{3} 0.5^n + \frac{1}{3}(-0.25)^n \right) u[n]$$

then

$$\mathbf{A}^n = \begin{bmatrix} p[n] - 0.25p[n-1] & p[n-1] \\ 0.125p[n-1] & p[n] \end{bmatrix},$$

which is used in finding the zero-input response

$$\begin{aligned} y[n] &= \begin{bmatrix} 1 & 0 \end{bmatrix} \begin{bmatrix} p[n] - 0.25p[n-1] & p[n-1] \\ 0.125p[n-1] & p[n] \end{bmatrix} \mathbf{v}[0] \\ &= (p[n] - 0.25p[n-1])v_1[0] + p[n-1]v_2[0]. \quad \square \end{aligned}$$

Canonical Realizations

Just like in the continuous case, discrete-time state-variable realizations have different canonical forms. In this section we will illustrate the process for obtaining a parallel realization to demonstrate the similarity.

The state-variable parallel realization is obtained by realizing each of the partial fraction expansion components of a transfer function. Consider the case of simple poles of the system, so that the transfer function is

$$H(z) = \frac{Y(z)}{X(z)} = \sum_{i=1}^{N} \underbrace{\frac{A_i}{1 - \alpha_i z^{-1}}}_{H_i(z)}$$

so that

$$Y(z) = \sum_{i=1}^{N} \underbrace{H_i(z)X(z)}_{Y_i(z)}.$$

The state variable for $Y_i(z)(1 - \alpha_i z^{-1}) = X(z)$ or $y_i[n] - \alpha_i y_i[n-1] = x[n]$ would be $v_i[n] = y_i[n-1]$ so that

$$v_i[n+1] = y_i[n] = \alpha_i v_i[n] + x[n], \qquad i = 1, \cdots, N.$$

For the whole system we then have

$$v_1[n+1] = \alpha_1 v_1[n] + x[n],$$
$$v_2[n+1] = \alpha_2 v_2[n] + x[n],$$
$$\cdots$$
$$v_N[n+1] = \alpha_N v_N[n] + x[n],$$

and the output is

$$y[n] = \sum_{i=1}^{N} y_i[n] = \sum_{i=1}^{N} \alpha_i v_i[n] + Nx[n]$$

or in matrix form

$$\mathbf{v}[n+1] = \mathbf{A}\mathbf{v}[n] + \mathbf{b}x[n],$$
$$y[n] = \mathbf{c}^T \mathbf{v}[n] + dx[n],$$

$$\mathbf{A} = \begin{bmatrix} \alpha_1 & 0 & \cdots & 0 \\ 0 & \alpha_2 & \cdots & 0 \\ \vdots & \vdots & \vdots & \vdots \\ 0 & 0 & \cdots & \alpha_N \end{bmatrix}, \quad \mathbf{b} = \begin{bmatrix} 1 \\ 1 \\ \vdots \\ 1 \end{bmatrix},$$

$$\mathbf{c}^T = \begin{bmatrix} 1 & 1 & \cdots & 1 \end{bmatrix}, \quad d = N.$$

In general, whenever the transfer function is factored in terms of first- and second-order systems, with real coefficients, a parallel realization is obtained by doing a partial fraction expansion in terms of first- and second-order components and realizing each of these components. The following example illustrates the procedure.

Example 10.32. Obtain a parallel state-variable realization from the transfer function

$$H(z) = \frac{z^3}{(z+0.5)[(z-0.5)^2 + 0.25]}.$$

Show a block realization of the corresponding state and output equations.

Solution: If we obtain a partial fraction expansion of the form

$$H(z) = \frac{1}{(1 + 0.5z^{-1})[(1 - 0.5z^{-1})^2 + 0.25z^{-2}]} = \frac{A_1}{1 + 0.5z^{-1}} + \frac{A_2 + A_3 z^{-1}}{1 - z^{-1} + 0.5z^{-2}}$$

where

$$A_1 = H(z)(1 + 0.5z^{-1})|_{z^{-1}=-2} = \frac{1}{1 - z^{-1} + 0.5z^{-2}}\Big|_{z^{-1}=-2} = \frac{1}{5}$$

and then

$$\frac{A_2 + A_3 z^{-1}}{1 - z^{-1} + 0.5z^{-2}} = \frac{1}{(1 + 0.5z^{-1})(1 - z^{-1} + 0.5z^{-2})} - \frac{1/5}{1 + 0.5z^{-1}}$$

$$= \frac{1 - 0.2(1 - z^{-1} + 0.5z^{-2})}{(1 + 0.5z^{-1})(1 - z^{-1} + 0.5z^{-2})} = \frac{0.8 - 0.2z^{-1}}{1 - z^{-1} + 0.5z^{-2}}$$

so that $A_2 = 0.8$ and $A_3 = -0.2$. This allows us to write the output as

$$Y(z) = \underbrace{\frac{0.2X(z)}{1 + 0.5z^{-1}}}_{Y_1(z)} + \underbrace{\frac{(0.8 - 0.2z^{-1})X(z)}{1 - z^{-1} + 0.5z^{-2}}}_{Y_2(z)}$$

or for two difference equations

$$y_1[n] + 0.5y_1[n - 1] = 0.2x[n],$$
$$y_2[n] - y_2[n - 1] + 0.5y[n - 2] = 0.8x[n] - 0.2x[n - 1],$$

for which we need to obtain state variables. Minimal realizations can be obtained as follows. The first system is a "constant-numerator" system, in terms of negative powers of z, and a minimal realization is obtained directly. The transfer function of the second difference equation can be written

$$Y_2(z) = \underbrace{\frac{X(z)}{1 - z^{-1} + 0.5z^{-2}}}_{W(z)} \times \underbrace{(0.8 - 0.2z^{-1})}_{Y_2(z)/W(z)},$$

from which we obtain the equations

$$w[n] - w[n - 1] + 0.5w[n - 2] = x[n],$$
$$y_2[n] = 0.8w[n] - 0.2w[n - 1].$$

Notice that the realization of these equations only require two delays. If we let

$$v_1[n] = w[n - 1],$$
$$v_2[n] = w[n - 2],$$

we get

$$v_1[n+1] = w[n] = v_1[n] - 0.5v_2[n] + x[n],$$
$$v_2[n+1] = v_1[n],$$
$$y_2[n] = 0.8(v_1[n] - 0.5v_2[n] + x[n]) - 0.2v_1[n] = 0.6v_1[n] - 0.4v_2[n] + 0.8x[n],$$

so that

$$\begin{bmatrix} v_1[n+1] \\ v_2[n+1] \end{bmatrix} = \begin{bmatrix} 1 & -0.5 \\ 1 & 0 \end{bmatrix} \begin{bmatrix} v_1[n] \\ v_2[n] \end{bmatrix} + \begin{bmatrix} 1 \\ 0 \end{bmatrix} x[n],$$

$$y_2[n] = \begin{bmatrix} 0.6 & -0.4 \end{bmatrix} \begin{bmatrix} v_1[n] \\ v_2[n] \end{bmatrix} + 0.8x[n].$$

The following script shows how to obtain the state-variable representation for the two components of $H(z)$ using the function *tf2ss*. The second part of the script shows that by using the function *ss2tf* we can get back the two components of $H(z)$. Notice that these are the same functions used in the continuous case. Fig. 10.17 shows the block diagram of the parallel realization. □

```
%%
% Example 10.32----State models
%%
clear all;
num1=[1 0]; den1=[1 0.5];
[A1,B1,C1,D1]=tf2ss(num1,den1)
num2=[0.8 -0.2 0]; den2=[1 -1 0.5];
[A2,B2,C2,D2]=tf2ss(num2,den2)
% verification
[nu,de]=ss2tf(A1,B1,C1,D1,1)
[nu1,de1]=ss2tf(A2,B2,C2,D2,1)


A1 = -0.5000
B1 =1
C1 = -0.5000
D1 =1


A2 = 1.0000   -0.5000
     1.0000        0
B2 = 1
     0
C2 = 0.6000   -0.4000
D2 = 0.8000
```

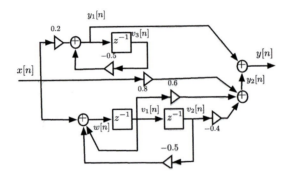

FIGURE 10.17

Block diagram of a parallel state-variable realization of the system with $H(z) = z^3/(z + 0.5)[(z - 0.5)^2 + 0.25]$. Notice this is a minimum realization with three delays corresponding to the third-order system.

10.7 TWO-DIMENSIONAL Z-TRANSFORM

The two-dimensional Z-transform is an extension of the one-dimensional transform. For a sequence $x[m, n]$ it is defined as

$$X(z_1, z_2) = \sum_{m=-\infty}^{\infty} \sum_{n=-\infty}^{\infty} x[m, n] z_1^{-m} z_2^{-n}, \qquad z_1, z_2 \in \text{ROC}, \qquad (10.52)$$

where ROC stands for the **region of convergence** of the double sum or the region where $X(z_1, z_2)$ is analytic—where it has no singularities of any kind.

Despite being an extension of the one-dimensional, the two-dimensional Z-transform is much more complex. In part, because it is defined in a four-dimensional space spanned by the two complex variables. Moreover, since the one-dimensional transform is a special case of the two-dimensional transform properties of the one-dimensional transform are not satisfied by the two-dimensional one as we will see. That is the case when we try to find the poles and zeros and the region of convergence of two-dimensional Z-transforms.

Example 10.33. Consider the two-dimensional Z-transform

$$X(z_1, z_2) = \frac{z_1^{-1} + z_2^{-1}}{1 + z_1^{-1} z^{-2}}.$$

Find the poles and zeros of $X(z_1, z_2)$.

Solution: Being functions of two variables the numerator $z_1^{-1} + z_2^{-1}$ and the denominator $1 + z_1^{-1} z^{-2}$ of $X(z_1, z_2)$ cannot be factored to obtain its 'zeros' and 'poles' like in one dimension. Letting the numerator equal zero, the zeros can be expressed as functions of either z_1 or z_2 and likewise for the poles. For $z_1 = r_1 e^{j\theta_1}$ and $z_2 = r_2 e^{j\theta_2}$ we have

$$\text{Zeros: } z_1^{-1} + z_2^{-1} = 0 \Rightarrow z_1 = -z_2 \Rightarrow r_1 = r_2 \text{ and } \theta_1 = \theta_2 + \pi,$$

$$\text{Poles: } 1 + z_1^{-1} z_2^{-1} = 0 \Rightarrow z_1 = -1/z_2 \Rightarrow r_1 = 1/r_2 \text{ and } \theta_1 = -\theta_2 + \pi.$$

The zero and pole loci can be obtained in different ways: (i) by mapping values of z_2 in the complex plane of z_1 as indicated above; (ii) by obtaining maps of r_1 vs. r_2 and θ_1 vs. θ_2, or (iii) by mapping the real and imaginary parts of $z_1 = \mathcal{R}e(z_1) + j\mathcal{I}m(z_1)$ and $z_2 = \mathcal{R}e(z_2) + j\mathcal{I}m(z_2)$. Notice also that the zero and the pole loci intersect when $z_1 = -1$ and $z_2 = 1$ (and when $z_1 = 1$ and $z_2 = -1$) which implies that the value $(-1, 1)$ (and $(1, -1)$) is at the same time a zero and a pole. Thus $X(-1, 1) = X(1, -1) = 0/0$ is undefined because of the division by zero, and it is a special singularity that cannot be canceled since the numerator and denominator cannot be factored.[1] □

Example 10.34. The definition of the region of convergence differs from that of the one-dimensional transform. Determine the Z-transform of $x[m, n] = \alpha^m u_1[m, m]$, find its region of convergence and plot it. If $\alpha = 0.5$ and $\alpha = 2$, for which of these two values is $X(e^{j\omega_1}, e^{j\omega_2})$ defined?

Solution: We have

$$X(z_1, z_2) = \sum_{k=0}^{\infty} \alpha^k z_1^{-k} z_2^{-k} = \frac{1}{1 - \alpha z_1^{-1} z_2^{-1}}, \quad \text{ROC: } |\alpha z_1^{-1} z_2^{-1}| < 1, \text{ or } |z_1| > \frac{|\alpha|}{|z_2|}. \quad (10.53)$$

The regions of convergence in the one-dimensional case were given in relation with the unit circle $|z| = 1$, but that cannot be done in the four-dimensional complex plane (z_1, z_2). Thus, the ROC is expressed in terms of $|z_1|$ and $|z_2|$ in the two-dimensional plane $(|z_1|, |z_2|)$ where the unit bidisc $|z_1| = 1$, $|z_2| = 1$ becomes a point. See Fig. 10.18. For $\alpha = 0.5$, the unit bidisc given by the point $(1, 1)$ is inside the region of convergence, which will permit us to determine $X(e^{j\omega_1}, e^{j\omega_2})$. If $\alpha = 2$, the ROC does not contain the unit bidisc and as such $X(e^{j\omega_1}, e^{j\omega_2})$ is not defined. □

The difficulty in obtaining poles explicitly complicates the inversion of two-dimensional Z-transforms. The partial fraction expansion method used to invert one-dimensional Z-transforms thus cannot be extended to the two-dimensional case. It is thus important to understand the mapping that exists between the sequence $x[m, n]$ and the transform $X(z_1, z_2)$. First, the two-dimensional Z-transform is a linear transform, i.e., if $X(z_1, z_2)$ and $Y(z_1, z_2)$ are the transforms of $x[m, n]$ and $y[m, n]$ then the

---

[1] This is a special singularity that does not occur in one dimension. Suppose that you have a Z-transform $X(z) = (1 - z)/(1 - z^2)$ having a zero and a pole at $z = 1$, causing an undetermined $X(1) = 0/0$. However, because $1 - z^2 = (1 - z)(1 + z)$ the expression for $X(z)$ can be simplified by canceling the pole and the zero at 1 giving $X(z) = 1/(1 + z)$ with no zeros and a pole at -1. The factorization of the numerator and denominator makes it possible to cancel the zero and the pole that are causing the undetermined $0/0$. That is not possible in most cases with functions of (z_1, z_2).

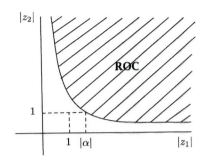

FIGURE 10.18

Region of convergence when $|\alpha| > 1$ in $X(z_1, z_2) = 1/(1 - \alpha z_1^{-1} z_2^{-1})$.

transform of the weighted sum $\alpha x[m, n] + \beta y[m, n]$ for nonzero constants α and β is

$$
\begin{aligned}
\mathcal{Z}_2(\alpha x[m, n] + \beta y[m, n]) &= \sum_{m=-\infty}^{\infty} \sum_{n=-\infty}^{\infty} (\alpha x[m, n] + \beta y[m, n]) z_1^{-m} z_2^{-n} \\
&= \alpha \sum_{m=-\infty}^{\infty} \sum_{n=-\infty}^{\infty} x[m, n] z_1^{-m} z_2^{-n} + \beta \sum_{m=-\infty}^{\infty} \sum_{n=-\infty}^{\infty} y[m, n] z_1^{-m} z_2^{-n} \\
&= \alpha \mathcal{Z}_2(x[m, n]) + \beta \mathcal{Z}_2(y[m, n]). \tag{10.54}
\end{aligned}
$$

Now, how to invert it. Suppose that $X(z_1, z_2)$, with its respective region of convergence,[2] is given and we wish to find the sequence $x[m, n]$ that generated it. According to the definition of $X(z_1, z_2)$ the mapping for every value $x[k, \ell] \delta[m - k, n - \ell]$ assigns a monomial $x[k, \ell] z_1^{-k} z_2^{-\ell}$, and by the linearity for any signal $x[m, n]$ represented in the generic way in terms of $\delta[m, n]$ we have

$$
\begin{aligned}
\mathcal{Z}_2(x[m, n]) &= \mathcal{Z}_2 \left(\sum_{k=-\infty}^{\infty} \sum_{\ell=-\infty}^{\infty} x[k, \ell] \delta[m - k, n - \ell] \right) \\
&= \sum_{k=-\infty}^{\infty} \sum_{\ell=-\infty}^{\infty} x[k, \ell] \underbrace{\mathcal{Z}_2(\delta[m - k, n - \ell])}_{z_1^{-k} z_2^{-\ell}} = X(z_1, z_2). \tag{10.55}
\end{aligned}
$$

Thus the shifting of $\delta[m, n]$ or $\delta[m - k, n - \ell]$ has a transform $\mathcal{Z}_2(\delta[m - k, n - \ell]) = z_1^{-k} z_2^{-\ell}$, and in particular $\mathcal{Z}_2(\delta[m, n]) = 1$. Using this connection it is possible to find the inverse transform, however, it might not be in a closed form.

---

[2] The ROC makes the inverse unique, without it there could be more than one sequence with the given Z-transform.

The inverse of (10.53) is easily obtained from the sum:

$$X(z_1, z_2) = \frac{1}{1 - \alpha z_1^{-1} z_2^{-1}} = \sum_{k=0}^{\infty} \alpha^k z_1^{-k} z_2^{-k}$$

$$= \underbrace{1}_{x[0,0]} + \underbrace{\alpha}_{x[1,1]} z_1^{-1} z_2^{-1} + \underbrace{\alpha^2}_{x[2,2]} z_1^{-2} z_2^{-2} + \cdots + \underbrace{\alpha^k}_{x[k,k]} z_1^{-k} z_2^{-k} + \cdots$$

so that the inverse two-dimensional Z-transform is

$$x[m, n] = \begin{cases} \alpha^m & 0 \leq m < \infty, \ n = m \\ 0 & \text{otherwise} \end{cases} = \alpha^m u_1[m, m].$$

Although in this case we have a closed form for $x[m, n]$, that is not always the case.

The application of the two-dimensional Z-transform to systems is very similar to the one-dimensional case.

Example 10.35. A system is represented by the difference equation

$$y[m, n] = x[m, n] + y[m - 1, n] + y[m, n - 1], \qquad m \geq 0, n \geq 0;$$

find the transfer function $H(z_1, z_2) = Y(z_1, z_2)/X(z_1, z_2)$ and connect it with the impulse response $h[m, n]$ of the system.

Solution: To obtain its impulse response $h[m, n]$ we let $x[m, n] = \delta[m, n]$ be the only input (i.e., the boundary conditions are assumed to be zero) of the difference equation, and thus the impulse response must satisfy

$$h[m, n] = \delta[m, n] + h[m - 1, n] + h[m, n - 1], \qquad m \geq 0, n \geq 0$$

with zero boundary conditions. If we use the linearity, the shifting properties of the two-dimensional Z-transform and let $H(z_1, z_2) = \mathcal{Z}_2(h[m, n])$, the transfer function,[3] we get

$$H(z_1, z_2) = 1 + H(z_1, z_2)z_1^{-1} + H(z_1, z_2)z_2^{-1} \quad \Rightarrow \quad H(z_1, z_2) = \frac{1}{1 - z_1^{-1} - z_2^{-1}}$$

after letting $\mathcal{Z}_2(\delta[m, n]) = 1$. The region of convergence is obtained by expressing

$$H(z_1, z_2) = \frac{1}{1 - (z_1^{-1} + z_2^{-1})} = \sum_{k=0}^{\infty} (z_1^{-1} + z_2^{-1})^k \qquad |z_1^{-1} + z_2^{-1}| < 1$$

thus the ROC is $|z_1^{-1} + z_2^{-1}| < 1$.   □

---

[3] As in one dimension, the transfer function

$$H(z_1, z_2) = \mathcal{Z}_2(h[m, n]) = \frac{Y(z_1, z_2)}{X(z_1, z_2)},$$

i.e., it is the two-dimensional Z-transform of the impulse response $h[m, n]$ and at the same time the ratio of the two-dimensional Z-transforms of the output $y[m, n]$ and the input $x[m, n]$ of an LSI system.

A valuable application of the two-dimensional Z-transform is in the two-dimensional convolution of finite-support sequences. In such cases, the convolution can be seen as a way to obtain the coefficients of a polynomial resulting from the multiplication of other two polynomials in z_1 and z_2.

Example 10.36. Let the input $x[m, n]$ and the impulse response $h[m, n]$ of a system be

$$x[m, n] = \delta[m, n] + 2\delta[m - 1, n] + 3\delta[m, n - 1],$$
$$h[m, n] = \delta[m, n] + \delta[m - 1, n].$$

Find the output $y[m, n]$ of the system.

Solution: The corresponding two-dimensional Z-transforms are

$$X(z_1, z_2) = 1 + 2z_1^{-1} + 3z_2^{-1},$$
$$H(z_1, z_2) = 1 + z_1^{-1},$$

and their product is

$$Y(z_1, z_2) = H(z_1, z_2)X(z_1, z_2) = 1 + 3z_1^{-1} + 2z_1^{-2} + 3z_2^{-1} + 3z_1^{-1}z_2^{-1}.$$

The coefficients of $Y(z_1, z_2)$ are the values obtained from the two-dimensional convolution

$$y[m, n] = (x * h)[m, n] = \delta[m, n] + 3\delta[m - 1, n] + 2\delta[m - 2, n] + 3\delta[m, n - 1] + 3\delta[m - 1, n - 1]$$

(verify it with *conv2* in MATLAB). Thus the coefficients of the product of two bivariate polynomials can be found using the two-dimensional convolution. □

Thus in general for an LSI system, if the input $x[m, n]$ has a transform $X(z_1, z_2)$ and the impulse response $h[m, n]$ has a transform $H(z_1, z_2)$ (transfer function) then the output $y[m, n]$ and its transform $Y(z_1, z_2)$ are related by the following equations in the space domain and in the Z-transform domains:

$$y[m, n] = (h * x)[m, n] \quad \Leftrightarrow \quad Y(z_1, z_2) = H(z_1, z_2)X(z_1, z_2). \tag{10.56}$$

Consider then the case when the input $x[m, n]$ and the impulse response $h[m, n]$ of a two-dimensional LSI system are separable and of finite or infinite size. In that case we see that the Z-transform of $x[m, n] = x_1[m]x_2[n]$ is given by

$$
\begin{aligned}
X(z_1, z_2) &= \sum_{k=-\infty}^{\infty} \sum_{\ell=-\infty}^{\infty} x_1[k]x_2[\ell]z_1^{-k}z_2^{-\ell} \\
&= \sum_{k=-\infty}^{\infty} x_1[k]z_1^{-k} \sum_{\ell=-\infty}^{\infty} x_2[\ell]z_2^{-\ell} = X_1(z_1)X_2(z_2).
\end{aligned} \tag{10.57}
$$

Likewise, if $h[m, n] = h_1[m]h_2[n]$ then

$$H(z_1, z_2) = H_1(z_1)H_2(z_2)$$

and

$$Y(z_1, z_2) = [H_1(z_1)X_1(z_1)][H_2(z_2)X_2(z_2)] \tag{10.58}$$

so that the output $y[m, n]$ is

$$y[m, n] = (h_1 * x_1)[m] * (h_2 * x_2)[n]. \tag{10.59}$$

If $h[m, n]$ is not separable, but $x[m, n]$ is, we then have for the output $w[m, n]$:

$$
\begin{aligned}
W(z_1, z_2) &= H(z_1, z_2)X_1(z_1)X_2(z_2) & (10.60) \\
w[m, n] &= (h * (x_1 * x_2))[m, n], & (10.61)
\end{aligned}
$$

i.e., we perform a one-dimensional convolution of $x_1[m]$ with $x_2[n]$ and the result is two-dimensionally convolved with $h[m, n]$. Thus the two-dimensional convolution can be done in many more forms than in one dimension.

The conditions for BIBO stability can be expressed using the two-dimensional Z-transform. As indicated before, a two-dimensional LSI system is bounded-input bounded-output (BIBO) if its impulse response $h[m, n]$ is absolutely summable, or

$$\sum_{k, \ell} |h[k, \ell]| < \infty.$$

This condition can be obtained in the frequency domain as follows. Computing the corresponding transfer function $H(z_1, z_2) = \mathcal{Z}_2(h[m, n])$ at $z_1 = e^{j\omega_1}$ and $z_2 = e^{j\omega_2}$, or on the unit bidisc, we have

$$|H(e^{j\omega_1}, e^{j\omega_2})| = \left| \sum_{k, \ell} h[k, \ell] e^{-jk\omega_1} e^{-j\ell\omega_2} \right| \le \sum_{k, \ell} |h[k, \ell]| < \infty, \tag{10.62}$$

which means that the system is BIBO stable if the bidisc $z_1 = e^{j\omega_1}$ and $z_2 = e^{j\omega_2}$ or $|z_1| = |z_2| = 1$, is in the region of convergence of $H(z_1, z_2)$ so that $H(e^{j\omega_1}, e^{j\omega_2})$ is defined.

Example 10.37. Determine the BIBO stability of a system represented by the difference equation

$$y[m, n] = x[m, n] + y[m - 1, n] + y[m, n - 1] \qquad m \ge 0, \ n \ge 0.$$

Solution: The impulse response corresponding to this system is

$$h[m, n] = \binom{m + n}{n} u_1[m, n]$$

and that it grows un-boundedly as m and n grow so it is not absolutely summable, and as such the system is not BIBO stable. In Example 10.35 we found the transfer function for this system to be

$$H(z_1, z_2) = \frac{Y(z_1, z_2)}{X(z_1, z_2)} = \frac{1}{1 - (z_1^{-1} + z_2^{-1})}, \qquad \text{ROC: } |z_1^{-1} + z_2^{-1}| < 1.$$

The unit bidisc, i.e., when $z_1 = e^{j\omega_1}$ and $z_2 = e^{j\omega_1}$, is not in the ROC. Indeed, when $\omega_1 = \omega_2 = 0$ we have $z_1 = e^{j0} = 1$ and $z_2 = e^{j0} = 1$ and $|z_1^{-1} + z_2^{-1}| = 1 + 1| = 2$, which is not less than 1. This confirms that this system is not BIBO stable. $\quad\square$

A non-recursive or FIR system is BIBO stable given that its impulse response is absolutely summable due to its finite support. Thus, any system with a transfer function

$$H(z_1, z_2) = \sum_{k=0}^{M-1} \sum_{\ell=0}^{N-1} h[k, \ell] z_1^{-k} z_2^{-\ell} \tag{10.63}$$

has an ROC consisting of the whole (z_1, z_2) plane except for values for which $|z_1| = |z_2| = 0$. Thus, the unit bidisc $|z_1| = |z_2| = 1$ is certainly included in the ROC and as such the system is BIBO stable. On the other hand, a recursive or IIR system has a more general transfer function as it is a ratio of two two-dimensional polynomials. The transfer function corresponding to an impulse response $h[m, n]$ with a first-quadrant support is given as

$$H(z_1, z_2) = \frac{B(z_1, z_2)}{A(z_1, z_2)} = \sum_{k=0}^{\infty} \sum_{\ell=0}^{\infty} h[k, \ell] z_1^{-k} z_2^{-\ell}. \tag{10.64}$$

Because the impulse response has an infinite length, establishing that it is absolutely summable is more difficult than in the non-recursive systems. However, it stills hold true that the filter is BIBO stable if the ROC of $H(z_1, z_2)$ contains the unit bidisc, which makes the corresponding impulse response absolutely summable. As you probably noticed, in Equation (10.64) we consider a system with an impulse response with a first-quadrant support. It so happens that two-dimensional IIR filters can have support in either of the four quadrants or in wedges with angles less of 180 degrees (also called 'asymmetric half-plane' supports). To simplify the stability testing conditions, we first consider the conditions for first-quadrant support filters and for filters with any other support we use linear mappings to converter them into first-quadrant filters.

The most general BIBO stability result is that a system with a transfer function

$$H(z_1, z_2) = \frac{1}{A(z_1, z_2)} \tag{10.65}$$

with impulse response $h[m, n]$ with first-quadrant support is BIBO stable if

$$A(z_1, z_2) \neq 0 \quad \text{for } |z_1| \geq 1, \ |z_2| \geq 1. \tag{10.66}$$

Recall that the one-dimensional BIBO stability condition requires that the poles of the transfer function $H(z)$ be inside the unit disc (this excludes the unit disc). Although the poles of $H(z_1, z_2)$ are surfaces, not points in the complex plane, the stability condition still requires that $H(z_1, z_2)$ be analytic or devoid of singularities in the 'outside' of the unit bidisc, i.e., when $|z_1| \geq 1, \ |z_2| \geq 1$.

Example 10.38. Consider a filter with transfer function

$$H(z_1, z_2) = \frac{1}{1 - \alpha z_1^{-1} z_2^{-1}} \qquad \text{ROC: } |\alpha||z_1^{-1}||z_2^{-1}| < 1.$$

For what values of α is this system BIBO stable?

Solution: The given transfer function can be written as

$$H(z_1, z_2) = \sum_{k=0}^{\infty} \alpha^k (z_1 z_2)^{-k}, \qquad |\alpha||z_1^{-1}||z_2^{-1}| < 1,$$

the impulse response is

$$h[m, n] = \sum_{k=0}^{\infty} \alpha^k \delta[m - k, n - k].$$

If $|\alpha| < 1$, $h[m, n]$ is absolutely summable and so the filter is BIBO stable. Moreover, the poles of $H(z_1, z_2)$ are the values that make $1 - \alpha z_1^{-1} z_2^{-1} = 0$ or

$$z_2^{-1} = \frac{1}{\alpha z_1^{-1}} \quad \text{or} \quad z_2 = \frac{\alpha}{z_1}$$

cannot be on the unit bidisc (if $|z_1| = 1$ then $|z_2| = |\alpha| \neq 1$) or on the outside of it (if $|z_1| = 1/|\alpha| > 1$ then $|z_2| = \alpha^2 < 1$). Thus, for $|\alpha| < 1$

$$A(z_1, z_2) = 1 - \alpha z_1^{-1} z_2^{-1} \neq 0 \quad \text{for } |z_1| \geq 1, |z_2| \geq 1. \quad \square$$

A simplified BIBO stability test for systems with first-quadrant support impulse response $h[m, n]$ and transfer function

$$H(z) = \frac{1}{A(z_1, z_2)}, \qquad \text{consists of two tests:}$$

1. $A(e^{j\omega_1}, z_2) \neq 0$ for all ω_1 and $|z_2| \geq 1$,
2. $A(z_1, 1) \neq 0$ for $|z_1| \geq 1$,

i.e., the roots of $A(e^{j\omega_1}, z_2)$ and $A(z_1, 1)$ must be inside their corresponding unit discs. Otherwise, the filter is not BIBO stable.

Example 10.39. Let $\alpha = 0.5$ in the previous example. Use the above result to verify that the system with the corresponding transfer function is BIBO stable.

Solution: The transfer function is then

$$H(z_1, z_2) = \frac{1}{1 - 0.5 z_1^{-1} z_2^{-1}},$$

and we have

$$A(e^{j\omega_1}, z_2) = 1 - 0.5 e^{-j\omega_1} z_2^{-1},$$
$$A(z_1, 1) = 1 - 0.5 z_1^{-1}.$$

The poles corresponding to these two polynomials are

$$z_2 = 0.5e^{-j\omega_1}, \quad -\pi \leq \omega_1 \leq \pi,$$
$$z_1 = 0.5,$$

with $|z_2| = |z_1| = 0.5 < 1$, i.e., inside their unit circles, so that $A(e^{j\omega_1}, z_2) \neq 0$ for $|z_2| > 1$ and likewise $A(z_1, 1) \neq 0$ for $|z_1| > 1$, and thus the filter is BIBO stable. $\qquad\square$

Example 10.40. Consider a system with transfer function

$$H(z_1, z_2) = \frac{1}{A(z_1, z_2)} = \frac{1}{1 - 0.5z_1^{-1} - 0.5z_2^{-1}}.$$

Determine if it is BIBO stable or not. Use the above result to check.

Solution: The stability conditions are

$$A(e^{j\omega_1}, z_2) = 1 - 0.5e^{-j\omega_1} - 0.5z_2^{-1},$$
$$A(z_1, 1) = 1 - 0.5z_1^{-1} - 0.5,$$

with corresponding poles

$$z_2 = \frac{0.5}{1 - 0.5e^{-j\omega_1}}, \quad -\pi \leq \omega_1 \leq \pi,$$
$$z_1 = 1.$$

The pole $z_1 = 1$ makes the second condition fail so the system is not BIBO stable. $\qquad\square$

Two characteristics of the above results are:

- the effect of the numerator $B(z_1, z_2)$ has not been considered (we have taken it to be unity), and
- the coefficients of the denominator $A(z_1, z_2)$ have first-quarter support.

The reason for considering $B(z_1, z_2) = 1$ is that with two-dimensional transfer functions is possible to obtain cases where the denominator changes the stability of the filter, that is, the filter $H(z_1, z_2) = B(z_1, z_2)/A(z_1, z_2)$ is stable, but the system with transfer function $H'(z_1, z_2) = 1/A(z_1, z_2)$ is unstable. This is due to "singularities of the second kind," which do not occur in one-dimensional transfer functions. This type of singularity occurs when the numerator and the denominator are both zero for some values of z_1 and z_2 and because of the lack of factorization they cannot be canceled. Although these singularities rarely occur, in determining the stability of two-dimensional systems we will ignore the numerator. The second characteristic of only considering system with denominators with first-quadrant support is done to unify the stability test. Wedge supports for $A(z_1, z_2)$ is connected with the recursive computability of the difference equation resulting from the given transfer function and the support of the corresponding impulse response. The case of filters with a wedge support is dealt by transforming the denominator into a first-quadrant polynomial and testing its stability and if stable the other is also stable. Such a linear transformation does not affect the stability of the system.

10.8 WHAT HAVE WE ACCOMPLISHED? WHERE DO WE GO FROM HERE?

Although the history of the Z-transform is originally connected with probability theory, for discrete-time signals and systems it can be connected with the Laplace transform. The periodicity in the frequency domain and the possibility of infinite number of poles and zeros, makes this connection operationally not very useful. Defining a new complex variable in polar form provides the definition of the Z-transform and the Z-plane. As with the Laplace transform, poles of the Z-transform characterize discrete-time signals by means of frequency and attenuation. One- and two-sided Z-transforms are possible, although the one-sided can be used to obtain the two-sided one. The region of convergence makes the Z-transform have a unique relationship with the signal, and will be useful in obtaining the discrete Fourier representations in the next chapter.

Dynamic systems characterized by difference equations use the Z-transform for representation by means of the transfer function. The one-sided Z-transform is useful in the solution of difference equations with nonzero initial conditions. As in the continuous-time case, filters can be represented by difference equations. However, discrete filters represented by polynomials are also possible. These non-recursive filters give significance to the convolution sum, and will motivate us to develop methods that efficiently compute it. For control of digital systems we use the Z-transform and transfer functions (see for instance [27]). Discrete-time systems can also be represented by state-variable models ([42]). The state represents the memory of the system and it is more general than the transfer function. Modern control is based on state-representation of continuous as well as discrete-time systems.

In this chapter we considered also the two-dimensional Z-transform that will be useful in the processing and representation of two-dimensional signals and systems. Because this new transform is more general than the one-dimensional Z-transform it is more difficult and many of the properties of the one-dimensional transform are not satisfied. However, as the one-dimensional transform it is useful in the representation of two-dimensional systems by the transfer function, in the solution of difference equations and the testing of BIBO stability.

10.9 PROBLEMS
10.9.1 BASIC PROBLEMS

10.1 The poles of the Laplace transform $X(s)$ of an analog signal $x(t)$ are $p_{1,2} = -1 \pm j1$, $p_3 = 0$, $p_{4,5} = \pm j1$, and there are no zeros. If we use the transformation $z = e^{sT_s}$ with $T_s = 1$,

    (a) Determine where the given poles are mapped into the Z-plane. Carefully plot the poles and zeros of the analog and the discrete-time signals in the Laplace and the Z-planes.

    (b) How would you determine if these poles are mapped inside, on, or outside the unit disk in the Z-plane? Explain.

    Answer: Poles $s_{1,2} = -1 \pm j1$ are mapped to $z_{1,2} = e^{-1}e^{\pm j1}$ inside the unit disc.

10.2 Given the anticausal signal $x[n] = -\alpha^n u[-n]$.

    (a) Determine its Z-transform $X(z)$, and carefully plot the ROC when $\alpha = 0.5$ and $\alpha = 2$. For which of the two values of α does $X(e^{j\omega})$ exist?

    (b) Find the signal that corresponds to the derivative $dX(z)/dz$. Express it in terms of α.

    Answer: If $\alpha = 2$, ROC: $|z| < 2$ includes unit circle so $X(e^{j\omega})$ exists.

10.3 Consider the signal $x[n] = 0.5(1 + [-1]^n)u[n]$.

   (a) Plot $x[n]$ and use the sum definition of the Z-transform to obtain its Z-transform, $X(z)$.

   (b) Use the linearity property and the Z-transforms of $u[n]$ and $[-1]^n u[n]$ to find the Z-transform $X(z) = \mathcal{Z}[x[n]]$. Compare this result with the previous one.

   (c) Determine and plot the poles and zeros of $X(z)$.

   Answer: $X(z) = z^2/(z^2 - 1)$, $|z| > 1$.

10.4 A LTI system is represented by the first-order difference equation

$$y[n] = x[n] - 0.5y[n-1] \qquad n \geq 0$$

where $y[n]$ is the output and $x[n]$ is the input.

   (a) Find the Z-transform $Y(z)$ in terms of $X(z)$ and the initial condition $y[-1]$.

   (b) Find an input $x[n] \neq 0$ and an initial condition $y[-1] \neq 0$ so that the output is $y[n] = 0$ for $n \geq 0$. Verify you get this result by solving the difference equation recursively.

   (c) For a zero initial condition, find the input $x[n]$ so that $y[n] = \delta[n] + 0.5\delta[n-1]$.

   Answer: $x[n] = \delta[n]$, $y[-1] = 2$ give $Y(z) = 0$.

10.5 A second-order system is represented by the difference equation $y[n] = 0.25y[n-2] + x[n]$ where $x[n]$ is the input and $y[n]$ the output.

   (a) For the zero-input case, i.e., when $x[n] = 0$, find the initial conditions $y[-1]$ and $y[-2]$ so that $y[n] = 0.5^n u[n]$.

   (b) Suppose the input is $x[n] = u[n]$, without solving the difference equation can you find the corresponding steady state $y_{ss}[n]$? Explain how and give the steady state output.

   (c) Find the input $x[n]$ so that, for zero initial conditions, the output is given as $y[n] = 0.5^n u[n]$.

   (d) If $x[n] = \delta[n] + 0.5\delta[n-1]$ is the input to the above difference equation, find the impulse response $h[n]$ of the system.

   Answers: $y[-1] = 2$, $y[-2] = 4$; $y_{ss}[n] = 4/3$; $h[n] = [(-1)^n + 1]0.5^{n+1}u[n]$.

10.6 The transfer function of a causal LTI discrete-time system is $H(z) = (1 + z^{-1})/(1 - 0.5z^{-1})$.

   (a) Find the poles and zeros of $H(z)$. Choose the correct region of convergence corresponding to $H(z)$ from the following and explain your choice:

$$(i)\ |z| < 0.5, \qquad (ii)\ 0.5 < |z| < 1, \qquad (iii)\ |z| > 0.5.$$

   (b) Write the difference equation characterizing this system.

   (c) Find the impulse response $h[n]$ of this system.

   (d) From the pole and zero information determine what type of filter is this system.

   Answers: ROC is (iii); $h[n] = \delta[n] + 3 \times 0.5^n u[n-1]$.

10.7 Suppose we cascade a differentiator and a smoother. The equations for the differentiator is $w[n] = x[n] - x[n-1]$ where $w[n]$ is the output and $x[n]$ the input, and for the smoother the equation is $y[n] = \frac{2}{3}y[n-1] + \frac{1}{3}w[n]$ where $y[n]$ is the output and $w[n]$ the input.

   (a) Obtain the difference equation relating the input $x[n]$ to the output $y[n]$ of the whole system.

(b) If $x[n] = u[n]$, calculate $y[n]$. Assume zero initial conditions.

(c) Determine the steady-state response $y[n]$ to an input $x[n] = u[n] + \sin(\pi n/2)u[n]$.

Answer: If $x[n] = u[n]$ then $y[n] = (1/3)(2/3)^n u[n]$.

10.8 An LTI discrete-time system is characterized by the difference equation

$$y[n] + ay[n-1] + by[n-2] = x[n].$$

Determine for which of the following sets of coefficients the system is BIBO stable:

$$(i) \ a = 2, \ b = 2, \quad (ii) \ a = 1, \ b = 0.5.$$

Answer: (ii) corresponds to a stable system.

10.9 The following problems relate to FIR and IIR systems.

(a) The input and the output of a causal LTI discrete-time system are

$$\text{Input } x[n] = (-1)^n u[n] \quad \text{Output } y[n] = \begin{cases} 0 & n < 0, \\ n+1 & n = 0, \ 1, \ 2, \\ 3(-1)^n & n \geq 3. \end{cases}$$

Determine the impulse response $h[n]$.

(b) The transfer function $H(z)$ of an LTI discrete-time system has a pole at $z = 0$, a double zero at $z = -1$ and a dc gain $H(1) = 2$.

   i. Determine the transfer function $H(z)$ and its region of convergence. Is this a FIR or an IIR filter?

   ii. If the input of the LTI system is $x[n] = (1 + \cos(\pi n/2) + (-1)^n) u[n]$, what is the steady-state response $y_{ss}[n]$ of this system?

(c) The transfer function of an LTI discrete-time system is

$$H(z) = \frac{z}{(z - 0.5)(z + 2)}.$$

   i. Find the poles and zeros of $H(z)$ and indicate its region of convergence if the impulse response $h[n]$ of the system is noncausal.

   ii. Assume $h[n]$ is causal, and find it from $H(z)$.

   iii. Determine whether the system is BIBO stable or not when the impulse response is noncausal or causal. Explain.

Answer: (a) $H(z) = 1 + 3z^{-1} + 5z^{-2}$; (b) $H(z) = 0.5(z+1)^2/z$, FIR; (c) $h(n)$ noncausal, ROC: $1/2 < |z| < 2$; causal: $h[n] = (2/5)(0.5^n - (-2)^n)u[n]$.

10.10 The Z-transform of the unit-step response of a causal LTI discrete-time system is

$$S(z) = \frac{3}{1 - z^{-1}} - \frac{1.5}{1 - 0.5z^{-1}}.$$

Determine the impulse response of the system.

Answer: $h[n] = s[n] - s[n-1]$ where $s[n] = 3(1 - 0.5^{n+1})u[n]$.

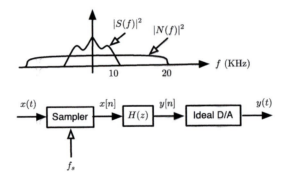

FIGURE 10.19

Problem 10.12.

10.11 The impulse response of a causal LTI discrete-time system is

$$h[n] = \begin{cases} 1 & n = 0,\ 1, \\ 0.25 & n = 2, \\ 0 & \text{otherwise.} \end{cases}$$

(a) If the input of the system is a pulse $x[n] = u[n] - u[n-3]$, determine the length of the output of the system $y[n]$. Graphically calculate the output of the system $y[n]$ as the convolution sum of $x[n]$ and $h[n]$.

(b) Determine the transfer function $H(z)$ of the system. Is this system BIBO stable? Explain.

(c) Suppose you want to recover $x[n]$ from the output $y[n]$ by cascading a system with transfer function $\hat{H}(z)$ to the given system with transfer function $H(z)$. What should $\hat{H}(z)$ be equal to? Is the cascaded system with $\hat{H}(z)$ as transfer function BIBO stable? Can you guarantee recovery of the original signal $x[n]$? Explain.

Answer: $H(z) = 1 + z^{-1} + 0.25z^{-2}$, so the system is BIBO stable.

10.12 We are given a noisy signal

$$x(t) = s(t) + \eta(t)$$

where $s(t)$ is the desired signal and $\eta(t)$ is additive noise. From experience, we know that the average power of the desired signal and the noise has finite support. That is

$$|S(f)|^2 = 0 \qquad f \geq 10 \text{ kHz}$$
$$|N(f)|^2 = 0 \qquad f \geq 20 \text{ kHz}$$

where $S(f)$ and $N(f)$ are the Fourier transforms of $s(t)$ and $\eta(t)$ in terms of the frequency f (see top figure in Fig. 10.19). Suppose we use the system shown at the bottom of Fig. 10.19 to process $x(t)$ to make it "look better", i.e., reduce the effects of the noise without distorting the desired signal.

We have four second-order filters (A to D) to choose from. The gain $K = 0.25$ for each and the poles $\{p_i\}$ and zeros $\{z_i\}$ are

A	$z_i = \pm 1, \quad p_i = (\sqrt{2}/2)\, e^{\pm j\pi/2}$,
B	$z_1 = -1, \; z_2 = 0, \quad p_i = (\sqrt{2}/2)\, e^{\pm j\pi/4}$,
C	$z_1 = -1, \; z_2 = 0, \quad p_i = \sqrt{2}\, e^{\pm j\pi/2}$,
D	$z_i = \pm 1, \quad p_i = \pm 0.5$,

where $i = 1, 2$.

(a) Suppose that for hardware reasons the sampling rate f_s you can use is limited to 10, 20, 30, 40 or 50 kHz. Select f_s from those available and explain the reason for your choice.

(b) Select an appropriate filter from those available and explain the reason for your choice.

(c) Determine $H(z) = KN(z)/D(z)$ for the filter of your choice, where $N(z)$ and $D(z)$ are polynomial in z with real coefficients.

(d) Determine and sketch $|H(e^{j\omega})|$, for $-\pi \le \omega \le \pi$. Evaluate carefully the values at $\omega = 0, \pm\pi$.

Answers: Select either $f_s = 40$ kHz or $f_s = 50$ kHz; select filter B that is stable and low-pass.

10.13 The transfer function of a discrete-time system is

$$H(z) = \frac{z^2 - 1}{z^2 - (\alpha + \beta)z + \alpha\beta}$$

with $\alpha = r_1 e^{j\theta_1}$ and $\beta = r_2 e^{j\theta_2}$, where $r_i > 0$ and θ_i are angles between 0 and 2π.

(a) Determine the range of values of r_1 and r_2 for which the impulse response $h[n]$ of the system tends to zero as $n \to \infty$.

(b) For what values of α and β would the frequency response be zero at $\omega = 0$ and π and ∞ at $\omega = \pi/2$?

Answer: For $h[n] \to 0$ as $n \to \infty$ we need $0 \le r_i < 1$ for $i = 1, 2$.

10.14 A model for echo generation is shown in Fig. 10.20.

(a) Calculate the transfer function $H(z) = Y(z)/X(z)$ of the echo system shown above.

(b) Suppose you would like to recover the original signal $x[n]$ from the output signal $y[n]$ by passing it through a LTI system with transfer function $R(z)$. What should be $R(z)$ so its corresponding output is $x[n]$?

(c) Suppose that $y[n] = u[n] - u[n-2]$; find the original signal $x[n]$.

Answer: $x[n] = \delta[n] + 0.5\delta[n-1] - 0.75\delta[n-2] - 0.25\delta[n-3]$.

10.15 The following are matrices for the state variable and the output equations of a LTI system:

$$\mathbf{A} = \begin{bmatrix} 0 & 1 \\ -1/3 & 1 \end{bmatrix}, \quad \mathbf{b} = \begin{bmatrix} 0 \\ 1/3 \end{bmatrix}, \quad \mathbf{c}^T = \begin{bmatrix} -\tfrac{1}{3} & 1 \end{bmatrix}, \quad d = \begin{bmatrix} \tfrac{1}{3} \end{bmatrix}.$$

Assume $v_i[n]$, $i = 1, 2$, are the state variables, and $x[n]$ the input and $y[n]$ the output. Use the following transformation matrix:

$$\mathbf{T} = \begin{bmatrix} 0 & 1 \\ 1 & 0 \end{bmatrix}$$

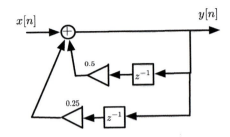

FIGURE 10.20

Problem 10.14.

to obtain a different set of state variables

$$\mathbf{w}[n] = \mathbf{T}\mathbf{v}[n].$$

Obtain the matrices $(\tilde{\mathbf{A}}, \tilde{\mathbf{b}}, \tilde{\mathbf{c}}^T)$ for the new state variables and a corresponding block diagram.
Answers: $w_1[n] = v_2[n]$, $w_2[n] = v_1[n]$; $\tilde{\mathbf{b}}^T = \begin{bmatrix} 1/3 & 0 \end{bmatrix}^T$.

10.16 Given the matrices corresponding to the state and output equations for a system with input $x[n]$ and output $y[n]$:

$$\mathbf{A} = \begin{bmatrix} -a_1 & -a_2 \\ 1 & 0 \end{bmatrix}, \quad \mathbf{b} = \begin{bmatrix} 1 \\ 0 \end{bmatrix}, \quad \mathbf{c}^T = \begin{bmatrix} 0 & b_2 \end{bmatrix}.$$

(a) Find the transfer function $H(z) = Y(z)/X(z)$ corresponding to the state and output equations.

(b) Obtain a minimum realization for $H(z)$. Draw a block diagram of the system. Indicate the state variables.
Answer: $H(z) = b_2 z^{-2}/(1 + a_1 z^{-1} + a_2 z^{-2})$.

10.17 Consider the following two state-variable representations:

$$\mathbf{A}_c = \begin{bmatrix} -a_1 & -a_2 \\ 1 & 0 \end{bmatrix}, \quad \mathbf{b}_c = \begin{bmatrix} 1 \\ 0 \end{bmatrix}, \quad \mathbf{c}_c^T = \begin{bmatrix} b_1 & b_2 \end{bmatrix}$$

$$\mathbf{A}_o = \begin{bmatrix} -a_1 & 1 \\ -a_2 & 0 \end{bmatrix}, \quad \mathbf{b}_o = \begin{bmatrix} b_1 \\ b_2 \end{bmatrix}, \quad \mathbf{c}_o^T = \begin{bmatrix} 1 & 0 \end{bmatrix}$$

where the first is the controller form and the second the observer form. Find the corresponding functions $H_c(z) = Y_c(z)/X_c(z)$ and $H_o(z) = Y_o(z)/X_o(z)$, for inputs $x_c[n]$, $x_o[n]$ and outputs $y_c[n]$ and $y_o[n]$. How are $H_c(z)$ and $H_o(z)$ related?
Answer: $H_c(z) = H_o(z)$.

10.18 Find an inverting transformation represented by the matrix

$$\mathbf{T} = \begin{bmatrix} t_1 & t_2 \\ t_3 & t_4 \end{bmatrix}$$

that changes the controller form into the observer form given in the previous problem. *Hint:* consider the transformation of $\mathbf{b}_c$ and $\mathbf{c}_c^T$ and then that of $\mathbf{A_c}$.
Answer: $t_1 = b_1$, $t_2 = t_3 = b_2$.

10.19 Find a state variable and output matrix equations corresponding to the transfer function

$$H(z) = \frac{0.8z^2 - 0.2z}{z^2 - z + 0.5}.$$

Answer: Let $Y(z) = (0.8 - 0.2z^{-1})W(z)$ and $X(z) = (1 - z^{-1} + 0.5z^{-2})W(z)$ for an additional variable $w[n]$.

10.20 Consider the separable two-dimensional Z-transform

$$X(z_1, z_2) = \frac{z_1 z_2}{1 - 0.5z_1 - 2z_2 + z_1 z_2} \qquad \text{ROC: } |z_1| > p_0, \ |z_2| > p_1$$

where p_0 and p_1 are poles of $X(z_1, z_2)$.
(a) Determine the poles and zeros of $X(z_1, z_2)$.
(b) Carefully indicate the ROC and the location of the unit bidisc in a $(|z_1|, |z_2|)$ plane. According to this, does $X(e^{j\omega_1}, e^{j\omega_2})$ exist?
(c) Find the inverse Z-transform of $X(z_1, z_2)$ or $x[m, n]$ and determine if it is absolutely summable.
Answers: The poles are $p_0 = 2$, $p_1 = 0.5$; $x[m, n]$ is not absolutely summable.

10.21 The transfer function of a two-dimensional system is

$$H(z_1, z_2) = \frac{1}{1 - z_1^{-1}z_2^{-1}}, \qquad \text{ROC: } |z_1| > 1, |z_2| > 1.$$

(a) Obtain the impulse response $h[m, n]$.
(b) Obtain the ROC by looking at the convergence of $H(z_1, z_2)$.
Answer: $h[m, n] = 1$ when $m = n$ and zero otherwise.

10.22 Use the binomial theorem

$$(x + y)^{n_1} = \sum_{n_2 = 0}^{n_1} \binom{n_1}{n_2} x^{n_2} y^{n_1 - n_2}$$

to express the transfer function $H(z_1, z_2) = 1/(1 - z_1^{-1} + z_2^{-1})$ as

$$H(z_1, z_2) = \sum_{m=0}^{\infty} \sum_{n=0}^{\infty} h[m, n] z_1^{-m} z_2^{-n}$$

and determine the impulse response of the system.
Answer: $h[m, n] = \binom{n+m}{m}$.

10.23 Consider a system represented by the convolution sum

$$y[m, n] = \sum_{k=0}^{\infty} \sum_{\ell=-k}^{\infty} h[k, \ell]x[m - k, n - \ell].$$

(a) Obtain the BIBO stability condition for this system, assuming that the input $x[m, n]$ is bounded.

(b) Use the variable transformation

$$\begin{bmatrix} m' \\ n' \end{bmatrix} = \begin{bmatrix} 1 & 0 \\ 1 & 1 \end{bmatrix} \begin{bmatrix} m \\ n \end{bmatrix}$$

to obtain the BIBO stability condition when $x[m, n]$ is exactly the same as before. Is this condition equivalent to the previous? Explain.

Answer: The variable transformation maps $h[m, n]$ with edge support to $h[m', n' - m']$ of quadrant support.

10.24 A filter has a transfer function

$$H(z_1, z_2) = \frac{1 - z_1^{-1}}{1 + 0.5z_1^{-1} + 1.5z_2^{-1}}.$$

(a) Find the poles and zeros of $H(z_1, z_2)$.

(b) For what values of $\hat{z}_1$ and $\hat{z}_2$ is $H(\hat{z}_1, \hat{z}_2) = 0/0$?

(c) Ignoring the numerator, i.e., letting

$$\hat{H}(z_1, z_2) = \frac{1}{1 + 0.5z_1^{-1} + 1.5z_2^{-1}}.$$

Is this filter BIBO stable? Show your work.

Answers: $(1, -1)$ is a singularity of second kind; the filter is not BIBO stable.

10.25 Show that the filter with transfer function

$$H(z_1, z_2) = \frac{1}{1 - (z_1^{-1}z_2^{-1})}$$

is not BIBO stable.

Answers: $A(z_1, 1) \neq 0, |z_1| \geq 1$ is not satisfied.

10.26 The impulse response of a system is

$$h[m, n] = \alpha^{m+n} \beta^{m-n} u_1[m, n].$$

(a) Determine the corresponding transfer function $H(z_1, z_2)$ and its region of convergence.

(b) For the following values of α and β determine if the systems with impulse responses $h[m, n]$ are BIBO stable using the region of convergence:

$$
\begin{array}{ll}
(i) & \alpha = 0.5, \ \beta = 1 \\
(ii) & \alpha = 0.5, \ \beta = 0.5 \\
(iii) & \alpha = 0.5, \ \beta = 2.
\end{array}
$$

(c) For $\alpha = 2$ and $\beta = 1$ determine if the system is BIBO stable using the stability conditions and the region of convergence.
Answer: ROC is $|z_1| > |\alpha\beta|$, $|z_2| > |\alpha/\beta|$.

10.9.2 PROBLEMS USING MATLAB

10.27 Convolution sum and product of polynomials—The convolution sum is a fast way to find the coefficients of the polynomial resulting from the multiplication of two polynomials.
  (a) Suppose $x[n] = u[n] - u[n - 3]$ find its Z-transform $X(z)$, a second-order polynomial in z^{-1}.
  (b) Multiply $X(z)$ by itself to get a new polynomial $Y(z) = X(z)X(z) = X^2(z)$. Find $Y(z)$.
  (c) Do graphically the convolution of $x[n]$ with itself and verify that the result coincides with the coefficients of $Y(z)$.
  (d) Use the *conv* function to find the coefficients of $Y(z)$.
  Answer: $Y(z) = X^2(z) = 1 + 2z^{-1} + 3z^{-2} + 2z^{-3} + z^{-4}$.

10.28 Inverse Z-transform—Use symbolic MATLAB to find the inverse Z-transform of

$$
X(z) = \frac{2 - z^{-1}}{2(1 + 0.25z^{-1})(1 + 0.5z^{-1})}
$$

and determine $x[n]$ as $n \to \infty$.
Answer: $x[n] = [-3(-0.25)^n + 4(-0.5)^n]u[n]$.

10.29 Transfer function, stability and impulse response—Consider a second-order discrete-time system represented by the following difference equation:

$$
y[n] - 2r\cos(\omega_0)y[n - 1] + r^2 y[n - 2] = x[n] \qquad n \geq 0
$$

where $r > 0$ and $0 \leq \omega_0 \leq 2\pi$, $y[n]$ is the output and $x[n]$ the input.
  (a) Find the transfer function $H(z)$ of this system.
  (b) Determine the values of ω_0 and of r that make the system stable. Use the MATLAB function *zplane* to plot the poles and zeros for $r = 0.5$ and $\omega_0 = \pi/2$ rad.
  (c) Let $\omega_0 = \pi/2$, find the corresponding impulse response $h[n]$ of the system. For what other value of ω_0 would get the same impulse response?
  Answer: $H(z) = z^2/[(z - re^{j\omega_0})(z - re^{j\omega_0})]$.

10.30 MATLAB partial fraction expansion—Consider finding the inverse Z-transform of

$$
X(z) = \frac{2z^{-1}}{(1 - z^{-1})(1 - 2z^{-1})^2} \qquad |z| > 2.
$$

(a) MATLAB does the partial fraction expansion:

$$X(z) = \frac{A}{1 - z^{-1}} + \frac{B}{1 - 2z^{-1}} + \frac{C}{(1 - 2z^{-1})^2}$$

while we do it in the following form:

$$X(z) = \frac{D}{1 - z^{-1}} + \frac{E}{1 - 2z^{-1}} + \frac{Fz^{-1}}{(1 - 2z^{-1})^2}.$$

Show that the two partial fraction expansions give the same result.

(b) Obtain $x[n]$ analytically using the two expansions and verify your answer with MATLAB.

Answer: MATLAB: $X(z) = 2/(1 - z^{-1}) + (-2 + 8z^{-1})/(1 - 2z^{-1})^2$.

10.31 State-variable representation—Two systems with transfer functions

$$H_1(z) = \frac{0.2}{1 + 0.5z^{-1}}, \quad H_2(z) = \frac{0.8 - 0.2z^{-1}}{1 - z^{-1} + 0.5z^{-2}}$$

are connected in parallel.

(a) Use MATLAB to determine the transfer function $H(z)$ of the overall system.

(b) Use the function *tf2ss* to obtain state-variable representations for $H_1(z)$ and $H_2(z)$, and the use *ss2tf* to verify these transfer functions are the transfer functions obtained from the models.

(c) Obtain a state-variable representation for $H(z)$ and compare it with the one you would obtain from the state-variable models for $H_1(z)$ and $H_2(z)$.

Answer: $H(z) = H_1(z) + H_2(z) = 1/(1 - 0.5z^{-1} - 0.25z^{-3})$.

DISCRETE FOURIER ANALYSIS

CONTENTS

Signals and Systems Using MATLAB®. https://doi.org/10.1016/B978-0-12-814204-2.00022-3

I am a great believer in luck, and I find the harder I work, the more I have of it.

President Thomas Jefferson (1743–1826)

11.1 INTRODUCTION

Similar to the connection between the Laplace and the Fourier transforms in continuous-time signals and systems, if the region of convergence of the Z-transform of a signal or of the transfer function of a discrete system includes the unit circle then the Discrete-Time Fourier Transform (DTFT) of the signal or the frequency response of the system are easily found from them. Duality in time and in frequency is used whenever signals and systems do not satisfy this condition. We can thus obtain the Fourier representation of most discrete-time signals and systems although, in general, the class of functions having Z-transforms is larger than the class of functions having DTFTs.

Two computational disadvantages of the DTFT are: the direct DTFT is a function of a continuously varying frequency, and the inverse DTFT requires integration. These disadvantages can be removed by sampling in frequency the DTFT resulting in the so-called Discrete Fourier Transform (DFT) (notice the difference in the naming of these two related frequency representations for discrete-time signals). An interesting connection between the time and the frequency domains for these transforms determines their computational feasibility: a *discrete-time* signal has a *periodic continuous-frequency* transform—the DTFT, while a *periodic discrete-time* signal has a *periodic and discrete-frequency* transform—the DFT. As we will discuss in this chapter, any periodic or aperiodic signal can be represented by the DFT, a computationally feasible transformation where both time and frequency are discrete and no integration is required, and very importantly: that can be implemented efficiently by the Fast Fourier Transform (FFT) algorithm.

A great deal of the Fourier representation of discrete-time signals and characterization of discrete systems can be obtained from our knowledge of the Z-transform. Interestingly, to obtain the DFT we will proceed in an opposite direction as we did in the continuous-time: first we consider the Fourier representation of aperiodic signals and then that of periodic signals, and finally use this representation to obtain the DFT.

Two-dimensional signals and systems can be represented and processed in the frequency domain using an analogous transformation, the two-dimensional discrete Fourier transform. Processing of images using linear systems can again be more efficiently done in the frequency domain than in the space domain. The two dimensions require a lot of more care in the processing, but the processing is very similar to the one-dimensional. Image data compression is a topic of great interest and one of the possible implementations is by means of transforms. Given that the Fourier transform gives a signal representation in terms of complex coefficients, alternate transform are used. One commonly used for images is the two-dimensional discrete cosine transform, which provides real-valued coefficients and is able to represent images efficiently.

11.2 THE DISCRETE-TIME FOURIER TRANSFORM (DTFT)

The Discrete-Time Fourier Transform (DTFT) of a discrete-time signal $x[n]$,

$$X(e^{j\omega}) = \sum_n x[n]e^{-j\omega n} \qquad -\pi \le \omega < \pi \tag{11.1}$$

converts the discrete-time signal $x[n]$ into a function $X(e^{j\omega})$ of the discrete frequency ω (rad), while the inverse transform

$$x[n] = \frac{1}{2\pi} \int_{-\pi}^{\pi} X(e^{j\omega})e^{j\omega n} d\omega \tag{11.2}$$

gives back $x[n]$ from $X(e^{j\omega})$.

Remarks

1. The DTFT measures the frequency content of a discrete-time signal.
2. That the DTFT $X(e^{j\omega})$ is periodic in frequency, of period 2π, can be seen from

$$X(e^{j(\omega+2\pi k)}) = \sum_n x[n]e^{-j(\omega+2\pi k)n} = X(e^{j\omega}) \qquad k \text{ integer,}$$

since $e^{-j(\omega+2\pi k)n} = e^{-j\omega n}e^{-j2\pi kn} = e^{-j\omega n}$. Thus, one can think of Equation (11.1) as the Fourier series in frequency of $X(e^{j\omega})$: if $\varphi = 2\pi$ is the period, the Fourier series coefficients are given by

$$x[n] = \frac{1}{\varphi} \int_{\varphi} X(e^{j\omega})e^{j2\pi n\omega/\varphi} d\omega = \frac{1}{2\pi} \int_{-\pi}^{\pi} X(e^{j\omega})e^{jn\omega} d\omega,$$

which coincides with the inverse transform given in (11.2). Because of the periodicity, we only need to consider $X(e^{j\omega})$ for $\omega \in [-\pi, \pi)$.

3. For the DTFT $X(e^{j\omega})$ to converge, being an infinite sum, it is necessary that

$$|X(e^{j\omega})| \le \sum_n |x[n]||e^{j\omega n}| = \sum_n |x[n]| < \infty$$

or that $x[n]$ be absolutely summable. Which means that only for absolutely summable signals, the direct (11.1) and the inverse (11.2) DTFT definitions are valid. We will see next how to obtain the DTFT of signals that are not absolutely summable.

11.2.1 SAMPLING, Z-TRANSFORM, EIGENFUNCTIONS AND THE DTFT

The connection of the DTFT with sampling, eigenfunctions and the Z-transform can be shown as follows:

1. **Sampling and the DTFT.** When sampling a continuous-time signal $x(t)$ the sampled signal $x_s(t)$ can be written as

$$x_s(t) = \sum_n x(nT_s)\delta(t - nT_s).$$

Its Fourier transform is then

$$\mathcal{F}[x_s(t)] = \sum_n x(nT_s)\mathcal{F}[\delta(t - nT_s)] = \sum_n x(nT_s)e^{-jn\Omega T_s}.$$

Letting $\omega = \Omega T_s$, the discrete frequency in radians, and $x[n] = x(nT_s)$ the above equation can be written

$$X_s(e^{j\omega}) = \mathcal{F}[x_s(t)] = \sum_n x[n]e^{-jn\omega}, \tag{11.3}$$

coinciding with the DTFT of the discrete-time signal $x(nT_s) = x(t)|_{t=nT_s}$ or $x[n]$.
At the same time, the spectrum of the sampled signal can be equally represented as

$$X_s(e^{j\Omega T_s}) = X_s(e^{j\omega}) = \sum_k \frac{1}{T_s} X\left(\frac{\omega}{T_s} - \frac{2\pi k}{T_s}\right) \qquad \omega = \Omega T_s, \tag{11.4}$$

which is a periodic repetition, with fundamental period $2\pi/T_s$ (rad/s) in the continuous frequency or 2π (rad) in the discrete frequency, of the spectrum of the continuous-time signal being sampled. Thus sampling converts the continuous-time signal into a discrete-time signal with a periodic spectrum in a continuous frequency.

2. **Z-transform and the DTFT.** If in the above we ignore T_s and consider $x(nT_s)$ a function of n, we can see that

$$X_s(e^{j\omega}) = X(z)|_{z=e^{j\omega}}, \tag{11.5}$$

i.e., the Z-transform computed on the unit circle. For the above to happen, $X(z)$ must have a region of convergence that includes the unit circle $z = 1e^{j\omega}$. There are discrete-time signals for which we cannot find their DTFTs from the Z-transform because they are not absolutely summable, i.e., their ROCs do not include the unit circle. For those signals we will use the duality properties of the DTFT.

3. **Eigenfunctions and the DTFT.** The frequency response of a discrete-time linear time-invariant (LTI) system is shown to be the DTFT of the impulse response of the system. Indeed, according to the eigenfunction property of LTI systems if the input of such a system is a complex exponential, $x[n] = e^{j\omega_0 n}$, the steady-state output, calculated with the convolution sum, is given by

$$y[n] = \sum_k h[k]x[n-k] = \sum_k h[k]e^{j\omega_0(n-k)} = e^{j\omega_0 n} H(e^{j\omega_0}) \tag{11.6}$$

where

$$H(e^{j\omega_0}) = \sum_k h[k]e^{-j\omega_0 k} \tag{11.7}$$

or the DTFT of the impulse response $h[n]$ of the system computed at $\omega = \omega_0$. Notice that the input exponential appears in the steady-state output, multiplied by the frequency response at the frequency of the exponential, i.e., the eigenfunction characteristic. As with continuous-time systems, the system in this case needs to be bounded input–bounded output (BIBO) stable, since without the stability of the system there is no guarantee that there will be a steady-state response.

Example 11.1. Consider the noncausal signal $x[n] = \alpha^{|n|}$ with $|\alpha| < 1$. Determine its DTFT. Use the obtained DTFT to find

$$\sum_{n=-\infty}^{\infty} \alpha^{|n|}.$$

Solution: The Z-transform of $x[n]$ is

$$X(z) = \sum_{n=0}^{\infty} \alpha^n z^{-n} + \sum_{m=0}^{\infty} \alpha^m z^m - 1 = \frac{1}{1 - \alpha z^{-1}} + \frac{1}{1 - \alpha z} - 1 = \frac{1 - \alpha^2}{1 - \alpha(z + z^{-1}) + \alpha^2}$$

where the first term has a ROC: $|z| > |\alpha|$, and the ROC of the second is $|z| < 1/|\alpha|$. Thus the region of convergence of $X(z)$ is

$$\text{ROC: } |\alpha| < |z| < \frac{1}{|\alpha|}$$

and it includes the unit circle. Thus, the DTFT is

$$X(e^{j\omega}) = X(z)|_{z=e^{j\omega}} = \frac{1 - \alpha^2}{(1 + \alpha^2) - 2\alpha \cos(\omega)}. \tag{11.8}$$

Now, according to the formula for the DTFT at $\omega = 0$ we have

$$X(e^{j0}) = \sum_{n=-\infty}^{\infty} x[n]e^{j0n} = \sum_{n=-\infty}^{\infty} \alpha^{|n|} = \frac{2}{1 - \alpha} - 1 = \frac{1 + \alpha}{1 - \alpha}$$

and according to Equation (11.8) equivalently we have

$$X(e^{j0}) = \frac{1 - \alpha^2}{1 - 2\alpha + \alpha^2} = \frac{1 - \alpha^2}{(1 - \alpha)^2} = \frac{1 + \alpha}{1 - \alpha}. \quad \square$$

11.2.2 DUALITY IN TIME AND IN FREQUENCY

There are many signals of interest that do not satisfy the absolute summability condition and as such we cannot find their DTFTs with the definition given in Equation (11.1). Duality in time and frequency permits us to obtain the DTFT of those signals.

Consider the DTFT of the signal $\delta[n-k]$ for some integer k. Since $\mathcal{Z}[\delta[n-k]] = z^{-k}$ with ROC the whole Z-plane except for the origin, the DTFT of $\delta[n-k]$ is $e^{-j\omega k}$. By duality, as in the continuous-time case, we would expect that the signal $e^{-j\omega_0 n}$, $-\pi \leq \omega_0 < \pi$, would have $2\pi\delta(\omega + \omega_0)$ (where $\delta(\omega)$ is the continuous-time delta function) as its DTFT. Indeed, the inverse DTFT of $2\pi\delta(\omega + \omega_0)$ gives

$$\frac{1}{2\pi}\int_{-\pi}^{\pi} 2\pi\delta(\omega + \omega_0)e^{j\omega n}d\omega = e^{-j\omega_0 n}\int_{-\pi}^{\pi}\delta(\omega + \omega_0)d\omega = e^{-j\omega_0 n}.$$

Using these results we have the following dual pairs:

$$\sum_{k=-\infty}^{\infty} x[k]\delta[n-k] \quad \Leftrightarrow \quad \sum_{k=-\infty}^{\infty} x[k]e^{-j\omega k},$$

$$\sum_{k=-\infty}^{\infty} X[k]e^{-j\omega_k n} \quad \Leftrightarrow \quad \sum_{k=-\infty}^{\infty} 2\pi X[k]\delta(\omega + \omega_k). \tag{11.9}$$

The top-left equation in (11.9) is the generic representation of a discrete-time signal $x[n]$ and the corresponding term on the right is its DTFT $X(e^{j\omega})$, one more verification of Equation (11.1). The pair in the bottom of (11.9) is a dual of the above.[1]

The following are two instances where for signals of interest we cannot find their DTFTs unless we use the duality discussed above.

- A constant signal $y[n] = A$, $-\infty < n < \infty$, is not absolutely summable; however, using the equations in (11.9), we have the following dual pairs as special cases:

$$x[n] = A\delta[n] \quad \Leftrightarrow \quad X(e^{j\omega}) = A,$$
$$y[n] = A, \quad -\infty < n < \infty \quad \Leftrightarrow \quad Y(e^{j\omega}) = 2\pi A\delta(\omega) \quad -\pi \leq \omega < \pi.$$

  The signal $y[n]$ does not change from $-\infty$ to ∞, so that its frequency is $\omega = 0$, thus its DTFT $Y(e^{j\omega})$ is concentrated in that frequency.
- Consider then a sinusoid $x[n] = \cos(\omega_0 n + \theta)$ which is not absolutely summable. According to the bottom pair in (11.9) we get

$$x[n] = \frac{1}{2}\left[e^{j(\omega_0 n + \theta)} + e^{-j(\omega_0 n + \theta)}\right] \quad \Leftrightarrow \quad X(e^{j\omega}) = \pi\left[e^{j\theta}\delta(\omega - \omega_0) + e^{-j\theta}\delta(\omega + \omega_0)\right].$$

The DTFT of the cosine indicates that its power is concentrated at the frequency ω_0.

---

[1]Calling this is a dual is not completely correct given that the ω_k are discrete values of frequency instead of continuous as expressed by ω, and that the delta functions are not the same in the continuous and the discrete domains, but a duality of some sort exists in these two pairs which we would like to take advantage of.

The "dual" pairs

$$\delta[n-k], \quad \text{integer } k \quad \Leftrightarrow \quad e^{-j\omega k}, \tag{11.10}$$

$$e^{-j\omega_0 n}, \quad -\pi \le \omega_0 < \pi \quad \Leftrightarrow \quad 2\pi\delta(\omega+\omega_0), \tag{11.11}$$

allow us to obtain the DTFT of signals that do not satisfy the absolutely summable condition. Thus in general, we have

$$\sum_k X[k]e^{-j\omega_k n} \quad \Leftrightarrow \quad \sum_k 2\pi X[k]\delta(\omega+\omega_k). \tag{11.12}$$

The linearity of the DTFT, and the above result shows that for a non-absolutely summable signal

$$x[n] = \sum_\ell A_\ell \cos(\omega_\ell n + \theta_\ell) = \sum_\ell 0.5 A_\ell (e^{j(\omega_\ell n + \theta_\ell)} + e^{-j(\omega_\ell n + \theta_\ell)}),$$

its DTFT is

$$X(e^{j\omega}) = \sum_\ell \pi A_\ell \left[e^{j\theta_\ell}\delta(\omega-\omega_\ell) + e^{-j\theta_\ell}\delta(\omega+\omega_\ell) \right] \quad -\pi \le \omega < \pi.$$

If $x[n]$ is periodic, the discrete frequencies are harmonically related, i.e., $\omega_\ell = \ell\omega_0$ for ω_0 the fundamental frequency of $x[n]$.

Example 11.2. The DTFT of a signal $x[n]$ is

$$X(e^{j\omega}) = 1 + \delta(\omega-4) + \delta(\omega+4) + 0.5\delta(\omega-2) + 0.5\delta(\omega+2).$$

The signal $x[n] = A + B\cos(\omega_0 n)\cos(\omega_1 n)$ is given as a possible signal that has $X(e^{j\omega})$ as its DTFT. Determine whether you can find A, B and ω_0 and ω_1 to obtain the desired DTFT. If not, provide a better $x[n]$.

Solution: Using trigonometric identities we have

$$\cos(\omega_0 n)\cos(\omega_1 n) = 0.5\cos((\omega_1+\omega_0)n) + 0.5\cos((\omega_1-\omega_0)n)$$

so that $x[n] = A + 0.5B\cos((\omega_0+\omega_1)n) + 0.5B\cos((\omega_1-\omega_0)n)$. Letting $\omega_2 = \omega_1 + \omega_0$ and $\omega_3 = \omega_1 - \omega_0$ the DTFT of $x[n]$ is

$$X(e^{j\omega}) = 2\pi A + 0.25\pi B\left[\delta(\omega-\omega_2) + \delta(\omega+\omega_2)\right] + 0.25\pi B\left[\delta(\omega-\omega_3) + \delta(\omega+\omega_3)\right].$$

Comparing this DTFT with the given one we find that

$$2\pi A = 1 \Rightarrow A = 1/(2\pi),$$

$$\omega_2 = \omega_1 + \omega_0 = 4, \quad \omega_3 = \omega_1 - \omega_0 = 2, \Rightarrow \omega_0 = 1, \; \omega_1 = 3,$$

$$0.25\pi B = 1, \; 0.5 \text{ thus there is no unique value for } B.$$

Although we find that for A, ω_0, and ω_1, there is no unique value for B, the given $x[n]$ is not the correct answer. The correct signal is

$$x[n] = \frac{1}{2\pi} + \frac{1}{0.5\pi}\cos(4n) + \frac{1}{\pi}\cos(2n),$$

which has the desired DTFT.   □

11.2.3 COMPUTATION OF THE DTFT USING MATLAB

According to the definition of the direct and the inverse DTFT, see Equations (11.1) and (11.2), their computation needs to be done for a continuous frequency $\omega \in [-\pi, \pi)$ and requires integration. In MATLAB the DTFT is computed for signals of finite support (if the signal has not finite support we need to window it) in a discrete set of frequency values, and a summation instead of integration is used. As we will see later, this can be done by sampling in frequency the DTFT to obtain the discrete Fourier transform or DFT, which in turn is efficiently implemented by an algorithm called the Fast Fourier Transform or FFT. We will introduce the FFT later in this chapter, and so for now consider the FFT as a black box capable of giving a discrete approximation of the DTFT.

To understand the use of the MATLAB function *fft* in the script below consider the following issues:

1. The command X=fft(x), where x is a vector with entries the sample values $x[n]$, $n = 0, \cdots, L - 1$, computes the FFT $X[k]$, $k = 0, \cdots, L - 1$, or the DTFT $X(e^{j\omega})$ at discrete frequencies

$$\omega_k = 2\pi k/L, \ k = 0, \cdots, L - 1.$$

2. The $\{k\}$ values correspond to the discretized frequencies $\{\omega_k = 2\pi k/L\}$ which go from 0 to $2\pi(L - 1)/L$ (close to 2π for large L). This is a discretization of the frequency $\omega \in [0, 2\pi)$.
3. To find an equivalent representation of the frequency, $\omega \in [-\pi, \pi)$, we simply subtract π from $\omega_k = 2\pi k/L$ to get a band of frequencies

$$\tilde{\omega}_k = \omega_k - \pi = \pi \frac{2k - L}{L}, \ k = 0, \cdots, L - 1 \ \text{or} \ -\pi \leq \tilde{\omega}_k < \pi.$$

The frequency $\tilde{\omega}$ can be normalized to $[-1, 1)$, with no units, by dividing it by π. This change in the frequency scale requires a corresponding shift of the magnitude and the phase spectra. This is done by means of the MATLAB function *fftshift*.
4. When plotting the signal, which is discrete in time, the function *stem* is more appropriate than *plot*. However, the *plot* function is more appropriate for plotting the magnitude and the phase frequency response functions which are supposed to be continuously varying with respect to frequency.
5. The function *abs* computes the magnitude and the function *angle* computes the phase of the frequency response. The magnitude and phase are even and odd symmetric, respectively, when plotted in $\omega \in [-\pi, \pi)$ or in the normalized frequency $\omega/\pi \in [-1, 1)$.

Example 11.3. Consider three discrete-time signals: a rectangular pulse, a windowed sinusoid and a chirp. Use the MATLAB functions *fft*, *fftshift*, *abs* and *angle* to compute an approximate DTFT $X(e^{j\omega})$.

Solution: In the script below, to process one of the signals you delete the corresponding comment % and keep it for the other two. The length of the FFT is set to $L = 256$, which is larger than or equal to the length of either of the three signals.

```
%%
% Example 11.3---DTFT of aperiodic signals
%%
% signals
L=256;                                          % length of FFT
% N=21;x=[ones(1,N) zeros(1,L-N)];              % pulse
% N=200; n=0:N-1;x=[cos(4*pi*n/N) zeros(1,L-N)]; % windowed sinusoid
n=0:L-1;x=cos(pi*n.^2/(4*L));                   % chirp
X=fft(x);
w=0:2*pi/L:2*pi-2*pi/L;w1=(w-pi)/pi;            % normalized freq.
n=0:length(x)-1;
subplot(311)
stem(n,x); axis([0 length(n)-1 1.1*min(x) 1.1*max(x)]); grid;
xlabel('n'); ylabel('x(n)')
subplot(312)
plot(w1,fftshift(abs(X)));axis([min(w1) max(w1) 0 1.1*max(abs(X))]);
ylabel('|X|'); grid
subplot(313)
plot(w1,fftshift(angle(X))); ylabel('<X'); xlabel('\omega/\pi'); grid
axis([min(w1) max(w1) 1.1*min(angle(X)) 1.1*max(angle(X))])
```

As expected, the magnitude spectrum for the rectangular pulse looks like a sinc. The windowed sinusoid has a spectrum that resembles that of the sinusoid but the rectangular window makes it broader. Finally, a chirp is a sinusoid with time varying frequency, thus its magnitude spectrum displays components over a range of frequencies. We will comment on the phase spectra later. The results are shown in Fig. 11.1. □

When computing the DTFT of a sampled signal, it is important to display the frequency in rad/s or in Hz rather than the discrete frequency in radians. The discrete frequency ω (rad) is converted into the continuous-time signal Ω (rad/s) according to the relation $\omega = \Omega T_s$ where T_s is the sampling period used. Thus

$$\Omega = \omega / T_s \text{ (rad/s)}. \tag{11.13}$$

If the signal is sampled using the Nyquist sampling rate condition, the discrete frequency range $\omega \in [-\pi, \pi)$ (rad) corresponds to $\Omega \in [-\pi/T_s, \pi/T_s)$ or $[-\Omega_s/2, \Omega_s/2)$, where $\Omega_s/2 \geq \Omega_{max}$ for Ω_s the sampling frequency in (rad/s) and Ω_{max} the maximum frequency in the signal being sampled.

Example 11.4. To illustrate the above, sample a signal $x(t) = 5^{-2t}u(t)$ with $T_s = 0.01$ s/sample, create a vector of 256 values from the signal (we approximate the infinite support of $x(t)$ by a finite support) and compute its FFT as before.

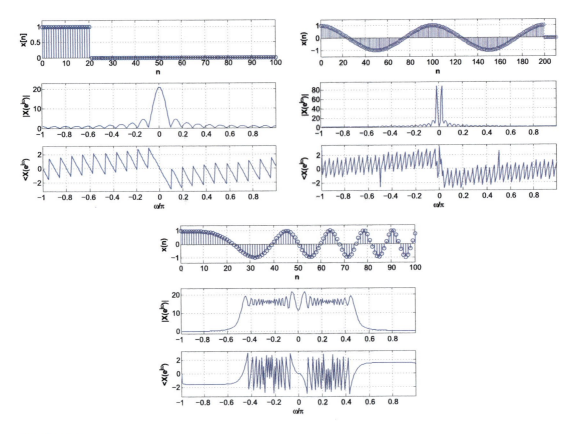

FIGURE 11.1

MATLAB computation of the DTFT of a pulse, a windowed sinusoid and a chirp: magnitude and phase spectra for each.

Solution: To solve this we modify the script in Example 11.3; the changes are

```
%%
% Example 11.4---DTFT of sampled signal
%%
L=256;Ts=0.01; t=0:Ts:(L-1)*Ts; x=5.^(-2*t);    % sampling of signal
X=fft(x);
w=0:2*pi/L:2*pi-2*pi/L;W=(w-pi)/Ts;             % analog  frequency
```

The results are shown in Fig. 11.2. Given that the signal is very smooth, most of the frequency components have low frequencies. □

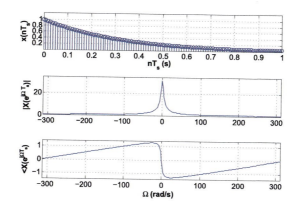

FIGURE 11.2

DTFT of a windowed-sampled signal (top), magnitude of DTFT and phase of DTFT (bottom).

11.2.4 TIME AND FREQUENCY SUPPORTS

The Fourier representation of a discrete-time signal gives a complementary characterization to its time representation, just like in the continuous-time case. The following examples illustrate the complementary nature of the DTFT of discrete-time signals.

> Just like with continuous-time signals, the frequency-support of the DTFT of a discrete-time signal is inversely proportional to the time support of the signal.

Example 11.5. Consider a discrete pulse

$$p[n] = u[n] - u[n - N].$$

Find its DTFT $P(e^{j\omega})$ and discuss the relation between its frequency support and the time support of $p[n]$ when $N = 1$ and when $N \to \infty$.

Solution: Since $p[n]$ has a finite support, its Z-transform has as region of convergence the whole Z-plane, except for $z = 0$, and we can find its DTFT from it. We have

$$P(z) = \sum_{n=0}^{N-1} z^{-n} = 1 + z^{-1} + \cdots + z^{-(N-1)} = \frac{1 - z^{-N}}{1 - z^{-1}}.$$

The DTFT of $p[n]$ is then given by

$$P(e^{j\omega}) = 1 + e^{-j\omega} + \cdots + e^{-j\omega(N-1)},$$

or equivalently

$$P(e^{j\omega}) = \frac{1 - e^{-j\omega N}}{1 - e^{-j\omega}} = \frac{e^{-j\omega N/2}[e^{j\omega N/2} - e^{-j\omega N/2}]}{e^{-j\omega/2}[e^{j\omega/2} - e^{-j\omega/2}]} = e^{-j\omega((N-1)/2)} \frac{\sin(\omega N/2)}{\sin(\omega/2)}.$$

The function $\sin(\omega N/2)/\sin(\omega/2)$ is the discrete counterpart of the sinc function in frequency. It can be shown that like the sinc function, this function is

- even function of ω, as both the numerator and denominator are odd functions of ω,
- $0/0$ at $\omega = 0$, so that using L'Hopital's rule

$$\lim_{\omega \to 0} \frac{\sin(\omega N/2)}{\sin(\omega/2)} = N \lim_{\omega \to 0} \frac{\cos(\omega N/2)}{\cos(\omega/2)} = N,$$

- zero at $\omega = 2\pi k/N$, for integer values $k \neq 0$, as $\sin(\omega N/2)|_{\omega = 2\pi k/N} = \sin(\pi k) = 0$,
- periodic of period 2π when N is odd, which can be seen from its equivalent representation,

$$\frac{\sin(\omega N/2)}{\sin(\omega/2)} = e^{j\omega((N-1)/2)} P(e^{j\omega}),$$

since $P(e^{j\omega})$ is periodic of period 2π and

$$e^{j(\omega + 2\pi)((N-1)/2)} = e^{j\omega((N-1)/2)} e^{j\pi(N-1)} = e^{j\omega((N-1)/2)}$$

when N is odd, e.g., $N = 2M + 1$ for some integer M. When N is even, for instance $N = 2M$ for some integer M, then $e^{j\pi(N-1)} = e^{j2\pi M} e^{-j\pi} = -1$, so $\sin(\omega N/2)/\sin(\omega/2)$ is not periodic of period 2π for N even.

For the discrete pulse when $N = 1$, i.e., $p[n] = u[n] - u[n-1] = \delta[n]$ or the discrete impulse, the DTFT is then $P(e^{j\omega}) = 1$. In this case, the support of $p[n]$ is one point, while the support of $P(e^{j\omega})$ is all discrete frequencies, or $[-\pi, \pi)$.

As we let $N \to \infty$, the pulse tends to a constant, 1, making $p[n]$ not absolutely summable. In the limit, we find that $P(e^{j\omega}) = 2\pi \delta(\omega)$, $-\pi \leq \omega < \pi$. Indeed, the inverse DTFT is found to be

$$\frac{1}{2\pi} \int_{-\pi}^{\pi} 2\pi \delta(\omega) e^{j\omega n} d\omega = \int_{-\pi}^{\pi} \delta(\omega) d\omega = 1.$$

The time support of $p[n] = 1$ is infinite, while $P(e^{j\omega}) = 2\pi \delta(\omega)$ exists at only one frequency, $\omega = 0$ radians. $\qquad \square$

11.2.5 DECIMATION AND INTERPOLATION

Although the expanding and contracting of discrete-time signals is not as obvious as in the continuous-time, the dual effects of contracting and expanding in time and frequency also occur in the discrete case. Contracting and expanding of discrete-time signals relate to down-sampling and up-sampling.

Down-sampling a signal $x[n]$ means getting rid of samples, i.e., contracting the signal. The signal down-sampled by an integer factor $M > 1$ is given by

$$x_d[n] = x[Mn]. \tag{11.14}$$

If $x[n]$ has a DTFT $X(e^{j\omega})$, $-\pi/M \leq \omega \leq \pi/M$, and zero otherwise in $[-\pi, \pi)$ (analogous to band-limited signals in the continuous-time), by replacing n by Mn in the inverse DTFT of $x[n]$ gives

$$x[Mn] = \frac{1}{2\pi} \int_{-\pi/M}^{\pi/M} X(e^{j\omega}) e^{jMn\omega} d\omega = \frac{1}{2\pi} \int_{-\pi}^{\pi} \frac{1}{M} X(e^{j\rho/M}) e^{jn\rho} d\rho$$

where we let $\rho = M\omega$. Thus the DTFT of $x_d[n]$ is $\frac{1}{M} X(e^{j\omega/M})$, i.e., an expansion of the DTFT of $x[n]$ by a factor of M.

Up-sampling a signal $x[n]$, on the other hand, consists in adding $L - 1$ zeros (for some integer $L > 1$) in between the samples of $x[n]$, i.e., the up-sampled signals is

$$x_u[n] = \begin{cases} x[n/L] & n = 0, \pm L, \pm 2L, \cdots, \\ 0 & \text{otherwise,} \end{cases} \tag{11.15}$$

thus expanding the original signal. The DTFT of the up-sampled signal, $x_u[n]$, is found to be

$$X_u(e^{j\omega}) = X(e^{jL\omega}) \qquad -\pi \leq \omega < \pi.$$

Indeed, the DTFT of $x_u[n]$ is

$$X_u(e^{j\omega}) = \sum_{n=0,\pm L,\cdots} x[n/L] e^{-j\omega n} = \sum_{m=-\infty}^{\infty} x[m] e^{-j\omega Lm} = X(e^{jL\omega}), \tag{11.16}$$

indicating a contraction of the DTFT of $x[n]$.

- A signal $x[n]$, band-limited to π/M in $[-\pi, \pi)$ or $|X(e^{j\omega})| = 0$, $|\omega| > \pi/M$ for an integer $M > 1$, can be **down-sampled** by a factor of M to generate a discrete-time signal

$$x_d[n] = x[Mn] \quad \text{with} \quad X_d(e^{j\omega}) = \frac{1}{M} X(e^{j\omega/M}), \tag{11.17}$$

  an expanded version of $X(e^{j\omega})$.
- A signal $x[n]$ is **up-sampled** by a factor of $L > 1$ to generate a signal $x_u[n] = x[n/L]$ for $n = \pm kL$, $k = 0, 1, 2, \cdots$ and zero otherwise. The DTFT of $x_u[n]$ is $X(e^{jL\omega})$ or a compressed version of $X(e^{j\omega})$.

Example 11.6. Consider an ideal low-pass filter with frequency response

$$H(e^{j\omega}) = \begin{cases} 1 & -\pi/2 \leq \omega \leq \pi/2, \\ 0 & -\pi \leq \omega < -\pi/2 \text{ and } \pi/2 < \omega \leq \pi, \end{cases}$$

which is the DTFT of an impulse response $h[n]$. Determine $h[n]$. Suppose that we down-sample $h[n]$ with a factor $M = 2$. Find the down-sampled impulse response $h_d[n] = h[2n]$ and its corresponding frequency response $H_d(e^{j\omega})$.

Solution: The impulse response $h[n]$ corresponding to the ideal low-pass filter is found to be

$$h[n] = \frac{1}{2\pi}\int_{-\pi/2}^{\pi/2} e^{j\omega n}\,d\omega = \begin{cases} 0.5 & n = 0, \\ \sin(\pi n/2)/(\pi n) & n \neq 0. \end{cases}$$

The down-sampled impulse response is given by

$$h_d[n] = h[2n] = \begin{cases} 0.5 & n = 0, \\ \sin(\pi n)/(2\pi n) = 0 & n \neq 0, \end{cases}$$

or $h_d[n] = 0.5\delta[n]$, with a DTFT $H_d(e^{j\omega}) = 0.5$ for $-\pi \leq \omega < \pi$, i.e., an all-pass filter. This agrees with the down-sampling theory, i.e.,

$$H_d(e^{j\omega}) = \frac{1}{2}H(e^{j\omega/2}) = 0.5, \qquad -\pi \leq \omega < \pi,$$

i.e., $H(e^{j\omega})$ multiplied by $1/M = 0.5$ and expanded by $M = 2$. □

Example 11.7.   A discrete pulse is given by $x[n] = u[n] - u[n-4]$. Suppose we down-sample $x[n]$ by a factor of $M = 2$, so that the length 4 of the original signal is reduced to 2, giving

$$x_d[n] = x[2n] = u[2n] - u[2n-4] = u[n] - u[n-2].$$

Find the corresponding DTFTs for $x[n]$ and $x_d[n]$, and determine how are they related.

Solution: The Z-transform of $x[n]$ is

$$X(z) = 1 + z^{-1} + z^{-2} + z^{-3}$$

with the whole Z-plane (except for the origin) as its region of convergence. Thus,

$$X(e^{j\omega}) = e^{-j(\frac{3}{2}\omega)}\left[e^{j(\frac{3}{2}\omega)} + e^{j(\frac{1}{2}\omega)} + e^{-j(\frac{1}{2}\omega)} + e^{-j(\frac{3}{2}\omega)}\right] = 2e^{-j(\frac{3}{2}\omega)}\left[\cos\left(\frac{\omega}{2}\right) + \cos\left(\frac{3\omega}{2}\right)\right].$$

The Z-transform of the down-sampled signal ($M = 2$) is $X_d(z) = 1 + z^{-1}$ and the DTFT of $x_d[n]$ is

$$X_d(e^{j\omega}) = e^{-j(\frac{1}{2}\omega)}\left[e^{j(\frac{1}{2}\omega)} + e^{-j(\frac{1}{2}\omega)}\right] = 2e^{-j(\frac{1}{2}\omega)}\cos\left(\frac{\omega}{2}\right).$$

Clearly this is not equal to $0.5X(e^{j\omega/2})$. This is caused by aliasing: the maximum frequency of $x[n]$ is not $\pi/M = \pi/2$ and so $X_d(e^{j\omega})$ is the sum of superposed $X(e^{j\omega})$.

Suppose we pass $x[n]$ through an ideal low-pass filter $H(e^{j\omega})$ with cut-off frequency $\pi/2$. Its output would be a signal $x_1[n]$ with a maximum frequency of $\pi/2$ and down-sampling it with $M = 2$ would give a signal with a DTFT $0.5X_1(e^{j\omega/2})$. □

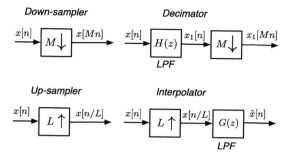

FIGURE 11.3

Down-sampler and decimator (top) and up-sampler and interpolator (bottom). The decimator is a down-sampler with no aliasing, the interpolator changes the zeros of the up-sampled signal to actual samples of the signal.

Cascading a low-pass filter to a down-sampler, to avoid aliasing, gives a **decimator**. Cascading a low-pass filter to an up-sampler, to smooth the signal, gives an **interpolator**.

Decimation—When down-sampling by a factor of M, to avoid aliasing (caused by the signal not being band-limited to $[-\pi/M, \ \pi/M]$) an antialiasing discrete low-pass filter is used before the down-sampler. The frequency response of the filter is

$$H(e^{j\omega}) = \begin{cases} 1 & -\pi/M \le \omega \le \pi/M, \\ 0 & \text{otherwise in } [-\pi, \pi). \end{cases} \tag{11.18}$$

Interpolation—When up-sampling by a factor of L, to change the zero samples in the up-sampler into actual samples, we smooth out the up-sampled signal using an ideal low-pass with frequency response

$$G(e^{j\omega}) = \begin{cases} L & -\pi/L \le \omega \le \pi/L, \\ 0 & \text{otherwise in } [-\pi, \pi). \end{cases} \tag{11.19}$$

Fig. 11.3 displays the down-sampler and the decimator, and the up-sampler and the interpolator.

Although it is clear that a low-pass filter $G(e^{j\omega})$ is needed to smooth out the up-sampled signal in the interpolator, it is not clear why the filter must have a gain of L. This can be seen by cascading an interpolator and a decimator with $L = M$, so that the input and the output of the cascade is exactly the same. The resulting cascade of the low-pass filters with frequency responses $G(e^{j\omega})$ and $H(e^{j\omega})$ gives an ideal low-pass filter with frequency response

$$F(e^{j\omega}) = G(e^{j\omega})H(e^{j\omega}) = \begin{cases} L & -\pi/L \le \omega \le \pi/L, \\ 0 & \text{otherwise in } [-\pi, \pi). \end{cases}$$

Suppose the input to the interpolator is $x[n]$ then the output of the filter with frequency response $F(e^{j\omega})$ is

$$X(e^{j\omega L})F(e^{j\omega}) = \begin{cases} LX(e^{j\omega L}) & -\pi/L \le \omega \le \pi/L, \\ 0 & \text{otherwise in } [-\pi, \pi), \end{cases}$$

and the output of the down-sampler is

$$X(e^{j\omega L/M})\frac{1}{M}F(e^{j\omega/M}) = \frac{L}{M}X(e^{j\omega L/M}) = X(e^{j\omega}) \qquad -\pi M/L = -\pi \leq \omega \leq \pi M/L = \pi,$$

since $L/M = 1$. Moreover, we have

$$\frac{1}{M}F(e^{j\omega/M}) = \underbrace{L/M}_{1} \qquad \underbrace{-\pi M/L}_{-\pi} \leq \omega \leq \underbrace{\pi M/L}_{\pi}$$

is an all-pass filter. Without the gain of L we would not find that the cascade of the interpolator and the decimator is an all-pass filter.

Example 11.8. Discuss the effects of down-sampling a discrete signal that is not band-limited versus the case of one that is. Consider a unit rectangular pulse of length $N = 10$, down-sample it by a factor $M = 2$, compute and compare the DTFTs of the pulse and its down-sampled version. Do a similar procedure to a sinusoid of discrete frequency $\pi/4$ and comment on the results. Explain the difference between these two cases. Use the MATLAB function *decimate* to decimate the signals and comment on the differences with down-sampling. Use the MATLAB function *interp* to interpolate the down-sampled signals.

Solution: As indicated above, when we down-sample a discrete-time signal $x[n]$ by a factor of M, in order not to have aliasing in frequency the signal must be band-limited to π/M. If the signal satisfies this condition, the spectrum of the down-sampled signal is an expanded version of the spectrum of $x[n]$. To illustrate the aliasing, consider the following script where we down-sample by a factor $M = 2$ first the pulse signal, which is not band-limited to $\pi/2$, and then the sinusoid, which is.

```
%%
% Example 11.8---Down-sampling, decimation and interpolation
%%
x=[ones(1,10) zeros(1,100)];
Nx=length(x); n1=0:19;                                  % first signal
% Nx=200;n=0:Nx-1; x=cos(pi*n/4);                       % second signal
y=x(1:2:Nx-1);                                          % down-sampling
X=fft(x);Y=fft(y);                                      % ffts
L=length(X); w=0:2*pi/L:2*pi-2*pi/L; w1=(w-pi)/pi;      % frequency range
z=decimate(x,2,'fir');                                  % decimation
Z=fft(z);                                               % fft
% interpolation
s=interp(y,2);
```

As shown in Fig. 11.4, the rectangular pulse is not band-limited to $\pi/2$, as it has frequency components beyond $\pi/2$, while the sinusoid is. The DTFT of the down-sampled rectangular pulse (a narrower pulse) is not an expanded version of the DTFT of the pulse, while the DTFT of the down-sampled sinusoid is. The MATLAB function *decimate* uses an FIR low-pass filter to smooth out $x[n]$ to a frequency of $\pi/2$ before down-sampling. In the case of the sinusoid, which satisfies the down-sampling

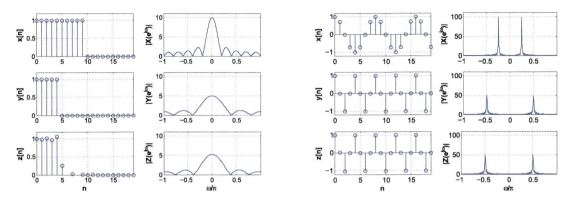

FIGURE 11.4

Down-sampling vs. decimation of non-band-limited (left) and band-limited (right) discrete-time signals. The signals $x[n]$ on left and right correspond to the original signals, while $y[n]$, $z[n]$ are their down-sampled and decimated signals. The corresponding magnitude spectra are shown.

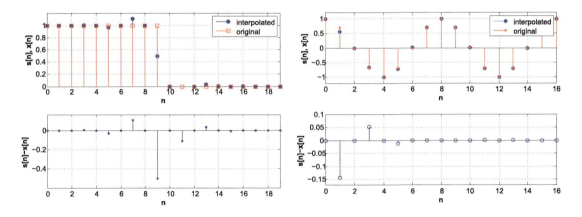

FIGURE 11.5

Interpolation of non-band-limited (left) and band-limited (right) discrete-time down-sampled signals. The interpolated signal is compared to the original signal, the interpolation error is shown.

condition, the down-sampling and the decimation provide the same results. But not for the rectangular pulse.

The original discrete-time signals can be reconstructed by interpolation of the down-sampled signals—very close to the original signal when the signal is band-limited. The MATLAB function *interp* is used to that effect. If we use the down-sampled signal as input to this function, we obtain better results for the sinusoid than for the pulse when comparing the interpolated to the original signals. The results are shown in Fig. 11.5. The error $s[n] - x[n]$ is shown also. The signal $s[n]$ is the interpolation of the down-sampled signal $y[n]$. □

11.2.6 ENERGY/POWER OF APERIODIC DISCRETE-TIME SIGNALS

As for continuous-time signals, the energy or power of a discrete-time signal $x[n]$ can be equivalently computed in time or in frequency.

Parseval's energy equivalence—If the DTFT of a finite-energy signal $x[n]$ is $X(e^{j\omega})$, the energy E_x of the signal is given by

$$E_x = \sum_{n=-\infty}^{\infty} |x[n]|^2 = \frac{1}{2\pi}\int_{-\pi}^{\pi} |X(e^{j\omega})|^2 d\omega. \tag{11.20}$$

Parseval's power equivalence—The power of a finite-power signal $y[n]$ is given by

$$P_y = \lim_{N\to\infty} \frac{1}{2N+1} \sum_{n=-N}^{N} |y[n]|^2 = \frac{1}{2\pi}\int_{-\pi}^{\pi} S_y(e^{j\omega}) d\omega$$

where $S_y(e^{j\omega}) = \lim_{N\to\infty} \frac{|Y_N(e^{j\omega})|^2}{2N+1}$,

$$Y_N(e^{j\omega}) = \mathcal{F}(y[n]W_{2N+1}[n]) \quad \text{DTFT of } y_N[n],$$

$$W_{2N+1} = u[n+N] - u[n-(N+1)] \quad \text{rectangular window.} \tag{11.21}$$

The Parseval's energy equivalence for finite energy $x[n]$ is obtained as follows:

$$
\begin{aligned}
E_x &= \sum_n |x[n]|^2 = \sum_n x[n]x^*[n] = \sum_n x[n]\left[\frac{1}{2\pi}\int_{-\pi}^{\pi} X^*(e^{j\omega})e^{-j\omega n} d\omega\right] \\
&= \frac{1}{2\pi}\int_{-\pi}^{\pi} X^*(e^{j\omega}) \underbrace{\sum_n x[n]e^{-j\omega n}}_{X(e^{j\omega})} d\omega = \frac{1}{2\pi}\int_{-\pi}^{\pi} |X(e^{j\omega})|^2 d\omega.
\end{aligned}
$$

The magnitude square $|X(e^{j\omega})|^2$ has the units of energy per radian, and so it is called an **energy density**. When $|X(e^{j\omega})|^2$ is plotted against frequency ω, the plot is called the **energy spectrum** of the signal, or how the energy of the signal is distributed over frequencies.

Now, if the signal $y[n]$ has finite power we have

$$P_y = \lim_{N\to\infty} \frac{1}{2N+1} \sum_{n=-N}^{N} |y[n]|^2$$

and windowing $y[n]$ with a rectangular window $W_{2N+1}[n]$

$$y_N[n] = y[n]W_{2N+1}[n] \quad \text{where } W_{2N+1}[n] = \begin{cases} 1 & -N \le n \le N, \\ 0 & \text{otherwise,} \end{cases}$$

we have

$$
\begin{aligned}
P_y &= \lim_{N\to\infty} \frac{1}{2N+1} \sum_{n=-\infty}^{\infty} |y_N[n]|^2 = \lim_{N\to\infty} \frac{1}{2N+1} \left[\frac{1}{2\pi} \int_{-\pi}^{\pi} |Y_N(e^{j\omega})|^2 d\omega \right] \\
&= \frac{1}{2\pi} \int_{-\pi}^{\pi} \underbrace{\lim_{N\to\infty} \frac{|Y_N(e^{j\omega})|^2}{2N+1}}_{S_y(e^{j\omega})} d\omega.
\end{aligned}
$$

Plotting $S_y(e^{j\omega})$ as a function of ω provides the distribution of the power over frequency. Periodic signals constitute a special case of finite-power signals and their power spectrum is much simplified by their Fourier series as we will see later in this chapter.

The significance of the above results is that for any signal, whether of finite energy or of finite power, we obtain a way to determine how the energy or power of the signal is distributed over frequency. The plots of $|X(e^{j\omega})|^2$ and $S_y(e^{j\omega})$ versus ω, corresponding to the finite-energy signal $x[n]$ and the finite-power signal $y[n]$ are called the energy spectrum and the power spectrum, respectively. If the signal is known to be infinite energy and finite power, the windowed computation of the power allows us to approximate the power and the power spectrum for a finite number of samples.

11.2.7 TIME AND FREQUENCY SHIFTS

Shifting a signal in time does not change its frequency content. Thus the magnitude of the DTFT of the signal is not affected, only the phase is. Indeed, if $x[n]$ has a DTFT $X(e^{j\omega})$, then the DTFT of $x[n-N]$ for some integer N is

$$
\mathcal{F}(x[n-N]) = \sum_n x[n-N]e^{-j\omega n} = \sum_m x[m]e^{-j\omega(m+N)} = e^{-j\omega N} X(e^{j\omega}).
$$

If $x[n]$ has a DTFT

$$
X(e^{j\omega}) = |X(e^{j\omega})|e^{j\theta(\omega)}
$$

where $\theta(\omega)$ is the phase, the shifted signal $x_1[n] = x[n-N]$ has a DTFT

$$
\begin{aligned}
X_1(e^{j\omega}) &= X(e^{j\omega})e^{-j\omega N} \\
&= |X(e^{j\omega})|e^{-j(\omega N - \theta(\omega))}.
\end{aligned}
$$

In a dual way, when we multiply a signal by a complex exponential $e^{j\omega_0 n}$, for some frequency ω_0, the spectrum of the signal is shifted in frequency. So if $x[n]$ has a DTFT $X(e^{j\omega})$ the modulated signal $x[n]e^{j\omega_0 n}$ has as DTFT $X(e^{j(\omega-\omega_0)})$. Indeed, the DTFT of $x_1[n] = x[n]e^{j\omega_0 n}$ is

$$
X_1(e^{j\omega}) = \sum_n x_1[n]e^{-j\omega n} = \sum_n x[n]e^{-j(\omega-\omega_0)n} = X(e^{j(\omega-\omega_0)}).
$$

The following pairs illustrate the duality in time and frequency shifts: if the DTFT of $x[n]$ is $X(e^{j\omega})$ then

$$x[n - N] \quad \Leftrightarrow \quad X(e^{j\omega})e^{-j\omega N},$$

$$x[n]e^{j\omega_0 n} \quad \Leftrightarrow \quad X(e^{j(\omega-\omega_0)}).$$

(11.22)

Remark

The signal $x[n]e^{j\omega_0 n}$ is called modulated because $x[n]$ modulates the complex exponential $e^{j\omega_0 n}$ or discrete-time sinusoids, as it can be written

$$x[n]e^{j\omega_0 n} = x[n]\cos(\omega_0 n) + jx[n]\sin(\omega_0 n).$$

Example 11.9. The DTFT of $x[n] = \cos(\omega_0 n)$, $-\infty < n < \infty$, cannot be found from the Z-transform or from the sum defining the DTFT as $x[n]$ is not a finite-energy signal. Use the frequency-shift and the time-shift properties to find the DTFTs of $x[n] = \cos(\omega_0 n)$ and $y[n] = \sin(\omega_0 n)$.

Solution: Using Euler's identity we have $x[n] = \cos(\omega_0 n) = 0.5(e^{j\omega_0 n} + e^{-j\omega_0 n})$, and so the DTFT of $x[n]$ is given by

$$X(e^{j\omega}) = \mathcal{F}[0.5e^{j\omega_0 n}] + \mathcal{F}[0.5e^{-j\omega_0 n}] = \mathcal{F}[0.5]_{\omega-\omega_0} + \mathcal{F}[0.5]_{\omega+\omega_0} = \pi\,[\delta(\omega - \omega_0) + \delta(\omega + \omega_0)]$$

where we used $\mathcal{F}[0.5] = 2\pi(0.5)\delta(\omega)$.

Since $y[n] = \sin(\omega_0 n) = \cos(\omega_0 n - \pi/2) = \cos(\omega_0(n - \pi/(2\omega_0))) = x[n - \pi/(2\omega_0)]$, we see that according to the time-shift property its DTFT is given by

$$
\begin{aligned}
Y(e^{j\omega}) &= X(e^{j\omega})e^{-j\omega\pi/(2\omega_0)} = \pi\left[\delta(\omega - \omega_0)e^{-j\omega\pi/(2\omega_0)} + \delta(\omega + \omega_0)e^{-j\omega\pi/(2\omega_0)}\right] \\
&= \pi\left[\delta(\omega - \omega_0)e^{-j\pi/2} + \delta(\omega + \omega_0)e^{j\pi/2}\right] = -j\pi\,[\delta(\omega - \omega_0) - \delta(\omega + \omega_0)].
\end{aligned}
$$

Thus the frequency content of the cosine and the sine is concentrated at the frequency ω_0.   □

11.2.8 SYMMETRY

When plotting or displaying the spectrum of a real-valued discrete-time signal it is important to know that it is only necessary to show the magnitude and the phase spectra for frequencies $[0\ \pi]$, since the magnitude and the phase of $X(e^{j\omega})$ are even an and odd functions of ω, respectively.

For a real-valued signal $x[n]$ with DTFT

$$
\begin{aligned}
X(e^{j\omega}) &= |X(e^{j\omega})|e^{j\theta(\omega)} \\
&= \mathcal{R}e[X(e^{j\omega})] + j\mathcal{I}m[X(e^{j\omega})]
\end{aligned}
$$

its magnitude is even with respect to ω, while its phase is odd with respect to ω or

$$|X(e^{j\omega})| = |X(e^{-j\omega})|,$$

$$\theta(e^{j\omega}) = -\theta(e^{-j\omega}).$$

(11.23)

The real part of $X(e^{j\omega})$ is an even function of ω, and its imaginary part is an odd function of ω:

$$\mathcal{R}e[X(e^{j\omega})] = \mathcal{R}e[X(e^{-j\omega})],$$
$$\mathcal{I}m[X(e^{j\omega})] = -\mathcal{I}m[X(e^{-j\omega})]. \tag{11.24}$$

This is shown by considering that a real-valued signal $x[n]$ has as inverse DTFT

$$x[n] = \frac{1}{2\pi} \int_{-\pi}^{\pi} X(e^{j\omega})e^{j\omega n} d\omega$$

and its complex conjugate is

$$x^*[n] = \frac{1}{2\pi} \int_{-\pi}^{\pi} X^*(e^{j\omega})e^{-j\omega n} d\omega = \frac{1}{2\pi} \int_{-\pi}^{\pi} X^*(e^{-j\omega'})e^{j\omega' n} d\omega'.$$

But since $x[n] = x^*[n]$, as $x[n]$ is real, comparing the above integrals we have

$$X(e^{j\omega}) = X^*(e^{-j\omega}),$$
$$|X(e^{j\omega})|e^{j\theta(\omega)} = |X(e^{-j\omega})|e^{-j\theta(-\omega)},$$
$$\mathcal{R}e[X(e^{j\omega})] + j\mathcal{I}m[X(e^{j\omega})] = \mathcal{R}e[X(e^{-j\omega})] - j\mathcal{I}m[X(e^{-j\omega})],$$

or that the magnitude is an even function of ω, and that the phase is an odd function of ω. Likewise, the real and the imaginary parts of $X(e^{j\omega})$ are also even and odd functions of ω.

Example 11.10. For the signal $x[n] = \alpha^n u[n], 0 < \alpha < 1$, find the magnitude and the phase of its DTFT $X(e^{j\omega})$.

Solution: The DTFT of $x[n]$ is

$$X(e^{j\omega}) = \frac{1}{1 - \alpha z^{-1}} \Big|_{z=e^{j\omega}} = \frac{1}{1 - \alpha e^{-j\omega}},$$

since the Z-transform of $x[n]$ has a region of convergence $|z| > \alpha$, which includes the unit circle. Its magnitude is

$$|X(e^{j\omega})| = \frac{1}{\sqrt{(1 - \alpha \cos(\omega))^2 + \alpha^2 \sin^2(\omega)}},$$

which is an even function of ω given that $\cos(\omega) = \cos(-\omega)$ and $\sin^2(-\omega) = (-\sin(\omega))^2 = \sin^2(\omega)$. The phase is given by

$$\theta(\omega) = -\tan^{-1}\left[\frac{\alpha \sin(\omega)}{1 - \alpha \cos(\omega)}\right],$$

which is an odd function of ω: as functions of ω, the numerator is odd and the denominator is even so that the argument of the inverse tangent is odd, which is in turn odd. □

Example 11.11. For a discrete-time signal $x[n] = \cos(\omega_0 n + \phi)$, $-\pi \le \phi < \pi$, determine how the magnitude and the phase of the DTFT $X(e^{j\omega})$ change with ϕ.

Solution: The signal $x[n]$ has a DTFT

$$X(e^{j\omega}) = \pi \left[e^{-j\phi} \delta(\omega - \omega_0) + e^{j\phi} \delta(\omega + \omega_0) \right].$$

Its magnitude is

$$|X(e^{j\omega})| = |X(e^{-j\omega})| = \pi \left[\delta(\omega - \omega_0) + \delta(\omega + \omega_0) \right]$$

for all values of ϕ. The phase of $X(e^{j\omega})$ is

$$\theta(\omega) = \begin{cases} \phi & \omega = -\omega_0, \\ -\phi & \omega = \omega_0, \\ 0 & \text{otherwise.} \end{cases}$$

In particular, if $\phi = 0$ the signal $x[n]$ is a cosine and its phase is zero. If $\phi = -\pi/2$, $x[n]$ is a sine and its phase is $\pi/2$ at $\omega = -\omega_0$ and $-\pi/2$ at $\omega = -\omega_0$. The DTFT of a sine is

$$X(e^{j\omega}) = \pi \left[\delta(\omega - \omega_0) e^{-j\pi/2} + \delta(\omega + \omega_0) e^{j\pi/2} \right].$$

The DTFTs of the cosine and the sine are only different in the phase. □

The symmetry property, like other properties, also applies to systems. If $h[n]$ is the impulse response of an LTI discrete-time system, and it is real-valued, its DTFT is

$$H(e^{j\omega}) = \mathcal{Z}(h[n])\big|_{z=e^{j\omega}} = H(z)\big|_{z=e^{j\omega}}$$

if the region of convergence of $H(z)$ includes the unit circle. As with the DTFT of a signal, the frequency response of the system, $H(e^{j\omega})$, has a magnitude that is an even function of ω, and a phase that is an odd function of ω. Thus, the **magnitude response** of the system is such that

$$|H(e^{j\omega})| = |H(e^{-j\omega})| \tag{11.25}$$

and the **phase response** is such that

$$\angle H(e^{j\omega}) = -\angle H(e^{-j\omega}). \tag{11.26}$$

According to these symmetries and that the frequency response is periodic, it is only necessary to give these responses in $[0, \pi]$ rather than in $(-\pi, \pi]$.

Computation of the Phase Spectrum Using MATLAB

Computation of the phase using MATLAB is complicated by the following three issues:

- Definition of the phase of a complex number: Given a complex number $z = x + jy = |z|e^{j\theta}$, its phase θ is computed using the inverse tangent function

$$\theta = \tan^{-1}\left(\frac{y}{x}\right).$$

  This computation is not well defined because the principal values of $\tan^{-1}$ are $[-\pi/2, \pi/2]$, while the phase can extend beyond those values. By adding in which quadrant x and y are, the principal values can be extended to $[-\pi, \pi)$. When the phase is linear, i.e., $\theta = -N\omega$ for some integer N, using the extended principal values is not good enough. Plotting a linear phase would show discontinuities when π, or $-\pi$, and their multiples are reached.
- Significance of magnitude when computing phase: Given two complex numbers $z_1 = 1 + j = \sqrt{2}e^{j\pi/4}$ and $z_2 = z_1 \times 10^{-16} = \sqrt{2} \times 10^{-16}e^{j\pi/4}$, they both have the same phase of $\pi/4$ but the magnitudes are very different $|z_2| = 10^{-16}|z_1|$. For practical purposes the magnitude of z_1 is more significant than that of z_2, which is very close to zero, so one could disregard the phase of z_2 with no effect on computations. When computing the phase of these numbers MATLAB cannot make this distinction, and so even very insignificant complex values can have significant phase values.
- Noisy measurements: Given that noise is ever present in actual measurements, even very small noise present in the signal can change the computation of phase.

Phase unwrapping. The problem with phase computation using MATLAB has to do with the way the phase is displayed as a function of frequency with values within $[-\pi, \pi)$ or the **wrapped phase**. If the magnitude of the DTFT of a signal is zero or infinite at some frequencies, the phase at those frequencies is not defined, as any value would be as good as any other because the magnitude is zero or infinity. On the other hand, if there are no zeros or poles on the unit circle to make the magnitude zero or infinite at some frequencies, the phase is continuous. However, because of the way the phase is computed and displayed with values between $-\pi$ to π it seems discontinuous. When the phase is continuous, these phase discontinuities are 2π wide, making the phase values identical. Finding the frequencies where these discontinuities occur and patching the phase, it is possible to obtain the continuous phase. This process is called **unwrapping of the phase**. The MATLAB function *unwrap* is used for this purpose.

Example 11.12. Consider a sinusoid $x[n] = \sin(\pi n/2)$ to which we add a Gaussian noise $\eta[n]$ generated by the MATLAB function *randn*, which theoretically can take any real value. Use the significance of the magnitude computed by MATLAB to estimate the phase.

Solution: The DTFT of $x[n]$ consists of two impulses, one at $\pi/2$ and the other at $-\pi/2$. Thus, the phase should be zero everywhere except at these frequencies. However, any amount of added noise (even in noiseless cases, given the computational precision of MATLAB) would give nonzero phases in places where it should be zero.

In this case, since we know that the magnitude spectrum is significant at the negative and positive values of the sinusoid frequency, we use the magnitude as a mask to indicate where the phase computation can be considered important given the significance of the magnitude. The following script is used to illustrate the masking using the significance of the magnitude:

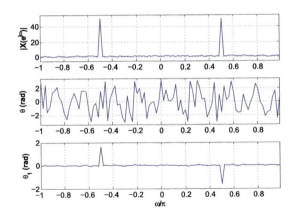

FIGURE 11.6

Phase spectrum of sinusoid in Gaussian noise using magnitude masking.

```
%%
% Example 11.12---Phase of sinusoid in noise
%%
n=0:99; x=sin(pi*n/2)+0.1*randn(1,100);        % sine plus noise
X=fftshift(fft(x));                            % fft of signal
X1=abs(X); theta=angle(X);                     % magnitude and phase
theta1=theta.*X1/max(X1);                      % masked phase
L=length(X);w=0:2*pi/L:2*pi-2*pi/L;w1=(w-pi)/pi; % frequency range
```

Using the mask, the noisy phase (middle plot in Fig. 11.6) is converted into the phase of the sine which occurs when the magnitude is significant (see top and bottom plots of Fig. 11.6).   □

Example 11.13. Consider two FIR filters with impulse responses

$$h_1[n] = \sum_{k=0}^{9} \frac{1}{10}\delta[n-k], \quad h_2[n] = 0.5\delta[n-3] + 1.1\delta[n-4] + 0.5\delta[n-5].$$

Determine which of these filters has linear phase, and use the MATLAB function *unwrap* to find their unwrapped phase functions. Explain the results.

Solution: The transfer function of the filter with $h_1[n]$ is

$$H_1(z) = \frac{1}{10}\sum_{n=0}^{9} z^{-n} = 0.1\frac{1-z^{-10}}{1-z^{-1}} = 0.1\frac{z^{10}-1}{z^9(z-1)} = 0.1\frac{\prod_{k=1}^{9}(z-e^{j2\pi k/10})}{z^9}.$$

Because this filter has nine zeros on the unit circle, its phase is not defined at the frequencies of the zeros (not continuous) and it cannot be unwrapped.

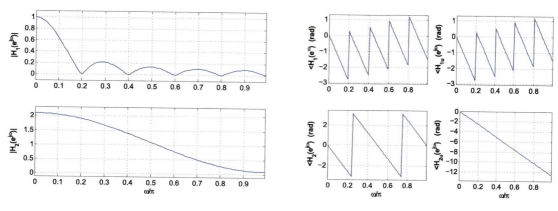

FIGURE 11.7

Unwrapping of the phase: (top) magnitude response, wrapped and unwrapped phase responses of filter with transfer function $H_1(z)$; (bottom) magnitude response, wrapped and unwrapped phase (linear phase) responses of filter with transfer function $H_2(z)$.

The impulse response of the second filter is symmetric about $n = 4$, thus its phase is linear and continuous. Indeed, the transfer function of this filter is

$$H_2(z) = 0.5z^{-3} + 1.1z^{-4} + 0.5z^{-5} = z^{-4}(0.5z + 1.1 + 0.5z^{-1}),$$

which gives the frequency response

$$H_2(e^{j\omega}) = e^{-j4\omega}(1.1 + \cos(\omega)).$$

Since $1.1 + \cos(\omega) > 0$ for $-\pi \leq \omega < \pi$, the phase $\angle H_2(e^{j\omega}) = -4\omega$, which is a line through the origin with slope -4, i.e., a linear phase.

The following script is used to compute the frequency responses of the two filters using *fft*, their wrapped phases using *angle* and then unwrapping them using *unwrap*. The results are shown in Fig. 11.7. □

```
%%
% Example 11.13---Phase unwrapping
%%
h1=(1/10)*ones(1,10);                      % fir filter 1
h2=[ zeros(1,3) 0.5 1.1 0.5 zeros(1,3)];   % fir filter 2
H1=fft(h1,256);                            % fft of h1
H2=fft(h2,256);                            % fft of h2
H1m=abs(H1(1:128));H1p=angle(H1(1:128));   % magnitude/phase of H1(z)
H1up=unwrap(H1p);                          % unwrapped phase of H1(z)
H2m=abs(H2(1:128));H2p=angle(H2(1:128));   % magnitude/phase of H2(z)
H2up=unwrap(H2p);                          % unwrapped phase of H1(z)
```

11.2.9 CONVOLUTION SUM

The computation of the convolution sum, just like the convolution integral in the continuous-time, is simplified in the Fourier domain.

If $h[n]$ is the impulse response of a stable LTI system, its output $y[n]$ can be computed by means of the convolution sum

$$y[n] = \sum_k x[k]\, h[n-k]$$

where $x[n]$ is the input. The Z-transform of $y[n]$ is the product

$$Y(z) = H(z)X(z) \qquad \text{ROC:} \mathcal{R}_Y = \mathcal{R}_H \cap \mathcal{R}_X.$$

If the unit circle is included in $\mathcal{R}_Y$ then

$$Y(e^{j\omega}) = H(e^{j\omega})X(e^{j\omega}) \qquad \text{or}$$
$$|Y(e^{j\omega})| = |H(e^{j\omega})||X(e^{j\omega})|,$$
$$\angle Y(e^{j\omega}) = \angle H(e^{j\omega}) + \angle X(e^{j\omega}). \tag{11.27}$$

Remarks

1. Since the system is stable, the ROC of $H(z)$ includes the unit circle and so if the ROC of $X(z)$ includes the unit circle, the intersection of these regions will also include the unit circle.
2. We will see later it is still possible for $y[n]$ to have a DTFT when the input $x[n]$ does not have a Z-transform with a region of convergence including the unit circle, as when the input is periodic. In this case the output is also periodic. These signals are not finite energy, but finite power, and they can be represented by DTFTs containing continuous-time delta functions.

Example 11.14. Let $H(z)$ be the cascade of first-order systems with transfer functions

$$H_i(z) = K_i \frac{z - 1/\alpha_i}{z - \alpha_i^*} \qquad |z| > |\alpha_i|, \quad i = 1, \cdots N - 1$$

where $|\alpha_i| < 1$ and $K_i > 0$. Such a system is called an **all-pass system** because its magnitude response is a constant for all frequencies. If the DTFT of the filter input $x[n]$ is $X(e^{j\omega})$, determine the gains $\{K_i\}$ so that the magnitude of the DTFT of the output $y[n]$ of the system coincides with the magnitude of $X(e^{j\omega})$.

Solution: Notice that if $1/\alpha_i$ is a zero of $H_i(z)$, a pole at the conjugate reciprocal α_i^* exists. To show that the magnitude of $H_i(e^{j\omega})$ is a constant for all frequencies, consider the magnitude squared function

$$
\begin{aligned}
|H_i(e^{j\omega})|^2 &= H_i(e^{j\omega})H_i^*(e^{j\omega}) = K_i^2 \frac{(e^{j\omega} - 1/\alpha_i)(e^{-j\omega} - 1/\alpha_i^*)}{(e^{j\omega} - \alpha_i^*)(e^{-j\omega} - \alpha_i)} \\
&= K_i^2 \frac{e^{j\omega}(e^{-j\omega} - \alpha_i)e^{-j\omega}(e^{j\omega} - \alpha_i^*)}{\alpha_i \alpha_i^*(e^{j\omega} - \alpha_i^*)(e^{-j\omega} - \alpha_i)} = \frac{K_i^2}{|\alpha_i|^2}.
\end{aligned}
$$

Table 11.1 Discrete-Time Fourier Transform (DTFT) properties		
Z-transform:	$x[n]$, $X(z)$, $\|z\| = 1 \in$ ROC	$X(e^{j\omega}) = X(z)\|_{z=e^{j\omega}}$
Periodicity:	$x[n]$	$X(e^{j\omega}) = X(e^{j(\omega+2\pi k)})$, k integer
Linearity:	$\alpha x[n] + \beta y[n]$	$\alpha X(e^{j\omega}) + \beta Y(e^{j\omega})$
Time-shifting:	$x[n - N]$	$e^{-j\omega N} X(e^{j\omega})$
Frequency shift:	$x[n]e^{j\omega_0 n}$	$X(e^{j(\omega-\omega_0)})$
Convolution:	$(x * y)[n]$	$X(e^{j\omega})Y(e^{j\omega})$
Multiplication:	$x[n]y[n]$	$\dfrac{1}{2\pi}\displaystyle\int_{-\pi}^{\pi} X(e^{j\theta})Y(e^{j(\omega-\theta)})d\theta$
Symmetry:	$x[n]$ real-valued	$\|X(e^{j\omega})\|$ even function of ω
		$\angle X(e^{j\omega})$ odd function of ω
Parseval's relation:	$\displaystyle\sum_{n=\infty}^{\infty} \|x[n]\|^2 = \dfrac{1}{2\pi}\displaystyle\int_{-\pi}^{\pi} \|X(e^{j\omega})\|^2 d\omega$	

Thus by letting $K_i = |\alpha_i|$, the above gives a unit magnitude. The cascade of the $H_i(z)$ gives a transfer function

$$H(z) = \prod_i H_i(z) = \prod_i |\alpha_i| \frac{z - 1/\alpha_i}{z - \alpha_i},$$

so that

$$H(e^{j\omega}) = \prod_i H_i(e^{j\omega}) = \prod_i |\alpha_i| \frac{e^{j\omega} - 1/\alpha_i}{e^{j\omega} - \alpha_i},$$

$$\Rightarrow \quad |H(e^{j\omega})| = \prod_i |H_i(e^{j\omega})| = 1, \quad \angle H(e^{j\omega}) = \sum_i \angle H_i(e^{j\omega}),$$

which in turn gives

$$Y(e^{j\omega}) = |X(e^{j\omega})|e^{j(\angle X(e^{j\omega})+\angle H(e^{j\omega}))},$$

so that the magnitude of the output coincides with that of the input; however, the phase of $Y(e^{j\omega})$ is the sum of the phases of $X(e^{j\omega})$ and $H(e^{j\omega})$. Thus the all-pass system allows all frequency components in the input to appear at the output with no change in the magnitude spectrum, but with a phase shift.   □

Table 11.1 shows some of the properties of the discrete-time Fourier transform.

11.3 FOURIER SERIES OF DISCRETE-TIME PERIODIC SIGNALS

Like in the continuous-time we are interested in finding the response of an LTI system to a periodic signal. As in that case, we represent the periodic signal as a combination of complex exponentials and use the eigenfunction property of LTI systems to find the response.

Notice that we are proceeding in the reverse order we followed in the continuous-time: we are considering the Fourier representation of periodic signals after that of aperiodic signals. Theoretically there is no reason why this cannot be done, but practically it has the advantage of ending with a representation that is discrete and periodic in both time and frequency and as such it can be implemented in a computer. This is the basis of the so-called Discrete Fourier Transform (DFT), which is fundamental in digital signal processing and which we will discuss in the next section. An algorithm called the Fast Fourier Transform (FFT) implements the DFT very efficiently.

Before finding the representation of periodic discrete-time signals recall that:

- A discrete-time signal $x[n]$ is periodic if there is a positive integer N such that $x[n + kN] = x[n]$, for any integer k. This value N is the smallest positive integer that satisfies this condition and it is called the fundamental period of $x[n]$. For the periodicity to hold we need that $x[n]$ be of infinite support, i.e., $x[n]$ must be defined in $-\infty < n < \infty$.
- According to the eigenfunction property of discrete-time LTI systems, whenever the input to such systems is a complex exponential $Ae^{j(\omega_0 n + \theta)}$ the corresponding output in the steady state is

$$y[n] = Ae^{j(\omega_0 n + \theta)} H(e^{j\omega_0}),$$

where $H(e^{j\omega_0})$ is the frequency response of the system at the input frequency ω_0. The advantage of this, as demonstrated in the continuous-time, is that for LTI systems if we are able to express the input signal as a linear combination of complex exponentials then superposition gives us a linear combination of the responses to each exponential. Thus, if the input signal is of the form

$$x[n] = \sum_k A[k] e^{j\omega_k n}$$

then the output will be

$$y[n] = \sum_k A[k] e^{j\omega_k n} H(e^{j\omega_k}).$$

This property is valid whether the frequency components of the input signal are harmonically related (when $x[n]$ is periodic), or not.
- We showed before that a signal $x(t)$, periodic of fundamental period T_0, can be represented by its Fourier series

$$x(t) = \sum_{k=-\infty}^{\infty} \hat{X}[k] e^{j\frac{2\pi kt}{T_0}}. \tag{11.28}$$

If we sample $x(t)$ using a sampling period $T_s = T_0/N$ ($\Omega_s = N\Omega_0$ thus satisfying the Nyquist sampling condition) where N is a positive integer, we get

$$x(nT_s) = \sum_{k=-\infty}^{\infty} \hat{X}[k] e^{j\frac{2\pi kn T_s}{T_0}} = \sum_{k=-\infty}^{\infty} \hat{X}[k] e^{j\frac{2\pi kn}{N}}.$$

The last summation repeats the frequencies between 0 to 2π. To avoid these repetitions, we let $k = m + rN$ where $0 \le m \le N - 1$ and $r = 0, \pm 1, \pm 2, \cdots$, i.e., we divide the infinite support of k

into an infinite number of finite segments of length N, we then have

$$
\begin{aligned}
x(nT_s) &= \sum_{m=0}^{N-1} \sum_{r=-\infty}^{\infty} \hat{X}[m+rN] e^{j\frac{2\pi(m+rN)n}{N}} = \sum_{m=0}^{N-1} \left[\sum_{r=-\infty}^{\infty} \hat{X}[m+rN] \right] e^{j\frac{2\pi mn}{N}} \\
&= \sum_{m=0}^{N-1} X[m] e^{j\frac{2\pi mn}{N}}.
\end{aligned}
$$

This representation is in terms of complex exponentials with frequencies $2\pi m/N$, $m = 0, \cdots, N-1$, from 0 to $2\pi(N-1)/N$. It is this Fourier series representation that we will develop next.

Circular Representation of Discrete-Time Periodic Signals

Since both the fundamental period N of a periodic signal $x[n]$, and the samples in a first period $x_1[n]$ completely characterize a periodic signal $x[n]$, a circular rather than a linear representation would more efficiently represent the signal. The circular representation is obtained by locating uniformly around a circle the values of the first period starting with $x[0]$ and putting in a clockwise direction the remaining terms $x[1], \cdots, x[N-1]$. If we were to continue in the clockwise direction we would get $x[N] = x[0]$, and $x[N+1] = x[1], \cdots, x[2N-1] = x[N-1]$ and so on. In general, any value $x[m]$ where m is represented as

$$
m = kN + r
$$

for integers k, the exact divisor of m by N, and the residue $0 \leq r < N$, equals one of the samples in the first period, that is,

$$
x[m] = x[kN + r] = x[r].
$$

This representation is called **circular** in contrast to the equivalent **linear** representation introduced before:

$$
x[n] = \sum_{k=-\infty}^{\infty} x_1[n + kN],
$$

which superposes shifted versions of the first period $x_1[n]$. The circular representation becomes very useful in the computation of the DFT as we will see later in this chapter. Fig. 11.8 shows the circular and the linear representations of a periodic signal $x[n]$ of fundamental period $N = 4$.

11.3.1 COMPLEX EXPONENTIAL DISCRETE FOURIER SERIES

Consider the representation of a discrete-time signal $x[n]$, periodic of fundamental period N, using the orthogonal basis functions $\{\phi[k, n] = e^{j2\pi kn/N}\}$ for $n, k = 0, \cdots, N-1$. Two important characteristics of these functions are:

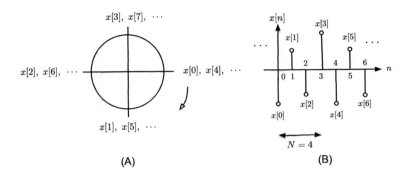

(A)

(B)

FIGURE 11.8

Circular (A) and linear (B) representations of a periodic discrete-time signal $x[n]$ of fundamental period $N = 4$. Notice how the circular representation shows the periodicity $x[0] = x[4], \cdots, x[3] = x[7], \cdots$ for positive and negative integers.

- The functions $\{\phi[k, n]\}$ are periodic with respect to k and to n, with fundamental period N. In fact,

$$\phi[k + \ell N, n] = e^{j \frac{2\pi (k + \ell N)n}{N}} = e^{j \frac{2\pi kn}{N}} e^{j 2\pi \ell n} = e^{j \frac{2\pi kn}{N}}$$

where we used $e^{j2\pi \ell n} = 1$. It can be equally shown that the functions $\{\phi(k, n)\}$ are periodic with respect to n with a fundamental period N.
- The functions $\{\phi(k, n)\}$ are orthogonal over the support of n, that is,

$$\sum_{n=0}^{N-1} \phi(k, n)\phi^*(\ell, n) = \sum_{n=0}^{N-1} e^{j \frac{2\pi}{N} kn} (e^{j \frac{2\pi}{N} \ell n})^* = \begin{cases} N & \text{if } k - \ell = 0, \\ 0 & \text{if } k - \ell \neq 0, \end{cases}$$

and can be normalized by dividing them by $\sqrt{N}$. So $\{\phi[k, n]/\sqrt{N}\}$ are orthonormal functions: indeed, when $k \neq \ell$

$$\frac{1}{N} \sum_{n=0}^{N-1} \phi(k, n)\phi^*(\ell, n) = \frac{1}{N} \sum_{n=0}^{N-1} e^{j \frac{2\pi}{N} (k - \ell)n} = \frac{1 - e^{j \frac{2\pi}{N} (k - \ell)N}}{1 - e^{j \frac{2\pi}{N} (k - \ell)}} = 0$$

or the basis functions are orthogonal. When $k = \ell$ the sum is N so that the basis functions are normal.

These two properties will be used in obtaining the Fourier series representation of periodic discrete-time signals.

The Fourier series representation of a periodic signal $x[n]$ of fundamental period N is

$$x[n] = \sum_{k=k_0}^{k_0+N-1} X[k]e^{j\frac{2\pi}{N}kn} \tag{11.29}$$

where the Fourier series coefficients $\{X[k]\}$ are obtained from

$$X[k] = \frac{1}{N} \sum_{n=n_0}^{n_0+N-1} x[n]e^{-j\frac{2\pi}{N}kn}. \tag{11.30}$$

The frequency $\omega_0 = 2\pi/N$ (rad) is the fundamental frequency, and k_0 and n_0 in (11.29) and (11.30) are arbitrary integer values. The Fourier series coefficients $X[k]$, as functions of frequency $2\pi k/N$, are periodic of fundamental period N.

Remarks

1. The connection of the above two equations can be verified by using the orthonormality of the basis functions. In fact, if we multiply $x[n]$ by $e^{-j(2\pi/N)\ell n}$ and sum these values for n changing over a period, we get using Equation (11.29):

$$\sum_n x[n]e^{-j2\pi n\ell/N} = \sum_n \sum_k X[k]e^{j2\pi(k-\ell)n/N} = \sum_k X[k] \sum_n e^{j2\pi(k-\ell)n/N} = NX[\ell],$$

since $\sum_n e^{j2\pi(k-\ell)n/N}$ is zero, except when $k - \ell = 0$ in which case the sum equals N. Dividing by N we obtain Equation (11.30) changing k by ℓ.

2. Both $x[n]$ and $X[k]$ are periodic with respect to n and k, of the same fundamental period N, as can be easily shown using the periodicity of the basis functions $\{\phi(k,n)\}$. Consequently, the sum over k in the Fourier series and the sum over n in the Fourier coefficients are computed over any period of $x[n]$ and $X[k]$. Thus the sum in the Fourier series is computed from $k = k_0$ to $k_0 + N - 1$ for any value of k_0. Likewise, the sum when computing the Fourier coefficients goes from $n = n_0$ to $n_o + N - 1$, which is an arbitrary period for any integer value n_0.

3. Notice that both $x[n]$ and $X[k]$ can be found using a computer since the frequency is discrete and only sums are needed to implement them. We will use these characteristics in the practical computation of the Fourier transform of discrete-time signals, or the Discrete Fourier Transform (DFT).

Example 11.15. Find the Fourier series of $x[n] = 1 + \cos(\pi n/2) + \sin(\pi n)$, $-\infty < n < \infty$.

Solution: The fundamental period of $x[n]$ is $N = 4$. Indeed,

$$x[n+4] = 1 + \cos(2\pi(n+4)/4) + \sin(\pi(n+4)) = 1 + \cos(2\pi n/4 + 2\pi) + \sin(\pi n + 4\pi) = x[n].$$

The frequencies in $x[n]$ are: a dc frequency, corresponding to the constant, and frequencies $\omega_0 = \pi/2$ and $\omega_1 = \pi = 2\omega_0$ corresponding to the cosine and the sine. No other frequencies are present in the signal. Letting the fundamental frequency be $\omega_0 = 2\pi/N = \pi/2$, the complex exponential Fourier series can be obtained directly from $x[n]$ using Euler's equation:

$$
\begin{aligned}
x[n] &= 1 + 0.5(e^{j\pi n/2} + e^{-j\pi n/2}) - 0.5j(e^{j\pi n} - e^{-j\pi n}) \\
&= X[0] + X[1]e^{j\omega_0 n} + X[-1]e^{-j\omega_0 n} + X[2]e^{j2\omega_0 n} + X[-2]e^{-j2\omega_0 n} \qquad \omega_0 = \frac{\pi}{2};
\end{aligned}
$$

thus the Fourier coefficients are $X[0] = 1$, $X[1] = X^*[-1] = 0.5$, and $X[2] = X^*[-2] = -0.5j$. $\qquad \square$

11.3.2 CONNECTION WITH THE Z-TRANSFORM

Recall that the Laplace transform was used to find the Fourier series coefficients. Likewise, for periodic discrete-time signals the Z-transform of a period of the signal—which always exists—can be connected with the Fourier series coefficients.

If $x_1[n] = x[n](u[n] - u[n - N])$ is the period of a periodic signal $x[n]$, of fundamental period N, its Z-transform

$$
\mathcal{Z}(x_1[n]) = \sum_{n=0}^{N-1} x[n]z^{-n}
$$

has the whole plane, except for the origin, as its region of convergence. The Fourier series coefficients of $x[n]$ are thus determined as

$$
\begin{aligned}
X[k] &= \frac{1}{N} \sum_{n=0}^{N-1} x[n]e^{-j\frac{2\pi}{N}kn} \\
&= \frac{1}{N} \mathcal{Z}(x_1[n]) \Big|_{z=e^{j\frac{2\pi}{N}k}}.
\end{aligned}
\qquad (11.31)
$$

Example 11.16. Consider a discrete pulse $x[n]$ with a fundamental period $N = 20$, and $x_1[n] = u[n] - u[n - 10]$ is the period between 0 and 19. Find the Fourier series of $x[n]$.

Solution: The Fourier series coefficients are found as ($\omega_0 = 2\pi/20$ rad)

$$
X[k] = \frac{1}{20} \mathcal{Z}(x_1[n]) \Big|_{z=e^{j\frac{2\pi}{20}k}} = \frac{1}{20} \sum_{n=0}^{9} z^{-n} \Big|_{z=e^{j\frac{2\pi}{20}k}} = \frac{1}{20} \frac{1 - z^{-10}}{1 - z^{-1}} \Big|_{z=e^{j\frac{2\pi}{20}k}}.
$$

A closed expression for $X[k]$ is obtained as follows:

$$
X[k] = \frac{z^{-5}(z^5 - z^{-5})}{20z^{-0.5}(z^{0.5} - z^{-0.5})} \Big|_{z=e^{j\frac{2\pi}{20}k}} = \frac{e^{-j\pi k/2} \sin(\pi k/2)}{20e^{-j\pi k/20} \sin(\pi k/20)} = \frac{e^{-j9\pi k/20}}{20} \frac{\sin(\pi k/2)}{\sin(\pi k/20)}. \qquad \square
$$

11.3.3 DTFT OF PERIODIC SIGNALS

A discrete-time periodic signal $x[n]$, of fundamental period N, with a Fourier series representation

$$x[n] = \sum_{k=0}^{N-1} X[k] e^{j2\pi nk/N} \tag{11.32}$$

has a DTFT transform

$$X(e^{j\omega}) = \sum_{k=0}^{N-1} 2\pi X[k] \delta(\omega - 2\pi k/N) \qquad -\pi \le \omega < \pi. \tag{11.33}$$

If we let $\mathcal{F}(.)$ be the DTFT operator, the DTFT of a periodic signal $x[n]$ is

$$
\begin{aligned}
X(e^{j\omega}) &= \mathcal{F}(x[n]) = \mathcal{F}\left(\sum_k X[k] e^{j2\pi nk/N} \right) = \sum_k \mathcal{F}(X[k] e^{j2\pi nk/N}) \\
&= \sum_k 2\pi X[k] \delta(\omega - 2\pi k/N) \qquad -\pi \le \omega < \pi
\end{aligned}
$$

where $\delta(\omega)$ is the continuous-time delta function since ω varies continuously.

Example 11.17. The periodic signal

$$\delta_M[n] = \sum_{m=-\infty}^{\infty} \delta[n - mM]$$

has a fundamental period M. Find its DTFT.

Solution: The DTFT of $\delta_M[n]$ is given by

$$\Delta_M(e^{j\omega}) = \sum_{m=-\infty}^{\infty} \mathcal{F}(\delta[n - mM]) = \sum_{m=-\infty}^{\infty} e^{-j\omega mM}. \tag{11.34}$$

An equivalent result can be obtained if before we find the DTFT we find the Fourier series of $\delta_M[n]$. The coefficients of the Fourier series of $\delta_M[n]$ are ($\omega_0 = 2\pi/M$ is its fundamental period):

$$\Delta_M[k] = \frac{1}{M} \sum_{n=0}^{M-1} \delta_M[n] e^{-j2\pi nk/M} = \frac{1}{M} \sum_{n=0}^{M-1} \delta[n] e^{-j2\pi nk/M} = \frac{1}{M}$$

so that the Fourier series of $\delta_M[n]$ is

$$\delta_M[n] = \sum_{k=0}^{M-1} \frac{1}{M} e^{j2\pi nk/M}$$

and its DTFT is then given by

$$\Delta_M(e^{j\omega}) = \sum_{k=0}^{M-1} \mathcal{F}\left(\frac{1}{M}e^{j2\pi nk/M}\right) = \frac{2\pi}{M}\sum_{k=0}^{M-1}\delta\left(\omega - \frac{2\pi k}{M}\right) \qquad -\pi \leq \omega < \pi, \qquad (11.35)$$

which is equivalent to Equation (11.34). Moreover, putting the pair $\delta_M[n]$ and $\Delta_M(e^{j\omega})$ together gives an interesting relation

$$\sum_{m=-\infty}^{\infty}\delta[n-mM] \quad \Leftrightarrow \quad \frac{2\pi}{M}\sum_{k=0}^{M-1}\delta\left(\omega - \frac{2\pi k}{M}\right) \qquad -\pi \leq \omega < \pi.$$

Both terms are discrete in time and in frequency, and both are periodic. The fundamental period of $\delta_M[n]$ is M and the one of $\Delta(e^{j\omega})$ is $2\pi/M$. Furthermore, the DTFT of an impulse train in time is also an impulse train in frequency. However, it should be observed that the delta functions $\delta[n-mM]$ on the left term are discrete, while the ones on the right term $\delta(\omega - 2\pi k/M)$ are continuous. □

Table 11.2 compares the properties of the Fourier Series and of the Discrete Fourier Transform (DFT).

Computation of the Fourier Series Using MATLAB

Given that periodic signals only have discrete frequencies, and that the Fourier series coefficients are obtained using summations, the computation of the Fourier series can be implemented with a frequency discretized version of the DTFT, or the Discrete Fourier Transform (DFT) which can be efficiently computed using the FFT algorithm.

Example 11.18. To illustrate the above using MATLAB, consider three different signals as indicated in the following script and displayed in Fig. 11.9.

Solution: Each of the signals is periodic and we consider 10 periods to compute their FFTs to obtain the Fourier series coefficients (only 3 periods are displayed in Fig. 11.9). A very important issue to remember is that one needs to input exactly one or more periods, and thus that the FFT lengths must be that of a period or of multiples of a period. Notice that the sign function *sign* is used to generate a periodic train of pulses from the cosine function. □

```
%%
% Example 11.18---Fourier series using FFT
%%
N=10; M=10; N1=M*N;n=0:N1-1;
x=cos(2*pi*n/N);                          % sinusoid
x1=sign(x);                               % train of pulses
x2=x-sign(x);                             % sinusoid minus train of pulses
X=fft(x)/M;X1=fft(x1)/M;X2=fft(x2)/M;     % ffts of signals
X=X/N;X1=X1/N;X2=X2/N;                     % FS coefficients
```

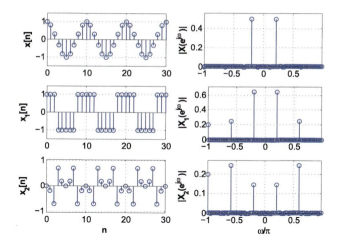

FIGURE 11.9

Computation of Fourier series coefficients of different periodic signals.

11.3.4 RESPONSE OF LTI SYSTEMS TO PERIODIC SIGNALS

Let $x[n]$, a periodic signal of fundamental period N, be the input of an LTI system with transfer function $H(z)$. If the Fourier series of $x[n]$ is

$$x[n] = \sum_{k=0}^{N-1} X[k]e^{jk\omega_0 n} \qquad \omega_0 = \frac{2\pi}{N} \text{ fundamental frequency}$$

then according to the eigenfunction property of LTI systems, the output is also periodic of fundamental period N with Fourier series

$$y[n] = \sum_{k=0}^{N-1} X[k]H(e^{jk\omega_0})e^{jk\omega_0 n} \qquad \omega_0 = \frac{2\pi}{N} \text{ fundamental frequency}$$

and Fourier series coefficients $Y[k] = X[k]H(e^{jk\omega_0})$, i.e., the coefficients $X[k]$ of the input changed by the frequency response of the system

$$H(e^{jk\omega_0}) = H(z)\big|_{z=e^{jk\omega_0}}$$

at the harmonic frequencies $\{k\omega_0, k = 0, \cdots, N-1\}$.

Remarks

1. Although the input $x[n]$ and the output $y[n]$ of the LTI system are both periodic of the same fundamental period, the Fourier series coefficients of the output are affected by the frequency response $H(e^{jk\omega_0})$ of the system at each of the harmonic frequencies.

2. A similar result is obtained by using the convolution property of the DTFT, so that if $X(e^{j\omega})$ is the DTFT of the periodic signal $x[n]$, then the DTFT of the output $y[n]$ is given by

$$
\begin{aligned}
Y(e^{j\omega}) &= X(e^{j\omega})H(e^{j\omega}) \\
&= \left[\sum_{k=0}^{N-1} 2\pi X[k]\delta(\omega - 2\pi k/N) \right] H(e^{j\omega}) \\
&= \sum_{k=0}^{N-1} 2\pi X[k]H(e^{j2\pi k/N})\delta(\omega - 2\pi k/N)
\end{aligned}
$$

and letting $X[k]H(e^{j2\pi k/N}) = Y[k]$ we get the DTFT of the periodic output $y[n]$.

Example 11.19. Consider how to implement a crude spectral analyzer for discrete-time signals using MATLAB. Divide the discrete frequencies $[0 \; \pi]$ in three bands: $[0 \; 0.1\pi]$, $(0.1\pi \; 0.6\pi]$ and $(0.6\pi \; \pi]$ to obtain low-pass, band-pass and high-pass components of the signal $x[n] = \text{sign}(\cos(0.2\pi n))$. Use the MATLAB function *fir1* to obtain the three filters. Plot the original signal, and its components in the three bands. Verify that the overall filter is an all-pass filter. Obtain the sum of the outputs of the filters, and explain how it relates to the original signal.

Solution: The script for this example uses several MATLAB functions that facilitate the filtering of signals, and for which you can get more information by using *help*.

The spectrum of $x[n]$ is found using the FFT algorithm as indicated before. The low-pass, band-pass and high-pass filters are obtained using *fir1*, and the function *filter* allows us to obtain the corresponding outputs $y_1[n]$, $y_2[n]$ and $y_3[n]$. The frequency responses $\{H_i(e^{j\omega})\}$, $i = 1, 2, 3$, of the filters are found using the function *freqz*.

The three filters separate $x[n]$ into its low-, middle-, and high-frequency band components from which we are able to obtain the power of the signal in these three bands, i.e., we have a crude spectral analyzer. Ideally we would like the sum of the filters outputs to be equal to $x[n]$ and so the sum of the frequency responses

$$
H(e^{j\omega}) = H_1(e^{j\omega}) + H_2(e^{j\omega}) + H_3(e^{j\omega})
$$

should be the frequency response of an all-pass filter. Indeed, that is the result shown in Fig. 11.10, where we obtain the input signal—delayed due to the linear phase of the filters—as the output of the filter with transfer function $H(z)$. ☐

```
%%
% Example 11.19---Spectral analyzer
%%
N=500;n=0:N-1; x=cos(0.2*pi*n); x=sign(x);      % pulse signal
X=fft(x)/50; X=X(1:250);                        % approximate DTFT
L=500;w1=0:2*pi/L:pi-2*pi/L;w1=w1/pi;           % frequencies
h1=fir1(30,0.1);                                % low-pass filter
h2=fir1(30,0.6,'high');                         % high-pass filter
h3=fir1(30,[0.1 0.6]);                          % band-pass filter
```

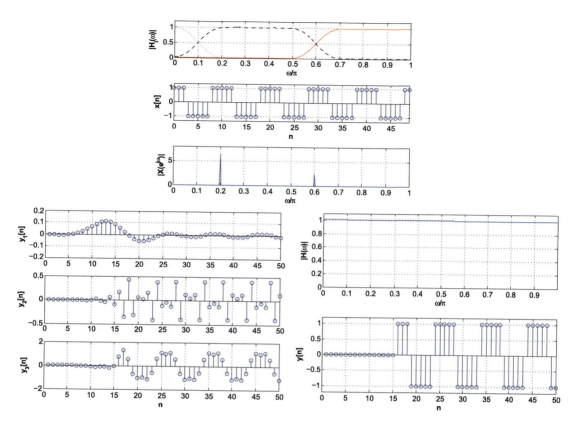

FIGURE 11.10

A crude spectrum analyzer: magnitude response of low-pass, band-pass and high-pass filters, input signal and its magnitude spectrum (top, center figure). Outputs of filters (bottom left), overall magnitude response of the bank of filters (all-pass filter) and overall response. Delay is due to linear phase of the bank of filters. Notice the difference in the amplitude scales used to display $y_i[n]$, $i = 1, 2, 3$.

```
y1=filter(h1,1,x); y2=filter(h2,1,x); y3=filter(h3,1,x);
y=y1+y2+y3;                                    % outputs of filters
[H1,w]=freqz(h1,1); [H2,w]=freqz(h2,1); [H3,w]=freqz(h3,1);
H=H1+H2+H3;                                    % frequency responses
```

11.3.5 CIRCULAR SHIFTING AND PERIODIC CONVOLUTION

Circular shifting. When a periodic signal $x[n]$, of fundamental period N, is shifted by M samples the signal is still periodic. The circular representation provides the appropriate visualization of this shift, as it concentrates on the period displayed by the representation. Values are rotated circularly.

The Fourier series of the shifted signal $x_1[n] = x[n - M]$ is obtained from the Fourier series of $x[n]$ by replacing n by $n - M$ to get

$$x_1[n] = x[n - M] = \sum_k X[k]e^{j2\pi(n-M)k/N} = \sum_k \left(X[k]e^{-j2\pi Mk/N} \right) e^{j2\pi nk/N},$$

so that the shifted signal and its Fourier series coefficients are related as

$$x[n - M] \quad \Leftrightarrow \quad X[k]e^{-j2\pi Mk/N}. \tag{11.36}$$

It is important to consider what happens for different values of M. This shift can be represented as

$$M = mN + r, \qquad m = 0, \pm 1, \pm 2, \cdots, \quad 0 \le r \le N - 1,$$

and as such

$$e^{-j2\pi Mk/N} = e^{-j2\pi(mN+r)k/N} = e^{-j2\pi rk/N} \tag{11.37}$$

for any value of M. So that shifting more than a period is equivalent to shifting the residue r of dividing the shift M by N.

Example 11.20. To visualize the difference between a linear shift and a circular shift consider the periodic signal $x[n]$ of fundamental period $N = 4$ with a first period $x_1[n] = n$, $n = 0, \cdots, 3$. Plot $x[-n]$, and $x[n - 1]$ as functions of n using circular representations.

Solution: In the circular representation of $x[n]$, the samples $x[0]$, $x[1]$, $x[2]$ and $x[3]$ of the first period are located in a clockwise direction in the circle (see Fig. 11.11). Considering the rightmost direction the origin of the representation, delaying by M in the circular representation corresponds to shifting circularly, or rotating, M positions in the clockwise direction. Advancing by M corresponds to shifting circularly in the counterclockwise direction M positions. Reflection corresponds to placing the entries of $x_1[n]$ counterclockwise starting with $x[0]$. For the periodic signal $x[n]$ of fundamental period 4, the different shifted versions, circular and linear, are displayed in Fig. 11.11. □

Periodic convolution. Consider then the multiplication of two periodic signals $x[n]$ and $y[n]$ of the same fundamental period N. The product $v[n] = x[n]y[n]$ is also periodic of fundamental period N, and its Fourier series coefficients are

$$V[k] = \sum_{m=0}^{N-1} X[m]Y[k - m] \qquad 0 \le k \le N - 1,$$

as we will show next. That $v[n]$ is periodic of fundamental period N is clear. Its Fourier series is found by letting the fundamental frequency be $\omega_0 = 2\pi/N$ and replacing the given Fourier coefficients we

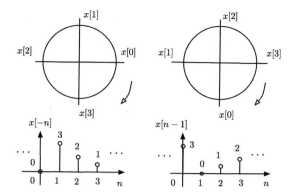

FIGURE 11.11

Circular representation of $x[-n]$ and $x[n-1]$.

have for $0 \leq n \leq N - 1$:

$$
\begin{aligned}
v[n] &= \sum_{m=0}^{N-1} V[m]e^{j\omega_0 nm} = \sum_{m=0}^{N-1}\sum_{k=0}^{N-1} X[k]Y[m-k]e^{j\omega_0 nm} \\
&= \sum_{k=0}^{N-1} X[k]\left(\sum_{m=0}^{N-1} Y[m-k]e^{j\omega_0 n(m-k)}\right)e^{j\omega_0 kn} = \sum_{k=0}^{N-1} X[k]y[n]e^{j\omega_0 kn} = y[n]x[n].
\end{aligned}
$$

Thus we see that the Fourier series coefficients of the product of two periodic signals, of the same fundamental period, gives the pair

$$
x[n]y[n], \quad \text{periodic of fundamental period } N \quad \Leftrightarrow \quad \sum_{m=0}^{N-1} X[m]Y[k-m], \quad 0 \leq k \leq N-1 \quad (11.38)
$$

and by duality

$$
\sum_{m=0}^{N-1} x[m]y[n-m], \quad 0 \leq n \leq N-1 \quad \Leftrightarrow \quad NX[k]Y[k], \quad \text{periodic of fundamental period } N.
$$

$$(11.39)$$

Although

$$
\sum_{m=0}^{N-1} x[m]y[n-m] \quad \text{and} \quad \sum_{m=0}^{N-1} X[m]Y[k-m]
$$

look like the convolution sums we had before, the periodicity of the sequences makes them different. These are called **periodic convolution sums**. Given the infinite support of periodic signals, the convolution sum of periodic signals does not exist—it would not be finite. The periodic convolution is done only for a period of periodic signals of the same fundamental period.

Remarks

1. As before, multiplication in one domain causes convolution in the other domain.
2. When computing a periodic convolution we need to remember that: (i) the convolving sequences must have the same fundamental period, and (ii) the Fourier series coefficients of a periodic signal share the same fundamental period with the signal.

Example 11.21. To understand how the periodic convolution results, consider the product of two periodic signals $x[n]$ and $y[n]$ of fundamental period $N = 2$. Find the Fourier series of their product $v[n] = x[n]y[n]$.

Solution: The multiplication of the Fourier series

$$x[n] = X[0] + X[1]e^{j\omega_0 n}$$
$$y[n] = Y[0] + Y[1]e^{j\omega_0 n} \qquad \omega_0 = 2\pi/N = \pi$$

can be seen as a product of two polynomials in complex exponentials $\zeta[n] = e^{j\omega_0 n}$ so that

$$
\begin{aligned}
x[n]y[n] &= (X[0] + X[1]\zeta[n])(Y[0] + Y[1]\zeta[n]) \\
&= X[0]Y[0] + (X[0]Y[1] + X[1]Y[0])\,\zeta[n] + X[1]Y[1]\zeta^2[n].
\end{aligned}
$$

Now $\zeta^2[n] = e^{j2\omega_0 n} = e^{j2\pi n} = 1$, so that

$$x[n]y[n] = \underbrace{(X[0]Y[0] + X[1]Y[1])}_{V[0]} + \underbrace{(X[0]Y[1] + X[1]Y[0])}_{V[1]}\,e^{j\omega_0 n} = v[n]$$

after replacing $\zeta[n]$. Using the periodic convolution formula we have

$$V[0] = \sum_{k=0}^{1} X[k]Y[-k] = X[0]Y[0] + X[1]Y[-1] = X[0]Y[0] + X[1]Y[2-1],$$

$$V[1] = \sum_{k=0}^{1} X[k]Y[1-k] = X[0]Y[1] + X[1]Y[0],$$

where in the upper equation we used the periodicity of $Y[k]$ so that $Y[-1] = Y[-1+N] = Y[-1+2] = Y[1]$. Thus, the multiplication of periodic signals can be seen as the product of polynomials in $\zeta[n] = e^{j\omega_0 n} = e^{j2\pi n/N}$, such that

$$\zeta^{mN+r}[n] = e^{j\frac{2\pi}{N}(mN+r)} = e^{j\frac{2\pi}{N}r} = \zeta^r[n] \qquad m = \pm 1, \pm 2, \cdots, \ 0 \le r \le N-1,$$

which ensures that the resulting polynomial is always of order $N-1$.

Graphically, we proceed in a manner analogous to the convolution sum (see Fig. 11.12): representing $X[k]$ and $Y[-k]$ circularly, and shifting $Y[-k]$ clockwise while keeping $X[k]$ stationary. The

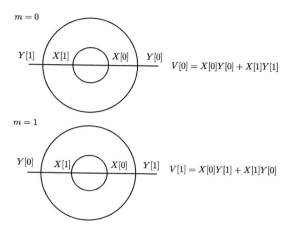

FIGURE 11.12

Periodic convolution of the Fourier series coefficients $\{X[k]\}$ and $\{Y[k]\}$.

circular representation of $X[k]$ is given by the internal circle with the values of $X[0]$ and $X[1]$ in the clockwise direction, while $Y[m - k]$ is represented in the outer circle with the two values of a period in the counterclockwise direction (corresponding to the reflection of the signal or $Y[-k]$ for $m = 0$). Multiplying the values opposite to each other and adding them we get $V[0] = X[0]Y[0] + X[1]Y[1]$. If we shift the outer circle 180 degrees clockwise, for $m = 1$, and multiply the values opposite to each other and add their product, we get $V[1] = X[0]Y[1] + X[1]Y[0]$. There is no need for further shifting as the results would coincide with the ones obtained before. □

For periodic signals $x[n]$ and $y[n]$, of fundamental period N, we have

1. **Duality in time and frequency circular shifts:** The Fourier series coefficients of the signals on the left are the terms on the right

$$x[n - M] \quad \Leftrightarrow \quad X[k]e^{-j2\pi Mk/N},$$
$$x[n]e^{j2\pi Mn/N} \quad \Leftrightarrow \quad X[k - M]. \tag{11.40}$$

2. **Duality in multiplication and periodic convolution sum:** The Fourier series coefficients of the signals on the left are the terms on the right

$$z[n] = x[n]y[n] \quad \Leftrightarrow \quad Z[k] = \sum_{m=0}^{N-1} X[m]Y[k - m], \ 0 \le k \le N - 1$$

$$v[n] = \sum_{m=0}^{N-1} x[m]y[n - m], \ 0 \le n \le N - 1 \quad \Leftrightarrow \quad V[k] = NX[k]Y[k] \tag{11.41}$$

where $z[n]$ and $V[k]$ are periodic of fundamental period N.

Example 11.22. A periodic signal $x_1[n]$ of fundamental period $N = 4$ has a period

$$x_1[n] = \begin{cases} 1 & n = 0, 1, \\ 0 & n = 2, 3. \end{cases}$$

Suppose that we want to find the periodic convolution of $x[n]$ with itself, call it $v[n]$. Let then $y[n] = x[n - 2]$, find the periodic convolution of $x[n]$ and $y[n]$, call it $z[n]$. How does $v[n]$ compare with $z[n]$?

Solution: The circular representation of $x[n]$ is shown in Fig. 11.13. To find the periodic convolution we consider a period $x_1[n]$ of $x[n]$, and represent the stationary signal by the internal circle, and the circularly shifted signal by the outside circle. Multiplying the values in each of the spokes, and adding them we find the values of a period of $v[n]$, which for its period $v_1[n]$ are:

$$v_1[n] = \begin{cases} 1 & n = 0, \\ 2 & n = 1, \\ 1 & n = 2, \\ 0 & n = 3. \end{cases} \tag{11.42}$$

Analytically, the Fourier series coefficients of $v[n]$ are $V[k] = N(X[k])^2 = 4(X[k])^2$. Using the Z-transform, $X_1(z) = 1 + z^{-1}$, the Fourier series coefficients of $x[n]$ are

$$X[k] = \frac{1}{N}(1 + z^{-1})|_{z = e^{j2\pi k/4}} = \frac{1}{4}(1 + e^{-j\pi 2k/4}) \qquad 0 \le k \le 3$$

and thus

$$V[k] = 4(X[k])^2 = \frac{1}{4}(1 + 2e^{-j\pi k/2} + e^{-j\pi k}).$$

This can be verified by using the period obtained from the periodic convolution equation (11.42):

$$V[k] = \frac{1}{N} \sum_{n=0}^{N-1} v[n]e^{-j2\pi nk/N} = \frac{1}{4}(1 + 2e^{-j2\pi k/4} + e^{-j2\pi 2k/4}),$$

which equals the above expression. Graphically, the circular convolution of $x[n]$ with itself is displayed in the middle of Fig. 11.13. The inner circle corresponds to the circular representation of $x[m]$ and the outside to that of $x[n - m]$ which is shifted clockwise. The resulting periodic signal $v[n]$ is shown on the side of the circular convolution.

Graphically the periodic convolution of $x[n]$ and $y[n]$ is shown in Fig. 11.13, where the stationary signal is chosen as $x[m]$, represented by the inner circle, and the circularly-shifted signal $y[n - m]$, represented by the outer circle. The result of the convolution is a periodic signal $z[n]$ of period

$$z_1[n] = \begin{cases} 1 & n = 0, \\ 0 & n = 1, \\ 1 & n = 2, \\ 2 & n = 3. \end{cases}$$

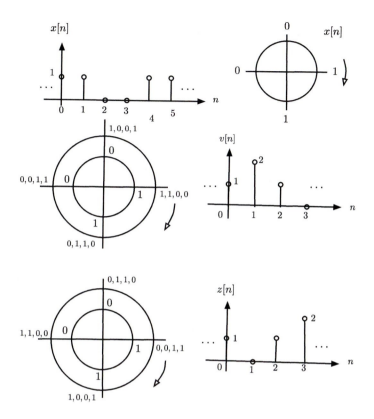

FIGURE 11.13

Linear and circular representations of $x[n]$ (top). Periodic convolution of $x[n]$ with itself to get $v[n]$, and linear representation of $v[n]$ (middle); periodic convolution of $x[n]$ and $y[n] = x[n-2]$, the result is $z[n] = v[n-2]$ represented linearly (bottom).

Analytically, the Fourier series coefficients of $z[n]$ are given by

$$
Z[k] = 4 \frac{X_1(z)Y_1(z)}{4 \times 4} \Big|_{z=e^{j2\pi k/4}} = \frac{z^{-2}X_1^2(z)}{4} = \frac{z^{-2} + 2z^{-3} + z^{-4}}{4} \Big|_{z=e^{j2\pi k/4}}
$$
$$
= \frac{1}{4}(e^{-j2\pi\,2k/4} + 2e^{-j2\pi\,3k/4} + e^{-j2\pi\,4k/4}) = \frac{1}{4}(1 + e^{-j2\pi\,2k/4} + 2e^{-j2\pi\,3k/4}),
$$

which coincides with the $Z[k]$ obtained by the periodic convolution:

$$
Z[k] = \frac{1}{N}\sum_{n=0}^{N-1} z[n]e^{-j2\pi nk/N} = \frac{1}{4}(1 + z^{-2} + 2z^{-3})\Big|_{z=e^{j2\pi k/4}} = \frac{1}{4}(1 + e^{-j2\pi\,2k/4} + 2e^{-j2\pi\,3k/4}). \quad \square
$$

11.4 THE DISCRETE FOURIER TRANSFORM (DFT)

Recall that the direct and the inverse DTFTs of a discrete-time signal $x[n]$ are

$$X(e^{j\omega}) = \sum_{n} x[n]e^{-j\omega n}, \qquad -\pi \le \omega < \pi,$$

$$x[n] = \frac{1}{2\pi} \int_{-\pi}^{\pi} X(e^{j\omega})e^{j\omega n} d\omega.$$

These equations have the following computational disadvantages:

- The frequency ω varies continuously from $-\pi$ to π, and as such computing $X(e^{j\omega})$ needs to be done for an uncountable number of frequencies.
- The inverse DTFT requires integration which cannot be implemented exactly in a computer.

To resolve these issues we consider the Discrete Fourier Transform or DFT (notice the name difference with respect to the DTFT) which is computed at discrete frequencies and its inverse does not require integration. Moreover, the DFT is efficiently implemented using an algorithm called the Fast Fourier Transform (FFT).

The development of the DFT is based on the representation of periodic discrete-time signals and taking advantage of the fact that both the signal and its Fourier coefficients are periodic of the same fundamental period. Thus the *representation of discrete-time periodic signals* is *discrete and periodic in both time and frequency*. We need then to consider how to extend aperiodic signals into periodic signals, using an appropriate fundamental period, to obtain their DFTs.

11.4.1 DFT OF PERIODIC DISCRETE-TIME SIGNALS

A periodic signal $\tilde{x}[n]$, of fundamental period N, is represented by N values in a period. Its discrete Fourier series is

$$\tilde{x}[n] = \sum_{k=0}^{N-1} \tilde{X}[k]e^{j\omega_0 nk} \qquad 0 \le n \le N-1 \tag{11.43}$$

where $\omega_0 = 2\pi/N$ is the fundamental frequency. The coefficients $\{\tilde{X}[k]\}$ correspond to harmonic frequencies $\{k\omega_0\}$, for $0 \le k \le N-1$, so that $\tilde{x}[n]$ has no frequency components at any other frequencies. Thus $\tilde{x}[n]$ and $\tilde{X}[k]$ are both discrete and periodic of the same fundamental period N. The Fourier series coefficients can be calculated using the Z-transform

$$\tilde{X}[k] = \frac{1}{N} \mathcal{Z}[\tilde{x}_1[n]]\Big|_{z=e^{jk\omega_0}} = \frac{1}{N} \sum_{n=0}^{N-1} \tilde{x}[n]e^{-j\omega_0 nk} \qquad 0 \le k \le N-1, \ \omega_0 = 2\pi/N \tag{11.44}$$

where $\tilde{x}_1[n] = \tilde{x}[n]W[n]$ is a period of $\tilde{x}[n]$ and $W[n]$ a rectangular window, i.e.,

$$W[n] = u[n] - u[n-N] = \begin{cases} 1 & 0 \le n \le N-1, \\ 0 & \text{otherwise.} \end{cases}$$

Moreover, $\tilde{x}[n]$ can be expressed linearly as

$$\tilde{x}[n] = \sum_{r=-\infty}^{\infty} \tilde{x}_1[n+rN].$$ (11.45)

Although one could call Equation (11.44) the **DFT of the periodic signal** $\tilde{x}[n]$ and Equation (11.43) the corresponding **inverse DFT**, traditionally the DFT of $\tilde{x}[n]$ is $N\tilde{X}[k]$ or

$$X[k] = N\tilde{X}[k] = \sum_{n=0}^{N-1} \tilde{x}[n]e^{-j\omega_0 nk} \qquad 0 \le k \le N-1, \quad \omega_0 = 2\pi/N$$ (11.46)

and the inverse DFT is then defined as

$$\tilde{x}[n] = \frac{1}{N} \sum_{k=0}^{N-1} X[k]e^{j\omega_0 nk} \qquad 0 \le n \le N-1.$$ (11.47)

Equations (11.46) and (11.47) show that the representation of periodic signals is completely discrete: summations instead of integrals and discrete rather than continuous frequencies. Thus the DFT and it inverse can be evaluated by computer.

Given a periodic signal $x[n]$, of fundamental period N, its DFT is given by

$$X[k] = \sum_{n=0}^{N-1} x[n]e^{-j2\pi nk/N} \qquad 0 \le k \le N-1.$$ (11.48)

Its inverse DFT is

$$x[n] = \frac{1}{N} \sum_{k=0}^{N-1} X[k]e^{j2\pi nk/N} \qquad 0 \le n \le N-1.$$ (11.49)

Both $X[k]$ and $x[n]$ are periodic of the same fundamental period N.

11.4.2 DFT OF APERIODIC DISCRETE-TIME SIGNALS

We obtain the DFT of an aperiodic signal $y[n]$ by sampling in frequency its DTFT $Y(e^{j\omega})$. Suppose we choose $\{\omega_k = 2\pi k/L, k = 0, \cdots, L-1\}$ as the sampling frequencies, where an appropriate value for the integer $L > 0$ needs to be determined. Analogous to the sampling in time in Chapter 8, sampling in frequency generates a periodic signal in time

$$\tilde{y}[n] = \sum_{r=-\infty}^{\infty} y[n+rL].$$ (11.50)

Now, if $y[n]$ is of finite length N, then when $L \geq N$ the periodic extension $\tilde{y}[n]$ clearly displays a first period equal to the given signal $y[n]$ (with some zeros attached at the end when $L > N$). On the other hand, if the length $L < N$ the first period of $\tilde{y}[n]$ does not coincide with $y[n]$ because of superposition of shifted versions of it—this corresponds to **time aliasing**, the dual of **frequency-aliasing** which occurs in time-sampling. Thus, for $y[n]$ of finite length N, and letting $L \geq N$, we have

$$\tilde{y}[n] = \sum_{r=-\infty}^{\infty} y[n+rL] \quad \Leftrightarrow \quad Y[k] = Y(e^{j2\pi k/L}) = \sum_{n=0}^{N-1} y[n]e^{-j2\pi nk/L} \qquad 0 \leq k \leq L-1. \quad (11.51)$$

The equation on the right above is the DFT of $y[n]$. The inverse DFT is the Fourier series representation of $\tilde{y}[n]$ (normalized with respect to L) or its first period

$$y[n] = \frac{1}{L} \sum_{k=0}^{L-1} Y[k]e^{j2\pi nk/L} \qquad 0 \leq n \leq L-1 \qquad (11.52)$$

where $Y[k] = Y(e^{j2\pi k/L})$.

In practice, the generation of the periodic extension $\tilde{y}[n]$ is not needed, we just need to generate a period that either coincides with $y[n]$ when $L = N$, or that is $y[n]$ with a sequence of $L - N$ zeros attached to it (i.e., $y[n]$ is **padded with zeros**) when $L > N$. To avoid time aliasing we do not consider choosing $L < N$.

If the signal $y[n]$ is a very long signal, in particular if $N \to \infty$, it does not make sense to compute its DFT, even if we could. Such a DFT would give the frequency content of the whole signal and since a large support signal could have all types of frequencies its DFT would just give no valuable information. A possible approach to obtain the frequency content of a signal with a large time support is to window it and compute the DFT of each of these segments. Thus when $y[n]$ is of infinite length, or its length is much larger than the desired or feasible length L, we use a window $W_L[n]$ of length L, and represent $y[n]$ as the superposition

$$y[n] = \sum_{m} y_m[n] \quad \text{where} \quad y_m[n] = y[n]W_L[n - mL], \qquad (11.53)$$

so that by the linearity of the DFT we have the DFT of $y[n]$ is

$$Y[k] = \sum_{m} \text{DFT}(y_m[n]) = \sum_{m} Y_m[k] \qquad (11.54)$$

where each $Y_m[k]$ provides a frequency characterization of the windowed signal or the local frequency content of the signal. Practically, this would be more meaningful than finding the DFT of the whole signal. Now we have frequency information corresponding to segments of the signal and possibly evolving over time.

The DFT of an aperiodic signal $y[n]$ of finite length N, is found as follows:

- Choose an integer $L \geq N$, the length of the DFT, to be the fundamental period of a periodic extension $\tilde{y}[n]$ having $y[n]$ as a period with padded zeros if necessary.
- Find the DFT of $\tilde{y}[n]$,

$$\tilde{y}[n] = \frac{1}{L} \sum_{k=0}^{L-1} \tilde{Y}[k] e^{j2\pi nk/L} \qquad 0 \leq n \leq L-1$$

and the inverse DFT

$$\tilde{Y}[k] = \sum_{n=0}^{L-1} \tilde{y}[n] e^{-j2\pi nk/L} \qquad 0 \leq k \leq L-1.$$

- Then,
  <u>DFT of $y[n]$:</u> $Y[k] = \tilde{Y}[k]$ for $0 \leq k \leq L-1$, and
  <u>Inverse DFT or IDFT of $Y[k]$:</u> $y[n] = \tilde{y}[n]W[n]$, $0 \leq n \leq L-1$, where $W[n] = u[n] - u[n-L]$ is a rectangular window of length L.

11.4.3 COMPUTATION OF THE DFT VIA THE FFT

Although the direct and the inverse DFT uses discrete frequencies and summations making them computational feasible, there are still several issues that should be understood when computing these transforms. Assuming that the given signal is finite length, or made finite length by windowing, we have:

1. **Efficient computation with the Fast Fourier Transform or FFT algorithm**—A very efficient computation of the DFT is done by means of the FFT algorithm, which takes advantage of some special characteristics of the DFT as we will discuss later. It should be understood that the FFT is not another transformation but an algorithm to efficiently compute DFTs. For now, we will consider the FFT a black box that for an input $x[n]$ (or $X[k]$) gives as output the DFT $X[k]$ (or IDFT $x[n]$).
2. **Causal aperiodic signals**—If the given signal $x[n]$ is causal of length N, i.e., samples

$$\{x[n], \ n = 0, 1, \cdots, N-1\},$$

we proceed to obtain $\{X[k], \ k = 0, 1, \cdots, N-1\}$ or the DFT of $x[n]$ by means of an FFT of length $L = N$. To compute an $L > N$ DFT we simply attach $L - N$ zeros at the end of the above sequence and obtain L values corresponding to the DFT of $x[n]$ of length L (why this could be seen as a better version of the DFT of $x[n]$ is discussed below in frequency resolution).
3. **Noncausal aperiodic signals**—When the given signal $x[n]$ is noncausal of length N, i.e., samples

$$\{x[n], \ n = -n_0, \cdots 0, 1, \cdots, N - n_0 - 1\}$$

are given, we need to recall that a periodic extension of $x[n]$ or $\tilde{x}[n]$ was used to obtain its DFT. This means that we need to create a sequence of N values corresponding to the first period of $\tilde{x}[n]$,

Table 11.2 Fourier series and discrete Fourier transform

	$x[n]$ periodic signal of period N	$X[k]$ periodic FS coefficients of period N
Z-transform	$x_1[n] = x[n](u[n] - u[n-N])$	$X[k] = \dfrac{1}{N} \mathcal{Z}(x_1[n]) \big|_{z=e^{j2\pi k/N}}$
DTFT	$x[n] = \displaystyle\sum_k X[k]e^{j2\pi nk/N}$	$X(e^{j\omega}) = \displaystyle\sum_k 2\pi X[k]\delta(\omega - 2\pi k/N)$
LTI response	input: $x[n] = \displaystyle\sum_k X[k]e^{j2\pi nk/N}$	output: $y[n] = \displaystyle\sum_k X[k]H(e^{jk\omega_0})e^{j2\pi nk/N}$
		$H(e^{j\omega})$ (frequency response of system)
Time-shift (circular shift)	$x[n-M]$	$X[k]e^{-j2\pi kM/N}$
Modulation	$x[n]e^{j2\pi Mn/N}$	$X[k-M]$
Multiplication	$x[n]y[n]$	$\displaystyle\sum_{m=0}^{N-1} X[m]Y[k-m]$ periodic convolution
Periodic convolution	$\displaystyle\sum_{m=0}^{N-1} x[m]y[n-m]$	$NX[k]Y[n]$
	$x[n]$ finite-length N aperiodic signal	$\tilde{x}[n]$ periodic extension of period $L \geq N$
	$\tilde{x}[n] = \dfrac{1}{N}\displaystyle\sum_{k=0}^{L-1} \tilde{X}[k]e^{j2\pi nk/L}$	$\tilde{X}[k] = \displaystyle\sum_{n=0}^{L-1} \tilde{x}[n]e^{-j2\pi nk/L}$
IDFT/DFT	$x[n] = \tilde{x}[n]W[n],$ $W[n] = u[n] - u[n-N]$	$X[k] = \tilde{X}[k]W[k],$ $W[k] = u[k] - u[k-N]$
Circular convolution	$(x \otimes_L y)[n]$	$X[k]Y[k]$
Circular and linear convolutions	$(x \otimes_L y)[n] = (x * y)[n], \quad L \geq M + K - 1$ $M = $ length of $x[n], \quad K = $ length of $y[n]$	

i.e.,

$$\underbrace{x[0]\, x[1]\, \cdots x[N-n_0-1]}_{\text{causal samples}} \; \underbrace{x[-n_0]\, x[-n_0+1] \cdots x[-1]}_{\text{noncausal samples}}$$

where, as indicated, the samples $x[-n_0]\, x[-n_0 + 1] \cdots x[-1]$ are the values that make $x[n]$ noncausal. If we wish to consider zeros after $x[N - n_0 - 1]$ to be part of the signal, so as to obtain a better DFT transform as we discuss later in frequency resolution, we simply attach zeros between the causal and noncausal components, that is,

$$\underbrace{x[0]\, x[1]\, \cdots x[N-n_0-1]}_{\text{causal samples}} 0\,0 \, \cdots \, 0\,0 \; \underbrace{x[-n_0]\, x[-n_0+1] \cdots x[-1]}_{\text{noncausal samples}}$$

to compute an $L > N$ DFT of the noncausal signal. The periodic extension $\tilde{x}[n]$ represented circularly instead of linearly would clearly show the above sequences.

4. **Periodic signals**—If the signal $x[n]$ is periodic of fundamental period N we will then choose $L = N$ (or a multiple of N) and calculate the DFT $X[k]$ by means of the FFT algorithm. If we use a multiple of the fundamental period, e.g., $L = MN$ for some integer $M > 0$, we need to divide the obtained DFT by the value M. For periodic signals we cannot choose L to be anything but a multiple of N

as we are really computing the Fourier series of the signal. Likewise, no zeros can be attached to a period (or periods when $M > 1$) to improve the frequency resolution of its DFT—by attaching zeros to a period we distort the signal.

5. **Frequency resolution**—When the signal $x[n]$ is periodic of fundamental period N, the DFT values are normalized Fourier series coefficients of $x[n]$ that only exist at the harmonic frequencies $\{2\pi k/N\}$, as no frequency components exist at other frequencies. On the other hand, when $x[n]$ is aperiodic, the number of possible frequencies depend on the length L chosen to compute its DFT. In either case, the frequencies at which we compute the DFT can be seen as frequencies around the unit circle in the Z-plane. In both cases one would like to have a significant number of frequencies in the unit circle so as to visualize well the frequency content of the signal. The number of frequencies considered is related to the **frequency resolution** of the DFT of the signal.

   - If the signal is aperiodic we can improve the frequency resolution of its DFT by increasing the number of samples in the signal without distorting the signal. This can be done by **padding the signal with zeros**, i.e., attaching zeros to the end of the signal. These zeros do not change the frequency content of the signal (they can be considered part of the aperiodic signal), but permit us to increase the available frequency components of the signal displayed by the DFT.
   - On the other hand, the harmonic frequencies of a periodic signal, of fundamental period N, are fixed to $2\pi k/N$ for $0 \leq k < N$. In the case of periodic signals we cannot pad the given period of the signal with an arbitrary number of zeros, because such zeros are not part of the periodic signal. As an alternative, to increase the frequency resolution of a periodic signal we consider several periods. The DFT values, or normalized Fourier series coefficients, appear at the harmonic frequencies independent of the number of periods considered, but by considering several periods zero values appear at frequencies in between the harmonic frequencies. To obtain the same values for one period, it is necessary to divide the DFT values by the number of periods used.

6. **Frequency scales**—When computing the N-length DFT of a signal $x[n]$ of length N, we obtain a sequence of complex values $X[k]$ for $k = 0, 1, \cdots, N - 1$. Since each of the k values corresponds to a discrete frequency $2\pi k/N$ the $k = 0, 1, \cdots, N - 1$ scale can be converted into a discrete frequency scale $[0 \quad 2\pi(N - 1)/N]$ (rad) (the last value is always smaller than 2π to keep the periodicity in frequency of $X[k]$) by multiplying the integer scale $\{0 \leq k \leq N - 1\}$ by $2\pi/N$. Subtracting π from this frequency scale we obtain discrete frequencies $[-\pi \quad \pi - 2\pi/N]$ (rad) where again the last frequency does not coincide with π in order to keep the periodicity of 2π of the $X[k]$. Finally, to obtain a normalized discrete frequency scale we divide the above scale by π so as to obtain a non-units normalized scale $[-1 \quad 1 - 2/N]$. If the signal is the result of sampling and we wish to display the continuous-time frequency, we then use the relation where T_s is the sampling period and f_s the sampling frequency:

$$\Omega = \frac{\omega}{T_s} = \omega f_s \text{ (rad/s)} \quad \text{or} \quad f = \frac{\omega}{2\pi T_s} = \frac{\omega f_s}{2\pi} \text{ (Hz)} \quad (11.55)$$

giving scales $[-\pi f_s \quad \pi f_s]$ (rad/s) and $[-f_s/2 \quad f_s/2]$ (Hz) and where according to the Nyquist sampling condition $f_s \leq f_{max}$, for f_{max} the maximum frequency in the signal.

Example 11.23. Determine the DFT, via the FFT, of the causal signal

$$x[n] = \sin(\pi n/32)(u[n] - u[n - 33])$$

and of its advanced version $x_1[n] = x[n + 16]$. To improve their frequency resolution compute FFTs of length $L = 512$. Explain the difference between computing the FFTs of the causal and the noncausal signals.

Solution: As indicated above, when computing the FFT of a causal signal the signal is simply inputted into the function. However, to improve the frequency resolution of the FFT we attach zeros to the signal. These zeros provide additional values of the frequency components of the signal, with no effect on the frequency content of the signal.

For the noncausal signal $x[n + 16]$, we need to recall that the DFT of an aperiodic signal was computed by extending the signal into a periodic signal with an arbitrary fundamental period L, which exceeds the length of the signal. Thus, the periodic extension of $x[n + 16]$ can be obtained by creating an input array consisting of $x[n]$ for $n = 0, \cdots, 16$, followed by $L - 33$ zeros (L being the length of the FFT and 33 the length of the signal) zeros to improve the frequency resolution, and $x[n]$, $n = -16, \cdots, -1$.

In either case, the output of the FFT is available as an array of length $L = 512$ values. This array $X[k]$, $k = 0, \cdots, L - 1$ can be understood as values of the signal spectrum at frequencies $2\pi k/L$, i.e., from 0 to $2\pi(L - 1)/L$ (rad). We can change this scale into other frequency scales, for instance if we wish a scale that considers positive as well as negative frequencies, to the above scale we subtract π, and if we wish a normalized scale $[-1 \quad 1)$, we simply divide the previous scale by π. When shifting to a $[-\pi \quad \pi)$ or $[-1 \quad 1)$ frequency scale, the spectrum also needs to be shifted accordingly—this is done using the MATLAB function *fftshift*. To understand this change recall that $X[k]$ is also periodic of fundamental period L.

The following script is used to compute the DFT of $x[n]$ and $x[n + 16]$. The results are shown in Fig. 11.14. □

```
%%
% Example 11.23---FFFT computation of causal and noncausal signals
%%
clear all; clf
L=512;                                          % order of the FFT
n=0:L-1;
% causal signal
x=[ones(1,33) zeros(1,L-33)]; x=x.*sin(pi*n/32);    % zero-padding
X=fft(x); X=fftshift(X);                         % shifting to [-1 1]
w=2*[0:L-1]./L-1;                                % normalized freq.
n1=[-9:40];                                      % time scale
% noncausal signal
xnc=[zeros(1,3) x(1:33) zeros(1,3)];             % noncausal signal
x=[x(17:33) zeros(1,N-33) x(1:16)];             % periodic extension
                                                 % and zero-padding

X=fft(x); X=fftshift(X);
n1=[-19:19];                                     % time scale
```

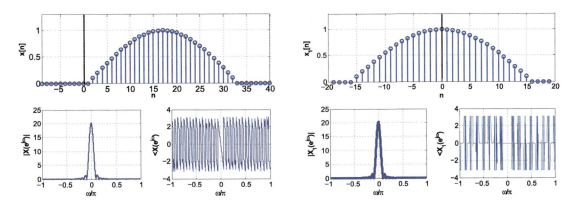

FIGURE 11.14

Computation of the FFT of causal and noncausal signals. Notice that as expected the magnitude responses are equal, only the phase responses change.

Example 11.24. Consider improving the frequency resolution of a periodic sampled signal

$$y(nT_s) = 4\cos(2\pi f_0 n T_s) - \cos(2\pi f_1 n T_s) \qquad f_0 = 100 \text{ Hz}, f_1 = 4f_0$$

where the sampling period is $T_s = 1/(3f_1)$ s/sample.

Solution: In the case of a periodic signal, the frequency resolution of its FFT cannot be improved by attaching zeros. The length of the FFT must be the fundamental period or a multiple of the fundamental period of the signal. The following script illustrates how the FFT of the given periodic signal can be obtained by using 4 or 12 periods. As the number of periods increases the harmonic components appear in each case at exactly the same frequencies, and only zeros in between these fixed harmonic frequencies result from increasing the number of periods. The magnitude frequency response is, however, increasing as the number of periods increases. Thus we need to divide by the number of periods used in computing the FFT.

Since the signal is sampled, it is of interest to have the frequency scale of the FFTs in Hz so we convert the discrete frequency ω (rad) into f (Hz) according to

$$f = \frac{\omega}{2\pi T_s} = \frac{\omega f_s}{2\pi}$$

where $f_s = 1/T_s$ is the sampling rate given in samples/s. The results are shown in Fig. 11.15. □

```
%%
% Example 11.24---Improving frequency resolution of FFT of periodic signals
%%%
f0=100; f1=4*f0;                    % frequencies in Hz of signal
Ts=1/(3*f1);                        % sampling period
t=0:Ts:4/f0;                        % time for 4 periods
```

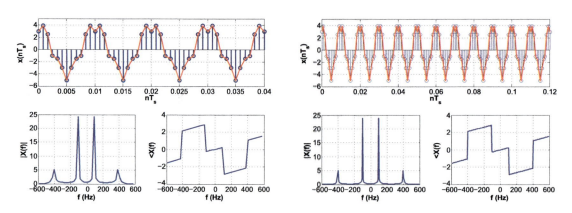

FIGURE 11.15

Computation of the FFT of a periodic signal using 4 and 12 periods to improve the frequency resolution of the FFT. Notice that both magnitude and phase responses look alike, but when we use 12 periods these spectra look sharper due to the increase in the number of frequency components added.

```
y=4*cos(2*pi*f0*t)-cos(2*pi*f1*t);      % sampled signal (4 periods)
M=length(y);
Y=fft(y,M); Y=fftshift(Y)/4;            % fft, shifting and normalizing
t1=0:Ts:12/f0;                          % time for 12 periods
y1=4*cos(2*pi*f0*t1)-cos(2*pi*f1*t1);   % sampled signal (12 periods)
Y1=fft(y1);Y1=fftshift(Y1)/12;          % fft, shifting and normalizing
w=2*[0:M-1]./M-1;f=w/(2*Ts);            % frequency scale (4 periods)
N=length(y1);
w1=2*[0:N-1]./N-1;f=w/(2*Ts);           % frequency scale (12 periods)
```

11.4.4 LINEAR AND CIRCULAR CONVOLUTION

The most important property of the DFT is the convolution property which permits the computation of the linear convolution sum very efficiently by means of the FFT.

Consider the convolution sum that gives the output $y[n]$ of a discrete-time LTI system with impulse response $h[n]$ and input $x[n]$:

$$y[n] = \sum_m x[m]h[n-m].$$

In frequency, $y[n]$ is the inverse DTFT of the product

$$Y(e^{j\omega}) = X(e^{j\omega})H(e^{j\omega}).$$

Assuming that $x[n]$ has a finite length M, and that $h[n]$ has a finite length K then $y[n]$ has a finite length $N = M + K - 1$. If we choose a period $L \geq N$ for the periodic extension $\tilde{y}[n]$ of $y[n]$, we would obtain the frequency-sampled periodic sequence

$$Y(e^{j\omega})|_{\omega=2\pi k/L} = X(e^{j\omega})H(e^{j\omega})|_{\omega=2\pi k/L}$$

or the DFT of $y[n]$ as the product of the DFTs of $x[n]$ and $h[n]$

$$Y[k] = X[k]H[k] \quad \text{for} \quad k = 0, 1, \cdots, L-1.$$

We then obtain $y[n]$ as the inverse DFT of $Y[k]$. It should be noticed that the L-length DFT of $x[n]$ and of $h[n]$ requires that we pad $x[n]$ with $L - M$ zeros, and $h[n]$ with $L - K$ zeros, so that both $X[k]$ and $H[k]$ have the same length L and can be multiplied at each k. Thus we have

---

Given $x[n]$ and $h[n]$ of lengths M and K, the linear convolution sum $y[n]$, of length $M + K - 1$, can be found by following the following three steps:

- Compute DFTs $X[k]$ and $H[k]$ of length $L \geq M + K - 1$ for $x[n]$ and $h[n]$.
- Multiply them to get $Y[k] = X[k]H[k]$.
- Find the inverse DFT of $Y[k]$ of length L to obtain $y[n]$.

Although it seems computationally more expensive than performing the direct computation of the convolution sum, the above approach implemented with the FFT can be shown to be much more efficient in the number of computations.

The above procedure could be implemented by a **circular convolution sum** in the time domain, although in practice it is not done due to the efficiency of the implementation with FFTs. A **circular convolution** uses circular rather than linear representation of the signals being convolved. The **periodic convolution sum** introduced before is a circular convolution of fixed length—the period of the signals being convolved. When we use the DFT to compute the response of an LTI system the length of the circular convolution is given by the possible length of the linear convolution sum. Thus if the system input is a finite sequence $x[n]$ of length M and the impulse response of the system $h[n]$ has a length K then the output $y[n]$ is given by a linear convolution of length $M + K - 1$. The DFT $Y[k] = X[k]H[k]$ of length $L \geq M + K - 1$ corresponds to a circular convolution of length L of the $x[n]$ and $h[n]$ (padded with zeros so that both have length L). In such a case the circular and the linear convolutions coincide. This can be summarized as follows:

If $x[n]$, of length M, is the input of an LTI system with impulse response $h[n]$ of length K, then

$$Y[k] = X[k]H[k] \quad \Leftrightarrow \quad y[n] = (x \otimes_L h)[n] \tag{11.56}$$

where $X[k]$, $H[k]$ and $Y[k]$ are DFTs of length L of the input, the impulse response and the output of the LTI system, and $\otimes_L$ stands for circular convolution of length L.

If L is chosen so that $L \geq M + K - 1$, the circular and the linear convolution sums coincide, i.e.,

$$y[n] = (x \otimes_L h)[n] = (x * h)[n]. \tag{11.57}$$

If we consider the periodic expansions of $x[n]$ and $h[n]$ with period $L = M + K - 1$, we can use their circular representations and implement the circular convolution as shown in Fig. 11.16. Since

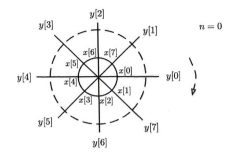

FIGURE 11.16

Circular convolution of length $L = 8$ of $x[n]$ and $y[n]$. Signal $x[k]$ is stationary with circular representation given by the inside circle, while $y[n-k]$ is represented by the outside circle and rotated in the clockwise direction. This figure corresponds to the case when $n = 0$.

the length of the linear convolution or convolution sum, $M + K - 1$, coincides with the length of the circular convolution, the two convolutions coincide. Given the efficiency of the FFT algorithm in computing the DFT, the convolution is typically done using the DFT as indicated above.

Example 11.25. To see the connection between the circular and the linear convolution, compute using MATLAB the circular convolution of a pulse signal $x[n] = u[n] - u[n-21]$, of length $N = 20$, with itself for different values of its length. Determine the length for which the circular convolution coincides with the linear convolution of $x[n]$ with itself.

Solution: We know that the length of the linear convolution, $z[n] = (x * x)[n]$ is $N + N - 1 = 2N - 1 = 39$. If we use our function *circonv2* to compute the circular convolution of $x[n]$ with itself with length $L = N < 2N - 1$ the result will not equal the linear convolution. Likewise, if the circular convolution is of length $L = N + 10 = 30 < 2N - 1$ only part of the result resembles the linear convolution. If we let the length of the circular convolution be $L = 2N + 9 = 49 > 2N - 1$, the result is identical to the linear convolution. (See Fig. 11.17.) The script is given below.

```
%%
% Example 11.25---Linear and circular convolution
%%
clear all; clf
N=20; x=ones(1,N);
% linear convolution
z=conv(x,x);z=[z zeros(1,10)];
% circular convolution
y=circonv(x,x,N);
y1=circonv(x,x,N+10);
y2=circonv(x,x,2*N+9);
Mz=length(z); My=length(y); My1=length(y1);My2=length(y2);
y=[y zeros(1,Mz-My)]; y1=[y1 zeros(1,Mz-My1)]; y2=[y2 zeros(1,Mz-My2)];
```

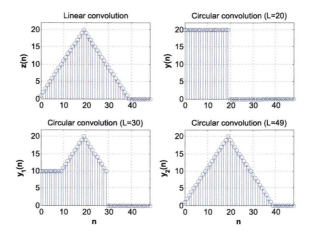

FIGURE 11.17

Circular vs. linear convolutions: Top-left plot corresponds to the linear convolution of $x[n]$ with itself. Top-right and bottom-left plots are circular convolutions of $x[n]$ with itself of length $L < 2N - 1$. Bottom-right plot is circular convolution of $x[n]$ with itself of length $L > 2N - 1$ coinciding with the linear convolution.

Our function *circonv* has as inputs the signals to be convolved and the desired length of the circular convolution. It computes and multiplies the FFTs of the signals and then finds the inverse FFT to obtain the circular convolution. If the desired length of the circular convolution is larger than the length of each of the signals, the signals are padded with zeros to make them of the length of the circular convolution.

```
function xy=circonv(x,y,N)
M=max(length(x),length(y))
if M>N
    disp('Increase N')
end
x=[x zeros(1,N-M)];
y=[y zeros(1,N-M)];
% circular convolution
X=fft(x,N);Y=fft(y,N);XY=X.*Y;
xy=real(ifft(XY,N));
```

Example 11.26. A significant advantage of using the FFT for computing the DFT is in filtering. Assume that the signal to filter consists of the MATLAB file *laughter.mat*, multiplied by 5, to which we add a signal that changes between -0.3 and 0.3 from sample to sampled. Recover the original *laughter* signal using an appropriate filter. Use the MATLAB function *fir1* to design the filter.

Solution: Noticing that the disturbance $0.3(-1)^n$ is a signal of frequency π, we need a low-pass filter with a wide bandwidth so as to get rid of the disturbance while trying to keep the frequency components

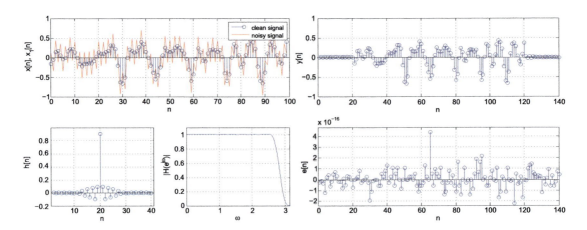

FIGURE 11.18

Finite impulse response (FIR) filtering of disturbed signal. Comparison of results using *conv* and *fft* functions. Top left: actual (samples with circles) and noisy signal (continuous line); bottom left: impulse response of FIR filter and its magnitude response. Top right: denoised signal; bottom right: error signal $\epsilon[n] = y[n] - y_1[n+20]$ between output from *conv* and *fft*-based filtering.

of the desired signal. The following script is used to design the desired low-pass filter, and to implement the filtering. To compare the results obtained with the FFT we use the function *conv* to find the output of the filter in the time domain. The results are shown in Fig. 11.18. Notice that the denoised signal is delayed 20 samples due to the linear phase of the filter of order 40. □

```
%%
% Example 11.26---Filtering using convolution and FFT
%%
clear all; clf
N=100; n=0:N-1;
load laughter
x=5*y(1:N)'; x1=x+0.3*(-1).^n;     % desired signal plus disturbance
h=fir1(40,0.9);[H,w]=freqz(h,1);   % low-pass FIR filter design
% filtering using convolution
y=conv(x,h);                       % convolution
 % computing using FFT
M=length(x)+length(h)-1;           % length of convolutions equal
X=fft(x,M);
 H=fft(h,M);
Y=X.*H;
y1=ifft(Y);                        % output of filtering
```

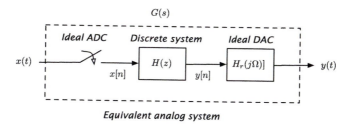

FIGURE 11.19

Discrete processing of analog signals using ideal analog-to-digital converter (ADC) and digital-to-analog converter (DAC). $G(s)$ is the transfer function of the overall system, while $H(z)$ is the transfer function of the discrete-time system.

11.4.5 THE FAST FOURIER TRANSFORM ALGORITHM

Given the advances in digital technologies and computers, processing of signals is being done mostly digitally. Early results in sampling, analog to digital conversion, and especially the fast computation of the output of linear systems using the Fast Fourier Transform (FFT) made it possible for digital signal processing to become a technical area on its own—the first textbooks in this area [68,61] come from the middle 1970s. Although the origins of the FFT have been traced back to the German mathematician Gauss in the early 1800s, the modern theory of the algorithm comes from the 1960s. It should be understood that the FFT is not yet another transform, but an efficient algorithm to compute the Discrete Fourier Transform (DFT).

In many applications, such as speech processing or acoustics, one would like to digitally process analog signals. In practice, this is possible by converting these signals into binary signals using an A/D converter, and if the output is desired in analog form a D/A converter is used to convert the binary signal into a continuous-time signal. Ideally, if sampling with no quantization is considered and if the discrete-time signal is converted into an analog signal by sinc interpolation the system can be visualized as in Fig. 11.19.

Viewing the whole system as a black box, with an analog signal $x(t)$ as input, and giving as output also an analog signal $y(t)$, the processing can be seen as a continuous-time system with a transfer function $G(s)$. Under the assumption of no quantization, the discrete-time signal $x[n]$ is obtained by sampling $x(t)$ using a sampling period determined by the Nyquist sampling condition. Likewise, considering the transformation of a discrete-time (or sampled signal) $y[n]$ into a continuous-time signal $y(t)$ by means of the sinc interpolation the ideal D/A converter is an analog low-pass filter that interpolates the discrete-time samples to obtain an analog signal. Finally, the discrete-time signal $x[n]$ is processed by a discrete-time system with transfer function $H(z)$, which depends on the desired transfer function $G(s)$.

Thus, one can process discrete- or continuous-time signals using discrete systems. A great deal of the computational cost of this processing is due to the convolution sum used to obtain the output of the discrete system. That is where the significance of the Fast Fourier Transform (FFT) algorithm lies. Although the Discrete Fourier Transform (DFT) allows us to simplify the convolution to a multiplication, it is the FFT that as an algorithm provides a very efficient implementation of this process. We thus introduce the FFT and provide some of the basics of this algorithm to understand its efficiency.

Comparing the equations for the DFT and the inverse DFT

$$X[k] \;=\; \sum_{n=0}^{N-1} x[n] W_N^{kn} \qquad k=0,\cdots,N-1, \tag{11.58}$$

$$x[n] \;=\; \frac{1}{N} \sum_{k=0}^{N-1} X[k] W_N^{-kn} \qquad n=0,\cdots,N-1, \tag{11.59}$$

where $W_N = e^{-j2\pi/N}$, one sees duality between the two transforms (more so if both the DFT and the IDFT had the term $1/\sqrt{N}$ instead of only the $1/N$ in the IDFT). Since $X[k]$ is typically complex, one can also see that if we assume $x[n]$ to be complex the same algorithm could be used to compute both the direct and the inverse DFTs. We thus consider $x[n]$ complex in the following.

Two issues used to assess the complexity of an algorithm are:

- **Total number of additions and multiplications:** Typically the complexity of a computational algorithm is assessed by the number of additions and multiplications it requires. The direct calculation of $X[k]$ using Equation (11.58) for $k=0,\cdots,N-1$ requires $N \times N$ complex multiplications, and $N \times (N-1)$ complex additions. Computing the number of real multiplications (4 real multiplications and 3 real additions for each complex multiplication) and real additions (2 real addiction for each complex addition) needed, it is found that the total number of these operations is of the order of N^2.

- **Storage:** Besides the number of computations, the required storage is also an issue of interest. Given that the $\{X[k]\}$ are complex, $2N^2$ locations in memory are required.

THE MODERN FFT

A paper by James Cooley, an IBM researcher, and Professor John Tukey from Princeton University [16] describing an algorithm for the machine calculation of complex Fourier series, appeared in *Mathematics of Computation* in 1965. Cooley, a mathematician, and Tukey, a statistician, had in fact developed an efficient algorithm to compute the Discrete Fourier Transform (DFT), which will be called the Fast Fourier Transform or FFT. Their result was a turning point in digital signal processing: the proposed algorithm was able to compute the DFT of a sequence of length N using $N \log N$ arithmetic operations, much smaller than the N^2 operations that had blocked the practical use of the DFT. As Cooley indicated in his paper "How the FFT Gained Acceptance" [15], his interest in the problem came from a suggestion from Tukey on letting N be a composite number, which would allow a reduction in the number of operations of the DFT computation. The FFT algorithm was a great achievement for which the authors received deserved recognition, but also benefited the new digital signal processing area, and motivated further research on the FFT. But as in many areas of research, Cooley and Tukey were not the only ones who had developed an algorithm of this class. Many other researchers before them had developed similar procedures. In particular, Danielson and Lanczos, in a paper published in the *Journal of the Franklin Institute* in 1942 [20], proposed an algorithm that came very close to Cooley and Tukey's results. Danielson and Lanczos showed that a DFT of length N could be represented as a sum of two $N/2$ DFTs proceeding recursively with the condition that $N = 2^\gamma$. Interestingly, they mention that (remember this was in 1942!):

Adopting these improvements the approximation times for Fourier analysis are: 10 minutes for 8 coefficients, 25 minutes for 16 coefficients, 60 minutes for 32 coefficients, and 140 minutes for 64 coefficients.

Radix-2 FFT Decimation-in-Time Algorithm

In the following, we assume that the FFT length is $N = 2^\gamma$ for an integer $\gamma > 1$. An excellent reference on the DFT and the FFT is [12].

The FFT algorithm:

- Uses the fundamental principle of **"Divide and Conquer,"** i.e., dividing a problem into smaller problems with similar structure, the original problem can be successfully solved by solving each of the smaller problems.
- Takes advantage of periodicity and symmetry properties of W_N^{nk}:

   1. **Periodicity:** W_N^{nk} is periodic of fundamental period N with respect to n, and with respect to k, i.e.,

   $$W_N^{nk} = \begin{cases} W_N^{(n+N)k}, \\ W_N^{n(k+N)}. \end{cases}$$

   2. **Symmetry:** The conjugate of W_N^{nk} is such that

   $$\left[W_N^{nk}\right]^* = W_N^{(N-n)k} = W_N^{n(N-k)}.$$

Applying the "divide and conquer" principle, we express $X[k]$ as

$$X[k] = \sum_{n=0}^{N-1} x[n] W_N^{nk} = \sum_{n=0}^{N/2-1} \left[x[2n] W_N^{k(2n)} + x[2n+1] W_N^{k(2n+1)} \right] \qquad k = 0, \cdots, N-1,$$

i.e., we gather the samples with even argument separately from those with odd arguments. From the definition of W_N^{nk} we have

$$W_N^{k(2n)} = e^{-j2\pi(2kn)/N} = e^{-j2\pi kn/(N/2)} = W_{N/2}^{kn}$$
$$W_N^{k(2n+1)} = W_N^k W_{N/2}^{kn},$$

which allows us to write

$$\begin{aligned} X[k] &= \sum_{n=0}^{N/2-1} x[2n] W_{N/2}^{kn} + W_N^k \sum_{n=0}^{N/2-1} x[2n+1] W_{N/2}^{kn} \\ &= Y[k] + W_N^k Z[k] \qquad k = 0, \cdots, N-1 \end{aligned} \tag{11.60}$$

where $Y[k]$ and $Z[k]$ are DFTs of length $N/2$ of the even-numbered sequence $\{x[2n]\}$ and of the odd-numbered sequence $\{x[2n+1]\}$, respectively. Although it is clear how to compute the values of $X[k]$ for $k = 0, \cdots, (N/2) - 1$ as

$$X[k] = Y[k] + W_N^k Z[k] \qquad k = 0, \cdots, (N/2) - 1, \tag{11.61}$$

it is not clear how to proceed for $k \geq N/2$. The $N/2$ periodicity of $Y[k]$ and $Z[k]$ allow us to find those values:

$$
\begin{aligned}
X[k + N/2] &= Y[k + N/2] + W_N^{k+N/2} Z[k + N/2] \\
&= Y[k] - W_N^k Z[k] \qquad k = 0, \cdots, N/2 - 1,
\end{aligned} \tag{11.62}
$$

where besides the periodicity of $Y[k]$ and $Z[k]$, we used

$$
W_N^{k+N/2} = e^{-j2\pi[k+N/2]/N} = e^{-j2\pi k/N} e^{-j\pi} = -W_N^k.
$$

Writing Equations (11.61) and (11.62) in a matrix form we have

$$
\mathbf{X}_N = \begin{bmatrix} \mathbf{I}_{N/2} & \boldsymbol{\Omega}_{N/2} \\ \mathbf{I}_{N/2} & -\boldsymbol{\Omega}_{N/2} \end{bmatrix} \begin{bmatrix} \mathbf{Y}_{N/2} \\ \mathbf{Z}_{N/2} \end{bmatrix} = \mathbf{A}_1 \begin{bmatrix} \mathbf{Y}_{N/2} \\ \mathbf{Z}_{N/2} \end{bmatrix} \tag{11.63}
$$

where $\mathbf{I}_{N/2}$ is a unit matrix and $\boldsymbol{\Omega}_{N/2}$ is a diagonal matrix with entries $\{W_N^k, \; k = 0, \cdots, N/2 - 1\}$, both of dimension $N/2 \times N/2$. The vectors $\mathbf{X}_N$, $\mathbf{Y}_{N/2}$ and $\mathbf{Z}_{N/2}$ contain the coefficients of $x[n]$, $y[n]$ and $z[n]$. This procedure is called **decimation in time FFT algorithm**.

Repeating the above computation for the $Y[k]$ and the $Z[k]$ we can express it in a similar matrix form until we reduce the process to 2×2 matrices. While performing these computations, the ordering of the $x[n]$ is changed. This scrambling of the $x[n]$ is obtained by a permutation matrix $\mathbf{P}_N$ (with 1 and 0 entries indicating the resulting ordering of the $x[n]$ samples).

If $N = 2^\gamma$, the $\mathbf{X}_N$ vector, containing the DFT terms $\{X[k]\}$, is obtained as the product of γ matrices $\mathbf{A}_i$ and the permutation matrix $\mathbf{P}_N$. That is

$$
\mathbf{X}_N = \left[\prod_{i=1}^{\gamma} \mathbf{A}_i \right] \mathbf{P}_N \, \mathbf{x} \qquad \mathbf{x} = [x[0], \cdots, x[N-1]]^T \tag{11.64}
$$

where T stands for the matrix transpose. Given the large number of 1s and 0s in the $\{\mathbf{A}_i\}$ and the $\mathbf{P}_N$ matrices, the number of additions and multiplications is much lower than those in the original formulas. The number of operations is found to be of the order of $N \log_2 N = \gamma N$, which is much smaller than the original number of order N^2. For instance if $N = 2^{10} = 1024$ the number of additions and multiplications for the computation of the DFT from its original formula is $N^2 = 2^{20} = 1.048576 \times 10^6$ while the FFT computation requires $N \log_2 N = 1024 \times 10 = 0.010240 \times 10^6$, i.e., the FFT requires about one percent of the number of operations required by the original formula for the DFT.

Example 11.27. Consider the decimation-in-time FFT algorithm for $N = 4$. Give the equations to compute the four DFT values $X[k]$, $k = 0, \cdots, 3$, in matrix form.

Solution: If we compute the DFT of $x[n]$ directly we have

$$
X[k] = \sum_{n=0}^{3} x[n] W_4^{nk} \qquad k = 0, \cdots, 3,
$$

which can be rewritten in the matrix form

$$
\begin{bmatrix} X[0] \\ X[1] \\ X[2] \\ X[3] \end{bmatrix} = \begin{bmatrix} 1 & 1 & 1 & 1 \\ 1 & W_4^1 & W_4^2 & W_4^3 \\ 1 & W_4^2 & 1 & W_4^2 \\ 1 & W_4^3 & W_4^2 & W_4^1 \end{bmatrix} \begin{bmatrix} x[0] \\ x[1] \\ x[2] \\ x[3] \end{bmatrix}
$$

where we used

$$
\begin{aligned}
W_4^4 &= W_4^{4+0} = e^{-j2\pi 0/4} = W_4^0 = 1, \\
W_4^6 &= W_4^{4+2} = e^{-j2\pi 2/4} = W_4^2, \\
W_4^9 &= W_4^{4+4+1} = e^{-j2\pi 1/4} = W_4^1,
\end{aligned}
$$

which requires 16 multiplications (8 if multiplications by 1 are not counted) and 12 additions. So 28 (or 20 if multiplications by 1 are not counted) total additions and multiplications. Since the entries are complex these are complex additions and multiplications. A complex addition requires two real additions, and a complex multiplication four real multiplications and two real additions. Indeed, for two complex numbers $z = a + jb$ and $v = c + jd$, $z + v = (a + c) + j(b + c)$ and $zv = (ac - bd + j(bc + ad))$. Thus the number of real multiplications is 16×4 and of real additions is $12 \times 2 + 16 \times 2$ giving a total of 120 operations.

Separating the even- and the odd-numbered entries of $x[n]$ we have

$$
\begin{aligned}
X[k] &= \sum_{n=0}^{1} x[2n] W_2^{kn} + W_4^k \sum_{n=0}^{1} x[2n+1] W_2^{kn} \\
&= Y[k] + W_4^k Z[k] \qquad k = 0, \cdots, 3,
\end{aligned}
$$

which can be written, considering that $Y[k+2] = Y[k]$ and $Z[k+2] = Z[k]$, i.e., periodic of fundamental period 2 and that $W_4^{k+2} = W_4^2 W_4^k = e^{-j\pi} W_4^k = -W_4^k$, as

$$
\begin{aligned}
X[k] &= Y[k] + W_4^k Z[k] \\
X[k+2] &= Y[k] - W_4^k Z[k] \qquad k = 0, 1.
\end{aligned}
$$

In matrix form the above equations can be written as

$$
\begin{bmatrix} X[0] \\ X[1] \\ \cdots \\ X[2] \\ X[3] \end{bmatrix} = \begin{bmatrix} 1 & 0 & \vdots & 1 & 0 \\ 0 & 1 & \vdots & 0 & W_4^1 \\ \cdots & \cdots & \cdots & \cdots & \cdots \\ 1 & 0 & \vdots & -1 & 0 \\ 0 & 1 & \vdots & 0 & -W_4^1 \end{bmatrix} \begin{bmatrix} Y[0] \\ Y[1] \\ \cdots \\ Z[0] \\ Z[1] \end{bmatrix} = \mathbf{A}_1 \begin{bmatrix} Y[0] \\ Y[1] \\ Z[0] \\ Z[1] \end{bmatrix},
$$

which is in the form indicated by Equation (11.63).

Now we have

$$Y[k] = \sum_{n=0}^{1} x[2n]W_2^{kn} = x[0]W_2^0 + x[2]W_2^k,$$

$$Z[k] = \sum_{n=0}^{1} x[2n+1]W_2^{kn} = x[1]W_2^0 + x[3]W_2^k \qquad k = 0, 1,$$

where $W_2^0 = 1$ $W_2^k = e^{-j\pi k} = (-1)^k$, giving the matrix

$$
\begin{bmatrix} Y[0] \\ Y[1] \\ \cdots \\ Z[0] \\ Z[1] \end{bmatrix}
=
\begin{bmatrix}
1 & 1 & \vdots & 0 & 0 \\
1 & -1 & \vdots & 0 & 0 \\
\cdots & \cdots & \cdots & \cdots & \cdots \\
0 & 0 & \vdots & 1 & 1 \\
0 & 0 & \vdots & 1 & -1
\end{bmatrix}
\begin{bmatrix} x[0] \\ x[2] \\ \cdots \\ x[1] \\ x[3] \end{bmatrix}
= A_2
\begin{bmatrix} x[0] \\ x[2] \\ x[1] \\ x[3] \end{bmatrix}.
$$

Notice the ordering of the $\{x[n]\}$. The scrambled $\{x[n]\}$ entries can be written

$$
\begin{bmatrix} x[0] \\ x[2] \\ x[1] \\ x[3] \end{bmatrix}
=
\begin{bmatrix}
1 & 0 & 0 & 0 \\
0 & 0 & 1 & 0 \\
0 & 1 & 0 & 0 \\
0 & 0 & 0 & 1
\end{bmatrix}
\begin{bmatrix} x[0] \\ x[1] \\ x[2] \\ x[3] \end{bmatrix}
= P_4
\begin{bmatrix} x[0] \\ x[1] \\ x[2] \\ x[3] \end{bmatrix},
$$

which finally gives

$$
\begin{bmatrix} X[0] \\ X[1] \\ X[2] \\ X[3] \end{bmatrix}
= A_1\, A_2\, P_4
\begin{bmatrix} x[0] \\ x[1] \\ x[2] \\ x[3] \end{bmatrix}.
$$

The count of multiplications is now much lower given the number of ones and zeros. The complex additions and multiplications is now 10 (2 complex multiplications and 8 complex additions) if we do not count multiplications by 1 or -1. Half of what it was when calculating the DFT directly! □

11.4.6 COMPUTATION OF THE INVERSE DFT

The FFT algorithm can be used to compute the inverse DFT without any changes in the algorithm. Assuming the input $x[n]$ is complex ($x[n]$ being real is a special case), the complex conjugate of the inverse DFT equation, multiplied by N, is

$$Nx^*[n] = \sum_{k=0}^{N-1} X^*[k]W^{nk}. \tag{11.65}$$

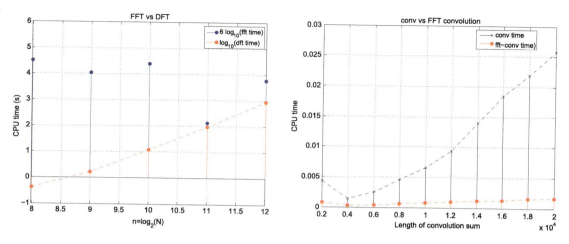

FIGURE 11.20

Left: execution times for the *fft* and the *dft* functions, in logarithmic scale, used in computing the DFT of sequences of ones of increasing length $N = 256$ to 4096 (corresponding to $n = 8, \cdots 12$). The CPU time for the FFT is multiplied by 10^6. Right: comparison of execution times of convolution of sequence of ones with itself using MATLAB's conv function and a convolution sum implemented with FFT.

Ignoring that the right-hand side term is in the frequency domain, we recognize it as the DFT of a sequence $\{X^*[k]\}$ and can be computed using the FFT algorithm discussed before. The desired $x[n]$ is thus obtained by computing the complex conjugate of Equation (11.65) and dividing it by N. As a result, the same algorithm, with the above modification, can be used to compute both the direct and the inverse DFTs.

Remark

In the FFT algorithm the $2N$ memory allocations for the complex input (one allocation for the real part and another for the imaginary part of the input) are the same ones used for the output. The results at each step uses the same locations. Since $X[k]$ is typically complex, to have identical allocation with the output, the input sequence $x[n]$ is assumed to be complex. If $x[n]$ is real, it is possible to transform it into a complex sequence and use properties of the DFT to obtain $X[k]$.

Example 11.28.   In this example we compare the efficiency of the FFT algorithm, in terms of computation time as measured by the functions *tic* and *toc*, when computing

- the DFT of a signal consisting of ones with increasing length $N = 2^r$, for $r = 8, \cdots , 12$, or 256 to 4096. We will compare the computation time of the FFT with our algorithm *dft* that computes the DFT using its definition;
- the convolution sum of a signal composed of ones with itself and of increasing lengths from 1000 to 10,000. The FFT-based convolution is compared with the MATLAB function *conv*.

Solution: To compare the algorithms we use the following script. Results of the two experiments are shown in Fig. 11.20. The left figure shows the results of comparing the execution time of the *fft* and the *dft* using a logarithmic scale. To make the resulting values comparable, the logarithmic values for the *fft* are multiplied by 6, indicating the FFT time is about one percent of the time for the *dft*. The figure on the right displays the execution times for *conv* and the FFT-based calculation of the convolution. Clearly, in both cases the FFT performance is superior to the DFT calculation using *dft* and to the convolution calculation using *conv* (showing on the side that this function is not based on the FFT). Our function *dft* is given after the script. □

```
%%
% Example 11.28
%%
% fft vs dft
clf; clear all
time=zeros(1,12-8+1); time1=zeros(1,12-8+1);
for r=8:12,
   N(r)=2^r;  i=r-7;
   % fft elapsed time
   t=tic;  X1=fft(ones(1,N(r)),N(r));  time(i)=toc(t);
   % dft elapsed time
   t=tic;  Y1=dft(ones(N(r),1),N(r)); time1(i)=toc(t);
end

% comparison of conv and fft
clear all
time1=zeros(1,10); time2=time1;
for i=1:10,
   NN=1000*i; x=ones(1,NN);
   % elapsed time for convolution using conv
   t1=tic;  y=conv(x,x); time1(i)=toc(t1);
   % elapsed time for convolution using fft
   x=[x zeros(1,NN)];  t2=tic; X=fft(x); X=fft(x); Y=X.*X; y=real(ifft(Y));
   time2(i)=toc(t2);
end

function X=dft(x,N)
n=0:N-1;
W=ones(1,N);
for k=1:N-1,
    W=[W; exp(-j*2*pi*n*k/N)];
end
X=W*x;
```

GAUSS AND THE FFT

Going back to the sources used by the FFT researchers it was discovered that many well known mathematicians had developed similar algorithms for different values of N. But that an algorithm similar to the modern FFT had been developed and used by Carl Gauss, the German mathematician, probably in 1805, predating even Fourier work on harmonic analysis in 1807, was an interesting discovery—although not surprising [35]. Gauss has been called the "Prince of Mathematicians" for his prodigious work in so many areas of mathematics, and for the dedication to his work. His motto was *Pauca sed matura* (few, but ripe); he would not disclose any of his work until he was very satisfied with it. Moreover, as it was customary in his time, his treatises were written in Latin using a difficult mathematical notation, which made his results not known or understood by modern researchers. Gauss' treatise describing the algorithm was not published in his lifetime, but appeared later in his collected works. He, however, deserves the paternity of the FFT algorithm.

The developments leading to the FFT, as indicated by Cooley, points out two important concepts in numerical analysis (the first of which applies to research in other areas): (i) *the divide-and-conquer approach*, i.e., it pays to break a problem into smaller pieces of the same structure, and (ii) *the asymptotic behavior of the number of operations*. Cooley's final recommendations in his paper are worth serious consideration by researchers in technical areas:

- *Prompt publication of significant achievements is essential.*
- *Review of old literature can be rewarding.*
- *Communication among mathematicians, numerical analysts, and workers in a wide range of applications can be fruitful.*
- *Do not publish papers in neoclassic Latin!*

11.5 TWO-DIMENSIONAL DISCRETE TRANSFORMS

Analogous to the one-dimensional case, a rectangular periodic discrete sequence $\tilde{x}[m, n]$ of period (N_1, N_2) has also a discrete and periodic (of the same period (N_1, N_2)) Fourier series[2]

$$\tilde{x}[m, n] = \frac{1}{N_1 N_2} \sum_{k=0}^{N_1-1} \sum_{\ell=0}^{N_2-1} \tilde{X}(k, \ell) e^{j2\pi(mk/N_1 + n\ell/N_2)} \tag{11.66}$$

where the discrete Fourier coefficients $\{\tilde{X}(k, \ell)\}$ are found using the orthogonality of the two-dimensional basis functions $\{e^{j2\pi(mk/N_1 + n\ell/N_2)}\}$ to be

$$\tilde{X}(k, \ell) = \sum_{m=0}^{N_1-1} \sum_{n=0}^{N_2-1} \tilde{x}[m, n] e^{-j2\pi(mk/N_1 + n\ell/N_2)}. \tag{11.67}$$

Thus a period of the infinite-support bisequence $\tilde{x}[m, n]$, $-\infty < m, n < \infty$ is sufficient to represent it. The coefficients $\{\tilde{X}(k, \ell)\}$ form an infinite, periodic or circular,[3] discrete set of values.

---

[2] Although the term $1/N_1 N_2$ could be attached to the Fourier series coefficients it conventionally appears in the Fourier series of $\tilde{x}[m, n]$. This allows one to define the 2D Discrete Fourier Transform in the conventional way.

[3] This means the coefficients $\tilde{X}(k + uN_1, \ell + vN_2) = \tilde{X}(k, \ell)$ given the periodicity of the complex exponentials or the circularity of the frequencies $(2\pi(k + uN_1)/N_1, 2\pi(\ell + vN_2)/N_2) = (2\pi k/N_1, 2\pi \ell/N_2)$.

These coefficients correspond to the harmonic frequencies $(2\pi k/N_1, 2\pi \ell/N_2)$ for $k = 0, \cdots, N_1$ and $\ell = 0, \cdots, N_2$.

Although the above definitions are not practical—periodic signals are not ever found—they form the basis for the definition of the Two-Dimensional Discrete Fourier Transform (2D-DFT) of non-periodic signals. Indeed, if we have a non-periodic signal $x[m, n]$ with a finite support[4] $\mathcal{S}$, it is possible to assume that $x[m, n]$ corresponds to the period of an extended periodic signal $\tilde{x}[m, n]$, to find the Fourier coefficients of $\tilde{x}[m, n]$ and to use them to represent $x[m, n]$.

Example 11.29. Let $x[m, n]$ have a finite support in the first quadrant $0 \leq m \leq M - 1, 0 \leq n \leq N - 1$. Consider how to obtain its 2D-DFT $X(k, \ell)$ and its inverse.

Solution: Taking $N_1 \geq M$ and $N_2 \geq N$ (adding zeros to $x[m, n]$ when $N_1 > M$ and $N_2 > N$) we consider an extended periodic signal $\tilde{x}[m, n]$ of rectangular period (N_1, N_2), which has as period

$$\tilde{x}[m, n] = \begin{cases} x[m, n] & 0 \leq m \leq M - 1, \ 0 \leq n \leq N - 1, \\ 0 & M \leq m \leq N_1 - 1, \ N \leq n \leq N_2 - 1. \end{cases} \qquad (11.68)$$

The discrete Fourier series of $\tilde{x}[m, n]$ has coefficients $\{\tilde{X}(k, \ell)\}$ calculated as in Equation (11.67) and will now be called the Two-Dimensional Discrete Fourier Transform (2D-DFT) $\{X(k, \ell)\}$ of the non-periodic signal $x[m, n]$. When processing two-dimensional signals with different supports it is important to recall this connection between $x[m, n]$ and $\tilde{x}[m, n]$.

Now, if we are given the DFT sequence $\{X(k, \ell)\}, 0 \leq k \leq N_1 - 1, 0 \leq \ell \leq N_2 - 1$ the inverse DFT will give the original signal padded with zeros shown in Equation (11.68). Notice that the main restriction is that we consider an extended periodic signal of periods bigger than the lengths of the signal. If the signal is periodic of period (M_1, M_2), this condition can be satisfied by taking $N_1 = uM_1$ and $N_2 = vM_2$, where u and v are positive integers. □

Analogous to the one-dimensional case, the resolution of the 2D-DFT is improved by making N_1 and N_2 much larger than the actual lengths M and N, and when the signal is periodic we need not only to take N_1 and N_2 to be multiples of the periods M_1 and M_2 but to scale the result by the product $1/uv$.

The computation of the 2D-DFT is done efficiently by means of the Two-dimensional Fast Fourier Transform (2D-FFT), which is not a transform but an algorithm. If the support of the signal is not the first quarter plane, we need to consider the periodic extension to generate the signal supported in the first quarter plane.

Two very important properties of the 2D-DFT are the convolution and the shifting properties. The two-dimensional convolution of $x[m, n]$, $0 \leq m \leq M - 1$, $0 \leq n \leq N - 1$ and $h[m, n]$, $0 \leq m \leq K - 1$, $0 \leq n \leq L - 1$ gives as result a signal $y[m, n]$, $0 \leq m \leq M + K - 1$, $0 \leq n \leq N + L - 1$. That is, the support of $y[m, n] = (x * h)[m, n]$ is $(M + K) \times (N + L)$. According to the convolution property, taking a 2D-DFT of size bigger or equal to $(M + K) \times (N + L)$ of $y[m, n]$ to get $Y(k, \ell)$ we should also take 2D-DFTs of the same size $(M + K) \times (N + L)$ for $x[m, n]$ and $h[m, n]$. Thus, the convolution property is

$$y[m, n] = (x * h)[m, n] \quad \Leftrightarrow \quad Y(k, \ell) = X(k, \ell)H(k, \ell) \qquad (11.69)$$

---

[4]If the support of the signal is not finite, it can be made finite by multiplying it by a finite mask.

where $X(k, \ell)$, $H(k, \ell)$ and consequently $Y(k, \ell)$ are 2D-DFTs of size bigger or equal to $(M + K) \times (N + L)$. As you can see, this is an extension of the one-dimensional convolution property of the one-dimensional DFT.[5]

The shifting property simply establishes that

$$2\text{D-DFT}(x[m - u, n - v]) = e^{-j2\pi uk/N_1} e^{-j2\pi v\ell/N_2} X(k, \ell) \tag{11.70}$$

where $N_1 \times N_2$ is the size of the DFT $X(k, \ell)$.

Example 11.30. Consider filtering the image *cameraman* with a filter with impulse response

$$h_1[m, n] = \frac{1}{3}\left(\delta[m, n] + 2\delta[m - 1, n] + \delta[m - 2, n] - \delta[m, n - 2]\right.$$
$$\left. - 2\delta[m - 1, n - 2] - \delta[m - 2, n - 2]\right)$$

and the result being filtered with a filter of impulse response

$$h_2[m, n] = \frac{1}{3}\left(\delta[m, n] - \delta[m - 2, n] + 2\delta[m, n - 1] - 2\delta[m - 2, n - 1]\right.$$
$$\left. - \delta[m, n - 2] - \delta[m - 2, n - 2]\right),$$

i.e., the original image is filtered by a cascade of filters with impulse responses $h_1[m, n]$ and $h_2[m, n]$. Use the 2D-FFT to find the output image. Make this image into a binary image by first making it a positive image and then thresholding it so that values bigger than 1.1 are converted into 1 and the rest are zero.

Solution: The dimension of the image is 256×256 pixels, so that the size of the 2D-FFTs for the first filtering should be greater or equal to $(256 + 3 - 1) \times (256 + 3 - 1) = 258 \times 258$ or the size of the image added to the one of the filter and subtracting one. Likewise, for the second filtering the minimum size of the 2D-FFTs should be $(258 + 3 - 1) \times (258 + 3 - 1) = 300 \times 300$. If we choose 300×300 as the size of the 2D-FFTs we satisfy both size conditions. The following MATLAB code is used.

```
%%
% Example 11.30
%%
clear all; clf
RGB = imread('cameraman.tif'); i = im2double(RGB);
h1=[1 2 1;0 0 0;-1 -2 -1];h1=h1/3; h2=h1';
H1=fft2(h1,300,300); H2=fft2(h2,300,300);
I=fft2(i,300,300);
Y=(H1.*I).*H2; y=ifft2(Y);y1=y;
y=-min(min(y))+y;y=y>1.1;
```

---

[5]Although this can also be related to the circular convolution for bi-sequences, it is better to avoid that and establish the property directly in terms of the linear convolution as we have just done.

```
w1=-1:2/300:1-2/300;w2=w1;
H=H1.*H2;
figure(1)
colormap('gray')
subplot(221)
imshow(i);title('Original')
subplot(222)
imshow(y1);title('Filtered')
subplot(223)
imshow(y); title('Binarized')
subplot(224)
mesh(w1,w2,fftshift(abs(H)));
title('Magnitude response of filter')
axis([-1,1,-1,1,0,3])
xlabel('\omega_1');ylabel('\omega_2');
zlabel('|H(\omega_1,\omega_2)|')
```

Notice that the binarization is done by making the filtered image positive and then thresholding with the value 1.1. The original and processed images and the magnitude frequency response of the filter are shown in Fig. 11.21. The filter as a result is found to be a pass-band filter.   □

In the image coding application, given the inverse relation of the support of a signal in the space domain and the support in the frequency domain[6] it is common practice to represent signals by their transforms in the frequency domain. This procedure is called the transform coding method. Although the 2D-DFT can be used in transform coders, the fact that the coefficients are complex requires sophisticated coding and as such transforms that give real coefficients and perhaps provide more compression are typically used. The Discrete Cosine Transform (DCT) is one of those transforms. Because the extension of the one-dimensional to the two-dimensional DCT is straightforward and to emphasize the connection of the one-dimensional DFT and the one-dimensional DCT we show next one possible way to obtain it.

Consider determining the DFT of an even real-valued sequence $y[n]$, $-(N-1) \leq n \leq N$. The length of this sequence is $2N$ and the even symmetry implies that $y[n] = y[-n]$ for $n = 1, \cdots, N-1$ and taking advantage of this symmetry its DFT is found to be

$$
\begin{aligned}
Y(k) &= \sum_{n=-(N-1)}^{N} y[n]W_{2N}^{-nk} = y[0] + \sum_{n=1}^{N-1} (y[n]W_{2N}^{-nk} + y[-n]W_{2N}^{nk}) + y[N]\cos(\pi k) \\
&= y[0] + \sum_{n=1}^{N-1} 2y[n]\cos\left(\frac{\pi nk}{N}\right) + y[N]\cos(\pi k), \qquad -(N-1) \leq k \leq N, \qquad (11.71)
\end{aligned}
$$

where we have defined

$$
W_{2N} = e^{j2\pi/2N}, \quad \text{and we used } W_{2N}^{Nk} = e^{j2\pi Nk/2N} = e^{j\pi k} = \cos(\pi k) + j\sin(\pi k) = \cos(\pi k).
$$

---

[6]A small support signal $x[m, n]$ has a large support for $X(k, \ell)$ in the frequency domain, and vice versa.

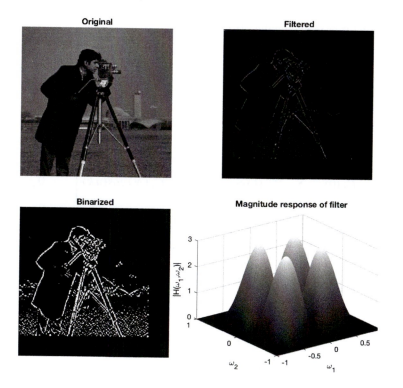

FIGURE 11.21

Cascade filtering and binarization of the image *cameraman*. Clockwise from top left: original and filtered images, magnitude response of cascaded filter and binarized image.

Notice that the values $y[n]$ for $-(N-1) \leq n \leq -1$ are not used, and that, since $Y(k) = Y(-k)$, due to the even symmetry of the cosines, the values $Y(k)$ for $-(N-1) \leq k \leq -1$ are redundant, so that we can see Equation (11.71) as a mapping of the two real vectors

$$(y[0], y[1], \cdots, y[N]) \quad \text{and} \quad (Y(0), Y(1), \cdots, Y(N)).$$

To take advantage of this result, consider an $N+1$ real sequence $x[n]$, $0 \leq n \leq N$, and zero outside this support, which as such is not even symmetric, connected with $y[n]$ as follows:

$$y[n] = \begin{cases} x[0] & n = 0, \\ x[n] & 1 \leq n \leq N-1, \\ x[-n] & -(N-1) \leq n \leq -1, \\ x[N] & n = N, \end{cases} \tag{11.72}$$

then we have, as indicated above, the mapping

$$(x[0], x[1], \cdots, x[N]) \quad \text{and} \quad (\tilde{C}_x(0), \tilde{C}_x(1), \cdots, \tilde{C}_x(N)),$$

where the transform (after replacing the $y[n]$s by $x[n]$s)

$$\tilde{C}_x(k) = x[0] + \sum_{n=1}^{N-1} 2x[n] \cos\left(\frac{\pi nk}{N}\right) + x[N] \cos(\pi k) \qquad 0 \le k \le N \qquad (11.73)$$

is the discrete cosine transform or DCT of $x[n]$. Its inverse is found by using the IDFT of $Y(k)$, which is real, and given by

$$x[n] = \frac{1}{2N} \left[\tilde{C}_x(0) + \sum_{k=1}^{N-1} 2\tilde{C}_x(k) \cos\left(\frac{\pi nk}{N}\right) + \tilde{C}_x(N) \cos(\pi k) \right] \qquad 0 \le n \le N. \qquad (11.74)$$

Not only this transform converts a real-valued sequence $\{x[n]\}$ into another real-valued sequence $\{\tilde{C}_x[n]\}$, but the DCT and the IDCT can be computed using the same algorithm up to a multiplicative factor.

The above is one possible form of the DCT another is given by

$$C_x(k) = \left(\frac{2}{N}\right)^{1/2} \sum_{n=0}^{N-1} \Lambda(n) x[n] \cos\left(\frac{\pi k}{2N}(2n+1)\right)$$

$$\Lambda(n) = \begin{cases} 1/\sqrt{2} & n = 0, \\ 1 & \text{otherwise.} \end{cases} \qquad (11.75)$$

The extension of this formula to 2D gives the 2D-DCT

$$C_x(k, \ell) = \left(\frac{4}{MN}\right)^{1/2} \sum_{m=0}^{M-1} \sum_{n=0}^{N-1} \Lambda(m, n) x[m, n] \cos\left(\frac{\pi k}{2M}(2m+1)\right) \cos\left(\frac{\pi \ell}{2N}(2n+1)\right) \qquad (11.76)$$

where $\Lambda(m, n) = \Lambda(m)\Lambda(n)$ and $\Lambda(m)$ and $\Lambda(n)$ are defined as in the one-dimensional case, given above. The inverse DCT for the one- and two-dimensional cases is very much like the DCT, except for a constant term.

Example 11.31. Consider the *cameraman* image for which we compute its two-dimensional DCT using the function *dct2*, and obtain a compressed image having 2D-DCT coefficients bigger than 0.1 in absolute value. Show the original and the compressed images and their corresponding 2D-DCT values. Calculate the square error between the original and the compressed image, to quantify the level of compression, or tell how close the original and the compressed images are.

Solution: The 2D-DCT values of the original image *cameraman* are within a range $[-16.6, 119.2]$, and when thresholded so that the DCT values are set to zero whenever in the range $[-0.1, 0.1]$ we obtain the compressed image shown in the middle. Fig. 11.22 illustrates the application of the 2D-DCT in data compression of images.

  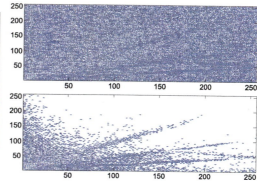

FIGURE 11.22

Compression using thresholded DCT values: original image (left), compressed image (center), support of original 2D-DCT values (top right) and thresholded 2D-DCT values (bottom right).

The following MATLAB code is used.

```
%%
% Example 11.31---2D-DCT compression
%%
clear all; clf
RGB = imread('cameraman.tif'); I = im2double(RGB);
J = dct2(I);
J0=J;
J(abs(J)<0.1) = 0;
J1=J;
K = idct2(J);
MM=size(J); M=MM(1);
figure(1)
subplot(221); imshow(I);title('Original')
subplot(222); imshow(K); title('Compressed')
subplot(223),contour(J0);title('DCT of original')
subplot(224),contour(J1);title('DCT of compressed')
% mean square error
esq=(I-K).*(I-K); error2=sum(esq(:))/M^2
```

The support of the 2D-DCT values for the original image and for those for the compressed image are shown in the figure on the right of Fig. 11.22. A much smaller number of coefficients are being used for the compressed image which is very similar to the original image. Indeed, the square error between the original and the compressed image was found to be 0.0013, indicating that the compressed image is quite close to the original image.   □

11.6 WHAT HAVE WE ACCOMPLISHED? WHERE DO WE GO FROM HERE?

In this chapter we considered the Fourier representation of discrete-time signals. Just as with the Laplace and the Fourier transforms in the continuous case, there is a large class of discrete-time signals for which we are able to find their discrete-time Fourier transform from their Z-transforms. For signals that are not absolutely summable, the time–frequency duality and other properties of the transform are used to find their DTFTs. Although theoretically useful, the DTFT is computationally not feasible, due to the continuity of the frequency variable and to the integration required in the inverse transformation. It is the Fourier series of discrete-time signals that makes the Fourier representation computationally feasible. The Fourier series coefficients constitute a periodic sequence of the same fundamental period as the signal, thus both are periodic. Moreover, the Fourier series and its coefficients are obtained as sums, and the frequency used is discretized. Thus they can be obtained by computer. To take advantage of this, the spectrum of an aperiodic signal resulting from the DTFT is sampled so that in the time domain there is a periodic repetition of the original signal. For finite-support signals we can then obtain a periodic extension which gives the discrete Fourier transform or DFT.

The significance of this result is that we have frequency representations of discrete-time signals that are computed algorithmically. However, what makes the DFT computationally efficient is the Fast Fourier Transform (FFT) algorithm. As shown, this algorithm not only efficiently computes the DFT but also makes the convolution sum feasible.

The two-dimensional discrete Fourier transform is obtained in a similar way to the one-dimension case, by considering the Fourier series representation of rectangular periodic signals. Once this is accomplished, its implementation is achieved by the two-dimensional FFT algorithm. The 2D-DFT is important particularly in the implementation of filtering problems. Data compression of images can be implemented using transforms by using the inverse relationship between the space and frequency supports of signals. Of particular interest in this problem is the use of the two-dimensional discrete cosine transform that not only gives real-valued coefficients, but that also provides a significant degree of compression.

11.7 PROBLEMS

11.7.1 BASIC PROBLEMS

11.1 From the direct and the inverse DTFT of $x[n] = 0.5^{|n|}$:

    (a) Determine the sum

$$\sum_{n=-\infty}^{\infty} 0.5^{|n|}.$$

    (b) Find the integral

$$\int_{-\pi}^{\pi} X(e^{j\omega})d\omega.$$

    (c) Find the phase of $X(e^{j\omega})$.

(d) Determine the sum

$$\sum_{n=-\infty}^{\infty} (-1)^n 0.5^{|n|}.$$

Answers: $\sum_n x[n] = 3$; $\sum_n x[n](-1)^n = 1/3$.

11.2 Consider the connection between the DTFT and the Z-transform in the following problems.

(a) Let $x[n] = u[n+2] - u[n-3]$.

    i. Can you find the DTFT $X(e^{j\omega})$ of $x[n]$ using the Z-transform? If so, what is it?

    ii. Is it true that $X(e^{j0}) = 5$? Explain.

(b) For the noncausal signal $x[n] = \alpha^n u[-n]$, $\alpha > 0$, give values of α for which $X(e^{j\omega}) = X(z)|_{z=e^{j\omega}}$.

(c) Consider the causal and anticausal signals

$$x_1[n] = \left(\frac{1}{2}\right)^n u[n], \quad x_2[n] = -\left(\frac{1}{2}\right)^n u[-n-1],$$

show that $X_1(z) = X_2(z)$, and find the regions of convergence. According to the ROCs, which of the two signals has a DTFT that can be found from its Z-transform?

Answer: (a) Yes, $X(e^{j0}) = 5$; (b) $\alpha > 1$; (c) $X_2(e^{j\omega})$ cannot be found from $X_2(z)$.

11.3 A triangular pulse is given by

$$t[n] = \begin{cases} 3+n & -2 \le n \le -1, \\ 3-n & 0 \le n \le 2, \\ 0 & \text{otherwise.} \end{cases}$$

Find a sinusoidal expression for the DTFT of $t[n]$

$$T(e^{j\omega}) = B_0 + \sum_{k=1}^{\infty} B_k \cos(k\omega).$$

Determine the coefficients B_0 and B_k.

Answers: $B_0 = 3$, $B_k = 2(3-k)$, $k = 1, 2$.

11.4 Find the DTFT of $x[n] = e^{j\theta}\delta[n+\tau] + e^{-j\theta}\delta[n-\tau]$, and use it to find the DTFT of $\cos(\omega_0 n + \theta)$, $-\infty < n < \infty$ according to the duality property.

(a) Verify the results coincide with those in the tables for $\theta = 0$ and $\theta = -\pi/2$.

(b) Determine the DTFT of a signal

$$x_1[n] = 1 + \sum_{k=1}^{5} A_k \cos(k\omega_0 n + \theta_k).$$

Answer: $X_1(e^{j\omega}) = 2\pi\delta(\omega) + \sum_{k=1}^{5} A_k \pi [e^{j\theta_k}\delta(\omega + k\omega_0) + e^{-j\theta_k}\delta(\omega - k\omega_0)]$.

11.5 Consider the application of the DTFT properties to filters.

(a) Let $h[n]$ be the impulse response of an ideal low-pass filter with frequency response

$$H(e^{j\omega}) = \begin{cases} 1 & -0.4\pi \le \omega \le 0.4\pi, \\ 0 & \text{otherwise in } (-\pi, \pi]. \end{cases}$$

If we let the impulse response of a new filter be $h_1[n] = [1 + (-1)^n] h[n]$, find the frequency response $H_1(e^{j\omega})$ in terms of $H(e^{j\omega})$. What type of filter is the new filter?

(b) Consider the frequency response of a filter

$$H(e^{j\omega}) = \frac{0.75}{1.25 - \cos(\omega)}.$$

i. From $H(e^{j\omega})$ find the sum $\sum_{n=-\infty}^{\infty} h[n]$.

ii. Given that $H(e^{j\omega}) = H(e^{-j\omega})$, i.e., it is real and an even function of ω, show that $h[n]$ is an even function of n. Use the inverse DTFT definition.

iii. Is it true that the phase response $\angle H(e^{j\omega})$ is zero for all discrete frequencies? Explain.

Answers: (a) $H_1(e^{j\omega})$ is a band-eliminating filter; (b) $H(e^{j0}) = 3$; zero phase.

11.6 Let $x[n] = u[n+2] - u[n-3]$.

(a) Find the DTFT $X(e^{j\omega})$ of $x[n]$ and sketch $|X(e^{j\omega})|$ vs. ω giving it at $\omega = \pm\pi, \pm\pi/2, 0$.

(b) If $x_1[n] = x[2n]$, i.e., $x[n]$ is down-sampled with $M = 2$, find its DTFT $X_1(e^{j\omega})$. Carefully sketch $x_1[n]$ and $|X_1(e^{j\omega})|$ indicating its values at $\omega = \pm\pi, \pm\pi/2, 0$. Is $X_1(e^{j\omega}) = 0.5X(e^{j\omega/2})$? If not, how would you process $x[n]$ so that when $x_1[n] = x[2n]$ you would satisfy this condition? Explain.

(c) Consider now the up-sampled signal

$$x_2[n] = \begin{cases} x[n/2] & n \text{ even}, \\ 0 & \text{otherwise}. \end{cases}$$

Find the DTFT $X_2(e^{j\omega})$ of $x_2[n]$, and carefully sketch both (in particular, when plotting $X_2(e^{j\omega})$ indicate the values at frequencies $\omega = \pm\pi, \pm\pi/2, 0$). Explain the differences between this case the down-sampling cases.

Answers: $X(e^{j\omega}) = 1 + 2\cos(\omega) + 2\cos(2\omega)$; $X_1(e^{j\omega}) \ne 0.5X(e^{j\omega/2})$; $X_2(e^{j\omega}) = X(e^{j2\omega})$.

11.7 Consider a LTI discrete-time system with input $x[n]$ and output $y[n]$. It is well known that the impulse response of the system is

$$h[n] = \frac{\sin(\pi(n-10)/3)}{\pi(n-10)}, \quad n \ne 10, \quad h[10] = 1/3.$$

(a) Determine the magnitude and phase responses $|H(e^{j\omega})|$ and $\angle H(e^{j\omega})$.

(b) Find the output $y[n]$ if $x[n] = \delta[n-1] + \cos(\pi n/5)$, $-\infty < n < \infty$.

Answers: $\angle H(e^{j\omega}) = -10\omega$, $-\pi < \omega \le \pi$; $y[n] = h[n-1] + \cos(\pi n/5)$, $-\infty < n < \infty$.

11.8 Consider the following problems related to the properties of the DTFT.

    (a) For the signal $x[n] = \beta^n u[n]$, $\beta > 0$, for what values of β you are able to find the DTFT of $x[n]$, $X(e^{j\omega})$, from $X(z)$?

    (b) Given the DTFT $X(e^{j\omega}) = \delta(\omega)$, i.e., a continuous-frequency delta function, what is

        i. the inverse DTFT $x[n]$?

        ii. the inverse DTFT of $X_1(e^{j\omega}) = X(e^{j(\omega-\pi)}) = \delta(\omega - \pi)$?

        iii. the inverse of $X_2(e^{j\omega}) = \delta(\omega) + \delta(-\omega)$?

    (c) Is it true that

$$1 + e^{-j\omega} + \cdots + e^{-j\omega(N-1)} = \frac{1 - e^{-j\omega N}}{1 - e^{-j\omega}}$$

    for any integer $N > 1$? What happens when $N = 1$?

    Answer: (a) $\beta < 1$; (b) $x_1[n] = (-1)^n/(2\pi)$; $x_2[n] = 2x[n]$.

11.9 The impulse response of an FIR filter is $h[n] = (1/3)(\delta[n] + \delta[n-1] + \delta[n-2])$.

    (a) Find the frequency response $H(e^{j\omega})$, and determine the magnitude and the phase responses for $-\pi < \omega \leq \pi$.

    (b) Find poles and zeros of the transfer function $H(z)$ of this filter and determine its region of convergence.

    (c) At what frequencies $\{\omega_i\}$ is $H(e^{j\omega_i}) = 0$?

    (d) Would unwrapping change the phase response? Explain.

    Answers: $H(e^{j\omega}) = e^{-j\omega}(1 + 2\cos(\omega))/3$; magnitude is zero at $\omega_0 = \cos^{-1}(0.5)$.

11.10 Determine the Fourier series coefficients $X_i[k]$, $i = 1, \cdots, 4$, for each of the following periodic discrete-time signals. Explain the connection between these coefficients and the symmetry of the corresponding signals.

    (a) $x_1[n]$ has a fundamental period $N = 5$ and in a period $x_1[n] = 1$ in $-1 \leq n \leq 1$ and $x_1[-2] = x_1[2] = 0$.

    (b) $x_2[n]$ has a fundamental period $N = 5$ and in a period $x_2[n] = 0.5^n$ in $-1 \leq n \leq 1$ and $x_2[-2] = x_2[2] = 0$.

    (c) $x_3[n]$ has a fundamental period $N = 5$ and in a period $x_3[n] = 2^n$ in $-1 \leq n \leq 1$ and $x_3[-2] = x_3[2] = 0$.

    (d) $x_4[n]$ has a fundamental period $N = 5$ and in a period $x_4[n] = n$ in $-1 \leq n \leq 1$ and $x_4[-2] = x_4[2] = 0$.

    (e) Consider a period of $x_1[n]$ starting at $n = 0$ and find the Fourier series coefficients. How do they compare with the ones found above?

    Answers: $X_1[k] = (1/5)(1 + 2\cos(2\pi k/5))$, $0 \leq k \leq 4$; $X_4[k] = -(2j/5)\sin(2\pi k/5)$, $0 \leq k \leq 4$.

11.11 Determine the Fourier series coefficients of the following periodic discrete-time signals:

    (a) $x_1[n] = 1 - \cos(2\pi n/3)$, $x_2[n] = 2 + \cos(8\pi n/3)$,
        $x_3[n] = 3 - \cos(2\pi n/3) + \cos(8\pi n/3)$, $x_4[n] = 2 + \cos(8\pi n/3) + \cos(2\pi n/3)$,
        $x_5[n] = \cos(\pi n/3) + (-1)^n$.

    (b) (i) $y[n]$ has a fundamental period $N = 3$ and a period $y_1[n] = (0.5)^n$ for $n = 0, 1, 2$;
        (ii) $v[n]$ has a fundamental period $N = 5$ and a period $v_1[n] = (0.5)^{|n|}$ for $-2 \leq n \leq 2$;
        (iii) $w[n]$ has a fundamental period $N = 4$ and a period $w_1[n] = 1 + (0.5)^n$ for $0 \leq n \leq 3$.

Indicate for each its fundamental frequency.

Answers: (a) $X_3[0] = 3$, $X_3[1] = X_3[-1] = 0$;   (b) $Y[k] = (1/3)[1 + 0.5e^{-j2\pi k/3} + 0.25e^{-j4\pi k/3}]$.

11.12 A continuous-time periodic signal $x(t)$ with fundamental period $T_0 = 2$ has a period $x_1(t) = u(t) - u(t-1)$.
  (a) Is $x(t)$ a band-limited signal? Find the Fourier coefficients X_k of $x(t)$.
  (b) Would $\Omega_{max} = 5\pi$ be a good value for the maximum frequency of $x(t)$? Explain.
  (c) Let $T_s = 0.1$ s/sample so that in a period the sampled signal is $x(nT_s) = u[n] - u[n-11]$. Obtain its discrete Fourier series coefficients of the periodic signal $x[n]$ and compare $X[0]$, $X[1]$ of $x[n]$ with X_0, X_1 of $x(t)$.

Answers: $x(t)$ not band-limited; $X_k = e^{j\pi k/2} \sin(\pi k/2)/(\pi k)$, $k \neq 0$, and $X_0 = 1/2$.

11.13 The output of an ideal low-pass filter is

$$y[n] = 1 + \sum_{k=1}^{2}(2/k)\cos(0.2\pi kn) \qquad -\infty < n < \infty.$$

  (a) Assume the filter input is a periodic signal $x[n]$. What is its fundamental frequency ω_0? What is the fundamental period N_0?
  (b) When the filter is low-pass with $H(e^{j0}) = H(e^{j\omega_0}) = H(e^{-j\omega_0}) = 1$ and $H(e^{j2\omega_0}) = H(e^{-j2\omega_0}) = 1/2$ determine the Fourier series for the input $x[n]$.

Answer: $x[n] = 1 + 2\cos(2\pi n/10) + 2\cos(4\pi n/10)$.

11.14 For the periodic discrete-time signal $x[n]$ with a period $x_1[n] = n$, $0 \leq n \leq 3$ use its circular representation to find

$$x[n-2], \quad x[n+2], \quad x[-n], \quad x[-n+k], \quad \text{for } 0 \leq k \leq 3$$

display several periods.

Answer: $x[n-2]$ has values $2, 3, 0, 1, 2, 3, 0, 1, \cdots$ starting at zero.

11.15 Let $x[n] = 1 + e^{j\omega_0 n}$ and $y[n] = 1 + e^{j2\omega_0 n}$ be periodic signals of fundamental period $\omega_0 = 2\pi/N$, find the Fourier series of their product $z[n] = x[n]y[n]$ by
  (a) calculating the product $x[n]y[n]$,
  (b) using the periodic convolution of length $N = 3$ of the Fourier series coefficients of $x[n]$ and $y[n]$. Is the periodic convolution equal to $x[n]y[n]$ when $N = 3$? Explain.

Answer: $x[n]y[n] = 1 + e^{j\omega_0 n} + e^{j2\omega_0 n} + e^{j3\omega_0 n}$ for $\omega_0 = 2\pi/N$; yes, the two results are equal.

11.16 The periodic signal $x[n]$ has a fundamental period $N_0 = 4$, and a period is given by $x_1[n] = u[n] - u[n-2]$. Calculate the periodic convolution of length $N_0 = 4$ of
  (a) $x[n]$ with itself and call it $v[n]$,
  (b) $x[n]$ and $x[-n]$ and call it $z[n]$.

Answers: $v[0] = 1$, $v[1] = 2$, $v[2] = 1$, $v[3] = 0$.

11.17 Consider the aperiodic signal

$$x[n] = \begin{cases} 1 & n = 0, 1, 2, \\ 0 & \text{otherwise.} \end{cases}$$

Find the DFT of length $L = 4$ of

(i) $x[n]$, (ii) $x_1[n] = x[n-3]$, (iii) $x_2[n] = x[-n]$, (iv) $x_3[n] = x[n+3]$.

Answers: $X[k] = 1 + e^{-j\pi k/2} + e^{-j\pi k}$; $X_3[k] = e^{-j2\pi k/4} + e^{-j\pi k} + e^{j2\pi k/4}$.

11.18 Given the signals $x[n] = 2^n(u[n] - u[n-3])$ and $y[n] = 0.5^n(u[n] - u[n-3])$ write a matrix equation to compute their circular convolution of lengths $N = 3, 4$ and 5. Call the results $z[n]$, $w[n]$ and $v[n]$, respectively. For which of these lengths does the circular and the linear convolutions coincide? Explain.
Answer: $v[n] = \delta[n] + 2.5\delta[n-1] + 5.25\delta[n-2] + 2.5\delta[n-3] + \delta[n-4]$, which is the same as for a linear convolution.

11.19 Consider the discrete-time signal $x[n] = u[n] - u[n-M]$ where M is a positive integer.
   (a) Let $M = 1$, calculate and sample the DTFT $X(e^{j\omega})$ in the frequency domain using a sampling frequency $2\pi/N$ with $N = M$ to obtain the DFT of length $N = 1$.
   (b) Let $N = 10$, still $M = 1$, sample the DTFT to obtain the DFT and carefully plot $X(e^{j2\pi k/N}) = X[k]$ and the corresponding signal. What would happen if N is made 1024, what would be the differences with this case and the previous one? Explain.
   (c) Let then $M = 10$, and $N = 10$ for the sampling of the DTFT. What does $X[k]$ imply in terms of aliasing? Comment on what would happen if $N \gg 10$, and when $N < 10$
   Answers: $M = 10$, $N = 10$, $X[k] = 10$ for $k = 0$ and $X[k] = 0$ for $k = 1, \cdots, 9$.

11.20 The convolution sum of a finite sequence $x[n]$ with the impulse response $h[n]$ of an FIR system can be written in a matrix form $\mathbf{y} = \mathbf{Hx}$ where $\mathbf{H}$ is a matrix, $\mathbf{x}$ and $\mathbf{y}$ are input and output values. Let $h[n] = (1/3)(\delta[n] + \delta[n-1] + \delta[n-2])$ and $x[n] = 2\delta[n] + \delta[n-1]$.
   (a) Write the matrix equation and compute the values of $y[n]$.
   (b) Use the DFT matrix representation $\mathbf{Y} = \mathbf{Fy}$, replace $\mathbf{y}$ to obtain an expression in terms of $\mathbf{H}$ and $\mathbf{x}$. Use the DFT matrix representation $\mathbf{X} = \mathbf{Fx}$ and the orthogonality properties of $\mathbf{F}$ to obtain an expression in the DFT domain for the convolution sum.
   Answers: $y[0] = 2/3$, $y[1] = y[2] = 1$, $y[3] = 1/3$, $y[4] = 0$.

11.21 The signal $x[n] = 0.5^n(u[n] - u[n-3])$ is the input of a LTI system with an impulse response $h[n] = (1/3)(\delta[n] + \delta[n-1] + \delta[n-2])$.
   (a) Determine the length of the output $y[n]$ of the system.
   (b) Calculate the output $y[n]$ using the convolution sum.
   (c) Use the Z-transform to determine the output $y[n]$.
   (d) Use the DTFT to determine the output $y[n]$.
   (e) Use the DFT to find the output $y[n]$.
   Answers: $y[n] = (x * h)[n]$ is of length 5; use $Y(z) = X(z)H(z)$ to find $y[n]$.

11.7.2 PROBLEMS USING MATLAB

11.22 Zero-phase—Given the impulse response

$$h[n] = \begin{cases} \alpha^{|n|} & -2 \leq n \leq 2 \\ 0 & \text{otherwise} \end{cases}$$

where $\alpha > 0$. Find values of α for which the filter has zero phase. Verify your results with MATLAB.

Answer: $0 < \alpha \leq 0.612$.

11.23 Eigenfunction property and frequency response—An IIR filter is characterized by the following difference equation: $y[n] = 0.5y[n-1] + x[n] - 2x[n-1]$, $n \geq 0$, where $x[n]$ is the input and $y[n]$ the output of the filter. Let $H(z)$ be the transfer function of the filter.

(a) The given filter is LTI, as such the eigenfunction property applies. Obtain the magnitude response $H(e^{j\omega})$ of the filter using the eigenfunction property.

(b) Compute the magnitude response $|H(e^{j\omega})|$ at discrete frequencies $\omega = 0$, $\pi/2$ and π radians. Show that the magnitude response is constant for $0 \leq \omega \leq \pi$ and as such this is an all-pass filter.

(c) Use the MATLAB function *freqz* to compute the frequency response (magnitude and phase) of this filter and to plot them.

(d) Determine the transfer function $H(z) = Y(z)/X(z)$, find its pole and zero and indicate how they are related.

Answers: $|H(e^{j0})| = |H(e^{j\pi})| = |H(e^{j\pi/2})| = 2$, all-pass filter.

11.24 Frequency transformation of low-pass to high-pass filters—You have designed an IIR low-pass filter with an input–output relation given by the difference equation

$$(i) \quad y[n] = 0.5y[n-1] + x[n] + x[n-1] \qquad n \geq 0$$

where $x[n]$ is the input and $y[n]$ the output. You are told that by changing the difference equation to

$$(ii) \quad y[n] = -0.5y[n-1] + x[n] - x[n-1] \qquad n \geq 0,$$

you obtain a high-pass filter.

(a) From the eigenfunction property find the frequency response of the two filters at $\omega = 0$, $\pi/2$ and π radians. Use the MATLAB functions *freqz* and *abs* to compute the magnitude responses of the two filters. Plot them to verify that the filters are low-pass and high-pass.

(b) Call $H_1(e^{j\omega})$ the frequency response of the first filter and $H_2(e^{j\omega})$ the frequency response of the second filter. Show that $H_2(e^{j\omega}) = H_1(e^{j(\pi-\omega)})$ and relate the impulse response $h_2[n]$ to $h_1[n]$.

(c) Use the MATLAB function *zplane* to find and plot the poles and zeros of the filters and determine the relation between the poles and zeros of the two filters.

Answer: The transformation shifts poles and zeros to their opposite side of the Z-plane.

11.25 Chirps for jamming—A chirp signal is a sinusoid of continuously changing frequency. Chirps are frequently used to jam communication transmissions. Consider the chirp $x[n] = \cos(\theta n^2)u[n]$, $\theta = \frac{\pi}{2L}$, $0 \le n \le L - 1$.

(a) A measure of the frequency of the chirp is the so-called instantaneous frequency which is defined as the derivative of the phase in the cosine, i.e., $IF(n) = d(\theta n^2)/dn$. Find the instantaneous frequency of the given chirp. Use MATLAB to plot $x[n]$ for $L = 256$.

(b) Let $L = 256$ and use MATLAB to compute the DTFT of $x[n]$ and to plot its magnitude. Indicate the range of discrete frequencies that would be jammed by the given chirp.

Answer: $IF(n) = 2\theta n = \pi n/L$ which as a function of n is a line of slope π/L.

11.26 Time-specifications for FIR filters—When designing discrete filters the specifications can be given in the time domain. One can think of converting the frequency domain specifications into the time domain. Assume you wish to obtain a filter that approximates an ideal low-pass filter with a cut-off frequency $\omega_c = \pi/2$ and that has a linear phase $-N\omega$. Thus, the frequency response is

$$H(e^{j\omega}) = \begin{cases} 1e^{-jN\omega} & -\pi/2 \le \omega \le \pi/2, \\ 0 & -\pi \le \omega < \pi/2 \ \text{and} \ \pi/2 < \omega \le \pi. \end{cases}$$

(a) Find the corresponding impulse response using the inverse DTFT of $H(e^{j\omega})$.

(b) If $N = 50$, plot $h[n]$ using the MATLAB function *stem* for $0 \le n \le 100$. Comment on the symmetry of $h[n]$. Assume $h[n] \approx 0$ for $n < 0$ and $n > 100$, plot it as well as the corresponding magnitude and phase of $H(e^{j\omega})$.

(c) Suppose we want a band-pass filter of center frequency $\omega_0 = \pi/2$, use the above impulse response $h[n]$ to obtain the impulse response of the desired band-pass filter.

Answer: $h[n] = \sin(\pi(n - N)/2)/(\pi(n - N))$ for $n \ne N$, $h[N] = 0.5$.

11.27 Down-sampling and DTFT—Consider the pulses $x_1[n] = u[n] - u[n - 20]$ and $x_2[n] = u[n] - u[n - 10]$, and their product $x[n] = x_1[n]x_2[n]$.

(a) Plot the three pulses. Could you say that $x[n]$ is a down-sampled version of $x_1[n]$? What would be the down-sampling rate? Find $X_1(e^{j\omega})$.

(b) Find directly the DTFT of $x[n]$ and compare it to $X_1(e^{j\omega/M})$, where M is the down-sampling rate found above. If we down-sample $x_1[n]$ to get $x[n]$, would the result be affected by aliasing? Use MATLAB to plot the magnitude DTFT of $x_1[n]$ and $x[n]$.

Answers: $x[n]$ is $x_1[n]$ down-sampled with a rate of $M = 2$; $\frac{1}{2}X_1(e^{j\omega/2}) \ne X(e^{j\omega})$.

11.28 Cascading of interpolator and decimator—Suppose you cascade an interpolator (an up-sampler and a low-pass filter) and a decimator (a low-pass filter and a down-sampler).

(a) If both the interpolator and the decimator have the same rate M, carefully draw a block diagram of the interpolator–decimator system.

(b) Suppose that the interpolator is of rate 3 and the decimator of rate 2, carefully draw a block diagram of the interpolator–decimator. What would be the equivalent of sampling the input of this system to obtain the same output?

(c) Use the MATLAB functions *interp* and *decimate* to process the first 100 samples of the test signal *handel* where the interpolator's rate is 3 and the decimator's 2. How many samples does the output have?

Answers: Cascade of interpolator and decimator of same rate M is equivalent to cascading up-sampler with low-pass filter and down-sampler.

11.29 Linear phase and phase unwrapping—Let $X(e^{j\omega}) = 2e^{-j4\omega}$, $-\pi \leq \omega < \pi$.

(a) Use the MATLAB functions *freqz* and *angle* to compute the phase of $X(e^{j\omega})$ and then plot it. Does the phase computed by MATLAB appear linear? What are the maximum and minimum values of the phase, how many radians separate the minimum from the maximum?

(b) Now, recalculate the phase but after using *angle* use the function *unwrapping* in the resulting phase and plot it. Does the phase appear linear?

Answers: $\angle X(e^{j\omega}) = -4\omega$, $-\pi \leq \omega \leq \pi$; a wrapped phase does not appear linear.

11.30 Windowing and DTFT—A window $w[n]$ is used to consider the part of a signal we are interested in.

(a) Let $w[n] = u[n] - u[n-20]$ be a rectangular window of length 20. Let $x[n] = \sin(0.1\pi n)$ and we are interested in a period of the infinite-length signal $x[n]$, or $y[n] = x[n]w[n]$. Compute the DTFT of $y[n]$ and compare it with the DTFT of $x[n]$. Write a MATLAB script to compute $Y(e^{j\omega})$.

(b) Let $w_1[n] = (1 + \cos(2\pi n/11))(u[n+5] - u[n-5])$ be a raised cosine window which is symmetric with respect to $n = 0$ (noncausal). Adapt the script in the previous part to find the DTFT of $z[n] = x[n]w_1[n]$ where $x[n]$ is the sinusoid given above.

Answer: $y[n] = 0.5w[n](e^{j0.1\pi n} - e^{-j0.1\pi n})/(2j)$.

11.31 Z-transform and Fourier series—Let $x_1[n] = 0.5^n$, $0 \leq n \leq 9$ be a period of a periodic signal $x[n]$.

(a) Use the Z-transform to compute the Fourier series coefficients of $x[n]$.

(b) Use MATLAB to compute the Fourier series coefficients using the analytic expression obtained above and the FFT. Plot the magnitude and phase line spectrum, i.e., $|X_k|$ and $\angle X_k$ versus frequency $-\pi \leq \omega \leq \pi$.

Answer: $X_k = 0.1(1 - 0.5^{10})/(1 - 0.5e^{-j0.2\pi k})$, $0 \leq k \leq 9$.

11.32 Operations on Fourier series—A periodic signal $x[n]$ of fundamental period N can be represented by its Fourier series

$$x[n] = \sum_{k=0}^{N-1} X_k e^{j2\pi nk/N} \qquad 0 \leq n \leq N-1.$$

If you consider this a representation of $x[n]$:

(a) Is $x_1[n] = x[n - N_0]$ for any value of N_0 periodic? If so use the Fourier series of $x[n]$ to obtain the Fourier series coefficients of $x_1[n]$.

(b) Let $x_2[n] = x[n] - x[n-1]$, i.e., the finite difference. Determine if $x_2[n]$ is periodic, and if so find its Fourier series coefficients.

(c) If $x_3[n] = x[n](-1)^n$, is $x_3[n]$ periodic? If so determine its Fourier series coefficients.

(d) Let $x_4[n] = \text{sign}[\cos(0.5\pi n)]$ where $\text{sign}(\xi)$ is a function that gives 1 when $\xi \geq 0$ and -1 when $\xi < 0$. Determine the Fourier coefficients of $x_4[n]$ if periodic.

(e) Let $x[n] = \text{sign}[\cos(0.5\pi n)]$, and $N_0 = 3$. Use MATLAB to find the Fourier series coefficients for $x_i[n]$, $i = 1, 2, 3$.

Answers: $x_1[n]$ is periodic of fundamental period N and FS coefficients $X_k e^{-j2\pi N_0 k/N}$; for $x_3[n]$ to be periodic of fundamental period N, N should be even.

11.33 Fourier series of even and odd signals—Let $x[n]$ be an even signal, and $y[n]$ an odd signal.

(a) Determine whether the Fourier coefficients X_k and Y_k corresponding to $x[n]$ and $y[n]$ are complex, real or imaginary.

(b) Consider $x[n] = \cos(2\pi n/N)$ and $y[n] = \sin(2\pi n/N)$ for $N = 3$ and $N = 4$, use the above results to find the Fourier series coefficients for the two signals with the different periods.

(c) Use MATLAB to find the Fourier series coefficients of the above two signals with the different periods, and plot their real and imaginary parts. Use 10 periods to compute the Fourier series coefficients using the *fft*. Comment on the results.

Answer: Independent of the period for $x_1[n]$, $X_1 = X_{-1} = 0.5$.

11.34 Response of LTI systems to periodic signals—Suppose you get noisy measurements

$$y[n] = (-1)^n x[n] + A\eta[n]$$

where $x[n]$ is the desired signal, and $\eta[n]$ is a noise that varies from 0 to 1 at random.

(a) Let $A = 0$, and $x[n] = \text{sign}[\cos(0.7\pi n)]$. Determine how to recover it from $y[n]$. Specify the type of filter you might need. Consider the first 100 samples of $x[n]$ and use MATLAB to find the spectrum of $x[n]$ and $y[n]$ to show that the filter you recommend will do the job.

(b) Use MATLAB function *fir1* to generate the kind of filter you decided to use above (choose an order $N \leq 40$ to get good results) and show that when filtering $y[n]$, for $A = 0$, you obtain the desired result.

(c) Consider the first 1000 samples of the MATLAB file *handel* a period of a signal that continuously replays these values over and over. Let $x[n]$ be the desired signal that results from this. Now let $A = 0.01$, and use the function *rand* to generate the noise, and come up with suggestions as to how to get rid of the effects of the multiplication by $(-1)^n$ and of the noise $\eta[n]$. Recover the desired signal $x[n]$

Answer: Demodulate $y[n]$ by multiplying it by $(-1)^n$, and then pass it through a low-pass filter.

11.35 DFT of aperiodic and periodic signals—Consider a signal $x[n] = 0.5n(0.8)^n(u[n] - u[n - 40])$.

(a) To compute the DFT of $x[n]$ we pad it with zeros so as to obtain a signal with length 2^γ, larger than the length of $x[n]$ but the closest to it. Determine the value of γ and use the MATLAB function *fft* to compute the DFT $X[k]$ of the padded-with-zeros signal. Plot its magnitude and phase using *stem*. Compute then an $N = 2^{10}$ FFT of $x[n]$, and compare its magnitude and phase DFTs by plotting using *stem*.

(b) Consider $x[n]$ a period of a periodic signal of fundamental period $N = 40$. Consider 2 and 4 periods and compute their DFTs using the *fft* algorithm and then plot its magnitude and phase. How do the magnitude responses compare? What do you need to make them equal? How do the phases compare after the magnitude responses are made equal?

Answers: $N = 2^6 = 64$; divide by number of periods to make magnitudes equal in periodic case.

11.36 Frequency resolution of DFT—When we pad an aperiodic signal with zeros, we are improving its frequency resolution, i.e., the more zeros we attach to the original signal the better the frequency resolution, as we obtain the frequency representation at a larger number of frequencies around the unit circle.

(a) Consider an aperiodic signal $x[n] = u[n] - u[n - 10]$, compute its DFT by means of the *fft* function padding it with 10 and then 100 zeros. Plot the magnitude response using *stem*. Comment on the frequency resolution of the two DFTs.

(b) When the signal is periodic, one cannot pad a period with zeros. When computing the FFT in theory we generate a periodic signal of period L equal to or larger than the length of the signal when the signal is aperiodic, but if the signal is periodic we must let L be the signal fundamental period or a multiple of it. Adding zeros to the period makes the signal be different from the periodic signal. Consider $x[n] = \cos(\pi n/5)$, $-\infty < n < \infty$ be a periodic signal, and do the following:

- Consider exactly one period of $x[n]$, and compute the FFT of this sequence.
- Consider 10 periods of $x[n]$ and compute the FFT of this sequence.
- Consider attaching 10 zeros to one period and compute the FFT of the resulting sequence.

If we consider the first of these cases giving the correct DFT of $x[n]$, how many harmonic frequencies does it show. What happens when we consider the 10 periods? Are the harmonic frequencies the same as before? What are the values of the DFT in frequencies in between the harmonic frequencies? What happened to the magnitude at the original frequencies. Finally, does the last FFT relate at all to the first FFT?

11.37 Circular and linear convolutions—Consider the circular convolution of two signals $x[n] = n$, $0 \leq n \leq 3$ and $y[n] = 1$, $n = 0, 1, 2$ and zero for $n = 3$.

(a) Compute the convolution sum or linear convolution of $x[n]$ and $y[n]$. Do it graphically and verify your results by multiplying the DFTs of $x[n]$ and $y[n]$.

(b) Use MATLAB to find the linear convolution. Plot $x[n]$, $y[n]$ and the linear convolution $z[n] = (x * y)[n]$.

(c) We wish to compute the circular convolution of $x[n]$ and $y[n]$ for different lengths $N = 4$, $N = 7$ and $N = 10$. Determine for which of these values the circular and the linear convolutions coincide. Show the circular convolution for the three cases. Use MATLAB to verify your results.

(d) Use the convolution property of the DFT to verify your result in the above part of the problem.

Answer: Circular and linear convolutions coincide when $N \geq 7$.

11.38 2D-DFT using 1D-DFT—To compute the 2D-DFT one can use 1D-DFT by separating the equation for the 2D-DFT as

$$
\begin{aligned}
X(k, \ell) &= \sum_{m=0}^{M-1} \left[\exp\left(\frac{-j2\pi km}{M}\right) \underbrace{\sum_{n=0}^{N-1} x[m,n] \exp\left(\frac{-j2\pi \ell n}{N}\right)}_{X(m,\ell)} \right] \\
&= \sum_{m=0}^{M-1} X(m, \ell) \exp\left(\frac{-j2\pi km}{M}\right).
\end{aligned}
$$

(a) Using the one-dimensional MATLAB function *fft* to implement the above result, and for the signal

$$
x[m, n] = \begin{cases} 1 & 0 \le m \le (M_1 - 1), \ 0 \le n \le (N_1 - 1), \\ 0 & M_1 \le m \le (M - 1), \ N_1 \le n \le (N - 1), \end{cases}
$$

with values $M = N = 10$. Compute $X(m, \ell)$ for every value of m and then find $X(k, \ell)$, the two-dimensional DFT of $x[m, n]$. Verify your result is the same as when you directly use the function *fft2* to compute $X(k, \ell)$.

(b) Is the MATLAB expression *fft(fft(x).').'* equivalent to *fft2(x)*? Try it and then explain.

(c) Use different values of M_1 and N_1 to verify that the support of $x[m, n]$ in the space domain is inversely proportional to the support of $X(k, \ell)$.

Answer: Results should coincide with those using the function *fft2*.

11.39 Image filtering using 2D-FFT—Consider the linear filtering of an image using the 2D-FFT. Load the image *clown* and use three different filters to process it given in different formats.

- Low-pass FIR filter with impulse response

$$
h_1[m, n] = \begin{cases} 1/100 & 0 \le m, n \le 10, \\ 0 & \text{otherwise.} \end{cases}
$$

- High-pass FIR filter with transfer function

$$
H_2(z_1, z_2) = 1 - z_1^{-1} - z_2^{-1} + z_1^{-1} z_2^{-1}.
$$

- Six-order separable IIR Butterworth filter with coefficients of one of the filters given by the MATLAB function *butter*:

$$
[b, a] = butter(3, [0.05, 0.95]).
$$

(a) Determine the impulse response corresponding to the high-pass FIR and the IIR filter (use *filter*).

(b) Find the magnitude of the DFT of the image (use a logarithmic scale to plot it) to determine the frequencies present in the image.

(c) Using the pixel array from the image as input and the convolution property of 2D-DFT to compute the output of the filter.

(d) Plot the magnitude response of the three filters (normalize the frequencies ω_1, ω_2 in the plots to get a support $[-1, 1] \times [-1, 1]$). Determine the type of filter that the IIR filter is.

Answer: Use 2D-FFTs of size 524×524.

11.40 Image blurring—An image can be blurred by means of a Gaussian filter which has an impulse response

$$h[m, n] = \frac{e^{-(m^2+n^2)}}{4\sigma^4} \qquad -3\sigma \leq m \leq 3\sigma, \ -3\sigma \leq n \leq 3\sigma.$$

(a) If $h[m, n] = h_1[m]h_1[n]$, i.e., separable determine $h_1[n]$. Find the DFT of $h[m, n]$ and plot its magnitude response. Use the functions *mesh* and *contour* to display the characteristics of the Gaussian filter (normalize the frequencies (ω_1, ω_2) to $[-1, 1] \times [-1, 1]$). What are these?

(b) To see the effect of blurring read the test image *cameraman.tif* and convolve it several times using two-dimensional FFTs and IFFTs. Use the function *imagesc* to display the resulting images. Why do the images besides being display the blurring seem to be moving down towards the right?

Answer: To generate Gaussian filters for $i = 1, \cdots, 10$ let $\sigma = i$ and $h_1[m] = \exp(-m)^2/(2\sigma^2)$ for values of m in $[-3\sigma, 3\sigma]$.

11.41 DCT image compression—The problem with thresholding the DCT coefficients of an image to compress it, is that the locations of the chosen coefficients are arbitrary and difficult to code. Consider then using a mask $W(k, \ell)$:

$$W(k, \ell) = \begin{cases} 1 & 1 \leq k \leq 4, 1 \leq \ell \leq 5 - k \\ 0 & \text{otherwise} \end{cases}$$

of the same dimension as the DCT array $\{C_x(k, \ell)\}$. Now the location of the chosen DCT coefficients is known. Consider the *cameraman* image, compute its DCT coefficients that form a matrix $\mathbf{C}_x$ and let

$$\hat{\mathbf{C}}_x = \mathbf{W}\mathbf{C}_x$$

be the DCT coefficients of a compressed image. Show the original and the compressed images, and the supports of the original and the masked DCT coefficients.

INTRODUCTION TO THE DESIGN OF DISCRETE FILTERS

CONTENTS

Signals and Systems Using MATLAB®. https://doi.org/10.1016/B978-0-12-814204-2.00023-5

When in doubt, don't.

Benjamin Franklin (1706–1790), printer, inventor, scientist and diplomat

12.1 INTRODUCTION

Filtering is an important application of linear time-invariant (LTI) systems. According to the eigenfunction property of discrete-time LTI systems, the steady-state response of a discrete-time LTI system to a sinusoidal input is also a sinusoid of the same frequency as that of the input, but with magnitude and phase affected by the response of the system at the frequency of the input. Since periodic as well as aperiodic signals have Fourier representations consisting of sinusoids of different frequencies, these signal components can be modified by appropriately choosing the frequency response of a LTI system, or filter. Filtering can thus be seen as a way to change the frequency content of an input signal.

The appropriate filter is specified using the spectral characterization of the input and the desired spectral characteristics of the output of the filter. Once the specifications of the filter are set, the problem becomes one of approximation, either by a ratio of polynomials or by a polynomial (if possible). After establishing that the filter resulting from the approximation satisfies the given specifications, it is then necessary to check its stability (if not guaranteed by the design method)—in the case of the filter being a rational approximation—and if stable, we need to figure out what would be the best possible way to implement the filter in hardware or in software. If not stable, we need either to repeat the approximation or to stabilize the filter before its implementation.

In the continuous-time domain, filters are obtained by means of rational approximation. In the discrete-time domain, there are two possible types of filters: one that is the result of a rational approximation—these filters are called recursive or Infinite Impulse Response (IIR) filters. The other type is the non-recursive or Finite Impulse Response (FIR) filter, which results from a polynomial approximation. As we will see, the discrete filter specifications can be in the frequency or in the time domain. For recursive or IIR filters, the specifications are typically given in the form of magnitude and phase specifications, while the specifications for non-recursive or FIR filters can be in the time domain as a desired impulse response. The discrete filter design problem then consists in: Given the specifications of a filter we look for a polynomial or rational (ratio of polynomials) approximation to the specifications. The resulting filter should be realizable, which besides causality and stability requires that the filter coefficients be real-valued.

There are different ways to attain a rational approximation for discrete IIR filters: by transformation of analog filters, or by optimization methods that include stability as a constraint. We will see that the classical analog design methods (Butterworth, Chebyshev, elliptic, etc.) can be used to design discrete filters by means of the bilinear transformation that maps the analog s-plane into the Z-plane. Given that the FIR filters are unique to the discrete domain, the approximation procedures for FIR filters are unique to that domain.

The difference between discrete and digital filters is in quantization and coding. For a discrete filter we assume that the input and the coefficients of the filter are represented with infinite precision, i.e., using an infinite number of quantization levels, and thus no coding is performed. The coefficients of a

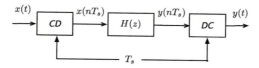

FIGURE 12.1

Discrete filtering of analog signals using ideal continuous-to-discrete (CD), or a sampler, and discrete-to-continuous (DC) converter, or a reconstruction filter.

digital filter are binary, and the input is quantized and coded. Quantization thus affects the performance of a digital filter, while it has no effect in discrete filters.

Considering continuous-to-discrete (CD) and discrete-to-continuous (DC) ideal converters simply as samplers and reconstruction filters, respectively, theoretically it is possible to implement the filtering of band-limited analog signals using discrete filters (Fig. 12.1). In such an application, an additional specification for the filter design is the sampling period. In this process it is crucial that the sampling period in the CD and DC converters be synchronized. In practice, filtering of analog signals is done using analog-to-digital (A/D) and digital-to-analog (D/A) converters together with digital filters.

Two-dimensional filtering finds an important application in image processing. Edge detection, de-blurring, de-noising, and compression of images are possible using two-dimensional linear shift-invariant filters. The overall structure of images is characterized by low-frequency components, while edges and texture are typically high-frequency components. Thus edge detection can be implemented using high-frequency filters capable of detecting abrupt changes, or edges, in the image. Likewise, denoising consists in smoothing or preserving the low-frequency components of the image. In many situations, non-linear filters perform better than linear filters. Decomposing a signal into components with specific frequency ranges is useful in image compression. Bank of filters are capable of separating an image into components in frequency sub-bands that when added give the whole spectrum of frequencies; each of the components can be represented perhaps more efficiently. MATLAB provides a comprehensive set of tools for image processing and the design of two-dimensional filters.

12.2 FREQUENCY SELECTIVE DISCRETE FILTERS

The principle behind discrete filtering is easily understood by considering the response of a linear time-invariant (LTI) system to sinusoids. If $H(z)$ is the transfer function of a discrete-time LTI system, and the input is

$$x[n] = \sum_k A_k \cos(\omega_k n + \phi_k) \tag{12.1}$$

(if the input is periodic the frequencies $\{\omega_k\}$ are harmonically related, otherwise they are not). According to the eigenfunction property of LTI systems the steady-state response of the system is

$$y_{ss}[n] = \sum_k A_k |H(e^{j\omega_k})| \cos(\omega_k n + \phi_k + \theta(\omega_k)) \tag{12.2}$$

where $|H(e^{j\omega_k})|$ and $\theta(\omega_k)$ are the magnitude and the phase of $H(e^{j\omega})$, the frequency response of the system, at the discrete frequency ω_k. The frequency response is the transfer function computed on the unit circle, i.e., $H(e^{j\omega}) = H(z)|_{z=e^{j\omega}}$. It becomes clear from Equation (12.2) that by judiciously choosing the frequency response of the LTI system we can select the frequency components of the input we wish to have at the output, and as such attenuate or amplify their amplitudes or change their phases. In general, for an input $x[n]$ with Z-transform $X(z)$, the Z-transform of the output of the filter is

$$Y(z) = H(z)X(z) \quad \text{or on the unit circle when } z = e^{j\omega}$$
$$Y(e^{j\omega}) = H(e^{j\omega})X(e^{j\omega}). \tag{12.3}$$

By selecting the frequency response $H(e^{j\omega})$ we allow some frequency components of $x[n]$ to appear in the output, and others to be filtered out. Although ideal frequency selective filters such as low-pass, high-pass, band-pass and stop-band cannot be realized, they serve as prototypes for the actual filters.

12.2.1 PHASE DISTORTION

A filter changes the spectrum of its input in magnitude as well as in phase. Distortion in magnitude can be avoided by using an all-pass filter with unit magnitude response for all frequencies. Phase distortion can be avoided by requiring the phase response of the filter to be linear (in particular zero).

For instance, when transmitting a voice signal in a communication system it is important that the signals at the transmitter and at the receiver be equal within a time delay and a constant attenuation factor. To achieve this, the transfer function of an ideal communication channel should be an all-pass filter with a linear phase. Indeed, if the output of an ideal discrete transmitter is a signal $x[n]$ and the recovered signal at an ideal discrete receiver is $\alpha x[n - N_0]$, for an attenuation factor α and a time delay N_0, the channel is represented by the transfer function of an all-pass filter:

$$H(z) = \frac{\mathcal{Z}(\alpha x[n - N_0])}{\mathcal{Z}(x[n])} = \alpha z^{-N_0}.$$

The constant gain of the all-pass filter permits all frequency components of the input to appear in the output. As we will see, the linear phase simply delays the signal, which is a very tolerable distortion that can be reversed.

To appreciate the effect of linear phase, consider the filtering of a signal

$$x[n] = 1 + \cos(\omega_0 n) + \cos(\omega_1 n) \quad \omega_1 = 2\omega_0 \quad n \geq 0$$

using an all-pass filter with transfer function $H(z) = \alpha z^{-N_0}$. The magnitude response of this filter is α, for all frequencies, and its phase is linear as shown in Fig. 12.2A. The steady-state output of the

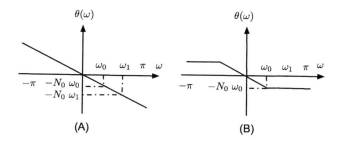

FIGURE 12.2

(A) Linear, (B) non-linear phase.

all-pass filter is

$$y_{ss}[n] = 1H(e^{j0}) + |H(e^{j\omega_0})|\cos(\omega_0 n + \angle H(e^{j\omega_0})) + |H(e^{j\omega_1})|\cos(\omega_1 n + \angle H(e^{j\omega_1}))$$
$$= \alpha[1 + \cos(\omega_0(n - N_0)) + \cos(\omega_1(n - N_0))] = \alpha x[n - N_0],$$

which is the input signal attenuated or amplified by α and delayed N_0 samples.

Suppose then that the all-pass filter has a phase function which is non-linear, for instance the one in Fig. 12.2B. The steady-state output would then be

$$y_{ss}[n] = 1H(e^{j0}) + |H(e^{j\omega_0})|\cos(\omega_0 n + \angle H(e^{j\omega_0})) + |H(e^{j\omega_1})|\cos(\omega_1 n + \angle H(e^{j\omega_1}))$$
$$= \alpha[1 + \cos(\omega_0(n - N_0)) + \cos(\omega_1(n - 0.5N_0))] \neq \alpha x[n - N_0].$$

In the case of the linear phase each of the frequency components of $x[n]$ is delayed N_0 samples, and thus the output is just a delayed version of the input. On the other hand, in the case of the non-linear phase the frequency component of frequency ω_1 is delayed less than the other two frequency components creating distortion in the signal so that the output is not a delayed version of the input and the phase effect cannot be reversed.

Group Delay

A measure of linearity of the phase $\theta(\omega)$ of a LTI system is obtained from the **group delay function** which is defined as

$$\tau(\omega) = -\frac{d\theta(\omega)}{d\omega}. \tag{12.4}$$

The group delay is constant when the phase is linear. Deviation of the group delay from a constant indicates the non-linearity of the phase.

In the above cases, when the phase is linear, i.e., for $0 \leq \omega \leq \pi$

$$\theta(\omega) = -N_0\omega \implies \tau(\omega) = N_0$$

and when the phase is non-linear

$$\theta(\omega) = \begin{cases} -N_0\omega & 0 < \omega \le \omega_0, \\ -N_0\omega_0 & \omega_0 < \omega \le \pi, \end{cases}$$

then we see that the group delay is

$$\tau(\omega) = \begin{cases} N_0 & 0 < \omega \le \omega_0 \\ 0 & \omega_0 < \omega \le \pi \end{cases} \qquad 0 \le \omega \le \pi,$$

which is not constant.

Remarks

1. Integrating (12.4) when $\tau(\omega)$ is a constant τ gives a general expression for the linear phase: $\theta(\omega) = -\tau\omega + \theta_0$. If $\theta_0 = 0$ the phase, as a function of ω, is a line through the origin with slope $-\tau$. Otherwise, as an odd function of ω the phase is

$$\theta(\omega) = \begin{cases} -\tau\omega - \theta_0 & -\pi \le \omega < 0, \\ 0 & \omega = 0, \\ -\tau\omega + \theta_0 & 0 < \omega \le \pi, \end{cases}$$

   i.e., it has a discontinuity at $\omega = 0$ but it is still considered linear.
2. The group delay τ of a linear-phase system is not necessarily an integer. Suppose we have an ideal analog low-pass filter with frequency response

$$H(j\Omega) = [u(\Omega + \Omega_0) - u(\Omega - \Omega_0)]e^{-j\zeta\Omega}, \quad \zeta > 0.$$

   Sampling the impulse response $h(t)$ of this filter using a sampling frequency $\Omega_s = 2\Omega_0$ gives an all-pass discrete filter with frequency response

$$H(e^{j\omega}) = 1e^{-j\zeta\omega/T_s} \quad -\pi < \omega \le \pi$$

   where the group delay ζ/T_s can be an integer or a real positive value.
3. Notice in the above example that when the phase is linear the group delay corresponds to the time delay in the output signal. This is due to the phase delay being proportional to the frequency of the signal component. When this proportionality is not present, i.e., the phase is not linear, the delay is different for different signal components causing the phase distortion.

12.2.2 IIR AND FIR DISCRETE FILTERS

- A discrete filter with transfer function

$$H(z) = \frac{B(z)}{A(z)} = \frac{\sum_{m=0}^{M-1} b_m z^{-m}}{1 + \sum_{k=1}^{N-1} a_k z^{-k}} = \sum_{n=0}^{\infty} h[n]z^{-n} \qquad (12.5)$$

is called **Infinite Impulse Response** or **IIR** since its impulse response $h[n]$ typically has infinite length. It is also called **recursive** because if the input of the filter $H(z) = Y(z)/X(z)$ is $x[n]$ and $y[n]$ its output, the input/output relationship is given by the following difference equation:

$$y[n] = -\sum_{k=1}^{N-1} a_k y[n-k] + \sum_{m=0}^{M-1} b_m x[n-m] \tag{12.6}$$

where the output recurs on previous outputs (i.e., the output is fed back).

- The transfer function of a **Finite Impulse Response** or **FIR** filter is

$$H(z) = B(z) = \sum_{m=0}^{M-1} b_m z^{-m}. \tag{12.7}$$

Its impulse response is $h[n] = b_n, n = 0, \ldots, M-1$, and zero otherwise, thus of finite length. The input/output relationship using $H(z) = Y(z)/X(z)$ is given by

$$y[n] = \sum_{m=0}^{M} b_m x[n-m] = (b * x)[n] \tag{12.8}$$

and coincides with the convolution sum of the filter coefficients (or impulse response) and the input. This filter is also called **non-recursive** as the output only depends on the input.

For practical reasons, these filters must be causal (i.e., $h[n] = 0$ for $n < 0$) and bounded input–bounded output (BIBO) stable (i.e., all the poles of $H(z)$ must be *inside the unit circle*). This guarantees that the filter can be implemented and used in real-time processing, and that the output remains bounded when the input is bounded.

Remarks

1. Calling the IIR filters recursive is more appropriate. Although it is traditional to refer to these filters as IIR, it is possible to have a filter with a rational transfer function that does not have an infinite length impulse response. For instance, consider a filter with transfer function

$$H(z) = \frac{1}{M} \frac{z^M - 1}{z^{M-1}(z-1)}$$

for some integer $M \geq 1$. This filter appears to be IIR, but if we express its transfer function in negative powers of z we obtain

$$H(z) = \frac{1}{M} \frac{1 - z^{-M}}{1 - z^{-1}} = \sum_{n=0}^{M-1} \frac{1}{M} z^{-n},$$

which is an Mth-order FIR filter with impulse response $h[n] = 1/M, n = 0, \cdots, M-1$.

2. When comparing the IIR and the FIR filters, neither has a definite advantage:

   - IIR filters are implemented more efficiently than FIR filters in terms of number of operations and required storage (having similar frequency responses, an IIR filter has fewer coefficients than an FIR filter).
   - The implementation of an IIR filter using the difference equation, resulting from its transfer function, is simple and computationally efficient, but FIR filters can be implemented using the computationally efficient Fast Fourier Transform (FFT) algorithm.
   - Since the transfer function of any FIR filter only has poles at the origin of the Z-plane, FIR filters are always BIBO stable, but for an IIR filter we need to check that the poles of its transfer function (i.e., the zeros of its denominator $A(z)$) be inside the unit circle if the design procedure does not guarantee stability.
   - FIR filters can be designed to have linear phase, while IIR filters usually have non-linear phases.

Example 12.1. The phase of IIR filters is always non-linear. Although it is possible to design FIR filters with linear phase, not all FIR filters have linear phase. Consider the following two filters with input/output equations:

$$(i) \ y[n] = 0.5y[n-1] + x[n], \quad (ii) \ y[n] = \tfrac{1}{3}(x[n-1] + x[n] + x[n+1]),$$

where $x[n]$ is the input and $y[n]$ the output. Determine the frequency phase response of the filters indicating if they are linear or not. Use MATLAB to compute and plot the magnitude and phase response of each of these filters.

Solution: The transfer functions of the given filters are

$$(i) \ H_1(z) = \frac{1}{1 - 0.5z^{-1}},$$

$$(ii) \ H_2(z) = \frac{1}{3}[z^{-1} + 1 + z] = \frac{1 + z + z^2}{3z} = \frac{(z - 1e^{j2.09})(z - 1e^{-j2.09})}{3z};$$

thus the first is an IIR filter and the second an FIR filter (notice that this filter is noncausal as it requires future values of the input to compute the present output). The phase response of the IIR filter is

$$H_1(e^{j\omega}) = \frac{1}{[1 - 0.5\cos(\omega)] + j0.5\sin(\omega)} \quad \text{then} \quad \angle H_1(e^{j\omega}) = -\tan^{-1}\left(\frac{0.5\sin(\omega)}{1 - 0.5\cos(\omega)}\right),$$

which is clearly non-linear. For the FIR filter we have

$$H_2(e^{j\omega}) = \frac{1}{3}(e^{-j\omega} + 1 + e^{j\omega}) = \frac{1 + 2\cos(\omega)}{3} \quad \text{then} \quad \angle H_2(e^{j\omega}) = \begin{cases} 0 & \text{when } 1 + 2\cos(\omega) \geq 0, \\ \pi & \text{when } 1 + 2\cos(\omega) < 0. \end{cases}$$

Since $H_2(z)$ has zeros at $z = 1e^{\pm j2.09}$, its magnitude response becomes zero at $\omega = \pm 2.09$ rads. The frequency response $H_2(e^{j\omega})$ is real-valued, and for $\omega \in [0, 2.09]$ the response $H_2(e^{j\omega}) \geq 0$ so the phase is zero, while for $\omega \in (2.09, \pi]$ the response $H_2(e^{j\omega}) < 0$ so the phase is π. The phase of the two filters is computed using the MATLAB function *freqz* and together with the corresponding magnitudes is shown in Fig. 12.3. ☐

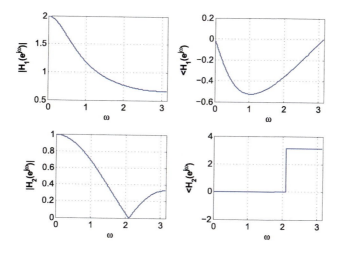

FIGURE 12.3

Magnitude and phase responses of an IIR filter with transfer function $H_1(z) = 1/(1 - 0.5z^{-1})$ (top), and of an FIR filter with transfer function $H_2(z) = (z - 1e^{j2.09})(z - 1e^{-j2.09})/3z$ (bottom). Notice the phase responses are non-linear.

Example 12.2. A simple model for the multi-path effect in the channel of a wireless system is

$$y[n] = x[n] - \alpha x[n - N_0], \qquad \alpha = 0.8, \ N_0 = 11,$$

i.e., the output $y[n]$ is a combination of the input $x[n]$, and of a delayed and attenuated version of the input, $\alpha x[n - N_0]$. Determine the transfer function of the filter representing the channel, i.e., that gives the above input/output equation. Use MATLAB to plot its magnitude and phase. If the phase is non-linear, how would you recover the input $x[n]$ (which is the message)? Let the input be $x[n] = 2 + \cos(\pi n/4) + \cos(\pi n)$. In practice, the delay N_0 and the attenuation α are not known at the receiver and need to be estimated. What would happen if the delay is estimated to be 12 and the attenuation 0.79?

Solution: The transfer function of the filter with input $x[n]$ and output $y[n]$ is

$$H(z) = \frac{Y(z)}{X(z)} = 1 - 0.8z^{-11} = \frac{z^{11} - 0.8}{z^{11}}$$

with a pole $z = 0$ of multiplicity 11, and zeros the roots of $z^{11} - 0.8 = 0$ or

$$z_k = (0.8)^{1/11} e^{j2\pi k/11} = 0.9799 e^{j2\pi k/11} \qquad k = 0, \cdots, 10.$$

Using the *freqz* function to plot its magnitude and phase responses (see Fig. 12.4), we find that the phase is non-linear and as such the output of $H(z)$, $y[n]$, will not be a delayed version of the input and

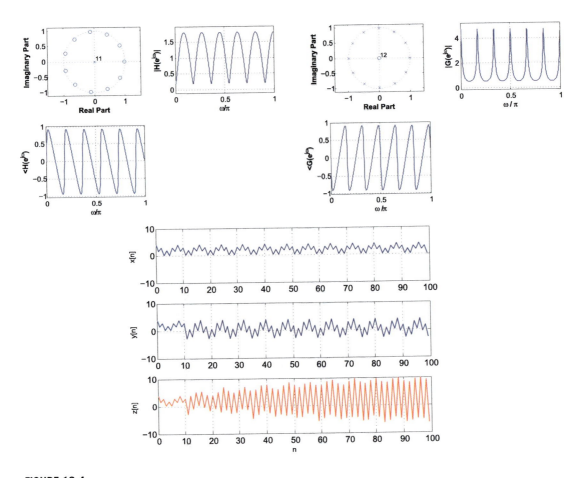

FIGURE 12.4

Poles and zeros, and frequency response of the FIR comb filter $H(z) = (z^{11} - 0.8)/z^{11}$ (top left) and the estimated inverse IIR comb filter $\hat{G}(z) = z^{12}/(z^{12} - 0.79)$ (top right). The bottom plot shows the message $x[n] = 2 + \cos(\pi n/4) + \cos(\pi n)$, the output $y[n]$ of channel $H(z)$ and output $z[n]$ of the estimated inverse filter $\hat{G}(z)$.

cannot be recovered by shifting it back. To recover the input, we use an **inverse filter** $G(z)$ such that cascaded with $H(z)$ the overall filter is an all-pass filter, i.e., $H(z)G(z) = 1$. Thus

$$G(z) = \frac{z^{11}}{z^{11} - 0.8}.$$

The poles and zeros, and the magnitude and phase response of $H(z)$ are shown in the top left plot in Fig. 12.4. The filters with transfer functions $H(z)$ and $G(z)$ are called **comb filters** because of the shape of their magnitude responses.

If the delay is estimated to be 11 and the attenuation 0.8, the input signal $x[n]$ (the message) is recovered exactly. However, if we have slight variations on these values the message might not be recovered. When the delay is estimated to be 12 and the attenuation 0.79, the inverse filter is

$$\hat{G}(z) = \frac{z^{12}}{z^{12} - 0.79},$$

having the poles, zeros, and magnitude and phase responses shown in the top right plot of Fig. 12.4. In the bottom figure in Fig. 12.4, the effect of these changes is illustrated. The output of the inverse filter $z[n]$ does not resemble the sent signal $x[n]$. The signal $y[n]$ is the output of the channel with transfer function $H(z)$.   □

12.3 FILTER SPECIFICATIONS

There are two ways to specify a discrete filter: in the frequency domain and in the time domain. Frequency specification of the magnitude and phase for the desired filter are more common in IIR filter design, while the time specification in terms of the desired impulse response of the filter is used in FIR filter design.

12.3.1 FREQUENCY SPECIFICATIONS

Conventionally, when designing IIR filters a prototype low-pass filter is obtained first and then converted into the desired filter using frequency transformations. The magnitude specifications of a discrete low-pass filter are given for frequencies $[0, \pi]$ due to the periodicity and the evenness of the magnitude. Typically, the phase is not specified but it is expected to be approximately linear.

For a low-pass filter, the desired magnitude $|H_d(e^{j\omega})|$ is to be close to unity in a passband frequency region, and close to zero in a stopband frequency region. To allow a smooth transition from the passband to the stopband a transition region where the filter is not specified is needed. Thus, the magnitude specifications are displayed in Fig. 12.5. The **passband** $[0, \omega_p]$ is the band of frequencies for which the attenuation specification is the smallest; the **stopband** $[\omega_{st}, \pi]$ is the band of frequencies where the attenuation specification is the greatest; and the **transition band** (ω_p, ω_{st}) is the frequency band where the filter is not specified. The frequencies ω_p, and ω_{st} are called the **passband** and the **stopband** frequencies.

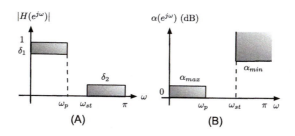

(A)   (B)

FIGURE 12.5

Low-pass magnitude specifications for an IIR filter: (A) linear scale, (B) loss scale.

Loss Function

As for analog filters, the linear-scale specifications shown in Fig. 12.5A, do not give the sense of the attenuation and thus the loss or log-specification in decibels (dB) is preferred. This logarithmic scale also provides a greater resolution of the magnitude.

The magnitude specifications of a discrete low-pass filter in a linear scale (Fig. 12.5A) are

$$\text{Passband:} \quad \delta_1 \leq |H(e^{j\omega})| \leq 1 \quad 0 \leq \omega \leq \omega_p,$$
$$\text{Stopband:} \quad 0 < |H(e^{j\omega})| \leq \delta_2 \quad \omega_{st} \leq \omega \leq \pi, \quad (12.9)$$

for $0 < \delta_2 < \delta_1 < 1$.
Defining the **loss function** for a discrete filter as

$$\alpha(e^{j\omega}) = -10\log_{10}|H(e^{j\omega})|^2 = -20\log_{10}|H(e^{j\omega})| \text{ dB} \quad (12.10)$$

equivalent magnitude specifications for a discrete low-pass filter (Fig. 12.5B) are

$$\text{Passband:} \quad 0 \leq \alpha(e^{j\omega}) \leq \alpha_{max} \quad 0 \leq \omega \leq \omega_p,$$
$$\text{Stopband:} \quad \alpha_{min} \leq \alpha(e^{j\omega}) < \infty \quad \omega_{st} \leq \omega \leq \pi, \quad (12.11)$$

where $\alpha_{max} = -20\log_{10}\delta_1$ and $\alpha_{min} = -20\log_{10}\delta_2$ (which are positive since both δ_1 and δ_2 are positive and smaller than 1). The frequency specification ω_p is the **passband frequency** and ω_{st} is the **stopband frequency**, both in rad. Increasing the loss 20 dB makes the filter attenuate the input signal by a factor of 10^{-1}.

Remarks

1. The dB scale is an indicator of attenuation: if we have a unit magnitude the corresponding loss is 0 dB and for every 20 dB in loss this magnitude is attenuated by 10^{-1}, so that when the loss is 100 dB the unit magnitude would be attenuated to 10^{-5}. The dB scale also has the physiological significance of being a measure of how humans detect levels of sound.
2. Besides the physiological significance, the loss specifications have intuitive appeal. It indicates that in the passband, where minimal attenuation of the input signal is desired, the "loss" is minimal as it is constrained to be below a maximum loss of α_{max} dB. Likewise, in the stopband where maximal attenuation of the input signal is needed, the "loss" is set to be larger than α_{min}.
3. When specifying a high-quality filter the value of α_{max} should be small, the α_{min} value should be large and the transition band as narrow as possible, i.e., approximating as much as possible the frequency response of an ideal low-pass filter. The cost of this is a large order for the resulting filter, making the implementation expensive computationally and requiring large memory space.

Example 12.3. Consider the following specifications for a low-pass filter:

$$0.9 \leq |H(e^{j\omega})| \leq 1.0 \quad 0 \leq \omega \leq \pi/2,$$
$$0 < |H(e^{j\omega})| \leq 0.1 \quad 3\pi/4 \leq \omega \leq \pi.$$

Determine the equivalent loss specifications.

Solution: The loss specifications are then

$$0 \leq \alpha(e^{j\omega}) \leq 0.92 \qquad 0 \leq \omega \leq \pi/2,$$
$$\alpha(e^{j\omega}) \geq 20 \qquad 3\pi/4 \leq \omega \leq \pi,$$

where $\alpha_{max} = -20\log_{10}(0.9) = 0.92$ dB, and $\alpha_{min} = -20\log_{10}(0.1) = 20$ dB. These specifications indicate that in the passband the loss is small, or that the magnitude would change between 1 and $10^{-\alpha_{max}/20} = 10^{-0.92/20} = 0.9$ while in the stopband we would like a large attenuation, at least α_{min}, or that the magnitude would have values smaller than $10^{-\alpha_{min}/20} = 0.1$. □

Magnitude Normalization

The specifications of the low-pass filter in Fig. 12.5 are **normalized in magnitude**: the dc gain is assumed to be unity (or the dc loss is 0 dB), but there are many cases where that is not so.

Not normalized magnitude specifications (Fig. 12.6): In general, the loss specifications are

$$\alpha_1 \leq \hat{\alpha}(e^{j\omega}) \leq \alpha_2 \qquad 0 \leq \omega \leq \omega_p,$$
$$\alpha_3 \leq \hat{\alpha}(e^{j\omega}) \qquad \omega_{st} \leq \omega \leq \pi.$$

Writing these loss specifications as

$$\hat{\alpha}(e^{j\omega}) = \alpha_1 + \alpha(e^{j\omega}) \tag{12.12}$$

we have:

- the normalized specifications are given as

$$0 \leq \alpha(e^{j\omega}) \leq \alpha_{max} \qquad 0 \leq \omega \leq \omega_p,$$
$$\alpha_{min} \leq \alpha(e^{j\omega}) \qquad \omega_{st} \leq \omega \leq \pi,$$

where $\alpha_{max} = \alpha_2 - \alpha_1$ and $\alpha_{min} = \alpha_3 - \alpha_1$.
- the dc loss of α_1 is achieved by multiplying a magnitude-normalized filter by a constant K such that

$$\hat{\alpha}(e^{j0}) = \alpha_1 = -20\log_{10} K \quad \text{or} \quad K = 10^{-\alpha_1/20}. \tag{12.13}$$

Example 12.4. Suppose the loss specifications of a low-pass filter are

$$10 \leq \hat{\alpha}(e^{j\omega}) \leq 11, \qquad 0 \leq \omega \leq \frac{\pi}{2},$$
$$\hat{\alpha}(e^{j\omega}) \geq 50, \qquad \frac{3\pi}{4} \leq \omega \leq \pi.$$

Determine the loss specifications that can be used to design a magnitude normalized filter. Find a gain K that when it multiplies the normalized filter the resulting filter satisfies the specifications.

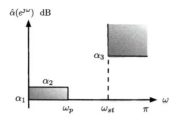

FIGURE 12.6

Not normalized loss specifications for a low-pass filter.

Solution: If we let

$$\hat{\alpha}(e^{j\omega}) = 10 + \alpha(e^{j\omega}),$$

the loss specifications for a normalized filter would be

$$0 \leq \alpha(e^{j\omega}) \leq 1, \qquad 0 \leq \omega \leq \frac{\pi}{2},$$

$$\alpha(e^{j\omega}) \geq 40, \qquad \frac{3\pi}{4} \leq \omega \leq \pi.$$

Then the dc loss is 10 dB and $\alpha_{max} = 11 - 10 = 1$ and $\alpha_{min} = 50 - 10 = 40$ dB. Suppose that we design a filter $H(z)$ that satisfies the normalized filter specifications. If we let $\hat{H}(z) = K H(z)$ be the filter that satisfies the given loss specifications, at the dc frequency we must have

$$-20 \log_{10} |\hat{H}(e^{j0})| = -20 \log_{10} K - 20 \log_{10} |H(e^{j0})| \quad \text{or} \quad 10 = -20 \log_{10} K + 0,$$

so that $K = 10^{-0.5} = 1/\sqrt{10}$. ☐

Frequency Scales

Given that discrete filters can be used to process continuous- and discrete-time signals, there are different equivalent ways the frequency of a discrete filter can be expressed (see Fig. 12.7).

In the discrete processing of continuous-time signals the sampling frequency (f_s in Hz or Ω_s in rad/s) is known, and so we have the following possible scales:

- the f (Hz) scale from 0 to $f_s/2$ (the fold-over or Nyquist frequency) derived from the sampling theory,
- the scale $\Omega = 2\pi f$ (rad/s), where f is the previous scale, the frequency range is then from 0 to $\Omega_s/2$,
- the discrete frequency scale $\omega = \Omega T_s$ (rad) ranging from 0 to π,
- a normalized discrete frequency scale ω/π (no units) ranging from 0 to 1.

If the specifications are in the discrete domain, the scales used are the ω (rad) or the normalized ω/π.

Other scales are possible, but less used. One of these consists in dividing by the sampling frequency either in Hz or in rad/s: the f/f_s (no units) scale goes from 0 to 0.5, and so does the Ω/Ω_s (no units) scale. It is clear that when the specifications are given in any scale, it can easily be transformed into

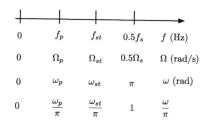

FIGURE 12.7

Frequency scales used in discrete filter design.

any other desired scale. If the filter is designed for use in the discrete domain only the scale is in rad and the normalized ω/π are the ones to use.

12.3.2 TIME DOMAIN SPECIFICATIONS

Time-domain specifications consist in giving a desired impulse response $h_d[n]$. For instance, when designing a low-pass filter with cut-off frequency ω_c and linear phase $\phi(\omega) = -N\omega$, the desired frequency response in $0 \leq \omega \leq \pi$ is

$$H_d(e^{j\omega}) = \begin{cases} 1e^{-j\omega N} & 0 \leq \omega \leq \omega_c, \\ 0 & \omega_c < \omega \leq \pi. \end{cases}$$

The desired impulse response for this filter is then found from

$$h_d[n] = \frac{1}{2\pi} \int_{-\pi}^{\pi} H_d(e^{j\omega})e^{j\omega n}d\omega = \frac{1}{2\pi} \int_{-\omega_c}^{\omega_c} 1e^{-j\omega N}e^{j\omega n}d\omega.$$

The resulting $h_d[n]$ will be used as the desired impulse response to approximate.

Example 12.5. Consider an FIR filter design with the following desired magnitude response:

$$|H_d(e^{j\omega})| = \begin{cases} 1 & 0 \leq \omega \leq \pi/4, \\ 0 & \pi/4 < \omega \leq \pi, \end{cases}$$

and zero phase. Find the desired impulse response $h_d[n]$ that we wish to approximate.

Solution: Since the magnitude response is even function of ω, the desired impulse response is computed as follows:

$$h_d[n] = \frac{1}{2\pi} \int_{-\pi}^{\pi} H_d(e^{j\omega})e^{j\omega n}d\omega = \frac{1}{2\pi} \int_{-\pi/4}^{\pi/4} e^{j\omega n}d\omega = \begin{cases} \sin(\pi n/4)/\pi n & n \neq 0, \\ 0.25 & n = 0, \end{cases}$$

which corresponds to the impulse response of a noncausal system. As we will see later, windowing and shifting of $h_d[n]$ are needed to make it into a causal, finite length filter. □

12.4 IIR FILTER DESIGN

Two possible approaches in the design of IIR filters are:

- using analog filter design methods and transformations between the s-plane and the Z-plane, and
- using optimization techniques.

The first is a **frequency transformation approach**. Using a mapping between the analog and the discrete frequencies we obtain the specifications for an analog filter from the discrete filter specifications. Applying well-known analog filter design methods, we then design the analog filter from the transformed specifications. The discrete filter is finally obtained by transforming the designed analog filter.

The **optimal approach designs** the filter directly, setting the rational approximation as a non-linear optimization. The added flexibility of this approach is diminished by the need to ensure stability of the designed filter. Stability is guaranteed, on the other hand, in the transformation approach.

12.4.1 TRANSFORMATION DESIGN OF IIR DISCRETE FILTERS

To take advantage of analog filter design, a common practice is to design discrete filters by means of analog filters and mappings of the s-plane into the Z-plane. Two mappings used are:

- the sampling transformation $z = e^{sT_s}$, and
- the bilinear transformation

$$s = K \frac{1 - z^{-1}}{1 + z^{-1}}.$$

Recall the transformation $z = e^{sT_s}$ was found when relating the Laplace transform of a sampled signal with its Z-transform. Using this transformation, we convert the analog impulse response $h_a(t)$ of an analog filter into the impulse response $h[n]$ of a discrete filter and obtain the corresponding transfer function. The resulting design procedure is called the **impulse invariant method**. Advantages of this method are:

- it preserves the stability of the analog filter, and
- given the linear relation between the analog frequency Ω and the discrete frequency ω, the specifications for the discrete filter can easily be transformed into the specifications for the analog filter.

Its drawback is possible frequency aliasing. Sampling of the analog impulse response requires that the analog filter be band-limited which might not be possible to satisfy in all cases. Due to this, we will concentrate on the approach based on the bilinear transformation.

The Bilinear Transformation

The bilinear transformation results from the **trapezoidal rule approximation of an integral**. Suppose that $x(t)$ is the input and $y(t)$ the output of an integrator with transfer function

$$H(s) = \frac{Y(s)}{X(s)} = \frac{1}{s}. \tag{12.14}$$

Sampling the input and the output of this filter using a sampling period T_s, we see that the integral at time nT_s is

$$y(nT_s) = \int_{(n-1)T_s}^{nT_s} x(\tau)d\tau + y((n-1)T_s)$$

where $y((n-1)T_s)$ is the integral at time $(n-1)T_s$. If T_s is very small, the integral between $(n-1)T_s$ and nT_s can be approximated by the area of a trapezoid with bases $x((n-1)T_s)$ and $x(nT_s)$ and height T_s (this is called the trapezoidal rule approximation of an integral):

$$y(nT_s) \approx \frac{[x(nT_s) + x((n-1)T_s)] T_s}{2} + y((n-1)T_s)$$

with a Z-transform given by

$$Y(z) = \frac{T_s(1+z^{-1})}{2(1-z^{-1})} X(z).$$

The discrete transfer function is thus

$$H(z) = \frac{Y(z)}{X(z)} = \frac{T_s}{2} \frac{1+z^{-1}}{1-z^{-1}}, \tag{12.15}$$

which can be obtained directly from $H(s)$ in Equation (12.14) by letting

$$s = \frac{2}{T_s} \frac{1-z^{-1}}{1+z^{-1}}. \tag{12.16}$$

Thinking of the above transformation as a transformation from the z to the s variable, solving for the variable z in that equation we obtain a transformation from the s to the z variable:

$$z = \frac{1 + (T_s/2)s}{1 - (T_s/2)s}. \tag{12.17}$$

The **bilinear transformation** (linear in the numerator and in the denominator) that transforms from the s-plane into the z-plane is

$$z = \frac{1 + s/K}{1 - s/K} \qquad K = \frac{2}{T_s}, \tag{12.18}$$

and it maps

- the $j\Omega$-axis in the s-plane into the unit circle in the Z-plane,
- the open left-hand s-plane $\mathcal{R}e[s] < 0$ into the inside of the unit circle in the Z-plane, or $|z| < 1$, and
- the open right-hand s-plane $\mathcal{R}e[s] > 0$ into the outside of the unit circle in the Z-plane, or $|z| > 1$.

The bilinear transformation from the z-plane to the s-plane is

$$s = K\frac{1-z^{-1}}{1+z^{-1}}. \tag{12.19}$$

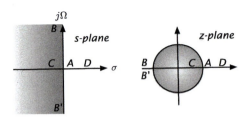

FIGURE 12.8

Bilinear transformation mapping of s-plane into Z-plane.

Thus as shown in Fig. 12.8, the point (A), $s = 0$ or the origin of the s-plane, is mapped into $z = 1$ on the unit circle; points (B) and (B'), $s = \pm j\infty$, are mapped into $z = -1$ on the unit circle; point (C), $s = -1$, is mapped into $z = (1 - 1/K)/(1 + 1/K) < 1$ which is inside the unit circle; finally, point (D), $s = 1$, is mapped into $z = (1 + 1/K)/(1 - 1/K) > 1$, located outside the unit circle.

In general, by letting $z = re^{j\omega}$ and $s = \sigma + j\Omega$ in (12.18) we obtain

$$r = \sqrt{\frac{(1 + \sigma/K)^2 + (\Omega/K)^2}{(1 - \sigma/K)^2 + (\Omega/K)^2}},$$

$$\omega = \tan^{-1}\left(\frac{\Omega/K}{1 + \sigma/K}\right) + \tan^{-1}\left(\frac{\Omega/K}{1 - \sigma/K}\right), \tag{12.20}$$

from which we have:

- On the $j\Omega$-axis of the s-plane, i.e., when $\sigma = 0$ and $-\infty < \Omega < \infty$, we obtain $r = 1$ and $-\pi \leq \omega < \pi$, which corresponds to the unit circle of the Z-plane.
- In the open left-hand s-plane, or equivalently when $\sigma < 0$ and $-\infty < \Omega < \infty$, we obtain $r < 1$ and $-\pi \leq \omega < \pi$, or the inside of the unit circle in the Z-plane.
- Finally, in the open right-hand s-plane, or equivalently when $\sigma > 0$ and $-\infty < \Omega < \infty$, we obtain $r > 1$ and $-\pi \leq \omega < \pi$, or the outside of the unit circle in the Z-plane.

The above transformation can be visualized by thinking of a giant who puts a nail in the origin of the s-plane, grabs the plus and minus infinity extremes of the $j\Omega$-axis and pulls them together to make them agree into one point, getting a magnificent circle, keeping everything in the left-hand s-plane inside, and keeping out the rest. If our giant lets go, we get back the original s-plane!

The bilinear transformation maps the whole s-plane into the whole Z-plane, differently from the transformation $z = e^{sT_s}$ that only maps a slab of the s-plane into the Z-plane (see Chapter 10 on the Z-transform). Thus a stable analog filter with poles in the open left-hand s-plane will generate a discrete filter that is also stable, as it has all its poles inside the unit circle.

Frequency Warping

A minor drawback of the bilinear transformation is the non-linear relation between the analog and the discrete frequencies. Such a relation creates a warping that needs to be taken care of when specifying the analog filter using the discrete filter specifications.

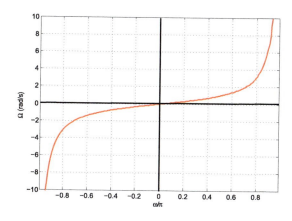

FIGURE 12.9

Relation between Ω and ω for $K = 1$.

The analog frequency Ω and the discrete frequency ω according to the bilinear transformation are related by

$$\Omega = K \tan(\omega/2), \qquad (12.21)$$

which when plotted displays a linear relation around the low frequencies but it warps as we get into larger frequencies (see Fig. 12.9).

The relation between the frequencies is obtained by letting $\sigma = 0$ in the second equation in (12.20). The linear relationship at low frequencies can be seen using the expansion of the tan(.) function

$$\Omega = K \left[\frac{\omega}{2} + \frac{\omega^3}{24} + \cdots \right] \approx \frac{\omega}{T_s}$$

for small values of ω or $\omega \approx \Omega T_s$. As the frequency increases the effect of the terms beyond the first one makes the relation non-linear. See Fig. 12.9.

To compensate for the non-linear relation between the frequencies, or the warping effect, the following steps to design a discrete filter are followed:

1. Using the frequency warping relation (12.21) the specified discrete frequencies ω_p, and ω_{st} are transformed into specified analog frequencies Ω_p and Ω_{st}. The magnitude specifications remain the same in the different bands—only the frequency is being transformed.
2. Using the specified analog frequencies and the discrete magnitude specifications an analog filter $H_N(s)$ that satisfies these specifications is designed.
3. Applying the bilinear transformation to the designed filter $H_N(s)$, the discrete filter $H_N(z)$ that satisfies the discrete specifications is obtained.

12.4.2 DESIGN OF BUTTERWORTH LOW-PASS DISCRETE FILTERS

Applying the warping relation between the continuous and the discrete frequencies

$$\Omega = K \tan(\omega/2) \tag{12.22}$$

to the magnitude squared function of the Butterworth low-pass analog filter

$$|H_N(j\Omega')|^2 = \frac{1}{1 + (\Omega')^{2N}}, \qquad \Omega' = \frac{\Omega}{\Omega_{hp}}, \tag{12.23}$$

gives the magnitude squared function for the Butterworth low-pass discrete filter

$$|H_N(e^{j\omega})|^2 = \frac{1}{1 + \left[\frac{\tan(0.5\omega)}{\tan(0.5\omega_{hp})}\right]^{2N}}. \tag{12.24}$$

As a frequency transformation (no change to the loss specifications) we directly obtain the minimum order N and the half-power frequency bounds by replacing

$$\frac{\Omega_{st}}{\Omega_p} = \frac{\tan(\omega_{st}/2)}{\tan(\omega_p/2)} \tag{12.25}$$

in the corresponding formulas for N and Ω_{hp} of the analog filter, giving

$$N \geq \frac{\log_{10}[(10^{0.1\alpha_{min}} - 1)/(10^{0.1\alpha_{max}} - 1)]}{2 \log_{10}\left[\frac{\tan(\omega_{st}/2)}{\tan(\omega_p/2)}\right]}, \tag{12.26}$$

$$2 \tan^{-1}\left[\frac{\tan(\omega_p/2)}{(10^{0.1\alpha_{max}} - 1)^{1/2N}}\right] \leq \omega_{hp} \leq 2 \tan^{-1}\left[\frac{\tan(\omega_{st}/2)}{(10^{0.1\alpha_{min}} - 1)^{1/2N}}\right]. \tag{12.27}$$

The normalized half-power frequency $\Omega'_{hp} = 1$ in the continuous domain is mapped into the discrete half-power frequency ω_{hp}, giving the constant in the bilinear transformation

$$K_b = \frac{\Omega'}{\tan(0.5\omega)}\bigg|_{\Omega'=1, \omega=\omega_{hp}} = \frac{1}{\tan(0.5\omega_{hp})}. \tag{12.28}$$

The bilinear transformation $s = K_b(1 - z^{-1})/(1 + z^{-1})$ is then used to convert the analog filter $H_N(s)$, satisfying the transformed specifications, into the desired discrete filter

$$H_N(z) = H_N(s)\big|_{s=K_b(1-z^{-1})/(1+z^{-1})}.$$

The basic idea of this design is to convert an **analog frequency-normalized** Butterworth magnitude squared function into a discrete function using the relationship (12.22). To understand why this is an efficient approach consider the following issues that derive from the application of the bilinear transformation to the Butterworth design:

- Since the discrete magnitude specifications are not changed by the bilinear transformation, we only need to change the analog frequency term in the formulas obtained before for the Butterworth low-pass analog filter.

- It is important to recognize that when finding the minimum order N and the half-power relation the value of K is not used. This constant is only important in the final step where the analog filter is transformed into the discrete filter using the bilinear transformation.
- When considering that $K = 2/T_s$ depends on T_s, one might think that a small value for T_s would improve the design, but that is not the case. Given that the analog frequency is related to the discrete frequency as

$$\Omega = \frac{2}{T_s} \tan\left(\frac{\omega}{2}\right) \qquad (12.29)$$

for a given value of ω if we choose a small value of T_s the specified analog frequency Ω is large, and if we choose a large value of T_s the analog frequency Ω decreases. In fact, in the above equation we can only choose either Ω or T_s. To avoid this ambiguity, we ignore the connection of K with T_s and concentrate on K.
- An appropriate value for K for the Butterworth design is obtained by connecting the normalized half-power frequency $\Omega'_{hp} = 1$ in the analog domain with the corresponding frequency ω_{hp} in the discrete domain. This allows us to go from the discrete domain specifications *directly* to the analog normalized frequency specifications. Thus we map the normalized half-power frequency $\Omega'_{hp} = 1$ into the discrete half-power frequency ω_{hp}, by means of K_b.
- Once the analog filter $H_N(s)$ is obtained, using the bilinear transformation with the K_b we transform $H_N(s)$ into a discrete filter

$$H_N(z) = H_N(s)\Big|_{s=K_b \frac{z-1}{z+1}}.$$

- The filter parameters (N, ω_{hp}) can also be obtained directly from the discrete loss function obtained from Equation (12.24) as

$$\alpha(e^{j\omega}) = 10\log_{10}\left[1 + (\tan(0.5\omega)/\tan(0.5\omega_{hp}))^{2N}\right] \qquad (12.30)$$

and the loss specifications

$$0 \le \alpha(e^{j\omega}) \le \alpha_{max} \qquad 0 \le \omega \le \omega_p,$$
$$\alpha(e^{j\omega}) \ge \alpha_{min} \qquad \omega \ge \omega_{st},$$

just as we did in the continuous case. The results coincide with those where we replace the warping frequency relation.

Example 12.6. The analog signal $x(t) = \cos(40\pi t) + \cos(500\pi t)$ is sampled using the Nyquist frequency and processed with a discrete filter $H(z)$ which is obtained from a second-order, high-pass analog filter

$$H(s) = \frac{s^2}{s^2 + \sqrt{2}s + 1}.$$

The discrete-time output $y[n]$ is then converted into an analog signal. Apply MATLAB's *bilinear* function to obtain the discrete filter with half-power frequencies $\omega_{hp} = \pi/2$. Use MATLAB to plot the poles and zeros of the discrete filters in the Z-plane and the corresponding magnitude and phase responses. Use the function *plot* to plot the sampled input and the filter output and consider these approximations to the analog signals. Change the frequency scale of the discrete filter into f in Hz and indicate what is the corresponding half-power frequency in Hz.

Solution: The coefficients of the numerator and denominator of the discrete filter are found from $H(s)$ using the MATLAB function *bilinear*. The input F_s in this function equals $K_b/2$, where K_b corresponds to the transformation of the discrete half-power frequency, ω_{hp}, into the normalized analog half-power frequency $\Omega_{hp} = 1$.

The following script is used.

```
%%
% Example 12.6---Bilinear transformation and high-pass filtering
%%
b=[1 0 0]; a=[1 sqrt(2) 1];     % coefficients of analog filter
whp=0.5*pi;                      % desired half-power frequency
Kb=1/tan(whp/2); Fs=Kb/2;
[num, den]=bilinear(b,a,Fs);    % bilinear transformation
Ts=1/500;                       % sampling period
n=0:499; x1=cos(2*pi*20*n*Ts)+cos(2*pi*250*n*Ts);   % sampled signal
zplane(num, den)                % poles/zeros of discrete filter
[H,w]=freqz(num,den);           % frequency response discr. filter
phi=unwrap(angle(H));           % unwrapped phase discr. filter
y=filter(num,den,x1);           % output of discr. filter, input x1
```

We find the transfer function of the discrete filter to be

$$H(z) = \frac{0.2929(1 - z^{-1})^2}{1 + 0.1715z^{-2}}.$$

The poles and zeros of $H(z)$ can be found with the MATLAB function *roots* and plotted with *zplane*. The frequency response is obtained using *freqz*. To have the frequency scale in Hz we consider that $\omega = \Omega T_s$, letting $\Omega = 2\pi f$, then

$$f = \frac{\omega}{2\pi T_s} = \left(\frac{\omega}{\pi}\right)\left(\frac{f_s}{2}\right)$$

so we multiply the normalized discrete frequency ω/π by $f_s/2 = 250$, resulting in a maximum frequency of 250 Hz. The half-power frequency in Hz is thus 125 Hz. The magnitude and phase responses of $H(z)$ are shown in Fig. 12.10. Notice that phase is approximately linear in the passband, despite the fact that no phase specifications are considered.

Since the maximum frequency of $x(t)$ is 250 Hz we choose the sampling period to be $T_s = 1/(2f_{max}) = 1/500$. As a high-pass filter, when we input $x(nT_s)$ into $H(z)$ its low-frequency component $\cos(40\pi nT_s)$ is attenuated. The input and corresponding output of the filter are shown in Fig. 12.10. Notice from the figures that the input signal $x(t)$ is a high-frequency signal riding over a low-frequency signal and the filtering gets rid of the low-frequency signal leaving only the high-frequency signal in the output $y(t)$. □

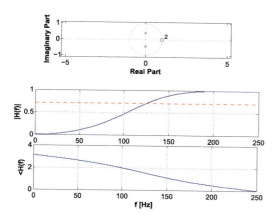

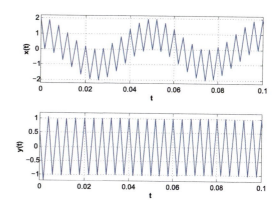

FIGURE 12.10

Bilinear transformation of a high-pass analog filter into a discrete filter with half-power frequencies $\omega_{hp} = \pi/2$ or $f_{hp} = 125$ Hz. Poles and zeros, and magnitude and phase response of the discrete filter are shown on left plots. The plots on the right show the analog input and output obtained using the MATLAB function *plot* to interpolate the sampled signal $x(nT_s)$ and the output of the discrete filter $y(nT_s)$ into $x(t)$ and $y(t)$.

Remarks

1. The Butterworth low-pass filter design is simplified when giving as a specification a desired half-power frequency ω_{hp}. We only need then to calculate the order of the filter by using the stopband constraint. In such a case, we could let $\alpha_{max} = 3$dB and $\omega_p = \omega_{hp}$ and use (12.26) to find N.

2. A very important consequence of using the bilinear transformation is that the resulting transfer function $H_N(z)$ is guaranteed to be BIBO stable. This transformation maps the poles of a stable filter $H_N(s)$ in the open left-hand s-plane into poles inside the unit circle corresponding to $H_N(z)$, making the discrete filter stable.

3. The bilinear transformation creates a pole and a zero in the Z-plane for each pole in the s-plane. Analytic calculation of the poles of $H_N(z)$ is not as important as in the analog case, the MATLAB function *zplane* can be used to plot its poles and zeros, and the function *roots* to find the values of the poles and zeros.

4. Applying the bilinear transformation by hand to filters of order higher than 2 is cumbersome. When doing so, $H_N(s)$ should be expressed as a product or a sum of first- and second-order transfer functions before applying the bilinear transformation to each, i.e.,

$$H_N(s) = \prod_i H_{N_i}(s) \quad \text{or} \quad H_N(s) = \sum_\ell H_{N_\ell}(s)$$

where $H_{Ni}(s)$ and $H_{N\ell}(s)$ are first- or second-order functions with real coefficients. Applying the bilinear transformation to each of the $H_{Ni}(s)$ or $H_{N\ell}(s)$ components to obtain $H_{Ni}(z)$ and $H_{N\ell}(z)$, the discrete filter becomes

$$H_N(z) = \prod_i H_{N_i}(z) \quad \text{or} \quad H_N(z) = \sum_\ell H_{N_\ell}(z).$$

5. The zeros of the discrete low-pass Butterworth filter are at $z = -1$; this is due to the rationalization of the transfer function of the transformed analog filter. If the analog low-pass Butterworth filter has a transfer function

$$H_N(s) = \frac{1}{a_0 + a_1 s + \cdots + a_N s^N}$$

letting $s = K_b(1 - z^{-1})/(1 + z^{-1})$ we obtain a discrete filter

$$
\begin{aligned}
H_N(z) &= \frac{1}{a_0 + a_1 K_b(1 - z^{-1})/(1 + z^{-1}) + \cdots + a_N K_b^N (1 - z^{-1})^N/(1 + z^{-1})^N} \\[2mm]
&= \frac{(1 + z^{-1})^N}{a_0(1 + z^{-1})^N + a_1 K_b(1 - z^{-1})(1 + z^{-1})^{N-1} + \cdots + a_N K_b^N (1 - z^{-1})^N}
\end{aligned}
$$

with N zeros at $z = -1$.

6. Since the resulting filter has normalized magnitude, a specified dc gain can be attained by multiplying $H_N(z)$ by a constant value G so that $|GH(e^{j0})|$ equals the desired dc gain.

Example 12.7. The specifications of a low-pass discrete filter are

$$
\begin{aligned}
\omega_p &= 0.47\pi \ \text{(rad)} & \alpha_{max} &= 2 \ \text{dB}, \\
\omega_{st} &= 0.6\pi \ \text{(rad)}, & \alpha_{min} &= 9 \ \text{dB}, \\
\alpha(e^{j0}) &= 0 \ \text{dB}.
\end{aligned}
$$

Use MATLAB to design a discrete low-pass Butterworth filter using the bilinear transformation.

Solution: Normalizing the frequency specifications to ω_p/π and ω_{st}/π we use these values directly in the MATLAB function *buttord* as inputs along with α_{max}, and α_{min}. The function gives the minimum order N and the normalized half-power frequency ω_{hp}/π of the filter. With these as inputs of the function *butter* we obtain the coefficients of the numerator and denominator of the designed filter $H(z) = B(z)/A(z)$. The function *roots* is used to find the poles and zeros of $H(z)$, while *zplane* plots them. The magnitude and phase responses are found using *freqz* aided by the functions *abs*, *angle* and *unwrap*. Notice that *butter* obtains the normalized analog filter and transform it using the bilinear transformation. The script used is

```
%%
% Example 12.7---Design of low-pass Butterworth filter
%%
% LP Butterworth
alphamax=2; alphamin=9;                  % loss specifications
wp=0.47;ws=0.6;                          % passband and stopband frequencies
[N,wh]=buttord(wp,ws,alphamax,alphamin)  % minimum order, half-power frequency
[b,a]=butter(N,wh);                      % coefficients of designed filter
[H,w]=freqz(b,a);w=w/pi;N=length(H);     % frequency response
spec1=alphamax*ones(1,N); spec2=alphamin*ones(1,N);  % specification lines
```

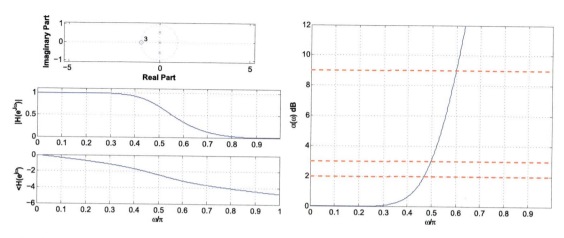

FIGURE 12.11

Design of low-pass Butterworth filter using MATLAB: poles and zeros, magnitude and phase response (left); verification of specifications using loss function $\alpha(\omega)$.

```
hpf=3.01*ones(1,N);                                  % half-power frequency line
disp('poles')                                        % display poles
roots(a)
disp('zeros')                                        % display zeros
roots(b)
alpha=-20*log10(abs(H));                             % loss in dB
```

The results of the design are shown in Fig. 12.11. The minimum order of the designed filter is $N = 3$ and the half-power frequency is $\omega_{hp} = 0.499\pi$. The poles are along the imaginary axis of the Z-plane ($K_b = 1$) and there are three zeros at $z = -1$. The transfer function of the designed filter is

$$H(z) = \frac{0.166 + 0.497z^{-1} + 0.497z^{-2} + 0.166z^{-3}}{1 - 0.006z^{-1} + 0.333z^{-2} - 0.001z^{-3}}.$$

Finally, to verify that the specifications are satisfied we plot the loss function $\alpha(e^{j\omega})$ along with three horizontal lines corresponding to $\alpha_{max} = 2$ dB, 3 dB for the half-power frequency, and $\alpha_{min} = 9$ dB. The crossings of these lines with the filter loss function indicate that at the normalized frequencies [0 0.47] the loss is less than 2 dB, at the normalized frequencies (0.6 1] the loss is bigger than 9 dB, as desired, and that the normalized half-power frequency is about 0.5.   □

Example 12.8. In this example we consider designing a Butterworth low-pass discrete filter for processing an analog signal. (See Fig. 12.12.) The filter specifications are:

$$f_p = 2250 \text{ Hz passband frequency,} \quad \alpha_1 = -18 \text{ dB dc loss,}$$
$$f_{st} = 2700 \text{ Hz stopband frequency,} \quad \alpha_2 = -15 \text{ dB loss in passband,}$$
$$f_s = 9000 \text{ Hz sampling frequency,} \quad \alpha_3 = -9 \text{ dB loss in stopband.}$$

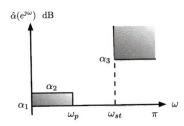

FIGURE 12.12

Loss specifications for a discrete low-pass filter used to process analog signals.

Solution: The specifications are not normalized. Normalizing them we have

$$\hat{\alpha}(e^{j0}) = -18 \text{ dB, dc gain,} \quad \alpha_{max} = \alpha_2 - \alpha_1 = 3 \text{ dB,} \quad \alpha_{min} = \alpha_3 - \alpha_1 = 9 \text{ dB,}$$

$$\omega_p = \frac{2\pi f_{hp}}{f_s} = 0.5\pi, \quad \omega_{st} = \frac{2\pi f_{st}}{f_s} = 0.6\pi.$$

Note that $\omega_p = \omega_{hp}$ since the difference in the losses at dc and at ω_p is 3 dB.
The sample period is

$$T_s = 1/f_s = (1/9) \times 10^{-3} \text{ s/sample} \quad \Rightarrow \quad K_b = \cot(\pi f_{hp} T_s) = 1.$$

Given that the half-power frequency is known, only the minimum order of the filter is needed. The loss function for the Butterworth filter is then

$$\alpha(e^{j\omega}) = 10\log_{10}\left(1 + \left[\frac{\tan(0.5\omega)}{\tan(0.5\omega_{hp})}\right]^{2N}\right) = 10\log_{10}(1 + (\tan(0.5\omega))^{2N})$$

since $0.5\omega_{hp} = \pi/4$. For $\omega = \omega_{st}$, letting $\alpha(e^{j\omega_{st}}) \geq \alpha_{min}$ solving for N we get, using the symbol $\lceil . \rceil$ for ceiling,

$$N = \left\lceil \frac{\log_{10}(10^{0.1\alpha_{min}} - 1)}{2\log_{10}(\tan(0.5\omega_{st}))} \right\rceil$$

or the integer that is larger than the argument of $\lceil . \rceil$, which by replacing α_{min} and ω_{st} gives as minimum order $N = 4$. The MATLAB script used is

```
%%
% Example 12.8---Filtering of analog signal
%%
wh=0.5*pi; ws=0.6*pi; alphamin=9; Fs=9000;    % filter specifications
N=log10((10^(0.1*alphamin)-1))/(2*log10(tan(ws/2)/tan(wh/2)));
```

```
N=ceil(N);
[b,a]=butter(N,wh/pi);
[H,w]=freqz(b,a);w=w/pi;N=length(H);f=w*Fs/2;
alpha0=-18;
G=10^(-alpha0/20);H=H*G;
spec2=alpha0+alphamin*ones(1,N);
hpf=alpha0+3.01*ones(1,N);
disp('poles'); p=roots(a); disp('zeros'); z=roots(b)
alpha=-20*log10(abs(H));
```

To ensure that the loss is -18 dB at $\omega = 0$, if the transfer function of the normalized loss (0 dB) filter is $H(z)$, we include a gain G in the numerator so that $H'(z) = GH(z)$ has the desired dc loss of -18 dB. The gain G is found from $\alpha(e^{j0}) = -18 = -20\log_{10} G$ to be $G = 7.94$. Thus the final form of the filter is

$$H'(z) = GH(z) = \frac{(z+1)^4}{(z^2+0.45)(z^2+0.04)},$$

which satisfies the loss specifications. Notice that when $K_b = 1$, the poles are imaginary.

Since this filter is used to filter an analog signal the frequency scale of the magnitude and phase response of the filter is given in Hz. See Fig. 12.13. To verify that the specifications are satisfied the loss function is plotted and compared with the losses corresponding to f_{hp} and f_{st}. The loss at $f_{hp} = 2250$ (Hz) coincides with the dc loss plus 3 dB, and the loss at $f_{st} = 2700$ (Hz) is above the specified value. □

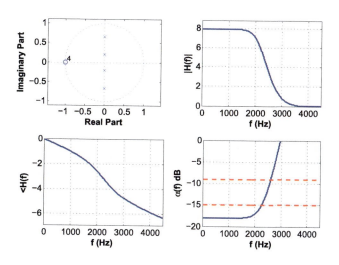

FIGURE 12.13

Low-pass filter for filtering of analog signal.

12.4.3 DESIGN OF CHEBYSHEV LOW-PASS DISCRETE FILTERS

The constant K_c of the bilinear transform for the Chebyshev filter is calculated by transforming the normalized pass frequency $\Omega'_p = 1$ into the discrete frequency ω_p giving

$$K_c = \frac{1}{\tan(0.5\omega_p)}. \tag{12.31}$$

Using the relation between frequencies in the bilinear transformation, we have

$$\frac{\Omega}{\Omega_p} = \frac{\tan(0.5\omega)}{\tan(0.5\omega_p)} \tag{12.32}$$

and replacing it in the magnitude squared function for the Chebyshev analog filter, the magnitude squared function of the discrete Chebyshev low-pass filter is

$$|H_N(e^{j\omega})|^2 = \frac{1}{1 + \varepsilon^2 C_N^2(\tan(0.5\omega)/\tan(0.5\omega_p))} \tag{12.33}$$

where $C(.)$ are the Chebyshev polynomials of the first kind encountered before in the analog design. The ripple parameter remains the same as in the analog design (since it does not depend on frequency)

$$\varepsilon = (10^{0.1\alpha_{max}} - 1)^{1/2}. \tag{12.34}$$

Replacing (12.32) in the analog formulas gives the order of the filter

$$N \geq \frac{\cosh^{-1}\left([(10^{0.1\alpha_{min}} - 1)/(10^{0.1\alpha_{max}} - 1)]^{1/2}\right)}{\cosh^{-1}[\tan(0.5\omega_{st})/\tan(0.5\omega_p)]} \tag{12.35}$$

and the half-power frequency

$$\omega_{hp} = 2\tan^{-1}\left[\tan(0.5\omega_p)\cosh\left(\frac{1}{N}\cosh^{-1}\left(\frac{1}{\varepsilon}\right)\right)\right]. \tag{12.36}$$

After calculating these parameters, the transfer function of the Chebyshev discrete filter is found by transforming the Chebyshev analog filter of order N into a discrete filter using the bilinear transformation:

$$H_N(z) = H_N(s)|_{s=K_c(1-z^{-1})/(1+z^{-1})}. \tag{12.37}$$

Remarks

1. Just as with the Butterworth filter, the equations for the filter parameters (N, ω_{hp}) can be obtained from the analog formulas by substituting

$$\frac{\Omega_{st}}{\Omega_p} = \frac{\tan(0.5\omega_{st})}{\tan(0.5\omega_p)}.$$

Since the analog ripple factor ε only depends on the magnitude specifications, it is not affected by the bilinear transformation—a frequency only transformation.

2. The filter parameters $(N, \omega_{hp}, \varepsilon)$ can also be found from the loss function obtained from the discrete Chebyshev squared magnitude:

$$\alpha(e^{j\omega}) = 10 \log_{10} \left[1 + \varepsilon^2 C_N^2 \left(\frac{\tan(0.5\omega)}{\tan(0.5\omega_p)} \right) \right]. \tag{12.38}$$

This is done by following a similar approach to the one in the analog case.

3. Like in the discrete Butterworth, for Chebyshev filters the dc gain (i.e., gain at $\omega = 0$) can be set to any desired value by allowing a constant gain G in the numerator such that

$$|H_N(e^{j0})| = |H_N(1)| = G \frac{|N(1)|}{|D(1)|} = \text{desired gain.} \tag{12.39}$$

4. MATLAB provides two functions to design Chebyshev filters. The function *cheby1* is for designing the filters covered in this section, while *cheby2* is to design filters with a flat response in the passband and with ripples in the stopband. The minimum order of the filter is found using *cheb1ord* and *cheb2ord*. The functions *cheby1* and *cheby2* give the filter coefficients.

Example 12.9. Consider the design of two low-pass Chebyshev filters. The specifications for the first filter are

$$\alpha(e^{j0}) = 0 \text{ dB},$$
$$\omega_p = 0.47\pi \text{ rad}, \quad \alpha_{max} = 2 \text{ dB},$$
$$\omega_{st} = 0.6\pi \text{ rad}, \quad \alpha_{min} = 6 \text{ dB}.$$

For the second filter change ω_p to 0.48π rad, keep the other specifications. Use MATLAB.

Solution: In Example 12.7, we obtained a third-order Butterworth low-pass filter that satisfies the specifications of the first filter. According to the results in this example, a second-order Chebyshev filter satisfies the same specifications. It is always so that a Chebyshev filter satisfies the same specifications as a Butterworth filter using a lower minimum order. For the second filter we narrow the transition band by 0.01π rad, and so the minimum order of the Chebyshev filter increases by one, as we will see. The following is the script for the design of the two filters.

```
%%
% Example 12.9---LP Chebyshev
%%
alphamax=2;   alphamin=9;                      % loss specs
figure(1)
for i=1:2,
  wp=0.47+(i-1)*0.01; ws=0.6;                   % normalized frequency specs
   [N,wn]=cheb1ord(wp,ws,alphamax,alphamin)
   [b,a]=cheby1(N,alphamax,wn);
  wp=wp*pi;
  % magnitude and phase
```

```
[H,w]=freqz(b,a); w=w/pi; M=length(H);H=H/H(1);
% to verify specs
spec0=zeros(1,M); spec1=alphamax*ones(1,M)*(-1)^(N+1);
spec2=alphamin*ones(1,M);
alpha=-20*log10(abs(H));
 hpf=(3.01+alpha(1))*ones(1,M);
 % epsilon and half-power frequency
 epsi=sqrt(10^(0.1*alphamax)-1);
 whp= 2*atan(tan(0.5*wp)*cosh(acosh(sqrt(10^(0.1*3.01)-1)/epsi)/N));
 whp=whp/pi
 % plotting
 subplot(221); zplane(b,a)
 subplot(222)
 plot(w,abs(H)); grid; ylabel('|H|');
 axis([0 max(w) 0 1.1*max(abs(H))])
 subplot(223)
 plot(w,unwrap(angle(H))); grid;
 ylabel('<H (rad)'); xlabel('\omega/\pi')
 subplot(224)
 plot(w,alpha); ylabel('\alpha(\omega) dB'); xlabel('\omega/\pi')
 hold on; plot(w,spec0,'r'); hold on; plot(w,spec1,'r')
 hold on; plot(w,hpf,'k'); hold on
 plot(w,spec2,'r'); grid;
 axis([0 max(w) 1.1*min(alpha) 1.1*(alpha(1)+3)]); hold off
 figure(2)
end
```

The transfer function of the first filter is

$$H_1(z) = \frac{0.224 + 0.449z^{-1} + 0.224z^{-2}}{1 - 0.264z^{-1} + 0.394z^{-2}},$$

and its half-power frequency is $\omega_{hp} = 0.493\pi$ rad. The second filter has a transfer function

$$H_2(z) = \frac{0.094 + 0.283z^{-1} + 0.283z^{-2} + 0.094z^{-3}}{1 - 0.691z^{-1} + 0.774z^{-2} - 0.327z^{-3}},$$

and a half-power frequency $\omega_{hp} = 0.4902\pi$. The poles and zeros as well as the magnitude and phase responses of the two filters are shown in Fig. 12.14. Notice the difference in the gain (or losses) in the passband of the two filters. In order for the dc gain to be unity, the magnitude response in the even-order filter $H_1(z)$ has values above 1, while the odd-order filter $H_2(z)$ does the opposite.

It is important to indicate that the output frequency ω_n given by *cheb1ord* and that *cheby1* uses as input is the passband frequency ω_p. Since the half power is not given by *cheb1ord* the half-power frequency of the filter can be calculated using the minimum order N, the ripple factor ε and the passband frequency ω_p in Equation (12.36). See the script. $\quad\square$

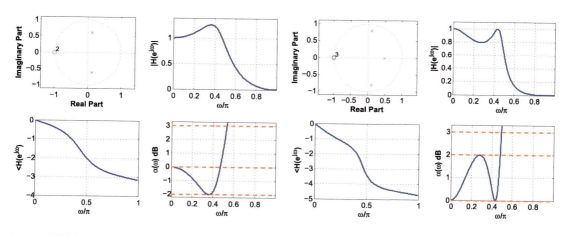

FIGURE 12.14

Two Chebyshev filters with different transition bands: even-order filter for $\omega_p = 0.47\pi$ on the left, and odd-order filter for $\omega_p = 0.48\pi$ (narrower transition band) on the right.

Example 12.10. Consider the following specifications of a filter that will be used to filter an acoustic signal:

$$\text{dc gain} = 10, \qquad\qquad \text{half-power frequency } f_{hp} = 4 \text{ kHz}$$
$$\text{band-stop frequency } f_{st} = 5 \text{ kHz}, \quad \alpha_{min} = 60 \text{ dB}$$
$$\text{sampling frequency } f_s = 20 \text{ kHz}.$$

Design a Butterworth and a Chebyshev low-pass filters of the same order and compare their frequency responses.

Solution: The specifications of the discrete filter are

$$\text{dc gain} = 10 \Rightarrow \alpha(e^{j0}) = -20 \text{ dB},$$
$$\text{half-power frequency: } \omega_{hp} = 2\pi f_{hp}(1/f_s) = 0.4\pi \text{ rad},$$
$$\text{band-stop frequency: } \omega_{st} = 2\pi f_{st}(1/f_s) = 0.5\pi \text{ rad}.$$

When designing the Butterworth filter we only need to find the minimum order N given that the half-power frequency is specified. We find that $N = 15$ satisfies the specifications. Using this value and the given discrete half-power frequency the function *butter* gives the coefficients of the filter. See Fig. 12.15 for results.

The design of the Chebyshev filter with order $N = 15$ and half-power frequency $\omega_{hp} = 0.4\pi$ cannot be done directly with the function *cheby1*, as we do not have the passband frequency ω_p. Using Equation (12.36) to compute the half-power frequency, we solve for ω_p after we give a value to α_{max} (arbitrarily chosen to be 0.001 dB) which allows us to compute the ripple factor ε (see Equation (12.34)). See the script part corresponding to the Chebyshev design. The function *cheby1* with inputs N, α_{max}, and ω_p gives the coefficients of the designed filter. Using these coefficients we plot the

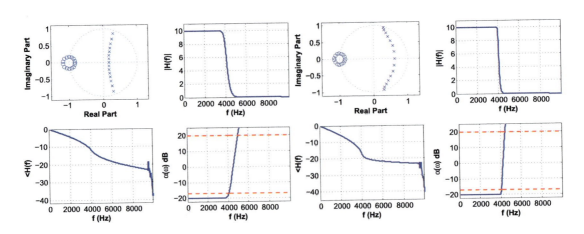

FIGURE 12.15

Equal-order ($N = 15$) Butterworth (left) and Chebyshev filters for filtering of acoustic signal.

poles/zeros, magnitude and phase responses, and the loss function as shown in Fig. 12.15. According to the loss function plots the Chebyshev filter displays a sharper response in the transition band than the Butterworth filter, as expected.

```
%%
% Example 12.10---Butterworth/Chebyshev filters for analog signal
%%
wh=0.4*pi;ws=0.5*pi; alphamin=40;Fs=20000;
% Butterworth
        %N=log10((10^(0.1*alphamin)-1))/(2*log10(tan(ws/2)/tan(wh/2))); N=ceil(N)
        % [b,a]=butter(N,wh/pi); % to get Butterworth filter get rid of ' % '
% Chebyshev
        alphamax=0.001;
        epsi=sqrt(10^(0.1*alphamax)-1);
        % computation of wp for Chebyshev design
        wp=2*atan(tan(0.5*wh)/(cosh(acosh(sqrt(10^(0.1*3.01)-1)/epsi)/N)));
        wp=wp/pi;  [b,a]=cheby1(N,alphamax,wp);
% magnitude and phase
        [H,w]=freqz(b,a);w=w/pi;M=length(H); f=w*Fs/2;
        alpha0=-20;H=H*10;
% verifying specs
        spec2=alpha0+alphamin*ones(1,M);
        hpf=alpha0+3.01*ones(1,M);
        alpha=-20*log10(abs(H));
        Ha=unwrap(angle(H));
```

12.4.4 RATIONAL FREQUENCY TRANSFORMATIONS

As indicated before, the conventional approach to filter design is to obtain first a prototype low-pass filter and then to convert it into different types of filters by means of frequency transformations. When using analog filters to design IIR discrete filters the frequency transformation could be done in two ways:

- transform a prototype low-pass analog filter into a desired analog filter which in turn is converted into a discrete filter using the bilinear, or other transformation, into a discrete filter,
- design a prototype low-pass discrete filter and then transform it into the desired discrete filter.

The first approach has the advantage that the analog frequency transformations are available and well understood. Its drawback appears when applying the bilinear transformation as it may cause undesirable warping in the higher frequencies. So the second approach will be used.

Given a prototype low-pass filter $H_{\ell p}(Z)$, we wish to transform it into a desired filter $H(z)$, typically another low-pass, band-pass, high-pass or band-stop filter. The transformation

$$G(z^{-1}) = Z^{-1} \qquad (12.40)$$

should preserve the rationality and the stability of the low-pass prototype. Accordingly, $G(z^{-1})$ should

- be rational, to preserve rationality,
- map the inside of the unit circle in the Z-plane into the inside of the unit circle in the z-plane, to preserve stability, and
- map the unit circle $|Z| = 1$ into the unit circle $|z| = 1$, so that the frequency response of the prototype filter is mapped into the frequency response of the desired filter.

If $Z = Re^{j\theta}$ and $z = re^{j\omega}$, the third condition on $G(z^{-1})$ corresponds to

$$G(e^{-j\omega}) = |G(e^{-j\omega})|e^{j\angle(G(e^{-j\omega}))} = \underbrace{1\,e^{-j\theta}}_{\text{unit circle in Z-plane}}, \qquad (12.41)$$

indicating that the frequency transformation $G(z^{-1})$ has the characteristics of an all-pass filter, with magnitude $|G(e^{-j\omega})| = 1$ and phase $\angle G(e^{-j\omega}) = -\theta$.

Using the general form of the transfer function of an all-pass filter (ratio of two equal-order polynomials with poles and zeros being the inverse conjugate of each other) the general form of the rational transformation is

$$Z^{-1} = G(z^{-1}) = A \prod_k \frac{z^{-1} - \alpha_k}{1 - \alpha_k^* z^{-1}} \qquad (12.42)$$

where the values of A and $\{\alpha_k\}$ are obtained from the prototype and the desired filters.

Low-Pass to Low-Pass Transformation

We wish to obtain the transformation $Z^{-1} = G(z^{-1})$ to convert a prototype low-pass filter into a different low-pass filter. The all-pass transformation should be able to expand or contract the frequency support of the prototype low-pass filter but keep its order. Thus it should be a ratio of linear transformations:

$$Z^{-1} = A \frac{z^{-1} - \alpha}{1 - \alpha z^{-1}} \tag{12.43}$$

for some parameters A and α. Since the zero frequency in the Z-plane is to be mapped into the zero frequency in the z-plane, if we let $Z = z = 1$ in the transformation we get $A = 1$. To obtain α, we let $Z = 1e^{j\theta}$ and $z = 1e^{j\omega}$ in (12.43) to obtain

$$e^{-j\theta} = \frac{e^{-j\omega} - \alpha}{1 - \alpha e^{-j\omega}}. \tag{12.44}$$

The value of α that maps the cut-off frequency θ_p of the prototype into the desired cut-off frequency ω_d is found as follows. First, from (12.44) we have

$$\begin{aligned}
\alpha &= \frac{e^{-j\omega} - e^{-j\theta}}{1 - e^{-j(\theta+\omega)}} = \frac{e^{-j\omega} - e^{-j\theta}}{2je^{-j0.5(\theta+\omega)} \sin((\theta + \omega)/2)} \\
&= \frac{e^{j0.5(\theta-\omega)} - e^{-j0.5(\theta-\omega)}}{2j \sin((\theta + \omega)/2)} = \frac{\sin((\theta - \omega)/2)}{\sin((\theta + \omega)/2)}
\end{aligned}$$

and then replacing θ and ω by θ_p and ω_d gives

$$\alpha = \frac{\sin((\theta_p - \omega_d)/2)}{\sin((\theta_p + \omega_d)/2)}. \tag{12.45}$$

Notice that if the prototype filter coincides with the desired filter, i.e., $\theta_p = \omega_d$, then $\alpha = 0$, and the transformation is $Z^{-1} = z^{-1}$. For different values of α between 0 and 1 the transformation shrinks the support of the prototype low-pass filter, and conversely for $-1 \leq \alpha < 0$ the transformation expands the support of the prototype. (In Fig. 12.16 the frequencies θ and ω are normalized to values between 0 and 1, i.e., both are divided by π.)

Remarks

1. The LP-LP transformation then consists in:

   - given θ_p and ω_d finding the corresponding α value (Equation (12.45)), and then
   - using the found α in the transformation (12.43) with $A = 1$.

2. Even in this simple low-pass to low-pass case the relation between the frequencies θ and ω is highly non-linear. Indeed, solving for $e^{-j\omega}$ in the transformation (12.44) we get

$$\begin{aligned}
e^{-j\omega} &= \left(\frac{e^{-j\theta} + \alpha}{1 + \alpha e^{-j\theta}} \right) \left(\frac{1 + \alpha e^{j\theta}}{1 + \alpha e^{j\theta}} \right) = \frac{e^{-j\theta} + 2\alpha + \alpha^2 e^{j\theta}}{1 + 2\alpha \cos(\theta) + \alpha^2} \\
&= \underbrace{\frac{2\alpha + (1 + \alpha^2)\cos(\theta)}{1 + 2\alpha \cos(\theta) + \alpha^2}}_{B} - j \underbrace{\frac{(1 - \alpha^2)\sin(\theta)}{1 + 2\alpha \cos(\theta) + \alpha^2}}_{C}
\end{aligned}$$

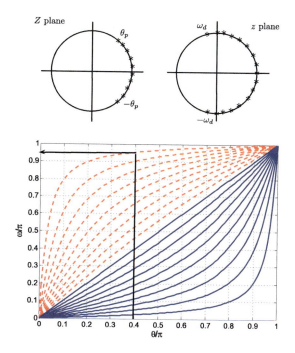

FIGURE 12.16

Frequency transformation from a prototype low-pass filter with cut-off θ_p into a low-pass filter with desired cut-off frequency ω_d (top figure). Mapping of θ/π into ω/π frequencies in low-pass to low-pass frequency transformation: the continuous lines correspond to $0 < \alpha \leq 1$, while the dashed lines correspond to values $-1 \leq \alpha \leq 0$. The arrow shows the transformation of $\theta_p = 0.4\pi$ into $\omega_d \approx 0.95\pi$ when $\alpha = -0.9$.

and since $e^{-j\omega} = \cos(\omega) - j\sin(\omega)$ we find that $\cos(\omega) = B$ and $\sin(\omega) = C$ so that $\tan(\omega) = C/B$ and thus

$$\omega = \tan^{-1}\left[\frac{(1-\alpha^2)\sin(\theta)}{2\alpha + (1+\alpha^2)\cos(\theta)}\right], \tag{12.46}$$

which when plotted for different values of α gives Fig. 12.16. These curves clearly show the mapping of a frequency θ_p into ω_d and the value of α needed to perform the correct transformation.

Low-Pass to High-Pass Transformation

The duality between low-pass and high-pass filters indicates that this transformation, like the LP-LP, should be linear in both numerator and denominator. Also notice that the prototype low-pass filter can be transformed into a high-pass filter with the same bandwidth, by changing Z^{-1} into $-Z^{-1}$. Indeed, complex poles or zeros $R_1 e^{\pm j\theta_1}$ of the low-pass filter are mapped into $-R_1 e^{\pm j\theta_1} = R_1 e^{j(\pi \pm \theta_1)}$

(i.e., let $\theta_1 \to \pi - \theta_1$) corresponding to a high-pass filter. For instance, a low-pass filter

$$H(Z) = \frac{Z + 1}{Z - 0.5}$$

with a zero at -1 and a pole at 0.5 becomes, when letting $Z^{-1} \to Z^{-1}$,

$$H_1(Z) = \frac{-Z + 1}{-Z - 0.5} = \frac{Z - 1}{Z + 0.5}$$

with a zero at 1 and a pole at -0.5 or a high-pass filter.

The LP-HP transformation is then

$$Z^{-1} = -\left(\frac{z^{-1} - \alpha}{1 - \alpha z^{-1}} \right)$$

and to obtain α we replace θ_p by $\pi - \theta_p$ in Equation (12.45):

$$\begin{aligned} \alpha &= \frac{\sin(-(\theta_p + \omega_d)/2 + \pi/2)}{\sin(-(\theta_p - \omega_d)/2 + \pi/2)} \\ &= \frac{\cos(-(\theta_p + \omega_d)/2)}{\cos(-(\theta_p - \omega_d)/2)} = \frac{\cos((\theta_p + \omega_d)/2)}{\cos((\theta_p - \omega_d)/2)}. \end{aligned} \tag{12.47}$$

As before, θ_p is the cut-off frequency of the prototype low-pass filter and ω_d the desired cut-off frequency of the high-pass filter.

The above transformation confirms that when the low-pass and the high-pass filters have the same bandwidth, i.e., $\omega_d = \pi - \theta_p$, we have $\theta_p + \omega_d = \pi$ and so $\alpha = 0$ giving as transformation $Z^{-1} = -z^{-1}$, which as we indicated converts a low-pass filter into a high-pass filter, both of the same bandwidth.

Low-Pass to Band-Pass and Low-Pass to Band-Stop Transformations

By being linear in both the numerator and the denominator, the LP-LP and LP-HP transformations preserve the number of poles and zeros of the prototype filter. To transform a low-pass filter into a band-pass or a band-stop filter, the number of poles and zeros must be doubled. For instance, if the prototype is a first-order low-pass filter (with real-valued poles/zeros) we need a quadratic, rather than a linear, transformation in both numerator and denominator to obtain band-pass or band-stop filters from the low-pass filter since band-pass and band-stop filters cannot be of first order.

The low-pass to band-pass (LP-BP) transformation is

$$Z^{-1} = -\left(\frac{z^{-2} - bz^{-1} + c}{cz^{-2} - bz^{-1} + 1} \right) \tag{12.48}$$

while the low-pass to band-stop (LP-BS) transformation is

$$Z^{-1} = \frac{z^{-2} - (b/k)z^{-1} - c}{-cz^{-2} - (b/k)z^{-1} + 1} \tag{12.49}$$

where

$$b = 2\alpha k/(k+1), \qquad\qquad\qquad c = (k-1)/(k+1),$$

$$\alpha = (\cos((\omega_{du} + \omega_{d\ell})/2))/(\cos((\omega_{du} - \omega_{d\ell})/2)), \qquad k = \cot((\omega_{du} - \omega_{d\ell})/2)\tan(\theta_p/2).$$

The frequencies $\omega_{d\ell}$ and ω_{du} are the desired lower and higher cut-off frequencies in the band-pass and band-stop filters. Notice that when $\omega_{du} + \omega_{d\ell} = \pi$ (i.e., there is symmetry of the magnitude response around $\pi/2$) the above equations become simpler because $\alpha = 0$.

12.4.5 GENERAL IIR FILTER DESIGN WITH MATLAB

The following function *buttercheby1* can be used to design low-pass, high-pass, band-pass and band-stop Butterworth and Chebyshev filters. One important thing to remember when designing band-pass and band-stop filters is that the order of the low-pass prototype is half that of the desired filter.

```
function [b,a,H,w]=buttercheby1(lp_order,wn,BC,type)
%
%  Design of frequency discriminating filters
%  using Butterworth and Chebyshev methods, the bilinear transformation and
%  frequency transformations
%
%  lp_order : order of lowpass filter prototype
%  wn : vector containing the cutoff normalized frequency(ies)
%  (entries must be normalized)
%  BC: Butterworth (0) or Chebyshev1 (1)
%  type : type of filter desired
%            1 = lowpass
%            2 = highpass
%            3 = bandpass
%            4 = stopband
%    [b,a] : numerator, denominator coefficients of designed filter
%    [H,w] : frequency response, frequency range
%  USE:
%    [b,a,H,w]=buttercheby1(lp_order,wn,BC,type)
  if BC==0;                          % Butterworth filter
     if type == 1
     [b,a]=butter(lp_order,wn);           % lowpas
     elseif type == 2
     [b,a]=butter(lp_order,wn,'high');  % highpass
     elseif type == 3
     [b,a]=butter(lp_order,wn);           % bandpass
     else
     [b,a]=butter(lp_order,wn,'stop');  % stopband
     end
```

```
    [H,w]=freqz(b,a,256);
    else % Chebyshev1 filter
    R=0.01;
    if type == 1,
        [b,a]=cheby1(lp_order,R,wn);          % lowpas
        elseif type == 2,
        [b,a]=cheby1(lp_order,R,wn,'high'); % highpass
        elseif type == 3,
        [b,a]=cheby1(lp_order,R,wn);          % bandpass
        else
        [b,a]=cheby1(lp_order,R,wn,'stop'); % stopband
    end
    [H,w]=freqz(b,a,256);
    end
```

Example 12.11. To illustrate the design of filters other than low-pass filters, consider the design of a Butterworth and Chebyshev band-stop filters of order $N = 30$ and half-power frequencies $[0.4\pi\ 0.6\pi]$.

Solution: For the filter design we use the following script. Fig. 12.17 shows the results.   □

```
%%
% Example 12.11---Band-stop filters
%%
clear all; clf
ind=input('Butterworth (1) or Chebyshev (2) band-stop?    ')
if ind==1,
[b,a]=buttercheby1(15,[0.4 0.6],0,4)
else
[b,a]=buttercheby1(15,[0.4 0.6],1,4)
end
% magnitude and phase
[H,w]=freqz(b,a); w=w/pi; M=length(H);
Ha=unwrap(angle(H));
% loss function
alpha=-20*log10(abs(H));

figure(1)
subplot(221)
zplane(b,a)
if ind==1,
    title(' Stop-band Butterworth')
else
    title(' Stop-band Butterworth')
end
```

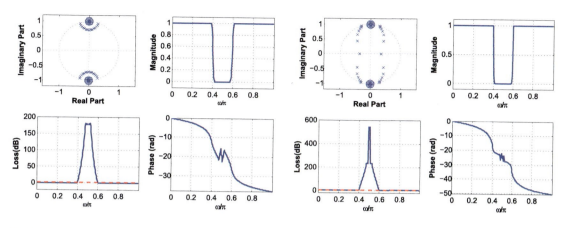

FIGURE 12.17

Bandstop Butterworth (left) and Chebyshev (right) filters: (clockwise for each side from top left) poles/zeros, magnitude, phase frequency responses, loss.

```
subplot(222)
plot(w,abs(H)); grid; axis([0 1 0 1.1*max(abs(H))]); ylabel('|H(e^{j\omega})|');
 xlabel('\omega/\pi')
subplot(223)
plot(w,Ha);grid; ylabel('<H(e^{j\omega}) (rad)'); xlabel('\omega/\pi')
axis([0 1 1.1*min(Ha) 1.1*max(Ha)]);
subplot(224)
plot(w,alpha); ylabel('\alpha(e^{j\omega}) dB'); xlabel('\omega/\pi');grid
axis([0 1 -20 250]);
```

Example 12.12. There are other filters that can be designed with MATLAB, following a procedure similar to the previous cases. For instance, to design a bandpass elliptic filter with cut-off frequencies $[0.45\pi \quad 0.55\pi]$ of order 20 and with losses specifications of 0.1 and 40 dB in the passband and in the stopband, respectively, we use the script shown below. Likewise, to design a high-pass filter using the *cheby2* function we specify the order 10, the loss in the stopband and the cut-off frequency 0.55π and indicate it is a high-pass filter. The results are shown in Fig. 12.18.

```
%%
% Example 12.12---Elliptic and Cheby2
%%
[b1,a1]=ellip(10,0.1,40,[0.45 0.55]);
[b2,a2]=cheby2(10,40, 0.55,'high');
```

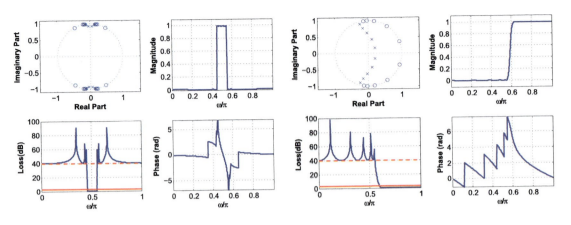

FIGURE 12.18

Elliptic band-pass filter (left) and high-pass using *cheby2*: (clockwise for each side from top left) poles/zeros, magnitude, phase frequency responses, loss.

12.5 FIR FILTER DESIGN

The design of FIR filters is typically discrete. The specification of FIR filters is usually given in the time rather than in the frequency domain. FIR filters have three definite advantages: (i) stability, (ii) possible linear phase, and (iii) efficient implementation. Indeed, the poles of an FIR filter are at the origin of the Z-plane, thus FIR filters are stable. An FIR filter can be designed to have linear phase, and since the input/output equation of an FIR filter is equivalent to a convolution sum, FIR filters are implemented using the Fast Fourier Transform (FFT). A minor disadvantage is the storage required, typically FIR filters have a large number of coefficients.

Example 12.13. A moving average filter has an impulse response $h[n] = 1/M$, $0 \leq n \leq M - 1$, and zero otherwise. The transfer function of this filter is

$$H(z) = \sum_{n=0}^{M-1} \frac{1}{M} z^{-n} = \frac{1 - z^{-M}}{M(1 - z^{-1})} = \frac{1}{M} \frac{z^M - 1}{z^{M-1}(z - 1)}.$$

Consider the stability of this filter, determine if the phase of this filter is linear and what type of filter it is.

Solution: The impulse response $h[n]$ is absolutely summable given its finite length M, thus the filter is BIBO stable. Notice that $H(1)$ is $0/0$, according to the final expression above, indicating that a pole and a zero at $z = 1$ might exist, but also from the sum $H(1) = 1$, so there are no poles at $z = 1$. The zeros of $H(z)$ are values that make $z^M - 1 = 0$, or $z_k = e^{j2\pi/M}$ for $k = 0, \cdots, M - 1$. When $k = 0$ the zero $z = 1$ cancels the pole at 1, thus

$$H(z) = \frac{(z - 1) \prod_{k=1}^{M-1}(z - e^{j2\pi/M})}{Mz^{M-1}(z - 1)} = \frac{\prod_{k=1}^{M-1}(z - e^{j2\pi/M})}{Mz^{M-1}}.$$

To convince yourself of the pole–zero cancellation let $M = 3$ for which

$$H(z) = \frac{1}{3} \frac{z^3 - 1}{z^2(z - 1)} = \frac{1}{3} \frac{(z^2 + z + 1)(z - 1)}{z^2(z - 1)} = \frac{z^2 + z + 1}{3z^2}$$

showing the pole–zero cancellation. Notice the two poles are at zero, confirming the filter is BIBO stable.

Since the zeros of the filter are on the unit circle, the phase of this filter is not linear, as the phase will display discontinuities at the frequencies of the zeros. Although the filter is considered a low-pass filter, it is of very poor quality in terms of its magnitude response. □

12.5.1 WINDOW DESIGN METHOD

The usual filter specifications of magnitude and linear phase can be translated into a time-domain specification (i.e., a desired impulse response) by means of the discrete-time Fourier transform. In this section, we will show how to design FIR filters using this specification with the **window method**. You will see that this is a trial-and-error method, as there is no measure of how close the designed filter is to the desired response, and that using different windows we obtain different results.

Let $H_d(e^{j\omega})$ be the desired frequency response of an ideal discrete low-pass filter. Assume that the phase of $H_d(e^{j\omega})$ is zero. The desired impulse response is given by the inverse discrete-time Fourier transform:

$$h_d[n] = \frac{1}{2\pi} \int_{-\pi}^{\pi} H_d(e^{j\omega}) e^{j\omega n} \, d\omega \qquad -\infty < n < \infty, \qquad (12.50)$$

which is noncausal and of infinite length. The filter

$$H_d(z) = \sum_{n=-\infty}^{\infty} h_d[n] z^{-n} \qquad (12.51)$$

is thus not an FIR filter.[1] To obtain an FIR filter that approximates $H_d(e^{j\omega})$ we need to window the impulse response $h_d[n]$ to get finite length, and then to delay the resulting windowed impulse response to achieve causality.

For an odd integer N, define

$$h_w[n] = h_d[n] w[n] = \begin{cases} h_d[n] & -(N-1)/2 \le n \le (N-1)/2, \\ 0 & \text{elsewhere}, \end{cases}$$

where $w[n]$ is a **rectangular window**

$$w[n] = \begin{cases} 1 & -(N-1)/2 \le n \le (N-1)/2, \\ 0 & \text{otherwise}, \end{cases} \qquad (12.52)$$

---

[1] To obtain a zero phase the desired frequency response $H_d(e^{j\omega})$ must be real, which in turns requires even symmetry in the impulse response, i.e., $h[n] = h[-n]$, and that it be noncausal. Moreover, $h[n]$ would be of infinite length in general; otherwise, it would be a noncausal FIR.

that causes the truncation of $h_d[n]$. The windowed impulse response $h_w[n]$ has a discrete-time Fourier transform

$$H_w(e^{j\omega}) = \sum_{n=-(N-1)/2}^{(N-1)/2} h_w[n]e^{-j\omega n}.$$

For a large value of N we have $H_w(e^{j\omega})$ must be a good approximation of $H_d(e^{j\omega})$, i.e., for a large N we could have

$$|H_w(e^{j\omega})| \approx |H_d(e^{j\omega})|, \quad \angle H_w(e^{j\omega}) = \angle H_d(e^{j\omega}) = 0. \tag{12.53}$$

It is not clear how the value of N should be chosen—this is what we meant by stating that this design is a trial-and-error method.

For the N value that makes possible (12.53), to convert $H_w(z)$ into a causal filter, we delay the impulse response $h_w[n]$ by $(N-1)/2$ samples to obtain

$$
\begin{aligned}
\hat{H}(z) &= H_w(z)z^{-(N-1)/2} = \sum_{m=-(N-1)/2}^{(N-1)/2} h_w[m]z^{-(m+(N-1)/2)} \\
&= \sum_{n=0}^{N-1} h_d[n-(N-1)/2]w[n-(N-1)/2]z^{-n}
\end{aligned}
$$

after letting $n = m + (N-1)/2$. For a large value of N, we have

$$
\begin{aligned}
|\hat{H}(e^{j\omega})| &= |H_w(e^{j\omega})e^{-j\omega(N-1)/2}| = |H_w(e^{j\omega})| \approx |H_d(e^{j\omega})| \\
\angle\hat{H}(e^{j\omega}) &= \angle H_w(e^{j\omega}) - \frac{N-1}{2}\omega = -\frac{N-1}{2}\omega
\end{aligned}
\tag{12.54}
$$

since $\angle H_w(e^{j\omega}) = \angle H_d(e^{j\omega}) = 0$, according to (12.53). That is, the magnitude response of the FIR filter $\hat{H}(z)$ is approximately (depending on the value of N) the desired response and its phase response is linear. These results can be generalized as follows.

General Window Method for FIR Design

- If the desired low-pass frequency response has a magnitude

$$|H_d(e^{j\omega})| = \begin{cases} 1 & -\omega_c \le \omega \le \omega_c, \\ 0 & \text{otherwise,} \end{cases} \tag{12.55}$$

and a linear phase

$$\theta(\omega) = -\omega M/2.$$

The corresponding impulse response is given by

$$h_d[n] = \begin{cases} \sin(\omega_c(n-M/2))/(\pi(n-M/2)) & n \neq M/2, \\ \omega_c/\pi & n = M/2 \text{ for } M \text{ even, else not defined.} \end{cases} \tag{12.56}$$

Using a window $w[n]$ of length M and centered at $M/2$, the windowed impulse response is

$$h[n] = h_d[n]w[n],$$

and the designed FIR filter is

$$H(z) = \sum_{n=0}^{M-1} h[n]z^{-n}.$$

• The design using windows is a trial-and-error procedure. Different tradeoffs can be obtained by using various windows and various lengths of the windows.

• The symmetry of the impulse response $h[n]$ with respect to $M/2$, independent of whether this is an integer or not, guarantees the linear phase of the filter.

12.5.2 WINDOW FUNCTIONS

In the previous section, the windowed impulse response $h_w[n]$ was written $h_w[n] = h_d[n]w[n]$, where

$$w[n] = \begin{cases} 1 & -(N-1)/2 \le n \le (N-1)/2, \\ 0 & \text{otherwise,} \end{cases} \tag{12.57}$$

is a **rectangular window** of length N. If we wish $H_w(e^{j\omega}) = H_d(e^{j\omega})$, we would need a rectangular window of infinite length so that the impulse responses $h_w[n] = h_d[n]$, i.e., no windowing. This ideal rectangular window has a discrete-time Fourier transform

$$W(e^{j\omega}) = 2\pi\delta(\omega) \qquad -\pi \le \omega < \pi. \tag{12.58}$$

Since $h_w[n] = w[n]h_d[n]$, a product of functions of n, then $H_w(e^{j\omega})$ is the convolution of $H_d(e^{j\omega})$ and $W(e^{j\omega})$ in the frequency domain, i.e.,

$$H_w(e^{j\omega}) = \frac{1}{2\pi}\int_{-\pi}^{\pi} H_d(e^{j\theta})W(e^{j(\omega-\theta)})d\theta = \int_{-\pi}^{\pi} H_d(e^{j\theta})\delta(\omega-\theta)d\theta = H_d(e^{j\omega}).$$

Thus, for $N \to \infty$, the result of this convolution is $H_d(e^{j\omega})$, but if N is finite the convolution in the frequency domain would give a distorted version of $H_d(e^{j\omega})$. Thus to obtain a good approximation of $H_d(e^{j\omega})$ using a finite window $w[n]$ the window must have a spectrum approximating that of the ideal rectangular window, an impulse in frequency in $-\pi \le \omega < \pi$ as in (12.58) with most of the energy concentrated in the low frequencies. The smoothness of the window makes this possible.

Examples of windows that are smoother than the rectangular window are:

Triangular or Barlett Window

$$w[n] = \begin{cases} 1 - \frac{2|n|}{N-1} & -(N-1)/2 \le n \le (N-1)/2, \\ 0 & \text{otherwise.} \end{cases} \tag{12.59}$$

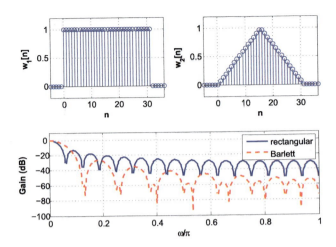

FIGURE 12.19

Rectangular (top left) and Bartlett (top right) causal windows and their spectra.

Hamming Window

$$w[n] = \begin{cases} 0.54 + 0.46\cos(2\pi n/(N-1)) & -(N-1)/2 \le n \le (N-1)/2, \\ 0 & \text{otherwise.} \end{cases} \tag{12.60}$$

Kaiser Window. This window has a parameter β, which can be adjusted. It is given by

$$w[n] = \begin{cases} \dfrac{I_0\left(\beta\sqrt{1-(n/(N-1)^2)}\right)}{I_0(\beta)} & -(N-1)/2 \le n \le (N-1)/2, \\ 0 & \text{otherwise.} \end{cases} \tag{12.61}$$

Here $I_0(x)$ is the zeroth-order Bessel function of the first kind which can be computed by the series

$$I_0(x) = 1 + \sum_{k=1}^{\infty} \left(\frac{(0.5x)^k}{k!}\right)^2. \tag{12.62}$$

When $\beta = 0$ the Kaiser window coincides with a rectangular window, since $I_0(0) = 1$. As β increases the window becomes smoother.

The above definitions are for windows symmetric with respect to the origin. Figs. 12.19 and 12.20 show the causal rectangular, Barlett, Hamming, and Kaiser windows, and their magnitude spectra. Given that the side lobes for the Kaiser window have the largest loss, the Kaiser window is considered the best of these four, followed by the Hamming, the Bartlett and the rectangular windows. Notice that the width of the first lobe is the widest for the Kaiser and the narrowest for the rectangular, indicating the Kaiser window is very smooth and most of its energy is in the low frequencies and the opposite for the rectangular window.

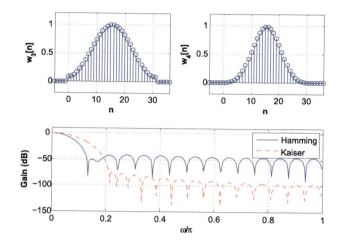

FIGURE 12.20

Hamming (top left) and Kaiser (top right) causal windows and their spectra.

12.5.3 LINEAR PHASE AND SYMMETRY OF THE IMPULSE RESPONSE

The linear phase is a result of the symmetry of the impulse response of the designed filter. As we show next, if the impulse response $h[n]$ of the FIR filter is even or odd symmetric with respect to a middle sample the filter has linear phase.

Consider the following cases for an FIR filter, of order $M-1$ or length M, with transfer function

$$H(z) = \sum_{n=0}^{M-1} h[n] z^{-n}. \tag{12.63}$$

1. M is odd, i.e., $M = 2N + 1$. Assume that the impulse response $h[n]$ is symmetric with respect to the N value:

$$h[n] = h[2N - n] \qquad n = 0, \cdots, N \tag{12.64}$$

or $h[0] = h[2N]$, $h[1] = h[2N - 1]$, $\cdots$, $h[N]$ equal to itself, then we rewrite (12.63) as

$$
\begin{aligned}
H(z) &= z^{-N} \sum_{n=0}^{2N} h[n] z^{N-n} = z^{-N} \left[\sum_{n=0}^{N-1} h[n] z^{N-n} + h[N] + \sum_{n=N+1}^{2N} h[n] z^{N-n} \right] \\
&= z^{-N} \left[\sum_{n=0}^{N-1} h[n] z^{N-n} + h[N] + \sum_{m=0}^{N-1} h[2N - m] z^{-N+m} \right] \\
&= z^{-N} \left[h[N] + \sum_{n=0}^{N-1} h[n] (z^{N-n} + z^{-(N-n)}) \right]
\end{aligned}
$$

where we let $m = 2N - n$ and used the symmetry of the impulse response. The frequency response is then

$$H(e^{j\omega}) = e^{-j\omega N} \left[h[N] + \sum_{n=0}^{N-1} 2h[n]\cos((N-n)\omega) \right] = e^{-j\omega N} A(e^{j\omega})$$

and since $A(e^{j\omega})$ is real,

$$\angle H(e^{j\omega}) = \begin{cases} -\omega N & A(e^{j\omega}) \geq 0, \\ -\omega N - \pi & A(e^{j\omega}) < 0, \end{cases} \tag{12.65}$$

that is, the group delay $\tau(\omega) = -d\angle H(e^{j\omega})/d\omega = N$ is constant, so the phase is linear.

2. M is even, i.e., $M = 2N$. Assume the impulse response in this case is symmetric with respect to the value in between samples $N - 1$ and N, or $N - 0.5$:

$$h[n] = h[2N - 1 - n] \qquad n = 0, \cdots, N - 1 \tag{12.66}$$

or $h[0] = h[2N - 1]$, $h[1] = h[2N - 2]$, $\cdots$, $h[N - 1] = h[N]$. Using a similar approach to the even case, we find the frequency response to be

$$H(e^{j\omega}) = e^{-j(N-0.5)\omega} \sum_{n=0}^{N-1} 2h[n]\cos((N - 0.5 - n)\omega) = e^{-j(N-0.5)\omega} B(e^{j\omega})$$

and since $B(e^{j\omega})$ is real, the phase of the FIR filter in this case is

$$\angle H(e^{j\omega}) = \begin{cases} -(N - 0.5)\omega & B(e^{j\omega}) \geq 0, \\ -(N - 0.5)\omega - \pi & B(e^{j\omega}) < 0. \end{cases} \tag{12.67}$$

The group delay $\tau(\omega) = -d\angle H(e^{j\omega})/d\omega = N - 0.5$ is constant so that the phase is considered linear.

3. It is possible to have odd symmetry in the impulse response for the above two cases:

- M is odd, i.e., $M = 2N + 1$. Assume that the impulse response $h[n]$ is odd symmetric with respect to the N value:

$$h[n] = -h[2N - n] \qquad n = 0, \cdots, N \tag{12.68}$$

or $h[0] = -h[2N]$, $h[1] = -h[2N - 1]$, $\cdots$, $h[N] = h[-N] = 0$. As above, the frequency response of the FIR filter is

$$H(e^{j\omega}) = -je^{-j\omega N} \left[\sum_{n=0}^{N-1} 2h[n]\sin((N-n)\omega) \right] = -je^{-j\omega N} C(e^{j\omega})$$

and since $C(e^{j\omega})$ is real,

$$\angle H(e^{j\omega}) = \begin{cases} -\omega N - \pi/2 & C(e^{j\omega}) \geq 0, \\ -\omega N - 3\pi/2 & C(e^{j\omega}) < 0, \end{cases} \tag{12.69}$$

that is, the phase of the filter is linear.

• M is even, i.e., $M = 2N$. Assume the impulse response in this case is odd symmetric with respect to the value in between samples $N - 1$ and N or $N - 0.5$:

$$h[n] = -h[2N - 1 - n] \qquad n = 0, \cdots, N - 1 \tag{12.70}$$

or $h[0] = -h[2N - 1]$, $h[1] = -h[2N - 2]$, $\cdots$, $h[N - 1] = -h[N]$. The frequency response is found to be

$$H(e^{j\omega}) = je^{-j(N-0.5)\omega} \sum_{n=0}^{N-1} 2h[n]\sin((N - 0.5 - n)\omega) = je^{-j(N-0.5)\omega}D(e^{j\omega}),$$

and since $D(e^{j\omega})$ is real, the phase of the FIR filter in this case is

$$\angle H(e^{j\omega}) = \begin{cases} -(N - 0.5)\omega + \pi/2 & D(e^{j\omega}) \geq 0, \\ -(N - 0.5)\omega - \pi/2 & D(e^{j\omega}) < 0, \end{cases} \tag{12.71}$$

or the phase is linear.

Example 12.14. Design a low-pass FIR filter of length $M = 21$ to be used in filtering analog signals and that approximates the following ideal frequency response:

$$H_d(e^{jf}) = \begin{cases} 1 & -125 \leq f \leq 125 \text{ Hz}, \\ 0 & \text{elsewhere in } -f_s/2 < f \leq f_s/2, \end{cases}$$

and $f_s = 1000$ Hz is the sampling rate. Use first a rectangular window, and then a Hamming window. Compare the designed filters.

Solution: Using $\omega = 2\pi f/f_s$, the discrete frequency response is given by

$$H_d(e^{j\omega}) = \begin{cases} 1 & -\pi/4 \leq \omega \leq \pi/4 \text{ rad}, \\ 0 & \text{elsewhere in } -\pi < \omega \leq \pi. \end{cases}$$

The desired impulse response is thus

$$h_d[n] = \frac{1}{2\pi}\int_{-\pi}^{\pi} H_d(e^{j\omega})e^{j\omega n}d\omega = \frac{1}{2\pi}\int_{-\pi/4}^{\pi/4} e^{j\omega n}d\omega = \begin{cases} \sin(\pi n/4)/(\pi n) & n \neq 0, \\ 0.25 & n = 0. \end{cases}$$

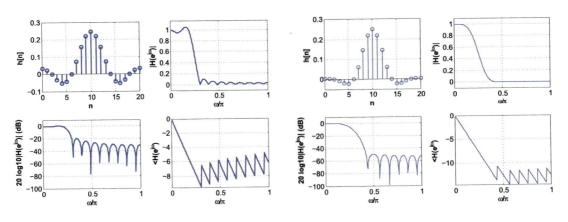

FIGURE 12.21

Low-pass FIR filters using rectangular (left) and Hamming (right) windows: impulse and magnitude responses on top and magnitude in dB and phase in rad responses on bottom.

Using a rectangular window, the FIR filter is then of the form (since $M = 2N + 1 = 21$, the delay is $N = 10$):

$$\hat{H}(z) = H_w(z)z^{-10} = \sum_{n=0}^{20} h_d[n-10]z^{-n} = 0.25z^{-10} + \sum_{n=0,n\neq 10}^{20} \frac{\sin(\pi(n-10)/4)}{\pi(n-10)} z^{-n}.$$

The magnitude and phase of this filter are shown in Fig. 12.21 when we use a rectangular (left) and a Hamming window (right).

The magnitude and phase responses of the filter designed using the Hamming window is much improved over the one obtained using the rectangular window. Notice that the second lobe in the stopband for the Hamming window design is at about -50 dB while for the rectangular window design is at about -20 dB, a significant difference. In both cases, the phase response is linear in the passband of the filter, corresponding to the impulse response $h[n]$ being symmetric with respect to the $n = 10$ sample.

The following is the script used.   □

```
%%
% Example 12.14---Filter from fir
%%
N=20;fc=125;Fs=1000; wc=2*fc/Fs;wo=0;
ind= input(' rectangular (1) or hamming (2) windows?  ')
if ind==1,
  wind=1;
else
  wind=3;
end
```

```
[b]=fir(N,wc,wo,wind);
disp(' impulse response')
h=b; [H,w]=freqz(b,1,256);
n=0:N;

figure(1)
subplot(221)
stem(n,b); axis([0 N 1.2*min(b) 1.2*max(b)]); ylabel('h[n]');xlabel('n'); grid
subplot(222)
plot(w/pi,abs(H));grid; xlabel('\omega/\pi'); ylabel('|H(e^{j\omega})|')
axis([0 1 -0.1 1.1*max(abs(H))])
subplot(223)
plot(w/pi,20*log10(abs(H)));axis([0 1 -100 10])
xlabel('\omega/\pi'); ylabel('20 log10|H(e^{j\omega})| (dB)'); grid
subplot(224)
phase=unwrap(angle(H));
plot(w/pi,phase); axis([0 1 1.1*min(phase) 1.1*max(phase)])
xlabel('\omega/\pi'); ylabel('<H(e^{j\omega})'); grid
```

Example 12.15.   Design a high-pass filter of order $M - 1 = 14$, and cut-off frequency 0.2π using the Kaiser window. Use MATLAB.

Solution: Let $h_{lp}[n]$ be the impulse response of an ideal low-pass filter

$$H_{lp}(e^{j\omega}) = \begin{cases} 1 & -\omega_c \leq \omega \leq \omega_c, \\ 0 & \text{otherwise in } [-\pi, \ \pi). \end{cases}$$

According to the modulation property of the DTFT, we have that

$$2h_{lp}[n]\cos(\omega_0 n) \quad \Leftrightarrow \quad H_{lp}(e^{j(\omega+\omega_0)}) + H_{lp}(e^{j(\omega-\omega_0)}).$$

If we let $\omega_0 = \pi$ then the right term gives a high-pass filter, and so $h_{hp}[n] = 2h_{lp}[n]\cos(\pi n) = 2(-1)^n h_{lp}[n]$ is the desired impulse response of the high-pass filter. The following script shows how to use our function *fir* (shown below) to design the high-pass filter.

```
%%
% Example 12.15---FIR filter from 'fir'
%%
No=14;wc=0.2;wo=1;wind=4;
[b]=fir(No,wc,wo,wind);
[H,w]=freqz(b,1,256);
```

The results are shown in Fig. 12.22. Notice the symmetry of the impulse response with respect to sample at $n = 7$ gives linear phase in the passband of the high-pass filter. The second lobe of the gain in dB is about -50 dB. $\square$

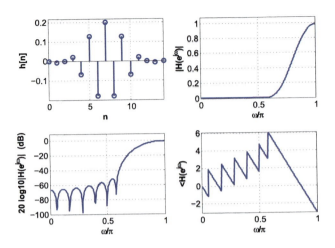

FIGURE 12.22

High-pass FIR filter design using Kaiser window: impulse and magnitude responses (top), magnitude in dB and phase responses on the bottom.

Our function *fir* can be used to design low-pass, high-pass and band-pass FIR filters using different types of windows. When designing high-pass and band-pass FIRs, *fir* first designs a prototype low-pass filter and then uses the modulation property to shift it in frequency to a desired center frequency.

```
function [b]=fir(N,wc,wo,wind)
%
% FIR filter design using window method and
% frequency modulation
%
%  N : order of the FIR filter
%  wc : normalized cutoff frequency (between 0 and 1)
%  of low-pass prototype
%  wo : normalized center frequency (between 0 and 1)
%  of highpass, bandpass filters
%  wind : type of window function
%      1 : rectangular
%      2 : hanning
%      3 : hamming
%      4 : kaiser
%  [b] : coefficients of designed filter
%
%  USE:
%  [b]=fir(N,wc,wo,wind)
%
```

```
n=0:N;
if wind ==1
window=boxcar(N+1);
disp(' ***** RECTANGULAR WINDOW  *****')
elseif wind ==2
window=hanning(N+1);
disp(' *****HANNING WINDOW *****')
elseif wind == 3
window=hamming(N+1);
disp(' *****   HAMMING WINDOW *****')
else
window=kaiser(N+1,4.55);
disp(' *****   KAISER WINDOW *****')
end
% calculation of ideal impulse response
den=pi*(n-N/2);
num=sin(wc*den);
% if N even, this prevents 0/0
if fix(N/2)==N/2,
num(N/2+1)=wc;
den(N/2+1)=1;
end
b=(num./den).*window';
% frequency shifting
[H,w]=freqz(b,1,256);          % low-pass
if wo>0 & wo<1,
b=2*b.*cos(wo*pi*(n-N/2))/H(1);
elseif wo==0,
    b=b/abs(H(1));
elseif wo==1;
    b=b.*cos(wo*pi*(n-N/2));
end
```

MATLAB provides the function *fir1* to design FIR filters with the window method. As expected, the results are identical for either *fir1* and *fir*. The reason for writing *fir* is to simplify the code and to show how the modulation property can be used in the design of filters different from the low-pass case.

12.6 REALIZATION OF DISCRETE FILTERS

The realization of a discrete filter can be done in hardware or in software. In either case, the implementation of the transfer function $H(z)$ of a discrete filter requires delays, adders and constant multipliers as actual hardware or as symbolic components. Fig. 10.13 in Chapter 10 depicts the operation of each of these components as block diagrams.

In choosing a structure over another to realize a filter, two factors to consider are:

1. **Computational complexity** which relates to the number of operations (mainly multiplications and additions), but more importantly to the number of delays used. The aim is to obtain minimal realizations.
2. **Quantization effects** or the representation of filter parameters using finite length registers. The aim is to minimize quantization effects on parameters and on operations.

We will consider the computational complexity of the structures for minimum realizations, i.e., optimizing the number of delays used. The quantization effects are not considered.

12.6.1 REALIZATION OF IIR FILTERS

The structures commonly used to realize IIR filters are:

1. Direct form
2. Cascade
3. Parallel

The direct form represents the difference equation resulting from the transfer function of the IIR filter while attempting to minimize the number of delays. The cascade and parallel structures are based on the product or sum of first- and second-order filters to express the filter transfer function, which are in turn implemented using a direct form.

Direct Form Realization

Given the transfer function of a causal IIR filter

$$H(z) = \frac{Y(z)}{X(z)} = \frac{\sum_{k=0}^{M-1} b_k z^{-k}}{1 + \sum_{k=1}^{N-1} a_k z^{-k}} \qquad M \le N \tag{12.72}$$

where $Y(z)$ and $X(z)$ are the z-transforms of the output $y[n]$ and the input $x[n]$, the input/output relationship is given by the difference equation

$$y[n] = -\sum_{k=1}^{N-1} a_k y[n-k] + \sum_{k=0}^{M-1} b_k x[n-k]. \tag{12.73}$$

The direct form attempts to realize this equation with no more than $N-1$ delays. Assuming the input $x[n]$ is available, then $M-1$ delays are needed to generate the delayed inputs $\{x[n-k]\}$, $k = 1, \cdots, M-1$, and realizing the output components requires additional $N-1$ delays. Thus a direct realization requires $M+N-2$ delays for an $(N-1)$th-order difference equation. Such a realization is not minimal.

The direct form provides minimal realizations—the number of delays coincides with the order of the system—for filters with a constant numerator transfer function. Indeed, if

$$H(z) = \frac{Y(z)}{X(z)} = \frac{b_0}{1 + \sum_{k=1}^{N-1} a_k z^{-k}} \tag{12.74}$$

the input/output relationship is given by the difference equation

$$y[n] = -\sum_{k=1}^{N-1} a_k\, y[n-k] + b_0 x[n], \tag{12.75}$$

which only requires $N-1$ delays for the output, and none for the input. This is a minimum realization of $H(z)$ as only $N-1$ delays are needed.

If the numerator of the transfer function $H(z)$ of the filter we wish to realize is not a constant, but a polynomial

$$B(z) = \sum_{k=0}^{M-1} b_k z^{-k}$$

as in Equation (12.72), we need to factor $H(z)$ into a transfer function $B(z)$ and a constant numerator transfer function $1/A(z)$, where

$$A(z) = \sum_{k=0}^{N-1} a_k z^{-k}$$

is the denominator of $H(z)$. We then have

$$Y(z) = H(z)X(z) = B(z)\left[\frac{X(z)}{A(z)}\right]. \tag{12.76}$$

Defining an output $w[n]$ with $W(z) = X(z)/A(z)$, corresponding to the second term in the last equation, we obtain a constant-numerator IIR filter with transfer function

$$\frac{W(z)}{X(z)} = \frac{1}{A(z)} \;\Rightarrow\; w[n] = -\sum_{k=1}^{N-1} a_k w[n-k] + b_0 x[n]$$

or a difference equation which only requires $N-1$ delays for the output, and none for the input. The output $y[n]$ is then obtained from

$$Y(z) = B(z)W(z) \;\Rightarrow\; y[n] = \sum_{k=0}^{M-1} b_k w[n-k],$$

an input–output equation which uses the delayed signals $\{w[n-k]\}$ from above. Since the filter is causal $M \leq N$ and no additional delays are required. Thus the number of delays used corresponds to the order of the denominator $A(z)$, which is the order of the filter.

Example 12.16. Consider the transfer function

$$H(z) = \frac{1 + 1.5z^{-1}}{1 + 0.1z^{-1}}$$

of a first-order IIR filter. Find a minimal direct realization for it.

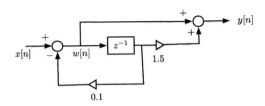

FIGURE 12.23

Minimal direct realization of $H(z) = (1 + 1.5z^{-1})/(1 + 0.1z^{-1})$.

Solution: The transfer function corresponds to a system with a first-order difference equation

$$y[n] = x[n] + 1.5x[n-1] - 0.1y[n-1],$$

so $M = N = 2$ and this equation can be realized with $M + N - 2 = 2$ delays—not a minimal realization.

To obtain a minimal realization we let

$$W(z) = \frac{X(z)}{1 + 0.1z^{-1}} \quad \Rightarrow \quad w[n] = x[n] - 0.1w[n-1],$$
$$Y(z) = (1 + 1.5z^{-1})W(z) \quad \Rightarrow \quad y[n] = w[n] + 1.5w[n-1],$$

which give the minimal direct realization shown in Fig. 12.23.

To obtain the transfer function from the minimal direct realization we need to obtain the transfer function corresponding to the constant numerator IIR filter first, then that of the FIR and use their Z-transforms. Thus,

$$w[n] = x[n] - 0.1w[n-1],$$
$$y[n] = w[n] + 1.5w[n-1].$$

If we replace the first equation into the second we obtain an expression containing $w[n]$ and $w[n-2]$ and $x[n]$ so that we cannot express $y[n]$ directly in terms of the input. Instead, the Z-transforms of the above equations are

$$(1 + 0.1z^{-1})W(z) = X(z),$$
$$Y(z) = (1 + 1.5z^{-1})W(z),$$

from which we obtain the transfer function $H(z)$. □

Since the cascade and the parallel realizations will connect first and second-order systems to realize a given transfer function, overall minimal realizations can be obtained by using minimal direct realiza-

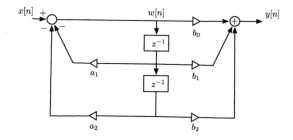

FIGURE 12.24

Minimal direct realization of first- and second-order filters (for the first-order filters let $a_2 = b_2 = 0$, eliminate the constant multipliers and the lower delay).

tions for first- and second-order filters. For a second-order filter with a general transfer function with constant coefficients

$$H_2(z) = \frac{b_0 + b_1 z^{-1} + b_2 z^{-2}}{1 + a_1 z^{-1} + a_2 z^{-2}} \tag{12.77}$$

a minimal direct realization is shown in Fig. 12.24. A minimal realization for a first-order filter is obtained from this by getting rid of the constant multipliers corresponding to b_2 and a_2 and the lower delay, corresponding to letting $b_2 = a_2 = 0$ when we obtain a general transfer function for a first-order system from $H_2(z)$.

To obtain the transfer function from a minimal direct realization, obtain the equations for the IIR and the FIR components and apply the Z-transform.

Cascade Realization

The cascade realization is obtained by representing the given transfer function $H(z) = B(z)/A(z)$ as a product of first- and second-order filters $H_i(z)$ with real coefficients:

$$H(z) = \prod_i H_i(z). \tag{12.78}$$

Each transfer function $H_i(z)$ is realized using the minimal direct form and cascaded. Different from the analog case, this cascade realization is not constrained by loading.

Example 12.17. Obtain a cascade realization of the filter with transfer function

$$H(z) = \frac{3 + 3.6z^{-1} + 0.6z^{-2}}{1 + 0.1z^{-1} - 0.2z^{-2}}.$$

Solution: The poles of $H(z)$ are $z = -0.5$ and $z = 0.4$ and the zeros $z = -1$ and $z = -0.2$ all of which are real. One way of obtaining the cascade realization is to express $H(z)$ as

$$H(z) = \left[\frac{3(1 + z^{-1})}{1 + 0.5z^{-1}} \right] \left[\frac{1 + 0.2z^{-1}}{1 - 0.4z^{-1}} \right].$$

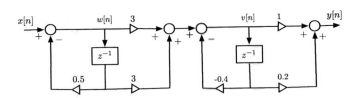

FIGURE 12.25

Cascade realization of $H(z) = (3 + 3.6z^{-1} + 0.6z^{-2})/(1 + 0.1z^{-1} - 0.2z^{-2})$.

If we let

$$H_1(z) = \frac{3(1 + z^{-1})}{1 + 0.5z^{-1}}, \quad H_2(z) = \frac{1 + 0.2z^{-1}}{1 - 0.4z^{-1}},$$

realizing each separately and then cascading them we obtain the realization shown in Fig. 12.25.
  It is also possible to express $H(z)$ as

$$H(z) = \underbrace{\left[\frac{1 + 0.2z^{-1}}{1 + 0.5z^{-1}}\right]}_{\hat{H}_1(z)} \underbrace{\left[\frac{3(1 + z^{-1})}{1 - 0.4z^{-1}}\right]}_{\hat{H}_2(z)},$$

which would give a different but equivalent realization of $H(z)$.
  Since loading is not applicable when cascading discrete filters, the product of the transfer functions always gives the overall transfer function. As LTI systems these realizations can be cascaded in different orders with the same result. □

Example 12.18. Obtain a cascade realization of

$$H(z) = \frac{1 + 1.2z^{-1} + 0.2z^{-2}}{1 - 0.4z^{-1} + z^{-2} - 0.4z^{-3}}.$$

Solution: The zeros of $H(z)$ are $z = -1$ and $z = -0.2$, while its poles are $z = \pm j$ and $z = 0.4$. We can thus rewrite $H(z)$ as

$$
\begin{aligned}
H(z) &= \left[\frac{1 + z^{-1}}{1 + z^{-2}}\right]\left[\frac{1 + 0.2z^{-1}}{1 - 0.4z^{-1}}\right] \\
&= \left[\frac{1 + 0.2z^{-1}}{1 + z^{-2}}\right]\left[\frac{1 + z^{-1}}{1 - 0.4z^{-1}}\right]
\end{aligned}
$$

where the complex conjugate poles give the denominator of the first filter. Realizing each of these components and cascading in any order would give different but equivalent representation of $H(z)$. Fig. 12.26 shows the realization of the top form of $H(z)$. □

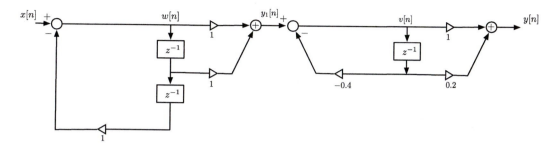

FIGURE 12.26

Cascade realization of $H(z) = [(1 + z^{-1})/(1 + z^{-2})][(1 + 0.2z^{-1})/(1 - 0.4z^{-1})]$.

Parallel Realization

In this case the given transfer function $H(z)$ is represented as a partial fraction expansion

$$H(z) = \frac{B(z)}{A(z)} = C + \sum_{i=1}^{r} H_i(z) \tag{12.79}$$

where C is a constant and the r filters $H_i(z)$ are first- or second-order with real coefficients that are implemented with the minimal direct realization.

The constant C in the expansion is needed when the numerator (in positive powers of z) is of larger or equal order than the denominator. If the numerator is of larger order than the denominator, the filter is noncausal. To illustrate this consider a first-order filter with a transfer function where the numerator is of second order (in terms of positive powers of z)

$$H(z) = \frac{Y(z)}{X(z)} = \frac{b_0 z^2 + b_1 z + b_2}{z + a_1} = \frac{b_0 z + b_1 + b_2 z^{-1}}{1 + a_1 z^{-1}}$$

where we multiplied the numerator and denominator by z^{-1} to be able to obtain the difference equation. The difference equation representing this system is

$$y[n] = -a_1 y[n-1] + b_0 x[n+1] + b_1 x[n] + b_2 x[n-1],$$

requiring a future input $x[n+1]$ to compute the present $y[n]$, i.e., corresponding to a noncausal filter. The cascade and parallel realizations are shown in Fig. 12.27.

Example 12.19. Let

$$H(z) = \frac{3 + 3.6z^{-1} + 0.6z^{-2}}{1 + 0.1z^{-1} - 0.2z^{-2}} = \frac{3z^2 + 3.6z + 0.6}{z^2 + 0.1z - 0.2}.$$

Obtain a parallel realization.

Cascade realization

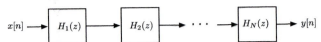

$H_i(z)$ *first- or second-order direct form II realization*

Parallel realization

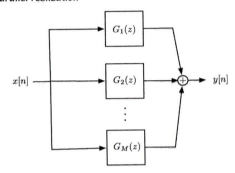

$G_i(z)$ *first- or second-order direct form II realization*

FIGURE 12.27

Cascade and parallel realizations of IIR filters.

Solution: The transfer function $H(z)$ is not proper rational, in either positive or negative powers of z, and the poles are $z = -0.5$ and $z = 0.4$. Thus the transfer function can be expanded as

$$H(z) = A_1 + \frac{A_2}{1 + 0.5z^{-1}} + \frac{A_3}{1 - 0.4z^{-1}}.$$

In this case we need A_1 because the numerator, in positive as well as in negative powers of z, is of the same order as the denominator. We then have

$$A_1 = H(z)|_{z=0} = -3,$$
$$A_2 = H(z)(1 + 0.5z^{-1})|_{z^{-1}=-2} = -1,$$
$$A_3 = H(z)(1 - 0.4z^{-1})|_{z^{-1}=2.5} = 7.$$

Letting

$$H_1(z) = \frac{-1}{1 + 0.5z^{-1}}, \quad H_2(z) = \frac{7}{1 - 0.4z^{-1}},$$

we obtain the parallel realization for $H(z)$ shown in Fig. 12.28. $\qquad\square$

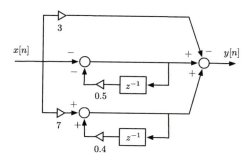

FIGURE 12.28

Parallel realization for $H(z) = (3 + 3.6z^{-1} + 0.6z^{-2})/(1 + 0.1z^{-1} - 0.2z^{-2})$.

12.6.2 REALIZATION OF FIR FILTERS

The realization of FIR filters can be done using direct and cascade forms. Since these filters are non-recursive, there is no way to implement FIR filters in parallel.

The direct realization of an FIR filter consists in realizing the input/output equation using delays, constant multipliers and summers. For instance, if the transfer function of an FIR filter is

$$H(z) = \sum_{k=0}^{M} b_k z^{-k}. \tag{12.80}$$

The Z-transform of the filter output can be written as $Y(z) = H(z)X(z)$ where $X(z)$ is the Z-transform of the filter input. In the time domain we have

$$y[n] = \sum_{k=0}^{M} b_k x(n-k),$$

which can be realized as shown in Fig. 12.29 for $M = 3$. Notice that M is the number of delays needed and that there are $M + 1$ taps, which has given the name of tapped filters to FIR filters realized this way.

The cascade realization of an FIR filter is based on the representation of $H(z)$ in (12.80) as a cascade of first- and second-order filters, i.e. we let

$$H(z) = \prod_{i=1}^{r} H_i(z),$$

where

$$
\begin{aligned}
H_i(z) &= b_{oi} + b_{1i}z^{-1} \quad \text{or} \\
H_i(z) &= b_{oi} + b_{1i}z^{-1} + b_{2i}z^{-2}.
\end{aligned}
$$

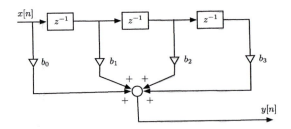

FIGURE 12.29

Direct form realization of FIR filter of order $M = 3$.

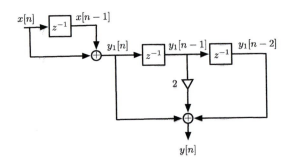

FIGURE 12.30

Cascade realization of FIR filter.

Example 12.20. Provide the cascade realization of an FIR filter with transfer function

$$H(z) = 1 + 3z^{-1} + 3z^{-2} + z^{-3}.$$

Solution: The transfer function is factored as

$$H(z) = (1 + 2z^{-1} + z^{-})(1 + z^{-1}),$$

which can be realized as the cascade of two FIR filters:

$$y_1[n] = x[n] + x[n-1],$$
$$y[n] = y_1[n] + 2y_1[n-1] + y_1[n-2],$$

which are realized as shown in Fig. 12.30.

12.7 TWO-DIMENSIONAL FILTERING OF IMAGES

The frequency content of an image is mostly composed of low-frequency components corresponding to the smooth structure of the image, while texture and edges correspond to the higher frequency components. The human visual system is sensitive to distortions caused by phase and by the smoothing of edges.

Because the high-frequency components in an image are related to the sharp details and edges of an object, low-pass filtering attenuates high-frequency components and passes the low-frequency components causing blurring of the image. Similarly, low-frequency components of an image corresponds to slowly-varying intensities of flat surfaces or slowly changing edges. As such, applying a high-pass filter to an image attenuates these low-frequency components enhancing the details and sharpening the edges corresponding to the high-frequency components of the image. Using a combination of low- and high-pass filters, one obtains "band-pass" filters that pass a selected frequency range between low and high frequencies, and "band-stop" filters that remove components in the selected frequency range.

There are significant differences in the approach to design and implement two-dimensional (2D) filters relative to one-dimensional (1D) filters:

- Different from the 1D design methods, including stability conditions in the design procedure for 2D filters is more difficult. This makes the design of non-recursive or FIR filters more attractive than that of recursive or IIR filters, despite the higher computational overhead due to the larger number of coefficients required by FIR filters.
- The design and the implementation of 1D filters are not coupled: the factorization of 1D polynomials permits different types of implementation for a given transfer function. In 2D, design and implementation must be coupled because of the lack of factorization. Thus, separable filters that are easy to design and implement are preferred, but the low-frequency concentration of most images makes circular filters (non-separable) more appropriate.
- The design methods for 2D IIR filters do not rely on the analog filter design methods, as in the 1D case.
- Because of the phase effects, linear or zero-phase filtering becomes an issue of interest in 2D more so that in 1D. This is another reason to prefer FIR filters which can be designed to have linear phase. A non-linear-phase filter distorts the proper registration of different frequency components that make up an image (lines, borders, etc.) thus distorting its shape in many ways.

12.7.1 SPATIAL FILTERING

Typical applications in image processing are de-noising, de-blurring and edge detection where commonly the processing is done in blocks using small size FIR filters. For the de-noising where the objective is to try to get rid of noise as much as possible to enhance the image[2] low-pass averaging filters are used. In the edge-detection application spatial filters that approximate a two-dimensional gradient to identify edges in the image are commonly used. It is important to notice that these filters have not been obtained via the conventional filter design used for one-dimensional filters. Such an approach will be covered later.

---

[2]The performance of the filtering depends on the type of noise present in the image. If it is additive noise with spectrum in a certain frequency band, deterministic linear filters in most cases do well, but whenever the noise has a broad spectrum—white noise—statistical approaches are needed. Likewise, non-linear filters are more effective than linear filters when the noise is a salt-and-pepper noise.

A 2D averaging filter is such that its output $y[m, n]$ is given by

$$y[m, n] = \frac{1}{MN} \sum_{k=0}^{M-1} \sum_{\ell=0}^{N-1} b_{k,\ell} x[m - k, n - \ell] \tag{12.81}$$

where $\{b_{k,\ell}\}$ are the coefficients of the filter and $x[m, n]$ is the input of the filter. If the coefficients are different from unity, the averaging is being done in a weighted way. A special case of a weighted averager is given by the Gaussian filter where the values of the impulse response are the samples of a bell-shaped curve

$$f(x, y) = A \exp\left(-\left[\frac{x^2}{2\sigma_x^2} + \frac{y^2}{2\sigma_y^2}\right]\right) \tag{12.82}$$

where σ_x and σ_y indicate the expansion of the curve about the origin.

Two special filters used for edge detection are the Sobel and the Prewitt filters. The Sobel filter is a second-order FIR filter which allows edge detection in either horizontal or vertical directions. The impulse responses for the horizontal, $h_{sh}[m, n]$, and the vertical $h_{sv}[m, n]$ are given by masks:

$$\mathbf{H_{sh}} = \begin{bmatrix} h_{sh}[0,0] & h_{sh}[0,1] & h_{sh}[0,2] \\ h_{sh}[1,0] & h_{sh}[1,1] & h_{sh}[1,2] \\ h_{sh}[2,0] & h_{sh}[2,1] & h_{sh}[2,2] \end{bmatrix} = \begin{bmatrix} 1 & 2 & 1 \\ 0 & 0 & 0 \\ -1 & -2 & -1 \end{bmatrix},$$

$$\mathbf{H_{sv}} = \begin{bmatrix} h_{sv}[0,0] & h_{sv}[0,1] & h_{sv}[0,2] \\ h_{sv}[1,0] & h_{sv}[1,1] & h_{sv}[1,2] \\ h_{sv}[2,0] & h_{sv}[2,1] & h_{sv}[2,2] \end{bmatrix} = \mathbf{H_{sh}}^T. \tag{12.83}$$

A similar second-order filter that approximates the gradient is the Prewitt filter, which has as impulse response masks

$$\mathbf{H_{ph}} = \begin{bmatrix} h_{ph}[0,0] & h_{ph}[0,1] & h_{ph}[0,2] \\ h_{ph}[1,0] & h_{ph}[1,1] & h_{ph}[1,2] \\ h_{ph}[2,0] & h_{ph}[2,1] & h_{ph}[2,2] \end{bmatrix} = \begin{bmatrix} 1 & 1 & 1 \\ 0 & 0 & 0 \\ -1 & -1 & -1 \end{bmatrix}, \quad \mathbf{H_{pv}} = \mathbf{H_{ph}}^T. \tag{12.84}$$

The parallel implementation of the horizontal and vertical filters giving outputs $y_h[m, n]$ and $y_v[m, n]$ can be combined to give

$$y[m, n] = \sqrt{y_h^2[m, n] + y_v^2[m, n]}, \tag{12.85}$$

an approximation of the magnitude of the gradient in two directions. To get the edges it might be necessary to threshold the filtered image with the $y[m, n]$ values.

An averaging filter smooths out or blurs an image. The blurring depends on the size and the type of filter, for instance using the MATLAB function *fspecial* an averaging and a Gaussian filter are available.

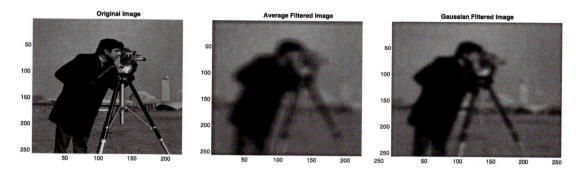

FIGURE 12.31

Filtering using average and Gaussian filters of size 15×15 and standard deviation of 3 for the Gaussian.

Example 12.21. Apply a 15×15 averaging filter and a 15×15 Gaussian filter, with a standard deviation of 3, to the *cameraman* image in MATLAB.

Solution: The MATLAB code for this example is

```
%%
% Example 12.21---Averaging and Gaussian low-pass filtering
%%
clear all;clf
h1=fspecial('average',15);          % averaging filter
h2=fspecial('gaussian',15,3);       % Gaussian filter
h1 = h1./ sum(h1(:));
h2 = h2./ sum(h2(:));
I = imread('cameraman.tif');I=double(I);
y1=filter2(h1,I);
y2=filter2(h2,I);
figure(1)
colormap('gray'); imagesc(I,[0 255]); title('Original Image')
figure(2)
colormap('gray'); imagesc(y1,[0 255]); title('Average Filtered Image')
figure(3)
colormap('gray'); imagesc(y2,[0 255]); title('Gaussian Filtered Image')
```

Fig. 12.31 shows the original *cameraman* image and the blurred images caused by an averaging filter of size 15×15, and a Gaussian filter of the same size with a standard deviation of 3. The de-blurring can be done by inverting the transfer function of the smoothing filter, giving a recursive filter typically of infinite dimensions. Approximating this inverse filter is done using a 2D IIR filter and dealing with the problem of stability.   □

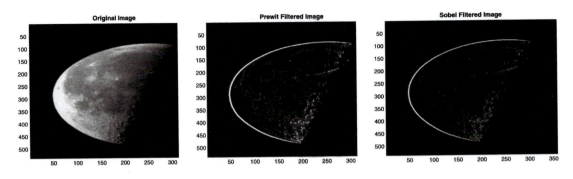

FIGURE 12.32

Filtering using Prewitt (middle) and Sobel (right) filters to detect edges of an image (left).

Example 12.22. Apply the Prewitt and Sobel horizontal and vertical filters for the edge detection of the *moon* image in MATLAB. Use Equation (12.85) to obtain the filtered image in the two directions.

Solution: We again use the function *fspecial* to obtain the Prewitt and the Sobel filters that approximate the gradient operator. The MATLAB code is given below:

```
%%
% Example 12.22---Edge Detection using Prewitt and Sobel
%%
clear all; clf
h1 =fspecial('prewit');     % Prewitt horizontal filter
h2 =fspecial('sobel');      % Sobel horizontal filter
h1 = h1./ max(h1(:));
h2 = h2./ max(h2(:));
I = imread('moon.tif');
y1h=filter2(h1,I); y1v=filter2(h1',I);
y2h=filter2(h2,I); y2v=filter2(h2',I);
y1=sqrt(y1h.^2+y1v.^2);
y2=sqrt(y2h.^2+y2v.^2);
figure(1)
colormap('gray'); imagesc(I,[0 255]); title('Original Image');
figure(2)
colormap('gray'); imagesc(y1,[0 255]); title('Prewitt Filtered Image');
figure(3)
colormap('gray'); imagesc(y2,[0 255]); title('Sobel Filtered Image');
```

Fig. 12.32 shows the original and the results of applying horizontally and vertically the Prewitt and Sobel filters. □

Depending on the type of noise, linear filters may not perform well. That is the case when *salt-and-pepper* noise[3] is added to an image. A non-linear filter such as the median filter will perform much better.

Example 12.23. Use the MATLAB function *imnoise* to add salt-and-pepper noise to the *peppers* image, and *medfilt2* to perform the median filtering on the noisy image. Compare the results of using 5×5 median and Gaussian filters.

Solution:

```
%%
% Example 12.23---Salt and Pepper noise reduction using median filtering
%%
clear all; clf
im1=imread('peppers.png');        % read in color image
im2=im2double(im1);               % convert into double precision
im3=rgb2gray(im2);                % convert color into gray level images
figure(1)
imshow(im3); title('Original image')      % show gray level image
im4=imnoise(im3,'salt & pepper', 0.3);    % add "salt and pepper" noise
figure(2)
imshow(im4); title('Noisy image')
h=fspecial('gaussian',[5 5],1);   % linear filter
im5=filter2(h,im4);               % filtering noisy image
figure(3)
imshow(im5); title('Filtered image with Gaussian filter')
im9=medfilt2(im4,[5 5]);          % 5x5 median filtering
figure(4)
imshow(im9); title('Filtered image with median')
```

Fig. 12.33 shows the original, the salt-and-pepper noise added images and the results obtained by using 5×5 Gaussian and medial filters. ☐

12.7.2 FREQUENCY DOMAIN FILTERING

Frequency selective filtering in two dimensions is an extension of the one-dimensional filtering. Given an input $x[m, n]$, for instance an image, and depending on the type of processing desired (e.g., denoising, edge detection, etc., of images) the filtering problem is to obtain a filter with an appropriate impulse response $h[m, n]$ or transfer function $H(z_1, z_2)$ to achieve the desired processing. Thus if we are interested in the smoothing of an image corrupted by noise, then we wish to low-pass filter the image so as to try to get rid of the noise on the flat or close-to-flat regions. Now, if we are interested in detecting the edges of an image, we could then consider designing a high-pass filter to enhance the

---

[3] Salt-and-pepper noise could have highest or lowest, or both, pixel values spread randomly on the image, giving the impression of white (salt) and black (pepper) specks.

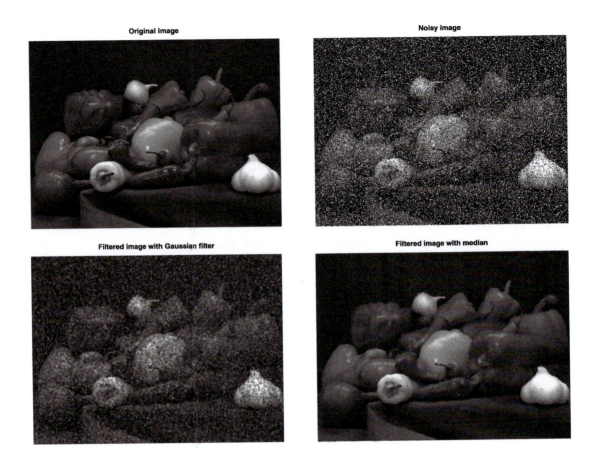

FIGURE 12.33

Comparison of linear and non-linear filtering for gray-level image (top left) distorted by salt-and-pepper noise (top right): Gaussian filtering (bottom left) and median filtering (bottom right).

edges and to filter out other components of the image. In a data-compression application, we would design a bank of filters[4] to separate an image into frequency-band components that could be represented more efficiently than the overall image. In all of these applications we are interested in designing stable filters that have real coefficients.

The lack of factorization of a desired transfer function $H_d(z_1, z_2)$ couples the design and the implementation of two-dimensional filters, thus the different implementations, e.g., cascade, parallel, available for 1D filters are not typically available for 2D filters. Likewise, the difficulty in imposing stability requirements in the design or in checking the stability of recursive filters, makes the non-recursive filter designs more practical than the design of recursive filters. In the following, we will

---

[4]A bank of filters is a combination of low-, band- and high-pass filters that has the magnitude response of an all-pass filter.

consider 2D filter design methods that guarantee stability and that can easily be implemented, thus putting emphasis on separable filter and non-recursive designs.

Separable filters, whether they are recursive or non-recursive, guarantee stability and implementation. However, they restrict the frequencies to those corresponding to vertical and horizontal filtering directions which are not necessarily the frequencies in an image. In separable filters with impulse response $h[m, n] = h_1[m]h_2[n]$, the output $y[m, n]$ of the filter with input $x[m, n]$ is obtained in two steps in the space domain

$$y_1[m, n] = h_1[m] * x[m, n] \quad \Rightarrow \quad y[m, n] = h_2[n] * y_1[m, n] \tag{12.86}$$

where $*$ is the convolution sum, or in two steps in the frequency domain computing the Z-transform of the above:

$$Y_1(z_1, z_2) = H_1(z_1)X(z_1, z_2) \quad \Rightarrow \quad Y(z_1, z_2) = H_2(z_2)Y_1(z_1, z_2). \tag{12.87}$$

Thus we have

$$y[m, n] = (h_1[m] * h_2[n]) * x[m, n] \quad \Rightarrow \quad Y(z_1, z_2) = \underbrace{H_1(z_1)H_2(z_2)}_{H(z_1, z_2)} X(z_1, z_2). \tag{12.88}$$

Given that $H_1(z_1)$ and $H_2(z_2)$ can be factored, they can be implemented as 1D filters and thus the filter with transfer function $H(z_1, z_2)$ is implementable. Also, if the two filters with transfer functions $H_1(z_1)$ and $H_2(z_2)$ are stable so is the two-dimensional filter with transfer function $H(z_1, z_2)$. The one-dimensional filters are specified and designed using the 1D methods covered before for recursive and non-recursive filters.

In general, the non-separable 2D FIR filter design is done using approaches similar to those in 1D. For instance, the 2D windowing method attempts to approximate a desirable impulse response $h_d[m, n]$, typically obtained from a desired frequency response

$$H_d(e^{j\omega_1}, e^{j\omega_2}) = |H_d(e^{j\omega_1}, e^{j\omega_2})|e^{j\Theta(\omega_1, \omega_2)} \tag{12.89}$$

by the inverse 2D discrete Fourier transform:

$$h_d[m, n] = \frac{1}{4\pi^2} \int_{-\pi}^{\pi} \int_{-\pi}^{\pi} H_d(e^{j\omega_1}, e^{j\omega_2})e^{-j(\omega_1 m + \omega_2 n)} d\omega_1 \, d\omega_2. \tag{12.90}$$

The desired magnitude response $|H_d(e^{j\omega_1}, e^{j\omega_2})|$ typically corresponds to one of the frequency selective types of filters, i.e., low-pass, band-pass, stop-band and high-pass, or a combination of these,[5] and $\Theta(\omega_1, \omega_2)$ is a desired linear or zero-phase response. Since it is quite possible that the resulting impulse response $h_d[m, n]$ does not have the desired finite support, we multiply it by real and symmetric windows $w[m, n]$. Just as in the 1D case, these windows are such that their frequency response

---

[5] In two dimensions it is possible to consider filters that are not related to these conventional filters, for instance a fan filter which hast the shape of a fan.

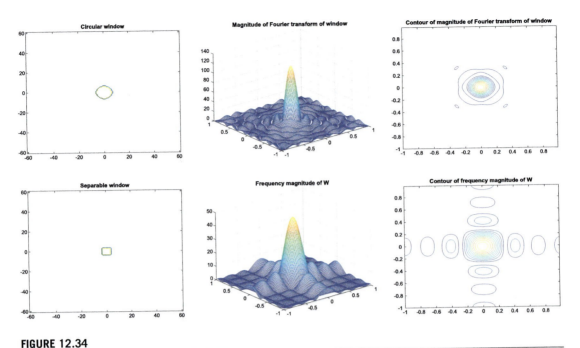

FIGURE 12.34

Magnitude responses (middle) and their contours (right) of circular (top) and separable (bottom) window with ideal contours shown on the left.

$W(e^{j\omega_1}, e^{j\omega_2})$ approximates an impulse function in the frequency domain. There are two typical ways to obtain these windows: as separable windows

$$w_R[m, n] = w_1[m]w_2[n] \tag{12.91}$$

or as circular windows

$$w_C[m, n] = w[\sqrt{m^2 + n^2}], \tag{12.92}$$

which are circular rotations of a one-dimensional window (see Fig. 12.34). The 1D windows used are those presented before in the one-dimensional filter design.

The frequency response of the resulting filter is given by

$$H(e^{j\omega_1}, e^{j\omega_2}) = \frac{1}{4\pi^2} \int_{-\pi}^{\pi} \int_{-\pi}^{\pi} H_d(e^{j\theta_1}, e^{j\theta_2}) W(e^{j(\omega_1-\theta_1)}, e^{j(\omega_2-\theta_2)}) d\theta_1 d\theta_2, \tag{12.93}$$

i.e., the convolution in the frequency domain of the desired frequency response $H_d(e^{j\omega_1}, e^{j\omega_2})$ and the discrete Fourier transform of the window $W(e^{j\omega_1}, e^{j\omega_2})$. Note that this discrete Fourier transform for the separable window is given by

$$W(e^{j\omega_1}, e^{j\omega_2}) = W_1(e^{j\omega_1})W_2(e^{j\omega_2}) \tag{12.94}$$

where $W_1(e^{j\omega_1})$ and $W_2(e^{j\omega_2})$ are the one-dimensional discrete Fourier transforms of the windows $w_1[m]$ and $w_2[n]$. If the frequency response of the desired filter $H_d(e^{j\omega_1}, e^{j\omega_2})$ is separable, the resulting filter will also be separable and in that case the filtering problem consists in designing two one-dimensional filters using the window method.

The window method, in either one dimension or two dimensions, is a trial-and-error method and several adjustment might be necessary to obtain the desirable filter. For instance, when choosing the window we need to consider that the discrete Fourier transform of a separable window has side lobes along the two frequencies (see bottom figure in Fig. 12.34) while the discrete Fourier transform of a circular window is circular with side lobes spreading circularly over all frequencies (see top figure in Fig. 12.34). Moreover, the smoothness of the window is given by the concentration of frequency components on the main lobe. Thus two-dimensional windows generated using Kaiser windows would be more appropriate than those generated using rectangular windows. The shape of the one-dimensional window affects the main-lobe and side-lobe behavior, while the size of the window affects mainly the energy concentration in the main lobe.

Example 12.24. Obtain using MATLAB:

a) a circular bandpass FIR filter with desired frequency response

$$H_d(e^{j\omega_1}, e^{j\omega_2}) = \begin{cases} 1 & 0.4 \leq \sqrt{\omega_1^2 + \omega_2^2} \leq 0.5, \\ 0 & \text{otherwise in } [-\pi, \pi] \times [-\pi, \pi], \end{cases}$$

b) a high-pass FIR circular filter with desired frequency response

$$H_d(e^{j\omega_1}, e^{j\omega_2}) = \begin{cases} 1 & \sqrt{\omega_1^2 + \omega_2^2} \geq 0.5, \\ 0 & \text{otherwise in } [-\pi, \pi] \times [-\pi, \pi]. \end{cases}$$

Use a Kaiser window with $\beta = 4$ and length of $N = 125$.

Solution: The following MATLAB code can be used to design different types of circular FIR filters using three different windows. In this case we will design band-pass and high-pass filters using the Kaiser window.

```
%%
% Example 12.24---Circular filter design using fwind1
%%
clear all; clf
% filter specifications
N=125; [f1,f2] = freqspace(N,'meshgrid');  Hd = ones(N);  r = sqrt(f1.^2 + f2.^2);
% type of filter
tf=input('type of filter, 1=lp,2=bp,3=hp >>>')
if tf==1,
    Hd(r>0.5)=0; %low-pass
elseif tf==2,
    Hd((r<0.4) | (r>0.5)) = 0; %band-pass
```

```
else
     Hd(r<0.5)=0; %high-pass
end
% type of window
tw=input('type of window 1=hamming, 2=kaiser >>>')
if tw==1,
        h1 = fwind1(Hd,hamming(N));
else
        h1=fwind1(Hd,kaiser(N,4));
end
H=freqz2(h1,124); w1=-1:2/124:1-2/124; w2=w1;
figure(1)
contour(f1,f2,Hd,20); grid; title('Frequency response of desired filter')
figure(2)
mesh(w1,w2,H); title('Frequency response of designed filter')
figure(3)
contour(w1,w2,H); grid; title('Contour of frequency response of designed filter')
ww=fftshift(fft2(Hd));
```

Magnitude responses of designed circular band- and high-pass filters are shown in Fig. 12.35.   □

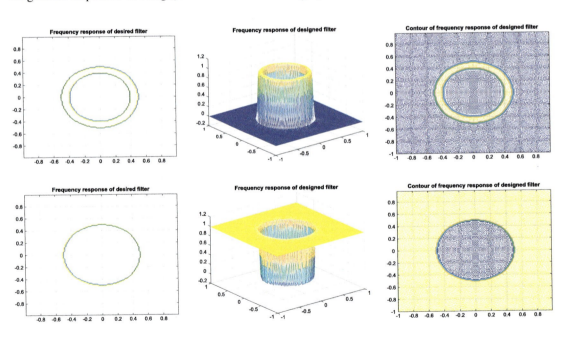

FIGURE 12.35

Contour of desired magnitude response of band- and high-pass filters (left). Magnitude response of designed 2D FIR filters using circular windows (middle) and their corresponding contours (left).

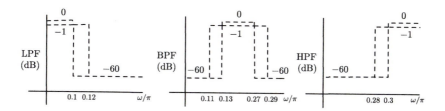

FIGURE 12.36

Magnitude specifications in dB for the three components of a 1D bank of filters: low-, band- and high-pass filters. Notice that the scale of the magnitude specification for the passband are distorted to be able to show it.

Example 12.25. In this example, we first obtain separable low-pass, band-pass, and high-pass filters. The magnitude specifications for the one-dimensional filters in dB are shown in Fig. 12.36. The three 1D filters cover the whole one-dimensional frequency range and thus any one-dimensional signal passed through these filters would give components that when added would closely approximate the original signal, that is, this 1D bank of filters is approximately an all-pass filter. The design of separable filters is done using the one-dimensional functions *designfilt* available in MATLAB. After designing the 1D FIR filters using the equiripple method, we implement the 2D separable filters and use them to separate the *cameraman* image into its frequency components. Show the magnitude response of the 2D bank of filters and the resulting data compressed image. Explain your results.

Solution: The following routine reads in the original image and designs the three separable filters. Although we will only show the final result, it would be of interest for you to see the different results.

```
%%
% Example 12.25---Design of bank of filter with separable filters
%%
% design of separable filters and filtering
clear all; clf
for in=1:3,
    I = imread('cameraman.tif');I=double(I);
    figure(1)
    colormap('gray'); imagesc(I,[0 255]); title('Original Image')
    if(in==1)
    lp = designfilt('lowpassfir', 'PassbandFrequency', 0.1,...
            'StopbandFrequency', 0.12, 'PassbandRipple',1, ...
            'StopbandAttenuation',60, 'DesignMethod', 'equiripple');
    fvtool(lp)
    h1=lp.Coefficients;
    h1=h1'*h1;
    figure(2)
    H1=fftshift(abs(fft2(h1,124,124)));
    contour(H1,10)
    y1=filter2(h1,I);
```

```
        figure(3)
        colormap('gray'); imagesc(y1,[0 255]); title('Low-pass Image')
        elseif(in==2)
        bp=designfilt('bandpassfir', 'PassbandFrequency1',0.13,...
                'StopbandFrequency1',0.11,'PassbandFrequency2', 0.27,...
                'StopbandFrequency2',0.29, 'PassbandRipple',1, ...
                'StopbandAttenuation1',60, 'DesignMethod', 'equiripple');
        fvtool(bp)
        h2=bp.Coefficients;
        h2=h2'*h2;
        figure(4)
        H2=fftshift(abs(fft2(h2,124,124)));
        contour(H2,20)
        y2=filter2(h2,I);
        figure(5)
        colormap('gray'); imagesc(y2,[0 255]); title('Band-pass Image')
        else
        hp= designfilt('highpassfir', 'PassbandFrequency', 0.30,...
                'StopbandFrequency', 0.28, 'PassbandRipple',1, ...
                'StopbandAttenuation',60, 'DesignMethod', 'equiripple');
        fvtool(hp)
        h3=hp.Coefficients;
        h3=h3'*h3;
           figure(6)
           H3=fftshift(abs(fft2(h3,124,124)));
           mesh(H3)
        y3=filter2(h3,I);
        figure(7)
        colormap('gray'); imagesc(y3,[0 255]); title('high-pass Image')
        end
end
% reconstructed image and magnitude response of bank of filters
figure(8)
colormap('gray'); imagesc(y1+y2+y3,[0 255]); title('reconstructed Image')
figure(9)
contour(H1+H2+H3)
```

The final results are shown in Fig. 12.37. It becomes clear, from the addition of the contour of the three filters that this bank of filters is not an all-pass filter, that it filters out components in certain bands so that the reconstructed image does not approximate the original (see middle figure in Fig. 12.37). A different approach is needed for the 2D bank of filters to approximate an all-pass filter. ☐

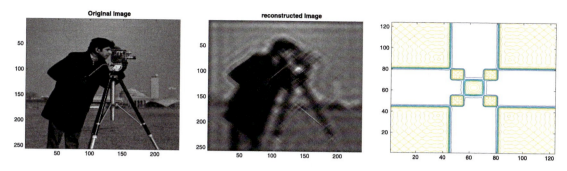

FIGURE 12.37

Original and reconstructed images, contour of the frequency response of the bank of filters composed of the low-, band- and high-pass filters specified; the white regions are filtered out.

12.8 WHAT HAVE WE ACCOMPLISHED? WHERE DO WE GO FROM HERE?

In Chapter 7 and in this chapter you have been introduced to the most important application of linear time-invariant systems: filtering. The design and realization of analog and discrete filters gathers many practical issues in signals and systems. If you pursue this topic, you will see the significance, for instance, of passive and active elements, feedback and operational amplifiers, reactance functions and frequency transformation in analog filtering. The design and realization of discrete filters brings together interesting topics such as quantization error and its effect on the filters, optimization methods for filter design, stabilization of unstable filters, finite register effects in the implementation of filters, etc.

Filtering is a very important tool in image processing, as you saw it can be used to de-noise, de-blur and to compress images. Due to complexity of recursive filters and the difficulty in satisfying stability conditions, 2D FIR filters are more commonly applied. The design of separable filters uses the one-dimensional theory, but circular filters require a new theory.

If you pursue filtering deeper, you will find that there is a lot more on filter design that what we have provided you in this chapter [8,3,19]. Also remember that MATLAB gives you a large number of functions to design and implement filters in one and two dimensions. It also has special functions for image processing.

12.9 PROBLEMS
12.9.1 BASIC PROBLEMS

12.1 Inputs to an ideal low-pass filter with frequency response

$$H(e^{j\omega}) = \begin{cases} 1 \, e^{-j10\omega} & -\frac{\pi}{2} \leq \omega \leq \frac{\pi}{2}, \\ 0 & \text{else in } -\pi \leq \omega \leq \pi, \end{cases}$$

are given below. Find the corresponding outputs $y_i[n]$, $i = 1, 2, 3$.

(a) $x_1[n] = 1 + \cos(0.3\pi n) + 23\sin(0.7\pi n)$, $-\infty < n < \infty$.

(b) $x_2[n] = \delta[n] + 2\delta[n-4]$.

(c) $x_3[n]$ has DTFT $X(e^{j\omega}) = H(e^{j\omega/2})$.

Answers: $y_1[n] = 1 - \cos(0.3\pi n)$; $y_2[n] = h[n] + 2h[n-4]$, $h[n]$ impulse response of filter.

12.2 The impulse response $h[n]$ of an FIR is given by $h[0] = h[3]$, $h[1] = h[2]$, the other values are zero.

(a) Find the Z-transform of the filter $H(z)$ and the frequency response $H(e^{j\omega})$.

(b) Let $h[0] = h[1] = h[2] = h[3] = 1$. Determine the zeros of the filter. Determine if the phase $\angle H(e^{j\omega})$ of the filter is linear? Explain.

(c) Under what conditions on the values of $h[0]$ and $h[1] - h[0]$ would the phase be linear.

(d) For linear phase, in general, where should the zeros of $H(z)$ be? Explain.

Answers: $H(e^{j\omega}) = e^{-j1.5\omega}(2h[0]\cos(1.5\omega) + 2h[1]\cos(0.5\omega))$; if $h[1] > 3h[0]$ phase is linear.

12.3 The impulse response of an FIR filter is $h[n] = \alpha\delta[n] + \beta\delta[n-1] + \alpha\delta[n-2]$, $\alpha > 0$ and $\beta > 0$.

(a) Determine the value of α and β for which this filter has a dc gain $|H(e^{j0})| = 1$, and linear phase $\angle H(e^{j\omega}) = -\omega$.

(b) For the smallest possible β and the corresponding α obtained above find the zeros of the filter, plot them in the Z-plane and indicate their relation. Generalize the relation of the zeros for all possible values of β and corresponding α, and find a general expression for the two zeros.

Answers: $H(e^{j\omega}) = e^{-j\omega}[\beta + 2\alpha\cos(\omega)]$; $\beta \geq 1/2$.

12.4 An FIR filter has a system function $H(z) = 0.05z^2 + 0.5z + 1 + 0.5z^{-1} + 0.05z^{-2}$.

(a) Find the magnitude $|H(e^{j\omega})|$ and phase response $\angle H(e^{j\omega})$ at frequencies $\omega = 0$, $\pi/2$ and π. Sketch each of these responses for $-\pi \leq \omega < \pi$, and indicate the type of filter.

(b) Determine the impulse response $h[n]$ and indicate if the filter is causal.

(c) If $H(z)$ is noncausal, how would you make it causal? What would be the effect of your procedure on the magnitude and the phase responses obtained before? Explain and plot the magnitude and phase of your causal filter.

Answers: $H(e^{j\omega}) = 1 + \cos(\omega) + 0.1\cos(2\omega)$; the filter is made causal by delaying $h[n]$ two samples.

12.5 The transfer function of an IIR filter is

$$H(z) = \frac{z+1}{z(z-0.5)}.$$

(a) Calculate the impulse response $h[n]$ of the filter.

(b) Would it be possible for this filter to have linear phase? Explain.

(c) Sketch the magnitude response $|H(e^{j\omega})|$ using a plot of the poles and zeros of $H(z)$ in the Z-plane. Use vectors to calculate the magnitude response.

Answers: $h[n] = 0.5^{n-1}u[n-1] + 0.5^{n-2}u[n-2]$; no linear phase possible.

12.6 The transfer function of an IIR filter is

$$H(z) = \frac{(z+2)(z-2)}{(z+0.5)(z-0.5)}.$$

Find the magnitude response of this filter at $\omega = 0$, $\omega = \pi/2$ and $\omega = \pi$. From the poles and the zeros of $H(z)$ find geometrically the magnitude response and indicate the type of filter.
Answer: $|H(e^{j\omega})| = 4$ for all ω, i.e., all-pass filter.

12.7 A second-order analog Butterworth filter has a transfer function

$$H(s) = \frac{1}{s^2 + \sqrt{2}s + 1}.$$

(a) Is the half-power frequency of this filter $\Omega_{hp} = 1$ rad/s?
(b) To obtain a discrete Butterworth filter we choose the bilinear transformation

$$s = K\frac{1 - z^{-1}}{1 + z^{-1}}, \quad K = 1,$$

what is the half-power frequency of the discrete filter?
(c) Find the transfer function $H(z)$ when we use the above bilinear transformation.
(d) Plot the poles and zeros of $H(s)$ and $H(z)$. Are both of these filters BIBO stable?
Answers: $|H(j1)| = \frac{1}{\sqrt{2}}|H(j0)|$; $\omega_{hp} = \pi/2$ rad.

12.8 A first-order low-pass analog filter has a transfer function $H(s) = 1/(s+1)$.
(a) If for this filter, the input is $x(t)$ and the output is $y(t)$ what is the ordinary differential equation representing this filter.
(b) Suppose that we change this filter into a discrete filter using the bilinear transformation

$$s = K\frac{1 - z^{-1}}{1 + z^{-1}}, \quad K = \frac{2}{T_s}.$$

Obtain the transfer function $H(z)$. If for the discrete filter, the input is $x[n]$ and the output $y[n]$ obtain the difference equation representing the discrete filter.
(c) Suppose $x(t) = u(t) - u(t - 0.5)$ find the output of the analog filter.
(d) Let $K = 1000$, so that $T_s = 2/K$ is used to sample $x(t)$ to get the discrete signal $x[n]$. Use the difference equation to solve for the output $y[n]$. Compare your result with the one obtained by solving the ordinary differential equation for the first three values.
Answers: $y(t) = (1 - e^{-t})u(t) + (1 - e^{-(t-0.5)})u(t - 0.5)$; $y[0] = 0.000999$, $y[1] = 0.0030$.

12.9 Given the discrete IIR filter realization shown in Fig. 12.38 where G is a gain value.
(a) Determine the difference equation that corresponds to the filter realization.
(b) Determine the range of values of the gain G so that the given filter is BIBO stable and has complex conjugate poles.
Answer: for complex conjugate poles and for the system to be BIBO stable $1/4 < G < 1$.

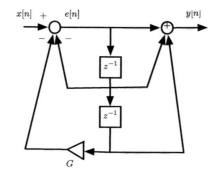

FIGURE 12.38

Problem 12.9.

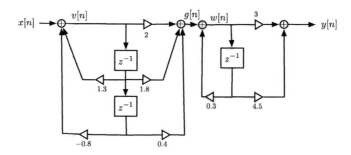

FIGURE 12.39

Problem 12.10: IIR realization.

12.10 Given the realization in Fig. 12.39. Obtain:
    (a) the difference equations relating $g[n]$ to $x[n]$ and $g[n]$ to $y[n]$,
    (b) the transfer function $H(z) = Y(z)/X(z)$ for this filter.
Answer: $g[n] = 1.3g[n-1] - 0.8g[n-2] + 2x[n] + 1.8x[n-1] + 0.4x[n-2]$.

12.9.2 PROBLEMS USING MATLAB

12.11 FIR and IIR filters: causality and zero phase—Let the filter $H(z)$ be the cascade of a causal filter with transfer function $G(z)$ and an anticausal filter with transfer function $G(z^{-1})$, so that

$$H(z) = G(z)G(z^{-1}).$$

(a) Suppose that $G(z)$ is an FIR filter with transfer function

$$G(z) = \frac{1}{3}(1 + 2z^{-1} + z^{-2}).$$

Find the frequency response $H(e^{j\omega})$ and determine its phase.

(b) Determine the impulse response of the filter $H(z)$. Is $H(z)$ a causal filter? If not, would delaying its impulse response make it causal? Explain. What would be the transfer function of the causal filter.

(c) Use MATLAB to verify the unwrapped phase of $H(z)$ you obtained analytically, and to plot the poles and zeros of $H(z)$.

(d) How would you use the MATLAB function *conv* to find the impulse response of $H(z)$.

(e) Suppose then that $G(z) = 1/(1 - 0.5z^{-1})$, find the filter $H(z) = G(z)G(z^{-1})$. Is this filter of zero-phase type? If so, where are its poles and zeros located? If you think of filter $H(z)$ as causal, is it BIBO stable?

Answers: $H(e^{j\omega})$ has zero phase; $h_1[n] = h[n - 2]$ corresponds to causal filter.

12.12 FIR and IIR filters: symmetry of impulse response and linear-phase—Consider two FIR filters with transfer functions

$$H_1(z) = 0.5 + 0.5z^{-1} + 2.2z^{-2} + 0.5z^{-3} + 0.5z^{-4},$$
$$H_2(z) = -0.5 - 0.5z^{-1} + 0.5z^{-3} + 0.5z^{-4}.$$

(a) Find the impulse responses $h_1[n]$ and $h_2[n]$ corresponding to $H_1(z)$ and $H_2(z)$. Plot them carefully and determine the sample with respect to which these impulse responses are even or odd.

(b) Show frequency response for $G(z) = z^2 H_1(z)$ is zero phase, and from it determine the phase of $H_1(e^{j\omega})$. Use MATLAB to find the unwrapped phase of $H_1(e^{j\omega})$ and confirm your analytic results.

(c) Find the phase of $H_2(e^{j\omega})$ by finding the phase of the frequency response for $F(z) = z^2 H_2(z)$. Use MATLAB to find the unwrapped phase of $H_2(e^{j\omega})$. Is it linear?

(d) If $H(z)$ were the transfer function of an IIR filter, according to the above arguments could it be possible for it to have linear phase? Explain.

Answer: $G(e^{j\omega}) = \cos(2\omega) + \cos(\omega) + 2.2$; $F(e^{j\omega}) = -j(\sin(2\omega) + \sin(\omega))$.

12.13 Butterworth vs. Chebyshev specifications—A Butterworth low-pass discrete filter of order N has been designed to satisfy the following specifications:

$$\text{Sampling period } T_s = 100 \ \mu s,$$
$$\alpha_{max} = 0.7 \ dB \quad \text{for } 0 \le f \le f_p = 1000 \ Hz,$$
$$\alpha_{min} = 10 \ dB \quad \text{for } f_{st} = 1200 \le f \le f_s/2 \ Hz.$$

What should be the new value of the stop-band frequency f_{st} so that an Nth-order Chebyshev low-pass filter satisfies the design specifications for $T_s, \alpha_{max}, \alpha_{min}$, and f_p?

Answer: If we choose $f_{st} = 1035$ we have $N_b = N_c = 10$.

12.14 Warping effect of the bilinear transformation—The non-linear relation between the discrete frequency ω (rad) and the continuous frequency Ω (rad/s) in the bilinear transformation causes warping in the high frequencies. To see this consider the following:

(a) Use MATLAB to design a Butterworth analog bandpass filter of order $N = 12$ and with half-power frequencies $\Omega_1 = 10$ and $\Omega_2 = 20$ (rad/s). Use MATLAB *bilinear*, with $K = 1$, to transform the resulting filter into a discrete filter. Plot the magnitude and the phase of the discrete filter. Plot the poles and zeros of the continuous and the discrete filters.

(b) Increase the order of the filter to $N = 14$ and keep the other specifications the same. Design an analog bandpass filter and use again *bilinear*, with $K = 1$, to transform the analog filter into a discrete filter. Plot the magnitude and the phase of the discrete filter. Plot the poles and zeros of the continuous and the discrete filters. Explain your results.

12.15 Warping effect of the bilinear transformation on phase—The warping effect of the bilinear transformation also affects the phase of the transformed filter. Consider a filter with transfer function $G(s) = e^{-5s}$.

(a) Find the transformed discrete frequencies ω (rad) corresponding to $0 \le \Omega \le 20$ (rad/s) using a bilinear transformation with $K = 1$. Plot Ω vs. ω.

(b) Discretize the continuous frequencies $0 \le \Omega \le 20$ (rad/s) to compute values of $G(j\Omega)$ and use MATLAB functions to plot the phase of $G(j\Omega)$.

(c) Find the function

$$H(e^{j\omega}) = G(j\Omega)|_{\Omega = \tan(\omega/2)},$$

and plot its unwrapped phase using MATLAB for the discrete frequencies corresponding to the analog frequencies to $0 \le \Omega \le 20$ (rad/s). Compare the phases of $G(j\Omega)$ and of $H(e^{j\omega})$.

Answer: $H(e^{j\omega}) = 1e^{-j5\tan(\omega)/2}$ has a non-linear phase.

12.16 All-pass IIR filter—Consider an all-pass analog filter

$$G(s) = \frac{s^4 - 4s^3 + 8s^2 - 8s + 4}{s^4 + 4s^3 + 8s^2 + 8s + 4}.$$

(a) Use MATLAB functions to plot the magnitude and phase responses of $G(s)$. Indicate whether the phase is linear.

(b) A discrete filter $H(z)$ is obtained from $G(s)$ by the bilinear transformation. By trial and error, find the value of K in the bilinear transformation so that the poles and zeros of $H(z)$ are on the imaginary axis of the Z-plane. Use MATLAB functions to do the bilinear transformation and to plot the magnitude and unwrapped phase of $H(z)$ and its poles. Is it an all-pass filter? If so, why?

(c) Let the input to the filter $H(z)$ be $x[n] = \sin(0.2\pi n)$, $0 \le n < 100$ and the corresponding output be $y[n]$. Use MATLAB functions to compute and plot $y[n]$. From these results would you say that the phase of $H(z)$ is approximately linear? Explain why or why not.

Answers: All-pass filter, $K = 1.4$, the phase is approximately linear.

12.17 Butterworth, Chebyshev, and elliptic filters—The gain specifications of a filter are

$$-0.1 \le 20\log_{10}|H(e^{j\omega})| \le 0 \text{ (dB)} \qquad 0 \le \omega \le 0.2\pi,$$
$$20\log_{10}|H(e^{j\omega})| \le -60 \text{ (dB)} \qquad 0.3\pi \le \omega \le \pi.$$

(a) Find the loss specifications for this filter.

(b) Design using MATLAB a Butterworth, a Chebyshev (using *cheby1*), and an elliptic filter. Plot in one plot the magnitude response of the three filters, compare them, and indicate which gives the lowest order.

Answers: The dc loss is 0 dB, $\alpha_{max} = 0.1$, and $\alpha_{min} = 15$ dB.

12.18 IIR comb filters—Consider a filter with transfer function

$$H(z) = K \frac{1 + z^{-4}}{1 + (1/16)z^{-4}}.$$

(a) Find the gain K so that this filter has unit dc gain. Use MATLAB to find and plot the magnitude response of $H(z)$, and its poles and zeros. Why is it called a comb filter?

(b) Use MATLAB to find the phase response of the filter $H(z)$. Why is it that the phase seems to be wrapped and it cannot be unwrapped by MATLAB?

(c) Suppose you wish to obtain an IIR comb filter that is sharper around the notches of $H(z)$ and flatter in between notches. Implement such a filter using the function *butter* to obtain two notch filters of order 10 and appropriate cut-off frequencies. Decide how to connect the two filters. Plot the magnitude and phase of the resulting filter, and its poles and zeros.

Answers: Zeros $z_k = e^{j(2k+1)\pi/4}$, poles $z_k = 0.5e^{jk\pi/4}$, $k = 0, \cdots, 3$.

12.19 Three-band discrete spectrum analyzer—To design a three-band discrete spectrum analyzer for audio signals, we need to design a low-pass, a band-pass and a high-pass IIR filters. Let the sampling frequency be $F_s = 10$ kHz. Consider the three bands, in kHz, to be $[0 \quad F_s/4]$, $(F_s/4 \quad 3F_s/8]$ and $(3F_s/8 \quad F_s/2]$. Consider the following two approaches:

(a) Let all the filters be of order $N = 20$, and choose the cut-off frequencies so that the sum of the tree filters is an all-pass filter of unit gain.

(b) Consider the MATLAB test signal *handel*, use the designed spectrum analyzer to obtain the spectrum in the three bands.

12.20 FIR filter design with different windows—Design a causal low-pass FIR digital filter with $N = 21$. The desired magnitude response of the filter is

$$|H_d(e^{jwT})| = \begin{cases} 1 & 0 \le f \le 250 \text{ Hz}, \\ 0 & \text{elsewhere in } 0 \le f \le (f_s/2), \end{cases}$$

and the phase is zero for all frequencies. The sampling frequency $f_s = 2000$ Hz.

(a) Use a rectangular window in your design. Plot magnitude and phase of the designed filter.

(b) Use a triangular window in the design and compare the magnitude and phase plots of this filter with those obtained in part (a).

Answers: $h_d[0] = 0.25$, $h_d[n] = \sin(\pi n/4)/(\pi n)$, $n \ne 0$.

12.21 FIR filter design—Design an FIR low-pass filter with a cut-off of $\pi/3$ and lengths $N = 21$ and $N = 81$,

(a) using a rectangular window,

(b) using Hamming and Kaiser ($\beta = 4.5$) windows. Compare the magnitudes of the resulting filters.

Answers: $h_d[0] = 0.33$, $h_d[n] = \sin(\pi n/3)/(\pi n)$, $n \ne 0$.

12.22 Down-sampling transformations—Consider down-sampling the impulse response $h[n]$ of a filter with transfer function $H(z) = 1/(1 - 0.5z^{-1})$.

(a) Use MATLAB to plot $h[n]$ and the down-sampled impulse response $g[n] = h[2n]$.

(b) Plot the magnitude responses corresponding to $h[n]$ and $g[n]$ and comment on the effect of the down-sampling. Is $G(e^{j\omega}) = 0.5H(e^{j\omega/2})$? Explain.

Answers: $g[n] = 0.25^n u[n]$; $G(e^{j\omega}) \neq 0.5H(e^{j\omega/2})$.

12.23 Modulation property transformation—Consider a moving average, low-pass, FIR filter

$$H(z) = \frac{1 + z^{-1} + z^{-2}}{3}.$$

(a) Use the modulation property to convert the given filter into a high-pass filter.

(b) Use MATLAB to plot the magnitude responses of the low-pass and the high-pass filters.

Answer: $G(e^{j\omega}) = e^{-j\omega}(2\cos(\omega) - 1)/3$.

12.24 Implementation of IIR rational transformation—Use MATLAB to design a Butterworth second-order low-pass discrete filter $H(Z)$ with half-power frequency $\theta_{hp} = \pi/2$, and dc gain of 1. Consider this low-pass filter a prototype that can be used to obtain other filters. Implement, using MATLAB, the frequency transformations $Z^{-1} = N(z)/D(z)$ using the convolution property to multiply polynomials to obtain:

(a) A high-pass filter with a half-power frequency $\omega_{hp} = \pi/3$ from the low-pass filter.

(b) A band-pass filter with $\omega_1 = \pi/2$ and $\omega_2 = 3\pi/4$ from the low-pass filter.

(c) Plot the magnitude of the low-pass, high-pass and band-pass filters.

Give the corresponding transfer functions for the low-pass and the high-pass and the band-pass filters.

Answer: Use the *conv* function to implement multiplication of polynomials.

12.25 Parallel connection of IIR filters—Use MATLAB to design a Butterworth second-order low-pass discrete filter with half-power frequency $\theta_{hp} = \pi/2$, and dc gain of 1, call it $H(z)$. Use this filter as a prototype to obtain a filter composed of a parallel combination of the following filters.

(a) Assume that we up-sample by $L = 2$ the impulse response $h(n)$ of $H(z)$ to get a new filter $H_1(z) = H(z^2)$. Determine $H_1(z)$, plot its magnitude using MATLAB and indicate the type of filter.

(b) Assume then that we shift $H(z)$ by $\pi/2$ to get a bandpass filter $H_2(z)$. Find the transfer function of $H_2(z)$ from $H(z)$, plot its magnitude and indicate the type of filter.

(c) If the filters $H_1(z)$ and $H_2(z)$ are connected in parallel, what is the overall transfer function $G(z)$ of the parallel connection? Plot the magnitude response corresponding to $G(z)$.

Answer: $H_1(z)$ is a band-eliminating filter and $H_2(z)$ is a band-pass filter.

12.26 Image blurring—An image can be blurred by means of a Gaussian filter, which has an impulse response

$$h[m, n] = \frac{e^{-(m^2 + n^2)}}{4\sigma^4} \qquad -3\sigma \leq m \leq 3\sigma, \ -3\sigma \leq n \leq 3\sigma.$$

(a) Is $h[m, n] = h_1[m]h_2[n]$, i.e., is this a separable case? If so, determine the 1D filters $h_1[m]$ and $h_2[n]$. Plot these functions for the parameter $\sigma = 1$ and $\sigma = 3$ what is the difference?

(b) Find the DFT of $h[m, n]$ for $\sigma = 1$ and $\sigma = 3$ and plot its magnitude response. Use the functions *mesh* and *contour* to display the characteristics of the Gaussian filter (normalize the frequencies (ω_1, ω_2) to $[-1, 1] \times [-1, 1]$). What type of filter is this?

(c) To see the effect of blurring read in the test image *cameraman.tif* and convolve it with $h[m, n]$ for integer values $1 \leq \sigma \leq 10$ using the 2D-FFT and 2D-IFFT. Use the function *imagesc* to display the resulting images. Pause the resulting image to see the effects of blurring. Why, besides the blurring, do the resulting images seem to be moving down towards the right? Explain.

Answers: The filter is separable, $h_1[m] = e^{-m^2}/(2\sigma^2)$ and $h_2[n] = h_1[n]$.

12.27 Edge detection—Consider the impulse response

$$
\begin{aligned}
h_u[m, n] &= \delta[m, n] - \delta[m - 2, n] + 2\delta[m, n - 1] - 2\delta[m - 2, n - 1] \\
&\quad + \delta[m, n - 2] - \delta[m - 2, n - 2], \\
h_c[m, n] &= h_u[n, m].
\end{aligned}
$$

For a given input $x[m, n]$ the output of an edge detector is

$$
y_1[m, n] = (h_u * x)[m, n], \quad y_2[m, n] = (h_c * x)[m, n],
$$

$$
y[m, n] = \sqrt{(y_1[m, n])^2 + (y_2[m, n])^2}.
$$

(a) Determine if $h_u[m, n]$ is the impulse response of a vertical or a horizontal Sobel filter. Is the proposed edge detector a linear filter?

(b) Use *imread* to read in the image *peppers.png*, and convert it into a gray image with double precision by means of the functions *rgb2gray* and *double*. Letting this image be input $x[m, n]$, implement the edge detection to obtain an image $y[m, n]$. Determine a threshold T such that whenever $y[m, n] > T$ we have a final image $z[m, n] = 1$ and zero otherwise. Display the images $z[m, n]$ and its complement.

(c) Repeat the above for the image *circuit.tif*, choose a threshold T so that you obtain a binarized image displaying the edges of the image.

Answers: The edge detector is not linear because of the squaring and the square root. Moreover, the thresholding makes it non-linear also.

12.28 Gaussian filtering—Consider a Gaussian filter with impulse response

$$
h[m, n] = \begin{cases} e^{-(m^2 + n^2)}/(4\sigma^4) & -3\sigma \leq m \leq 3\sigma, \ -3\sigma \leq n \leq 3\sigma, \\ 0 & \text{otherwise,} \end{cases}
$$

for two values $\sigma = 1$ and $\sigma = 6$.

(a) Determine the masks corresponding to the two values of σ, and call them H_1 and H_6. Find the 2D FFT of size 256×256 of H_1 and H_6 and use *mesh* to plot the masks and their magnitude responses together.

(b) Blur the image *cameraman.tif* and determine the mean square error for the two blurring systems to find out which blurs more.

Answer: The system with $\sigma = 6$ blurs more.

12.29 Bank of filters—Consider the design of a bank of filters using circular low-pass, band-pass and high-pass FIR filters using the following magnitude specifications where $r = \sqrt{f_1^2 + f_2^2} < 0.5$ for normalized frequencies $-1 \le f_1 \le 1$ and $-1 \le f_2 \le 1$ and $f_i = \omega_i / \pi$, $i = 1, 2$:

$$H_{lp}(e^{j\omega_1}, e^{j\omega_2}) = \begin{cases} 1 & r < 0.5, \\ 0 & \text{otherwise,} \end{cases}$$

$$H_{bp}(e^{j\omega_1}, e^{j\omega_2}) = \begin{cases} 1 & 0.5 < r < 0.8, \\ 0 & \text{otherwise,} \end{cases}$$

$$H_{hp}(e^{j\omega_1}, e^{j\omega_2}) = \begin{cases} 1 & r > 0.8, \\ 0 & \text{otherwise.} \end{cases}$$

Determine the magnitude frequency response of each of the filters and of the bank of filters. Plot them using *contour*.

Answers: The original image can be reconstructed almost exactly.

USEFUL FORMULAS

TRIGONOMETRIC RELATIONS

Reciprocal

$$\csc(\theta) = \frac{1}{\sin(\theta)} \qquad \sec(\theta) = \frac{1}{\cos(\theta)}$$

$$\cot(\theta) = \frac{1}{\tan(\theta)}$$

Pythagorean

$$\sin^2(\theta) + \cos^2(\theta) = 1$$

Sum and difference of angles

$$\sin(\theta \pm \phi) = \sin(\theta)\cos(\phi) \pm \cos(\theta)\sin(\phi)$$
$$\sin(2\theta) = 2\sin(\theta)\cos(\theta)$$
$$\cos(\theta \pm \phi) = \cos(\theta)\cos(\phi) \mp \sin(\theta)\sin(\phi)$$
$$\cos(2\theta) = \cos^2(\theta) - \sin^2(\theta)$$

Multiple-angle

$$\sin(n\theta) = 2\sin((n-1)\theta)\cos(\theta) - \sin((n-2)\theta)$$
$$\cos(n\theta) = 2\cos((n-1)\theta)\cos(\theta) - \cos((n-2)\theta)$$

Products

$$\sin(\theta)\sin(\phi) = \frac{1}{2}[\cos(\theta - \phi) - \cos(\theta + \phi)]$$

$$\cos(\theta)\cos(\phi) = \frac{1}{2}[\cos(\theta - \phi) + \cos(\theta + \phi)]$$

$$\sin(\theta)\cos(\phi) = \frac{1}{2}[\sin(\theta + \phi) + \sin(\theta - \phi)]$$

$$\cos(\theta)\sin(\phi) = \frac{1}{2}[\sin(\theta + \phi) - \sin(\theta - \phi)]$$

Euler's identity (θ in rad)

$$e^{j\theta} = \cos(\theta) + j\sin(\theta) \qquad j = \sqrt{-1}$$

$$\cos(\theta) = \frac{e^{j\theta} + e^{-j\theta}}{2} \qquad \sin(\theta) = \frac{e^{j\theta} - e^{-j\theta}}{2j}$$

$$\tan(\theta) = -j\left[\frac{e^{j\theta} - e^{-j\theta}}{e^{j\theta} + e^{-j\theta}}\right]$$

Hyperbolic trigonometry relations

Hyperbolic cosine $\qquad \cosh(\alpha) = \dfrac{1}{2}(e^{\alpha} + e^{-\alpha})$

Hyperbolic sine $\qquad \sinh(\alpha) = \dfrac{1}{2}(e^{\alpha} - e^{-\alpha})$

$$\cosh^2(\alpha) - \sinh^2(\alpha) = 1$$

CALCULUS

Derivatives
u, v functions of x; α, β constants

$$\frac{duv}{dx} = u\frac{dv}{dx} + v\frac{du}{dx}$$

$$\frac{du^n}{dx} = nu^{n-1}\frac{du}{dx}$$

Integrals

$$\int \phi(y)dx = \int \frac{\phi(y)}{y'}dy \qquad y' = \frac{dy}{dx}$$

$$\int u\,dv = uv - \int v\,du$$

$$\int x^n dx = \frac{x^{n+1}}{n+1} \qquad n \neq 1, \text{ integer}$$

$$\int x^{-1}dx = \log(x)$$

$$\int e^{ax}dx = \frac{e^{ax}}{a} \qquad a \neq 0$$

$$\int xe^{ax}dx = \frac{e^{ax}}{a^2}(ax - 1)$$

$$\int \sin(ax)dx = -\frac{1}{a}\cos(ax)$$

$$\int \cos(ax)dx = \frac{1}{a}\sin(ax)$$

$$\int \frac{\sin(x)}{x}dx = \sum_{n=0}^{\infty}(-1)^n \frac{x^{2n+1}}{(2n+1)(2n+1)!} \qquad \text{integral of sinc function}$$

$$\int_0^{\infty} \frac{\sin(x)}{x}dx = \int_0^{\infty} \left[\frac{\sin(x)}{x}\right]^2 dx = \frac{\pi}{2}$$

Bibliography

[1] L. Ahlfors, Complex Analysis, McGraw-Hill, New York, 1979.
[2] Analog, Op-Amp History, http://www.analog.com/media/en/training-seminars/design-handbooks/Op-Amp-Applications/SectionH.pdf. (Accessed 2018).
[3] A. Antoniou, Digital Filters, McGraw-Hill, New York, 1993.
[4] S. Attaway, MATLAB – A Practical Introduction to Programming and Problem Solving, Butterworth-Heinemann, Amsterdam, 2016.
[5] E.T. Bell, Men of Mathematics, Simon and Schuster, 1965.
[6] M. Bellis, The invention of radio, https://www.thoughtco.com/invention-of-radio-1992382. (Accessed 2017).
[7] J. Belrose, Fessenden and the early history of radio science, http://www.ewh.ieee.org/reg/7/millennium/radio/radio_radioscientist.html. (Accessed 2017).
[8] N.K. Bose, Digital Filters, Elsevier Science Pub., Amsterdam, 1985.
[9] C. Boyer, A History of Mathematics, John Wiley & Sons, 1991.
[10] R. Bracewell, The Fourier Transform and Its Application, McGraw-Hill, Boston, 2000.
[11] M. Brain, How CDs work, http://electronics.howstuffworks.com/cd.htm. (Accessed 2017).
[12] W. Briggs, V. Henson, The DFT, Society for Industrial and Applied Mathematics (SIAM), Philadelphia, 1995.
[13] O. Brigham, The Fast Fourier Transform and Its Applications, Prentice Hall, Englewood Cliffs, NJ, 1988.
[14] L. Chua, C. Desoer, E. Kuh, Linear and Nonlinear Circuits, McGraw-Hill, New York, 1987.
[15] J. Cooley, How the FFT gained acceptance, in: S. Nash (Ed.), A History of Scientific Computing, ACM (Association for Computing Machinery, Inc.) Press, 1990, pp. 133–140.
[16] J.W. Cooley, J.W. Tukey, An algorithm for the machine calculation of complex Fourier series, Mathematics of Computation 19 (Apr. 1965) 297.
[17] L.W. Couch, Digital and Analog Communication Systems, Pearson/Prentice Hall, Upper Saddle River, NJ, 2007.
[18] E. Craig, Laplace and Fourier Transforms for Electrical Engineers, Holt, Rinehart and Winston, New York, 1966.
[19] E. Cunningham, Digital Filtering, Houghton Miffling, Boston, 1992.
[20] G. Danielson, C. Lanczos, Some improvements in practical Fourier analysis and their applications to X-ray scattering from liquids, Journal of the Franklin Institute (1942) 365–380.
[21] C. Desoer, E. Kuh, Basic Circuit Theory, McGraw-Hill, New York, 1969.
[22] R. Dorf, R. Bishop, Modern Control Systems, Prentice Hall, Upper Saddle River, NJ, 2005.
[23] R. Dorf, J. Svoboda, Introduction to Electric Circuits, Wiley, Hoboken, NJ, 2014.
[24] D. Dudgeon, R. Mersereau, Multidimensional Digital Signal Processing, Prentice-Hall, Englewood Cliffs, NJ, 1984.
[25] Engineering and Technology History Wiki, Harry Nyquist, http://ethw.org/Harry_Nyquist. (Accessed 2017).
[26] R. Fessenden, Inventing the wireless telephone and the future, https://www.ieee.ca/millennium/radio/radio_wireless.html. (Accessed 2017).
[27] G. Franklin, J. Powell, M. Workman, Digital Control of Dynamic Systems, Addison-Wesley, Reading, MA, 1998.
[28] R. Gabel, R. Roberts, Signals and Linear Systems, Wiley, New York, 1987.
[29] Z. Gajic, Linear Dynamic Systems and Signals, Prentice-Hall, Upper Saddle River, NJ, 2003.
[30] S. Goldberg, Introduction to Difference Equations, Dover, New York, 1958.
[31] R. Graham, D. Knuth, O. Patashnik, Concrete Mathematics, Addison-Wesley, Reading, MA, 1994.
[32] S. Haykin, M. Moher, Modern Wireless Communications, Pearson/Prentice Hall, Upper Saddle River, NJ, 2005.
[33] S. Haykin, M. Moher, Introduction to Analog and Digital Communications, Wiley, Hoboken, NJ, 2007.
[34] S. Haykin, B. Van Veen, Signals and Systems, Wiley, New York, 2003.
[35] M. Heideman, D. Johnson, S. Burrus, Gauss and the history of the Fast Fourier Transform, IEEE ASSP Magazine (Oct. 1984) 14–21.
[36] K.B. Howell, Principles of Fourier Analysis, Chapman & Hall, CRC Press, Boca Raton, 2001.
[37] IAC Acoustics, Comparative examples of noise levels, http://www.industrialnoisecontrol.com/comparative-noise-examples.htm. (Accessed 2017).
[38] Intel, Moore's law, http://www.intel.com/technology/mooreslaw/index.htm. (Accessed May 2013).

[39] L. Jackson, Signals, Systems, and Transforms, Addison-Wesley, Reading, MA, 1991.

[40] G. Johnson, The New York Times: Claude Shannon, Mathematician, dies at 84, http://www.nytimes.com/2001/02/27/nyregion/claude-shannon-mathematician-dies-at-84.html. (Accessed 2013).

[41] E.I. Jury, Theory and Application of the Z-transform Method, J. Wiley, New York, 1964.

[42] T. Kailath, Linear Systems, Prentice-Hall, Englewood Cliffs, NJ, 1980.

[43] E. Kamen, B. Heck, Fundamentals of Signals and Systems, Pearson Prentice Hall, Upper Saddle River, NJ, 2007.

[44] S. Kuo, B. Lee, W. Tian, Real-time Digital Signal Processing: Fundamentals, Implementations and Applications, Wiley, New York, 2013.

[45] B.P. Lathi, Modern Digital and Analog Communication Systems, Oxford University Press, New York, 1998.

[46] B.P. Lathi, Linear Systems and Signals, Oxford University Press, New York, 2002.

[47] E. Lee, P. Varaiya, Structure and Interpretation of Signals and Systems, Addison Wesley, Boston, 2003.

[48] W. Lehr, F. Merino, S. Gillet, Software radio: implications for wireless services, industry structure, and public policy, http://itc.mit.edu. (Accessed 30 August 2002).

[49] J. Lim, Two-dimensional Signal and Image Processing, Prentice Hall, Englewood Cliffs, NJ, 1990.

[50] D. Luke, The origins of the sampling theorem, IEEE Communications Magazine (Apr. 1999) 106–108.

[51] D. Manolakis, V. Ingle, Applied Digital Signal Processing, Cambridge University Press, New York, NY, 2011.

[52] J. Marsden, M. Hoffman, Basic Complex Analysis, W. H. Freeman, New York, 1998.

[53] D. McGillem, G. Cooper, Continuous and Discrete Signals and System Analysis, Holt, Rinehart, and Winston, New York, 1984.

[54] U. Meyer-Baese, Digital Signal Processing with Field Programmable Gate Arrays, Springer-Verlag, Berlin, 2001.

[55] S. Mitra, Digital Signal Processing, McGraw-Hill, New York, 2006.

[56] C. Moler, The origins of MATLAB, https://www.mathworks.com/company/newsletters/articles/the-origins-of-matlab.html. (Accessed 2017).

[57] P.J. Nahin, Dr. Euler's Fabulous Formula: Cures Many Mathematical Ills, Princeton U. Press, 2006.

[58] F. Nebeker, Signal Processing – The Emergence of a Discipline, 1948 to 1998, IEEE History Center, New Brunswick, NJ, 1998. This book was especially published for the 50th anniversary of the creation of the IEEE Signal Processing Society in 1998.

[59] MIT News Office, MIT Professor Claude Shannon dies; was founder of digital communications, http://web.mit.edu/newsoffice/2001/shannon.html. (Accessed 2017).

[60] K. Ogata, Modern Control Engineering, Prentice Hall, Upper Saddle River, NJ, 1997.

[61] A. Oppenheim, R. Schafer, Digital Signal Processing, Prentice-Hall, Englewood Cliffs, NJ, 1975.

[62] A. Oppenheim, R. Schafer, Discrete-time Signal Processing, Prentice Hall, Upper Saddle River, NJ, 2010.

[63] A. Oppenheim, A. Willsky, Signals and Systems, Prentice Hall, Upper Saddle River, NJ, 1997.

[64] A. Papoulis, Signal Analysis, McGraw-Hill, New York, 1977.

[65] PBS.org, Tesla – master of lightning: who invented radio? http://www.pbs.org/tesla/ll/ll_whoradio.htm. (Accessed 2017).

[66] C. Phillips, J. Parr, E. Riskin, Signals, Systems and Transforms, Pearson/Prentice Hall, Upper Saddle River, NJ, 2003.

[67] J. Proakis, M. Salehi, Communication Systems Engineering, Prentice Hall, Upper Saddle River, NJ, 2002.

[68] L. Rabiner, B. Gold, Theory and Application of Digital Signal Processing, Prentice-Hall, Englewood Cliffs, NJ, 1975.

[69] GNU Radio, The GNU software radio, http://gnuradio.org/redmine/projects/gnuradio/wiki. (Accessed 2017).

[70] D. Slepian, On bandwidth, Proceedings of the IEEE 64 (Mar. 1976) 292–300.

[71] S. Soliman, M. Srinath, Continuous and Discrete Signals and Systems, Prentice Hall, Upper Saddle River, NJ, 1998.

[72] J. Soni, R. Goodman, A Mind at Play: How Claude Shannon Invented the Information Theory, Simon and Schuster, New York, 2017.

[73] A. Stanoyevitch, Introduction to Numerical Ordinary and Partial Differential Equations Using MATLAB, John Wiley, New York, 2005.

[74] H. Stern, S. Mahmoud, Communication Systems – Analysis and Design, Pearson, Upper Saddle River, NJ, 2004.

[75] M. Unser, Sampling – 50 years after Shannon, Proceedings of the IEEE 88 (2000) 569–587, https://doi.org/10.1109/5.843002.

[76] M. Van Valkenburg, Analog Filter Design, Oxford University Press, New York, 1982.

[77] Wikipedia, Harold S. Black, http://en.wikipedia.org/wiki/Harold_Stephen_Black. (Accessed 2017).

[78] Wikipedia, Nikola Tesla, http://en.wikipedia.org/wiki/Nikola_Tesla. (Accessed 2017).

[79] Wikipedia, Oliver Heaviside, http://en.wikipedia.org/wiki/Oliver_Heaviside. (Accessed 2017).

[80] M.R. Williams, A History of Computing Technologies, Wiley-IEEE Computer Soc. Press, 1997.

[81] J. Woods, Multidimensional Signal, Image, and Video Processing and Coding, Academic Press, Amsterdam, 2012.

Index